Bioremediation of Toxic Metal(loid)s

Editors

Anju Malik
Department of Energy and Environmental Sciences
Chaudhary Devi Lal University
Sirsa, Haryana
India

Mohd. Kashif Kidwai
Department of Energy and Environmental Sciences
Chaudhary Devi Lal University
Sirsa, Haryana
India

Vinod Kumar Garg
Department of Environmental Science and Technology
School of Environment and Earth Sciences
Central University of Punjab
Bathinda, Punjab
India

CRC Press is an imprint of the
Taylor & Francis Group, an **informa** business
A SCIENCE PUBLISHERS BOOK

First edition published 2023
by CRC Press
6000 Broken Sound Parkway NW, Suite 300, Boca Raton, FL 33487-2742

and by CRC Press
4 Park Square, Milton Park, Abingdon, Oxon, OX14 4RN

CRC Press is an imprint of Taylor & Francis Group, LLC

Library of Congress Cataloging-in-Publication Data (applied for)

ISBN: 978-1-032-13577-9 (hbk)
ISBN: 978-1-032-13579-3 (pbk)
ISBN: 978-1-003-22994-0 (ebk)

DOI: 10.1201/9781003229940

Typeset in Times New Roman
by Radiant Productions

Preface

The environment is being incessantly contaminated by a large number of toxic pollutants resulting from various technological advancements. Among them, metal(loid)s are an exclusive group of toxicants, that are persistent, do not degrade, tend to bioaccumulate, and are difficult to transform into non-toxic forms. The ever-increasing concentrations of metal(loid)s in the various spheres of the environment are serious threat to human beings, animals, plants, and other living beings. Even at trace concentrations, metal(loid)s have pernicious repercussions on living beings in particular carcinogenic, genotoxic, teratogenic, mutagenic, and other sub-lethal effects. Metal(loid)s pollution is a challenge continuously threatening the environment over the globe. Consequently, a pressing need has arisen for metal(loid)s remediation in an eco-friendly and cost-effective manner. In the past few decades, metal(loid)s remediation has drawn considerable attention from the scientific community and policymakers. Many conventional techniques based on physical, chemical, and biological strategies have been employed to remove, clean, transform, or sequester these toxic pollutants from the environment. However, most of these techniques have innate limitations including their low sustainability and they are non-economic, labor-intensive, potentially introduce secondary contaminants, disturb the native microflora and microfauna, and are often disquieting. Because of the limitations associated with these conventional techniques, there has been a paradigm shift towards a bio-based, environment-friendly approach recognized as *bioremediation.* Bioremediation has a potential advantageous edge over other conventional remediation technologies as it is simple, cost-intensive, environmental friendly, efficient, eco-friendly, eco-sustainable, and fast-emerging new technology for remediating toxic metal(loid)s and other pollutants. Bioremediation utilizes a vast array of biological materials, especially plants (phytoremediation), bacteria (microbial remediation), algae (phycoremediation), fungi (mycoremediation), biochar, nano-bio materials, nano-enzymes etc. In the processes of bioremediation, biodiversity acts as a toolbox, by which the metal(loid)s are transformed into less toxic species or detoxified in an environment-friendly way.

The present book, *Bioremediation of Toxic Metal(loid)s*, has a compilation of the available comprehensive knowledge of the fundamentals and advancements in the field of bioremediation of toxic metal(loid)s. The mechanisms, applications, and current advancements of various bioremediation strategies used for metal(loid)s have been described in 21 chapters contributed by leading experts from different institutes, universities, and research laboratories from various countries across the globe including Argentina, Canada, Chile, Colombia, France, India, Japan, Republic of Korea, United Kingdom, and Unites States of America. To reflect the theme of the book, it has been divided into five sections:

Section I: Fundamentals of Bioremediation of Toxic Metal(loid)s

Section II: Bioremediation of Specific Toxic Metal(loid)s

Section III: Biotechnological Strategies for Remediation of Toxic Metal(loid)s

Section IV: Nanotechnology and Metal(loid)s Remediation

Section V: General Aspects/Case Studies on Bioremediation of Metal(oid)s

This book describes the state-of-the-art and potential of emerging technologies on bioremediation of toxic metal(loid)s. In Section I, a comprehensive view of the mechanisms adopted by a vast array of biological materials, especially higher plants, mosses, lichens, bacteria, algae, fungi, lignocellulosic waste, etc., for bioprecipitating, accumulating, stabilizing, transforming, and removing metal(loid)s have been given in nine chapters. The introductory chapter gives a bird's eye view of the sources and toxic effects of metal(loid)s pollution, and various bioremediation approaches used for metal(loid) s decontamination. Some chapters in this section also emphasize post-bioremediation production of bioenergy from biomass. Bioremediation of specific metal(loid)s such as Arsenic, Chromium, Mercury, Uranium and other radionuclides has been addressed in Section II. The recent tools of genetic, molecular, protein, and microbial engineering to modify plants for enhanced metal(loid) s uptake, transport, and sequestration have been described in Section III. Section IV encompasses the chapters on the nano-technological perspective of metal(loid)s remediation. Section V includes chapters focussing on the restoration of old mining sites, bioremediation of mining waste and other copper-containing effluents, and remediation of contaminated chromite mine spoil by biochar application. All the chapters are comprehensive, can stand alone, and have relevant illustrations in the form of tables, figures, and pictures providing additional help for clarity. On the other hand, all the chapters have a unifying theme addressing bioremediation of toxic metal(loid)s.

We sincerely hope that this book will instil the current status, practicality, and implications of bioremediation of toxic metal(loid)s to the researchers, academicians, environmentalists, agriculturalists, scientists, extension workers, industrialists, students at undergraduate and postgraduate levels, practicing engineers, policymakers, and other enthusiastic people who are wholeheartedly devoted to the fields of Environmental Science, Microbiology, Biotechnology, Public Health, Civil Engineering, Chemistry, Biochemistry, Agriculture, Life Sciences, etc.

The editors would like to express their sincere gratitude to the multidisciplinary team of authors having expertise in this field of research for readily accepting our invitation, submitting their innovative, high quality and valuable chapters in a timely manner, and their help in making this voluminous and high-quality outcome a successful endeavor. Their commitments are greatly appreciated. We all have strived hard to ensure that the book is free from any erroneous or misleading information and any such mistake is inadvertent. Finally, this book is dedicated to our respective families for their patience, co-operation, and understanding during this year-long journey. Last but not the least, we thank our family members from the core of our hearts.

Anju Malik

Mohd. Kashif Kidwai

Vinod Kumar Garg

Contents

Section II: Bioremediation of Specific Toxic Metal(loid)s

Section III: Biotechnological Strategies for Remediation of Toxic Metal(loid)s

Section IV: Nanotechnology and Metal(loid)s Remediation

Section V: General Aspects/Case Studies on Bioremediation of Metal(loid)s

Section I
Fundamentals of Bioremediation of Toxic Metal(loid)s

CHAPTER 1

Metal(loid)s
Sources, Toxicity and Bioremediation

Mohd. Kashif Kidwai,[1] *Anju Malik*[1,*] and *Vinod Kumar Garg*[2]

1.1 Introduction

Global expansion in industrial sectors and urbanization for development have occurred at the cost of the environment and its quality. Pollutants such as phthalates, PAH, metal(loid)s, dioxins, etc., are generated as a byproduct of various industrial and other activities surfacing environmental pollution as a major global issue affecting both economy as well as ecology (Singh et al. 2021). Heavy metals and metal(loid)s are widely known as environmental pollutants or potentially toxic elements (PTE) due to their toxic properties affecting the overall biotic components of different ecosystems (Sharma et al. 2015, Akhtar et al. 2020). Heavy metals occur naturally in the Earth's crust. Higher demand for the use of heavy metals in various industrial processes resulted in an increase in exposure of various metal(loid)s in different ecosystems. Heavy metal pollution has emerged due to unsustainable human induced activities such as smelting, mining of different metals, foundries, leaching of metals in ground water, landfills, indiscriminate disposal of industrial waste, and sewage waste water, excretion, runoffs, automobiles exhaust, construction activities, etc. (Fig. 1.1). Due to the use of agrochemicals, improper sewage and industrial sludge disposal, discharge of untreated industrial waste water, etc., are considered as some of the major sources of pollution due to metal(loid)s. Geogenic and natural events like volcanic eruptions, weathering of rocks, floods, ground water, wind erosion, soil erosion, forest fires, etc., are also some of the natural sources of metal(loid)s in environment (Rotkittikhun et al. 2007, Jaishankar et al. 2014, Sharma et al. 2015, Mosa et al. 2016, Muthusaravanan et al. 2018, Akhtar et al. 2020, Briffa et al. 2020, Shah and Daverey 2020, Tarekegn et al. 2020, Arora and Chauhan 2021, Goswami et al. 2021, Kaur and Roy 2021, Li et al. 2021, Manori et al. 2021, Raffa et al. 2021, Poonia et al. 2021, Sharma and Kumar 2021, Thakare et al. 2021, Velez et al. 2021, Xiang et al. 2021, Zaynab et al. 2022). Both developing and developed nations are experiencing various environmental challenges due to improper disposal of metal(loid)s (Kaur and Roy 2021).

Heavy metals are inorganic elements with high atomic weight and density (more than 5 gcm^{-3}) and pose serious health hazards and ecological risks all over the globe even at low concentrations (Witkowksa et al. 2021) as exhibited in Fig. 1.2, Tables 1.1 and 1.2. In this chapter, the term

[1] Department of Energy and Environmental Sciences, Chaudhary Devi Lal University, Sirsa, Haryana, India.
[2] Department of Environmental Science and Technology, School of Environment and Earth Sciences, Central University of Punjab, Bathinda, Punjab, India.
* Corresponding author: anjumalik27@yahoo.com, anjumalik@cdlu.ac.in

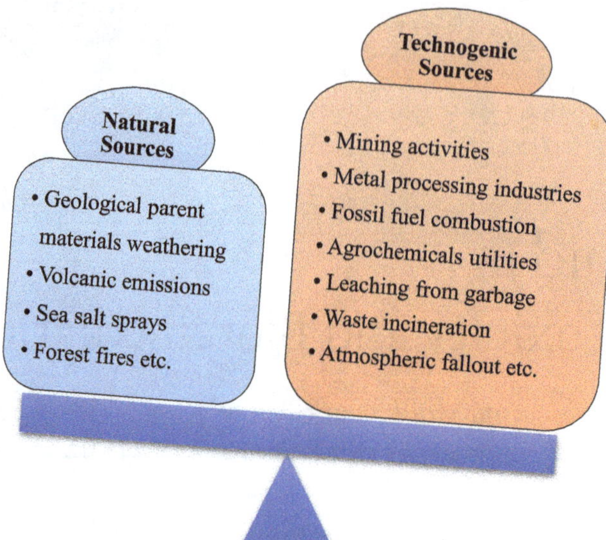

Fig. 1.1. Sources of metal(loid)s in the environment.

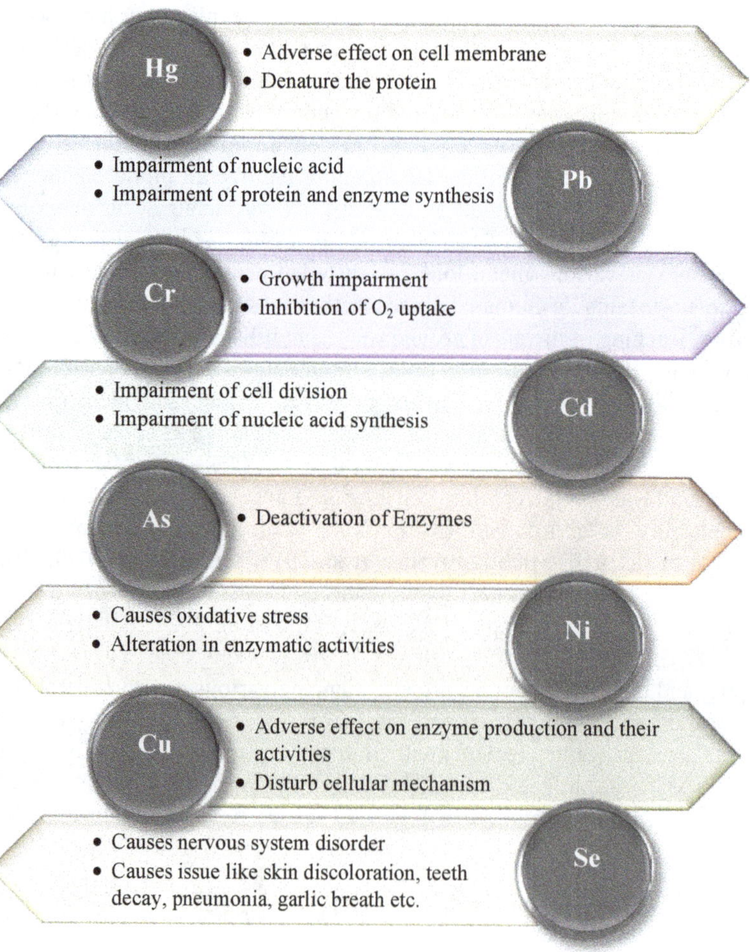

Fig. 1.2. Health hazards associated with metal(loid)s.

Table 1.1. Sources and health issues associated with metal(loid)s.

Sr. No.	Metal(oids)	Health issues	Source	References
1	Arsenic (As)	Skin allergies, Respiratory issues, Gastrointestinal problems, Infertility, Brain damage, DNA damage, Endocrinal, and hepatic disorder	Paints, Preservatives for wood, Thermal power plants, Fuel burning, Smelting, Doping agent in semiconductors, Color cosmetics	Khan et al. 2009, Jaishankar et al. 2014, Gupta and Kumar 2017, Sher and Rehman 2019, Briffa et al. 2020, Shah and Daveray 2020, Tarekegn et al. 2020, Cui et al. 2021, Guerra Sierra et al. 2021, Ozturk et al. 2022, Raffa et al. 2021, Thakare et al. 2021, Witkowksa et al. 2021
2	Cadmium (Cd)	Calcium metabolic alterations, CNS damage	Phosphatic fertilizer, Ni-Cd batteries waste, Glass paints, Incinerations, Stabilizers in plastic production, Seafood	Jaishankar et al. 2014, Gupta and Kumar 2017, Halwani et al. 2019, Zhou et al. 2019, Briffa et al. 2020, Shah and Daveray 2020, Guerra Sierra et al. 2021, Nandi and Chowdhuri 2021, Raffa et al. 2021, Thakare et al. 2021, Witkowksa et al. 2021
3	Chromium (Cr) (IV)	Apoptosis, Allergic dermatitis, Toxic nephritis, Bronchitis, Nausea, Gastrointestinal problems, liver and kidney diseases. Cause cancer via inhalation and ingestion	Electroplating industry, Tanneries, Coloring in glass, Dyes, Mining, Grains	Zhitkovich 2011, Jaishankar et al. 2014, Yaman 2020, Akhtar et al. 2020, Briffa et al. 2020, Shah and Daveray 2020, Tarekegn et al. 2020, Goswami et al. 2021, Guerra Sierra et al. 2021, Poonia et al. 2021, Raffa et al. 2021, Thakare et al. 2021, Witkowksa et al. 2021
4	Lead (Pb)	Increased risk of Thrombosis, Infertility, Renal impairment, Kidney damage, Disturbance in the synthesis of hemoglobin, Impairment in the development of the brain in children, Reduction of intelligence in children, Neurogenerative damage, Oxidative damage to DNA and proteins	Cottage industries, Potteries, Ceramics, Pesticides, Sports equipments, Lead piping, E-waste, Paints, Smelting operations, Toys, Glass industry	Wasi et al. 2013, Jaishankar et al. 2014, Kumar et al. 2017, Boskabady et al. 2018, Briffa et al. 2020, Shah and Daveray 2020, Tarekegn et al. 2020, Goswami et al. 2021, Guerra Sierra et al. 2021, Raffa et al. 2021, Thakare et al. 2021, Velez et al. 2021, Wang et al. 2021a, Witkowksa et al. 2021
5	Nickel (Ni)	Asthma, allergic skin diseases, Pulmonary fibrosis, Respiratory disorders, Impairment of heart functioning, Cardiovascular damage, Kidney disorder	Thermal power plants, Metallurgical industries, Welding, Rocket engines, Jewellery, Nichrome alloy, Steel making, Dark chocolates, Soy products, Smoking	Pasha et al. 2010, Sun et al. 2016, Briffa et al. 2020, Shah and Daveray 2020, Arora and Chauhan 2021, Dudek-Adamska et al. 2021, Infante et al. 2021, Goswami et al. 2021, Raffa et al. 2021, Thakare et al. 2021, Witkowksa et al. 2021
6	Mercury (Hg)	Visual disorders, Mental retardation, Renal issues, Genotoxic, Memory loss, Speech defects, Down syndrome, Teratogenic, Neurological issues	Thermometers, Cosmetics, Fluorescent bulbs, Barometers, Fumigants, Thermal power plants, Hospital waste, Electrical appliances	Jaishankar et al. 2014, Kumar et al. 2017, Budnik and Ludwine 2019, Briffa et al. 2020, Shah and Daveray 2020, Mergler 2021, Guerra Sierra et al. 2021, Raffa et al. 2021, Thakare et al. 2021, Witkowksa et al. 2021

Table 1.1 contd. ...

...Table 1.1 contd.

Sr. No.	Metal(oids)	Health issues	Source	References
7	Copper (Cu)	Liver damage, Hepatocellular degeneration, Liver cirrhosis, Insomnia, Wilson disease, Jaundice, Brain damage, Kidney damage	Timber, Paint, Pesticides, Electroplating, Copper alloys, Smelting, Fertilizer, Preservatives for wood and fabric	Briffa et al. 2020, Costa et al. 2020, Shah and Daveray 2020, Taylor et al. 2020, Arora and Chauhan 2021, Goswami et al. 2021, Guerra Sierra et al. 2021, Raffa et al. 2021, Thakare et al. 2021, Witkowksa et al. 2021
8	Selenium (Se)	Gastrointestinal issues, Discoloration of the skin, Decay of teeth, Garlic breath, Nervous system disorders, Pneumonia	Naturally in water, and soil, Volcanic rocks, Shale rocks, Coal, Glassware, Ceramics, Plastics, Colorant in glass, Red coloration in glass, Nuclear waste, Food products such as nuts, meat, etc.	Vinceti et al. 2001, Zwolak and Zaporowska 2012, Gebreeyessus and Zewge 2018, Natasha et al. 2018, Liang et al. 2019, Paul and Saha 2019, Briffa et al. 2020, Hasanuzzaman et al. 2020, Sabuda et al. 2020, Arora and Chauhan 2021, Raffa et al. 2021, Li et al. 2022

Table 1.2. Carcinogenic potential of different metal(loid)s to humans.

Sr. No.	Group	Carcinogenic levels in humans	Heavy metals
1	Group 1	Carcinogenic	Arsenic (Inorganic), Cadmium, Nickel, Chromium VI
2	Group 2A	Probably Carcinogenic	Lead (Inorganic Compounds)
3	Group 2B	Possibly Carcinogenic	Lead, Methyl Mercury, Cobalt, Nickel (Alloys and Metallic), Molybdenum trioxide, Vanadium pentaoxide
4	Group 3	Carcinogenicity not classifiable	Copper, Chromium III compounds, Mercury, Arsenic (Organic), Selenium
5	Group 4	Probably carcinogenic	Silver, Zinc, Manganese

(Source: Briffa et al. 2020)

metal(loid)s has been used as a convenient collective term representing both metals and metalloids. The presence of metals such as copper, cobalt, manganese, nickel, zinc, etc., is due to mineral weathering. Soil represents a medium for different heavy metals (Elgarahy et al. 2021). According to Zaynab et al. (2022), metal(loid)s ions are further categorized into three groups. The first group has mercury, cadmium, and lead reported for toxicity even at very low concentrations. The second group is represented by comparatively less toxic metals which include arsenic, bismuth, thallium, and antimony, whereas metals in the third group are zinc, cobalt, copper, iron, and selenium, which contribute to diverse biochemical processes and induce toxicities when present in high concentrations.

Metals are recognized as good conductor of electricity having a metallic luster. They are malleable and ductile and have basic oxides. Metals have diverse applications in various industrial sectors. Some metals also play a significant role in various biological processes. Both deficiency and excess of metals in living systems cause metabolic disorders and diseases. Some metals and metalloids are investigated to be essentially required for different life forms as they are present in enzymes, which influence biochemical reactions in diverse living organisms (Jaishankar et al. 2014, Ali et al. 2019, Kidwai and Dhull 2021).

Heavy metals are reported to react with nuclear proteins along with DNA causing onsite site-specific damages classified as direct damages and indirect damages. Heavy metals are reported to alter the signaling pathways (Nagajyoti et al. 2010, Jaishankar et al. 2014, Wilk et al. 2017, Briffa et al. 2020).

1.2 Essential and Non-essential Metalloids

Heavy metals are divided into two broad categories, i.e., essential and nonessential heavy metals. Essential heavy metals are significant for the biological system and required in a specifically lesser quantity viz. cobalt, iron, copper, manganese, etc., for various physiological and biochemical functions, whereas heavy metals like lead, cadmium, mercury, nickle, etc., are non-essential heavy metals and induce various toxicities including contact dermatitis, kidney diseases, lung fibrosis, cardiovascular, lung and nasal cancers, etc. (Maurya et al. 2019). However, the quantity of essential heavy metals may vary among different organisms- humans, plants and microorganisms (Nagajyoti et al. 2010, Anju 2017, Fu et al. 2017, Ali et al. 2019, Arora and Chauhan 2021).

1.3 Heavy Metals and Metal(loid)s

Heavy metals adversely affect nuclear proteins and DNA, resulting in causing specific damage. Direct damages due to heavy metals are the negative effects at the level of biomolecules. Furthermore, indirect effects occur with the generation of Reactive Oxygen Species (ROS) such as hydroxyl ions, hydrogen peroxide ions, superoxide ion, etc., causing oxidative stress (Valko et al. 2005, Briffa et al. 2020).

The heavy metal-induced toxicity induces the generation of free radicals causing damaged DNA along with the impairment of sulphydryl homeostasis along with lipid peroxidation. Calcium homeostasis is reported to be altered due to heavy metals. The induction of free radicals is associated with different heavy metal induced toxicities such as copper, nickel, cadmium, etc. (Briffa et al. 2020).

1.3.1 Arsenic (As)

Arsenic is a toxic metalloid. It enters the ecological cycles such as the food chain through dust, rainwater, open waters such as rivers and ponds, and groundwater. It is transferred to soil and aquatic ecosystems through various natural sources such as volcanic eruptions, hot springs, geysers, etc., as exhibited in Table 1.1. Some of the anthropogenic sources are mining, smelting, leather industry, petrochemical industries, agrochemicals, paints, cosmetics, glass industry, dyes, food, etc. The arsenic contaminated groundwater affects agriculture and poses grave health issues for humans and other organisms (Sher and Rehman 2019, Nath et al. 2021). Miners are more exposed to this metalloid because of their occupation. The human population residing near mining areas is constantly facing risk due to arsenic induced pollution. More than 140 million people from sixty countries including India, Bangladesh, Nepal, Cambodia, the USA, Chile, China, Vietnam, Hungary, Mexico, Argentina, New Zealand, Philippines, Ghana, Taiwan, etc., encounter the risk of arsenic contamination (Jaishankar et al. 2014, Zwolak 2020, Ozturk et al. 2022, Patel et al. 2021).

1.3.2 Cadmium (Cd)

Cadmium is a toxic metal and widely recognized as a major pollutant. It is a non-essential metal reported to develop cadmium induced toxicity even at very low quantity 0.001–0.1 mg L^{-1}. Cd is widely used in various industries, e.g., electroplating, pulp and paper industry, batteries, chlor-alkali, copper alloys, paint industry, mining operations, synthetic fertilizer, etc., as exhibited in Table 1.1 (Jaishankar et al. 2014, Chellaiah 2018, Halwani et al. 2019, Zwolak 2020, Li et al. 2021)

1.3.3 Chromium (Cr)

Chromium (Cr) is identified as one of the 129 priority pollutants and noxious heavy metals. Inhalation of low levels of chromium(VI) is reported to interact with DNA, thereby causing cancer. The toxic and carcinogenic properties of hexavalent chromium are due to chemical reactivity. Chromium(III) is reported to be noncarcinogenic for cells as it does not enter into the cell whereas Cr(VI) enters into the cell as anion transporter and is reduced to different oxidation states by cellular reductants affecting DNA (Yaman 2020). Cr and Ni are released through the persistent weathering of serpentine rocks. Cr (III) is an essential element for humans and beneficial nutrient for animals in trace amounts, but in its hexavalent form [Cr(VI)], Cr is investigated to be a probable carcinogen for human as exhibited in Table 1.1 and Table 1.2 (Jaishankar et al. 2014, Arora and Chauhan 2021, Infante et al. 2021, Poonia et al. 2021, Raffa et al. 2021).

1.3.4 Copper (Cu)

Copper is a metal and causes pollution in different spheres of the environment when present in excess quantities. Copper is a micronutrient required by a variety of living organisms. In plants, it is metabolically active in different processes such as photosynthesis, electron transport chains, cell wall metabolism, etc. (Rehman et al. 2019). Some of the anthropogenic sources of copper are paint and electronic industries, domestic waste water, landfill, mining activity, fossil fuels, etc. Copper based pollution develops various toxicities in humans after ingestion in high concentration as exhibited in Table 1.1 (Rehman et al. 2019, Arora and Chauhan 2021, Raffa et al. 2021, Thakare et al. 2021, Witkowksa et al. 2021).

1.3.5 Lead (Pb)

According to the World Health Organization, lead is a toxic metal posing negative health effects along with severe complications as exhibited in Table 1.1 (Velez et al. 2021). It is a soft, malleable, and heavy post transition metal. Lead as metal appears bluish-white, and later turns into dull greyish color when exposed to air. It gives shiny silver lustre when melted into liquid. Various human health issues are associated with lead such as neurotoxicity, hepatotoxicity, negatively affecting the development of brain with reduction of intelligence in children, neurogenerative damages as exhibited in Table 1.1 (Jaishankar et al. 2014, Kumar et al. 2017, Briffa et al. 2020, Shah and Daveray 2020, Tarekegn et al. 2020, Goswami et al. 2021, Guerra Sierra et al. 2021, Raffa et al. 2021, Thakare et al. 2021, Velez et al. 2021, Wang et al. 2021a).

1.3.6 Mercury (Hg)

Mercury is identified as one of the most toxic metals (Budnik and Ludwine 2019). Mercury exists in different oxidation states having different properties which makes it difficult to assess the contaminated sites in both soil and water. Mercury in organic state is highly hazardous detected as methyl mercury, a potent neurotoxin, formed due to the microbial activity in water from elemental mercury as exhibited in Table 1.1. Methyl mercury toxicity is reported to occur in humans through intake of contaminated food sources. The occurance of Minimata disease is one of the popular examples of mercury-based toxicity due to the intake of seafood by the local population residing near Minimata bay, Japan. Mercury is known to be relatively immobile in soil and binds strongly with soil constituents. The inorganic mercury gets accumulated in kidneys to cause renal damage as given in Table 1.1 (Wang et al. 2012, Jaishankar et al. 2014, Kumar et al. 2017, Budnik and Ludwine 2019, Liu et al. 2019, Briffa et al. 2020, Shah and Daveray 2020, Mergler 2021, Guerra Sierra et al. 2021, Raffa et al. 2021, Thakare et al. 2021, Witkowksa et al. 2021).

1.3.7 Nickel (Ni)

Nickel is identified as one of the heavy metal having the potential to cause health issues in humans. Nickel gets combined with metals to form alloy, e.g., iron, zinc, chromium, copper, etc. It is reported to induce carcinogenic effects in humans as exhibited in Table 1.1 and Table 1.2. Nickel carbonyl $Ni(CO)_4$, an airborne pollutant produced in refining activity, enters the body through inhalation causing diverse health issues, i.e., pneumonia, pulmonary edema, respiratory failure, etc. (Pasha et al. 2010, Jaishankar et al. 2014, Sun et al. 2016, Briffa et al. 2020, Shah and Daveray 2020, Tarekegn et al. 2020, Arora and Chauhan 2021, Dudek-Adamska et al. 2021, Infante et al. 2021, Goswami et al. 2021, Raffa et al. 2021, Thakare et al. 2021, Witkowksa et al. 2021).

1.3.8 Selenium (Se)

Selenium is a metalloid and an essential mineral occurring in soil and water naturally. Selenium is required as an essential dietary element for humans as well as animals. Selenium is termed a dual-edged sword as its deficiency and toxicity both cause health issues as presented in Table 1.1. Selenium has a narrow gap between its benefits and toxicity to humans. Selenium has been ranked as the 145th toxic element among the hazardous substances by ATSDR (2012). Sedimentary rocks based soils with high organic matter are reported for its potential toxic content of selenium in comparison to magmatic rocks based soils having low selenium concentration (Natasha et al. 2018). Different forms of selenium such as selenate and selenite occur in the natural aqueous environment. Both selenate and selenite can be transformed to elemental selenium enabling the removal of selenium due to its lower solubility (Li et al. 2022). Food supplement of selenium has several forms such as selenocysteine, selenite, selenomethionine and selenate (Vinceti et al. 2001, Zwolak and Zaporowska 2012, Jaishankar et al. 2014, Gebreeyessus and Zewge 2018, Liang et al. 2019, Paul and Saha 2019, Briffa et al. 2020, Hasanuzzaman et al. 2020, Sabuda et al. 2020, Arora and Chauhan 2021, Raffa et al. 2021, Li et al. 2022).

1.4 Remediation of Meta(loid)s

The sustainable elimination of heavy metals is a serious concern as several disadvantages are associated with conventional remediation methods. Conventional remediation methods are basically physical and chemical remediation methods that are uneconomical and pose various environmental hazards. Efficient sustainable bioremediation methods are required for the elimination of heavy metals (Kaur and Roy 2021). Various biotechnological strategies using diverse microorganisms and plants have emerged as a sustainable and efficient strategy in remediation of metal(loid)s (Zaynab et al. 2022).

1.5 Conventional Methods for Remediation of Meta(loid)s

Physicochemical methods are commonly used in the remediation of metal(loid)s as exhibited in Fig. 1.3. Some of the physical methods for remediation of metal(loid)s are magnetic separation, hydrodynamic separation, electrokinetic method, adsorption, activated carbon method, membrane filtration method, photocatalysis method, electrodialysis, vitrification, thermal treatment, etc., whereas chemical methods for remediation of metal(loid)s include chemical stabilization, solidification, ion exchange method, flotation, coagulation, flocculation, chemical washing of soil, etc. Several disadvantages associated with the application of physicochemical methods are the generation of sludge, non-availability of technical work force, energy-intensive, rusting, heavy investment of capital along with high operational cost, etc. (Ruchitha et al. 2015, Kaushal and Singh 2017, Sher and Rehman 2019, Akhtar et al. 2020, Kaur and Roy 2020, Goswami et al. 2021, Raffa

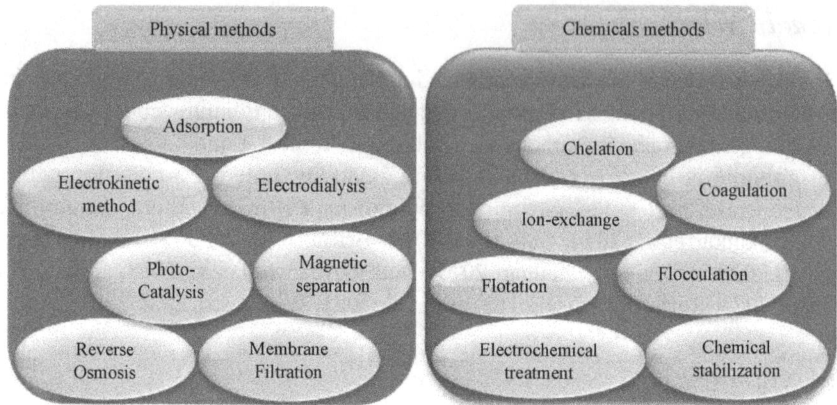

Fig. 1.3. Physicochemical methods commonly used in the remediation of metal(loid)s.

et al. 2021). Considering these limitations associated with conventional methods, novel, sustainable and cost effective methods are needed for the remediation of metal(loid)s.

1.6 Bioremediation

Bioremediation is an eco-friendly, non-invasive strategy, which is comparatively cost-effective than other conventional methods and provides a sustainable remedy with degradation or transformation of various environmental contaminants such as pesticides, polycyclic aromatic hydrocarbons, metal(loid)s, etc., into comparatively less toxic forms (Dzionek et al. 2016, Sharma and Kumar 2021). Diverse organisms are identified, explored, and applied in bioremediation to detoxify different pollutants including metal(loid)s in the environment due to their economic efficiency and promising results (Mosa et al. 2016). Bioremediation involves both natural and recombinant organisms for the *in situ* and *ex situ* remediation of various pollutants occuring in different spheres of the environment (Mosa et al. 2016, Bano et al. 2018). Different life forms such as fungi, bacteria, algae, higher and lower plants, etc., possess the potential to degrade or detoxify hazardous contaminants in various ecosystems. Efficient microorganisms degrade the pollutants through enzymes converting them into less harmful or harmless products. However, environmental factors influence the growth and metabolic activities of plants and diverse microorganisms (Kang et al. 2016, Bano et al. 2018). Various strategies used by the organisms for the bioremediation of the metal(loid)s are given in Fig. 1.4 for bioremediation of metal(loid)s. Bioremediation is applied *in situ* as well as *ex situ*.

1.6.1 Ex situ Bioremediation

Ex situ bioremediation involves the excavation of pollutants from site and transfer to other location for further processing. *Ex situ* bioremediation strategies depend on the cost of treatment, chemical features of pollutants, geological parameters of the polluted site, etc. (Azubuike et al. 2016). Some of the *ex situ* bioremediation strategies are biopile, windrows, bioreactor, etc. (Anusha and Natrajan 2020).

1.6.2 In situ Bioremediation

This category of bioremediation techniques is used in treating various types of pollutants such as hydrocarbons, dyes, metal(loid)s, etc., at the site of pollution. This type of bioremediation is comparatively cost feasible than *ex situ* bioremediation strategies. It is pertinent to mention that the abiotic factors such as soil porosity, pH, tempreture, moisture, etc. influence the success of *in situ*

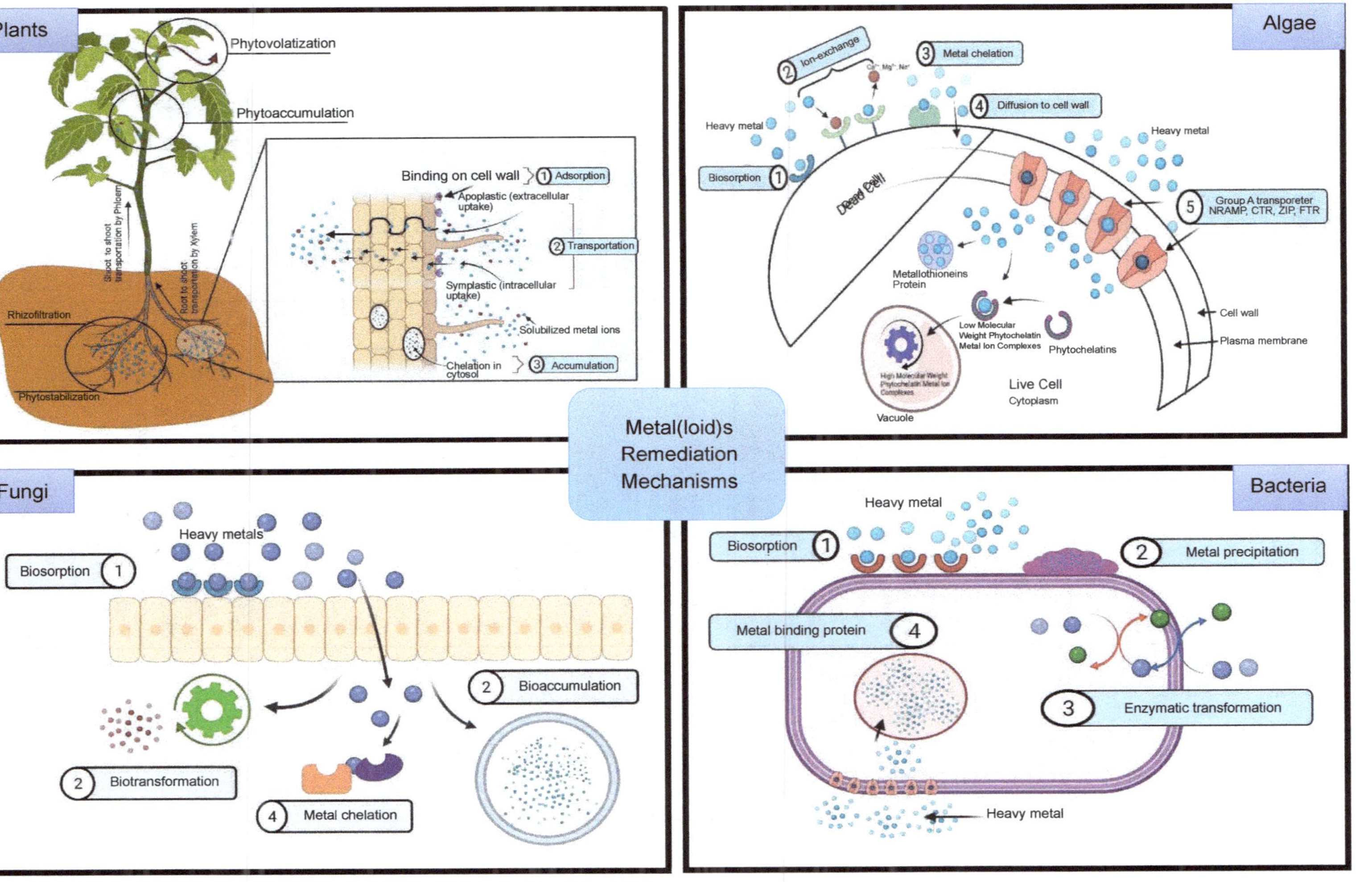

Fig. 1.4. Various strategies used by the organisms for metal(loid)s bioremediation.

Table 1.3. Bioremediation of metal(loid)s using bacteria.

Sr. No.	Metal(loid)s	Organism	References
1	Arsenic	*Alcaligenes faecalis* *Agrobacterium tumefaciens* *Bacillus aryabhattai* *Bacillus arsenooxydans* *Klebsiella pneumonia* *Microbacterium lacticum* *Pseudomonas extremorientalis* *Pseudomonas arsenitoxidans*	Satyapal et al. 2018, Paul and Saha 2019, Sher and Rehman 2019
2	Chromium	*Acinetobacter haemolyticus* *Acidithiobacillus ferrooxidans* *Acidithiobacillus thiooxidans* *Bacillus lentus* *Bacillus cereus* *Bacillus subtilis* *Escherichia coli* *Micrococcus roseus* *Enterobacter aerogenes* *Enterobacter cloaceae* *Streptococcus species* *Pseudomonas aeruginosa* *Pseudomonas putida* *Staphylococcus aureus* *Staphylococcus xylosus* *Escherichia coli* *Acidithiobacillus ferrooxidans*	Iyer et al 2004, Ziagova et al. 2007, Naik et al. 2011, Neha et al. 2013, Oves et al. 2013, Balamurugan et al. 2014, Nayan et al. 2018, Zakaria et al. 2018, Anusha and Natarajan 2020, Tarekegn et al. 2020, Wang et al. 2010
3	Copper	*Acidithiobacillus ferrooxidans* *Acinetobacter junii* *Bacillus thuringiensis* *Bacillus subtilis* *Bacillus cereus* *Bacillus lentus* *Bacillus tequilensis* *Bacillus endophyticus* *Bacillus pumilus* *Escherichia coli* *Micrococcus roseus* *Micrococcus luteus* *Microbacterium arborescens* *Enterobacter aerogenes* *Pseudomonas aeruginosa* *Pseudomonas putida* *Paenibacillus polymyxa* *Staphylococcus aureus* *Sinorhizobium meliloti* *Sulfolobus solfataricus* *Enterobacteria cloacae*	Chang et al. 1997, Daughney and Fein 1998, Pardo et al. 2003, Acosta et al. 2005, Wang et al. 2010, Puyen et al. 2012, Hou et al. 2013, Oves et al. 2013, Anusha and Natarajan 2020, Thakare et al. 2021
4	Cadmium	*Acidithiobacillus ferrooxidans* *B. thuringiensis* *Bacillus subtilis* *Geobacillus stereothermophilus* *Geobacillus thermocatenulatus* *Enterobacter cloacae* *Pseudomonas putida* *Pseudomonas aeruginosa* *Staphylococcus xylosus* *Zooloaeramigera*	Kuhnand Pfister 1990, Chang et al. 1997, Wang et al. 1997, Daughney and Fein 1998, Pardo et al. 2003, Hetzer et al. 2006, Shi et al. 2016, Ziagova et al. 2007, Oves et al. 2013, Banerjee et al. 2015, Chellaiah 2018, Wang et al. 2021

...Table 1.3 contd.

...Table 1.3 contd.

Sr. No.	Metal(loid)s	Organism	References
5	Mercury	*Bacillus licheniformis* *Escherichia coli* *Enterobacter cloacae* *Pseudomonas aeruginosa* *Vibrio anguillarum* *Vibrio parahaemolyticus*	Summers and Silver 1972, Deng and Wilson 2001, Golding et al. 2002, Jafari et al. 2015, Tarekegn et al. 2020
6	Nickel	*B. thuringiensis* *Bacillus subtilis* *Cloacibacterium normanense* *Enterobacter cloacae* *Pseudomonas aeruginosa* *Pseudomonas fluorescens* *Acidithiobacillus ferrooxidans*	Lopez et al. 2001, Zaidi et al. 2006, Gabr et al. 2008, Oves et al. 2013, Banerjee et al. 2015, Nouha et al. 2016, Wang et al. 2021
7	Lead	*Agrobacterium fabrum* *Bacillus subtilis* *Bacilluscereus* *Micrococcus luteus* *Methylobacterium organophilum* *Streptomyces rimosus* *Enterobacter cloacae* *Pseudomonas aeruginosa* *Pseudomonas alcaligenes* *Pseudomonas nitroreducens* *Pseudomonas putida* *Paenibacillus polymyxa* *Rhizobium Radiobacter* *Acidithiobacillus ferrooxidans* *Streptomyces rimosus*	Chang et al. 1997, Daughney and Fein 1998, Pardo et al. 2003, Selatnia et al. 2004, Gabr et al. 2008, Puyen et al. 2012, Wang et al. 2013, Wasi et al. 2013, Hassiba et al. 2014, Banerjee et al. 2015, Tiwari et al. 2017, Anusha and Natarajan 2020, Velez et al. 2021, Wang et al. 2013
8	Selenium	*Bacillus cereus* *Bacillus mycoides* *Pseudomonas putida* *Pseudomonas moraviensis*	Staicu et al. 2015, Paul and Saha 2019

bioremediation strategies. Some of the *in situ* bioremediation strategies are biosparging, bioventing, phytoremediation, etc. (Azubuike et al. 2016).

1.7 Bacterial Remediation

Among all classes of microorganisms, bacterial species are one of the most investigated type of microorganism for detoxification of metalloids. Several bacterial species are applied in the process of bioremediation against various metalloids as presented in Table 1.3. Bacteria employ various mechanisms such as biosorption, bioaccumulation, biotransformation, bioleaching, biomineralization for successful bioremediation of metalloids (Sher and Rehman 2019). Some bacterial species have the ability to transform the oxidation state of metal(loid)s, whereas some induce the solubility and immobility of metal(loid)s making them less toxic. Some of the bacterial species used for bioremediation of metal(loid)s are *Bacillus cereus, Bacillus mycoides, Bacillus lentus, Bacillus subtilis, Acinetobacter haemolyticus, Acidithiobacillus ferrooxidans, Acidithiobacillus thiooxidans, Pseudomonas putida, Pseudomonas moraviensis* (Iyer et al. 2004, Zakaria et al. 2007, Ziagova et al. 2007, Naik et al. 2011, Neha et al. 2013, Oves et al. 2013, Staicu et al. 2015, Paul and Saha 2019, Sher and Rehman 2019). Both gram positive and gram negative bacterial endophytes such as *Bacillus* species, *Pseudomonas* species, *Acinetobacter* species, *Achromobacter* species, etc., play significant role in bioremediation of metal(loid)s (Sharma and Kumar 2021).

1.8 Mycoremediation

The use of fungi in remediation of different pollutants is known as mycoremediation as presented in Table 1.4. Various fungal species are reported for remediation of various metalloids. Some of the fungi applied in bioremediation of metal(loid)s are *Aspergillus penicillioides, Aspergillus flavus, Exophiala pisciphila, Mucor circinelloides, Fusarium equiseti, Gaeumannomyces cylindrosporus, Penicillium aculeatum, Phenarochaete chrysosporium, Pleurotus ostreatus, Alternaria alternate, Aureobasidium pullulans, Mortierella humilis, Trichoderma harzianum, Trichoderma atroviride, Trichoderma reesei, Trichoderma asperellum, Phoma glomerata,* etc. (Garcia-Hernandez et al. 2017, Kidwai and Nehra 2017, Kumar and Dwivedi 2019, Liang et al. 2019, Sabuda et al. 2020, Tarekegn et al. 2020, Khalid et al. 2021). Specific fungi such as *Aspergillus* and *Penicillium* species are applied in various biohydrometallurgical processes. Fungi employ various mechanisms for the bioremediation of metal(loid)s viz. solubulization, biosorption, intracellular and extracellular sequestration, bioaccumulation, bioprecipitation, biotransformation which includes reduction, biomethylation, dealkylation, etc. They produce various organic acids, enzymes, metabolites, siderophores, proteins, metallothioneins, etc., which facilitate the mobilization and biotransformation of metal(loid)s.

1.9 Phycoremediation

The application of different algal species in remediation of different pollutants is known as phycoremediation. Live algal cells have the potential to remove metalloids from water, waste water, etc., through various mechanisms such as biosorption, etc. (Danouche et al. 2021, Wang et al. 2021b) or by using dead algal biomass as biochar (Singh et al. 2021) as exhibited in Table 1.5. Various functional groups, such as amide, carbonyl, carboxylic acid, hydroxyls, etc., contribute in the process of biosorption of metalloids by algal biomass. Some of the algal species used in bioremediation of metal(loid)s are *Chlorella vulgaris, Scenedesmus almeriensis, Hizikia fusiformis Ulva lactuca, Euglena gracilis, Halimedagracilis, Scenedesmus quadricauda, Laminaria digitata,* etc. (Devars et al. 2000, Jayakumar et al. 2014, Ibrahim et al. 2016, Wang et al. 2021b).

1.10 Bioremediation Strategies in Microorganisms

1.10.1 Biosorption

The uptake of different heavy metals by various microorganisms at cellular level is known as biosorption. Biosorption is further classified into two types (i) metabolism-independent biosorption, occurring in the exterior region of the microbial cell, whereas other (ii) metabolism-dependent bioaccumulation, which includes redox reaction, sequestration methods, etc. Both dead and living microbial cells are used in this process. Chemical, physical and biological mechanisms influence the process of bioaccumulation (Garcia-Hernandez et al. 2017, Igiri et al. 2018, Raffa et al. 2021).

1.10.2 Biosequestration

In intracellular sequestration, the metal ions get accumulated in the cytoplasm of microorganisms. Metals present in the cells of the microorganisms interact with the ligands present on the surface. The accumulation of cadmium, copper, etc., in *Pseudomonas putida* takes place through intracellular sequestration. Another type of sequestration is known as extracellular sequestration in which metal(loid)s get accumulated and gets precipitated in the periplasm; *G. sulfurreducens* and *G. metallireducens* are reported to transform chromium (Cr) from highly toxic Cr(VI) to least toxic

Table 1.4. Fungal species used for bioremediation of metal(loid)s.

Sr. No.	Metal(loid)s	Organism	References
1	Arsenic	*Aspergillus flavus* *Penicillium aculeatum* *Piriformospora indica* *Trichoderma harzianum* *Trametes versicolor*	Arriagada et al. 2009, Paul and Saha 2019, Khalid et al. 2021
2	Chromium	*Aspergillus niger* *Aspergillus flavus* *Cladosporium perangustum* *Fusarium solani* *Trichoderma harzianum* *Trichoderma gamsii* *Trichoderma inhamatum* *Rhizopus nigricans* *Termitomycesclypeatus* *Pichia guilliermondii* *Penicillium commune* *Mucor racemosus* *Termitomyces clypeatus* *Saccharomyces cerevisiae* *Fusarium equiseti* *Fusarium oxysporum*	Bai and Abraham 2001, Dursun et al. 2003, Ksheminska et al. 2003, Park et al. 2005, Ahmad et al. 2006, Liu et al. 2007, Morales and Cristiani 2008, Das and Guha 2009, Ramrakhiani et al. 2011, Sen and Dastidar 2011, Kavita and Keharia 2012, Abubacker and Kirthiga 2013, Soumik 2013, Sharma and Malaviya 2016, Garcia-Hernandez et al. 2017, Kumar and Dwivedi 2019, Tarekegn et al. 2020, Khalid et al. 2021, Thakare et al. 2021
3	Copper	*Aspergillus niger* *Aspergillus awamori* *Aspergillus brasiliensis* *Aspergillus versicolor* *Trichoderma viride* *Trichoderma reesei* *Rhizopus oryzae* *Phenarochaete chrysosporium* *Phialocephala fortinii* *Saccharomyces cerevisiae*	Price et al. 2001, Say et al. 2001, Dursun et al. 2003, Kim et al. 2003, Anand et al. 2006, Fu et al. 2012, Nascimento et al. 2019, Khalid et al. 2021
4	Cadmium	*Aspergillus niger* *Aspergillus sydowii* *Exophialapisciphila* *Fusarium oxysporum* *Trichoderma atroviride* *Trichoderma harzianum,* *Trichoderma reesei* *Saccharomyces cerevisiae* *Penicillium janthinellum* *Piriformospora indica* *Pleurotusostreatus*	Kim et al. 2003, Park et al. 2003, Ahmad et al. 2006, Cao et al. 2008, Freitas et al. 2011, Tay et al. 2011, Yaghoubian et al. 2019, Zhang et al. 2019, Khalid et al. 2021
5	Mercury	*Candida parapsilosis* *Trametes versicolor* *Pleurotus sajur-caju*	Arica et al. 2004, Tarekegn et al. 2020
6	Nickel	*Aspergillus niger* *Aspergillus versicolor* *Aspergillus flavus* *Trichoderma atroviride* *Pleurotusostreatus* *Phanerochaete chrysosporium*	Ahmad et al. 2006, Cao et al. 2008, Noormohamadi et al. 2019, Tarekegn et al. 2020, Khalid et al. 2021, Ozdemir et al. 2021, Thakare et al. 2021

...Table 1.4 contd.

...Table 1.4 contd.

Sr. No.	Metal(loid)s	Organism	References
7	Lead	*Aspergillus penicillioides* *Aspergillus flavus* *Exophiala pisciphila* *Mucor circinelloides* *Fusarium equiseti* *Gaeumannomyces cylindrosporus* *Penicillium aculeatum* *Phenarochaete chrysosporium* *Pleurotus ostreatus* *Paecilomyces lilacinus*	Say et al. 2001, Sun et al. 2017, Khalid et al. 2021, Ozdemir et al. 2021, Paria et al. 2021, Thakare et al. 2021 Wang et al. 2021
8	Selenium	*Alternaria alternate* *Aureobasidium pullulan* *Mortierella humilis* *Trichoderma harzianum* *Phoma glomerata*	Liang et al. 2019, Sabuda et al. 2020

Table 1.5. Algal species used in bioremediation of metal(loid)s.

Sr. No.	Metal(loid)s	Organism	References
1	Arsenic	*Chlamydomonas reinhardtii* *Chlorella vulgaris* *Scenedesmus almeriensis*	Saavedra et al. 2018, Goswami et al. 2021
2	Chromium	*Chlorella vulgaris* *Halimeda gracilis* *Scenedesmus quadricauda* *Laminaria digitata* *Ulva lactuca*	Dittert et al. 2014, Jayakumar et al. 2014, Ibrahim et al. 2016, Daneshvar et al. 2019, Goswami et al. 2021
3	Copper	*Arthrospira platensis* *Chlorella pyrenoidosa* *Chlorella vulgaris* *Chlorella sorokiniana* *Chlamydomonas reinhardtii* *Codium fragile* *Eucheuma denticulatum* *Hizikia fusiformis* *Nannochloropsis oculata* *Green gracilaria* *Scenedesmus acuminatus* *Scenedesmus incrassatulus* *Ulva lactuca*	Pena-Castro et al. 2004, Abboud and Wilkinson 2013, Ibrahim et al. 2016, Rahman and Sathasivam 2016, Hamed 2017, Moreira et al. 2019, Martınez-Macias et al. 2019, Piccini et al. 2019, Goswami et al. 2021, Wang et al. 2021a
4	Cadmium	*Chlorella vulgaris* *Arthrospira platensis* *Chlamydomonas reinhardtii* *Codium fragile* *Green gracilaria* *Hizikia fusiformis* *Ulva lactuca*	Abboud and Wilkinson 2013, Ibrahim et al. 2016, Piccini et al. 2019, Ibuot et al. 2020, Goswami et al. 2021, Wang et al. 2021a
5	Nickel	*Chlorella vulgaris* *Arthrospira platensis* *Codium fragile* *Green gracilaria* *Hizikia fusiformis*	Piccini et al. 2019, Goswami et al. 2021
6	Lead	*Pseudochlorococcum typicum* *Chlamydomonas reinhardtii* *Eucheuma denticulatum* *Codium fragile* *Green gracilaria,* *Hizikia fusiformis* *Ulva lactuca*	Shanab et al. 2012, Abboud and Wilkinson 2013, Ibrahim et al. 2016, Rahman and Sathasivam 2016,
7	Selenium	*Chlorella vulgaris*	Wang et al. 2021b
8	Mercury	*Euglena gracilis*	Devars et al. 2000

Cr(III) species (Igiri et al. 2018). Cr(VI) is reduced to (III) by the reducing mechanisms present in fungal cells (Garcia-Hernandez et al. 2017).

1.10.3 Biomethylation

Methylalation increases the lipophilicity of the metal(loid)s ions, thereby helping in enhancing the permeability through cell membranes. Methylated compounds such as mercury (II) can be biomethylated by *Escherichia* spp., *Bacillus* spp., and *Clostridium* spp. into methyl mercury in gaseous form. Biomethylation by microorganisms helps in remediation of selenium (Se) to volatile dimethyl selenide and arsenic As(V) into methylated arsenic compounds viz. dimethyl arsenic acid (DMAV), Monomethyl arsenic acid (MMA), Trimethyl arsine (TMA), etc. (Tripathi et al. 2017) as well as lead (Pb) to dimethyl lead (Igiri et al. 2018).

1.10.4 Bioreduction

The cells of microorganisms have the potential to change from one oxidation state to another, reducing the toxicity of metal(loid)s. Microorganisms, mainly bacteria and fungi, utilize different metals and metal(loid)s as electron donors or acceptors for the energy. In case of bacteria, metals in oxidized forms accept the electrons in anaerobic respiration. The reduction of metal ions through an enzymatic activity transforms the metal(loid)s into less toxic form of metal(loid)s such as in case of mercury, chromium, arsenic, etc. (Igiri et al. 2018, Sher and Rehman 2019).

1.10.5 Bioleaching

In the process of bioleaching, the insoluble pollutants such as metal(loid)s are converted into a solubulized form by microbial activity. Bacteria and fungi secrete specialised metabolites such as lipids, lipopeptides, organic acids, i.e., citric acid, oxalic acid, gloconic acid, etc., for enhancing the solubility, which results in lowering of bioavalibility and toxicity of metal(loids)s. Bioleaching of metals such as arsenic, chromium, lead, cadmium, copper, nickel, etc., was reported by the application of bacterial species *Acidithiobacillus thiooxidans, Acidithiobacillus ferrooxidans, Sulfobacillus acidophilus* and fungal species such as *Aspergillus niger*, *Penicillium* species, etc. (Zheng et al. 2015, Igiri et al. 2018, Sher et al. 2019, Raffa et al. 2021). Bioleaching is comparatively more efficient than chemical leaching but requires more time.

1.11 Phytoremediation

Phytoremediation is a plant-based detoxification mechanism utilizing plants's metabolic activities in the polluted environment to remediate various types of pollutants including metal(loid)s (Mosa et al. 2016, Sher and Rehman 2019). Plants are grouped in three categories based on their resistance mechanisms against metal(loid)s, i.e., metal excluder plants, metal indicator plants and hyperacuumulator plants. Phytoremediation is an energy driven process, which utilizes solar energy and innate metabolic processes in plants along with their symbiotic associations with different microorganisms such as endophytes for the sustainable detoxification of various types of pollutants including metal(loids)s (Sharma and Manchanda 2015, Mosa et al. 2016, Muthusaravanan et al. 2018, Sher and Rehman 2019, Shah and Daverey 2020, Shikha and Singh 2021, Guerra Sierra et al. 2021, Kaur and Roy 2021, Raffa et al. 2021, Thakare et al. 2021).

Diverse plant species including fern species having potential for removal or transformation of toxic metalliods are identified and applied as per their potential to remediate different metal(loid)s and other types of pollutants (Table 1.6, Table 1.7). Native plants adapted to the local environmental conditions of the area develop roots helping their growth and development, etc. Plants such as

Amaranthus, Indian mustard, Sunflower, Alfa alfa, Cowpea, Clover, etc., have been reported for their potential role in phytoremediation (Mosa et al. 2016, Shah and Daverey 2020, Kaur and Roy 2021, Thakare et al. 2021).

Phytoremediation is broadly classified into two categories: the first category is direct phytoremediation in which the absorption of various pollutants takes place through roots which are translocated to the shoot of the plants, whereas the second category is phytobial remediation, in which plant species secrete diverse metabolites and enzymes which induce the microbial growth for accumulation or coprecipitation of various pollutants (Muthusaravanan et al. 2018, Guerra Sierra et al. 2021).

Phytoremediation uses the application of specific plants known as hyperaccumulator plants along with the rhizospheric microorganisms for the transfer and degradation of various pollutants in the environment. Based on soil properties, types of pollutants, and plant species, different types of phytoremediation strategies are applied such as hydraulic barrier, phytoextraction, phytovolatilization, phytodegradation, phytofiltration and phytostabilization (Sharma et al. 2014, Shah and Daverey 2020, Guerra Sierra et al. 2021). Plants are natural organisms governed by various limiting factors such as certain abiotic factors of the local environment which adversely affect their metabolic activites responsible for bioremediation.

1.11.1 Phytostabilization

Phytostabilization or phytosequestration refers to the accumulation of metallic pollutants in the immobilized form in the roots of the plant, after the complex formation followed by the precipitation in roots. Pollutants are immobilized by the release of various chelating substances such as organic acids to form complexes with the metals reducing their bioavalibility. Absorption, adsorption, and accumulation also take place in roots by vacuole sequestration or cellular binding preventing the metal ions leaching into the groundwater with no translocation of pollutants. *Brachiaria decumbens* plants, when grown in heavy metal-contaminated soil, accumulates metal(loid)s in roots and are popularly used in phytostabilization, with better results in organically rich soils (Sharma et al. 2014, Mosa et al. 2016, Muthusaravanan et al. 2018, Shah and Daverey 2020)

1.11.2 Phytoextraction

Phytoextraction involves the role of roots of plant to extract the pollutants such as arsenic, cadmium, etc., and transfer them to the upper region of the plant. Later on, the plants are harvested followed by safe disposal of heavy metal enriched plants. Hyperaccumulator plant species, specially trees, are used in this technique (Sharma et al. 2014, Mosa et al. 2016, Muthusaravanan et al. 2018).

1.11.3 Phytovolatilization

Phytovolatilization is a type of phytotansformation in which the plants volatilize various metal(loid)s, i.e., mercury, arsenic, selenium, etc. Plants take various volatile contaminants through roots and transform them, followed by volatilizing them in atmosphere through shoots, leaves, etc. Chinese brake (*Pteris vittata*) extracts arsenic from soil in the elemental form. Further, absorbed metal gets converted into gaseous form by the biological processes and released into the atmosphere (Sharma et al. 2014, Mosa et al. 2016, Muthusaravanan et al. 2018, Shah and Daverey 2020).

Selenium is reported to be removed from waste by phytovolatilization. Inorganic Se is converted to the volatile form, i.e., dimethyl selenide (DMSe) by enzymatic activity of plant. Dimethyl diselenide (DMDSe), dimethyl selenone, dimethyl selenylsulfide, etc., are the volatile forms of selenium released by plants (Sharma et al. 2014). Mercury is reported to be remediated by phytovolatilization. The methylated form of Hg, i.e., MeHg gets biomagnified (Kumar et al.

2017, Liu et al. 2019). Methyl mercury gets absorbed through roots; further, the vascular system of plant translocates them to leaves for purging through transpiration. Enzymes help in transformation of metal(loid)s into volatile forms, like in case of selenium phytovolatilization, an enzyme S-methyltransferase catalyzes the methylation reaction related to transformation (Sharma et al. 2014, Kumar et al. 2017).

1.11.4 Phytotransformation

In the process of phytotransformation, plants convert metal(oids) to another chemical form by absorbing and metabolizing them, like in the case of chromium, using halophytic esturine plant *Halimione portulacoides*, highly toxic Cr(VI) was reduced to less toxic Cr(III) (Duarte et al. 2012). Plant mediates the biotransformation of the metal ions through the secretion of various enzymes at the cellular level (Shah and Daverey 2020).

1.11.5 Phytofiltration

Phytofiltration is a strategy which uses plant roots for absorbtion and precipitation of hazardous contaminants including heavy metals in various aquatic ecosytems. The pollutants get adsorbed on the surface of the roots. This strategy is basically applied in wastewater treatment. Plants with dense roots are applied in phytofiltration which accumulate high amount of contaminant in roots. Various plants species are applied in phytofiltration. Floating plants are efficient in phytofiltration, especially metal(loid)s in aquatic ecosystems (Sharma et al. 2014, Mosa et al. 2016, Muthusaravanan et al. 2018, Shah and Daverey 2020).

1.11.6 Hydraulic Barriers

Hydraulic barriers or hydraulic control is a process that uses hydrophilic plants or tree species that can uptake a large volume of water influencing the movement of ground water along with pollutants. The ground water resources are contaminated by leaching of heavy metal through soil or water. Use of deep-rooted tree species, having the capacity of abstracting polluted underground water, is recommended in this process. Roots play a significant role in helping the plant to avert groundwater contamination (Muthusaravanan et al. 2018).

1.12 Nanobioremediation: A Novel Strategy

Nanotechnology is an emerging concept of molecular biotechnology having economic viablility for the sustainable management of polluted environment with various applications at nano-scale by utilizing nanomaterials for environmental management (Kaur and Roy 2021). Nanoparticles are developed from different sources including organisms applied in remediation of polluted environment. Nanomaterials are of various types such as nanocrystals, nanoparticles, nanopowder, nanomembrane, nanotubes, etc. Nanoparticles are reported to enhance the bioavailability of toxic meta(loid)s due to their smaller size, enhancing the surface area and have affinity for toxic metal(loid)s (Rai et al. 2008). Nanomaterials like quantum dots have the ability to bind toxic metals such as lead, copper, etc. Nanomaterials are widely applied to enhance the efficiency with reduction of cost as in the case of water purification (Kaur and Roy 2021). Bionanomaterials have been developed from living organisms, i.e., plants, algae, bacteria, fungi, etc. Bionanoparticles are considered to be more effective than metallic nanoparticles. Nano-bioremediation is a promising strategy for sustainable management of various metal(loid)s by the application of bionanoparticles. Bionanoparticles are considered to be stable in nature and are applied in various detoxification processes (Yadav et al. 2017, Akhtar et al. 2020).

Table 1.6. Plant species used for phytoioremediation of meta(loid)s.

Sr. No.	Plant	Metal(loid)s	References
1	*Acacia nilotica* L.	Cadmium	Shabir et al. 2018
2	*Agrostis castellana* L	Copper, Lead	Muthusaravanan et al. 2018
3	*Amaranthus hybridus* L.	Cadmium, Lead	Guerra Sierra et al. 2021
4	*Amaranthus hypochondriacus* L.	Cadmium	Khalid et al. 2021
5	*Amaranthus spinosus* L.	Copper. Lead, Cadmium, Chromium	Sharma et al. 2014
6	*Amaranthus paniculatus* L.	Nickel	Azubuike et al. 2016
7	*Arundo donax* L.	Arsenic	Muthusaravanan et al. 2018
8		Lead	Zhao et al. 2016
9	*Bidens pilosa*L.	Cadmium	Manori et al. 2021
10	*Brassica campestris* L.	Chromium, Copper	Khalid et al. 2021
11	*Brassica juncea* (L.) Czern	Mercury, Arsenic, Cadmium, Lead Nickel, Selenium, Copper, Chromium	Morenoet al. 2005, Zaidi et al. 2006, Cao et al. 2008, Ko et al. 2008, Chauhan and Rai 2009, Koptsik 2014, Sharma et al. 2014, Kaur et al. 2018, Muthusaravanan et al. 2018, Natasha et al. 2018, Hasanuzzaman et al. 2020, Guerra Sierra et al. 2021
12	*Brassica napus*	Cadmium, Lead	Zhang et al. 2011, Shi et al. 2017, Khalid et al. 2021
13	*Brassica alboglabra*	Cadmium, Lead	Khalid et al. 2021
14	*Brassica oleracia*	Selenium	Natasha et al. 2018
15	*Canna glauca* L.	Arsenic	Muthusaravanan et al. 2018
16	*Cannabis sativa* L.	Nickel	Meers et al. 2005
17	*Carex pendula*	Lead	Azubuike et al. 2016
18	*Clethra barbinervis* L.	Copper, Lead	Shah and Daverey 2020
19	*Colocasia esculenta* L.	Arsenic	Muthusaravanan et al. 2018
20	*Cyperus papyrus* L.	Arsenic	Muthusaravanan et al. 2018
21	*Spinacia oleracea* L.	Cadmium, Chromium, Copper, Nickel	Pandey 2006
22	*Helianthus annuus* L.	Copper, Lead, Nickel, Mercury, Arsenic, Cadmium	Turgut et al. 2004, Sharma et al. 2014, Forte and Mutiti 2017, Ma et al. 2019, Shah and Daverey 2020, Guerra Sierra et al. 2021, Khalid et al. 2021
23	*Helianthus tuberosus* L.	Mercury	Mahar et al. 2016
24	*Miscanthus sinensis*	Arsenic, Copper, Lead, Cadmium	Shah and Daverey 2020

Table 1.6 contd. ...

...Table 1.6 contd.

Sr. No.	Plant	Metal(loid)s	References
25	*Nicotiana tabacum* L.	Cadmium, Copper	Evangelou et al. 2006, Khalid et al. 2021
26	*Ocimumratissimum*	Cadmium	Shah and Daverey 2020
27	*Linum usitatissimum*	Nickel	Khalid et al. 2021
28	*Lycopersicon esculentum* Mill	Mercury	Jagatheeswari et al. 2013
29	*Liriodendron tulipifera*	Mercury	Muthusaravanan et al. 2018
30	*Raphanus sativus* L.	Cadmium, Chromium, Copper, Nickel	Pandey 2006
31	*Lactuca sativa* L.	Copper	Shams et al. 2019
32	*Lemna minor* L.	Chromium, Copper, Lead, Nickel	Dirilgen and Inel 1994, Hurd and Sternberg 2008, Uysal and Taner 2009, Uysal 2013
33	*Lemnagibba* L.	Arsenic, Cadmium, Copper	Mkandawire et al. 2004, Megateli et al. 2009
34	*Panicum virgatum* L.	Cadmium, Lead	Khalid et al. 2021
35	*Zea mays* L.	Chromium, Cadmium, Arsenic, Lead, Selenium	Gupta et al. 2009, Sarma 2011, Chang et al. 2018, Natasha et al. 2018, Wang et al. 2018
36	*Triticum aestivum* L.	Selenium	Natasha et al. 2018, Hasanuzzaman et al. 2020
37	*Oryza sativa* L.	Selenium	Natasha et al. 2018, Hasanuzzaman et al. 2020
38	*Berkheya coddii*	Nickel	Mahar et al 2016
39	*Baccharis trimera*	Copper	Guerra Sierra et al. 2021
40	*Pisum sativum* L.	Lead, Copper	Sharma et al. 2014
41	*Halimione portulacoides*	Chromium	Duarte et al. 2012
42	*Achillea millefolium* L.	Mercury	Wang et al 2012
43	*Ipomoea alpina* L	Copper	Sarma 2011
44	*Horedeum vulgare* L.	Arsenic	Mains et al. 2006
45	*Solanum nigrum* L.	Lead, Cadmium, Copper	Sun et al. 2006, Sun et al. 2017, Shah and Daverey 2020
46	*Stanleya pinnata*	Selenium	Natasha et al. 2018
47	*Salix dasyclados* L.	Cadmium, Lead	Shah and Daverey 2020
48	*Salix fragilis*	Cadmium, Copper	Thakare et al. 2021
49	*Salix viminalis* L.	Cadmium, Copper, Lead	Sharma et al. 2014, Thakare et al. 2021
50	*Vetiveria zizanioides* L.	Lead, Cadmium, Chromium, Copper, Nickel	Chiu et al. 2006, Rotkittikhun et al. 2007, Sharma et al. 2014, Nayak et al. 2018, Shah and Daveray 2020

Table 1.6 contd. ...

...*Table 1.6 contd.*

Sr. No.	Plant	Metal(loid)s	References
51	*Thysanolaena maxima*	Lead	Rotkittikhun et al. 2007
52	*Cicer arietinum* L.	Mercury	Wang et al. 2021
53	*Phragmites australis* L.	Cadmium, Copper, Lead	Chiu et al. 2006, Rocha et al. 2014
54	*Phraghmites communis* L.	Cadmium	Shah and Daveray 2020
55	*Nymphaea spontanea*	Chromium	Choo et al. 2006
56	*Sedum alfredii*	Lead	Huang et al. 2012, Shah and Daveray 2020
57	*Detarium senegalense*	Chromium	Amaku et al. 2021
58	*Astragalus racemosus*	Selenium	White 2015
59	*Astragalus pattersonii*	Selenium	White 2015
60	*Heliconia psittacorum* L.	Cadmium, Lead	Madera-Parra et al. 2014
61	*Hydrocotyle umbellate* L.	Arsenic, Lead, Cadmium, Copper	Thakare et al. 2021
62	*Cecropia peltata* L.	Mercury	Guerra Sierra et al. 2021
63	*Jatropha curcas* L.	Mercury, Cadmium, Lead	Sharma et al. 2014, Guerra Sierra et al. 2021
64	*Juncus effusus*	Lead, Copper, Chromium, Cadmium, Nickel	Thakare et al. 2021
65	*Chrysopogon zizanioides* L.	Lead, Chromium, Cadmium	Guerra Sierra et al. 2021
66	*Ricinus communis* L.	Cadmium	Lu and He 2005
67	*Stanleya pinnata*	Selenium	Staicu et al. 2015
68	*Sesbania drummondii*	Lead	Sharma et al. 2014
69	*Trifolium repens* L.	Lead, Cadmium	Sharma et al. 2014
70	*Vigna radiata* (L.) R. Wilczek	Lead, Nickel	Sharma et al. 2014
71	*Typha latifolia* L.	Copper, Nickel	Yoon et al. 2006, Sharma et al. 2014
72	*Thlaspi caerulescens*	Cadmium	Zhao et al. 2003, Sharma et al. 2014
73	*Wolffia globosa*	Cadmium	Xie et al. 2013
74	*Tagetes erecta*	Copper, Cadmium, Lead	Sinhal et al. 2010
75	*Solanum nigrum* L.	Cadmium	Shi et al. 2016, Khalid et al. 2021
76	*Stanleya pinnata*	Selenium	Harris et al. 2014
77	*Spartina maritima*	Arsenic, Copper, Lead	Azubuike et al. 2016
78	*Eichhorina crassipes*	Cadmium, Copper, Chromium	Azubuike et al. 2016, Zhou et al. 2019
79	*Kyllinga brevifolia*	Lead, Nickel	Thakare et al. 2021
80	*Gratiolabogotensis*	Arsenic, Lead, Cadmium, Copper	Thakare et al. 2021

Nanobioremediation involves both adsorption as well as absorption. In adsorption, the interface between the metal(loid)s and the sorbent occurs at a surface level. On the contrary, in absorption the pollutant gets entered deep in sorbent. Specialised absorption such as chemisorption and physisorption are distinguished because in chemisorption, chemical reaction occurs, whereas in

Table 1.7. Fern species used for phytoremediation of metal(loid)s.

Sr. No.	Metal(loid)s	Fern	References
1	Arsenic	*Athyrium wardii* L. *Pteris biaurita* L. *Pteris. quadriaurita Retz* *Pteris ryukyuensis* L. *Pteris cretica* L. *Pteris vittata* L. *Pteris umbrosa* L.	Zhao et al. 2002, Srivastava et al. 2006, Shoji et al. 2008, Zhao et al. 2016, Muthusaravanan et al. 2018, Lampis et al. 2015 Kanwar et al. 2020,
2	Chromium	*Azolla species,* *Salvinia minima,* *Salvinia rotundifolia* *Pteris vittata* L.	Nichols et al. 2000, Arora et al. 2006, Kanwar et al. 2020,
3	Copper	*Salvinia natans*	Sen and Mondal 1990
4	Cadmium	*Azolla pinnata*	Rai 2008
5	Mercury	*Azolla pinnata,* *Azolla caroliniana*	Rai 2008, Muthusaravanan et al. 2018
6	Nickel	*Pteridium aquilinum*	Kubicka et al. 2015
7	Lead	*Salvinia natans*	Polechońska et al. 2019
8	Selenium	*Azolla caroliniana* *Pteris vittata* L.	Hasanuzzaman et al. 2020

physiosorption physical forces are involved, in immobilizing, sequestering and concentrating the pollutants in diverse forms (Vazquez-Nunez et al. 2020). In recent past, several studies concluded that application of nanotechnology along with phytoremediation is a synergistic strategy for the bioremediation of metal(loid)s (Shikha and Singh 2021).

Bionanoparticles are reported for their potential for removal of metal(loid)s from different ecosystems such as the use of silver based nanoparticles for removal of mercury, cadmium, chromium, copper, etc., in water bodies. Silver based nanoparticles have been developed with leaf extract of *Ficus Benjamina* for effective removal of cadmium. The removal efficiency is increased with the increase in the quantity of nanoparticles. Plant extracts and gums of *Piliostigma thonningii, Azadirachta indica, Araucaria heterophylla, Prosopis chilensis, Prosopis juliflora,* etc., have been reported to be used to develop silver nanoparticles (Samrot et al. 2019, Kaur and Roy 2021, Thakare et al. 2021). The integrated use of nanomaterials helps to avert and reduce the toxicity issues in microorganisms that enhances the efficacy of microorganisms for bioremediation of various pollutants including metal(loid)s.

1.13 Factors Influencing Bioremediation

Plants, fungi, bacteria and algae involved in different strategies of bioremediation are dependent on certain factors. Some of the abiotic factors such as soil, pH, tempreture, humidity, texture, redox potential, xenobiotic chemicals, climate change, etc., and biotic factors such as habitat destruction, negative population interactions viz. competition with exotic species, alien species, allelopathy, onset of disease causing pathogens, etc., determine the fate of bioremediation in different ecosystems.

1.14 Future Prospects

Genetically engineered organisms or genetically modified organisms, which include both microorganisms and plants, can play an important role for high efficiency in detoxification of metal(loids) present in different environments. Genetic engineering alters the properties of organisms including both plants as well as microorganisms to gain desirable traits, such as fast growth, survival in harsh abiotic conditions, and having less cost. Intensive research is required to investigate various stratagies of phytoremediation at molecular level for the commercial success at large scale. Use of different nanomaterials including bionanoparticles are reported for degradation of metal(loid)s but still limited literature is available regarding the synergetic effect of nanoparticles and bioremediation process. More investigations and studies are required to explore more efficient and less time consuming methods for bioremediation of toxic metal(loid)s. Plants having phytoremediation potential for toxic metal(loid)s may be explored for extraction of economically important metals by the process of phytomining.

1.15 Conclusion

Heavy metals and metalloids are potential polluting agents and pose serious health hazards to different life forms including humans. Toxicity, persistence and bioaccumulation of metal(loid)s are the characteristic features of metal(loid)s causing environmental problems and human health issues. Bioremediation is a sustainable concept to treat, transform and detoxify the metal(loid)s by using plants and microorganisms, but for successful bioremediation both the selection of organisms and identification of pollutants are required. Microorganisms and their interaction with metal(loid)s is a complex process and requires more intensive investigations. Potential organisms are necessary for more effective bioremediation, and novel strategies such as nanobioremediation can be used to treat contaminated sites with modification in existing bioremediation procedures for metal(loid)s remediation. Technological advancements may be integrated with existing bioremediation strategies to eliminate various toxic metal(loid)s.

Future research should focus on various molecular taxonomic tools, i.e., metagenomics, metatranscriptomics, etc., for identification and characterization of various naturally occurring organisms in different ecosystems, symbiotic associations among microorganisms along with the development and application of genetically modified microorganisms and genetically modified plants for application in various strategies of bioremediation. Both in present as well as in future, genetic engineering will provide avenues for improvement in the efficiency of bioremediation by identifying and using their specific target gene having metabolic potential for bioremediation in an organism. Application of endophytes for detoxification is a workable plan in remediation of different metal(loid)s, along with the involvement of plant species cultivated for addressing food security issues. Intensive and focused research may be undertaken in diverse areas for the upgradation of bioremediation technology such as the use of bioflocculant and its viability for the onsite treatment of different waste having metal(loid)s in large scale and provide environmental security to various organisms in different ecosystems.

References

Abboud, P. and K. J. Wilkinson. 2013. Role of metal mixtures (Ca, Cu, and Pb) on Cd bioaccumulation and phytochelatin production by *Chlamydomonas reinhardtii*. Environ. Pollut. 179: 33–38.

Abubacker, M. N. and B. Kirthiga. 2013. Biosorption of *Aspergillus flavus* NCBT 102 biomass in hexavalent chromium. Biosci. Biotechnol. Res. Asia. 10: 767–773.

Acosta, M. P., E. Valdman, S. Leite, F. Battaglini and S. Ruzal. 2005. Biosorption of copper by *Paenibacillus polymyxa* cells and their exopolysaccharide. World J. Microbiol. Biotechnol. 21: 1157–1163.

Ahmad, I., M. I. Ansari and F. Aqil. 2006. Biosorption of Ni, Cr, and Cd by metal tolerant *Aspergillus niger* and *Penicillium* sp. using single and multi-metal solution. Indian J. Exp. Biol. 44: 73–76.

Akhtar, F. Z., K. M. Archana, V. G. Krishnaswamy and R. Rajagopal. 2020. Remediation of heavy metals (Cr, Zn) using physical, chemical and biological methods: a novel approach. SN Appl. Sci. 2: 267.

Ali, H., E. Khan and I. Ikram. 2019. Environmental chemistry and ecotoxicology of hazardous heavy metals: environmental persistence, toxicity, and bioaccumulation. J. Chem. 1–14.

Amaku, J. F., C. M. Ngwu, S. A. Ogundare, K. G. Akpomie, O. I. Edozie and J. Conradie. 2021. Thermodynamics, kinetics and isothermal studies of chromium (VI) biosorption onto *Detariumsenegalense* stem bark extract coated shale and the regeneration potentials. Int. J. Phytoremediat. 1–11. doi:10.1080/15226514.2021.191399.

Anand, P., J. Isar, S. Saran and R. K. Saxena. 2006. Bioaccumulation of copper by *Trichoderma viride*. Bioresour. Technol. 97: 1018–1025.

Anju, M. 2017. Biotechnological strategies for remediation of toxic metal(loid)s from environment. pp. 315–359. *In*: Gahlawat, S., R. Salar, P. Siwach, J. Duhan, S. Kumar and P. Kaur (eds.). Plant Biotechnology: Recent Advancements and Developments. Springer, Singapore. 16.

Anusha, P. and D. Natarajan. 2020. Bioremediation potency of multi metal tolerant native bacteria *Bacillus cereus* isolated from bauxite mines, Kolli hills, Tamilnadu - A lab to land approach. Biocatal. Agric. Biotechnol. 25: 101581.

Arica, M. Y., C. Arpa, B. Kaya, S. Bektas,, A. Denizli and O. Genc. 2004. Comparative biosorption of mercuric ions from aquatic systems by immobilized live and heat inactivated *Trametes versicolor* and *Pleurotussajur-caju*. Bioresour. Technol. 89: 145–154.

Arora, A., S. Saxena and D. K. Sharma. 2006. Tolerance and phytoaccumulation of chromium by three *Azolla* species. World J. of Microbiol. Biotechnol. 22: 97–100.

Arora, N. K. and R. Chauhan. 2021. Heavy metal toxicity and sustainable interventions for their decontamination. Environ. Sustain. 4: 1–3.

Arriagada, C., E. Aranda, I. Sampedro, I. Garcia-Romera and J. A. Ocampo. 2009. Contribution of the saprobic fungi *Trametes ver*sicolor and *Trichoderma harzianum* and the arbuscular mycorrhizal fungi *Glomus deserticola* and *G. claroideum* to arsenic tolerance of Eucalyptus globules. Bioresour. Technol. 100: 6250–6257.

ATSDR. 2012. Toxicological profile for Cadmium. https://www.atsdr .cdc.gov/toxprofles/tp5.pdf. Accessed 28 May 2018.

Azubuike, C. C., C. B. Chickere and G. C. Okpokwasili. 2016. Bioremediation techniques–classification based on site of application: principles, advantages, limitations and prospects. World J. Microbiol. Biotechnol. 32: 180.

Bai, R. S. and T. E. Abraham. 2001. Biosorption of Cr(VI) from aqueous solution by *Rhizopus nigricans*. Bioresour. Technol. 79: 73–81.

Balamurugan, D., C. Udayasooriyan and B. Kamaladevi. 2014. Chromium (VI) reduction by *Pseudomonas putida* and *Bacillus subtilis* isolated from contaminated soils. Int. J. Environ. Sci. 5: 522–529.

Banerjee, G., S. Pandey. A. K. Ray and R. Kumar. 2015. Bioremediation of heavy metals by a novel bacterial strain *Enterobacter cloacae* and its antioxidant enzyme activity, flocculant production, and protein expression in presence of lead, cadmium, and nickel. Water Air Soil Pollut. 226: 91.

Bano A., J. Hussain, A. Akbar, K. Mehmood, M. Anwar, M. S. Hasni, S. Ullah, S. Sajid and I. Ali. 2018. Biosorption of heavy metals by obligate halophilic fungi. Chemosphere. 199: 218–222.

Boskabady, M., N. Marefati, T. Farkhondeh, F. Shakeri, A. Farshbaf and M. H. Boskabady. 2018. The effect of environmental lead exposure on human health and the contribution of inflammatory mechanisms, a review. Environ. Int. 120: 404–420.

Briffa, J., E. Sinagra and R. Blundell. 2020. Heavy metal pollution in the environment and their toxicological effects on humans. Heliyon. 6: e04691.

Budnik, L. T. and C. Ludwine. 2019. Mercury pollution in modern times and its socio-medical consequences. Sci. Total Environ. 654: 720–734.

Evangelou, M. W. H., M. Ebel and A. Schaefer. 2006. Evaluation of the effect of small organic acids on phytoextraction of Cu and Pb from soil with tobacco *Nicotiana tabacum*. Chemosphere 63: 996–1004.

Cao, L., M. Jiang, Z. Zeng, A. Du, H. Tan and Y. Liu. 2008. *Trichoderma atroviride* F6 improves phytoextraction efficiency of mustard (*Brassica juncea* (L.) Coss. var. foliosa Bailey) in Cd, Ni contaminated soils. Chemosphere. 71(9): 1769–1773.

Chang, J. S., R. Law and C. C. Chang. 1997. Biosorption of lead, copper and cadmium by biomass of *Pseudomonas aeruginosa* PU21. Water Res. 31: 1651–1658.

Chang, Q., F. W. Diao, Q. F. Wang, L. Pan, Z. H. Dang and W. Guo. 2018. Effects of arbuscular mycorrhizal symbiosis on growth, nutrient and metal uptake by maize seedlings (*Zea mays* L.) grown in soils spiked with lanthanum and cadmium. Environ. Pollut. 241: 607–615.

Chauhan, J. S. and J. P. N. Rai. 2009. Phytoextraction of soil cadmium and zinc by microbes inoculated Indian mustard (*Brassica juncea*). J. Plant Interact. 4(4): 279–287.

Chellaiah, A. H. 2018. Cadmium (heavy metals) bioremediation by *Pseudomonas aeruginosa*: a minireview. Appl. Water Sci. 8: 154.

Chiu, K. K., Z. H. Ye, and M. H. Wong. 2006. Growth of *Vetiveria zizanioides* and *Phragmities austra*lis on Pb/Zn and Cu mine tailings amended with manure compost and sewage sludge: a greenhouse study. Bioresour. Technol. 97: 158–170.

Choo, T. P., C.K. Lee, K. S. Low and O. Hishamuddin. 2006. Accumulation of chromium (VI) from aqueous solutions using water lilies (*Nymphaea spontanea*). Chemosphere. 62: 961–967.

Costa, M., B. Henriques, J. Pinto, E. Fabre, T. Viana, N. Ferreira, J. Amaral, C. Vale, J. Pinheiro-Torres and E. Pereira. 2020. Influence of salinity and rare earth elements on simultaneous removal of Cd, Cr, Cu, Hg, Ni and Pb from contaminated waters by living macroalgae. Environ. Pollut. 266: 115374.

Cui, D., P. Zhang, H. Li, Z. Zhang, Y. Song and Z. Yang. 2021. The dynamic changes of arsenic biotransformation and bioaccumulation in muscle of freshwater food fish crucian carp during chronic diet borne exposure. J. Environ. Sci. 100: 74–81.

Daneshvar, E., M. Javad and M. Kousha. 2019. Hexavalent chromium removal from water by microalgal-based materials: adsorption, desorption and recovery studies. Bioresour. Technol. 293: 122064.

Danouche, M., N. Ghachtouli and H. Arroussi. 2021. Phycoremediation mechanisms of heavy metals using living green microalgae: physicochemical and molecular approaches for enhancing selectivity and removal capacity. Heliyon. 7: e07609.

Das, S. K. and A. K. Guha. 2009. Biosorption of hexavalent chromium by *Termitomycesclypeatus* biomass: kinetics and transmission electron microscopic study. J. Hazard. Mater. 167: 685–691.

Daughney, C. J. and J. B. Fein. 1998. The effect of ionic strength on the adsorption of H^+, Cd^{2+}, Pb^{2+}, and Cu^{2+} by *Bacillus subtilis* and *Bacillus licheniformis*: a surface complexation. J. Colloid Interface Sci. 198: 53–77.

Deng, X. and D. B. Wilson. 2001. Bioaccumulation of mercury from wastewater by genetically engineered *Escherichia coli*. Appl. Microbiol. Biotechnol. 56: 276–279.

Devars, S., C. Aviles, C. Cervantes and R. Moreno-Sanchez. 2000. Mercury uptake and removal by *Euglena gracilis*. Arch. Microbiol. 174: 175–180.

Dirilgen, N. and Y. Inel. 1994. Effects of zinc and copper on growth and metal accumulation in duckweed, *Lemna minor*. Bull. Environ. Contam. Toxicol. 53: 442–448.

Dittert, I. M., H. de Lima Brandão, F. Pina, E. A. da Silva, S. M. G. U. de Souza, A. A. U. de Souza et al. 2014. Integrated reduction/oxidation reactions and sorption processes for Cr (VI) removal from aqueous solutions using *Laminaria digitata* macro-algae. Chem. Eng. J. 237: 443–454.

Duarte, B., V. Silva and I. Cacador. 2012. Hexavalent chromium reduction, uptake and oxidative biomarkers in *Halimioneportulacoides*. Ecotoxicol. Environ. Saf. 83: 1–7.

Dudek-Adamska, D., T. Lech, T. Konopka and P. Koscielniak. 2021. Nickel content in human internal organs. Biol. Trace Element Res. 199: 2138–2144.

Dursun, A. Y., G. Uslu, Y. Cuci and Z. Aksu. 2003. Bioaccumulation of Cu(II), Fe(II) and Cr(VI) by growing *Aspergillus niger*. Process Biochem. 38: 1647–1651.

Dzionek, A., D. Wojcieszynska and U. Guzik. 2016. Natural carriers in bioremediation: a review. Electronic J. Biotechnol. 23: 28–36.

Elgarahy, A. M., K. Elwakeel, S. H. Mohammad and G. A. Elshoubaky. 2021. A critical review of biosorption of dyes, heavy metals and metalloids from wastewater as an efficient and green process. Clean. Eng. Technol. 4: 100209.

Freitas, A. D. L., G. F. De Moura and M. A. B. De Lima. 2011. Role of the morphology and polyphosphate in *Trichoderma harzianum*related to cadmium removal. Mol. 16(3): 2486–2500.

Forte, J. and S. Mutiti. 2017. Phytoremediation potential of *Helianthus annuus* and *Hydrangea paniculata* in copper and lead contaminated soil. Water Air Soil Pollut. 228: 77.

Fu, Y. Q., Li S., H. Y. Zhu, R. Jiang and L. F. Yin. 2012. Biosorption of copper (II) from aqueous solution by mycelial pellets of *Rhizopus oryzae*. Afr. J. Biotechnol. 11: 1403–1411.

Fu, Z., W. Guo, Z. Dang et al. 2017. Refocusing on nonpriority toxic metals in the aquatic environment in China. Environ. Sci. Technol. 51(6): 3117–3118.

Gabr, R. M., S. H. Hassan and A. M. Shoreit. 2008. Biosorption of lead and nickel by living and non-living cells of *Pseudomonas aeruginosa* ASU 6a. Int. Biodeterior. Biodegrad. 62: 195–203.

Garcia-Hernandez, M. A., J. F. Villarreal-Chiu and M. T. Garza-Gonzalez. 2017. Metallophilic fungi research: an alternative for its use in the bioremediation of hexavalent chromium. Int. J. Environ. Sci. Technol. 14: 2023–2038.

Gebreeyessus, G. D. and F. Zewge. 2018. A review on environmental selenium issues. SN Appl. Sci. 1: 55.

Golding, G. R., C. A. Kelly, R. Sparling, P. C. Loewen, J. W. M. Rudd and T. Barkay. 2002. Evidence for facilitated uptake of Hg(II) by *Vibrio anguillarum* and *Escherichia coli* under anaerobic and aerobic conditions. Limnol. Oceanogr. 47: 967–975.

Goswami, R. K., K. Agarwal, M. P. Shah and P. Verma. 2021. Bioremediation of heavy metals from wastewater: a current perspective on microalgae-based future. Lett. Appl. Microbiol. 13564.

Guerra Sierra, B., J. M. Guerrero and S. Spkolski. 2021. Phytoremediation of heavy metals in tropical soils an overview. Sustain. 13: 2574.

Gupta, D. K., F. T. Nicoloso, M. R. C. Schetinger, L. V. RossatoPereira, L. B. Castro et al. 2009. Antioxidant defense mechanism in hydroponically grown *Zea mays* seedlings under moderate lead stress. J. Hazard. Mater. 172: 479–484.

Gupta, P. and V. Kumar. 2017. Value added phytoremediation of metal stressed soils using phosphate solubilizing microbial consortium. World J. Microbiol. Biotechnol. 33(1): 9 http://dx.doi.org/10.1007/s11274-016-2176-3.

Hamed, S. M. 2017. Ecotoxicology and environmental safety sensitivity of two green microalgae to copper stress: growth, oxidative and antioxidants analyses. Ecotoxicol. Environ. Saf. 144: 19–25.

Hasanuzzaman, M., M. H. M. Bhuyan, A. Raza, B. Hawrylak-Nowak, R. Matraszek-Gawron, K. Nahar et al. 2020. Selenium toxicity in plants and environment: biogeochemistry and remediation possibilities. Plants. 9(12): 1711.

Hassiba, M., A. Naima, K. Yahia and S. Zahra. 2014. Study of lead adsorption from aqueous solutions on agar beads with EPS produced from *Paenibacillus polymyxa*. Chem. Eng. Trans. 38: 31–36.

Halwani, D., M. Jurdi, F. Salem, M. Jafa, N. Amacha, R. Habib and H. R. Dhaini. 2019. Cadmium health risk assessment and anthropogenic sources of pollution in Mountlebanon springs. Exp. Health. 12: 163–178. https://doi.org/10.1007/s12403-019-00301-3.

Harris, J., K. A. Schneberg and E. A. H. Pilon-Smits. 2014. Sulfur–selenium–molybdenum interactions distinguish selenium hyperaccumulator *Stanleya pinnata* from non-hyperaccumulator *Brassica juncea* (Brassicaceae). Planta. 239: 479–491.

Hetzer, A., C. J. Daughney and H. W. Morgan. 2006. Cadmium ion biosorption by thermophilic bacteria *Geobacillus stereothermophilus* and *G. thermocatenulatus*. Appl. Environ. Microbiol. 72: 4020–4027.

Hou, W., Ma, Z., Sun, L., Han, M., Lu, J., Li, Z., O. Mohamad, and Z. Wei. 2013. Extracellular polymeric substances from copper-tolerance *Sinorhizobiummeliloti* immobilize Cu^{2+}. J. Hazard. Mater. 261: 614–620.

Huang, H., D. K. Gupta, S. Tian, X. E. Yang and T. Li. 2012. Lead tolerance and physiological adaptation mechanism in roots of accumulating and nonaccumulating ecotypes of *Sedum alfredii*. Environ. Sci. Pollut. Res. 19: 1640–1651.

Hurd, N. A. and S. P. K. Sternberg. 2008. Bioremoval of aqueous lead using *Lemna minor*. Int. J. Phytoremediat. 10: 278–288.

Ibrahim, W. M., A. F. Hassan and Y. A. Azab. 2016. Biosorption of toxic heavy metals from aqueous solution by *Ulva lactuca* activated carbon. Egypt. J. Basic Appl. Sci. 3(3): 241–249.

Ibuot, A., R. E. Webster, L. E. William and J. K. Pittman. 2020. Increased metal tolerance and bioaccumulation of zinc and cadmium in *Chlamydomonas reinhardtii* expressing a AtHMA4 C - terminal domain protein. Biotechnol. Bioeng. 117: 2996–3005.

Igiri, B. E., S. Okoduwa, G. O. Idoko, E. P. Akabuogu, A. O. Adeyi and I. K. Ejiogu. 2018. Toxicity and bioremediation of heavy metals contaminated ecosystem from tannery wastewater: a review. J. Toxicol. 2568038. doi.org/10.1155/2018/2568038.

Infante, E. F., C. P. Dulfo, G. P. Dicen, Z. Y. Hseu and I. A. Navarrete. 2021. Bioaccumulation and human health risk assessment of chromium and nickel in paddy rice grown in serpentine soils. Environ. Sci. Pollut. Res. 28(14): 17146–17157.

Iyer, A., K. Mody and B. Jha. 2004. Accumulation of hexavalent chromium by an exopolysaccharide producing marine *Enterobacter cloaceae*. Marine Pollut. Bull. 49: 974–977.

Jafari, S. A., S. Cheraghi, M. Mirbakhsh, R. Mirza, and A. Maryamabadi. 2015. Employing response surface methodology for optimization of mercury bioremediation by *Vibrio parahaemolyticus* PG02 in coastal sediments of Bushehr, Iran. CLEAN- Soil Air Water. 43(1): 118–126.

Jagatheeswari, D., P. Vedhanarayanan and P. Ranganathan. 2013. Phytoaccumulation of mercuric chloride polluted soil using tomato plants (*Lycopersicon esculentum* Mill.). Int J. Bot. Res. 3(2): 30–33.

Jaishankar, M., T. Tseten, N. Anbalagan, B. Mathew and K. N. Beeregowda. 2014. Toxicity, mechanism and health effects of some heavy metals. Interdiscip. Toxicol. 7(2): 60–72.

Jayakumar, R., M. Rajasimman and C. Karthikeyan. 2014. Sorption of hexavalent chromium from aqueous solution using marine green algae *Halimedagracilis*: optimization, equilibrium, kinetic, thermodynamic and desorption studies. J. Environ. Chem. Eng. 2(3): 1261–1274.

Kang, C., Y. Kwon and J. So. 2016. Bioremediation of heavy metals by using bacterial mixtures. Ecol. Eng. 89: 64–69.

Kanwar, V. S., A. Sharma, A. L. Srivastav and L. Rani. 2020. Phytoremediation of toxic metals present in soil and water environment: a critical review. Environ. Sci. Pollut. Res. 27: 44835–44860.

Kaur, P., S. Bali, A. Sharma, A. P. Vig and R. Bhardwaj. 2018. Role of earthworms in phytoremediation of cadmium (Cd) by modulating the antioxidative potential of *Brassica juncea* L. Appl. Soil Ecol. 124: 306–316.

Kaur, S. and A. Roy. 2021. Bioremediation of heavy metals from wastewater using Nanomaterials. Environ. Dev. Sustain. 23: 9617–9640.

Kaushal, A. and S. K. Singh. 2017. Removal of heavy metals by nanoadsorbents: a review. J. Environ. Biotechnol. Res. 6(1): 96–104.

Kavita, B. and H. Keharia. 2012. Biosorption Potential of *Trichoderma gamsii* biomass for Removal of Cr (VI) from Electroplating Industrial Effluent. Int. J. Chem. Eng. 1–7.

Khalid, M., S. Rahman, D. Hassani, K. Hayat, P. Zhou and N. Hui. 2021. Advances in fungal-assisted phytoremediation of heavy metals: a review. Pedosphere. 31(3): 475–495.

Khan, N. I., G. Owens, D. Bruce and R. Naidu. 2009. Human arsenic exposure and risk assessment at the landscape level: a review. Environ. Geochem. Health. 31(1): 143.

Kidwai, M. K. and M. Nehra. 2017. Biotechnological applications of *Trichoderma* species for environmental and food security. *In*: Gahlawat, S., R. Salar, P. Siwach, J. Duhan, S. Kumar and P. Kaur (eds.). Plant Biotechnology: Recent Advancements and Developments. Singapore. Springer Publication.

Kidwai, M. K. and S. B. Dhull. 2021. Heavy metals induced stress and metabolic responses in fenugreek (*Trigonella foenum-graecum* L.) Plants. *In*: Naeem, M. et al. (eds.). Fenugreek: Biology and Applications. Singapore. Springer Nature, Singapore.

Kim, S. K., B. P. Chun, M.K. Yoon and S. Y. Hyun. 2003. Biosorption of cadmium and copper by *Trichodermareesei* RUT C30. J. Ind. Eng. Chem. 9: 403–406.

Ko, B. G., C. W. N. Anderson, N. S. Bolan, K. Y. Huh and I. Vogeler. 2008. Potential for the phytoremediation of arsenic-contaminated mine tailings in Fiji. Aust. J. Soil Res. 46: 493–501.

Koptsik, G. N. 2014. Problems and prospects concerning the phytoremediation of heavy metal polluted soils: a review. Eurasian Soil Sci. 47(9): 923–939.

Ksheminska, H., A. Jaglarz, D. Fedorovych, L. Babyak, D. Yanovych, P. Kaszycki and H. Koloczek. 2003. Bioremediation of chromium by the yeast *Pichia guilliermondii*: toxicity and accumulation of Cr (III) and Cr (VI) and the influence of riboflavin on Cr tolerance. Microbiol. Res. 158: 59–67.

Kubicka, K., A. Samecka-Cymerman, K. Kolon, P. Kosiba and A. J. Kemper. 2015. Chromium and nickel in *Pteridium aquilinum* from environments with various levels of these metals. Environ. Sci. Pollut. Res. 22: 527–534.

Kuhn, S. P. and R. M. Pfister. 1990. Accumulation of Cadmium by immobilized *Zooloaeramigera*. Indian Microbiol. 115(6): 123–128.

Kumar, B., K. Smita and L. C. Flores. 2017. Plant mediated detoxification of mercury and lead. Arab. J. Chem. 10(2): s2335–s2342.

Kumar, V. and S. Dwivedi. 2019. Hexavalent chromium reduction ability and bioremediation potential of *Aspergillus flavus* CR500 isolated from electroplating wastewater. Chemosphere. 237: 124567.

Lampis, S., C. Santi, A. Ciurli, M. Andreolli and G. Vallini. 2015. Promotion of arsenic phytoextraction efficiency in the fern *Pteris vittata* by the inoculation of As-resistant bacteria: as oil bioremediation perspective. Front. Plant Sci. 6: 80.

Li, L., B. Zhang, L. Li and A. G. L. Brothwick. 2022. Microbial selenate detoxification linked to elemental sulfur oxidation: independent and synergic pathways. J. Hazard. Mater. 422: 126932.

Li, P., D. Karunanidhi, T. Subramani and K. Srinivasamoorthy. 2021. Sources and consequences of groundwater contamination. Arch. Environ. Contam. Toxicol. 80: 1–10.

Liang, X., M. Perez, K. C. Nwoko, P. Egbers, J. Feldmann and G. Gadd. 2019. Fungal formation of selenium and tellurium nanoparticles. Appl. Microbiol. Biotechnol. 103: 7241–7259.

Liu, J., W. Jianxu, N. Yongqiang, Y. Shaochen, W. Pengcong, S. M. Shaheen, X. Feng and J. Rinklebe. 2019. Methylmercury production in a paddy soil and its uptake by rice plants as affected by different geochemical mercury pools. Environ. Int. 129: 461–469.

Liu, T., H. Li, Z. Li, X. Xiao, L. Chen and L. Deng. 2007. Removal of hexavalent chromium by fungal biomass of *Mucor racemosus*: influencing factors and removal mechanism. World J. Microbiol. Biotechnol. 23: 1685–1693.

Lopez, A., N. Larao, J. Priergo and A. Marques. 2001. Effect of pH on the biosorption of nickel and other heavy metals by *Pseudomonas fluorescens* 4F39. J. Ind. Microbiol. Biotechnol. 24: 146–151.

Lu, X. Y. and C. Q. He. 2005. Tolerance uptake and accumulation of cadmium by *Ricinus communis* L. J. Agro-Environ. Sci. 24: 674–677.

Ma, Y., M. Rajkumar, R. S. Oliveira, C. Zhang and H. Freitas. 2019. Potential of plant beneficial bacteria and arbuscular mycorrhizal fungi in phytoremediation of metal-contaminated saline soils. J. Hazard. Mater. 379: 120813.

Madera-Parra, C., E. J. Peña-Salamanca, M. R. Pena, D. P. L. Rousseau and P. N. L. Lens. 2014. Phytoremediation of landfill leachate with *Colocasia esculenta*, *Gynerumsagittatum* and *Heliconia psittacorum* in constructed wetlands. Int. J. Phytoremediat. 17: 16–24.

Mahar, A., P. Wang, A. Ali, M. K. Awasthi, A. H. Lahori, Q. Wang and Z. Zhang. 2016. Challenges and opportunities in the phytoremediation of heavy metals contaminated soils: a review. Ecotoxicol. Environ. Saf. 126: 111–121.

Mains, D., D. Craw, C. Rufaut and C. Smith. 2006. Phytostabilization of gold mine tailings, New Zealand. Part 1: plant establishment in alkaline saline substrate. Int. J. Phytoremediat. 8: 131–147.

Manori, S., V. Shah, V. Soni, K. Dutta and A. Daverey. 2021. Phytoremediation of cadmium-contaminated soil by *Bidens pilosa* L.: impact of pine needle biochar amendment. Environ. Sci. Pollut. Res. 28(42): 58872–58884.

Martınez-Macias, M. D. R., M. A. Correa-Murrieta, Y. Villegas-Peralta, G. E. Devora-Isiordia, J. Alvarez-Sanchez, J. Saldivar-Cabrales and R. G. Sanchez-Duarte. 2019. Uptake of copper from acid mine drainage by the microalgae *Nannochloropsisoculata*. Environ. Sci. Pollut. Res. 26: 6311–6318.

Meers, E., A. Ruttens, M. Hopgood, E. Lesage and F. M. G. Tack. 2005. Potential of *Brassicrapa*, *Cannabis sativa*, *Helianthus annuus* and *Zea mays* for phytoextraction of heavy metals from calcareous dredged sediment derived soils. Chemosphere. 61: 561–572.

Megateli, S., S. Semsari and M. Couderchet. 2009. Toxicity and removal of heavy metals (cadmium, copper, and zinc) by *Lemnagibba*. Ecotoxicol. Environ. Saf. 6: 1774–1780.

Mergler, D. 2021. Ecosystem approaches to mercury and human health: A way toward the future. Ambio. 50: 527–531.

Mkandawire, M., B. Taubert and E. Dudel. 2004. Capacity of *Lemnagibba* L. (Duckweed) for uranium and arsenic phytoremediation in mine tailing waters. Int. J. Phytoremediat. 6(4): 347–362.

Morales, B. L. and U. E. Cristiani. 2008. Hexavalent chromium removal by a *Trichoderma inhamatum* fungal strain isolated from tannery effluent. Water Air Soil Pollut. 187: 327–336.

Moreira, V. R., Y. A. R. Lebron, S. J. Freire, L. V. S. Santos, F. Palladino and R. S. Jacob. 2019. Biosorption of copper ions from aqueous solution using *Chlorella pyrenoidosa*: optimization, equilibrium and kinetics studies. Microchem. J. 145: 119–129.

Moreno, F. N., C. W. N. Anderson, R. B. Stewart and B. H. Robinson. 2005. Mercury volatilisation and phytoextraction from base-metal mine tailings. Environ. Pollut. 136: 341–352.

Mosa, K. A., I. Saadoun, K. Kumar, M. Helmy and O. P. Dhankher. 2016. Potential Biotechnological Strategies for the cleanup of heavy metals and metalloids. Front. Plant Sci. 7: 303.

Muthusaravanan, S., N. Sivarajasekar, J. S. Vivek, T. Paramasivam, M. Naushad, J. Prakashraman, V. Gayathri and O. K. Al Douij. 2018. Phytoremediation of heavy metals: mechanisms, methods and enhancements. Environ. Chem. Lett. 16: 1339–1359.

Nagajyoti, P. C., K. D. Lee and T. V. M. Sreekanth. 2010. Heavy metals, occurrence and toxicity for plants: a review. Environ. Chem. Lett. 8(3): 199–216.

Naik, U. C., S. Srivastava and I. S. Thakur. 2011. Isolation and characterization of *Bacillus cereus* IST105 from electroplating effluent for detoxification of hexavalent chromium. Environ. Sci. Pollut. Res. 19: 3005–3014.

Nandi, A. and D. K. Chowdhuri. 2021. Cadmium mediated redox modulation in germline stem cells homeostasis affects reproductive health of *Drosophila* males. J. Hazard. Mater. 402: 123737.

Nascimento, J., J. Oliveira, A. Rizzo and S. Leite. 2019. Biosorption Cu (II) by the yeast *Saccharomyces cerevisiae*. Biotechnol. Rep. 21: e00315.

Natasha, M. Shahid, N. K. Niazi, S. Khalid, B. Murtaza, I. Bibi and M. I. Rashid. 2018. A critical review of selenium biogeochemical behavior in soil-plant system with an inference to human health. Environ. Pollut. 234: 915–934.

Nath, A., S. Samanta, S. Banerjee, A. Danda and S. Hazra. 2021. Threat of arsenic contamination, salinity and water pollution in agricultural practices of Sundarban Delta, India and mitigation strategies. S N Appl. Sci. 3: 560.

Nayak, A. K., Panda, S. S., Basu, A. and N. K. Dhal. 2018. Enhancement of toxic Cr (VI), Fe, and other heavy metals phytoremediation by the synergistic combination of native *Bacillus cereus* strain and *Vetiveriazizanioides* L. Int. J. Phytoremediat. 20: 682–691.

Nayan, A. K., S. S. Panda, A. Basu and N. K. Dhal. 2018. Enhancement of toxic Cr(VI), Fe, and other heavy metals phytoremediation by the synergistic combination of native *Bacillus cereus* strain and *Veltiveria* of phytoremediation. J. of Phytoremediation. 20(7): 682–691.

Neha, S., V. Tuhina and G. Rajeeva. 2013. Detoxification of hexavalent chromium by an indigenous facultative anaerobic *Bacillus cereus* strain isolated from tannery effluent. Afr. J. Biotechnol. 12(10): 1091–1103.

Nichols, P. B., J. D. Couch and S. H. Al Hamdani. 2000. Selected physiological responses of *Salvinia minima* to different chromium concentrations. Aquat. Bot. 68: 313–319.

Noormohamadi, H. R., M. R. Fat'hi, M. Ghaedi and G. R. Ghezelbash. 2019. Potentiality of white-rot fungi in biosorption of nickel and cadmium: modeling optimization and kinetics study. Chemosphere. 216: 124–130.

Nouha, K., R. S. Kumar and R. Tyagi. 2016. Heavy metals removal from wastewater using extracellular polymeric substances produced by *Cloacibacterium normanense* in wastewater sludge supplemented with crude glycerol and study of extracellular polymeric substances extraction by different methods. Bioresour. Technol. 212: 120–129.

Oves, M., S. Khan and A. Zaidi. 2013. Biosorption of heavy metals by *Bacillus thuringiensis* strain OSM29 originating from industrial effluent contaminated north Indian soil. Saudi J. Biol. Sci. 20: 121–129.

Ozdemir, S., Y. M. Serkan and E. Kılınç. 2021. Preconcentrations of Ni(II) and Pb(II) from water and food samples by solid-phase extraction using *Pleurotusostreatus* immobilized iron oxide nanoparticles. Food Chem. 336: 127675.

Ozturk, M., M. Metin, V. Altay, Rouf Ahmad Bhat, M. Ejaz, A. Gu et al. 2022. Arsenic and human health: genotoxicity, epigenomic effects, and cancer signaling. Biol. Trace Elem. Res. 200: 988–1001.

Pandey, S. N. 2006. Accumulation of heavy metals (Cd, Cr, Cu, Ni and Zn) in *Raphanus sativus* L. and *Spinacia oleracea* L. plants irrigated with industrial effluent. J. Environ. Biol. 27: 381–384.

Pardo, R., M. Herguedas, E. Barrado and M. Vega. 2003. Biosorption of cadmium, copper, lead and zinc by inactive biomass of *Pseudomonas putida*. Anal. Bioanal. Chem. 376: 26–32.

Park, J. K., J. W. Lee and J. Y. Jung. 2003. Cadmium uptake capacity of two strains of *Saccharomyces cerevisiae* cells. Enzyme Microb. Technol. 33: 371–378.

Park, D., Y. S. Yun, J. H. Jo and J. M. Park. 2005. Mechanism of hexavalent chromium removal by dead fungal biomass of *Aspergillus niger*. Water Res. 39: 533–540.

Paria, K., S. Pyne and S. K. Chakraborty. 2021. Optimization of heavy metal (lead) remedial activities of fungi *Aspergillus penicillioides* (F12) through extra cellular polymeric substances. Chemosphere. 286: 131874.

Pasha, Q., S. A. Malik, N. Shaheen and M. H. Shah. 2010. Investigation of trace metals in the blood plasma and scalp hair of gastrointestinal cancer patients in comparison with controls. Clin. Chim Acta. 411(7-8): 531–539.

Patel, A. K., A. Singh, N. Das and M. Kumar. 2021. Health risk associated with consumption of arsenic contaminated groundwater in the Ganga and the Brahmaputra floodplain of India. Case Stud. Chem. Environ. Eng. 3: 100103.

Paul, T. and N. C. Saha. 2019. Environmental Arsenic and Selenium contamination and approaches towards Its bioremediation through the exploration of Microbial Adaptations: A Review. Pedosphere. 29(5): 554–568.

Pena-Castro, J. M., F. Martınez-Jeronimo, F. Esparza-Garcıa and R. O. Canizares-Villanueva. 2004. Heavy metals removal by the microalga *Scenedesmus incrassatulus* in continuous cultures. Bioresour. Technol. 94: 219–222.

Piccini, M., S. Raikova, M. J. Allen and C. J. Chuck. 2019. A synergistic use of microalgae and macroalgae for heavy metal bioremediation and bioenergy production through hydrothermal liquefaction. Sustain. Ener. Fuels 3: 292–301.

Polechońska, L., A. Klink and M. Dambiec. 2019. Trace element accumulation in *Salvinia natans* from areas of various land use types. Environ. Sci. Pollut. Res. 26: 30242–30251.

Poonia, T., N. Singh and M. C. Garg. 2021. Contamination of Arsenic, Chromium and Fluoride in the Indian groundwater: a review, metaanalysis and cancer risk assessment. Int. J. Environ. Sci. Technol. 18: 2891–2902. doi.org/10.1007/s13762-020-03043-x.

Price, M. S., J. J. Classen and G. A. Payne. 2001. *Aspergillus niger* absorbs copper and zinc from swine wastewater. Bioresour. Technol. 77: 41–49.

Puyen, Z. M., E. Villagrasa, J. Maldonado, E. Diestra, I. Esteve and A. Sol´e. 2012. Biosorption of lead and copper by heavy-metal tolerant *Micrococcus luteus* DE2008. Bioresour. Technol. 126: 233–237.

Raffa, C. M., F. Chiampo and S. Shanthakumar. 2021. Remediation of metal/metalloid-polluted soils: a short review. Appl. Sci. 11: 4134.

Rahman, M. S. and K. V. Sathasivam. 2016. Heavy metal biosorption potential of a Malaysian Rhodophyte (*Eucheuma denticulatum*) from aqueous solutions. Int. J. Environ. Sci. Technol. 13: 1973–1988.

Rai, P. K. 2008. Phytoremediation of Hg and Cd from industrial effluents using an aquatic free floating macrophyte *Azolla pinnata*. Int. J. Phytoremediat. 10: 430–439.

Ramrakhiani, L., R. Majumder and S. Khowala. 2011. Removal of hexavalent chromium by heat inactivated fungal biomass of *Termitomycesclypeatus*: surface characterization and mechanism of biosorption. Chem. Eng. J. 171: 1060–1068.

Rehman, M., L. Liu, Q. Wang, M. Saleem, S. Basheer, S. Ullah and D. Peng. 2019. Copper environmental toxicology, recent advances, and future outlook: a review. Environ. Sci. Pollut. Res. 26: 18003–18016.

Rocha, A. C. S., C. M. Almeida, M. C. P. Basto and M. T. S. Vasconcelos. 2014. Antioxidant response of *Phragmites australis* to Cu and Cd contamination. Ecotoxicol. Environ. Saf. 109: 152–160.

Rotkittikhun, P., R. Chaiyarat, M. Kruatrachue, P. Pokethitiyook and A. J. M. Baker. 2007. Growth and lead accumulation by the grasses *Vetiveriazizanioides* and *Thysanolaena maxima* in lead-contaminated soil amended with pig manure and fertilizer: a glasshouse study. Chemosphere. 66: 45– 53.

Ruchitha, D., E. Wasiullah, D. Malaviya, K. Pandiyan, U. Singh and A. Sahu. 2015. Bioremediation of heavy metals from soil and aquatic environment: an overview of principles and criteria of fundamental processes. Sustain. 7: 2189–2221.

Saavedra, R., R. Mu~noz, M. E. Taboada, M. Vega and S. Bolado. 2018. Comparative uptake study of arsenic, boron, copper, manganese and zinc from water by different green microalgae. Bioresour. Technol. 263: 49–57.

Sabuda, Mary C., C. E. Rosenfeld, T. DeJournett, K. Schroeder, K. Wuolo-Journey and C. Santelli. 2020. Fungal bioremediation of selenium-contaminated industrial and municipal wastewaters. Front. Microbiol. 11: 2105.

Samrot, A. V., J. L. Angalene, S. Roshini, P. Raji, S. Stefi, R. Preethi and A. Madankumar. 2019. Bioactivity and heavy metal removal using plant gum mediated green synthesized silver nanoparticles. J. Cluster Sci. 30(6): 1599–1610.

Sarma, H. 2011. Metal hyperaccumulation in plants: a review focusing on phytoremediation technology. J. Environ. Sci. Technol. 4(2): 118–138.

Satyapal, G. K., S. K. Mishra, A. Srivastava, R. K. Rajan, K. Prakash, R. Haque and N. Kumar. 2018. Possible bioremediation of arsenic toxicity by isolating indigenous bacteria from the middle Gangetic plain of Bihar, India. Biotechnol. Rep. 17: 117–125.

Say, R., A. Denizil and Y. Arica. 2001. Biosorption of cadmium (II), Lead (II) and copper (II) with filamentous fungus *Phenarochaetechrysosporium*. Bioresour. Technol. 76: 67–70.

Selatnia, A., A. Boukazoula, N. Kechid, M. Z. Bakhti, A. Chergui and Y. Kerchich. 2004. Biosorption of lead (II) from aqueous solution by a bacterial dead *Streptomyces rimosus* biomass. Biochem. Eng. J. 19: 127–135.

Sen, K. A. and N. G. Mondal. 1990. Removal and uptake of copper (II) by *Salvinia natans* from waste water. Water Air Soil Pollut. 49: 1–6.

Sen, M. and M. G. Dastidar. 2011. Biosorption of Cr(VI) by resting cells of *Fusarium solani*. Iran J. Environ. Health Sci. Eng. 8(2): 153–158.

Shabir, R., G. Abbas, M. Saqib, M. Shahid, G. M. Shah, M. Akram, N. K. Niazi, M. A. Naeem, M. Hussain and F. Ashraf. 2018. Cadmium tolerance and phytoremediation potential of acacia (*Acacia nilotica* L.) under salinity stress. Int. J. Phytoremediat. 20: 739–746.

Shah, V. and A. Daveray. 2020. Phytoremediation: A multidisciplinary approach to clean up heavy metal contaminated soil. Environ. Technol. Innov. 12: 100774.

Shams, M., M. Ekinci, M. Turan, A. Dursun, R. Kul and E. Yildirim. 2019. Growth, nutrient uptake and enzyme activity response of lettuce (*Lactuca sativa* L.) to excess copper. Environ. Sustain. 2: 67–73.

Shanab, S., A. Essa, E. Shalaby, S. Shanab, A. Essa and E. Shalaby. 2012. Bioremoval capacity of three heavy metals by some microalgae species (Egyptian Isolates). Plant Sig. Behav. 7: 392–399.

Sharma, P. and S. Kumar. 2021. Bioremediation of heavy metals from industrial effluents by endophytes and their metabolic activity: Recent advances. Bioresour. Technol. 339: 125589.

Sharma, S., B. Singh and V. K. Manchanda. 2015. Phytoremediation: role of terrestrial plants and aquatic macrophytes in the remediation of radionuclides and heavy metal contaminated soil and water. Environ. Sci. Pollut. Res. 22(2): 946–962.

Sharma, S. and P. Malaviya. 2016. Bioremediation of tannery wastewater by chromium resistant novel fungal consortium. Ecol. Eng. 91: 419–425.

Sher, S. and A. Rehman. 2019. Use of heavy metals resistant bacteria—a strategy for arsenic bioremediation. Appl. Microbiol. Biotechnol. 103(15): 6007–6021. doi.org/10.1007/s00253-019-09933-6.

Shi, P., K. Zhu, Y. Zhang and T. Chai. 2016. Growth and cadmium accumulation of *Solanum nigrum* L. seedling were enhanced by heavy metal-tolerant strains of *Pseudomonas aeruginosa*. Water Air Soil Pollut. 227: 459.

Shi, Y. N., H. R. Xie, L. X. Cao, R. D. Zhang, Z. C. Xu, Z. Y. Wang and Z. J. Deng. 2017. Effects of Cd- and Pb-resistant endophytic fungi on growth and phytoextraction of *Brassica napus* in metal-contaminated soils. Environ. Sci. Pollut. Res. 24: 417–426.

Shikha, D. and P. K. Singh. 2021. *In situ* phytoremediation of heavy metal–contaminated soil and groundwater: a green inventive approach. Environ. Sci. Pollut. Res. 28: 4104–4124.

Shoji, R., R. Yajima and Y. Yano. 2008. Arsenic speciation for the phytoremediation by the Chinese brake fern, *Pteris vittata*. J. Environ. Sci. 20: 1463–1468.

Singh, A., R. Sharma, D. Pant and P. Malaviya. 2021. Engineered algal biochar for contaminant remediation and electrochemical applications. Sci. Total Environ. 774: 145676.

Sinhal, V. K., A. Srivastava and V. P. Singh. 2010. EDTA and citric acid mediated phytoextraction of Zn, Cu, Pb and Cd through marigold (*Tagetes erecta*). J. Environ. Biol. 31: 255–259.

Soumik, S. 2013. Hexavalent Chromium (Cr(VI)) removal by live mycelium of a *Trichoderma harzianum* Strain. Mol. Soil Biol. 4: 1–6.

Srivastava, M., L. Q. Ma and J. Santos. 2006. Three new arsenic hyperaccumulating ferns. Sci Total Environ. 364(1–3): 24–31.

Staicu, L. C., C. J. Ackerson, P. Cornelis, L. Ye, R. L. Berendsen, W. J. Hunter et al. 2015. *Pseudomonas moraviensis* sub sp. stanleyae, a bacterial endophyte of hyperaccumulator *Stanleya pinnata*, is capable of efficient selenite reduction to elemental selenium under aerobic conditions. J. Appl. Microbiol. 119: 400–410.

Summers, A. O. and S. Silver. 1972. Mercury resistance in a plasmid-bearing strain of *Escherichia coli*. J. Bacteriol. 112: 1228–1236.

Sun, Hongxia, W. Wu, J. Guo, R. Xiao, F. Jiang, L. Zheng and Z. Guren. 2016. Effects of nickel exposure on testicular function, oxidative stress, and male reproductive dysfunction in Spodoptera *litura Fabricius*. Chemosphere. 148: 178–187.

Sun, L. Q., X. F. Cao, M. Li, X. Zhang, X. Li and Z. J. Cui. 2017. Enhanced bioremediation of lead-contaminated soil by *Solanum nigrum* L. with *Mucor circinelloides*. Environ. Sci. Pollut. Res. 24: 9681–9689.

Sun, R. L. Q.-X. Zhou and C.-X Jin. 2006. Cadmium accumulation in relation to organic acids in leaves of *Solanum nigrum* L. as a newly found cadmium hyperaccumulator. Plant Soil. 285: 125–134.

Tarekegn, M. M., F. Z. Salillih and A. I. Ishetu. 2020. Microbes used as a tool for bioremediation of heavy metal from the environment. Cogent Food Agric. 6: 1783174.

Tay, C. C., H. H. Liew, C. Y. Yin, S. Abdul-Talib, S. Surif, A. A. Suhaimi and S. K. Yong. 2011. Biosorption of cadmium ions using *Pleurotusostreatus*: growth kinetics, isotherm study and biosorption mechanism. Kor. J. Chem. Eng. 28: 825–830.

Taylor, A., J. S. Tsuji, M. R. Garry, M. E. McArdle, W. L. Goodfellow, Jr., W. J. Adams and A. Menzie. 2020. Critical review of exposure and effects: implications for setting regulatory health criteria for ingested copper. Environ. Manag. 65: 131–159.

Thakare, M., H. Sharma, S. Datar, A. Roy, P. Pawar, K. Gupta, S. Pandit and R. Prasad. 2021. Understanding the holistic approach to plant-microbe remediation technologies for removing heavy metals and radionuclides from soil. Curr. Res. Biotechnol. 3: 84–98.

Tiwari, S., A. Hasan and L. M. Pandey. 2017. A novel bio-sorbent comprising encapsulated *Agrobacterium fabrum* (SLAJ731) and iron oxide nanoparticles for removal of crude oil co-contaminant, lead Pb (II). J. Environ. Chem. Eng. 5(1): 442–452.

Tripathi, P., P. Singh, A. Mishra, S. Srivastava, R. Chauhan, S. Awasthi, S. Mishra et al. 2017. Arsenic tolerant *Trichoderma* sp. reduces arsenic induced stress in chickpea (*Cicer arietinum*). Environ. Pollut. 223: 137–145.

Turgut, C., M. Katie Pepe and T. J. Cutright. 2004. The effect of EDTA and citric acid on phytoremediation of Cd, Cr, and Ni from soil using *Helianthus annuus*. Environ. Pollut. 131: 147–154.

Uysal, Y. and F. Taner. 2009. Effect of pH, temperature, and lead concentration on the bioremoval of lead from water using *Lemna minor*. Int. J. Phytoremediat. 11(7): 591–608.

Uysal, Y. 2013. Removal of chromium ions from wastewater by duckweed, *Lemna minor* L. by using a pilot system with continuous flow. J. Hazard. Mater. 263: 486–492.

Valko, M., H. Morris and M. Cronin. 2005. Metals, toxicity and oxidative stress. Curr. Med. Chem. 12: 1161–1208.

Vazquez-Nunez, C. E. Molina-Guerrero, J. M. Pena-Castro, F. Fernandez-Luqueno and M. G. Rosa-Alvarez. 2020. Use of nanotechnology for the bioremediation of contaminants: a review. Processes. 8: 826.

Velez, J. B., J. G. Martinez, J. T. Ospina and S. O. Agudelo. 2021. Bioremediation potential of *Pseudomonas* genus isolates from residual water, capable of tolerating lead through mechanisms of exopolysaccharide production and biosorption. Biotechnol. Rep. 32: e00685.

Vinceti, M., E. Wei, C. Malagoli, M. Bergomi and G. Vivoli. 2001. Adverse health efects of selenium in humans. Rev. Environ. Health. 16(4): 233–251.

Wang, C. L., P. C. Michels, S. C. Dawson, S. Kitisakkul, J. A. Baross, J. D. Keasling and D. S. Clark. 1997. Cadmium removal by a new strain of *Pseudomonas aeruginosa* in aerobic culture. Appl. Environ. Microbiol. 63: 4075–4078.

Wang, H., P. Huang, R. Zhang, X. Feng, Q. Tuing, S. Liu et al. 2021a. Efect of lead exposure from electronic waste on haemoglobin synthesis in children. Int. Arch. Occup. Environ. Health. 94: 911–918.
Wang, J., X. Feng, C. W. Anderson, Y. Xing and L. Shang. 2012. Remediation of mercury contaminated sites—a review. J. Hazard. Mater. 221: 1–18.
Wang, L., J. Yang, Z. Chen, X. Liu and F. Ma. 2013. Biosorption of Pb (II) and Zn (II) by extracellular polymeric substance (Eps) of *Rhizobium Radiobacter*: equilibrium, kinetics and reuse studies. Arch. Environ. Prot. 39: 129–140.
Wang, P. C., T. Mori, K. Komori, M. Sasatsu, K. Toda and H. Ohtaka. 2010. Isolation and characterization of an *Enterobacteria cloacae* strain that reduces hexavalent chromium under anaerobic conditions. Appl. Microbiol. Biotechnol. 55: 1665–1669.
Wang, S., S. Pan, G. M. Shah, Z. Zhang, L. Yang and S. Yang. 2018. Enhancement in arsenic remediation by maize (*Zea mays* L.) using EDTA in combination with arbuscular mycorrhizal fungi. Appl. Ecol. Environ. Res. 16: 5987–5999.
Wang, Y., Z.-J. Wang, J.-C. Huang, C. Zhou, H. Zou, S. He and V. Y. Chen. 2021b. Feasibility of using *Chlorella vulgaris* for the removal of selenium and chromium in water: Competitive interactions with sulfur, physiological effects on algal cells and its resilience after treatment. J. Clean. Prod. 313: 127939.
Wasi, S., S. Tabrez and M. Ahmad. 2013. Use of *Pseudomonas* spp. for the bioremediation of environmental pollutants: a review. Environ. Monit. Assess. 185: 8147–8155.
White, P. J. 2015. Selenium accumulation by plants. Ann. Bot. 1–19.
Wilk, A., E. Kalisinska, D. I. Kosik-Bogacka et al. 2017.Cadmium, lead and mercury concentrations in pathologically altered human kidneys. Environ. Geochem. Health. 39(4): 889–899.
Witkowksa, D., J. Slowik and K. Chilicka. 2021. Heavy metals and human health: possible exposure pathways and the competition for protein binding sites. Mol. 26: 6060.
Xiang, M., Y. Li, J. Yang, K. Lei, Y. Li, F. Li, D. Zheng, X. Fang and Y. Cao. 2021. Heavy metal contamination risk assessment and correlation analysis of heavy metal contents in soil and crops. Environ. Pollut. 278: 116911.
Xie, W., Q. Huang, G. Li, C. Rensing and Y. Zhu. 2013. Cadmium accumulation in the rootless macrophyte *Wolffia globosa* and its potential for phytoremediation. Int. J. Phytoremediat. 15: 385–397.
Yadav, K. K., J. K. Singh, N. Gupta and V. Kumar. 2017. A review of nanobioremediation technologies for environmental cleanup: a novel biological approach. J. Mater. Environ. Sci. 8(2): 740–757.
Yaghoubian, Y., S. A. Siadat, M. R. Telavat, H. Pirdashti and I. Yaghoubia. 2019. Bio-removal of cadmium from aqueous solutions by filamentous fungi: *Trichoderma* spp. and *Piriformospora indica*. Environ. Sci. Pollut. Res. 26: 7863–7872.
Yaman, B. 2020. Health efects of chromium and its concentrations in cereal foods together with sulfur. Exp. Health. 12: 153–161. https://doi.org/10.1007/s12403-019-00298-9.
Yoon, J., X. Cao, Q. Zhou and L. Q. Ma. 2006. Accumulation of Pb, Cu, and Zn in native plants growing on a contaminated Florida site. Sci. Total Environ. 368: 456–464.
Zaidi, S., S. Usmani, B. R. Singh and J. Musarrat. 2006. Significance of *Bacillus subtilis* SJ101 as a bioinoculant for concurrent plant growth promotion and nickel accumulation in *Brassica juncea*. Chemosphere. 64: 991–997.
Zakaria, Z. A., Z. Zakaria, S. Surif and W. A. Ahmad. 2018. Hexavalent chromium reduction by *Acinetobacter haemolyticus* isolated from heavy-metal contaminated wastewater. J. Hazard. Mater. 146(1-2): 30–38.
Zaynab, M., R. Al-yahyai, A. Ameen, Y. Sharif, L. Ali, M. Fatima, K. Khan and S. Li. 2022. Health and environmental effects of heavy metals. J. King Saud Univ. – Sci. 34: 101653.
Zhang, C., Y. Tao, S. Li, J. Tian, T. Ke, S. Wei, P. Wang and L. Chen. 2019. Simultaneous degradation of trichlorfon and removal of Cd (II) by *Aspergillus sydowii* strain PA F- 2. Environ. Sci. Pollut. Res. Int. 26: 26844–26854.
Zhao, F. J., S. J. Dunham and S. P. McGrath. 2002. Arsenic hyperaccumulation by different fern species. New Phytol. 156: 27–31.
Zhao, F. J., E. Lombi and S. P. McGrath. 2003. Assessing the potential for zinc and cadmium phytoremediation with the hyperaccumulator *Thlaspicaerulescens*. Plant Soil. 249: 37–43.
Zhao, L., T. Li, X. Zhang, G. Chen, Z. Zheng and H. Yu. 2016. Pb uptake and phytostabilization potential of the mining ecotype of *Athyrium wardii*(Hook.) grown in Pb-contaminated soil. Clean Soil Air Water. 44: 1184–1190.
Zhang, Y., L. He, Z. Chen, Q. Wang, M. Qian and X. Sheng. 2011. Characterization of ACC deaminase-producing endophytic bacteria isolated from copper-tolerant plants and their potential in promoting the growth and copper accumulation of *Brassica napus*. Chemosphere. 83: 57–62.
Zheng, X., S. Wei, L. Sun, D. A. Jacques, J. Tang, M. Lian, Z. Ji, J. Wang, J. Zhu and Xu, Zixiang. 2015. Bioleaching of heavy metals from contaminated sediments by the *Aspergillusniger* strain SY1. J. Soils Sediments. 15(4): 1029–1038.
Zhitkovich, A. 2011. Chromium in drinking water: sources, metabolism, and cancer risks. Chem. Res. Toxicol. 24: 1617–1629.
Zhou, R., M. Zhang, J. Zhou and J. Wang. 2019. Optimization of biochar preparation from the stem of *Eichhornia crassipes* using response surface methodology on adsorption of Cd^{2+}. Sci. Rep. 9: 17538.
Ziagova, M., G. Dimitriadis, D. Aslanidou, X. Papaioannou, E. Tzannetaki and M. L. Kyriakides. 2007. Comparative study of Cd(II) and Cr(VI) biosorption on *Staphylococcus xylosus* and *Pseudomonas* sp. in single and binary mixtures. Bioresour. Technol. 98: 2859–2865.
Zwolak, I. and H. Zaporowska. 2012. Selenium interactions and toxicity: a review. Cell Biol. Toxicol. 28: 31–46.
Zwolak, I. 2020. The role of selenium in arsenic and cadmium toxicity: an updated review of scientific literature. Biol. Trace Element Res. 193: 44–63.

CHAPTER 2

Bioprecipitation as a Remediation Technique for Metal(loid)s Contamination from Mining Activities

Samantha M. Wilcox,[1] *Catherine N. Mulligan*[1,*] and *Carmen Mihaela Neculita*[2]

2.1 Introduction

Bioprecipitation is a green and sustainable multidisciplinary science and engineering technique. The process requires an understanding of chemistry, biology, environmental engineering, and geotechnical engineering principles. As the world strives to achieve sustainability (the reduction of environmental and social impacts due to anthropogenic activities), the field of environmental engineering should also adapt to consider more green approaches to decrease the consumption of energy and natural resources, as well as reduce pollution of water, soil, and air during remediation processes.

Precipitation in mining is particularly important for wastewater treatment, metal recovery, and metal recycling. Metal resources are being depleted at alarming rates; however, as society moves to a green and sustainable future, their necessity is imperative. Green technology is highly reliant on metals, such that a 30% increase in demand is expected between 2025 and 2050 (Levett et al. 2021). Metal reuse from wastewater precipitation, therefore, offers an innovative solution mitigating the need or extent of additional mining and mineral excavation.

Bioprecipitation is an enhancement to chemical precipitation. The process transitions the state of soluble inorganics in a solution to insoluble metal(loid)s. This is done via metal precipitation, where microorganisms act as a catalyst speeding up the oxidative-reductive reactions taking place. A carbon or energy source is required to act as an electron donor during the oxidation-reduction reactions. The metal ions become trapped in the carbon matrix, precipitate to the solid phase, becoming less mobile in soil and/or groundwater (Jegatheesan et al. 2016). Bioprecipitation can

[1] Dept. Bldg, Civil and Environ. Eng., Concordia University, Montreal, Canada.
[2] Research Institute on Mines and the Environment (RIME), University of Quebec in Abitibi-Témiscamingue, Rouyn-Noranda, Canada.
* Corresponding author: mulligan@civil.concordia.ca

be useful at multiple phases throughout mining and processing; however, it is especially useful during remediation. This is particularly true of reclamation of acidic mine wastewater from mine wastes (e.g., tailings, waste rock, water treatment residuals) and exposed rocks (e.g., open pits, underground tunnels) referred to as acid rock drainage (ARD) or acid mine drainage (AMD), which will be the focus of this chapter.

2.2 Background

Precipitation is a chemical reaction in which a reagent is added to alter the liquid-solid equilibrium of a solution. At supersaturation, the solute concentration is greater than the solubility of the products causing precipitates to form. Thermodynamics, specifically the Gibbs phase rule, determines the number of precipitates capable of forming from a solute (Lewis 2017, Yong et al. 2014). This transition occurs as three separate kinetic mechanisms: nucleation, particle growth, and agglomeration. Nucleation is classified as either primary or secondary depending on the degree of saturation. At high saturation, primary nucleation will occur as either homogenous nucleation (spontaneously from a solution) or heterogeneous nucleation (by the addition of a reagent), whereas at lower saturation, secondary nucleation will occur in the presence of existing crystals (Lewis 2017). The growth stage is the adsorption of crystals on existing crystal surfaces. Again, this is governed by the supersaturation level, which, when high, will create rough crystals and, when low, will create smooth crystals. The particle size distribution of the crystal is governed by the nucleation rate with respect to the growth rate. Therefore, when nucleation is high compared to the growth rate, small grain crystals will form. Agglomeration is the collision of crystal particles, which form interparticle bridges over time. Once again, when nucleation is high, agglomeration is likely high, creating less pure crystals due to fluid entrapment during collision (Lewis 2017, Mihelcic and Zimmerman 2014). Factors impacting precipitation include the soil-water system, the type and concentration of metal(loid)s, the existence of organic and inorganic ligands (Yong et al. 2014), and the presence of co-contaminants. However, the main factor controlling precipitation is pH, whereby alkaline solutions are more likely to precipitate metal(loid)s, and acidic solutions will redissolve precipitates (Yong et al. 2014). In water and wastewater treatment, coagulation (charge neutralization by the addition of a reagent to form precipitates) and flocculation (aggregation and growth of particles) are methods to enhance metal(loid) precipitation and settling (Davis 2010, Mihelcic and Zimmerman 2014). In soil treatment, the precipitate will form in pore water, which will be adsorbed onto soil particle surfaces (Yong et al. 2014).

Bioprecipitation has gained recent traction in science and engineering as a remediation technique for metal(loid) contamination. In 1895, Beijerinck discovered the first signs of sulfate-reducing bacteria (SRB), noting the reduction of sulfate to sulfide in sediments by anaerobic respiration (Hao et al. 2014). However, the study of bioprecipitation dates to 1969, when Tuttle et al. (1969) discovered heterotrophic bacteria downstream from a log-cutting mill. Included in this microflora were dissimilatory SRB, which used wood dust as substrate (food and energy source) to reduce sulfate. In this study and those that followed, the neutrophilic bacteria were found in highly acidic environments ($pH \leq 3$). However, they were not active at reduced pH conditions ($pH \leq 5.5$) in laboratory settings (Johnson and Santos 2020). This was attributed to micro-sites of high pH where bacteria maintained favorable conditions ($pH \leq 6$) for sulfate reduction. In 1999, Sen and Johnson (1999) found the first acidophilic bacteria in mine wastes, meaning some bacterial strains were thriving in these low pH environments due to their ferric iron respiration ability in anaerobic sites. In 2014, there were 40 defined genera of SRB (Table 2.3) (Hao et al. 2014), and as of 2020, there are four acidophilic SRB species (Table 2.4), which are used to improve bioprecipitation as a form of sulfate-based remediation (Johnson and Santos 2020).

2.3 Methodology

2.3.1 Bioprecipitation in Mining

ARD or AMD are typically synonymous terms used to define acidic leachate from mine wastes (often the tailings) contaminating soil, groundwater, and surface water sources. The ARD/AMD is contaminated water, which is usually characterized by low pH with high concentrations of metal(loid)s and sulfates. This contamination derives from sulfide ores that, when extracted from bedrock and then stockpiled, are oxidized by the water instead of air in the geosphere. There are approximately 7800×10^{18} g of sulfur reservoirs in rocks and sediments, which predominately reside in iron sulfides minerals (Sánchez-Andrea et al. 2014), with specifically pyrite (FeS_2) as the most abundant sulfide mineral (Egiebor and Oni 2007).

The chemical reactions for ARD/AMD formation are shown in Eqs. 2.1 and 2.2 (illustrating pyrite oxidation). As pyrite oxidizes, ferrous iron (Fe^{2+}) is formed, which again oxidizes to form ferric iron (Fe^{3+}). The Fe^{3+} ion undergoes hydrolysis and precipitates ferric hydroxide ($Fe(OH)_3$, an orange precipitate), and acidity is formed as an aqueous sulfuric acid (SO_4^{2-} and H^+). The oxidized Fe^{3+} ions will also react with pyrite to speed up the reaction and increase ARD/AMD formation. Other ARD/AMD acidity producing reactions are shown in Table 2.1 (Taylor et al. 2005). It should be noted that the presence of acidophilic microorganisms in ARD/AMD can significantly increase the rate of reaction, acting as a catalyst for the oxidation of Fe^{2+} ions to Fe^{3+} ions and the sulfuric acid produced (Egiebor and Oni 2007). Three categories of microorganisms responsible for accelerating ARD/AMD formation include iron-oxidizing prokaryotes, acid-generating prokaryotes, and carbon-degrading prokaryotes (Johnson and Santos 2020).

ARD/AMD Formation (Part 1)

$$FeS_2 + 3.75O_2 + 3.5H_2O \leftrightarrow Fe(OH)_3 + 2SO_4^{2-} + 4H^+ \quad \text{(Eq. 2.1)}$$

ARD/AMD Formation (Part 2)

$$FeS_2 + 14Fe^{3+} + 8H_2O \leftrightarrow 15Fe^{2+} + 2SO_4^{2-} + 16H^+ \quad \text{(Eq. 2.2)}$$

Table 2.1. ARD/AMD formation from sulfide bearing minerals (Adapted from Taylor et al. 2005).

Mineral	Chemical Formula	Reactions
Sphalerite	ZnS	$ZnS + (3/2)O_2 + H_2O \rightarrow Zn^{2+} + SO_4^{2-} + 2H^+$
Galena	PbS	$PbS + (3/2)O_2 + H_2O \rightarrow Pb^{2+} + SO_4^{2-} + 2H^+$
Chalcocite	Cu_2S	$Cu_2S + (5/2)O_2 + 2H^+ \rightarrow 2Cu^{2+} + SO_4^{2-} + H_2O$
Covellite	CuS	$CuS + 2O_2 \rightarrow Cu^{2+} + SO_4^{2-}$

The rate of these reactions is dependent on the following (Taylor et al. 2005):

- Morphology
- Oxygen content
- Wetting/drying cycles
- Thermal and tectonics
- Presence of bacteria.

Precipitation is a key process involved in the remediation of ARD/AMD contamination. The addition of an alkaline reagent is intended to neutralize and precipitate metals to form insoluble compounds (Egiebor and Oni 2007). The process aims to increase pH and alkalinity, decrease salinity, and decrease sulfate and toxic metal concentrations. The reaction minimizes leachability and transforms the toxic sulfuric acid into a less toxic and mobile form, i.e., calcium sulfate known as gypsum ($CaSO_4$*$2H_2O$), as shown in Eq. 2.3 and 2.4 (Taylor et al. 2005). It is achieved by active or passive treatments, requiring engineered reactors or wetlands, respectively (further defined in Section 2.3.4 below) (Johnson and Hallberg 2005, Skousen et al. 2000). The addition of microorganisms can either ameliorate or alter these processes as a remediation technique. The biological processes involved in bioprecipitation can generate alkalinity to increase pH and precipitate metals, thereby decreasing ARD/AMD contamination (Sahinkaya et al. 2017).

ARD/AMD Mitigating using a Limestone Reagent (Part 1)

$$CaCO_{3(s)} + 2H^+ \rightarrow Ca^{2+} + H_2O + CO_2 \quad \text{(Eq. 2.3)}$$

ARD/AMD Mitigating using a Limestone Reagent (Part 2)

$$CaCO_{3(s)} + SO_4^{2-} + 2H^+ \rightarrow CaSO_{4(s)} + H_2O + CO_2 \quad \text{(Eq. 2.4)}$$

Microorganisms can be used to improve all aspects of the mining industry from exploration, surface/underground mining, mineral processing, and remediation. While this paper is focused on remediation, it is important to note that the metal-microbe interactions discussed are applicable to other mine related issues. For example, bacteria can be used as bioindicators for exploration, biohydrometallurgy for mineral leachate then subsequent precipitation and recovery, bioreactors for lixiviation and recovery, and biocementation for stabilization and/or solidification of mine tailings (Levett et al. 2021). Bioprecipitation plays an integral role within sustainable remediation and showcases innovative mining solutions to longstanding mine-related environmental problems.

2.3.2 Chemical Processes

The two primary mechanisms of bioprecipitation are sulfate-based metal bioprecipitation and iron-based bioprecipitation. These processes rely on oxidative-reductive reactions and the generation of alkalinity to precipitate metals (Johnson and Santos 2020).

Sulfate-based bioprecipitation precipitates metals, reduces sulfate, and increases pH using SRB (Willis and Donati 2017). Equations 2.5 and 2.6 demonstrate a generic sulfate-based metal bioprocess. An organic electron donor (simply expressed as CH_2O, formaldehyde) reacts with sulfate (SO_4^{2-}) from the toxic sulfuric acid of ARD/AMD to form hydrogen sulfide (H_2S), which reacts with metals to form low solubility metal sulfide precipitates. These metal precipitates (MS) can include the following divalent cations: nickel (Ni^{2+}), copper (Cu^{2+}), zinc (Zn^{2+}), ferrous iron (Fe^{2+}), cobalt (Co^{2+}), etc. However, this is an anaerobic process where oxygen should be avoided (Sánchez-Andrea et al. 2016). Equation 2.7 represents the neutralization of H^+ generated from Eq. 2.6 (Sahinkaya et al. 2017).

Sulfate-base metal bioprecipitation (Part 1)

$$2CH_2O + SO_4^{2-} \rightarrow H_2S + 2HCO_3^- \quad \text{(Eq. 2.5)}$$

Sulfate-base metal bioprecipitation (Part 2)

$$H_2S + M^{2+} \rightarrow MS_{(s)} + 2H^+ \quad \text{(Eq. 2.6)}$$

Sulfate-base metal bioprecipitation (Part 3)

$$HCO_3^- + H^+ \rightarrow CO_2 + H_2O \quad \text{(Eq. 2.7)}$$

The chemical process involved in ARD/AMD neutralization using microorganisms is illustrated in Fig. 2.1. The schematic details the four major steps occurring in ARD/AMD formation and mitigation via bioprecipitation. These steps include the oxidation of sulfide bearing minerals to form sulfuric acid, which subsequently solubilizes metals in the sulfidic ores and waste rock. Microorganisms then facilitate the reduction of SO_4^{2-} in sulfuric acid by dissimilatory sulfate reduction or assimilatory sulfate reduction, with the initial being most common. This figure details the specific oxidation reactions occurring during the transformation of biological sulfur. In these reactions, sulfate is the electron acceptor. However, sulfate must be activated for the reduction process to occur. The activation requires the consumption of adenosine triphosphate (ATP, responsible for storing and transferring energy), and the ATP molecules consumed per sulfate depend on the degree of reduction, i.e., two electrons for reduction to sulfite and eight electrons for reduction to hydrogen sulfide.

Further, sulfate as an electron acceptor has a low redox potential at neutral pH, indicating the poor energy yield experienced during these reactions (Johnson and Santos 2020). For remediation processes and to facilitate these reactions, appropriate electron donors or energy sources must be selected, which can be implemented into the influent wastewater *in situ* or *ex situ* (defined in Section 2.3.4). As the last step, the H_2S formed from the reduction of SO_4^{2-} reacts with the solubilized divalent cations to form insoluble sulfide precipitates.

Iron-oxidizing precipitation is prevalent in iron-rich mine wastewaters. The bioprecipitation of iron arises via "biologically controlled mineralization" and "biologically induced mineralization", defined either by the independent mechanism controlled by relevant organisms or the dependent mechanisms revolving around external environmental conditions, respectively (Kiskira et al. 2017). Goethite (-FeOOH), jarosite ($MFe_3(SO_4)_2(OH)_6$), and schwertmannite ($Fe_8O_8(OH)_6(SO_4)$) are prominent iron precipitates, and are shown in Eqs. 2.8, 2.9, and 2.10 (Sahinkaya et al. 2017). Oxidation of ferrous iron (Fe^{2+}) to ferric iron (Fe^{3+}) can lead to the precipitation of several Fe^{3+}

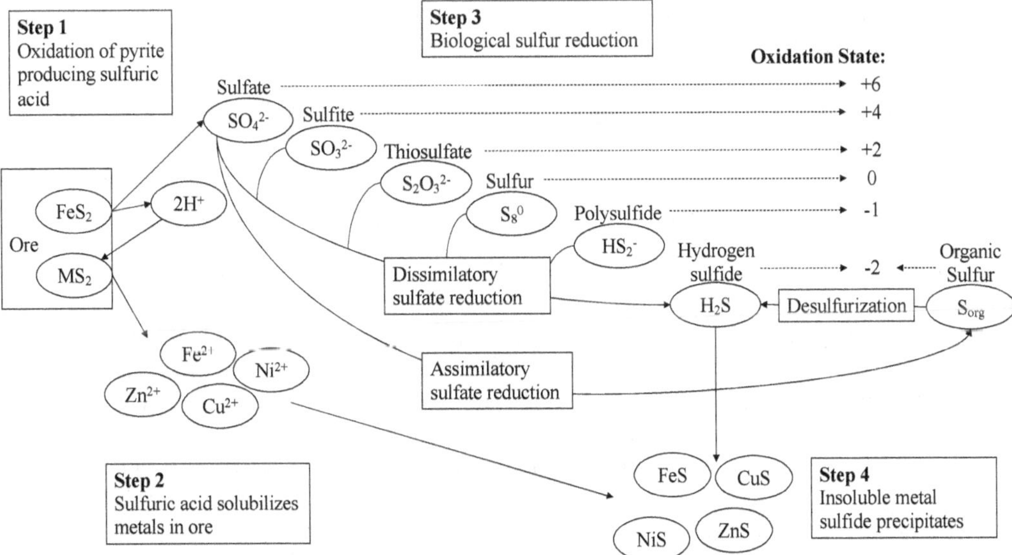

Fig. 2.1. Biological sulfur cycle related to ARD/AMD formation and remediation via bioprecipitation (Adapted from figures in Sánchez-Andrea et al. 2014, 2016).

species, including ferrihydrite ($Fe_5HO_8 \cdot 4H_2O$), goethite (-FeOOH), and hematite (Fe_2O_3). Further, since Fe^{3+} in ARD/AMD acts as a catalyst for the formation of sulfuric acid, the reduction of Fe^{3+} to Fe^{2+} can mitigate the formation of ARD/AMD and enhance co-precipitation with other metals (Eq. 2.6). However, this is based on the solubility of ferric iron (Fe^{3+}) (Kiskira et al. 2017).

Goethite Precipitation

$$Fe^{3+} + 2H_2O \rightarrow \alpha - FeOOH + 3H^{+} \quad \text{(Eq. 2.8)}$$

Jarosite Precipitation

$$3Fe^{3+} + M^{+} + 2HSO_4^{-} + 6H_2O \rightarrow MFe_3\,(SO_4)_2\,(OH)_6 + 8H^{+} \quad \text{(Eq. 2.9)}$$

Schwertmannite Precipitation

$$8Fe^{3+} + SO_4^{2-} + 14H_2O \rightarrow Fe_8\,O_8\,(OH)_6\,(SO_4) + 22\,H^{+} \quad \text{(Eq. 2.10)}$$

Other less common forms of biological precipitation during ARD/AMD removal include the reduction of uranium, the reduction of chromium, and the reduction of arsenic. These metallic elements can be found in high concentrations from mine wastewaters, either alone or together. Microorganisms can facilitate the biological reduction of these toxic metal(loid)s species by reducing them to their less toxic and mobile forms. For example, uranium (VI) (U^{6+}) can be reduced to uranium (IV) (U^{4+}), and chromium (VI) (Cr^{6+}) can be reduced to chromium (III) (Cr^{3+}). This allows for subsequent precipitation with other metals. The following are metal(loid) precipitated species often found after biological reduction and precipitation (Sahinkaya et al. 2017):

- Uranium: carboxyl, phosphate, and amide group precipitation; UO_2
- Chromium: CrO(OH); Cr_2S_3
- Arsenic: A_2S_3; FeAsS

In theory, sulfate-based metal precipitation is advantageous to iron-based and hydr(oxide) bioprecipitation due to the fast reaction rates, the precipitate settling abilities, the lower metal sulfide solubilities, potential precipitate reuse (Lewis 2010), and sludge reduction (Willis and Donati 2017). However, there are discrepancies between scientific and industrial research. In practice, the metal precipitates in the industry are amorphous as opposed to crystalline in structure, meaning their solubility and bioprecipitation ability are less effective. Other issues pertain to the odor and corrosiveness of the process (Lewis 2010).

2.3.3 Biological Processes

The selection of appropriate microorganisms to facilitate the sulfate-based metal bioprecipitation and/or the iron-based bioprecipitation is essential to the process efficacy. This also requires selecting an appropriate energy source to act as an electron donor to enable the chemical process.

SRB are required for sulfate-based metal bioprecipitation. There are two types of SRB: heterotrophic and autotrophic. The heterotrophic organisms metabolize organic compounds to carbon dioxide (CO_2) or acetate, while reducing the sulfur compound to sulfide (Hao et al. 2014). Autotrophic organisms use CO_2 as the energy source and receive electrons from hydrogen (H_2) oxidation. Table 2.2 provides examples of frequently used electron donors for sulfate-based bioprecipitation. The table demonstrates the chemical reactions, free energy, and potential rate of sulfate reduction for each electron donor example. As evident from the table, acetate (CH_3COO^-) is a highly effective electron donor used for sulfate-based bioprecipitation. However, based on the high value of Gibbs free energy change, all represent potential electron donors for oxidation with sulfate (as an electron acceptor). Further, enhanced sulfate reduction has been shown with multiple electron donors as opposed to the use of a single carbon source. An appropriate energy source should be selected based on its ability to completely reduce sulfate, reduce toxic by-products, and minimize cost (Liamleam and Annachhatre 2007).

As mentioned previously, there are 40 genera of SRB capable of sulfate-based bioprecipitation. The SRB are characterized as either incomplete oxidizers, complete oxidizers, or capable of both reactions (Hao et al. 2014). Table 2.3 provides a list of these SRB species and their oxidation ability, as well as their temperature and pH ranges for optimal growth. The *Desulfovibrio* genera is the most documented SRB found in sulfidogenic bioreactors (Kiran et al. 2017). Additional and thorough information pertaining to each of the organisms listed can be found in "Bergey's Manual of Systematics of Archaea and Bacteria". The book provides detailed material regarding the shape, metabolic activity, and electron donor reactions, all of which are useful tools for selecting an appropriate SRB species. In addition to this, Table 2.4 provides specifications for the acidophilic SRB ideal for acidic ARD/AMD wastewater remediation. These acid tolerant SRB species are all from the family *Peptococcaceae* (Johnson and Sánchez-Andrea 2019).

The acidophilic SRB require non-acidic organic substrates as energy sources, such as glycerol and sugars (Willis and Donati 2017). The reactions with glycerol as an electron donor with sulfuric acid are provided in Eqs. 2.11 and 2.12, showcasing the consumption of acidity from the ARD/AMD wastewater. However, the use of incomplete SRB oxidizers with glycerol can be acidity producing and form acetic acid (Eq. 2.13). This can be mitigated by the syntrophy of two acidophiles, which eliminate the production of acetic acid and increase sulfate removal (Johnson and Sánchez-Andrea 2019).

Glycerol as an electron donor for ARD/AMD remediation (1)

$$4C_3H_8O_3 + 7SO_4^{2-} + 14H^+ \rightarrow 12CO_2 + 7H_2S + 16H_2O \qquad \text{(Eq. 2.11)}$$

Glycerol as an electron donor for ARD/AMD remediation (2)

$$4C_3H_8O_3 + 3SO_4^{2-} + 6H^+ \rightarrow 4\,CH_3COOH + 4CO_2 + 3H_2S + 8H_2O \qquad \text{(Eq. 2.12)}$$

Acidity generating reaction with glycerol as an electron donor

$$4C_3H_8O_3 + 3SO_4^{2-} + H_2O \rightarrow 4\,CH_3COO^- + 4\,HCO_3^- + 3HS^- + 5H^+ + 5H_2O \qquad \text{(Eq. 2.13)}$$

A key element in predicting the effectiveness of the biological design, including SRB and electron donor selection, is the ratio of chemical oxygen demand (COD) per sulfate (SO_4^{2-}). This ratio relates the metabolic rate of the SRB in relation to the energy source and electron acceptor.

Table 2.2. Electron Donors for SRB Reactions (Adapted from Hao et al. 2014, Liamleam and Annachhatre 2007).

Electron Donor	**Chemical Formula**	**Reaction**	**Free Energy ΔG°' (kJ/reaction)**	**Sulfate Reduction Rate (SO_4^{2-} g/L/d)**
Formate	CH_2O_2	$4CH_2O_2^-+SO_4^{2-}\ HS^-+4HCO_3^-+3H^+$	–146.7	≤ 29
Methanol	CH_3OH	$4CH_3OH+3SO_4^{2-}\ 3HS^-+4HCO_3^-+4H_2O+H^+$ $2CH_3OH+SO_4^{2-}\ HS^-+2CH_2O_2^-+H^++2H_2O$	–361.7 –108.3	0.4 – 20.5
Ethanol	C_2H_5OH	$2C_2H_5OH+SO_4^{2-}\ 2CH_3COOH^-+HS^-+2H_2O+H^+$	–132.7	0.45 – 21
Lactate	$C_3H_6O_3$	$2C_3H_6O_3^-+SO_4^{2-}\ HS^-+2CH_3COOH^-+2HCO_3^-+H^+$	–159.6	0.36 – 5.76
Glucose	$C_6H_{12}O_6$	$C_6H_{12}O_6+SO_4^{2-}\ HS^-+2CH_3COOH^-+2HCO_3^-+H^+$ $C_6H_{12}O_6+3SO_4^{2-}\ 3HS^-+6HCO_3^-+3H^+$	–358.2 –452.5	0.9 – 2.2
Acetate	CH_3COOH	$CH_3COOH^-+SO_4^{2-}\ HS^-+2HCO_3^-+H^+$	–47.3	≤ 65

Table 2.3. Qualified sulfate reducing bacteria (Adapted from Hao et al. 2014, Whitman and Bergey's Manual Trust 2021).

Oxidation of Organics	**Genera**	**Temperature (°C)**	**pH**
Incomplete Oxidizers	*Desulfovibrio*	30–38	-
	Desulfomicrobium	25–37	7.2
	Desulfohalobium	37–40	-
	Desulfonatronum	30–40	8.0–9.5
	Desulfobotulus	32–34	-
	Desulfocella	Mesophilic	Neutrophilic
	Desulfofaba	7–20	-
	Desulforegula	25–30	-
	Desulfobulbus	25–40	6.6–7.5
	Desulfocapsa	20–30	7.3–7.5
	Desulfofustis	28	-
	Desulforhopalus	18–30	6.8–7.2
	Desulfotalea	10–18	7.2–7.9
	Thermodesulfobacterium	Thermophilic	-
	Thermodesulfovibrio	65	6.8–7.0
	Desulfosporosinus[1]	-	-
Complete Oxidizers	*Desulfothermus*	50–65	6.5
	Desulfobacter	28–34	6.5–7.4
	Desulfobacterium	28	-
	Desulfobacula	28	-
	Desulfococcus	28–35	6.7–7.6
	Desulfofrigus	10–18	7.0–7.5
	Desulfonema	30	-
	Desulfosarcina	28–34	-
	Desulfospira	Mesophilic	-
	Desulfotignum	Mesophilic	-
	Desulfatibacillum	28–35	6.8–7.5
	Desulfarculus	35–39	-
	Desulforhabdus	37	7.2–7.6
	Desulfovirga	35	-
	Desulfobacca	36–40	7.1–7.5
	Desulfacinum	60	-
	Desulfonauticus	45–58	7.0–7.8
	Desulfonatronovibrio	30–37	8.0–10.0
	Thermodesulforhabdus	60	6.9
	Thermodesulfobium[1]	50–55	4.0–6.5
	Archaeoglobus	60–95	5.5–7.5
Complete/Incomplete Oxidizers	*Desulfotomaculum*	-	-
	Desulfomonile	37	-

[1] Included in Table 2.4

Table 2.4. Validated acidophilic sulfate-reducing bacteria (Adapted from information in Johnson and Santos 2020).

Bacteria Type Strain	pH	Temperature	References
Desulfosporosinus acididurans	3.8–7.0	Mesophilic	Sánchez-Andrea et al. 2015
Desulfosporosinus acidiphilus	3.6–5.5	Mesophilic	Alazard et al. 2010
Thermodesulfobium narugense	4.0–6.5	Thermophilic	Mori et al. 2003
Thermodesulfobium acidiphilum	3.7–6.5	37–65°C	Frolov et al. 2017

To be considered a viable design process, a COD/SO_4^{2-} ratio of at least 0.67 is required (Kiran et al. 2017). Preliminary testing should be conducted to assess an optimal inoculum for a specific remediation scenario.

Iron-based metal bioprecipitation can be classified as either Fe^{2+} oxidation or Fe^{3+} reduction. Table 2.5 provides more generalized information relating to the conditions and microorganisms capable of iron-metal precipitation. Anaerobic nitrate-reducing bacteria capable of Fe^{2+} oxidation can play a significant role in ARD/AMD remediation. Pyrite oxidation (Eqs. 2.1 and 2.2), prevalent in ARD/AMD generation, acts as an electron donor during denitrification. Microorganisms oxidize Fe^{2+} to Fe^{3+}, and then Fe^{3+} (hydr)oxides precipitate along with other metal solutes. Equation 2.14 shows the general reaction (Kiskira et al. 2017).

Pyrite as an electron donor for denitrification

$$2FeS_2 + 6NO_3^- + 4H_2O \rightarrow 2Fe(OH)_3 + 3N_{2(g)} + 4SO_4^{2} + 2H^+ \quad \text{(Eq. 2.14)}$$

Table 2.5. Iron-based metal bioprecipitation, organisms, and conditions (Adapted from information in Kiskira et al. 2017).

Precipitation Mechanism	Bacteria Type	Specifications	Energy Source and Electron Donor
Fe^{2+} oxidation	Aerobic neutrophile	- Require oxygen - Neutral pH - Mesophilic	- Organic carbon source
	Aerobic acidophile	- Require oxygen - Acidic pH	- Solar energy - Organic carbon source - Inorganic carbon source
	Anaerobic phototroph	- Require lack of oxygen - Require carbon dioxide and light - Neutral pH	- Light - Fe^{2+}
	Anaerobic nitrate-reducing	- Require lack of oxygen - Require nitrate, perchlorate, or chlorate - Neutral pH	- Acetate - Lactate - Benzoate - Inorganic source (zero-valent Fe, FeS_2, H_2, reduced sulfur compounds, As^{3+}, CO_2, HCO_3^-)
Fe^{3+} reduction	Anaerobic/anoxic acidophile	- Require lack of oxygen - Neutral pH	- Reduced sulfur compounds and hydrogen - Simple fatty acids - Ethanol
	Anaerobic/anoxic neutrophile	- Require lack of oxygen - Acidic pH	- Reduced sulfur compounds and hydrogen - Simple fatty acids - Ethanol

Table 2.6 provides a summary of microorganisms involved in the biological metal reduction of other heavy metals in ARD/AMD from mining operations. These microorganisms facilitate the reduction and subsequent metal(loid) bioprecipitation leading to metal recovery.

Table 2.6. Biological metal reduction and precipitation of metal(loid)s in ARD/AMD (Adapted from information in Sahinkaya et al. 2017).

Metal Reduction	Microorganisms
$U^{6+}U^{4+}$	- Denitrifiers - SRB - Fe^{3+} reducing bacteria - Hyperthermophilic archaea - Thermophilic bacteria - Fermentative bacteria - Acidotolerant bacteria - Myxobacteria
$Cr^{6+}Cr^{3+}$	- SRB - Aerobic bacteria - Denitrifiers
$As^{5+}As^{3+}$	- SRB

2.3.4 Engineered Processes

Engineered systems should be implemented to optimize bioprecipitation as a remediation strategy for mining related soil and groundwater contamination. These systems can be operated *in situ* or *ex situ*. *In situ* treatment occurs at the contaminated site, while for *ex situ* treatment, the soil is extracted from the site and subsequently processed at a separate location/facility. Processes can be further categorized into passive treatment methods or active treatment methods. A passive operation is designed to focus on the enhancing of natural processes occurring and requires minimal initial investment and maintenance, whereas an active operation involves the continual addition of chemical and/or biological reagents (Egiebor and Oni 2007, Taylor et al. 2005).

In situ processes are typically passive in nature and include the fabrication of engineered wetlands. There are three main types of wetlands which include (Johnson and Santos 2020):

- Aerobic: geared toward iron-oxidation and metal (hydr)oxide precipitation. These processes can be acid-producing.
- Anaerobic: geared toward metal sulfide precipitation via iron and sulfate reduction. These processes are alkalinity-producing.
- Reducing and alkalinity-producing systems (RAPS): geared toward sulfate and ferric iron reduction via vertical flow into an organic substrate layer to consume oxygen and create anoxic conditions. These processes are acid-neutralizing.

Other *in situ* passive treatment methods include anoxic ponds, substrate injection into subsurface, infiltration beds, and permeable reactive barriers. These processes are considered more cost-effective both in terms of capital cost and maintenance. They also require less skilled labor. However, these processes are less reliable and are difficult to control. Metal precipitation and metal recovery can be challenging (Kiran et al. 2017).

Ex situ processes often use reactor configurations as an active treatment method. Soil/groundwater contamination is typically excavated from the site and transported to a reactor. The influent is engineered to optimize the temperature, pH, energy source, organisms, etc., to enhance metal precipitation and recovery. This is the main advantage of bioreactor processes, which can be easily manipulated, and more simply controlled. Further, the processes produce high metal precipitation under acidic conditions and a stable, less voluminous sludge (Neculita et al. 2007, 2021). Bioreactors can be operated in different modes and can incorporate different stages. The flow mode refers to how the influent is added to the reactor. For example, it is either added as a batch or applied continuously or semi-continuously. Furthermore, the stages refer to the separation of method processes. A one-stage reactor will allow the oxidation-reduction reactions and metal precipitation to occur together. However, a two-stage system will oxidize/reduce the influent in one reactor and precipitate the metals in another (Sánchez-Andrea et al. 2014).

Currently, there are numerous configurations of bioreactors. The following are some of these reactors and noteworthy characteristics relating to bioprecipitation (Kiran et al. 2017):

- Continuous stirred tank reactor (CSTR): the retention of biomass should be maximized via loading rates and/or reactor size to mitigate washout.
- Gas-lift bioreactor (GLB): gas moves upward from the bottom and has good mixing and mass transfer rates.
- Membrane bioreactor (MBR): membrane can be located at the side stream, immersed in the bioreactor, or extractive. The system has good biomass retention, but backwashing is required.
- Fluidized bed reactor (FBR): carrier material moves in the upflow or downflow (inverse) direction. Fluidized carrier material forms large surface areas of biofilm with high retention and mass transfer rates; however, biomass can be lost due to shear force.
- Anaerobic contact process (ACP): uses *in situ* sedimentation or flocculation to enhance biomass retention and recovery; however, sludge and flocs can break down due to shear force.
- Anaerobic filter reactor (AFR): carrier material moves in the upflow or downflow direction. The system has long sludge retention times and low shear force, but precipitates can cause flow channeling and clogging.
- Up-flow anaerobic sludge blanket (UASB) bioreactor: high sludge quality due to biomass retention and good settling; however, biomass washout can occur.
- Anaerobic hybrid reactor (AHR): a compilation of an AFR and UASB system. The system mitigates clogging and improves sludge removal.
- Anaerobic baffled reactor (ABR): modified USAB reactor with improved sludge retention time and high loading rates.
- Anaerobic fixed-film reactor: carrier material can operate in the upflow or downflow direction. Biomass growth occurs on the surface or within the pores space of support material.
- Anaerobic packed bed (APB) reactor: can operate in the upflow or downflow direction. Slow biomass growth rates where flow channeling and clogging can occur.

Many factors influence the effectiveness of the bioreactor design. As mentioned in Section 2.3.2, temperature and pH play an integral role in the growth of microorganisms. Species are classified as psychrophiles, mesophiles, thermophiles, or hyperthermophiles (with increased temperature ranges, respectively), dependent on their ability to grow under certain temperature conditions. However, pH not only impacts the growth of microorganisms but also impacts metal precipitation. In the instance of sulfate-based bioprecipitation in a bioreactor, SRB typically require a high temperature and pH to prevail over methanogen microorganism species (Kiran et al. 2017). Further, a higher pH is also preferable for metal precipitation. Other noteworthy factors that impact the bioreactor design include salinity, surface area, flocculants, H_2S concentration, hydraulic retention time (HRT), anaerobic conditions, nutrient levels, stirring and flushing conditions, and sludge retention time (Kiran et al. 2017).

Each reactor configuration should be selected based on the influent composition and the optimal metal recovery desired via bioprecipitation. As mentioned in Section 2.2, during precipitation, high nucleation rates will favor the formation of new crystals over the growth of existing crystals, meaning the overall crystal size will be small. In a reactor, it's preferable to have large crystal growth to promote settling and metal recovery. For example, a one-stage reactor with SRB will reach a homogeneous state with a lower sulfide concentration, which will result in the formation of larger crystal size. Not only does this improve metal recovery, but it is a cost-saving and simpler design (Costa et al. 2021, Sánchez-Andrea et al. 2014).

Sulfidogenic bioreactors are specific to sulfate-based bioprecipitation. They are characterized by the usage of SRB species to perform the dissimilatory sulfate reduction of SO_4^{2-} to H_2S using an electron donor (Sánchez-Andrea et al. 2014). Biofilms develop in these sulfidogenic bioreactors through the sorption of bacteria and extracellular polymeric substances (EPS). The biofilm plays an integral role in the success of these reactors and significantly influences the sorption and precipitation capability of metal(loid)s (Kiran et al. 2017); they are even considered nucleation seeds for crystal growth (Bijmans et al. 2009b, Villa-Gomez et al. 2011). The biofilm process entails adhesion to a bio-support material, colonization of SRB species, the proliferation of EPS, subsequent SRB detachment, and regeneration. The biofilm can also influence the surface area of microbial sites, which impacts the efficacy of bioprecipitation, such that a higher surface area means increased sulfate reduction and metal precipitation (Kiran et al. 2017). The location of metal sulfide precipitates is dependent on the sulfide concentrations in the reactor. High sulfide concentrations in the reactor produce metal precipitates outside the biofilm, while low sulfide concentrations in the reactor predominately produce metal precipitates inside the biofilm due to its local supersaturation (Villa-Gomez et al. 2011). Again, high saturation creates high nucleation, which creates small sized crystals with poor settling ability, which affects metal removal and recovery. Agglomeration, dependent on residence time, is a key factor essential to the increase in precipitate size and subsequent settling ability. An optimal saturation level is necessary to ensure that nucleation and agglomeration are maximized (Villa-Gomez et al. 2011, 2012).

Metal removal and metal recovery are essential to the recycling and reuse of precipitated metal(loid)s. Metal removal is typically representative of all metal(loid)s precipitated from the influent (including fines, i.e., small sized crystals), whereas metal recovery is specific to the settled precipitates (Villa-Gomez et al. 2011). Fines (within the reactor and biofilm) are, therefore, difficult on the recovery systems and require energy-intensive processes to extract and separate the metal sulfide precipitates (Costa et al. 2021, Villa-Gomez et al. 2012). Further, the quality of metal(loid) precipitation (i.e., shape, texture, purity) influences metal(loid) removal and recovery from the engineered design, which will likely require further processing for metal recycling and reuse.

2.3.5 Implementation procedures

All operations are specific to the site conditions, land usage requirements, and the legal regulations and requirements of a given region. This means each scenario is different. An assessment of the contaminated site is necessary to evaluate the extent, concentration, type, and fate transport of contamination. During the site assessment, an analysis of soil properties should be conducted to attain information relating to the soil characteristics and composition, grain size and morphology, porosity, hydraulic conductivity, etc. This will allow scientists and engineers to develop a better understanding of the contaminated site and necessary measures to consider for reclamation. Land usage is determined to assess the future land requirement desired by governmental agencies or industry. The land should be repurposed to meet societal needs while obliging to governmental laws and regulations. Every reclamation project should aim to leave the contaminated site at conditions better than or equal to that experienced prior to anthropogenic activities. To evaluate bioprecipitation as a remediation strategy, the appropriate mixture of microorganisms and a suitable

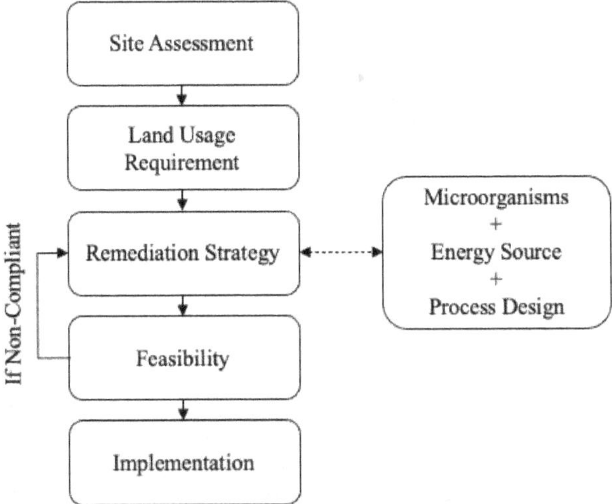

Fig. 2.2. Flowchart of the bioprecipitation process, design, and configuration.

carbon/energy source should be selected for that particular site. Laboratory experiments can be conducted to perform feasibility tests on consortia to predict the optimal performance (following parametric studies) for that specific site. The design and implementation of the strategy should review *in situ* vs. *ex situ* processes. For this step, project requirements such as timeframe and costs are considered. If the project is feasible, it can be implemented at the site. However, if there are any non-compliances, then remediation strategies should be re-evaluated to find a more optimal solution. A summary of the process is depicted in Fig. 2.2.

While each scenario is unique and should be designed accordingly, there are some noteworthy design parameters. Again, the types of microorganisms selected are based on the temperature and pH of the influent and the reaction. Specific species thrive under different conditions (Tables 2.3 and 2.4), and the design should be optimized to accommodate microbial growth. High metal removal is typically achieved at neutral to slightly alkaline pH values and mesophilic temperatures. Frequently for ARD/AMD remediation, there will be an overall increase in pH during the bioprecipitation process. Since ARD/AMD is extremely acidic, it is common to adjust the pH prior to inoculation. The pH can further be manipulated for the selective precipitation and recovery of explicit metal(loid)s. The influent sulfide concentration should also be monitored, as excess dissolved sulfide in the reactor can cause the dissolution of the formed precipitates. Furthermore, the selection of a mixture of electron donors shows promise for sulfate-based bioprecipitation (Liamleam and Annachhatre 2007), and a conglomeration of different microorganisms can improve the metabolic capability and enhance the biological process compared to pure cultures (Sahinkaya et al. 2017), as well as improve culture resistance (Kiran et al. 2017).

The engineered design is largely chosen based on the time required for remediation and the available costs for the project. HRT can play a role, altering the time and cost of the design, as well as impacting its success. This design parameter is essential to optimal nucleation rates and agglomeration. It also impacts hydraulic conductivity and microorganism contact (Vasquez et al. 2018). Therefore, the residence time is paramount to precipitate settling, and subsequent metal recovery. However, HRT can also impact sulfide toxicity when high and biomass washout when low (Kiran et al. 2017). An optimal HRT can be calculated and modified based on the volumetric flow rate, reactor size, and porosity in the bioreactor (Viggi et al. 2010). During laboratory experimentation, varying HRTs should be tested to find optimal conditions.

The design should further consider the environmental conditions of the region. Many mining operations are located in extreme climate conditions. These regions can have low temperatures, freeze/thaw cycles, high salinity, and changing climate conditions (i.e., dryness/wetness), which can impact bioprecipitation. While studies show that sulfate-based bioprecipitation is capable of performing in low temperature conditions (Nielsen et al. 2018, Virpiranta et al. 2022) and/or highly saline environments (Ben Ali et al. 2020), the effectiveness is mediocre. Further investigations are necessary to develop optimal operations. These specifications are necessary for scaling-up to field conditions for industry.

2.4 Case Studies

Tables 2.7 and 2.8 provide case studies showcasing bioprecipitation as a remediation technique for heavy metals from mining activities. Table 2.7 illustrates sulfate-based bioprecipitation using SRB, while Table 2.8 demonstrates iron-oxidation-based bioprecipitation with iron-oxidizer cultures. Both tables provide evidence of bioprecipitation as a promising and viable strategy for ARD/AMD remediation. The reactor configuration, energy source, HRT, temperature, and pH are provided to illustrate an overview of various process designs and their respective efficacies (i.e., metal removal).

To assess SRB capabilities to remediate mine tailings contamination, Martins et al. (2011) tested ARD/AMD from the Portuguese mines São Domingos and Tinoca, both located in the Alentejo region. They used two laboratory scaled up-flow anaerobic pack bed (UAPB) reactors to treat the tailings, which incorporated a column of calcite tailing to neutralize the high acidity of the samples. A mixture of SRB species was used with ethanol as an energy source. Also, ARD/AMD from the Tinoca mine was diluted prior to being incorporated into the reactor due to its severely low pH and high metal (specifically Cu, Fe, and Zn) and sulfate concentrations. Samples from the reactors were tested weekly, assessing the redox potential, pH, heavy metal, and sulfate concentrations. The heavy metal and sulfate removal was greater than 99% and 72%, respectively. However, significant removal is attributed to the calcite column in the bioreactor used in conjunction with the sulfate-based bioprecipitation process.

It should be noted that the study conducted by Martins et al. (2009) found insignificant sulfate reduction from mining samples. They sampled sediments at two locations at the São Domingos mining area in Alentejo, South Portugal. The samples were analyzed to evaluate their respective elemental compositions, focusing on Fe, Cu, and Zn. Anaerobic batch experiments were performed in glass bottles using lactose and ethanol as an energy source and SRB as microorganisms. However, SRB species were not found in these samples due to the highly acidic conditions, unconducive to bacterial growth. They, therefore, did not experience any significant bioprecipitation.

These studies demonstrate the necessity for preliminary testing specific to the ARD/AMD at the site prior to field remediation. The operation must be optimized for specified contamination considering all design components. Pre-treatment of the ARD/AMD may be required to effectively remediate contamination via bioprecipitation.

2.5 Conclusions and Recommendations

Bioprecipitation is a viable remediation technique to extract soluble metal(loid)s from contaminated soils and groundwater. Sulfate-based bioprecipitation acts as an optimal method for bioprecipitation, especially for ARD/AMD wastewater from mining related activities. Oxidation of sulfide bearing ores creates ARD/AMD that is highly acidic and contains a low oxygen content. Therefore, acidophilic SRB paired with a non-acidic electron donor shows promise as an optimal method to better remediate these sites via bioprecipitation. Further, a one-stage reactor design can enhance settling of the precipitated metal(loid)s ameliorating metal recovery and potential metal recycling and reuse. Site characterization should be conducted for each site to assess the contamination and

Table 2.7. Case studies using SRB metal bioprecipitation in mining remediation.

Metal Influent	Site Location	Lab vs. Field	Bioreactor	Energy Source	HRT	Temperature	pH	Metal Removal	References
Cu, Zn, Ni, Fe, Al, Mg	Woodcutters mine site, Australia	Bench-scale	UAPB with SRB	Lactate	16.16 h	25°C	Increased from 4.5–7.0	> 97.5% Cu, Zn, Ni > 77.5% As > 82% Fe Mg, Al unchanged	Jong and Parry 2003
Cu, Zn, Pb	Bolivar mine, Oruro, Bolivia	Bench-scale	UAPB with SRB	Papaya, apple, banana	20 days	20°C	Increased 3.0–7.5	100% Cu > 94% Zn > 92% Pb	Alvarez et al. 2007
Cu, Al, Mn, Co, Ni, Pb, Zn, Fe^{3+}, Fe^{2+}, SO_4^{2-}	Chessy-les-Mines mine site	Lab. pilot scale	Mechanically & magnetically stirred reactor with SRB	Hydrogen gas	Max sulfate reduction at 0.9 days	30°C	Increased from 2.55 – (pH_{Zn} = 3.5 pH_{Cu} = 2.8 $pH_{Ni, Fe(II)}$ = 6.0)	Zn Cu Ni Fe^{2+}	Foucher et al. 2001
Cu, Fe	Polymetallic Sulfide mine, Southwest Spain	Lab.	Batch tests with SRB	-	27 days	35°C	Increased from 5.0–7.0	100% Cu 97% Fe	García et al. 2001
Cu, Zn	Acid mine drainage	Lab.	ABR with SRB	Ethanol	2 days	20–25°C	Increased from 3.3–(6–7.5)	~ 100% Cu 83–98% Zn	Sahinkaya et al. 2009
Fe, Cu, Zn, Al	Mining area of Huelva Province, Spain	Lab.	Anaerobic reactor with 30% biogas	Cheese whey	8 days	25°C	Increased to 5.5	91.3% Fe 96.1% Cu 79% Zn 99% Al	Jiménez-Rodríguez et al. 2009
Zn	Synthetic replica of AMD	Lab.	Single-stage gas-lift bioreactor	Hydrogen & carbon dioxide	Varied	30°C	5.5	82.2–99.99% Zn	Bijmans et al. 2009a
Na, K, Mg, U, Ra, As, NO_3^-	Wastewater from a hydrometallurgical uranium mine	Lab.	Three-stage system with recirculation & SRB	Molasses	27 h	30–31°C	Increased from 5.2–7.7	92% U 99% Ra > 95% As 100% NO_3^-	Somlev and Banov 1998

Table 2.7 contd. ...

...Table 2.7 contd.

Metal Influent	Site Location	Lab vs. Field	Bioreactor	Energy Source	HRT	Temperature	pH	Metal Removal	References
Fe^{3+}, Zn, Mg, Cu, Cd, As, Pb, Cl, SO_4^{2-}, Na, CO_3^{2-}, OH^-	Synthetic replica of AMD	Lab.	Anaerobic filter reactor with SRB	Acetic acid	Varied	30°C	7.2	> 99% metal removal	Steed et al. 2000
Zn	Cueva de la Mora & Mina Esperanza Iberian Pyrite Belt mining districts	Lab.	Batch tests with SRB	-	35 days	21°C	7	100% Zn	Castillo et al. 2012
Fe, Zn, Cu, Mg	Simulated SMD	Lab.	Column reactors with SRB	Organic waste	20 days	30°C	6.8	99.8% Cu 99.9 % Zn > 80% Fe Mg incomplete	Chang et al. 2000
Fe, Zn, Cu	Synthetic AMD	Bench-scale	UASB with SRB	Ethanol	Varied	30°C	Increased from 4.0–6.5	88% Fe 98% Zn 99% Cu	Vieira et al. 2016
Fe, Ca, Mg, Ni, Zn, Na, K, Mn, Cd, SO_4^{2-}	Synthetic AMD	Lab.	Batch reactor with SRB	Organic waste	-	22°C	Increased from 3.9–(5–9)	100% Fe 99% Mn 99% Cd 99% Ni 94% Zn	Zagury et al. 2006
Fe, Cu, Zn, SO_4^{2-}	S. Domingos mine, South-east Portugal (2 samples)	Lab.	Batch and anaerobic reactor with SRB	Ethanol and lactate	-	21°C	3.7 & 4.5	Insignificant reduction	Martins et al. 2009
Cr^{6+}, SO_4^{2-}	Curilo mine district, Sophia	Lab.	Fixed bed column reactor with SRB	Ethanol	SO_4^{2-}: 9 days Cr: 2 days	25°C	7.5–8	65% SO_4^{2-} 95% Cr	Pagnanelli et al. 2012

As, Ca, Cd, Co, Cr, Cu, Fe, Hg, Mn, Mo, Ni, Pb, Sb, Se, Sn, Sr, Zn, SO_4^{2-}	Portuguese mines São Domingos and Tinoca	Lab.	UAPB with SRB	Ethanol	Varied	21°C	Increased from 2.8–6.5	> 99% metals removal > 72% SO_4^{2-}	Martins et al. 2011
Zn, Cu, Cd	Synthetic AMD	Lab.	Anaerobic sulfidogenic reactor with SRB	Rape straw	24 h	30°C	Increased from 5.0–(6.5–7.0)	> 95% metals removal	Wang et al. 2012
Al, As, Cd, Cu, Cr, Fe, Mn, Pb, Zn	Tailings from abandoned mine sites	Lab.	Column reactors with SRB and Fe-reducing bacteria	Food waste compost	-	-	Increased from 3.0–8.1	91% Zn 76% Mn 53% Cr 44% Fe 43% Cd 24% Al	Hwang and Neculita 2013
Zn, Cu, SO_4^{2-}	Outokumpu Oyj sulphide mine, Pyhäsalmi, Finland	Lab.	UASB	Glucose	Varied	20°C	Increased from 3.–7.1	> 99.9% Zn 23–72% SO_4^{2-}	Tuppurainen et al. 2002
Al, Ca, Cd, Co, Cr, Cu, Fe, K, Mg, Mn, Na, Ni, Pb, Zn	Copper mine, Elazığ, Turkey	Lab.	FBR with SRB	Ethanol	Varied	35°C	Increased from 2.3–(6–8)	> 99.9% Al 94% Mn	Sahinkaya et al. 2011
Cu, Fe	Synthetic AMD	Lab.	Three-stage FBR with SRB	Ethanol	1 day	35°C	Increased from 3.0- Neutral	> 99%	Ucar et al. 2011
Ur, Ra, Th, Cu, Zn, Cd	Curilo mine district, Sophia, Bulgaria	Lab.	Continuous fixed bed column reactor with SRB	Mixture of leaves, compost, Fe(0), silica sand, and limestone	SO_4^{2-}: 8 days Cd: 2.2 days Zn: 0.52 days	37°C	8.8	50% SO_4^{2-} 99% As 99% Cd 99% Cr^{6+} 98% Cu 99% Zn	Viggi et al. 2010

Table 2.7 contd. ...

...Table 2.7 contd.

Metal Influent	Site Location	Lab vs. Field	Bioreactor	Energy Source	HRT	Temperature	pH	Metal Removal	References
Fe, Cu	Synthetic AMD	Lab.	Downflow anaerobic packed bed reactor with SRB	Glycerol, lactate, sucrose, and a blend	15.9 h	30°C	4.5	> 90%	Costa et al. 2021
Fe, Cu, Cd, Zn	Synthetic AMD	Lab.	Upflow anaerobic packed bed reactor with SRB	Maize straw & sodium lactate	Varied	30°C	Increased from 2.7–(7.3–7.8)	> 99.9%	Zhang et al. 2016
Cu, Zn, Fe, Al	90% AMD from Wallenberg mine and 10% from Gammelgruva mine	Bench-scale	Batch reactor with SRB	Whey & cow manure	-	15°C	Increased from 3.4– (4–6)	> 90%	Christensen et al. 1996
Al, As, B, Ba, Be, Cd, Co, Cu, Fe, Pb, Mn, Ni, Sr, U, Zn	Finnish mine site AMD	Lab.	Continuous up-flow biofilm reactor with SRB	Succinate	10 h	11.7°C	Increased from 4.2–7.25	77% Co 60% Ni	Virpiranta et al. 2022
Fe^{2+}, Ni^{2+}, Cu^{2+}, SO_4^{2-}	Synthetic AMD	Lab.	Column passive biochemical reactors with SRB	Mixture of wood chips/ sawdust, leaf compost, ash, sand, and municipal sludge	2.5 and 5 days	22°C & 5°C	Increased from (2.8–3.5)–(6.8–8.4)	95–99% Cu 74–99% Ni 75–90% Fe 96–99% SO_4^{2-}	Ben Ali et al. 2020
SO_4^{2-}, Fe^{2+}, Mn^{2+}, Zn^{2+}, Ca^{2+}, Mg^{2+}	Artificial AMD	Lab.	Column biochemical passive reactors with SRB	Mixture of cow manure, mushroom compost, sajo sawdust, gravel, limestone, sediment	Varied	-	Increased from (3.0–3.7) – (> 7.0)	> 60% SO_4^{2-} > 95% Fe^{2+} & Zn^{2+}	Vasquez et al. 2018

Cd, Zn	Silver King mine site, Yukon Territory, Canada	Field	Anaerobic bioreactors with SRB	Molasses	2 weeks	5°C	6.5–7.1	20.9–89.3% Zn 39–90.5% Cd	Nielsen et al. 2018
Fe, Al, Ni, Cu, Zn	Leviathan Mine, Alpine California, U.S.A	Pilot scale	Column bioreactor with SRB	Ethanol and methanol	5 h and 6.6 h	6°C	2.5	> 93% Fe 42% SO_4^{2-}	Tsukamoto et al. 2004
Ca, Mg, Na, K, Cl, Fe, Mn, Al, Pb, Cu, Zn, SO_4^{2-}	Synthetic AMD replicating mine discharge from Cwm Rheidol, Wales, UK	Lab.	Column bioreactor with SRB	Methanol	12–14 h	-	Increased from 2.7–(6.6–7.6)	> 99% Pb & Cu 99% Zn	Mayes et al. 2011
Cu, Zn, Fe, Mn, Co, Ni	Simulated AMD	Lab.	Bioreactor with SRB	Cow manure, buffalo manure and goat manure	10 days	-	Increased from 2.7–(6.25–7.5)	51.5–99.3% Cu 35.1–99.8% Zn 17.9–99.1% Ni 63.6–99% Co 12.7–73.9% Mn	Choudhary and Sheoran 2012
Cu, Mn, Zn, Fe, SO_4^{2-}	Copper Mine, China	Bench-scale	Continuous up-flow anaerobic multiple-bed reactor with SRB	Lactate	Varied	25°C	Increased from 2.7–6.2	> 61% SO_4^{2-} 99% Cu 86% Fe^{2+} Mn^{2+} insignificant	Bai et al. 2013

Table 2.8. Case studies using iron oxidation metal bioprecipitation in mining related remediation.

Metal Influent	Site Location	Lab vs. Field	Bioreactor	HRT	Temperature	pH	Metal Removal	References
Fe^{2+}, Mn^{2+}, Al^{3+}, Na^{+}, Ca^{2+}	Simulated heap leaching solution	Lab.	Continuous flow FBR with *Leptospirillum ferriphilum*	Varied (max oxidation at 2 hrs)	37°C	2.0	98.5% Fe	Ozkaya et al. 2007
Fe^{2+}, Fe^{3+}, H^{+}, Cu^{2+}	Pyhäsalmi FeS_2-$CuFeS_2$- -(Zn,Fe)S mine	Lab.	Continuous flow FBR with *Leptospirillum ferriphilum*	Varied	35°C	< 1	> 99% Fe	Kinnunen and Puhakka 2005
Fe, S, Na, K, Mg, Mn, Ca, Cu, Zn, Ni, Co	Multimetal heap leaching pilot plant, Talvivaara, Finland	Bench-scale	FBR and gravity settler with *Leptospirillum ferriphilum*	2 h	37°C	3.5	99% Fe^{3+}	Nurmi et al. 2010
Fe^{2+}, Cu, Ni	Synthetic hydrometallurgical solution	Lab.	Two-stage CSTR with mixed culture mesophilic iron oxidisers	Varied	20–25°C	1.0–1.9	54% Fe 33% S 2.5% Cu 0.26% Ni	Kaksonen et al. 2014b
Fe^{2+}, Cu, Ni, K, Ca	Synthetic hydrometallurgical solution	Lab.	Continuous flow two-stage ALBR with mixed culture mesophilic iron oxidisers	Varied	23°C	1.5	30.9% Fe 16.7% S 1.1% Cu 0.2% Ni	Kaksonen et al. 2014a
Al, Cu, Mg, Zn, Fe	Synthetic mine water	Lab.	Packed bed reactor with *Ferrovum myxofaciens*	-	21–23°C	Increased from 2.0–3.5	~ 100% Fe	Hedrich and Johnson 2012
As	Synthetic arsenic-containing acidic solution (applicable to mining wastewater treatment)	Lab.	FBR with mixed cultures	208 h	27°C	3.0	99% As	Ahoranta et al. 2016
Fe, Zn, Al, Cd, Pb, SO_4^{2-}	Cwm Rheidol mine water	Lab.	Passive VFR with microbial Fe^{2+} oxidation	Varied	6.74–19.67°C	Increased from 2.9–4.94	67% Fe	Florence et al. 2016

predict the success of the engineered strategy prior to site implementation. Additional ARD/AMD treatment may be required prior to the active or passive treatment to increase overall process efficacy.

The following are noteworthy gaps in the literature regarding bioprecipitation:

- Studies assessing the thermodynamics, kinetics, and hydrodynamics of metal precipitation occurring during bioprecipitation with respect to different passive or active designs.
- Long-term studies evaluating dissolution of metal precipitates over time.
- Chemical and biological effects of biofilm during the lifecycle of various bioreactors.
- Field test and case studies evaluating SRB-based and iron-based bioprecipitation for ARD/AMD wastewaters at large scale and/or applied to industry. These studies should also analyze extreme climatic conditions and their impact on bioprecipitation.

To date, the application of bioprecipitation as a remediation technique at full scale is lacking. Additional experimentation is required to minimize the noted gaps and expand on current literature. Detailed cost-benefit analyses should be researched to scale-up operations and assess their viability for the industry. While technical differences have been noted between research and industry (Section 2.3.2), collaborations will lead to better optimization of the system at full scale and promote bioprecipitation as a feasible remediation technology.

References

Ahoranta, S. H., M. E. Kokko, S. Papirio, B. Özkaya and J. A. Puhakka. 2016. Arsenic removal from acidic solutions with biogenic ferric precipitates. J. Hazard. Mater. 306: 124–132. https://doi.org/10.1016/j.jhazmat.2015.12.012.

Alazard, D., M. Joseph, F. Battaglia-Brunet, J.-L. Cayol and B. Ollivier. 2010. *Desulfosporosinus acidiphilus* sp. *nov.*: A moderately acidophilic sulfate-reducing bacterium isolated from acid mining drainage sediments. Extremophiles 14(3): 305–312. https://doi.org/10.1007/s00792-010-0309-4.

Alvarez, M. T., C. Crespo and B. Mattiasson. 2007. Precipitation of Zn(II), Cu(II) and Pb(II) at bench-scale using biogenic hydrogen sulfide from the utilization of volatile fatty acids. Chemosphere 66(9): 1677–1683. https://doi.org/10.1016/j.chemosphere.2006.07.065.

Bai, H., Y. Kang, H. Quan, Y. Han, J. Sun and Y. Feng. 2013. Treatment of acid mine drainage by sulfate reducing bacteria with iron in bench scale runs. Bioresour. Technol. 128: 818–822. https://doi.org/10.1016/j.biortech.2012.10.070.

Ben Ali, H. E., C. M. Neculita, J. W. Molson, A. Maqsoud and G. J. Zagury. 2020. Salinity and low temperature effects on the performance of column biochemical reactors for the treatment of acidic and neutral mine drainage. Chemosphere 243(125303). https://doi.org/10.1016/j.chemosphere.2019.125303.

Bijmans, M. F. M., P.-J. van Helvoort, C. J. N. Buisman and P. N. L. Lens. 2009a. Effect of the sulfide concentration on zinc bio-precipitation in a single stage sulfidogenic bioreactor at pH 5.5. Sep. Purif. Technol. 69(3): 243–248. https://doi.org/10.1016/j.seppur.2009.07.023.

Bijmans, M. F. M., P.-J. van Helvoort, S. A. Dar, M. Dopson, P. N. L. Lens and C. J. N. Buisman. 2009b. Selective recovery of nickel over iron from a nickel–iron solution using microbial sulfate reduction in a gas-lift bioreactor. Water Res. 43(3): 853–861. https://doi.org/10.1016/j.watres.2008.11.023.

Castillo, J., R. Pérez-López, M. A. Caraballo, J. M. Nieto, M. Martins, M. C. Costa et al. 2012. Biologically-induced precipitation of sphalerite–wurtzite nanoparticles by sulfate-reducing bacteria: Implications for acid mine drainage treatment. Sci. Total Environ. 423: 176–184. https://doi.org/10.1016/j.scitotenv.2012.02.013.

Chang, I. S., P. K. Shin and B. H. Kim. 2000. Biological treatment of acid mine drainage under sulphate-reducing conditions with solid waste materials as substrate. Water Res. 34(4): 1269–1277. https://doi.org/10.1016/S0043-1354(99)00268-7.

Choudhary, R. P. and A. S. Sheoran. 2012. Performance of single substrate in sulphate reducing bioreactor for the treatment of acid mine drainage. Miner. Eng. 39: 29–35. https://doi.org/10.1016/j.mineng.2012.07.005.

Christensen, B., M. Laake and T. Lien. 1996. Treatment of acid mine water by sulfate-reducing bacteria; results from a bench scale experiment. Water Res. 30(7): 1617–1624. https://doi.org/10.1016/0043-1354(96)00049-8.

Costa, R. B., L. A. G. Godoi, A. F. M. Braga, T. P. Delforno, and D. Bevilaqua. 2021. Sulfate removal rate and metal recovery as settling precipitates in bioreactors: Influence of electron donors. J. Hazard. Mater. 403(123622). https://doi.org/10.1016/j.jhazmat.2020.123622.

Davis, M. L. 2010. Water and Wastewater Engineering: Design Principles and Practice. The McGraw-Hill Companies, Inc.

Egiebor, N. O. and B. Oni. 2007. Acid rock drainage formation and treatment: A review. Asia-Pacific Asia-Pac. J. Chem. Eng. 2(1): 47–62. https://doi.org/10.1002/apj.57.

Florence, K., D. J. Sapsford, D. B. Johnson, C. M. Kay and C. Wolkersdorfer. 2016. Iron-mineral accretion from acid mine drainage and its application in passive treatment. Environ. Technol. 37(11): 1428–1440. https://doi.org/10.1080/09593330.2015.1118558.

Foucher, S., F. Battaglia-Brunet, I. Ignatiadis and D. Morin. 2001. Treatment by sulfate-reducing bacteria of Chessy acid-mine drainage and metals recovery. Chem. Eng. Sci. 56(4): 1639–1645. https://doi.org/10.1016/S0009-2509(00)00392-4.

Frolov, E. N., I. V. Kublanov, S. V. Toshchakov, N. I. Samarov, A. A. Novikov, A. V. Lebedinsky et al. 2017. *Thermodesulfobium acidiphilum* sp. *nov*., a thermoacidophilic, sulfate-reducing, chemoautotrophic bacterium from a thermal site. Int. J. Syst. Evol. Microbiol. 67(5): 1482–1485. https://doi.org/10.1099/ijsem.0.001745.

García, C., D. A. Moreno, A. Ballester, M. L. Blázquez and F. González. 2001. Bioremediation of an industrial acid mine water by metal-tolerant sulphate-reducing bacteria. Miner. Eng. 14(9): 997–1008. https://doi.org/10.1016/S0892-6875(01)00107-8.

Hao, T., P. Xiang, H. R. Mackey, K. Chi, H. Lu, H. Chui et al. 2014. A review of biological sulfate conversions in wastewater treatment. Water Res. 65: 1–21. https://doi.org/10.1016/j.watres.2014.06.043.

Hedrich, S. and D. B. Johnson. 2012. A modular continuous flow reactor system for the selective bio-oxidation of iron and precipitation of schwertmannite from mine-impacted waters. Bioresour. Technol. 106: 44–49. https://doi.org/10.1016/j.biortech.2011.11.130.

Hwang, T. and C. M. Neculita. 2013. *In situ* immobilization of heavy metals in severely weathered tailings amended with food waste-based compost and zeolite. Water Air Soil Pollut. 224(1): 1388. https://doi.org/10.1007/s11270-012-1388-x.

Jegatheesan, V., H. Ravishankar, L. Shu and J. Wang. 2016. Application of green and physico-chemical technologies in treating water polluted by heavy metals. pp. 579–614. *In*: Ngo, H. H., W. Guo, Rao. Y. Surampalli and T. C. Zhang (eds.). Green Technologies for Sustainable Water Management. American Society of Civil Engineers. http://ascelibrary.org/doi/abs/10.1061/9780784414422.ch16.

Jiménez-Rodríguez, A. M., M. M. Durán-Barrantes, R. Borja, E. Sánchez, M. F. Colmenarejo and F. Raposo. 2009. Heavy metals removal from acid mine drainage water using biogenic hydrogen sulphide and effluent from anaerobic treatment: Effect of pH. J. Hazard. Mater. 165(1-3): 759–765. https://doi.org/10.1016/j.jhazmat.2008.10.053.

Johnson, D. B. and K. B. Hallberg. 2005. Acid mine drainage remediation options: A review. Sci. Total Environ. 338(1): 3–14. https://doi.org/10.1016/j.scitotenv.2004.09.002.

Johnson, D. B. and Sánchez-Andrea, I. 2019. Chapter Six-Dissimilatory reduction of sulfate and zero-valent sulfur at low pH and its significance for bioremediation and metal recovery. pp. 205–231. *In*: Poole, R. K. (ed.). Advances in Microbial Physiology (Vol. 75). Academic Press. https://doi.org/10.1016/bs.ampbs.2019.07.002.

Johnson, D. B. and A. L. Santos. 2020. Biological removal of sulfurous compounds and metals from inorganic wastewaters. pp. 215–246. *In*: P. Lens (ed.). Environmental Technologies to Treat Sulfur Pollution: Principles and Engineering (Second Edition). IWA Publishing.

Jong, T. and D. L. Parry. 2003. Removal of sulfate and heavy metals by sulfate reducing bacteria in short-term bench scale upflow anaerobic packed bed reactor runs. Water Res. 37(14): 3379–3389. https://doi.org/10.1016/S0043-1354(03)00165-9.

Kaksonen, A. H., C. Morris, F. Hilario, S. M. Rea, J. Li, K. M. Usher et al. 2014a. Iron oxidation and jarosite precipitation in a two-stage airlift bioreactor. Hydrometall. 150: 227–235. https://doi.org/10.1016/j.hydromet.2014.05.020.

Kaksonen, A. H., C. Morris, S. Rea, J. Li, J. Wylie, K. M. Usher et al. 2014b. Biohydrometallurgical iron oxidation and precipitation: Part I—Effect of pH on process performance. Hydrometall. 147–148: 255–263. https://doi.org/10.1016/j.hydromet.2014.04.016.

Kinnunen, P. H.-M. and J. A. Puhakka. 2005. High-rate iron oxidation at below pH 1 and at elevated iron and copper concentrations by a *Leptospirillum ferriphilum* dominated biofilm. Process Biochem. 40(11): 3536–3541. https://doi.org/10.1016/j.procbio.2005.03.050.

Kiran, M. G., K. Pakshirajan and G. Das. 2017. An overview of sulfidogenic biological reactors for the simultaneous treatment of sulfate and heavy metal rich wastewater. Chem. Eng. Sci. 158: 606–620. https://doi.org/10.1016/j.ces.2016.11.002.

Kiskira, K., S. Papirio, E. D. van Hullebusch and G. Esposito. 2017. Fe(II)-mediated autotrophic denitrification: A new bioprocess for iron bioprecipitation/biorecovery and simultaneous treatment of nitrate-containing wastewaters. Int. Biodeterior. Biodegrad. 119: 631–648. https://doi.org/10.1016/j.ibiod.2016.09.020.

Levett, A., S. A. Gleeson and J. Kallmeyer. 2021. From exploration to remediation: A microbial perspective for innovation in mining. Earth-Sci. Rev. 216: 103563. https://doi.org/10.1016/j.earscirev.2021.103563.

Lewis, A. 2017. Precipitation of heavy metals. pp. 101–120. *In*: Rene, E. R., E. Sahinkaya, A. Lewis and P. N. L. Lens (eds.). Sustainable Heavy Metal Remediation: Volume 1: Principles and Processes. Springer International Publishing. https://doi.org/10.1007/978-3-319-58622-9_4.

Lewis, A. E. 2010. Review of metal sulphide precipitation. Hydrometall. 104(2): 222–234. https://doi.org/10.1016/j.hydromet.2010.06.010.

Liamleam, W. and A. P. Annachhatre. 2007. Electron donors for biological sulfate reduction. Biotechnol. Adv. 25(5): 452–463. https://doi.org/10.1016/j.biotechadv.2007.05.002.

Martins, M., M. L. Faleiro, R. J. Barros, A. R. Veríssimo, M. A. Barreiros and M. C. Costa. 2009. Characterization and activity studies of highly heavy metal resistant sulphate-reducing bacteria to be used in acid mine drainage decontamination. J. Hazard. Mater. 166(2): 706–713. https://doi.org/10.1016/j.jhazmat.2008.11.088.

Martins, M., E. S. Santos, M. L. Faleiro, S. Chaves, R. Tenreiro, R. J. Barros et al. 2011. Performance and bacterial community shifts during bioremediation of acid mine drainage from two Portuguese mines. Int. Biodeterior. Biodegrad. 65(7): 972–981. https://doi.org/10.1016/j.ibiod.2011.07.006.

Mayes, W. M., J. Davis, V. Silva and A. P. Jarvis. 2011. Treatment of zinc-rich acid mine water in low residence time bioreactors incorporating waste shells and methanol dosing. J. Hazard. Mater. 193: 279–287. https://doi.org/10.1016/j.jhazmat.2011.07.073.

Mihelcic, J. R. and J. B. Zimmerman. 2014. Environmental Engineering: Fundamentals, Sustainability, Design (Second Edition). John Wiley & Sons, Inc.

Mori, K., H. Kim, T. Kakegawa and S. Hanada, 2003. A novel lineage of sulfate-reducing microorganisms: *Thermodesulfobiaceae fam. nov., Thermodesulfobium narugense, gen. nov.*, sp. *nov.*, a new thermophilic isolate from a hot spring. Extremophiles. 7(4): 283–290. https://doi.org/10.1007/s00792-003-0320-0.

Neculita, C. M., G. J. Zagury and B. Bussière. 2007. Passive Treatment of Acid Mine Drainage in Bioreactors using Sulfate-Reducing Bacteria. J. Environ. Qual. 36(1): 1–16. https://doi.org/10.2134/jeq2006.0066.

Neculita, C. M., G. J. Zagury and B. Bussière. 2021. Passive treatment of acid mine drainage at the reclamation stage. pp. 271–296. *In*: Bussière, B. and M. Guittonny (eds.). Hard Rock Mine Reclamation: From Prediction to Management of Acid Mine Drainage. CRC Press. https://doi.org/10.1201/9781315166698/-11.

Nielsen, G., I. Hatam, K. A. Abuan, A. Janin, L. Coudert, J. F. Blais et al. 2018. Semi-passive in-situ pilot scale bioreactor successfully removed sulfate and metals from mine impacted water under subarctic climatic conditions. Water Res. 140: 268–279. https://doi.org/10.1016/j.watres.2018.04.035.

Nurmi, P., B. Özkaya, K. Sasaki, A. H. Kaksonen, M. Riekkola-Vanhanen, O. H. Tuovinen et al. 2010. Biooxidation and precipitation for iron and sulfate removal from heap bioleaching effluent streams. Hydrometall. 101(1): 7–14. https://doi.org/10.1016/j.hydromet.2009.11.004.

Ozkaya, B., E. Sahinkaya, P. Nurmi, A. H. Kaksonen and J. A. Puhakka. 2007. Iron oxidation and precipitation in a simulated heap leaching solution in a Leptospirillum ferriphilum dominated biofilm reactor. Hydrometall. 88(1): 67–74. https://doi.org/10.1016/j.hydromet.2007.02.009.

Pagnanelli, F., C. Cruz Viggi, A. Cibati, D. Uccelletti, L. Toro and C. Palleschi. 2012. Biotreatment of Cr(VI) contaminated waters by sulphate reducing bacteria fed with ethanol. J. Hazard. Mater. 199–200: 186–192. https://doi.org/10.1016/j.jhazmat.2011.10.082.

Sahinkaya, E., M. Gungor, A. Bayrakdar, Z. Yucesoy and S. Uyanik. 2009. Separate recovery of copper and zinc from acid mine drainage using biogenic sulfide. J. Hazard. Mater. 171(1–3): 901–906. https://doi.org/10.1016/j.jhazmat.2009.06.089.

Sahinkaya, E., F. M. Gunes, D. Ucar and A. H. Kaksonen. 2011. Sulfidogenic fluidized bed treatment of real acid mine drainage water. Bioresour. Technol. 102(2): 683–689. https://doi.org/10.1016/j.biortech.2010.08.042.

Sahinkaya, E., D. Uçar and A. H. Kaksonen. 2017. Bioprecipitation of metals and metalloids. pp. 199–231. *In*: Rene, E. R., E. Sahinkaya, A. Lewis, and P. N. L. Lens (eds.). Sustainable Heavy Metal Remediation: Volume 1: Principles and Processes. Springer International Publishing. https://doi.org/10.1007/978-3-319-58622-9_7.

Sánchez-Andrea, I., J. L. Sanz, M. F. M. Bijmans and A. J. M. Stams. 2014. Sulfate reduction at low pH to remediate acid mine drainage. J. Hazard. Mater. 269: 98–109. https://doi.org/10.1016/j.jhazmat.2013.12.032.

Sánchez-Andrea, I., A. J. M. Stams, S. Hedrich, I. Ňancucheo and D. B. Johnson. 2015. *Desulfosporosinus acididurans* sp. *nov.*: An acidophilic sulfate-reducing bacterium isolated from acidic sediments. Extremophiles 19(1): 39–47. https://doi.org/10.1007/s00792-014-0701-6.

Sánchez-Andrea, I., A. J. M. Stams, J. Weijma, P. Gonzalez Contreras, H. Dijkman, R. A. Rozendal et al. 2016. A case in support of implementing innovative bio-processes in the metal mining industry. FEMS Microbiol. Lett. 363(11): 106. https://doi.org/10.1093/femsle/fnw106.

Sen, A. M. and B. Johnson. 1999. Acidophilic sulphate-reducing bacteria: Candidates for bioremediation of acid mine drainage. pp. 709–718. *In*: Amils, R. and A. Ballester (eds.). Process Metallurgy (Vol. 9). Elsevier. https://doi.org/10.1016/S1572-4409(99)80073-X.

Skousen, J. G., A. Sexstone and P. F. Ziemkiewicz. 2000. Acid mine drainage control and treatment. pp. 131–168. *In*: Barnhisel, R. I., R. G. Darmody and W. Lee Daniels (eds.). Reclamation of Drastically Disturbed Lands (Volume 41). John Wiley & Sons, Ltd. https://doi.org/10.2134/agronmonogr41.c6.

Somlev, V. and M. Banov. 1998. Three stage process for complex biotechnological treatment of industrial wastewater from uranium mining. Biotechnol. Tech. 12(8): 637–639.

Steed, V. S., M. T. Suidan, M. Gupta, T. Miyahara, C. M. Acheson and G. D. Sayles. 2000. Development of a sulfate-reducing biological process to remove heavy metals from acid mine drainage. Water Environ. Res. 72(5): 530–535. https://doi.org/10.2175/106143000X138102.

Taylor, J., S. Pape and N. Murphy. 2005. A Summary of Passive and Active Treatment Technologies for Acid and Metalliferous Drainage (AMD). 49.

Tsukamoto, T. K., H. A. Killion and G. C. Miller. 2004. Column experiments for microbiological treatment of acid mine drainage: Low-temperature, low-pH and matrix investigations. Water Res. 38(6): 1405–1418. https://doi.org/10.1016/j.watres.2003.12.012.

Tuppurainen, K. O., A. O. Väisänen and J. A. Rintala. 2002. Sulphate-reducing laboratory-scale high-rate anaerobic reactors for treatment of metal- and sulphate-containing mine wastewater. Environ. Technol. 23(6): 599–608. https://doi.org/10.1080/09593332308618382.

Tuttle, J. H., P. R. Dugan, C. B. Macmillan and C. I. Randles. 1969. Microbial dissimilatory sulfur cycle in acid mine water. J. Bacteriol. 97(2): 594–602. https://doi.org/10.1128/jb.97.2.594-602.1969.

Ucar, D., O. K. Bekmezci, A. H. Kaksonen and E. Sahinkaya. 2011. Sequential precipitation of Cu and Fe using a three-stage sulfidogenic fluidized-bed reactor system. Miner. Eng. 24(11): 1100–1105. https://doi.org/10.1016/j.mineng.2011.02.005.

Vasquez, Y., M. C. Escobar, J. S. Saenz, M. F. Quiceno-Vallejo, C. M. Neculita, Z. Arbeli et al. 2018. Effect of hydraulic retention time on microbial community in biochemical passive reactors during treatment of acid mine drainage. Bioresour. Technol. 247: 624–632. https://doi.org/10.1016/j.biortech.2017.09.144.

Vieira, B. F., P. T. Couto, G. P. Sancinetti, B. Klein, D. van Zyl and R. P. Rodriguez. 2016. The effect of acidic pH and presence of metals as parameters in establishing a sulfidogenic process in anaerobic reactor. J. Environ. Sci. Health, Part A 51(10): 793–797. https://doi.org/10.1080/10934529.2016.1181433.

Viggi, C. C., F. Pagnanelli, A. Cibati, D. Uccelletti, C. Palleschi and L. Toro. 2010. Biotreatment and bioassessment of heavy metal removal by sulphate reducing bacteria in fixed bed reactors. Water Res. 44(1): 151–158. https://doi.org/10.1016/j.watres.2009.09.013.

Villa-Gomez, D., H. Ababneh, S. Papirio, D. P. L. Rousseau and P. N. L. Lens. 2011. Effect of sulfide concentration on the location of the metal precipitates in inversed fluidized bed reactors. J. Hazard. Mater. 192(1): 200–207. https://doi.org/10.1016/j.jhazmat.2011.05.002.

Villa-Gomez, D. K., S. Papirio, E. D. van Hullebusch, F. Farges, S. Nikitenko, H. Kramer et al. 2012. Influence of sulfide concentration and macronutrients on the characteristics of metal precipitates relevant to metal recovery in bioreactors. Bioresour. Technol. 110: 26–34. https://doi.org/10.1016/j.biortech.2012.01.041.

Virpiranta, H., V.-H. Sotaniemi, T. Leiviskä, S. Taskila, J. Rämö, D. B. Johnson et al. 2022. Continuous removal of sulfate and metals from acidic mining-impacted waters at low temperature using a sulfate-reducing bacterial consortium. Chem. Eng. J. 427: 132050. https://doi.org/10.1016/j.cej.2021.132050.

Wang, J., S. Li, T. Chen, H. Zhang, N. Zhang and Z. Yue. 2012. Effects of heavy metals on the performance of anaerobic sulfidogenic reactor using rape straw as carbon source. Environ. Earth Sci. 67(7): 2161–2167. http://dx.doi.org.qe2a-proxy.mun.ca/10.1007/s12665-012-1657-4.

Whitman, W. B. and Bergey's Manual Trust. 2021. Bergey's Manual of Systematics of Archaea and Bacteria. John Wiley & Sons, Inc. https://onlinelibrary.wiley.com/browse/book/10.1002/9781118960608/title.

Willis, G. and E. R. Donati. 2017. Heavy metal bioprecipitation: Use of sulfate-reducing microorganisms. *In*: Donati, E. R. (ed.). Heavy Metals in the Environment: Microorganisms and Bioremediation. CRC Press. https://doi.org/10.1201/b22013.

Yong, R. N., C. N. Mulligan and M. Fukue. 2014. Sustainable Practices in Geoenvironmental Engineering (2nd ed.). CRC Press. https://doi.org/10.1201/b17443.

Zagury, G. J., V. I. Kulnieks and C. M. Neculita. 2006. Characterization and reactivity assessment of organic substrates for sulphate-reducing bacteria in acid mine drainage treatment. Chemosphere 64(6): 944–954. https://doi.org/10.1016/j.chemosphere.2006.01.001.

Zhang, M., H. Wang and X. Han. 2016. Preparation of metal-resistant immobilized sulfate reducing bacteria beads for acid mine drainage treatment. Chemosphere 154: 215–223. https://doi.org/10.1016/j.chemosphere.2016.03.103.

CHAPTER 3

Bioaccumulation of Metals in Lichens and Mosses

Understanding Atmospheric Deposition, Metal-induced Modifications and their Suitability as Biomonitors and Bioremediators

Sharfaa Hussain, Parijat Bharali, Barnali Koushik and *Raza R. Hoque**

3.1 Introduction

The heavy metal (HM) pollution in the environment can be traced back to the discovery of mining and metallurgical activities by human beings. Metals are a part of the natural geochemical systems whence nature can be considered the primary source of HM and human activities only as one of the main causes of their elevated levels (Martin 2012). As anthropogenic activities like industry, mining, traffic, power generation, incineration, construction, waste disposal, and agriculture attained pace, increased concentrations of HM began to be detected in the environment across the world. As such, the industrial revolution was one of the biggest contributors to the sudden rise of HM in the environment. The effects of HM pollution are far-reaching and have subverted several natural biogeochemical cycles of natural ecosystems (Nriagu 1979). Due to the long-distance transmission of pollutants, heavy metal deposition occurs on the saline and freshwater aquatic systems, alpine forests, rainforests, and polar ice sheets.

More so, HMs are a threat to the environment for their persistent nature. Once released into the environment, they enter the food chains through direct or indirect uptake and magnify in the higher trophic levels. Heavy metals are known to cause serious medical conditions in human beings and cause physiological damage to plants and animals (sensitive living components). To control or minimize HM toxicity, it is essential to monitor the levels of HM in the environment. This need for monitoring gave rise to the vast field of biological monitoring of HM using vascular plants, mosses, lichens, and animal tissues (Jeddi et al. 2021, Bartholomew et al. 2020, Turkyilmaz

Department of Environmental Science, Tezpur University, Tezpur 784028 (India).
* Corresponding author: rrh@tezu.ernet.in

et al. 2018, Malikova et al. 2019). Biological monitors are organisms that respond to the changes in pollution level or accumulate and hold the pollutants progressively over time. Apart from being more affordable, readily available, and safe against risks of vandalism and theft, biological monitors can account for actual responses of the biological system under study. With a long history of research in this area, lichens and mosses can be safely considered potential biomonitors of atmospheric pollutants (Richardson and Nieboer 1981, Daimari et al. 2020).

Mosses and lichens have a high capacity of HM accumulation (Kuik and Wolterbeek 1995, Bačkor and Loppi 2009, Cecconi et al. 2021). This allows for their excellent role as biomonitors, enabling a better understanding of the source, concentration, adaptation and tolerance mechanisms, and effect on community diversity due to pollution stress. Some of the most commonly used moss species are *Hylocomium splendens*, *Pleurozium schreberi* and *Hypnum cupressiforme* of the order Hypnales (Tyler 1970, Rühling and Tyler 2004). Mosses from order Hypnales, Hypnobryales, and Sphagnales are often seen to be used as biomonitors (Aničić et al. 2009). As described by Varela et al. (2016), the commonly used order of moss follows the sequence—Hypnales, Sphagnales, Pottiales, Hypnobryales, Bryales, Funariales, and Dicranales.

In the case of lichens, foliose and fruticose, lichens are mostly used in biomonitoring studies. Foliose lichens like *Flavoparmelia caperata*, *Parmelia sulcata*, and *Xanthoria parietina* are commonly chosen for *in situ* studies (Brunialti and Frati 2007, Nimis et al. 2000), whereas fruticose species like *Pseudevernia furfuracea* and *Evernia prunastri* are used in transplantation studies (Frati et al. 2005, Bari et al. 2001).

There has been a difference of opinion amongst researchers about which one is a better bioaccumulator or biomonitor between lichen and moss. It is seen that results vary based on the species of lichen or moss, type of pollutant, and environmental factors of their habitat. Bargagli et al. (2002) observed that epigeic mosses accumulate lithophilic metals, whereas lichens are efficient accumulators of atmophile metals. It is difficult to declare one of the two as a better biomonitor because of the variation in results found in different parts of the world (Chakrabortty and Paratkar 2006). However, both moss and lichen are irreplaceable by the other because of their difference in accumulation mechanisms.

Lichens and mosses are also used in the bioremediation of metals. Like biomonitoring, the technique of bioremediation is also age-old and has been utilized using the various classes of lower organisms. Considering the present-day environmental contamination, remediation techniques are important to mitigate the harmful effects of heavy metals. The demand of the situation is a robust methodology that can be at par with other technological methods and are also cost-effective and environment friendly (Volesky and Holan 1995, Sesli and Denchev 2008). Researchers have used varied organisms like *Arabidopsis thaliana* (Chaney et al. 1997), *Schizosaccharomyces pombe* (Ha et al. 1999), microalgae (Perales-Vela et al. 2005), model bacteria species (Mullen et al. 1989), and rhizospheric bacteria (Macek et al. 2007) to control environmental pollution using bioremediation. Mosses and lichens make potential contenders for bioremediation of heavy metals because of their biological design, and several successful studies support this idea (Pesch and Schroeder 2006, Kulkarni et al. 2014). Their ease of bioaccumulation naturally makes them good at bioremediation as well.

3.2 History of Bioaccumulation Research

3.2.1 Bioaccumulation in Mosses

Research on metal bioaccumulation in moss became frequent with the use of these organisms as a biomonitor for atmospheric deposition. Towards the end of 1960, large-scale research was executed in Scandinavia for the first time and later national level surveys in the 1970s and 1980s in countries like Sweden, Norway, and Denmark (Steinnes 1977). Since feasible techniques of biomonitoring studies evolved in Sweden, biomonitoring studies became frequent (Tyler 1970). The International Cooperative Programme on Effects of Air Pollution on Natural Vegetation and Crops (ICP Vegetation) of Europe did a routine evaluation of moss bioaccumulation for atmospheric monitoring after an interval of every five years since 1990 (Rühling and Steinnes 1998, Buse et al. 2003, Harmens et al. 2010). The latest evaluation by ICP Vegetation was conducted in 2010–2011 at 4500 sites of 25 countries that took part in the survey (Harmens et al. 2013). Most elevated levels of metal accumulation in mosses were recorded in 1990. Further, as a result of growing concern for environmental quality, metal concentration began to decline in all the countries. Since 1990, the metal accumulation declined most (45–72%) for arsenic, cadmium, iron, lead, and vanadium, followed by a 20–30% decline for copper, nickel, and zinc. Chromium and mercury showed the least percentage decline with 2% and 12%, respectively (Harmens and Norris 2008). Many regional surveys evaluating moss bioaccumulation for atmospheric studies were also carried out fundamentally in North America (Pott and Turpin 1996). *Sphagnum russowii* was evaluated using the moss-bag technique in the USA to trace atmospheric pollutants (Makholm and Mladenoff 2005). Countries like Romania, Russia, and Bulgaria also performed similar surveys using *Sphagnum girgensohnii* to investigate metal accumulation (Culicov et al. 2005). Research on moss biomonitoring and bioaccumulation thus stretched extensively to other parts of the world, and now we can uncover spatial and temporal differences, compare among contamination levels in geographically different areas, understand the influence of long-range transport and identify sources of emissions (Alvarez Montero et al. 2006, Charakrabortty and Paratkar 2006, Lee et al. 2004). Popular moss species considered in most research works are *Hypnum cupressiforme*, *Hylocomium splendens*, and *Pleurozium schreberi*. Kohler and Peichl (1993) investigated *H. cupressiforme* and *P. schreberi* and found that the former species was a better bio-accumulator. Many other researchers supported *H. cupressiforme* as a better accumulator (Suchara and Sucharova 1998, Carballeira et al. 2008, Schroder et al. 2008). Conversely, Galsomies et al. (2003) found *P. schreberi* to be a better accumulator as compared to *H. cupressiforme*. This difference in results can be explained by the influence of habitat on moss's accumulative ability. The immediate environment and factors like precipitation, temperature, etc., can influence the bioaccumulation in mosses, and hence a universal observation cannot be expected.

The first study of using moss as a biomonitor in India was carried out by Gupta (1995) in Shillong, India, where metal accumulation was studied in three moss species viz. *Plagiothecium denticulatum*, *Bryum argenteum* and *Sphagnum* sp. Studies carried out in the vicinity of Mumbai have recognized *Pinnatella alopccuroides, Pterobryopsis flexiceps*, and *Bryum* sp. as potential species for biomonitoring studies (Chakrabortty and Paratkar 2006). Saxena et al. (2008a) carried out a retrospective metal accumulation study from 1895–1999 using herbarium voucher specimens of *Barbula* sp. and reported a continuous percentage increase in Pb, Fe, Zn, Cd, and Ni. Saxena et al. (2008b) used *Rhodobyrum giganteum* and *Hypnum cupressiforme* to evaluate seasonal and temporal differences of atmophile metals (Cd, Zn, Cu, and Pb). Singh et al. (2017) evaluated *R. giganteum* of Kumaon and Garhwal hills of the Himalayas and observed high concentrations of Cu, Cd, Zn, and Pb in areas with high traffic density. Srivastava et al. (2014) studied bioaccumulation in *Barbula constricta* J. Linn to evaluate the seasonal depositional pattern in Mussoorie hills of the Himalayas and found the highest concentration of metals in summer than in winter and monsoon. In India, moss bioaccumulation investigations are mainly focused on understanding deposition patterns, seasonal and temporal trends, and identifying sources of emission. Other research articles using mosses as biomonitors for air pollutants (e.g., Saxena and Saxena 1999, Saxena et al. 2000, Chakrabortty et al. 2004, 2006, Saxena and Saiful-Arfeen 2006, Saxena and Afreen 2010, Saxena and Saiful-Arfeen 2010, Saxena et al. 2013, Mahapatra et al. 2019) have also been carried out in India. Still, studies are scarce in comparison to the rest of the world.

3.2.2 Bioaccumulation in Lichens

The application of lichens to the study of air pollution can be traced with Nylander's observations in Paris of 1866, but it was not until the 1960s when sulfur dioxide was identified as a major influence on lichen growth that biomonitoring studies for lichens took off. Since then, lichens have become popular in the application of air quality assessment (Ferry et al. 1973, Nimis et al. 1990).

Due to the physiological peculiarities, the lichens tend to be at risk from stressful conditions like anthropogenic emissions, climate change, or changes in forest structure (Giordani 2019). Extensive research in this field has been conducted in the European nations. The chosen research areas always leaned more towards urban, industrial, or waste disposal sites rather than natural ecosystems (Abas 2021).

In earlier works, the focus was more on the assessment of elemental concentration, but now the focus is seen to have concentrated towards understanding the physiochemical changes in lichens due to external elements and effects on lichen diversity and distribution.

A study by Chettri et al. (1997) using lichens *Cladonia convoluta* and *Cladonia rangiformis* indicated higher Cu and Pb deposition in dead thalli of *Cladonia*, while contradictory results were recorded for Zn. It can be inferred that passive uptake processes have an impact on the elemental accumulation in transplanted lichen as an overall trend of higher element accumulation in dead thalli (Cecconi et al. 2021). Chettri et al. (1997) also noted elevated deposition of Cu and Pb over Zn. Elevated levels of Zn content were observed inside the cells of living thalli (Fortuna et al. 2017). In the presence of a given element in excess, *Cladonia cariosa* species showed a defense against excessive intracellular uptake giving an insight into how pioneer species may probably colonize highly toxic environments (Rola 2020). Metal tolerant species have adaptive photobionts, so an increase in metal accumulation causes deceleration of chlorophyll degradation. Although pigment degradation is observed, no negative impact on photosynthetic ability was observed (Rola et al. 2019).

Pre-exposure and post-exposure studies for two foliose lichens, *Flavopunctelia soredica* and *Rhizoplaca chrysoleuca*, by Zhao et al. (2019) showed element concentrations to be species-specific and element-specific under similar observations. Generally, the metal concentrations in crustose lichens were higher than in foliose or fruticose lichens, as observed in an extensive comparative study by Chiarenzelli et al. (1997). Dependence on overall morphology has been speculated as a reason for the greater entrapment capacity of crustose lichens (Chiarenzelli et al. 1997).

In the Indian context, several lichenological studies have been conducted in the decades-spanning a diverse landscape and against a variety of pollution sources using both passive (on-site) and active (transplant) monitoring (Kumari 2019). The lichen bioaccumulation data in the country has been widely used to assess the air quality (quantification of metal pollutant) (Bajpai et al. 2004, 2010) of the studied areas, but lesser focus on regional patterns of pollution and its dispersal effects have been afforded. Bajpai et al. (2022) used the lichen *Phaeophyscia hispidula* to investigate spatial deposition trends for metals between the mountains and plains of north-western India using spatial interpolation technique. It was observed that along with local sources, wind direction and other meteorological conditions had an impact on the spatial behavior of metallic pollutants in the lichens. *Pyxine cocoes*, *Pyxine subcineria*, and *Phaeophyscia hispidula* are examples of some common lichen species found in the literature in the studies based in the Indian subcontinent. A study by Singh et al. (2018) in Srinagar and surrounding Garhwal hills of Western Himalaya to assess heavy metal accumulation by epiphytic foliose lichen species (*Canopermalia texana*, *Pyxine subcineria*, *Phaeophyscia hispidula*) showed *C. texana* to be sensitive and found to be only in minimum exposure areas, while *P. hispidula* and *P. subcineria* were regarded as tolerant species with a high accumulation capacity. Shukla and Upreti (2007) reported that metal accumulation affected the physiology of *P. hispidula*. Singh et al. (2018) also demonstrated the effect of traffic pollution on tree bark, which influences epiphytic lichen species composition of the area as many vulnerable species are rendered unviable in those conditions.

Extensive lichen biomonitoring research has been conducted in the Northern region of India with substantial work in the Garhwal hills and northern plains (Bajpai et al. 2004, Saxena et al. 2007, Shukla and Upreti 2008, 2009, Gupta et al. 2015, Kumar et al. 2021), followed by some studies in the Brahmaputra valley (Daimari et al. 2020, Singh et al. 2019, Chetia et al. 2021). Although past lichenology study in Assam leans more toward diversity assessment, Das et al. (2012, 2013) and Singh et al. (2019) are some bioaccumulation monitoring works (near paper mill factories) using lichens in the region. Daimari et al. (2021) also attempted to understand trace metal accumulation and its resulting anatomical, physiological and chemical alterations in lichens of Brahmaputra valley. According to Shukla and Upreti (2007), Bajpai et al. (2010), Danesh et al. (2013), Singh et al. (2019), *Pyxine cocoes* is one of the most toxicity resilient lichen species found in India. The study by Kumari (2019) demonstrated *Physcia dilata* as a viable species that can tolerate emissions from coal-burning. Furthermore, physiological parameters like total chlorophyll, carotenoid, and protein content have been reported to be maximum at the least polluted (control) site showing a direct effect of pollutant accumulation on lichen physiology.

In recent years, the impact of climate change on the biota has gained much attention. It has been reported that the cause of dominance of some species could be due to climate change and consider them as climate-resilient species (Singh et al. 2019), like members of Caliciaceae and Physicaceae families, more specifically *Phaeophyscia* and *Pyxine* (van Herk et al. 2002).

3.3 The Concept of Biomonitoring

The idea of biomonitoring dates back to the 20th century and uses the bioaccumulative nature of organisms to monitor atmospheric metallic pollutants. At times, there arises a conflict of understanding between the two terms "bioindicator" and "biomonitor". Bioindicator broadly speaks about any organism that can respond to a changing environment, whereas biomonitors provide quantitative data to understand the quality of the environment (Markert et al. 2003). The biomonitoring technique uses the responses of an organism as a function of its external environment to evaluate changes in the environment. Researchers have used vascular plants, mosses, lichens, and even animal tissues to study the bioaccumulation of metallic pollutants (Jeddi et al. 2021, Bartholomew et al. 2020, Turkyilmaz et al. 2018, Malikova et al. 2019). With the selection of a suitable biomonitor, monitoring atmospheric pollution is feasible almost anywhere from urban to rural habitations. The major advantages of the biomonitoring technique lie in replacing the long-term use of expensive samplers, cost-friendly and easy sample collection methods, and the capacity of biomonitor organisms to absorb atmospheric deposition over a more extended period (Chakrabortty and Paratkar 2006). Of biomonitor organisms, mosses and lichens stand out as an outstanding integrative tool (Rühling and Tyler 1968, Szczepaniak and Biziuk 2003). Biomonitors can be sensitive type (where a visual change in the organism due to pollutant stress is noted) or accumulative type (where the concentration of pollutants stored in its body is measured), but lichens are a great example of biomonitors that can act as both sensitive and accumulative type. Mosses can withstand vulnerable conditions, and their large surface area aids easy absorption. The biomonitoring technique has been frequently used for evaluating atmospheric bioaccumulation in different regions like industrial (Loppi and Bonini 2000, Frati et al. 2007), urban (Saeki et al. 1977, Gombert et al. 2004), and forests (Rühling and Tyler 1973, Steinnes 1995, Loppi and Pirintsos 2003). Anthropogenic emissions are an ongoing and never-ending issue. This has caused serious deterioration of environmental quality by accumulation into groundwater, soil system, and organisms, thereby interfering with the food chain. This necessitates careful monitoring of atmospheric deposition, for which biomonitoring is the best-suited method (Markert et al. 1999). A networked and global biomonitoring research would facilitate us with many insights, but the monitoring network for metallic pollutants remains scarce.

3.4 Selecting A Biomonitor

Selecting a good biomonitor is vital for smooth sample collection and achieving accurate results. The first crucial criterion considered for atmospheric deposition research is that the biomonitor should specifically accumulate from the atmosphere only (Rühling 1994). Other requirements to keep in mind while selecting the monitor organism are: easy availability in the area of interest, occurrence during all seasons, tolerant to relevant levels of pollution, element uptake remains unaffected by local conditions, mechanisms of elemental uptake should be known to better understand results, and the biomonitor should have low background concentrations (Chakrabortty and Paratkar 2006). An ideal biomonitor facilitates both continuous and retrospective monitoring of atmospheric depositions and is also cost-effective. When the goal is to achieve time-averaged trace-element concentrations, such non-mobile monitors are preferred.

In biomonitoring studies, large-scale mapping experiments normally use indigenous (i.e., naturally growing) mosses. In areas where the desired moss species are scarce or unavailable during all seasons, transplantation techniques are used (Goodman and Roberts 1971). Out of the two kinds of moss, epigeic mosses (mosses that grow on the ground) and epiphytic (growing in tree trunks), the former is mostly preferred for biomonitoring surveys (Steinnes 2000). It is clever not to use epiphytic mosses because the wet or dry deposition is interfered with by the canopy cover before reaching the moss, and elemental leaching from trees, specifically from leaves, might settle onto moss surface, thus creating bias. However, in cold regions that might typically lack suitable epigeic moss species, epiphytic species can be used, but one should be careful to collect samples from areas with the lowest possible canopy interference. In a warm climate, epigeic mosses directly grow on mineral soil, and hence soil particles are firmly attached to the rhizoids. In such a case, care has to be taken in the sample cleaning step to separate soil particles from the samples carefully.

Two biomonitoring methods utilizing indigenous lichen flora or transplantation from clean or relatively clean areas to the study site that lack indigenous colonization or expected species are performed to study the level and source of airborne heavy metal pollutants either for large-scale monitoring or individual pollution sources. A study on selected spontaneous lichen species collected *in situ* by Sawidis et al. (1995) found that epilithic lichens have the highest atmospheric heavy metal accumulation due to direct deposition of atmospheric pollution, unlike in epigeic lichen, in which nutrients are absorbed from the substrate. They also indicated the absence of epiphytic lichen near the highly polluted areas in response to SO_2 pollution. Many pieces of literature support the absence of epiphytic lichens in urban and highly polluted areas as these lichens are highly sensitive towards atmospheric pollution (Sugiyam et al. 1976, Bennett and Wetmore 1999, Begum and Harikrishna 2010). Epiphytic fruticose lichen *Pseudevernia furfuracea* (L) Zopf has been widely employed for bioaccumulation studies for a very long time (Lounamaa 1965, Folkeson 1979, Garty and Ammann 1987, Cardarelli et al. 1993, Caniglia et al. 1993, Bari et al. 2001). Epiphytic lichens are preferable over other ecological groups of lichen to evaluate atmospheric heavy metal pollutants because there is less chance of metal uptake from the soil, removing uncertainty in the interpretation of the potency of atmospheric precipitation (Sawidis et al. 1995). Single host tree species are selected for sample collection to prevent a lack of compatibility amongst samples. Substrate influences the elemental composition of epiphytic lichen depending on species of lichen, properties of bark, and the pollution condition of the atmosphere. To reduce these effects, selected lichen species for biomonitoring are preferred to have a loose attachment to the substrate. It is also necessary to select tree species, which are relatively unaffected by stem flow to collect samples (Osyczka and Rola 2019). The difference in age and exposure time develops a zonation pattern in concentrations of certain metals between the central and the peripheral region of lichen thalli (Bargagli et al. 1987, Hale and Lawrey 1985, Schutte 1977, Schwartzmann et al. 1987). When collecting lichen samples, canopy cover and throughfall should be avoided as far as possible. Several studies highlighted that elemental accumulation in lichen concerning air pollution is species-specific (Nimis et al. 2001, Cercasov et al. 2002, Minganti et al. 2003) and element-specific (Zhao et al. 2019). Therefore it is preferable to use the same species throughout the sampling process or carry out interspecies calibration in case more than one species is used (Minganti et al. 2003). Samples collected for a particular biomonitoring study should be of the same age.

3.5 Biological Design

Bioaccumulation studies in lichens and mosses date back to 1866 and 1970, respectively. The extensive use of mosses and lichens as biomonitors for atmospheric pollution is attributed to their "biological design".

3.5.1 Biological Design of Moss

Mosses bear several suitable features that make them a potential biomonitor of atmospheric pollution (Zeichmeister et al. 2003, Hussain and Hoque 2021): they depend solely on wet and dry deposition for nutrients, lack real roots, and hence prevent nutrient uptake from the soil, lack or have weakly developed cuticle allowing easy inflow of metals, high levels of polyuronic acid in their cell wall catalyzes ion exchange, large surface-to-weight ratio increases adsorption, can accumulate pollutants for a more extended period because their rate of growth is slow, and are perennial and widely distributed. Their perenniality allows studies to be conducted throughout the year, thereby making it easier for researchers to track the temporal changes in metal profiles (Migaszewski et al. 2009).

3.5.2 Biological Design of Lichen

The majority of lichen biomass is essentially made up of the fungal component, and it is primarily responsible for the uptake of nutrients accompanied by environmental toxins. The unique morphology, anatomy, and chemistry of lichens, like an absence of a selective outer barrier and the ability to retain trace metals, allow massive storage of pollutants in lichen thalli (Tessier and Boisvert 1999). There are no roots, specialized absorptive organs, and well-developed cuticles in lichens; therefore, both toxic elements and essential nutrients in the surrounding environment and substrate end up accumulating in lichens (Tyler 1989, Bačkor and Loppi 2009). High surface-to-volume ratio and wide intercellular spaces facilitate the high capacity of bioaccumulation (Zhao et al. 2019), which is further aided by the fact that lichens are slow-growing organisms with long life spans (Bačkor and Loppi 2009).

3.6 Monitoring Techniques

3.6.1 Application of Moss

In biomonitoring, different techniques have been used by different researchers owing to what the species or sample site demands. Researchers have used *in situ* collections from sampling points, transplantation of samples, and moss bag techniques. In *in situ* collection, mosses are directly collected from the sample site. However, care has to be taken that samples are collected from areas with the least interference of canopy and throughfall. Most of the national and international studies use the *in situ* technique for biomonitoring studies (Rühling et al. 1987, Rühling and Steinnes 1998, Saxena and Afreen 2010). In the transplantation method, the moss species are transplanted along with the substratum to the desired locations and given exposure for the desired period before analyzing for results (Goodman and Roberts 1971, Markert 1993). However, the transplanted moss is subjected to sudden changes in environment and habitat factors like light, humidity, precipitation, etc. (Tyler 1990, Ceburnis and Valiulis 1999). This technique has certain disadvantages because the moss samples might dry off, leading to loss of their original bioaccumulative property. In the moss bag technique, dried or fresh moss samples are kept in nylon bags and placed in the desired location for a definite time. Most moss bags are made up of nylon mesh in 20 × 20 or 20 × 30 cm sizes. The mesh size may vary from 0.07 cm^{-1} to 1 mesh cm^{-1} (Kelly et al. 1987). Mesh size has been seen to have little effect on metal accumulation; however, a mesh size larger than 0.07 mesh cm^{-1} is recommended (Kelly et al. 1987). The moss bag technique is adopted when an adequate amount of moss is hard to find, especially in urban and industrial areas (Goodman and Roberts 1971, Tyler 1990, Yurukova and Ganeva 1997, Ceburnis and Valiulis 1999, Saxena et al. 2013). It is of utmost importance to record the metal concentration of moss in the transplantation and bag method before exposing it to the desired area.

3.6.2 Application of Lichen

Approaches to monitoring heavy metal pollution using lichens are divided into two possible methods: *in situ* collection and transplantation method. The most common assessment techniques used in lichen transplant are namely bag method and board method. The *in situ* technique is based on sample collection from native lichens. These samples were analyzed to obtain information about the variable exposure time of atmospheric deposition of heavy metals. As lichen has a long life span and slow growth rate, analyzing the whole thallus can give information on long time exposure of lichen to trace elements, while analyzing 2–5 mm of peripheral lobes, the information is obtained for limited time exposure (Bargagli 1998). Transplantation techniques are used where due to air pollution, native lichens are diminished or expected lichen species are not available. In this technique, lichens from little or negligible air pollution regions are transplanted to monitoring sites applying different exposure methods. Lichen is kept in nylon bags with size 8 × 8 cm and mesh size 4 mm attached to plastic coated rope hanged 2 m above the soil by Bari et al. (2001). In another study, following the modified methods of Bari et al. (2001) and Nannoni et al. (2015), lichens were exposed in an 8 × 10 cm nylon net bag and anchored to trees at the height of 2–3 m above ground level (Zhao et al. 2019). The alternative technique is to use aboard. Daimari et al. (2021) fixed lichen thalli with the substrate using glue on a rectangular wooden board, and Bari et al. (2001) attached lichen thalli without substrate using nylon thread to a 50 × 50 cm Teflon board. Lichen thalli with substrate glued to a plastic tube were used as lichen transplants for monitoring by Ayrault et al. (2007). The transplants are mostly hung above 1.5–3 cm above the ground. In the transplantation method, exposed-to-control ratio or EC ratio is used to determine the temporal trends of heavy metal accumulation. The period of exposure of lichen samples in the study site varies from 15 days (Aprile et al. 2010) to 12 months (Bari et al. 2001, Nimis et al. 2001, Cercasov et al. 2002, Conti et al. 2004). It is observed that brief exposure of 1 to 3 months is suitable for detecting atmospheric fallout since long-time exposure may lose some biomass or become saturated with metals deposited from the atmosphere. Also, transplanted lichens exposed for a long time to air pollution significantly alter lichen morphological and physiological performances. A non-metallic roof is advised to install above lichen transplants to protect the lichen from the rain without interrupting air circulation (Ayrault et al. 2007). The significant correlation between most metals deposited in the lichen thalli exposed in a bag and the atmospheric concentration of these metals indicates the efficiency of the bag method to represent the atmospheric concentration (Cardarelli et al. 1993, Caniglia et al. 1993, Levin and Pignata 1995). Compared to the board method, exposure in bags is more effective with a significant correlation with atmospheric trace-metals concentration and higher accumulation capacity, greater water retention, large surface exposure, and better protection against external agents (Bari et al.

2001). Exposure on board seems to negatively correlate metal concentration with exposure time, probably due to stress and depressed accumulation capacity (Bari et al. 2001).

3.7 Mechanism of Bioaccumulation and Factors Affecting Particle Entrapment

3.7.1 Moss

The atmospheric deposition can enter the moss system in three forms: liquid suspension, gaseous mixture, or adhered particles. The size of the particle and the surface of the moss has an impact on particle entrapment. Once the particles have rested on the moss surface, bioaccumulation can happen in different ways, such as entrapment on the cell surface, attaching to the outermost cell wall via the mechanism of ion exchange, via diffusion, and transport channels or proteins to intercellular spaces (Brown and Bates 1990). A schematic representation of the uptake mechanism is shown in Fig. 3.1. Out of all processes, bioaccumulation in moss mostly takes place via cation exchange sites of the cell wall. Mosses possess a strong cation exchange capacity (CEC), which is explained by the number of protons exchangeable with metals ions binding to the tissue (Soudzilovskaia et al. 2010). At these cation exchange sites, metal adsorption is affected by the nature of the element (Rühling and Tyler 1970) and other cations present at exchange sites that can either inhibit or enhance the uptake (Gjengedal and Steinnes 1990). Mosses have a high content of polygalacturonic acid, functional groups like amine, phosphodiester, etc., and negatively charged groups like carboxyl and phosphoryl in their cell wall, which acts as exchangers for metal ions (Wiersma et al. 1990, Popper and Fry 2003). During the exchange of ions, metallic ions attach to functional groups by a process called chelation (Rao 1984). The high counter-gradient of mosses doesn't allow them to avert ions from entering their system (Shakya et al. 2008, Chakrabortty et al. 2006). Elements like Pb, Cd, Co, and Cu accumulate in higher quantities via wet deposition (Ross 1990). Mosses accumulate deposition from both wet and dry deposition, but a large portion is contributed from wet deposition only. The quantity and period of the precipitation affect the deposition and leaching of metals in moss (Berg et al. 1995). Dry deposition starts to contribute more in arid regions (Couto et al. 2004).

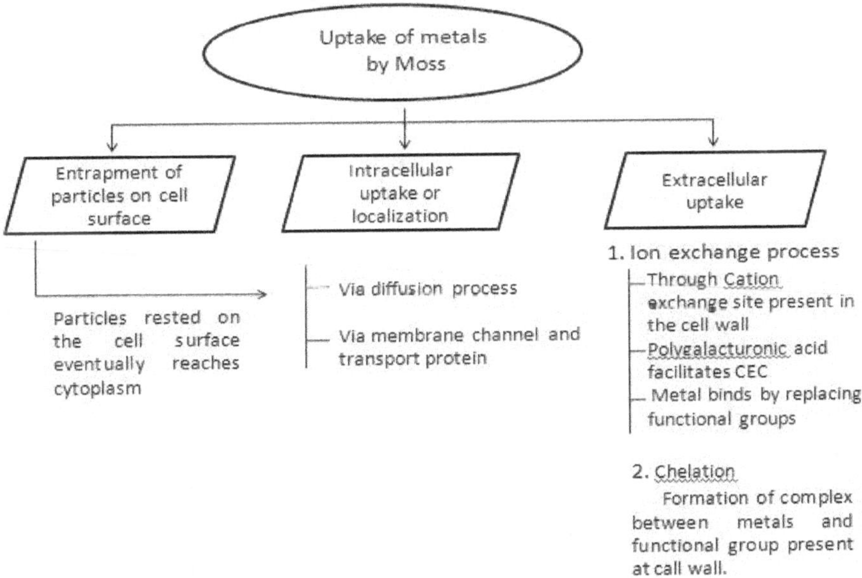

Fig. 3.1. Schematic representation of uptake of metals by Moss.

Besides physiochemical factors, the metal uptake also depends on natural factors. The habitat and environmental conditions of the moss also influence metal accumulation. Thöni et al. (1996) reported physiological differences between moss of the same species grown in different habitats. These naturally imposed differences in the morphology and physiology of the moss also, in turn, affect bioaccumulation. In areas with sparse vegetation or mineral-rich soil, Fe, Cr, Al, and Ti concentrations naturally increase in the moss (Mäkinen 1994). Some habitat-dependent factors influencing bioaccumulation include throughfall from canopies, leaching from vegetation layers, the amount of precipitation, interference of street dust, and long-range transport (Steinnes 1993, Zechmeister 1995, Gerdol et al. 2002).

3.7.2 Lichen

Lichens usually take up nutrients from the environment through three mechanisms—adsorbing particles in the surface of the thalli, binding of ions via extracellular exchange sites, or intracellular solubilization of ions. A schematic representation of the uptake mechanism is shown in Fig. 3.2. The larger part of pollutants accumulated in lichens is through extracellular trapping (Tretiach et al. 2011), which helps prevent toxicity at the cellular level. This is because the cell walls contain many compounds like chitin, glucans, polyketides having polyanionic functional groups (Cecconi et al. 2021) that can bind metal ions (Sarret et al. 1998) or form metal oxalates (Chisholm et al. 1987), and metal-lichenic acid complexes (Pawlik-Skowrońska and Bačkor 2011). Toxins having a greater affinity towards such functional groups keep continuously accumulating even in dead thalli (Chettri et al. 1997). The binding process might speed up with the death of the cells since previously masked exchange sites are exposed. Mechanisms like the extracellular binding of metals with chitin, organic acid, or secondary metabolites are vital for self-protection as well as protection of the photobiont. The photobiont has also been observed to have a high accumulation capacity, so it can be expected to play some role in the cation binding as well. Extracellular binding of cation is a fast and passive

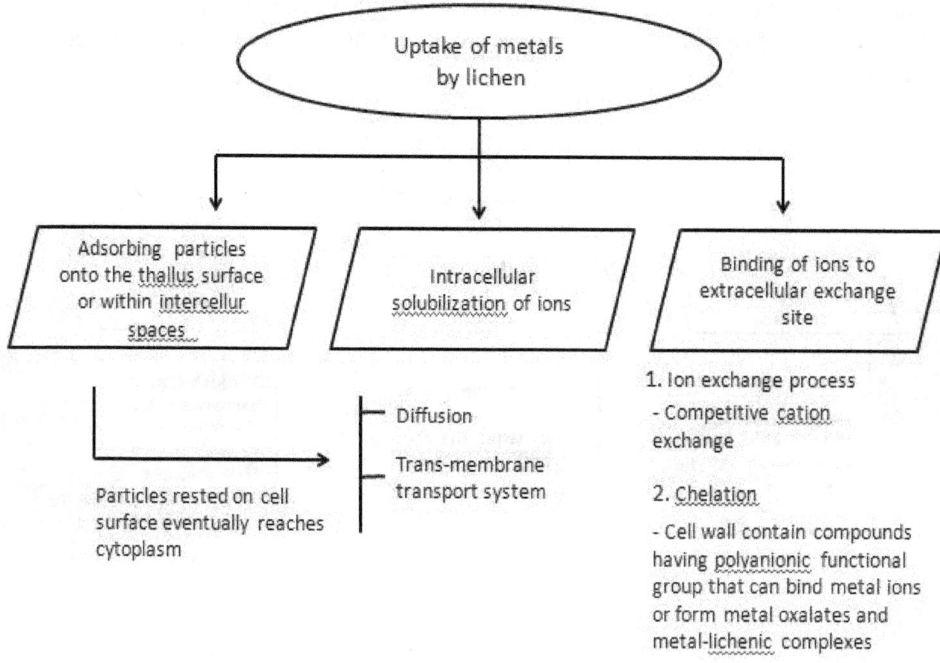

Fig. 3.2. Schematic representation of uptake of metals by Lichen.

mechanism that occurs together with the release of protons. Cations at the exchange site can be displaced by other cations with higher binding affinity with the specific sites, and hence the process is reversible in nature. On the contrary, the process of intracellular uptake is slow, irreversible, requires energy, and is selectively controlled by the plasma membrane. It can be suggested that lichens found in more toxic environments display more effective detoxification mechanisms than less polluted sites. Excess concentrations of Zn, Cd, and Pb have been seen to defend against the synthesis of phytochelatins from glutathione by lichen photobionts (Pawlik-Skowrońska et al. 2006, Bačkor et al. 2007). *Diploschistes muscorum*, a metal hyperaccumulator, was observed by Sarret et al. (1998) to tolerate abnormally high heavy metal concentrations through metal-oxalate production or the complex formation of metals with diploschistesic and lecanoric acid.

Referring to Nieboer et al. (1978), temperature may have an impact on metal accumulation as it has been observed to increase with temperature rise. Dron et al. (2021) observed mild temperature and high humidity to be more favorable for accumulation. Placement or location influences the availability of HM for lichen entrapment, most likely concerning different degrees of exposure to atmospheric depositions. Accumulation can be expected to be higher at high altitudes (Král et al. 1989). Low pH conditions allow greater solubility of metal species in the environment for lichens to take up. Many species can entrap more particles due to having extensive branched thalli systems or large intercellular spaces within the thalli (Bargagli 2016). Rainwater may dislodge accumulated substances from the surface of the thalli (Brown and Brown 1991), showing lower concentration levels of elements during rainy seasons. Factors like the location of the element trapped in the lichen (Brown and Brown 1991), outside meteorological conditions, the pollution level of the surrounding area (Rola 2020), varying accumulation capacity of thalli, and even the growth rate of the lichen (Osyczka and Rola 2019) may result in spatial and temporal variations in element entrapment observed in field studies. Chemical and physical factors like chemical structure, solubility, acidity, temperature, and the accumulated amount of entrapped particles also determine the toxicity of the metals to lichen.

3.8 Heavy Metal-Induced Modifications

3.8.1 Modifications in Moss

Bioaccumulation of metals into moss systems causes various physiochemical changes in their system. Heavy metal accumulation triggers reactive oxygen species (ROS) formation like superoxide radicals (O_2), hydrogen peroxide (H_2O_2), hydroxyl free radicals (OH), singlet and alpha oxygen, etc., that further create oxidative stress in the cell leading to damage of macromolecules (Sun et al. 2010, Aydogan et al. 2017). The ROS production increases lipid peroxidation in cells releasing products like Malondialdehyde (MDA) that can be quantified from cell protoplast to measure the degree of stress (Cobbett and Goldsbrough 2002, Ogunkunle et al. 2016). Panda and Choudhury (2005a) observed that on the application of Cu, Zn, and Cr in *Polytrichum commune*, MDA levels increased. This result was backed by observations made by Sun et al. (2010) in mosses *Hypnum plumaeforme* and *Brachythecium piligerum* under different concentrations of HM.

Another indication of metal toxicity is the increased concentration of carotenoids (Vajpayee et al. 2001). However, the nature of metal and moss species may create both an increase or decrease of carotenoids. Aydogan et al. (2017) observed a decline in carotenoid in *Pleurochaete squarrosa* with Ni application, but in *Timmiella barbuloides*, carotenoid increased with Cr and Pb application. Another indicator for metal toxicity is total chlorophyll concentration and chlorophyll fluorescence (Chen et al. 2019). High Cr concentration has been seen to disrupt the ultrastructure of chloroplast, thus disturbing the photosynthesis process (Panda and Choudhury 2005b, Oliveira 2012). More studies have observed changes in the ultrastructure of thylakoid under metal stress

(Basile et al. 2009, 2017, Esposito et al. 2018). Copper stress on moss *Rhytidiadelphus squarrosus* has been seen to decrease the count of total chlorophyll (Wells and Brown 1987, Ceburnis et al. 1997). Sun et al. (2009) observed a decline in chlorophyll content under Pb and Ni treatment. A decline in chlorophyll content was reported by Shakya et al. (2008) as well in *Thuidium sparsifolium* and *T. delicatulum* under mixed concentrations of heavy metals. Over the years, studies have established a linear relationship between the decline in chlorophyll content and intracellular heavy metal concentration, regardless of the exposure duration. Heavy metals Pb and Cu can disrupt the biochemical and physiological processes by lowering nitrogen uptake efficiency, finally leading to biological membrane damage. The decrease in nitrate reductase activity was confirmed by Panda and Choudhury (2005a) in moss *Polytrichum commune* under Cu, Zn, and Cr treatment. Thus, a decrease in nitrate uptake efficiency serves as another indicator of metal accumulation in moss (Choudhury and Panda 2004, Kapusta and Godzik 2013). Copper is part of enzymes that help in the functioning of the electron transfer chain (ETC) and also help in the reduction of ROS at suitable concentrations (Abreu and Cabelli 2010). Besides, high Cu concentration can alter or block the primary role of the enzyme and cause an increase in ROS concentration, degradation of the membrane, and further cause cell toxicity (Yruela 2005, Sharma 2009). It is also worth mentioning that the effects of metal accumulation are based on exposure time, quantity, and metal species (Carginale et al. 2004, Rau et al. 2007). The degree of toxicity can also change depending on whether the metals are attached to the cell wall or stored inside the cells (Brune et al. 1995, Carginale et al. 2004, Rau et al. 2007).

3.8.2 Modifications in Lichen

Lichen's response to pollution is documented in the literature, employing both field observation and laboratory experiments. Historically, qualitative observation of lichen, like change at the community level and morphological alteration of particular species, served as an indicator of pollutants. Quantifiable measurement of atmospheric pollutants and modification in the physiology of lichens has been used to indicate the presence of pollution (Osyczka and Rola 2019). Among the first visible lichen alterations towards the heavy metal polluted environment are bleaching and change in coloration, chlorotic and necrotic patches, convolution, impaired thallus development, detachment from the substratum, and change in layer thickness in cross-section (Scott and Hutchinson 1989, Otnyukova 2007). Some metal pollutants also cause modification of external morphology, induce calcium oxalate accumulation and reduce the efficiency of sexual reproduction at the intraspecific level (Nash and Gris 1995, Otnyukova 1997, Cuny et al. 2004, Paoli et al. 2015, Mateos and Gonzalez 2016). Goyal and Seaward (1982) showed that the morphological modifications in lichen are affected by high levels of metal accumulation and translocation from the external medium to and within the thallus when exposed to metal-polluted air. They reported that the accumulation of metal cations was the highest at the rhizinae from which metals can move freely to other parts of the thallus. These accumulated metals cause a reduction in rhizinal length and thallial size, dense rhizinal growth, hypertrophy of the medulla, and malformed veins. Munzi et al. (2009) reported membrane damage in lichen due to intercellular deposition of HM. Garty et al. (1985) found chlorophyll degradation in lichen under heavy metal exposure, which was experimentally established in later researches (Bačkor and Zetekova 2003, Bajpai et al. 2013). The lichen species *Cladonia pleurota* gathered from areas having Cu, and Zn contamination had less concentration of photosynthetic pigments, thus less efficiency of photosynthesis (Bačkor and Fahselt 2004). The biomass of the algal partner in lichen symbiosis, which is responsible for photosynthesis, is usually much lower than the fungal partner. Heavy metal contamination induces a tremendous decline in chlorophyll and carotenoid concentration in areas with heavy metal pollution (Bačkor et al. 2003, 2009, Vantová et al. 2013) and also cause disorder in chlorophyll synthesis (Rola et al. 2019). Heavy metals like Pb and Cd hold a powerful affinity for cell wall ligands (Osyczka and Rola 2019). In lichens, essential microelements like Zn and Cu are part of several enzymes (VanAssche and Clijsters 1990), and thus can readily pass through the plasma membrane (Bačkor and Loppi 2009), resulting in greater accumulation in the cell. Heavy metal pollution creates oxidative stress, increases lipid peroxidation, influences antioxidant activity, mutilates integrity of cell wall and chlorophyll pigment, reduces photosynthetic efficiency, and changes the concentration of assimilation pigments (Pisani et al. 2009, Carreras et al. 2005, Monnet et al. 2006, Munzi et al. 2009, Bajpai et al. 2010, 2012, Paoli et al. 2010, Karakoti el al. 2014, Bačkor et al. 2010, Bajpai et al. 2010, Gonzalez and Pignata 2000). In lichens, analyzing electric conductivity is the most sensitive parameter to detect environmental stress on physiological parameters (Mulgrew and Williams 2000). Many pieces of literature have evaluated deterioration in transplanted lichen such as degradation of chlorophyll (Garty et al. 1985), decrease in ATP content, rate of respiration, malondialdehyde (MDA) content, and increase in stress-ethylene concentration (Garty et al. 1985) exposed to heavy metal pollution. Transplantation around landfills showed discoloration, peroxidation of lipid, necrosis, and decreased ergosterol content due to increased heavy metal concentration in air (Paoli et al. 2015). Heavy metals like Cd, Cr, Cu, and Ni found around landfills have a toxic impact on the integrity of chlorophyll (Bačkor et al. 2009, 2010, Unal et al. 2010). A significant correlation of ergosterol with the quantity of metabolically active fungal cells (Ekblad et al. 1998) established it as a potential indicator of mycobiont. The activity of respiratory dehydrogenase is considered a viability indicator for heavy metal exposure (Bačkor and Fahselt 2005) as heavy metal concentrations negatively correlate with decreased activity of respiratory dehydrogenase (Bačkor 2011, Pisani et al. 2011).

3.9 Lichen vs. Moss

There has been a difference of opinion among the research fraternity about which one is a better bioaccumulator or biomonitor between lichen and moss. Results vary based on the species of lichen or moss, type of pollutant, and environmental factors of their habitat. Lithophilic elements have a predominant tendency to accumulate more in mosses as compared to lichens (Adamo et al. 2008, Kłos et al. 2018, Loppi and Bonini 2000). This is more common in arid areas and places with scanty vegetation, where moss accumulates lithophilic metals more easily than lichens (Loppi and Bonini 2000). However, surveys conducted in Finland do not demonstrate the same because there is adequate vegetation cover and hence the quantity of dry deposition is comparatively very less (Berg et al. 2001). Bargagli et al. (2002) observed that epigeic mosses are more prone to accumulate lithophilic elements due to their prolonged exposure to soil and street dust, whereas lichens are efficient accumulators of atmophilic metals. This observation can be supported by extensive studies conducted on epiphytic lichens having higher Pb and Zn concentrations as compared to mosses (Bargagli et al. 2002). Leaching of metals like Cu, Pb, and Zn from leaves and barks of trees into lichens may also be another cause for recording high concentrations of these metals (Steinnes 1993, Adamo et al. 2008). Basile et al. (2008) studied metal accumulation in different species of lichens and mosses and found that different species had different accumulative capacities; however, mosses had comparatively higher elemental accumulation and displayed continuous and linear accumulation, unlike lichens, pointing towards mosses being better bioaccumulators. Studies have also shown epiphytic lichens to be better accumulators of HM than mosses (Kansanen and Venetvaara 1991).

In Finland, the concentration of metals was reported to be higher in moss samples closer to the source of emission and lower in background areas, as compared to epiphytic lichens (Chakrabortty and Paratkar 2006). In Northern Europe, higher concentrations of HMs were found in epiphytic lichens than in mosses because lichens are exposed to atmospheric pollutants around the year, whereas mosses are covered in snow for a considerable period of time (Chakrabortty and Paratkar 2006). Wolterbeek et al. (1996) recommended moss over lichen because they easily depict regional changes in the accumulation of metals. According to a 2018 study in Poland (Kłos et al. 2018), there appeared to be a difference in uptake by lichens and mosses depending on the metal species. The effect of seasonality on mosses was not very clear, but in lichens, seasonal changes in metal uptake could be supposed. The difference in nutrient uptake mechanisms between mosses and lichens could be accountable for this dissimilarity. The variability in uptake efficiency can also be attributed to varied deposition states and the impact of throughfall on epiphytic lichens (Steinnes 1993). Mosses and lichens are seen to reflect wet deposition in different ways. Studies have led to the conclusion that just like lichens have an effect of throughfall or leaching in their metal concentrations, the concentration profile of mosses may be interfered with by precipitation. A study on *Hylocomium splendens* observed fast cation exchange capacity (CEC), which led to the conclusion that this species reflects immediate precipitation profile instead of accumulation occurring over time (Brown and Brûmelis 1996). Reimann et al. (1999) reported that the concentration of elements in mosses correlated with rainwater composition rather than with yearly accumulation depicted by terricolous lichens (Bargagli 1998).

It is difficult to declare one of the two as a better biomonitor due to the variation in results observed across the globe. However, both mosses and lichens are irreplaceable by the other because of their difference in accumulation mechanisms. Epiphytic lichens serve as a better accumulator than mosses in dry environments like cold and hot deserts (Chakrabortty and Paratkar 2006), whereas epigeic mosses are free from interference of throughfall and leaching. Thus, choosing one of the two will depend on the area of interest and the environmental conditions.

3.10 Bioremediation Potential of Mosses

Mosses particularly use the phytochelatin pathway for heavy metal metabolism. In this pathway, once the metal enters into the moss system through the surface of contact, Phytochelatin Synthase (PCS) enzyme gets activated. PCS cleaves Glutathione (GSH), a tripeptide (Glu-Cys-Gly) at C-terminal (Gly), conjugating the remaining Glu-Cys dipeptide to other Glu-Cys dipeptides. This leads to the production of metal-chelating peptides termed phytochelatins that bonds with metal ions. In this conjugated form, metal ions are no longer reactive or harmful to plants or humans (Hanquier 2020). The confirmation that mosses used the common phytochelatin plant pathway for heavy metal metabolism and sufficient knowledge on the mechanism of the pathway suggested that we can induce genetic modification into mosses for even better bioremediation (Petraglia et al. 2014). In a study by Pradhan et al. (2017), two moss species *Funaria hygrometrica* and *Physcomitrella patens* were studied to investigate their bioremediation potential. The study reported that the substratum of moss accumulates more heavy metal concentration as compared to shoot particles. Also, *P. patens* was found to be more sensitive and a better accumulator of metals than *F. hygrometrica.* Hanquier (2020) studied and confirmed *Hypnum* spp., A*mblystegium* spp., and *Mnium* spp. to have bioremediation capacity. Saxena and Saxena (2012) studied the bioremediation potential of *Sphagnum squarrosum* to Cd and Cu with external application of GSH and found that on application, bioaccumulation increased. This increase was attributed to the detoxifying and metal binding capacity of GSH. This study also supported the hypothesis that GSH acts as a substrate for PC synthesis (Leopold et al. 1999). Mosses also use surface complexation reactions (Ho and McKay 1999) to immobilize heavy metals from contaminated soil. *Sphagnum* is a fine example of low-cost adsorbent moss for heavy metal immobilization (Babel and Kurniawan 2003, Bailey et al. 1999, Lee et al. 2013). There are a large number of studies using *Sphagnum* for wastewater bioremediation, but only limited articles have reported their potential to bioremediate contaminated soil (Crist et al. 1996, Fine et al. 2005). Aquatic moss species, *Warnstorfia fluitans,* was reported to reduce arsenic (As) content in water by 36–56% (Sandhi and Gregor 2018). This reduction in As from water also reduced the concentration in vegetables irrigated with the treated water. This is an excellent example of how, if we remediate metals from soil and water, it stops the spread chain of metals and prevents them from entering the food chain and further into humans and other animals.

3.11 Bioremediation Potential of Lichen

Heavy metal remediation using physico-chemical methods can end up causing high energy consumptions, require expensive synthetic chemicals, generate huge amounts of toxic sludge, or even be ineffective at low concentrations of metal ions (Abraham 2001, Bingol et al. 2004). This creates an urgent need for innovative, low-cost, environmentally friendly options for the removal of heavy metals. Lichens can extracellularly sequester and accumulate high concentrations of heavy metals as crystals of oxalate or by making a complex with lichen acids. It has been reported that the non-living biomass of lichens can be of good use in removing heavy metals from wastewater, and biomass modification can substantially enhance its biosorption efficiency (Bingol et al. 2009). Bingol et al. (2009) used cationic surfactant modified *Cladonia rangiformis* (L.) to remove chromate ions from wastewater. Some other studies using lichens as biosorbents are the use of *Cladonia furcata* to remove nickel and lead by Sari et al. (2007), use of *Hypogymnia physodes* to remove cobalt by Pipíška et al. (2007), and an attempt to remove lead and chromium using *Parmelina titiaceae* by Uluozlu et al. (2008). Bioremediation of textile dyes is an enzyme-driven process involving laccase, veratryl alcohol oxidase, lignin peroxidase, manganese peroxidase, etc. Bioremediation and activity of these enzymes have been studied (Jadhav and Govindwar 2006, Waghmode et al. 2011), but the presence and activity of these enzymes and the bioremediation potential of lichen were first demonstrated by Kulkarni et al. (2014). They used *Permelia perlata* and observed that it could decolorize SR24 dye (Solvent Red 24 dye) completely within 24 hours, thus marking the onset of lichens as potential bioremediators. *Cladonia substellata* was investigated for its bioremediation potential for Fluvic Neosols, salinized by unsupervised irrigation. Hyperproduction of phenolics and salt concentration in the substrate was observed, pointing towards its bioremediation potential. A study by Koyuncu and Kul (2020) talks about the production of activated carbon from non-living lichen *Centraria islandica* (L.) Ach for application in wastewater remediation. The study also attempts to compare and comprehend the biosorption capability of activated carbon produced and the non-living lichen in removing the malachite green dye in wastewater treatment. This study was a novel step in the indirect use of lichens for remediation purposes. More research is needed in this area to explore similar treatments of environments contaminated by other compounds and metals.

3.12 Conclusion

Mosses and lichens are suitable bioaccumulators used in biomonitoring studies because of their biological design, physiology, and feedback towards environmental pressure at the molecular level allowing the research fraternity to use them as biomonitors. However, a feasible framework or systematized model for biomonitoring using moss and lichen is vital.

References

Abas, A. 2021. A systematic review on biomonitoring using lichen as the biological indicator: A decade of practices, progress and challenges. Ecol. Indic. 121: 107197.

Abraham, T. E. 2001. Biosorption of Cr (VI) from aqueous solution by *Rhizopus nigricans*. Bioresour. Technol. 79(1): 73–81.

Abreu, I. A. and D. E. Cabelli. 2010. Superoxide dismutases-a review of the metal-associated mechanistic variations. Biochim. Biophys. Acta Proteins Proteomics. 1804: 263–274.

Adamo, P., R. Bargagli, S. Giordano, P. Modenesi, F. Monaci, E. Pittao et al. 2008. Natural and pre-treatments induced variability in the chemical composition and morphology of lichens and mosses selected for active monitoring of airborne elements. Environ. Pollut. 152(1): 11–19.

Alvarez Montero, A., J. R. Estévez Alvarez, H. Iglesias Brito, O. Pérez Arriba, D. López Sánchez and H. T. Wolterbeek. 2006. Lichen based biomonitoring of air quality in Havana City west side. J. Radioanal. Nucl. Chem. 270 (1): 63–67.

Aničić, M., M. Tasić, M. V. Frontasyeva, M. Tomašević, S. Rajšić, Z. Mijić et al. 2009. Active moss biomonitoring of trace elements with *Sphagnum girgensohnii* moss bags in relation to atmospheric bulk deposition in Belgrade, Serbia. Environ. Pollut. 157(2): 673–679.

Aprile, G. G., M. Di Salvatore, G. Carratù, A. Mingoand and A. M. Carafa. 2010. Comparison of the suitability of two lichen species and one higher plant for monitoring airborne heavy metals. Environ. Monit. Assess. 162: 291–299.

Aydogan, S., B. Erdag and L. YIildiz Aktas. 2017. Bioaccumulation and oxidative stress impact of Pb, Ni, Cu, and Cr heavy metals in two bryophyte species, *Pleurochaete squarrosa* and *Timmiella barbuloides*. Turk. J. Bot. 41: 464–475.

Ayrault, S., R. Clochiatti, F. Carrot, L. Daudin and J. P. Bennett. 2007. Factors to consider for trace element deposition biomonitoring surveys with lichen transplants. Sci. Total Environ. 372: 717–727.

Babel, S. and T. A. Kurniawan. 2003. Low-cost adsorbents for heavy metals uptake from contaminated water: a review. J. Hazard. Mater. 97(1-3): 219–243.

Bačkor, M. and J. Zeteikova´. 2003. Effects of copper, cobalt and mercury on the chlorophyll content of lichens *Cetrariais landica* and *Flavocetraria cucullata*. J. Hattori Bot. Lab. 93: 175–187.

Bačkor, M., D. Fahselt, R. D. Davidson and C. T. Wu. 2003. Effects of copper on wild and tolerant strains of the lichen photobiont *Trebouxia erici* (Chlorophyta) and possible tolerance mechanisms. Arch. Environ. Contam. Toxicol. 45: 159–167.

Bačkor, M. and D. Fahselt. 2004. Physiological attributes of the lichen *Cladonia pleurota* in heavy metal-rich and control sites near Sudbury (Ontario, Canada). Environ. Exp. Bot. 52: 149–159.

Bačkor, M. and D. Fahselt. 2005. Tetrazolium reduction as an indicator of environmental stress in lichens and isolated bionts. Environ. Exp. Bot. 53: 125–133.

Bačkor, M., B. Pawlik-Skowrońska, J. Bud'ová, and T. Skowroński. 2007. Response to copper and cadmium stress in wild-type and copper tolerant strains of the lichen alga *Trebouxia erici*: metal accumulation, toxicity and non-protein thiols. Plant Growth Regul. 52(1): 17–27.

Bačkor, M. and S. Loppi. 2009. Interactions of lichens with heavy metals. Biol. Plant. 53: 214–222.

Bačkor, M., J. Kováčik, A. Dzubaj and M. Bačkorová. 2009. Physiological comparison of copper toxicity in the lichens Peltigerarufescens (Weis) Humb. and *Cladoina arbuscula* subsp. mitis (Sandst.) Ruoss. J. Plant Growth Regul. 58: 279–286.

Bačkor, M., J. Kováčik, J. Piovár, T. Pisani and S. Loppi. 2010. Physiological aspects of cadmium and nickel toxicity in the lichens *Peltigera rufescens* and *Cladina arbuscula* subsp. mitis. Water Air Soil Pollut. 207: 253–262.

Bačkor, M. 2011. Lichens and Heavy Metals: Toxicity and Tolerance. PavolJozefŠafárik University in Košice.

Bailey, S. E., T. J. Olin, R. M. Bricka and D. D. Adrian. 1999. A review of potentially low-cost sorbents for heavy metals. Water Res. 33(11): 2469–2479.

Bajpai, R., D. K. Upreti and S. K. Mishra. 2004. Pollution monitoring with the help of lichen transplant technique at some residential sites of Lucknow city, Uttar Pradesh. J. Environ. Biol. 25(2): 191–195.

Bajpai, R., D. K. Upreti, S. Nayaka and B. Kumari. 2010. Biodiversity, Bioaccumulation and Physiological changes in lichens growing in the vicinity of coal- based thermal power plant of Raebareli district North India. J. Hazard. Mater. 174: 429–436.

Bajpai, R., A. K. Pandey, F. Deeba, D. K. Upreti, S. Nayaka and V. Pandey. 2012. Physiological effects of arsenate on transplant thalli of the lichen *Pyxine cocoes* (Sw) Nyl. Environ. Sci. Pollut. Res. 19: 1494–1502.

Bajpai, R., N. Karakoti and D. K. Upreti. 2013. Performance of a naturally growing Parmelioid lichen *Remototrachyna awasthii* against organic and inorganic pollutants. Environ. Sci. Pollut. Res. 20(8): 5577–5592.

Bajpai, R., V. Shukla, C. P. Singh, A. Raju and D. K. Upreti. 2022. A geostatistical approach to compare metal accumulation pattern by lichens in plain and mountainous regions of northern and central India. Environ. Earth Sci. 81: 203.

Bargagli, R. 1998. Trace Element in Terrestrial Plants. A Ecophysiological Approach to Biomonitoring and Biorecovery. Springer Verlag, Berlin, New York, pp. 324.

Bargagli, R., F. Monaci, F. Borghini, F. Bravi and C. Agnorelli. 2002. Mosses and lichens as biomonitors of trace metals. A comparison study on *Hypnum cupressiforme* and *Parmeliacaperata* in a former mining district. Environ. Pollut. 116: 279–287.

Bargagli, R. 2016. Moss and lichen biomonitoring of atmospheric mercury: a review. Sci. Total Environ. 572: 216–231.

Bari, A., A. Rosso, M. R. Minciardi, F. Troiani and R. Piervittori. 2001. Analysis of heavy metals in atmospheric particulates in relation to their bioaccumulation in explanted *Pseudevernia furfuracea* thalli. Environ. Monit. Assess. 69: 205–220.

Bartholomew, C. J., N. Li, Y. Li, W. Dai, D. Nibagwire and T. Guo. 2020. Characteristics and health risk assessment of heavy metals in street dust for children in Jinhua, China. Environ. Sci. Pollut. Res. 27(5): 5042–5055.

Basile, A., S. Sorbo, G. Aprile, B. Conte and R. C. Cobianchi. 2008. Comparison of the heavy metal bioaccumulation capacity of an epiphytic moss and an epiphytic lichen. Environ. Pollut. 151(2): 401–407.

Basile, A., S. Sorbo, G. Aprile, B. Conte, R. C. Cobianchi, T. Pisani et al. 2009. Heavy metal deposition in the Italian "triangle of death" determined with the moss *Scorpiurum circinatum*. Environ. Pollut. 157(8-9): 2255–2260.

Basile, A., S. Loppi, M. Piscopo, L. Paoli, A. Vannini, F. Monaci et al. 2017. The biological response chain to pollution: A case study from the "Italian Triangle of Death" assessed with the liverwort *Lunularia cruciata*. Environ. Sci. Pollut. Res. 24: 26185–26193.

Begum, A. and S. HariKrishna. 2010. Monitoring air pollution using lichens species in South Bangalore, Karnataka. Int. J. Chemtech. Res. 2: 255–260.

Bennett, J. P. and C. M. Wetmore. 1999. Changes in element contents of selected lichens over 11 years in northern Minnesota, USA. Environ. Exp. Bot. 41: 75–82.

Berg, T., O. Røyset and F. Steinnes. 1995. Moss *Hylocomium splendens* used as biomonitor of atmospheric trace element deposition: estimation of uptake efficiencies. Atmos. Environ. 29: 353–360.

Berg, T., A. G. Hjebrekke and R. Larseen. 2001. Heavy metals and POPs within the EMEP region 1999. EMEP/CCC Report 9/2001. Norwegian Institute for Air Research.

Bingol, A., H. Ucun, Y. K. Bayhan, A. Karagunduz, A. Cakici and B. Keskinler. 2004. Removal of chromate anions from aqueous stream by a cationic surfactant-modified yeast. Bioresour. Technol. 94(3): 245–249.

Bingol, A., A. Aslan and A. Cakici. 2009. Biosorption of chromate anions from aqueous solution by a cationic surfactant-modified lichen (*Cladonia rangiformis* (L.)). J. Hazard. Mater. 161(2-3): 747–752.

Brown, D. H. and J. W. Bates. 1990. Bryophyte and nutrient cycling. Bot. J. Linn. Soc. 104: 129–147.

Brown, D. H. and R. M. Brown. 1991. Mineral cycling and lichens: the physiological basis. Lichenologist. 23(3): 293–307.

Brown, D. H. and G. Brumelis. 1996. A biomonitoring method using the cellular distribution of metals in moss. Sci. Total Environ. 187(2): 153–161.

Brune, A., W. Urbach and K. J. Dietz. 1995. Differential toxicity of heavy metals is partly related to a loss of preferential extraplasmic compartmentation: a comparison of Cd, Mo, Ni, and Zn-stress. New Phytol. 129: 403–409.

Brunialti, G. and L. Frati. 2007. Biomonitoring of nine elements by the lichen *Xanthoria parietina* in Adriatic Italy: A retrospective study over a 7-year time span. Sci. Total Environ. 387(1-3): 289–300.

Buse, A., D. Norris, H. Harmens, P. Büker, T. Ashenden and G. Mills. 2003. Heavy metals in European mosses: 2000/2001 survey. ICP Vegetation Programme Coordination Centre. Centre for Ecology and Hydrology, Bangor, UK.

Caniglia, G., C. Laveder, C. Zocca, I. Calliari and R. Zorer. 1993. Bioaccumulation of elements in thalli of *Pseudevernia furfuracea* exposed in urban and rural sites. G. Bot. Ital. 127(3): 621.

Carballeira, C. B., J. R. Aboal, J. A. Fernandez and A. Carballeira. 2008. Comparison of the accumulation of elements in two terrestrial moss species. Atmos. Environ. 42: 4904–4917.

Cardarelli, E., M. Achilli, L. Campanella and A. Bartoli. 1993. Monitoraggiodell'inquinamento da metallipesantimediantel'uso di licheninellacittà di Roma. Inquinamento. 6: 56–63.

Carginale, V., S. Sorbo, C. Capasso, F. Trinchella, G. Cafiero and A. Basile. 2004. Accumulation, localisation, and toxic effects of cadmium in the liverwort *Lunularia cruciata*. Protoplasma. 223: 53–61.

Carreras, H. A., E. D. Wannaz, C. A. Perez and M. L. Pignata. 2005. The role of urban air pollutants on the performance of heavy metal accumulation in *Usnea amblyoclada*. Environ. Res. 97: 50–57.

Ceburnis, D., A. Ruhling and K. Kvietkus. 1997. Extended study of atmospheric heavy metal deposition in Lithuania based on moss analysis. Environ. Monit. Assess. 47: 135–152.

Ceburnis, D. and D. Valiulis. 1999. Investigation of absolute metal uptake efficiency from precipitation in moss. Sci. Total Environ. 226: 247–253.

Cecconi, E., L. Fortuna, M. Peplis and M. Tretiach. 2021. Element accumulation performance of living and dead lichens in a large-scale transplant application. Environ. Sci. Pollut. Res. 28(13): 16214–16226.

Cercasov, V., A. Pantelica, M. Salagean, G. Caniglia and A. Scarlat. 2002. Comparative study of the suitability of three lichen species to trace-element air monitoring. Environ. Pollut. 119(1): 129–139.

Chakrabortty, S., S. K. Jha, G. T. Paratkar and V. D. Puranik. 2004. Distribution of trace elements in moss biomonitors near mumbai. Evansia 21(4): 180–188.

Chakrabortty, S. and G. T. Paratkar. 2006. Biomonitoring of trace element air pollution using mosses. Aerosol Air Qual. Res. 6(3): 247–258.

Chakrabortty, S., S. K. Jha, V. D. Puranik and G. T. Paratkar. 2006. Use of mosses and lichens as biomonitors in the study of air pollution near mumbai. Evansia 23: 1–8.

Chaney, R., M. Malik, Y. Li, S. Brown, E. Brewer, S. Angle et al. 1997. Phytoremediation of soil metals. Environ. Biotechnol. 8: 279–284.

Chen, Y. E., N. Wu, Z. W. Zhang, M. Yuan and S. Yuan. 2019. Perspective of monitoring heavy metals by moss visible chlorophyll fluorescence parameters. Front. Plant. Sci. 10: 35.

Chetia, J., N. Gogoi, R. Gogoi and F. Yasmin. 2021. Impact of heavy metals on physiological health of lichens growing in differently polluted areas of central Assam, North East India. Plant Physiol. Rep. 26(2): 210–219.

Chettri, M. K., T. Sawidis, G. A. Zachariadis and J. A. Stratis. 1997. Uptake of heavy metals by living and dead *Cladonia* thalli. Environ. Exp. Bot. 37(1): 39–52.

Chiarenzelli, J. R., L. B. Aspler, D. L. Ozarko, G. E. M. Hall, K. B. Powis and J. A. Donaldson. 1997. Heavy metals in lichens, southern district of Keewatin, Northwest Territories, Canada. Chemosphere 35(6): 1329–1341.

Chisholm, J. E., G. C. Jones and O. W. Purvis. 1987. Hydrated copper oxalate, moolooite, in lichens. Mineral. Mag. 51(363): 715–718.

Choudhury, S. and S. K. Panda. 2004. Induction of oxidative stress and ultrastructural changes in moss *Taxithelium nepalense* (Schwaegr.) Broth. under lead and arsenic phytotoxicity. Curr. Sci. 87: 342–348.

Cobbett, C. and P. Goldsbrough. 2002. Phytochelatins and metallothioneins: roles in heavy metal detoxification and homeostasis. Annu. Rev. Plant Biol. 53: 159–182.

Conti, M. E., M. Tudino, J. Stripeikis and G. Cecchetti. 2004. Heavy metal accumulation in the lichen *Evernia prunastri* transplanted at urban, rural and industrial sites in central Italy. J. Atmos. Chem. 49: 83–94.

Couto, J. A., J. Fernandez, J. R. Aboal and A. Carballeira. 2004. Active biomonitorng of element uptake with terrestrial mosses: a comparison of bulk and dry deposition. Sci. Total Environ. 324: 211–222.

Crist, R. H., J. R. Martin, J. Chonko and D. R. Crist. 1996. Uptake of metals on peat moss: an ion-exchange process. Environ. Sci. Technol. 30(8): 2456–2461.

Culicov, O. A., R. Mocanu, M.V. Frontasyeva, L. Yurukova and E. Steinnes. 2005. Active moss biomonitoring applied to an industrial site in romania: relative accumulation of 36 elements in moss-bags. Environ. Monit. Assess. 108: 22.

Cuny, D., C. van Haluwyn, P. Shirali, F. Zerimech, L. Jérôme and J. M. Haguenoer. 2004. Cellular impact of metal trace elements in terricolous lichen *Diploschistes muscorum* (Scop.) R. Sant.—identification of oxidative stress biomarker. Water Air Soil Pollut. 152: 55–69.

Daimari, R., P. Bhuyan, S. Hussain, S. Nayaka, M. J. Mazumder and R. R. Hoque. 2020. Biomonitoring by epiphytic lichen species—*Pyxine cocoes* (Sw.) Nyl.: Understanding characteristics of trace metal in ambient air of different landuses in mid-Brahmaputra Valley. Environ. Monit. Assess. 192(1): 1–11.

Daimari, R., P. Bhuyan, S. Hussain, S. Nayaka, M. J. Mazumder and R. R. Hoque. 2021. Anatomical, physiological, and chemical alterations in lichen (*Parmotrema tinctorum* (Nyl.) Hale) transplants due to air pollution in two cities of Brahmaputra Valley, India. Environ. Monit. Assess. 193: 101.

Danesh, N., E. T. Puttaiah and B. E. Basavarajappa. 2013. Studies on diversity of lichen, *Pyxine cocoes* to air pollution in Bhadravathi town, Karnataka, India. J. Environ. Biol. 34: 579–584.

Das, P., S. Joshi, J. Rout and D. K. Upreti. 2012. Impact of a paper mill on surrounding epiphytic lichen communities using multivariate analysis. Indian J. Ecol. 39(1): 38–43.

Das, P., S. Joshi, J. Rout and D. K. Upreti. 2013. Lichen diversity for environmental stress study: application of index of atmospheric purity (IAP) and mapping around a paper mill in Barak Valley, Assam, northeast India. Trop. Ecol. 54(3): 355–364.

dos Santos Lima, D. N., A. K. de Oliveira Silva, N. H. da Silva and E. C. Pereira. 2020. Bioremediation of salinized soils by the lichen *Cladonia substellata* fomented by a nitrogen source and gamma radiation. Raega-O Espaço Geográfico em Análise 49: 78–93.

Dron, J., A. Ratier, A. Austruy, G. Revenko, F. Chaspoul and E. Wafo. 2021. Effects of meteorological conditions and topography on the bioaccumulation of PAHs and metal elements by native lichen (*Xanthoria parietina*). J. Environ. Sci. 109: 193–205.

Ekblad, A., H. Wallander and T. Nasholm. 1998. Chitin and ergosterol combined to measure total and living biomass in ectomycorrhizae. New Phytol. 138: 143–149.

Esposito, S., S. Loppi, F. Monaci, L. Paoli, A. Vannini, S. Sorbo et al. 2018. In-field and *in-vitro* study of the moss *Leptodictyum riparium* as bioindicator of toxic metal pollution in the aquatic environment: Ultrastructural damage, oxidative stress and HSP70 induction. PLoS One. 13(4): e0195717.

Ferry, B. W., M. S. Baddeley and D. L. Hawksworth. 1973. Air pollution and lichens. *In*: Ferry, B. W., M. S. Baddeley and D. L. Hawksworth (eds.). Athlone Press of the University of London.

Fine, P., A. Scagnossi, Y. Chen and U. Mingelgrin. 2005. Practical and mechanistic aspects of the removal of cadmium from aqueous systems using peat. Environ. Pollut. 138(2): 358–367.

Folkeson, L. 1979. Interspecies calibration of heavy metal concentrations in nine mosses and lichens. Water Air Soil Pollut. 11: 253–260.

Fortuna, L., E. Baracchini, G. Adami and M. Tretiach. 2017. Melanization affects the content of selected elements in parmelioid lichens. J. Chem. Ecol. 43(11): 1086–1096.

Frati, L., G. Brunialti and S. Loppi. 2005. Problems related to lichen transplants to monitor trace element deposition in repeated surveys: a case study from central italy. J. Atmos. Chem. 52: 221–230.

Frati, L., S. Santoni, V. Nicolardi, C. Gaggi, G. Brunialti, A. Guttova et al. 2007. Lichen biomonitoring of ammonia emission and nitrogen deposition around a pig stockfarm. Environ. Pollut. 146: 311–316.

Galsomies, L., S. Ayrault, F. Carrot, C. Deschamps and M. Letrouit-Galinou. 2003. Interspecies calibration in mosses at regional scale—heavy metal and trace elements results from Ile-de-France. Atmos. Environ. 37: 241–251.

Garty, J., R. Ronen and M. Galun. 1985. Correlation between chlorophyll degradation and the amount of some elements in the lichen *Ramalinaduriaei* (De Not.). Jatta. Environ. Exp. Bot. 25: 67–74.

Garty, J. and K. Ammann. 1987. The amounts of Ni, Cr, Zn, Pb, Cu, Fe and Mn in some lichens growing in Switzerland. Environ. Exp. Bot. 27: 127–138.

Gerdol, R., L. Bragazza and R. Marchesini. 2002. Element concentrations in the forest moss *Hylocomium splendens*: variations associated with altitude, net primary production and soil chemistry. Environ. Pollut. 116: 129–135.

Gjengedal, E. and E. Steinnes. 1990. Uptake of metal ions in moss from artificial precipitation. Environ. Monit. Assess. 14(1): 77–87.

Giordani, P. 2019. Lichen diversity and biomonitoring: a special issue. Diversity. 11(9): 171.

Gombert, S., J. Asta and M. R. D. Seaward. 2004. Assessment of lichen diversity by index of atmospheric purity (IAP), index of human impact (IHI) and other environmental factors in an urban area (Grenoble, southeast France). Sci. Total Environ. 324: 183–199.

Gonzalez, C. M. and M. L. Pignata. 2000. Chemical response of transplanted lichen to different emission sources of air pollutants. Environ. Pollut. 110: 235–242.

Goodman, G. T. and T. M. Roberts. 1971. Plants and soils as indicators of metals in the air. Nature 231: 287–292.

Goyal, R. and M. R. D. Seaward. 1982. Metal uptake in terricolous lichens. II. Effects on the morphology of *Peltigera canina* and *Peltigera rufescens*. New Phytol. 90: 73–84.

Gupta, A. 1995. Heavy metal accumulation by three species of mosses in Shillong, North-Eastern India. Water Air Soil Pollut. 82(3): 751–6.

Gupta, N., V. Gupta, S. K. Dwivedi and D. K. Upreti. 2015. Comparative bioaccumulation potential of *Pyxine cocoes* and *Bacidia submedialis* in and around Faizabad city, Uttar Pradesh, India. *G*-J. Environ. Sci. Tech. 2(6): 86–92.

Ha, S., A. P. Smith, R. Howden, W. M. Dietrich, S. Bugg, M. J. O'Connell et al. 1999. Phytochelatin Synthase Genes from *Arabidopsis* and the Yeast *Schizosaccharomyces pombe*. Plant Cell. 11: 1153–1163.

Hale, M. E. and J. D. Lawrey. 1985. Annual rate of lead accumulation in the lichen *Pseudoparmelia baltimorensis*. Bryologist. 8: 5–7.

Hanquier, Z. 2020. Characterization of heavy metal bioremediation pathways in local moss species. Undergraduate honours thesis, Butler University, Indianapolis, United States.

Harmens, H. and D. Norris. 2008. Spatial and temporal trends in heavy metal accumulation in mosses in Europe (1990–2005). Surveydata is from the centre for Ecology & Hydrology, United Kingdom, pp. 51.

Harmens, H., D. A. Norris, E. Steinnes, E. Kubin, J. Piispanen, R. Alber et al. 2010. Mosses as biomonitors of atmospheric heavy metal deposition: spatial patterns and temporal trends in Europe. Environ. Pollut. 158: 3144–3156.

Harmens, H., D. A. Norris and G. Mills. 2013. Heavy metals and nitrogen in mosses: spatial patterns in 2010/2011 and long-term temporal trends in Europe. Centre for Ecology & Hydrology, Bangor, UK, pp. 63.

Ho, Y. S. and G. McKay. 1999. Pseudo-second order model for sorption processes. Process Biochem. 34(5): 451–465.

Hussain, S. and R. R. Hoque. 2021. Biomonitoring of metallic air pollutants in unique habitations of the Brahmaputra Valley using moss species-*Atrichum angustatum*: spatiotemporal deposition patterns and sources. Environ. Sci. Pollut. Res. 1–18.

Jadhav, J. P. and S. P. Govindwar. 2006. Biotransformation of malachite green by *Saccharomyces cerevisiae* MTCC 463. Yeast 23(4): 315–323.

Jeddi, K., M. Fatnassi, M. Chaieb and K. H. Siddique. 2021. Tree species as a biomonitor of metal pollution in arid Mediterranean environments: case for arid southern Tunisia. Environ. Sci. Pollut. Res. 1–8.

Kansanen, P. and J. Venetvaara. 1991. Comparison of biological collectors of airborne heavy metals near ferrochrome and steel works. Water Air Soil Pollut. 60: 337–359.

Kapusta, P. and B. Godzik. 2013. Does heavy metal deposition affect nutrient uptake by moss *Pleurozium schreberi*?. E3S Web of Conferences, EDP Sciences. 1: 29005.

Karakoti, N., R. Bajpai, D. K. Upreti, G. K. Mishra, A. Srivastava and S. Nayaka. 2014. Effect of metal content on chlorophyll fluorescence and chlorophyll degradation in lichen *Pyxine cocoes* (Sw.) Nyl.: a case study from Uttar Pradesh, India. Environ. Earth Sci. 71(5): 2177–2183.

Kelly, M. G., C. Girton and B. A. Whitton. 1987. Use of moss bags for monitoring heavy metals in rivers. Water Res. 21: 1429–1435.

Kłos, A., Z. Ziembik, M. Rajfur, A. Dołhańczuk-Śródka, Z. Bochenek, J.W. Bjerke et al. 2018. Using moss and lichens in biomonitoring of heavy-metal contamination of forest areas in southern and north-eastern Poland. Sci. Total. Environ. 627: 438–449.

Kohler, J. and L. Peichl. 1993. Moose alsBioindikatoren fur Schwermetalle. Bayerisches Staatsministerium fur Landesentwicklung und Umweltfragen: Munchen.

Koyuncu, H. and A. R. Kul. 2020. Synthesis and characterization of a novel activated carbon using nonliving lichen *Cetraria islandica* (L.) ach. and its application in water remediation: Equilibrium, kinetic and thermodynamic studies of malachite green removal from aqueous media. Surf. Interfaces 21: 100653.

Král, R., L. Krýžová and J. Liška. 1989. Background concentrations of lead and cadmium in the lichen *Hypogymnia physodes* at different altitudes. Sci. Total. Environ. 84: 201–209.

Kuik, P. and H. T. Wolterbeek. 1995. Factor analysis of atmospheric trace-element deposition data in the Netherlands obtained by moss monitoring. Water, Air, and Soil Pollut. 84(3): 323–346.

Kulkarni, A. N., A. A. Kadam, M. S. Kachole and S. P. Govindwar. 2014. Lichen Permelia perlata: A novel system for biodegradation and detoxification of disperse dye Solvent Red 24. J. Hazard. Mater. 276: 461–468.

Kumar, D., A. Pandey, S. Rawat, M. Joshi, R. Bajpai, D. K. Upreti et al. 2021. Predicting the distributional range shifts of *Rhizocarpon geographicum* (L.) DC. in Indian Himalayan Region under future climate scenarios. Environ. Sci. Pollut. Res. 1–15.

Kumari, A. 2019. Heavy metals accumulation and physiological changes in the lichens growing in the vicinity of coal-based thermal power of Kanti (Muzaffarpur), Bihar, India. Int. J. Plant Environ. 5(3): 165–169.

Lee, S. J., M. E. Lee, J. W. Chung, J. H. Park, K. Y. Huh and G. I. Jun. 2013. Immobilization of lead from Pb-contaminated soil amended with peat moss. J. Chem. 1–6.

Lee, Y., P. Johnson-Green and E. J. Lee. 2004. Correlation between environmental conditions and the distribution of mosses exposed to urban air Pollutants. Water Air Soil Pollut. 153: 293–305.

Leopold, I., D. Gunther, J. Schinidt and D. Neumann. 1999. Phytochelatins and heavy metal tolerance. Phytochemistry 50: 1323–1325.

Levin, A. G. and M. L. Pignata. 1995. *Ramalina ecklocnii* as a bioindicator of atmospheric pollution in Argentina. Can. J. Botany. 73: 1196–1202.

Loppi, S. and I. Bonini. 2000. Lichens and mosses as biomonitors of trace elements in areas with thermal springs and fumarole activity (Mt. Amiata, central Italy). Chemosphere 41: 1333–1336.

Loppi, S. and S. A. Pirintsos. 2003. Epiphytic lichens as sentinels for heavy metal pollution at forest ecosystems (central Italy). Environ. Pollut. 121: 327–332.

Lounamaa, K. J. 1965. Studies on the content of iron, manganese and zinc in macrolichens. Annales Botanici Fennici. 2: 127–137.

Macek, T., P. Kotrba, A. Svatos, N. Novakova, K. Demnerova and M. Mackova. 2007. Novel roles for genetically modified plants in environmental protection. Trends Biotechnol. 26(3): 146–152.

Mahapatra, B., N. K. Dhal, A. K. Dash, B. P. Panda, K. C. S. Panigrahi and A. Pradhan. 2019. Perspective of mitigating atmospheric heavy metal pollution: using mosses as biomonitoring and indicator organism. Environ. Sci. Pollut. Res. 26(29): 29620–29638.

Makholm, M. M. and D. J. Mladenoff. 2005. Efficacy of a Biomonitoring (moss bag) technique for determining element deposition on a Mid-Range (375) Km Scale. Environ. Monit. Assess. 104(1-3): 1–18.

Mäkinen, A. 1994. Biomonitoring of atmospheric deposition in the kola peninsula (Russia) and finnish lapland, based on the chemical analysis of mosses. Ministry of the Environment Rapport 4: 1–83.

Malikova, I. N., V. D. Strakhovenko and B. L. Shcherbov. 2019. Distribution of radionuclides in moss-lichen cover and needles on the same grounds of landscape-climatic zones of Siberia. J. Environ. Radioact. 198: 64–78.

Markert, B. 1993. Plants as biomonitors-indicators for heavy metals in terrestrial environment. VCH VerlagsgesellschaftmbH, Weinheim.

Markert, B., O. Wappelhorst, V. Weckert, U. Herpin, U. Siewers, K. Friese et al. 1999. The use of bioindicators for monitoring the heavy-metal status of the environment. J. Radioanal. Nucl. Chem. 240(2): 425–429.

Markert, B. A., A. M. Breure and H. G. Zechmeister. 2003. Definitions, strategies, and principles for bioindication/biomonitoring of the environment. pp. 3–39. *In*: Markert, B. A., A. M. Breure and H. G. Zechmeister (eds.). Elsevier, Oxford.

Martin, M. H. 2012. Biological monitoring of heavy metal pollution: land and air. Springer Science & Business Media, pp. 293.

Mateos, A. C. and C. M. González. 2016. Physiological response and sulfur accumulation in the biomonitor *Ramalina celastri* in relation to the concentrations of SO_2 and NO_2 in urban environments. Microchem. J. 125: 116–123.

Migaszewski, Z. M., A. Gałuszka, J. G. Crock, P. J. Lamothe and S. Dołęgowska. 2009. Interspecies and interregional comparisons of the chemistry of PAHs and trace elements in mosses Hylocomium splendens (Hedw.) BSG and Pleurozium schreberi (Brid.) Mitt. from Poland and Alaska. Atmos. Environ. 43(7): 1464–73.

Minganti, V., R. Capelli, G. Drava, R. De Pellegrini, G. Brunialti, P. Giordani et al. 2003. Biomonitoring of trace metals by different species of lichens (Parmelia) in North-West Italy. J. Atmos. Chem. 45: 219–229.

Monnet, F., F. Bordas, V. Deluchat and M. Baudu. 2006. Toxicity of Cu excess on the lichen *Dermatocarpon luridum*, antioxidant enzymes activities. Chemosphere. 65: 1806–1813.

Mulgrew, A. and P. Williams. 2000. Air Hygiene Report 10. WHO Collaborating Centre for Air Quality Management and Air Pollution Control, Berlin. WHO Report, pp. 165.

Mullen, M. D., D. C. Wolf, F. G. Ferris, T. J. Beveridgem, C. A. Flemming and G. W. Bailey. 1989. Bacterial sorption of heavy metals. Appl. Environ. Microbiol. 55(12): 3143–3149.

Munzi, S., T. Pisani and S. Loppi. 2009. The integrity of lichen cell membrane as a suitable parameter for monitoring biological effects of acute nitrogen pollution. Ecotoxicol. Environ. Saf. 72: 2009–2012.

Nannoni, F., R. Santolini and G. Protano. 2015. Heavy element accumulation in *Evernia prunastri* lichen transplants around a municipal solid waste landfill in central Italy. J. Waste Manage. 43: 353–362.

Nash, T. H. III and C. Gries. 1995. The response of lichens to atmospheric deposition with an emphasis on the Arctic. Sci. Total Environ. 160/161: 737–747.

Nieboer, E., D. H. S. Richardson and F. D. Tomassini. 1978. Mineral uptake and release by lichens: an overview. Bryologist 226–246.

Nimis, P. L., M. Castello and M. Perotti. 1990. Lichens as biomonitors of sulphur dioxide pollution in La Spezia (Northern Italy). Lichenologist 22(3): 333–344.

Nimis, P. L., G. Lazzarin, A. Lazzarin and N. Skert. 2000. Biomonitoring of trace elements with lichens in Veneto (NE Italy). Sci. Total Environ. 255(1-3): 97–111.

Nimis, P. L., S. Andreussi and E. Pittao. 2001. The performance of two lichen species as bioaccumulators of trace metals. Sci. Total Environ. 275(1–3): 43–51.

Nriagu, J. O. 1979. Global inventory of natural and anthropogenic emissions of trace metals to the atmosphere. Nature 279 (5712): 409–411.

Ogunkunle, C. O., A. M. Ziyath, S. S. Rufai and P. O. Fatoba. 2016. Surrogate approach to determine heavy metal loads in a moss species—*Barbula lambaranensis*. J. King Saud. Univ. – Sci. 28: 193–197.

Oliveira, H. 2012. Chromium as an environmental pollutant: insights on induced plant toxicity. J. Bot. 1–8.

Osyczka, P. and K. Rola. 2019. Integrity of lichen cell membranes as an indicator of heavy-metal pollution levels in soil. Ecotoxicol. Environ. Safe. 174: 26–34.

Otnyukova, T. 2007. Epiphytic lichen growth abnormalities and element concentrations as early indicators of forest decline. Environ. Pollut. 146: 359–365.

Otnyukova, T. N. 1997. Morphological conditions of *Cladina stellaris* (Cladoniacea, lichens) as a diagnostic feature of atmospheric pollution. *Botanicheskii zhurnal-moskva then sankt-peterburg* 82: 57–66.

Panda, S. K. and S. Choudhury. 2005a. Changes in nitrate reductase (NR) activity and oxidative stress response in the moss *Polytrichum commune* subjected to chromium, copper and zinc phytotoxicity. Braz. J. Plant. Physiol. 17: 191–197.

Panda, S. K. and S. Choudhury. 2005b. Chromium stress in plants. Braz. J. Plant Physiol. 17: 95–102.

Paoli, L., S. A. Pirintsos, K. Kotzabasis, T. Pisani, E. Navakoudis and S. Loppi. 2010. Effects of ammonia from livestock farming on lichen photosynthesis. Environ. Pollut. 158: 2258–2265.

Paoli, L., A. Guttová, A. Grassi, A. Lackovičová, D. Senko, S. Sorbo et al. 2015. Ecophysiological and ultrastructural effects of dust pollution in lichens exposed around a cement plant (SW Slovakia). Environ. Sci. Pollut. Res. 20: 15891–15902.

Pawlik-Skowrońska, B., O. W. Purvis, J. Pirszel and T. Skowroński. 2006. Cellular mechanisms of Cu-tolerance in the epilithic lichen *Lecanorapolytropa* growing at a copper mine. Lichenologist 38(3): 267–275.

Pawlik-Skowrońska, B. and M. Bačkor. 2011. Zn/Pb-tolerant lichens with higher content of secondary metabolites produce less phytochelatins than specimens living in unpolluted habitats. Environ. Exp. Bot.72(1): 64–70.

Perales-Vela, H. V., J. M. Peña-Castro and R. O. Cañizares-Villanueva. 2005. Heavy metal detoxification in eukaryotic microalgae. Chemosphere 64: 1–10.

Pesch, R. and W. Schroeder. 2006. Mosses as bio-indicators for metal accumulation: Statistical aggregation of measurement data to exposure indices. Ecol. Indic. 6: 137–152.

Petraglia, A., M. Benedictis, F. Degola, G. Pastore, M. Calcagno, R. Ruotolo et al. 2014. The capability to synthesize phytochelatins and the presence of constitutive and functional phytochelatin synthase are ancestral (plesiomorphic) characters for basal land plants. J. Exp. Bot. 65: 1153–1163.

Pipíška, M., M. Horník, L. U. Vrtoch, J. Augustín and J. Lesný. 2007. Biosorption of Co^{2+} ions by lichen *Hypogymnia physodes* from aqueous solutions. Biologia 62(3): 276–282.

Pisani, T., S. Munzi, L. Paoli, M. Bacˇkor and S. Loppi. 2009. Physiological effects of a geothermal element: Boron excess in the epiphytic lichen *Xanthoria parietina* (L) thfr. Chemospere 76: 921–926.

Pisani, T., S. Munzi, L. Paoli, M. Bacˇkor and S. Loppi. 2011. Physiological effects of arsenic in the lichen *Xanthoria parietina* (L.) Th. Fr. Chemosphere 82: 963–969.

Popper, Z. A. and S. Fry. 2003. Primary cell wall composition of bryophytes and charophytes. Ann. Bot. 91: 1–12.

Pott, U. and D. Turpin. 1996. Changes in Atmospheric Trace Element Deposition in Fraser Valley, B.C., Canada from 1960–1993 Measured by Moss Monitoring with *Isothecium stoloniferum*. Can. J. Bot. 74: 1345–1353.

Pradhan, A., S. Kumari, S. Dash, D. P. Biswal, A. K. Dash and K. C. Panigrahi. 2017. Heavy metal absorption efficiency of two species of Mosses (*Physcomitrella patens* and *Funaria hygrometrica*) studied in mercury treated culture under laboratory condition. In IOP Conference Series: Materials Science and Engineering 255(1): 012225.

Rao, D. N. 1984. Response of bryophytes to air pollution. pp. 445–471. *In*: Smith, A. J. E. (ed.). Chapman and Hall, London.

Rau, S., J. Miersch, D. Neumann, E. Weber and G. J. Krauss. 2007. Biochemical responses of the aquatic moss *Fontinalis antipyretica* to Cd, Cu, Pb and Zn determined by chlorophyll fluorescence and protein levels. Environ. Exp. Bot. 59: 299–306.

Reimann, C., J. H. Halleraker, G. Kashulina and I. Bogatyrev. 1999. Comparison of plant and precipitation chemistry in catchments with different levels of pollution in kola peninsula, russia. Sci. Total Environ. 243: 169–191.

Richardson, D. H. S. and E. Nieboer. 1981. Lichens and pollution monitoring. Endeavour 5(3): 127–133.

Rola, K., E. Latkowska, B. Myśliwa-Kurdziel and P. Osyczka. 2019. Heavy-metal tolerance of photobiont in pioneer lichens inhabiting heavily polluted sites. Sci. Total. Environ. 679: 260–269.

Rola, K. 2020. Insight into the pattern of heavy-metal accumulation in lichen thalli. J. Trace. Elem. Med. Bio. 61: 126512.

Ross, H. B. 1990. On the use of the Mosses *Hylocomium Splendens* and *Pleurozium schreberi* for estimating atmospheric trace metal deposition. Water Air Soil Pollut. 50: 63–76.

Rühling, A. 1994. Atmospheric heavy metal deposition in europe-estimations based on moss analysis. pp. 9. Nordic Council of Ministers. [Ed.]. AKA Print, A/S Arhus.

Rühling, A. and E. Steinnes. 1998. Atmospheric heavy metal deposition in Europe 1995–1996. Nord 1998, 15. Nordic Council of Ministry, Copenhagen.

Rühling, A. and G. Tyler. 1968. An ecological approach to the lead problem. Bot. Notiser. 121: 321–342.

Rühling, Å. and G. Tyler. 1973. Heavy metal deposition in Scandinavia. Water Air Soil Pollut. 2: 445–455.

Rühling, Å. and G. Tyler. 2004. Changes in the atmospheric deposition of minor and rare elements between 1975 and 2000 in south Sweden, as measured by moss analysis. Environ. Pollut. 131: 417–423.

Rühling, Å. and G. Tyler. 1970. Sorption and retention of heavy metals in the woodland moss *Hylocomium splendens* (Hedw.) Br. et Sch. *Oikos*, 92–97.

Rühling, Å., L. Rasmussen, K. Pilegaard, A. Mäkinen and E. Steinnes. 1987. Survey of atmospheric heavy metal deposition in the Nordic countries in 1985-monitored by moss analyses. 21–44.

Saeki, M., K. Kunii, T. Seki, K. Sugiyama and T. Suzuki. 1977. Metal burden of urban lichens. Environ. Res. 13: 256–266.

Sandhi, A. and M. Greger 2018. Moss based constructed wetland system: is it possible to use aquatic moss (*Warnstorfia fluitans*) for removal of as in an eco-friendly approach?. In 11th LNU Ecotech SWEDEN, Linnaeus university. pp. 17.

Sari, A., M. Tuzen, Ö. D. Uluözlü and M. Soylak. 2007. Biosorption of Pb (II) and Ni (II) from aqueous solution by lichen (*Cladonia furcata*) biomass. Biochem. Eng. J. 37(2): 151–158.

Sarret, G., A. Manceau, D. Cuny, C. V. Haluwyn, S. Déruelle, J. L. Hazemann et al. 1998. Mechanisms of lichen resistance to metallic pollution. Environ. Sci. Technol. 32(21): 3325–3330.

Sawidis, T., M. K. Chettri, G. A. Zachariadis, J. A. Stratis and M. R. D. Seaward. 1995. Heavy metal bioaccumulation in lichens from macedonia in northern greece. Toxicol Environ Chem. 50: 157–166.

Saxena, A. and A. Saxena. 2012. Bioaccumulation and glutathione-mediated detoxification of copper and cadmium in *Sphagnum squarrosum Crome Samml.* Environ Monit. Asses. 184(7): 4097–4103.

Saxena, D. K. and A. Saxena. 1999. Bio-monitoring of SO2 phytotoxicity on *Sphagnum squarrosum* cram. samml. J. Indian Bot. Soc. 78(3-4): 367–374.

Saxena, D. K., A. Saxena and H. S. Srivastava. 2000. Biomonitoring of metal precipitation at petrol pumps and their effects on moss *Sphagnum cuspidatum* Hoffm. J. Environ. Stud. Policy 3(2): 95–102.

Saxena, D. K. and M. Saiful-Arfeen. 2006. Biomonitoring and inter species comparison of metal precipitation through bryophytes at petrol pump on Kumaon hill. Environ. Conserv. J. 7: 69–77.

Saxena, D. K., K. Srivastava and S. Singh. 2008a. Retrospective metal data of the last 100 years deduced by moss, Barbulasp. from Mussoorie city, Garhwal Hills, India. Curr. Sci. 901–904.

Saxena, D. K., S. Singh and K. Srivastava. 2008b. Atmospheric heavy metal deposition in Garhwal hill area (India): Estimation based on native moss analysis. Aerosol Air Qual. Res. 8(1): 94–111.

Saxena, D. K. and M. Saiful-Arfeen. 2010. Active biomonitoring of atmospheric metal deposition by Bryum species around Almora, Nainital and Pithoragarh of Kumaon Hills India. Nat. Environ. Pollut. Technol. 9: 1–12.

Saxena, D. K. and S. Arfeen. 2010. Metal deposition pattern in Kumaon hills (India) through active monitoring using moss *Racomitrium crispulum*. Iranian J. Environ. Health. Sci. Eng. 7: 103–114.

Saxena, D. K., P. S. Hooda, S. Singh, K. Srivastava, H. M. Kalaji and D. Gahtori. 2013. An assessment of atmospheric metal deposition in Garhwal Hills, India by moss *Rhodobryum giganteum* (Schwaegr.) Par. Geophytology 43(1): 17–28.

Saxena, S., D. K. Upreti and N. Sharma. 2007. Heavy metal accumulation in lichens growing in north side of Lucknow city, India. J. Environ. Biol. 28(1): 49–51.

Schroder, W., R. Pesch, C. Englert, H. Harmens, I. Suchara, H. G. Zechmeister et al. 2008. Metal accumulation in mosses across national boundaries: uncovering and ranking causes of spatial variation. Environ. Pollut. 151: 377–388.

Schutte, J. A. 1977. Chromium in Two Corticolous Lichens from Ohio and West Virginia. Bryologist 80: 279–283.

Schwartzman, D., M. Kasiml, L. Stieff and J. H. Johnson. 1987. Quantitative monitoring of airborne lead pollution by a foliose lichen. Water, Air, And Soil Pollut. 32: 363–378.

Scott, M. G. and T. C. Hutchinson. 1989. Experiments and observations on epiphytic lichens as early warning sentinels of forest decline. pp. 205–216. *In*: Grossblatt, N. (ed.). Biologic markers of air pollution stress and damage in forests. National Academy Press, Washington, DC.

Sesli, E. and C. M. Denchev. 2008. Checklists of the myxomycetes, larger ascomycetes, and larger basidiomycetes in Turkey. Mycotaxon. 106: 65–67.

Shakya, K., M. K. Chettri and T. Sawidis. 2008. Impact of heavy metals (copper, zinc, and lead) on the chlorophyll content of some mosses. Arch. Environ. Contam. Toxicol. 54: 412–421.

Sharma, S. 2009. Study on impact of heavy metal accumulation in *Brachythecium populeum* (Hedw.) B.S.G. Ecol Indic 9: 807–811.

Shukla, V. and D. K. Upreti. 2007. Physiological response of the lichen *Phaeophyscia hispidula* (Ach.) Essl., to the urban environment of Pauri and Srinagar (Garhwal), Himalayas, India. Environ. Pollut. 150(3): 295–299.

Shukla, V. and D. K. Upreti. 2008. Effect of metallic pollutants on the physiology of lichen, *Pyxinesubcinerea* Stirton in Garhwal Himalayas. Environ. Monit. Assess. 141(1): 237–243.

Shukla, V. and D. K. Upreti. 2009. Polycyclic aromatic hydrocarbon (PAH) accumulation in lichen, *Phaeophyscia hispidula* of Dehradun City, Garhwal Himalayas. 149(1): 1–7.

Singh, P., P. K. Singh, P. K. Tondon and K. P. Singh. 2018. Heavy metals accumulation by epiphytic foliose lichens as biomonitors of air quality in Srinagar city of Garhwal hills, Western Himalaya (India). Curr. Res. Environ. Appl. Mycol. J. Fungal Biol 8: 282–289.

Singh, P. K., P. Bujarbarua, K. P. Singh and P. K. Tandon. 2019. Report on the bioaccumulation of heavy metals by foliose lichen (*Pyxine cocoes*) from air polluted area near Nagaon Paper Mill in Marigaon, Assam, North-East India.

Singh, S., K. Srivastava, D. Gahtori and D. K. Saxena. 2017. Bryomonitoring of atmospheric elements in *Rhodobryum giganteum* (Schwaegr.) Par., growing in Uttarakhand region of Indian Himalayas. Aerosol Air Qual. Res. 17(3): 810–820.

Soudzilovskaia, N. A., J. H. C. Cornelissen, H. J. During, R. S. P. Van Logtestijn, S. I. Lang and R. Aerts. 2010. Similar cation exchange capacities among bryophyte species refute a presumed mechanism of peatland acidification. Ecology 9: 2716–2726.

Srivastava, K., S. Singh and D. K. Saxena. 2014. Monitoring of metal deposition by moss *Barbulaconstricta* J. Linn., from Mussoorie Hills in the India. Environ. Res. Eng. Manage. 67(1): 54–62.

Steinnes, E. 1977. Atmospheric deposition of trace elements in norway studied by means of moss analysis. Kjeller Report, KR 154, Institute for Atomenegri, Kjeller, Norway.

Steinnes, E. 1993. Some aspects of biomonitoring of air pollutants using mosses as illustrated by the 1976 Norwegian Survey. pp. 381–394. *In*: Markert, B. (ed.). Plant as Biomonitors VHC, Weinheim.

Steinnes, E. 1995. A critical evaluation of the use of naturally growing moss to monitor the deposition of atmospheric metals. Sci. Total Environ. 160: 243–249.

Steinnes, E. 2000. Use of mosses as biomonitors of atmospheric deposition of trace elements. BioMap Proc. of an International Workshop Organized by the International Atomic Energy Agency in Co-operation with the Instituto Technologico Nuclear, Lisbon, Portugal, 100–107.

Suchara, I. and J. Sucharova. 1998. Atmospheric deposition levels of chosen elements in the Czech Republic determined in the framework of the International Bryomonitoring Program 1995. Sci. Total Environ. 223: 37–52.

Sugiyama, K., S. Kurokawa and G. Okada. 1976. Studies of lichens as a bioindicator of air pollution. I. Correlation of distribution of *Parmelia tinctorum* with SO_2 air pollution. Jpn. J. Ecol. 26: 209–212.

Sun, S. Q., M. He, T. Cao, Y. C. Zhang and W. Han. 2009. Response mechanisms of antioxidants in bryophyte (*Hypnum plumaeforme*) under the stress of single or combined Pb and/or Ni. Environ. Monit. Assess. 149: 291–302.

Sun, S. Q., M. He, G. X. Wang and T. Cao. 2010. Heavy metal-induced physiological alterations and oxidative stress in the moss *Brachythecium piligerum* chad. Environ. Toxicol. 26: 453–458.

Szczepaniak, K. and M. Biziuk. 2003. Aspects of the biomonitoring studies using mosses and lichens as indicators of metal pollution. Environ. Res. 93: 221–230.

Tessier, L. and J. L. Boisvert. 1999. Performance of terrestrial bryophytes as biomonitors of atmospheric pollution. A review. Toxicol. Environ. Chem. 68 (1-2): 179–220.

Thöni, L., N. Schnyder, N. and F. Krieg. 1996. Comparison of metal concentrations in three species of mosses and metal freights in bulk precipitations. Fresenius' J. Anal. Chem. 354(5): 703–708.

Tretiach, M., F. C. Carniel, S. Loppi, A. Carniel, A. Bortolussi, D. Mazzilis et al. 2011. Lichen transplants as a suitable tool to identify mercury pollution from waste incinerators: a case study from NE Italy. Environ. Monit. Assess. 175(1): 589–600.

Turkyilmaz, A., H. Sevik, M. Cetin and E. A. Ahmaida Saleh. 2018. Changes in heavy metal accumulation depending on traffic density in some landscape plants. Pol. J. Environ. Stud. 27(5): 2277–2284.

Tyler, G. 1970. Moss analysis-a method for surveying heavy metal deposition. pp. 129–132. *In*: Englaund, H. M. and W. T. Berry (eds.). Academic Press, New York.

Tyler, G. 1989. Uptake, retention, and toxicity of heavy metals in lichens. Water Air Soil Pollut. 47: 321–333.

Tyler, G. 1990. Bryophyte and heavy metals: A Literature Review. Bot. J. Linn. Soc. 104: 231–253.

Uluozlu, O. D., A. Sari, M. Tuzen and M. Soylak. 2008. Biosorption of Pb (II) and Cr (III) from aqueous solution by lichen (*Parmelina tiliaceae*) biomass. Bioresour. Technol. 99(8): 2972–2980.

Unal, D., N. O. Isik and A. Sukatar. 2010. Effects of chromium VI stress on photosynthesis, chlorophyll integrity, cell viability, and proline accumulation in lichen *Ramalina farinacea*. Russ. J. Plant Physiol. 57: 664–669.

Vajpayee, P., U. N. Rai, M. B. Ali, R. D. Tripathi, V. Yadav, S. Sinha et al. 2001 Chromium-induced physiologic changes in *Vallisneria spiralis* L. and its role in phytoremediation of tannery effluent. Bull Environ. Contam. Toxicol. 67: 246–256.

Van Assche, F. and H. Clijsters. 1990. Effects of metals on enzyme activity in plants. Plant Cell Environ. 13: 195–206.

van Herk, C. M., A. Aptroot and H. F. Van Dobben. 2002. Long-term monitoring in the Netherlands suggests that lichens respond to global warming. Lichenologist 34(2): 141–154.

Vantová, I., M. Bačkor, B. Klejdus, M. Bačkorová and J. Kováčik. 2013. Copper uptake and copper-induced physiological changes in the epiphytic lichen *Evernia prunastri.* Plant Growth Regul. 69: 1–9.

Varela, Z., R. García-Seoane, M. Arróniz-Crespo, A. Carballeira, J. A. Fernández and J. R. Aboal. 2016. Evaluation of the use of moss transplants (*Pseudoscleropodium purum*) for biomonitoring different forms of air pollutant nitrogen compounds. Environ. Pollut. 213: 841–849.

Volesky, Z. and S. Holan. 1995. Biosorption of heavy metal. J. Biotechnol. Microbial. 3: 11235–11250.

Waghmode, T. R., M. B. Kurade and S. P. Govindwar. 2011. Time dependent degradation of mixture of structurally different azo and non azo dyes by using *Galactomyces geotrichum* MTCC 1360. Int. Biodeter. Biodegrad. 65: 4.

Wells, J. M. and D. H. Brown. 1987. Factors affecting the kinetics of intra- and extracellular cadmium uptake by the moss *Rhytidiadelphus squarrosus*. New Phytol. 105: 123–137.

Wiersma, G. B., D. A. Bruns, C. Boelcke, C. Whitworth and L. McAnulty. 1990. Elemental composition of mosses from a remote Nothofagus forest site in southern Chile. Chemosphere 20(5): 569–583.

Wolterbeek, H. T., P. Bode and T. G. Verburg. 1996. Assessing the quality of biomonitoring via signal-to-noise ratio analysis. Sci. Total Environ. 180: 107–116.

Yruela, I. 2005. Copper in plants. Braz. J. Plant. Physiol. 17: 145–156.

Yurukova, L. and A. Ganeva. 1997. Active biomonitoring of atmospheric element deposition with Sphagnum species around a copper smelter in Bulgaria. J. Appl. Bot. 71: 14–20.

Zechmeister, H. G. 1995. Correlation between altitude and heavy metal deposition in Alps. Environ. Pollut. 89: 73–80.

Zeichmeister, H. G., K. Grodzinska and G. Szarek-Lukaszewska. 2003. Bryophytes. pp. 329–375. *In*: Markert B.A., A. M. Breure and H. G. Zeichmeister (eds.). Elsevier. Oxford.

Zhao, L., C. Zhang, S. Jia, Q. Liu, Q. Chen and X. Li. 2019. Element bioaccumulation in lichens transplanted along two roads: The source and integration time of elements. Ecol. Indic. 99: 101–107.

CHAPTER 4

Phytoremediation
A Green Technology for Treating Heavy Metal Contaminated Soil

Huijuan Shao,[1,*] *Guangyu Cui*[2] and *Sartaj Ahmad Bhat*[3]

4.1 Introduction

Due to the acceleration of industrialization and urbanization, heavy metal (HM) contamination of soil has become a serious environmental problem worldwide. Geological and human activities, including industrial discharge, mining, smelting, hazardous waste disposal, extensive use of agrochemicals, and transportation, are the primary sources of HMs in the soil (Liu et al. 2018a, Shah and Daverey 2020, Ali et al. 2019).

Once HMs get deposited in the soil, they can remain in the soil for several years due to their non-biodegradable nature. The mobility and availability of HMs are controlled by various physicochemical processes, including adsorption, precipitation, and mineralization, depending on soil and plant characteristics (He et al. 2015). HM contamination has adverse effects on the ecosystem balance, soil quality, and agricultural yield (Mao et al. 2015). Mobile HMs can be taken up by plant roots from soil interstitial water and then transported to plant shoots, leading to the high bioavailability of HMs, and eventually threatening human health. Considering these serious consequences of HM contamination, various soil remediation techniques such as soil excavation, landfill, electro-kinetic extraction, soil washing, and vitrification have been studied and carried out in the contaminated soil. Conventional physicochemical remediation techniques usually have relatively high efficiencies and can be applied to severely contaminated soils but are only suitable at small scales (Dada et al. 2015). Meanwhile, they also have many disadvantages such as high energy requirement and cost, destruction of soil structure and quality, possible secondary contamination caused by the released chemical substances applied during the remediation processes, etc. (Cristaldi et al. 2017). Therefore, it is necessary to develop low-cost, effective, and eco-friendly remediation strategies for HM contaminated soils.

[1] College of Resources and Environment, Shandong Agricultural University, Tai'an 271000, PR China.
[2] State Key Laboratory of Pollution Control and Resource Reuse, Tongji University, Shanghai 200092, PR China.
[3] River Basin Research Center, Gifu University, 1-1 Yanagido, Gifu 501-1193, Japan.
* Corresponding author: sharehui@foxmail.com

Bioremediation is biologically removing or degrading the contaminants to harmless forms or low concentrations (Kumar et al. 2011). Compared with physicochemical methods, bioremediation is an environment-friendly, safe, and cheap technology for solving HM contamination problems and soil reclamation (Xiao et al. 2019). Among bioremediation technologies, phytoremediation is a commonly used approach because it is easily operated, requires little manpower, and beautifies the environment. It uses vegetation to remove contaminants from the soil or decrease the availability of contaminants, thereby restoring the soil to a healthy level (Lajayer et al. 2019).

In order to provide the theoretical basis for developing phytoremediation technology that can be a tool to establish good environmental status, this chapter describes the current knowledge on soil contamination by HMs, phytoremediation strategies, improving approaches for phytoremediation efficiency, and the prospects of this remediation technology.

4.2 Heavy Metal Classification and Toxic Effects on the Environment

HMs are a kind of metallic chemical element with relatively high density, atomic weight, and number. Naturally, HMs are produced in the Earth's crust, but many of them are mainly released from agricultural and industrial activities (Tchounwou et al. 2012). HMs cannot be degraded by any physicochemical or biochemical processes; therefore, they can exist in the soil for a long time and pose severe threats to the environment and human health (Suman et al. 2018).

HMs in the soil can be divided into the following four types according to their contamination sources: (i) HMs contained in the parent materials of soil; (ii) HMs deposited from the atmosphere due to industrial production, mining, and vehicular exhausts; (iii) HMs induced by waste disposal; (iv) HMs added through agricultural activities. HMs can also be classified as extremely poisonous (Cd, As, Hg, Se, Pb, Zn), moderately poisonous (Ni, Co, Cr, Cu, Mo), and relatively less poisonous (Ba, Mn, Sr) according to their level of toxicity (Ashraf et al. 2019).

HMs are divided into essential and nonessential based on their role in the ecosystem. Essential HMs including Cu, Zn, Fe, Mn, and Ni are needed in minute quantities for the plants; however, excess levels have adverse impacts on the organisms via bioaccumulation and bioconcentration (Alirzayeva et al. 2017, Mortensen et al. 2018). Nonessential HMs like Pb, Cd, As, and Hg are usually very toxic, resulting in environmental contamination and negative effects on plant physiology and biochemistry, as well as plant production (Fasani et al. 2018, Luo et al. 2019, Ghori et al. 2019). Table 4.1 summarizes the toxicity of selected HMs to humans. For example, Cd is present as a free cation in liquid water or precipitated complexes with compounds. It has high bioavailability due to its high solubility. It is accumulated mainly in the kidney of the human body, endangering urinary system function. Zn is an essential human trace element, but excessive Zn can inactivate the soil enzyme, reduce the bacteria number, weaken the microbial action, and harm the crop growth. The

Table 4.1. Several common HMs along with their toxicity to human beings (Shah and Daverey 2020).

HMs	Toxicity
Cadmium (Cd)	Degenerative bone disease, kidney function harm, renal lesions, carcinogenic
Chromium (Cr)	DNA harm, carcinogen, irritant dermatitis, anaphylaxis, gastrointestinal bleeding
Lead (Pb)	Damage to fetal brain and kidney, anemia, neurological disorder
Copper (Cu)	Liver damage, Wilson disease, insomnia
Zinc (Zn)	Depression, lethargy, increased thirst
Nickel (Ni)	Nausea, chronic asthma, cardiovascular damage
Manganese (Mn)	Affects the heart and blood system, central nervous system, and breathing system
Mercury (Hg)	Blindness, hearing impairment, brain and kidney harm, indigestion, mental deficiency
Arsenic (As)	Skin and vascular diseases, carcinogenic, gastrointestinal problem, breathing system harm

mobility of Zn can be affected by soil properties such as soil pH, cation exchange capacity, and the contents of other existing elements (Broadley et al. 2007). Cu can bind with soil organic compounds, which can also be affected by soil pH, redox potential, and the existence of other anions (Cristaldi et al. 2017). Ni is a component widely distributed in soil and seawater, and it is retained by the soil depending on the soil pH, moisture, texture, organic matter, and hydroxide contents (Cristaldi et al. 2017). Arsenic (As) is a metalloid usually included in HMs and commonly exists in the Earth's crust and a place with water. The environmental contamination by As (As^{3+} and As^{5+} are the most dangerous forms) is the result of natural occurrence as well as human activities (Tchounwou et al. 2012).

4.3 Phytoremediation of Heavy Metal Contaminated Soil

Phytoremediation is a novel plant-based technology for the large-scale removal of low-level contaminants from the environment (Yan et al. 2021). There are several advantages of phytoremediation, including: (i) easy and low cost for application and management; (ii) eco-friendly and can improve soil fertility; (iii) applicable over a large-scale field; (iv) can prevent soil erosion and metal leaching through stabilization (Yan et al. 2020).

4.3.1 Efficient Plants for Accumulating Different Heavy Metals

Effective phytoremediation generally needs fast-growing plants with high biomass and prominent remediation ability for contaminants (Rascio and Navari-Izzo 2011, Tozser et al. 2017). Plants use rhizosphere ecosystems to adsorb and accumulate HMs, and regulate their bioavailability, thereby remediating the contaminated soil (Yan et al. 2020). Meanwhile, the selection of plants should consider not only their accumulation ability for HMs but also their adaptive capacity to the local natural and climate conditions; thus, native plants are usually selected for good growth rates and strong heavy metal accumulation characteristics (Wei et al. 2021). A few examples of hyperaccumulators are shown in Table 4.2.

4.3.2 Role of Plant Physiology

All plants are capable of taking-up essential HMs (Cu, Zn, Mn, Fe, Ni, etc.) from the soil; however, certain plants can also accumulate highly toxic HMs such as As, Cd, Pb, and Cr (Dai et al. 2017, Shah and Daverey 2020). The accumulation capability of a plant to HMs is affected by not only its genotype but also its rhizospheric microorganism, soil properties, and the bioavailability of HMs (Ma et al. 2003). The transport of HMs from plant roots to above-ground parts occurs with plant transpiration. The plant roots take in HMs by symplastic translocation (through the plasma membrane of endodermal cells) or apoplastic translocation (through cell spaces) (Ling et al. 2017, Thakur et al. 2016). The transport of HM ions from plant root to shoot is mainly through the xylem with membrane transporter proteins (Yan et al. 2021). Metal specific transporter proteins help HMs to cross the plasma membrane; the HMs are then chelated by plant-produced chelators, the chelated HMs stay in the vacuoles or cell wall; some of them could be transported to plant shoot through xylem (Pinto et al. 2014). Meier et al. (2018) revealed that the iron and Zn regulated transporter proteins were expressed in hyperaccumulators, whereas they were absent in non-hyperaccumulating species of *Senecio coronatus*. In addition, hyperaccumulators have a strong ability to detoxify and immobilize HMs, allowing them to accumulate plenty of HMs in the aboveground parts of plants safely (Shah and Daverey 2020). It was found that hyperaccumulators sequestered toxic HMs in the aboveground plant tissues, while non-hyperaccumulators in the underground plant tissues (Sharma et al. 2016).

Table 4.2. Representative hyperaccumulator plants for phytoremediation of HMs contaminated soils.

Plant species	HMs	Phytoremediation potential	References
Nerium oleander	Pb, Cd, Zn	Pb was mainly accumulated in the plant root. Cd and Zn were concentrated in the above-ground parts of the plant.	Ibrahim and El Afandi 2020
Trifolium repens L.	Cd, Cr, Pb	*Trifolium repens* L. accumulated both anions ($Cr_2O_7^{2-}$) and cations ($Cd^{2+}$and Pb^{2+}). The concentration in roots is higher than that in shoots.	Lin et al. 2021
Hydrangea paniculate and *Helianthus annuus*	Cu, Pb	The two plant species preferentially accumulated Cu in the leaves but Pb in the stems.	Forte and Mutiti 2017
Calendula officinalis L.	Cu	Cu was accumulated in both plant's shoots and roots at high concentrations.	Goswami and Das 2016
Tagetes erecta L.	Zn, Cd, Pb	This plant was revealed to be effective for Zn and Cd accumulation but not for Pb.	Minisha et al. 2020
Schimasuperba	Mn	Removal rate was 62412.3 mg kg^{-1}.	Yang et al. 2008
Alyssum murale and *Leptoplaxemarginata*	Ni	Both plants have high biomass production and Ni accumulation ability.	Pardo et al. 2018
Mirabilis jalapa	Cr	This plant showed high accumulation ability and was tolerant to high Cr concentration.	Miao and Yan 2013
Sedum plumbizincicola	Cd, Zn	In the plant shoots, the Cd concentration was much higher than 100 mg kg^{-1}, and the Zn concentration was around 10,000 mg kg^{-1}.	Hu et al. 2015
Abelmoschus manihot L. and *Bidens Pilosa* L.	Cd	The two plants are Cd hyperaccumulators. *Abelmoschus manihot* L. tolerated Cd by accumulating it in the vacuoles and cell wall.	Wu et al. 2018, Dai et al. 2017
Arundo donax L. and *Miscanthus* sp.	Zn	Both plants have high biomass production and high efficiencies for Zn phytoextraction.	Barbosa et al. 2015
Salix spp.	Cd, Zn	High concentrations of Cd and Zn were found in the plant leaves, and the bioaccumulation factors were 27 for Cd and 3 for Zn.	Wieshammer et al. 2007

4.3.3 Role of Soil Characteristics

The sorption and desorption capabilities of soil determine the mobility and availability of HMs. Various physicochemical properties of soil, such as pH, texture, and organic matter content, can influence the availability of HMs to plants (Fig. 4.1a).

Soil pH is a very important factor determining the availability of HMs to plants. Low soil pH (pH value < 5.0) signifies the high mobility and bioavailability of HMs (Shah and Daverey 2020). Dai et al. (2020) reported that when the soil pH value decreased, more Cd and Pb could be accumulated in the plant, generally. Hydrogen ion has high activity at low pH value and can compete with HM cations for binding sites of humic or fulvic acid, thereby reducing the formation of organometallic ligands and improving the mobility of HMs in soil (Yoon et al. 2015). On the contrary, at high pH levels, HM cations tend to be adsorbed onto soil particles, and their mobility and availability will be decreased; the reverse is true for metal anions (Kader et al. 2016).

The organic matter (OM) of soil governs the plant uptake of HMs because the HM ions tend to strongly bind to the organic ligands, thereby reducing the bioavailability of HMs (Shaheen et al. 2013). Cambier et al. (2014) reported that long-term application of OM affected the HM immobilization in the soil and decreased the availability of HM. Khan et al. (2018) reported that the application of several organic amendments, including hardwood biochar, rice husk, bagasse, and

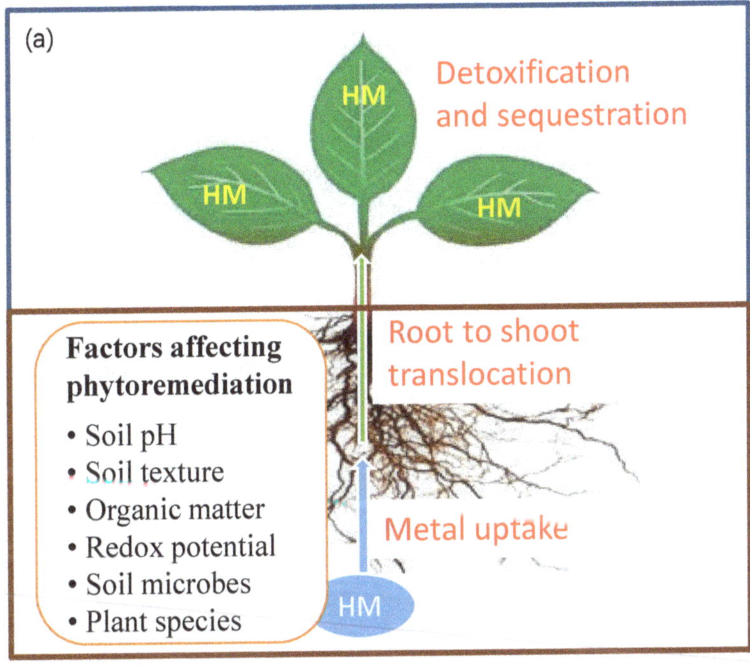

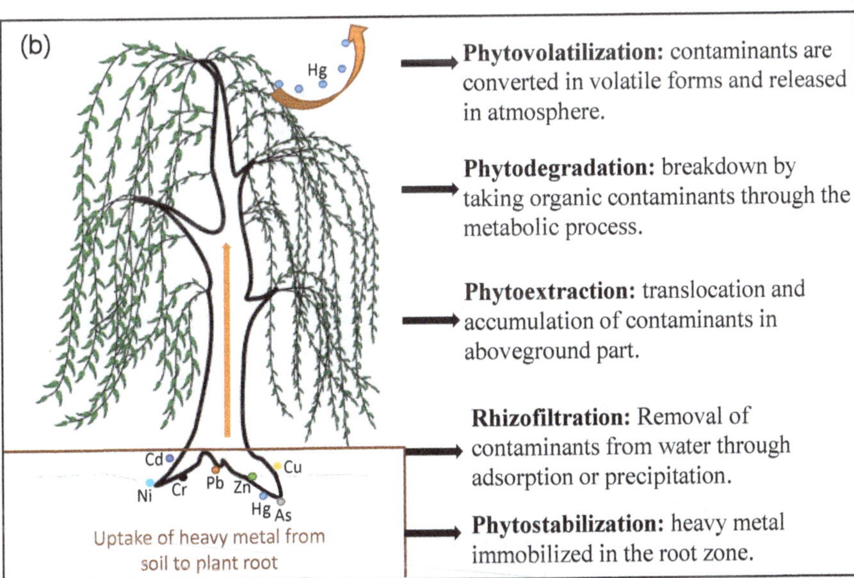

Fig. 4.1. Phytoremediation (a). Factors affecting (b). Strategies.

maize cob, negatively affected the bioavailability of Cd in the soil and eventually decreased the Cd accumulation in tomato and cucumber vegetables.

Soil texture (particle size distribution) also has important effects on the bioavailability and phytoextraction efficiency of HMs (Liu et al. 2018b). Lotfy and Mostafa (2014) reported that fine soil particles (size < 100 μm) had relatively large surface areas and could contain most HMs; therefore, the bioavailability of HM was higher in fine soil than that in coarse soil. Similarly, Volk and Yerokun (2016) observed that in clay loam soil, the bioavailable concentrations of Co and Cu were 2.7 times as high as that in sandy loam soil.

Another crucial factor for phytoremediation is related to soil microorganisms, including siderophores, biosurfactants, and organic acid-producing microorganisms. Various biochemical processes, including translocation, transformation, volatilization, precipitation, chelation, and complexation of HMs, as well as plant-microbe interaction, can affect phytoremediation (Rajkumar et al. 2012). Biosorption of HMs by soil microorganisms also influences the bioavailability of HMs; the biosorption process depends on several factors, including the HM species and concentrations, plant species, soil pH, climatic conditions, and so on (Shah and Daverey 2020).

4.4 Phytoremediation Strategies

Phytoremediation is a green technology, usually using fast-growing plants to remove, regulate, or detoxify contaminants in the environment (Mahar et al. 2016). Phytoremediation of HMs is grouped into several subclasses, usually including phytofiltration, phytoextraction, phytostabilization, phytodegradation, and phytovolatilization (Fig. 4.1b) (Parmar and Singh 2015, Ashraf et al. 2019, Anju 2017).

4.4.1 Phytofiltration

Phytofiltration is to remove HMs from water by using the vast filtration and absorption functions of the tolerant plants that accumulate HMs, and to harvest and properly treat these plants, thereby remediating the HM contamination in the water (Mukhopadhyay and Maiti 2010). It can be accomplished by plant roots, seedlings, and shoots (Da Conceiçao Gomes et al. 2016). Plants need to develop extensive root systems in uncontaminated water. The plants are then transferred to the contaminated site for rhizofiltration of HMs; the roots are finally disposed of once they become saturated with HMs (Wuana and Okieimen 2011). Plants suitable for rhizofiltration include aquatic plants, semi-aquatic plants, and some terrestrial plants such as sunflower, Indian mustard, cattail, and tobacco. They have strong accumulation ability and tolerance for HMs, high biomass, and fast growth (Hooda 2007). Magdalena et al. (2020) demonstrated that the bryophyte *Monosoleum tenerum* had a high potential for removing toxic HMs (Zn, Cu, Mn, Ni, and Fe) from solution via accumulation in plant tissue, indicating the feasibility of using *M. tenerum* for purifying HM polluted wastewater. Some land plants such as Indian mustard and sunflower also have high abilities to accumulate HMs since they usually have more extended and more vigorous root systems than water plants (Yan et al. 2020).

4.4.2 Phytoextraction

Among the phytoremediation strategies, phytoextraction is one of the most studied and promising (Vara Prasad and de Oliveira Freitas 2003). Phytoextraction or phytoaccumulation includes several processes: HM mobilization in the rhizosphere, root uptake of HM from the soil, transportation of HM from root to aboveground part, storage and compartmentalization of HM in plant tissue, harvesting, and continuous planting until soil contamination levels become acceptable (Kotrba et al. 2009, Ali et al. 2013, Bhargava et al. 2012). The phytoextraction efficiency depends on several factors, including the physiological and biochemical characteristics of plants, mobility and bioavailability of HMs, and soil characteristics. Since the HM-accumulating capacity and plant biomass are two important factors determining the phytoextraction efficiency, plants should be selected according to two principles: (i) selecting hyperaccumulators that can efficiently absorb and accumulate HMs; (ii) using plants with high biomass, which have an overall high accumulation ability for HMs (Yan et al. 2020, Anju 2017).

Hyperaccumulators usually possess characteristics including (Wei et al. 2008, Mahar et al. 2016):

(1) Strong accumulation and translocation capacity, compared to roots, the concentration of HMs in plant shoots should be higher in principle;
(2) Strong tolerance to the contamination, especially the plant growth should not show obvious toxic symptoms;
(3) Adaptation to the local environmental conditions;
(4) Resistance to diseases and pests;
(5) Developed root system, fast growth, and large biomass production. The fast growing plant is harvested to extract HMs from soil continuously, thereby reducing the phytoextraction time;
(6) The value of transfer factor (TF) (HM concentration ratio in plant to soil) should be larger than one when the concentration of HMs absorbed by hyperaccumulator plants is saturated.

In addition, the developed root system of the plant not only increases the contact areas between plants and HMs, thereby benefiting the absorption or adsorption of HMs, but also prevents soil erosion and HM leaching to the surrounding areas and groundwater (Yan et al. 2021). However, the shortcoming of phytoremediation is that most plants have a narrow growth adaptation range, small biomass, and low accumulation capability so that the repair cycle is long.

In the treatment of contaminated sites with complex HMs, the selection of polymetallic hyperaccumulators has become a new focus. It is necessary to enlarge the scope of plant species selection and application in order to give full play to the functions of phytoremediation technology in the reconstruction of vegetation and landscape, control of soil erosion, enrichment of biodiversity, improvement of land use-value, and so on. In addition, we should avoid using vegetation often eaten by animals for phytoextraction since the risks of HM contamination on human health will be raised through easily entering the food chain. For example, woody species are more proper than herbs and crops for phytoextraction because they are non-edible, have a high amount of biomass and deep root system, thereby effectively promoting the accumulation of high concentrations of HMs and preventing the spread of contamination (Suman et al. 2018). It is expected to be a new trend in developing phytoremediation technology to construct composite vegetation communities by utilizing the advantages of the rapid growth of suitable native plants and combining the functions of ecology-economic plants such as wood, industrial raw materials, medicine, energy, and landscape.

4.4.3 Phytostabilization

Phytostabilization is the use of plant absorption and some special substances in the plants' rhizosphere to convert soil HMs into relatively harmless forms, to reduce the mobility and bioavailability of HMs in soil, and reduce the possibility of further pollution through metal leaching into groundwater or air diffusion (Pulford and Watson 2003). Phytostabilization does not remove contaminants from the contaminated soil; conversely, the used vegetation stabilizes HMs in the soil mainly through precipitation or reduction in metal valence in the rooting zone, adsorption onto root cell walls, or sequestration in the root (Gerhardt et al. 2017, Yan et al. 2021). Unlike phytoextraction, phytostabilization has a specific advantage: the harvested contaminated plant does not need to be disposed of (Wuana and Okieimen 2011).

Effective plant species for phytostabilization should be tolerant to the HM conditions and accumulate HMs in their underground parts, which can inhibit the release of HMs to the food chain (Mahar et al. 2016). Similar to that for phytoextraction, the effective vegetation for phytostabilization should have developed root systems, a large amount of biomass, and fast growth (Yan et al. 2020). Moreover, the rhizospheric microorganisms can facilitate phytostabilization of HMs by promoting precipitation or adsorption of HMs onto root cell walls, generating chelators for HM complexation, developing plant root system, and inhibiting the transport of HM ions from root to aboveground parts of the plant (Göhre and Paszkowski 2006).

4.4.4 Phytodegradation

Phytodegradation is the degradation of organic contaminants by plant metabolism and its symbiotic microbial activities or the breakdown of contaminants outside the plant through enzymes secreted by the plant. The contributing enzymes in phytodegradation processes include dehalogenase, peroxidase, nitroreductase, nitrilase, and phosphatase (Susarla et al. 2002, Cristaldi et al. 2017). Phytodegradation is usually applied to remediate organic contamination, such as herbicides and PAHs (Yin et al. 2011, Cristaldi et al. 2017). This strategy is not suitable for HM-contaminated soil since HMs cannot be degraded.

The rhizodegradation process occurs due to the actions of bacteria, fungi, and yeasts, which obtain nutrients from the plant root exudates. It has advantages, including (i) it can be done on-site; (ii) the transport of contaminants to the aboveground part of the plant or to the air is less than other phytoremediation strategies; (iii) the implementation and maintenance costs are low. On the other hand, it also has disadvantages; for example, the process is slow and efficient only on the surface of contaminated soil since the root depth can be restricted by soil structure (Cristaldi et al. 2017). The contaminants with strong hydrophobicity cannot be migrated easily since they are usually closely bound to the root surface or soil particle; therefore, phytostabilization is more suitable for such contaminants.

4.4.5 Phytovolatilization

Phytovolatilization is using plants to convert toxic contaminants into less toxic gaseous substances. Plants then discharge the volatile contaminants into the atmosphere through plant transpiration, while the contaminants are limited to certain organic compounds and metals like Hg and Se (Kumar and Gunasundari 2018, Yan et al. 2020). In order to reduce environmental hazards, phytovolatilization requires that the toxicity of the transformed substances should be less than the original contaminants. For example, as the only metallic element that is liquid at room temperature, Hg can be absorbed by plant root or leaf, then be transformed from more toxic methyl-Hg to less toxic elemental mercury, which is volatilized, and released into the air (Bizily et al. 2000, Anju 2017).

Compared with other phytoremediation strategies, the advantage of phytovolatilization is that there is no need for plant harvesting and disposal during the remediation period (Yan et al. 2020). However, the toxic volatile compounds can contaminate the atmosphere and may be redeposited to the soil through rainfall; therefore, it is necessary to evaluate the risks before their practical application (Vangronsveld et al. 2009).

4.5 Challenges of Phytoremediation

Phytoremediation has got more and more attention recently because it is a low-cost, feasible, and eco-friendly remediation technology. At the same time, phytoremediation also has some challenges:

i. The remediation progress costs a long time (months or even years).
ii. The remediation efficiency depends on many factors, including soil contamination level, soil characteristics, plant growth rate, plant biomass, and root depth (Cameselle and Gouveia 2019). HMs which are tightly fixed on the soil are tough to be absorbed by plants due to their low bioavailability.

iii. Favorable weather and climatic conditions are needed for phytoremediation (Mahar et al. 2016). The accumulation ability of plants could be affected by pests or disease attacks.
iv. Proper disposal of contaminated plants is of substantial concern. Accumulated HMs could be transferred into the food chain if the harvested plant is not treated correctly. Considering economic and sustainable development, the contaminated plant can be used for gas, liquid, and solid biofuel production through several conversion processes, including anaerobic digestion, pyrolysis, hydrolysis, gasification, and esterification (Dastyar et al. 2019, He et al. 2019).
v. It is a challenge to introduce hyperaccumulators in the contaminated areas with different geographical and climatic conditions from their places of origin because of the limited adaptability of these hyperaccumulator plants; in addition, the production of hyperaccumulators is also a challenge with the gradual increase in global temperatures (Sarma et al. 2021).
vi. Phytoremediation is only applicable for medium to low degrees of soil contamination because plant growth can be restricted under high contamination degrees.

4.6 Phytoremediation Enhancing Approaches

Phytoremediation efficiency can be promoted usually by increasing the bioavailability of HM, increasing plant biomass, or shortening the remediation cycle (Ashraf et al. 2019). At present, chemical-assisted phytoremediation enhancement techniques, microorganism-assisted phytoremediation technology, genetic engineering technology, agronomic regulation technology, nanomaterials, and soil amendments with biochar and compost are the main strengthening measures studied.

Chemical-assisted technologies mainly change the chemical properties of soil by adding foreign substances or directly combining with HMs to alter their chemical species and bioavailability, and finally strengthen the absorption of HMs by plants. Common additives mainly include chelating agents, surfactants, acid-base regulators, organic materials, etc. For example, ethylene diamine tetra-acetic acid (EDTA) is the most common and effective chelating agent for promoting HM uptake and accumulation in plants (Sarwar et al. 2016). The application of chelating agents usually occurs a week before harvest when the biomass is at its maximum stage (Mahar et al. 2016). The chelating agents viz. EDTA (Ethylene diamine tetra-acetic acid), DTPA (diethylene triamine penta-acetic acid), and EGTA (ethylene glycol tetra-acetic acid) are generally used to enhance HMs uptake by plants (Sarwar et al. 2016). They can enter plant roots and form soluble complexes with HMs, which can enhance the availability of HMs (Jiang et al. 2019); the effect was more significant for Pb and Cu, while the pattern of plant uptake of HMs was similar to that without EDTA application (Dipu et al. 2012). However, EDTA could have adverse effects on plants, soil properties, and underground water quality. The chelating agents also activate other elements such as Fe, Mn, Ca in the soil, resulting in the nutrients leaching and lack of plant nutrition. Residual chelating agents may also cause new contamination in the soil. Activation of the chelating agent can increase the mobility of HMs in soil, which may cause contamination to surface and groundwater (Lu et al. 2017, Sarwar et al. 2016). In addition, reducing the soil pH can promote the dissolution of some HMs such as Pb, Zn, and Cd into soil solution and can be easily absorbed by plants. While in the case of arsenic (As), an increase in soil pH will allow more As to enter the soil solution, thereby increasing its bioavailability.

Plant biomass is an important factor in phytoremediation. Therefore, phytoremediation efficiency can be improved by applying nutrients, fertilizers, and a proper irrigation system (Nie et al. 2010). Bian et al. (2021) reported that intercropping of Moso bamboo and *Sedum plumbizincicola* increased the removal efficiency of Cu, Zn, and Cd from the soil due to the increased biomass and decreased soil pH. For heavily contaminated soils, excessive soil contaminants may threaten plant life because of the toxicity of contaminants; therefore, various amendments have been added to severely contaminated soils to promote plant growth and improve the physicochemical and biological properties of soil (Cui et al. 2020, Mahar et al. 2016). Biochar, produced from organic feedstocks under thermal and limited oxygen conditions, has typical properties such as high pH value, large specific surface area, high cation exchange capacity, and high carbon content (Sarwar et al. 2016). Unlike chelators, biochar application usually decreases the bioavailability of metals due to its high sorption capacity and the increase of soil pH (Shao et al. 2021). The application of limestone was reported to change soil physicochemical characteristics such as pH increase, promote plant growth, and effectively reduce the availability of HMs (Cui et al. 2020). The compost was also reported to be able to promote the plant accumulation of HMs by influencing the number of functional groups and redistributing HMs in the soil (Huang et al. 2020). Helmisaari et al. (2007) found that the application of compost and sawdust to soil effectively enhanced the phytoremediation of HMs. Another enhancing way is applying nanoparticles in size range of 1 ~ 100 nm as catalysts. Nanomaterials can improve the plant protein level, soil fertility and promote the absorption of HMs due to the interaction with the rhizospheric microorganisms (Zhu et al. 2019). Nanotechnology is also used to improve the surface area of biosolids, thereby improving their fixation ability on HMs.

Plant-microbial combined remediation technology makes full use of the coexistence relationship among soil, microorganisms, and plant, thereby improving the phytoremediation efficiency of soil contamination. For example, Khan et al. (2009) revealed that proper rhizobia application that promotes plant growth and HM tolerance could enhance the phytoremediation efficiency. Chen et al. (2008) found that the HM uptake ability of the plants increased obviously after the inoculation of an indigenous metal-resistant rhizobial strain, indicating that the strain may play an important role in accelerating HM uptake of the nodulated plant. Jin et al. (2019) isolated and identified a new soil fungal strain, *Simplicilliumchinense* QD10, which showed a strong adsorption capability and significantly enhanced the phytoremediation of Cd or Pb potentially attributed to the rhizospheric enrichment of HMs. Liu et al. (2021) found that the endophyte *Pseudomonas putida* enhanced *Trifolium repens* L. growth and HM uptake, and this could be a promising phytoremediation method of nonferrous metallic tailing.

Phytoremediation based on hyperaccumulator plants is limited by their small biomass, short growth cycle, and limited distribution. Applying modern agricultural techniques is a shortcut to improving the efficiency of phytoremediation. For example, crop breeding technology can be used to improve the performance of hyperaccumulator plants. The plant growth and absorption capacity of HMs can be enhanced by foliage application and reasonable water and fertilizer supply. In addition, genetic engineering technology has opened a new field of phytoremediation by improving the ability of plants to absorb and accumulate HMs and their tolerance to HMs (Sarma et al. 2021). It is worth noting that transgenic plants have potential ecological risks in practical application, which may threaten the local biological community. Therefore, safety evaluation of transgenic plants must be carried out before practical application. On the other hand, electro-kinetic technology can effectively activate soil HMs, enhance the accumulation and transport of HMs in plants, and improve the efficiency of phytoremediation (Chirakkara et al. 2016). The enhancing mechanisms generally include: increasing soil HM bioavailability, enhancing plant growth and metabolism, and affecting soil microbial activities.

4.7 Conclusion and Prospects

HM contamination in soil has potential long-range risks on the environment and human health, and it has become a severe concern worldwide. Phytoremediation is expected to be a commercially feasible technology for solving the contamination problem. Moderately contaminated soils are the most adaptable for phytoremediation. Phytoremediation shows other benefits: the harvested plant biomass could be reused to generate energy and produce biofuels; after plant incineration, the recovered HMs could be ingredients for industrial production.

In the future, more researches are needed to understand better the interactions among HMs, soil, microbes, and plant roots. Research is necessary for finding low-cost, environmentally safe, and effective catalysts to promote phytoremediation efficiency. Long-term *in situ* field studies are also necessary for the all-sided application of phytoremediation in soils. In addition, soil HM remediation is a systematic project, and it is difficult for a single remediation technology to achieve the desired effects. In order to increase the bioavailability of HMs and improve remediation efficiency, phytoremediation should be supplemented by other physical, chemical, and microbial methods. Therefore, combined remediation technology will be the main research direction for soil remediation in the future.

References

Ali, H., E. Khan and M. A. Sajad. 2013. Phytoremediation of heavy metals-concepts and applications. Chemosphere 91: 869–881.

Ali, H., E. Khan and I. Ilahi. 2019. Environmental chemistry and ecotoxicology of hazardous heavy metals: Environmental persistence, toxicity, and bioaccumulation. J. Chem. 2019: 6730305.

Alirzayeva, E., G. Neumann, W. Horst, Y. Allahverdiyeva, A. Specht and V. Alizade. 2017. Multiple mechanisms of heavy metal tolerance are differentially expressed in ecotypes of Artemisia fragrans. Envrion. Pollut. 220: 1024–1035.

Anju, M. 2017. Biotechnological strategies for remediation of toxic metal(loid)s from environment. pp. 315–360. *In*: Gahlawat, S. K. et al. (eds.). Plant Biotechnology: Recent Advancements and Developments, Springer, ISBN: 978-981-10-4731-2.

Ashraf, S., Q. Ali, Z. A. Zahir, S. Ashraf and H. N. Asghar. 2019. Phytoremediation: environmentally sustainable way for reclamation of heavy metal polluted soils. Ecotox. Environ. Safe. 174: 714–727.

Barbosa, B., S. Boleo, S. Sidella, J. Costa, M. P. Duarte, B. Mendes, et al. 2015. Phytoremediation of heavy metal-contaminated soils using the perennial energy crops *Miscanthus* spp. and *Arundo donax* L. Bioenerg. Res. 8: 1500–1511.

Bhargava, A., F. F. Carmona, M. Bhargava and S. Srivastava. 2012. Approaches for enhanced phytoextraction of heavy metals. J. Environ. Manag. 105: 103–120.

Bian, F., Z. Zhong, C. Li, X. Zhang, L. Gu, Z. Huang et al. 2021. Intercropping improves heavy metal phytoremediation efficiency through changing properties of rhizosphere soil in bamboo plantation. J. Hazard. Mater. 416: 125898.

Bizily, S. P., C. L. Rugh and R. B. Meagher. 2000. Phytodetoxification of hazardous organomercurials by genetically engineered plants. Nat. Biotechnol. 18: 213–217.

Broadley, M. R., P. J. White, J. P. Hammond, I. Zelko and A. Lux. 2007. Zinc in plants. New Phytol. 173(4): 677–702.

Cambier, P., V. Pot, V. Mercier, A. Michaud, P. Benoit, A. Revallier et al. 2014. Impact of long-term organic residue recycling in agriculture on soil solution composition and trace metal leaching in soils. Sci. Total Environ. 499: 560–573.

Cameselle, C. and S. Gouveia. 2019. Phytoremediation of mixed contaminated soil enhanced with electric current. J. Hazard. Mater. 361: 95–102.

Chen, W. M., C. H. Wu, E. K. James and J. S. Chang. 2008. Metal biosorption capability of *Cupriavidus taiwanensis* and its effects on heavy metal removal by nodulated *Mimosa pudica*. J. Hazard. Mater. 151 (2-3): 364–371.

Chirakkara, R. A., C. Cameselle and K. R. Reddy. 2016. Assessing the applicability of phytoremediation of soils with mixed organic and heavy metal contaminants. Rev. Environ. Sci. Biotechnol. 15: 299–326.

Cristaldi, A., G. O. Conti, E. H. Jho, P. Zuccarello and M. Ferrante. 2017. Phytoremediation of contaminated soils by heavy metals and PAHs. A brief review. Environ. Technol. Inno. 8: 309–326.

Cui, H., H. Li, S. Zhang, Q. Yi, J. Zhou, G. Fang et al. 2020. Bioavailability and mobility of copper and cadmium in polluted soil after phytostabilization using different plants aided by limestone. Chemosphere 242: 125252.

Da Conceiçao Gomes, M. A., R. A. Hauser-Davis, A. N. de Souza and A. P. Vitoria. 2016. Metal phytoremediation: general strategies, genetically modified plants and applications in metal nanoparticle contamination. Ecotoxicol. Environ. Saf. 134: 133–147.

Dada, E. O., K. I. Njoku, A. A. Osuntoki and M. O. Akinola. 2015. A review of current techniques of physico-chemical and biological remediation of heavy metals polluted soil. Ethiop. J. Environ. Stud. Manag. 8(5): 606–615.

Dai, H., S. Wei, I. Twardowska, R. Han and L. Xu. 2017. Hyperaccumulating potential of Bidens pilosa L. for Cd and elucidation of its translocation behavior based on cell membrane permeability. Environ. Sci. Pollut. Res. 24(29): 23161–23167.

Dai, H., S. Wei and L. Skuza. 2020. Effects of different soil pH and nitrogen fertilizers on *Bidens pilosa* L. Cd accumulation. Environ. Sci. Pollut. Res. 27: 9403–9409.

Dastyar, W., A. Raheem, J. He and M. Zhao. 2019. Biofuel production using thermo-chemical conversion of heavy metal-contaminated biomass (HMCB) harvested from phytoextraction process. Chem. Eng. J. 358: 759–785.

Dipu, S., A. A. Kumar and S. G. Thanga. 2012. Effect of chelating agents in phytoremediation of heavy metals. Remediat. J. 22: 133–146.

Fasani, E., A. Manara, F. Martini, A. Furini and G. DalCorso. 2018. The potential of genetic engineering of plants for the remediation of soils contaminated with heavy metals. Plant Cell Environ. 41: 1201–1232.

Forte, J. and S. Mutiti. 2017. Phytoremediation Potential of *Helianthus annuus* and *Hydrangea paniculata* in copper and lead-contaminated soil. Water Air Soil Pollut. 228: 77.

Gerhardt, K. E., P. D. Gerwing and B. M. Greenberg. 2017. Opinion: taking phytoremediation from proven technology to accepted practice. Plant Sci. 256: 170–185.

Ghori, N. H., T. Ghori, M. Q. Hayat, S. R. Imadi, A. Gul, V. Altay et al. 2019. Heavy metal stress and responses in plants. Int. J. Environ. Sci. Technol. 16: 1807–1828.

Goswami, S. and S. Das. 2016. Copper phytoremediation potential of *Calandula officinalis* L. and the role of antioxidant enzymes in metal tolerance. Ecotox. Environ. Safe. 126: 211–218.

Göhre, V. and U. Paszkowski. 2006. Contribution of the arbuscular mycorrhizal symbiosis to heavy metal phytoremediation. Planta 223: 1115–1122.

He, J., V. Strezov, R. Kumar, H. Weldekidan, S. Jahan, B. H. Dastjerdi et al. 2019. Pyrolysis of heavy metal contaminated Avicennia marina biomass from phytoremediation: characterisation of biomass and pyrolysis products. J. Clean. Prod. 234: 1235–1245.

He, S., Z. He, X. Yang, P. J. Stoffella and V. C. Baligar. 2015. Soil biogeochemistry, plant physiology, and phytoremediation of cadmium-contaminated soils. Adv. Agron. 134: 135–225.

Helmisaari, H. S., M. Salemaa, J. Derome, O. Kiikkila and T. M. Nieminen. 2007. Remediation of heavy metal-contaminated forest soil using recycled organic matter and native woody plants. J. Environ. Qual. 36(4): 1145–1153.

Hooda, V. 2007. Phytoremediation of toxic metals from soil and waste water. J. Environ. Biol. 28(2 Suppl.): 367–376.

Hu, P., Y. Wang, W. J. Przybyłowicz, Z. Li, A. Barnabas, L. Wu et al. 2015. Elemental distribution by cryo-micro-PIXE in the zinc and cadmium hyperaccumulator *Sedum plumbizincicola* grown naturally. Plant Soil 388: 267–282.

Huang, H. L., L. Luo, L. H. Huang, J. C. Zhang, P. Gikas and Y. Y. Zhou. 2020. Effect of manure compost on distribution of Cu and Zn in rhizosphere soil and heavy metal accumulation by *Brassica juncea*. Water Air Soil Pollut. 231: 195.

Ibrahim, N. and G. El Afandi. 2020. Phytoremediation uptake model of heavy metals (Pb, Cd and Zn) in soil using *Nerium oleander*, Heliyon 6(7): e04445.

Jiang, M., S. Liu, Y. Li, X. Li, Z. Luo, H. Song et al. 2019. EDTA-facilitated toxic tolerance, absorption and translocation and phytoremediation of lead by dwarf bamboos. Ecotoxicol. Environ. Saf. 170: 502–512.

Jin, Z., S. Deng, Y. Wen, Y. Jin, L. Pan, Y. Zhang et al. 2019. Application of *Simplicilliumchinense* for Cd and Pb biosorption and enhancing heavy metal phytoremediation of soils. Sci. Total Environ. 697: 134148.

Kader, M., D. T. Lamb, M. Megharaj and R. Naidu. 2016. Sorption parameters as a predictor of arsenic phytotoxicity in Australian soils. Geoderma 265: 103–110.

Khan, M. S., A. Zaidi, P. A. Wani and M. Oves. 2009. Role of plant growth promoting rhizobacteria in the remediation of metal contaminated soils. Environ. Chem. Lett. 7(1): 1–19.

Khan, M. A., X. Ding, S. Khan, M. L. Brusseau, A. Khan and J. Nawab. 2018. The influence of various organic amendments on the bioavailability and plant uptake of cadmium present in mine-degraded soil. Sci. Total Environ. 636: 810–817.

Kotrba, P., J. Najmanova, T. Macek, T. Ruml and M. Mackova. 2009. Genetically modified plants in phytoremediation of heavy metal and metalloid soil and sediment pollution. Biotechnol. Adv. 27: 799–810.

Kumar, A., B. S. Bisht, V. D. Joshi and T. Dhewa. 2011. Review on bioremediation of polluted environment: A management tool. Int. J. Environ. Sci. 1(6): 2011.

Kumar, P. S. and E. Gunasundari. 2018. Bioremediation of heavy metals. pp. 165–195. *In*: Varjani, S. J., A. K. Agarwal, E. Gnansounou and B. Gurunathan (eds.). Bioremediation: Applications for Environmental Protection and Management. Springer, Singapore.

Lajayer, B. A., N. K. Moghadam, M. R. Maghsoodi, M. Ghorbanpour and K. Kariman. 2019. Phytoextraction of heavy metals from contaminated soil, water and atmosphere using ornamental plants: mechanisms and efficiency improvement strategies. Environ. Sci. Pollut. Res. 26: 8468–8484.

Lin, H., C. Liu, B. Li and Y. Dong. 2021. *Trifolium repens* L. regulated phytoremediation of heavy metal contaminated soil by promoting soil enzyme activities and beneficial rhizosphere associated microorganisms. J. Hazard. Mater. 402: 123829.

Ling, T., Q. Gao, H. Du, Q. Zhao and J. Ren. 2017. Growing, physiological responses and Cd uptake of corn (*Zea mays* L.) under different Cd supply. Chem. Speciat. Bioavailab. 29: 216–221.

Liu, C., H. Lin, B. Li, Y. Dong and E. Menzembere. 2021. Endophyte Pseudomonas putida enhanced *Trifolium repens* L. growth and heavy metal uptake: A promising in-situ non-soil cover phytoremediation method of nonferrous metallic tailing. Chemosphere 272: 129816.

Liu, J., Y. J. Liu, Y. Liu, Z. Liu and A. N. Zhang. 2018a. Quantitative contributions of the major sources of heavy metals in soils to ecosystem and human health risks: a case study of Yulin, China. Ecotoxicol. Environ. Saf. 164: 261–269.

Liu, L., W. Li, W. Song and M. Guo. 2018b. Remediation techniques for heavy metal-contaminated soils: Principles and applicability. Sci. Total Environ. 633: 206–219.

Lotfy, S. M. and A. Z. Mostafa. 2014. Phytoremediation of contaminated soil with cobalt and chromium. J. Geochem. Explor. 144: 367–373.
Lu, Y., D. Luo, A. Lai, G. Liu, L. Liu and J. Long. 2017. Leaching characteristics of EDTA-enhanced phytoextraction of Cd and Pb by *Zea mays* L. in different particle-size fractions of soil aggregates exposed to artificial rain. Environ. Sci. Pollut. Res. 24(2): 1845–1853.
Luo, T., M. Shen, J. Zhou, X. Wang, J. Xia, Z. Fu et al. 2019. Chronic exposure to low doses of Pb induces hepatotoxicity at the physiological, biochemical, and transcriptomic levels of mice. Environ. Toxicol. 34(4): 521–529.
Ma, Y., N. M. Dickinson and M. H. Wong. 2003. Interactions between earthworms, trees, soil nutrition and metal mobility in amended Pb/Zn mine tailings from Guangdong, China. Soil Biol. Biochem. 35: 1369–1379.
Magdalena, S. L., J. Jerzy and R. Thomas. 2020. Phytofiltration of chosen metals by aquarium liverwort (*Monosoleumtenerum*). Ecotoxicol. Environ. Saf. 188: 109844.
Mahar, A., P. Wang, A. Ali, M. K. Awasthi, A. H. Lahori, Q. Wang et al. 2016. Challenges and opportunities in the phytoremediation of heavy metals contaminated soils: A review. Ecotoxicol. Environ. Saf. 126: 111–121.
Mao, X., R. Jiang, W. Xiao and J. Yu. 2015. Use of surfactants for the remediation of contaminated soils: A review. J. Hazard. Mater. 285: 419–435.
Meier, S. K., N. Adams, M. Wolf, K. Balkwill, A. M. Muasya, C. A. Gehring et al. 2018. Comparative RNA-seq analysis of nickel hyperaccumulating and non-accumulating populations of *Senecio coronatus* (Asteraceae). Plant J. 95: 1023–1038.
Miao, Q. and J. Yan. 2013. Comparison of three ornamental plants for phytoextraction potential of chromium removal from tannery sludge. J. Mater. Cycles. Waste. 15: 98–105.
Minisha, T. M., I. K. Shah, G. K. Varghese and R. K. Kaushal. 2020. Application of Aztec Marigold (*Tagetes erecta* L.) for phytoremediation of heavy metal polluted lateritic soil. Environ. Chem. Ecotoxicol. 3: 17–22.
Mortensen, L. H., R. Rønn and M. Vestergård. 2018. Bioaccumulation of cadmium in soil organisms–With focus on wood ash application. Ecotoxicol. Environ. Saf. 156: 452–462.
Mukhopadhyay, S. and S. K. Maiti. 2010. Phytoremediation of metal enriched mine waste: a review. Glob. J. Environ. Res. 4: 135–150.
Nie, S. W., W. S. Gao, Y. Q. Chen, P. Sui and A. E. Eneji. 2010. Use of life cycle assessment methodology for determining phytoremediation potentials of maize-based cropping systems in fields with nitrogen fertilizer over-dose. J. Clean. Prod. 18(15): 1530–1534.
Pardo, T., B. Rodríguez-Garrido, R. F. Saad, J. L. Soto-Vázquez, M. Loureiro-Vi˜ nas, A. Prieto-Fernández et al. 2018. Assessing the agromining potential of Mediterranean nickel-hyperaccumulating plant species at field-scale in ultramafic soils under humid-temperate climate. Sci. Total Environ. 630: 275–286.
Parmar, S. and V. Singh. 2015. Phytoremediation approaches for heavy metal pollution: a review. J. Plant Sci. Res. 2(2): 139.
Pinto, E., A. A. R. M. Aguiar and Isabel M. P. L. V. O. Ferreira. 2014. Influence of soil chemistry and plant physiology in the phytoremediation of Cu, Mn, and Zn. Crit. Rev. Plant Sci. 33: 351–373.
Pulford, I. and C. Watson. 2003. Phytoremediation of heavy metal-contaminated land by trees—a review. Environ. Int. 29: 529–540.
Rajkumar, M., S. Sandhya, M. N. V. Prasad and H. Freitas. 2012. Perspectives of plant-associated microbes in heavy metal phytoremediation. Biotechnol. Adv. 30: 1562–1574.
Rascio, N. and F. Navari Izzo. 2011. Heavy metal hyperaccumulating plants: how and why do they do it? And what makes them so interesting? Plant Sci. 180: 169–181.
Sarma, H., N. F. Islam, R. Prasad, M. N. V. Prasad, L. Q. Ma and J. Rinklebe. 2021. Enhancing phytoremediation of hazardous metal(loid)s using genome engineering CRISPR–Cas9 technology, J. Hazard. Mater. 414: 125493.
Sarwar, N., M. Imran, M. R. Shaheen, W. Ishaq, A. Kamran, A. Matloob et al. 2016. Phytoremediation strategies for soils contaminated with heavy metals: modifications and future perspectives. Chemosphere 171: 710–721.
Shah, V. and A. Daverey. 2020. Phytoremediation: A multidisciplinary approach to clean up heavy metal contaminated soil. Environ. Technol. Inno. 18: 100774.
Shaheen, S. M., C. D. Tsadilas and J. Rinklebe. 2013. A review of the distribution coefficients of trace elements in soils: Influence of sorption system, element characteristics, and soil colloidal properties. Adv. Colloid Interface Sci. 201–202, 43–56.
Shao, H., Y. Wei, C. Wei, F. Zhang and F. Li. 2021. Insight into cesium immobilization in contaminated soil amended with biochar, incinerated sewage sludge ash and zeolite. Environ. Technol. Inno. 23: 101587.
Sharma, S. S., K. J. Dietz and T. Mimura. 2016. Vacuolar compartmentalization as indispensable component of heavy metal detoxification in plants. Plant Cell Environ. 39: 1112–1126.
Suman, J., O. Uhlik, J. Viktorova and T. Macek. 2018. Phytoextraction of heavy metals: a promising tool for clean-up of polluted environment? Front Plant Sci. 9: 1476.
Susarla, S., V. F. Medina and S. C. McCutcheon. 2002. Phytoremediation: An ecological solution to organic chemical contamination. Ecol. Eng. 18: 647–658.

Tchounwou, P. B., C. G. Yedjou, A. K. Patlolla and D. J. Sutton. 2012. Heavy metals toxicity and the environment. pp. 133–164. *In*: Luch, A. (ed.). Molecular, Clinical and Environmental Toxicology. Experientia Supplementum, vol. 101. Springer, Basel. https://doi.org/10.1007/978-3-7643-8340-4_6.

Thakur, S., L. Singh, Z. A. Wahid, M. F. Siddiqui, S. M. Atnaw and M. F. M. Din. 2016. Plant-driven removal of heavy metals from soil: uptake, translocation, tolerance mechanism, challenges, and future perspectives. Environ. Monit. Assess. 188.

Tozser, D., T. Magura and E. Simon. 2017. Heavy metal uptake by plant parts of willow species: A meta-analysis. J. Hazard. Mater. 336: 101–109.

Vangronsveld, J., R. Herzig, N. Weyens, J. Boulet, K. Adriaensen, A. Ruttens et al. 2009. Phytoremediation of contaminated soils and groundwater: lessons from the field. Environ. Sci. Pollut. Res. 16: 765–794.

Vara Prasad, M. N. and H. M. de Oliveira Freitas. 2003. Metal hyperaccumulation in plants: biodiversity prospecting for phytoremediation technology. Electron. J. Biotechnol. 6: 285–321.

Volk, J. and O. Yerokun. 2016. Effect of application of increasing concentrations of contaminated water on the different fractions of Cu and Co in Sandy Loam and Clay Loam Soils. Agriculture 6(4): 64.

Wei, S. H., J. A. Teixeira da Silva and Q. X. Zhou. 2008. Agro-improving method of phytoextracting heavy metal contaminated soil. J. Hazard. Mater. 150: 662–668.

Wei, Z., Q. Van Le, W. Peng, Y. Yang, H. Yang, H. Gu et al. 2021. A review on phytoremediation of contaminants in air, water and soil. J. Hazard. Mater. 403: 123658.

Wieshammer, G., R. Unterbrunner, G. T. Baares, M. F. Zivkovic, M. Puschenreiter and W. W. Wenzel. 2007. Phytoextraction of Cd and Zn from agricultural soils by *Salix* ssp. and intercropping of *Salix caprea* and *Arabidopsis halleri*. Plant Soil 298: 255–264.

Wu, M., Q. Luo, Y. Zhao, Y. Long, S. Liu and Y. Pan. 2018. Physiological and biochemical mechanisms preventing Cd toxicity in the new hyperaccumulator *Abelmoschus manihot*. J. Plant Growth Regul. 37: 709–718.

Wuana, R. A. and F. E. Okieimen. 2011. Heavy metals in contaminated soils: a review of sources, chemistry, risks and best available strategies for remediation. Isrn Ecology 2011: 402647.

Xiao, R., A. Ali, P. Wang, R. Li, X. Tian and Z. Zhang. 2019. Comparison of the feasibility of different washing solutions for combined soil washing and phytoremediation for the detoxification of cadmium (Cd) and zinc (Zn) in contaminated soil. Chemosphere 230: 510–518.

Yan, A., Y. Wang, S. N. Tan, M. L. Mohd Yusof, S. Ghosh and Z. Chen. 2020. Phytoremediation: a promising approach for revegetation of heavy metal-polluted land. Front. Plant Sci. 11: 359.

Yan, L., Q. V. Le, C. Sonne, Y. Yang, H. Yang, H. Gu et al. 2021. Phytoremediation of radionuclides in soil, sediments and water. J. Hazard. Mater. 407: 124771.

Yang, S. X., H. Deng and M. S. Li. 2008. Manganese uptake and accumulation in a woody hyperaccumulator, Schima superba. Plant Soil Environ. 54(10): 441–446.

Yin, H., Q. Tan, Y. Chen, G. Lv, D. He and X. Hou. 2011. Polycyclic aromatic hydrocarbons (PAHs) pollution recorded in annual rings of gingko (*Gingko biloba* L.): Translocation, radial diffusion, degradation and modeling. Microchem. J. 97: 131–137.

Yoon, R. K. J., T. K. Jae, E. Y. Gary and K. Kim. 2015. Bioavailability of heavy metals in soils: definitions and practical implementation—a critical review. Environ. Geochem. Health 1041–1061.

Zhu, Y., F. Xu, Q. Liu, M. Chen, X. L. Liu, Y. Y. Wang et al. 2019. Nanomaterials and plants: Positive effects, toxicity and the remediation of metal and metalloid pollution in soil. Sci. Total Environ. 662: 414–421.

CHAPTER 5

Water Hyacinth (*Eichhornia crassipes*)
A Sustainable Strategy for Heavy Metals Removal from Contaminated Waterbodies

Apurba Koley,[1] *Douglas Bray,*[2] *Sandipan Banerjee,*[3] *Sudeshna Sarkar,*[1] *Richik Ghosh Thakur,*[1] *Amit Kumar Hazra,*[4] *Narayan Chandra Mandal,*[3] *Shibani Chaudhury,*[1] *Andrew B. Ross,*[2] *Miller Alonso Camargo-Valero*[5,6] and *Srinivasan Balachandran*[1,*]

5.1 Introduction

Overpopulation and the rapid increase of urbanization and industrialization have resulted in the mass generation of pollutants and subsequent release to the environment. In aquatic systems, the main sources of HM pollution include industrial, petrochemical, urban, mining, e-waste, and agricultural wastewaters (Dhote and Dixit 2009). The increase in heavy metal and metalloids (HM) concentrations imposes severe negative impacts that threaten aquatic species and the environment (Singh et al. 2008). Heavy Metal is a term describing any "metal and metalloid element with a relatively high density" (Sharma 2014), of ≥ 5 g cm^{-3}, encompassing ~ 53 elements (Majumdar et al. 2018). HM often play a vital role in plant functionality, e.g., Cu, Zn, Fe, provided they do not exceed certain concentrations, whereas HM like As, Pb, and Hg are toxic at low concentrations. HM are non-degradable and persistent in the environment; therefore, the increasing levels due to anthropogenic activities are a serious cause for concern (Ansari et al. 2016). There are several methods to treat contaminated water; however, in the case of the underdeveloped and developing countries, the huge expenditure of modern technologies to remove the metals and metalloids may be a limiting factor (Rai 2010, Rai and Tripathi 2007).

[1] Bioenergy Laboratory, Department of Environmental Studies, Institute of Science, Visva-Bharati, Santiniketan-731235, India.
[2] School of Chemical and Process Engineering, University of Leeds, Leeds, LS2 9JT, United Kingdom.
[3] Mycology and Plant Pathology Laboratory, Department of Botany, Visva-Bharati University, Santiniketan-731235, India.
[4] Socio-Energy Lab, Department of Lifelong Learning and Extension, Visva-Bharati, Santiniketan.
[5] BioResource Systems Research Group, School of Civil Engineering, University of Leeds, Leeds LS2 9JT, United Kingdom.
[6] Departamento de Ingeniería Química, Universidad Nacional de Colombia, Campus La Nubia, Manizales, Colombia.
* Corresponding author: s.balachandran@visva-bharati.ac.in
ORCID ID: 0000-0003-4247-408X

Phytoremediation is a plant-based approach to remove different pollutants or contaminants from water or soil; the HM uptake rate or concentration varies depending on the plant and conditions of the system. The advantages of phytoremediation include its economic feasibility, being environmentally and eco-friendly, its broad applicability, and ability to prevent metal leaching/spread (Yan et al. 2020). Accumulation of HM in plants can result from a number of processes, i.e., mobilization, uptake by the roots, xylem loading, root to shoot transport, and sequestration. Phytoremediation can be achieved through phytoextraction/accumulation, rhizofiltration, phytostabilisation, phytotransformation/degradation, and phytovolatilization (Hakeem et al. 2014, Saha et al. 2017, Prasad 2004). Some plants are able to accumulate HM that are not bioavailable by releasing exudates to increase solubility (Dalvi and Bhalerao 2013).

Water hyacinth is an invasive free-floating aquatic macrophyte in India, once noted for its beautiful flowers, and is generally regarded as a nuisance in the waterbody and the entire aquatic ecosystem. The dense mats formed by water hyacinth interfere with the irrigation, drainage channels, fishing, transport, and various domestic activities (Luu and Getsinger 1990). These mats also provide breeding grounds for various vectors of disease such as malaria, encephalitis, and filariasis due to the increased number of mosquitos resulting in an increased spread in areas where water bodies are used in domestic activities (Sucharit et al. 1981, Patnaik et al. 2022). However, the plant has been noted to demonstrate positive impacts associated with phytoremediation: the invasive nature of water hyacinth resulting in a rapid growth rate, particularly in nutrient-rich water. Water hyacinth has been reported as one of the most productive aquatic plants on the planet. Therefore, the use of water hyacinth offers great potential for both phytoremediation and the production of bioenergy and bioproducts. Some of the different uses of water hyacinth are described in Table 5.1.

Water hyacinth can play a key role in removing pollutants from waste effluents (Gopal 1987, Basu et al. 2021). Several studies emphasize the removal of HM and organic matter from water bodies and wastewater streams (Madikizela 2021, Sharma et al. 2021, Li et al. 2021). High water hyacinth yields result in the efficient uptake and removal of heavy metals and nutrients, providing opportunities for wastewater treatment worldwide (Rezania et al. 2013, Ilo et al. 2020).

The primary purpose of this study is to understand the effectiveness of water hyacinth in the removal of HM and the potential use of biomass for the sustainable production of energy.

Table 5.1. Different benefits of Water Hyacinth.

Use of Water Hyacinth	References
Use of water hyacinth waste to produce fiber-reinforced polymer composites for concrete confinement	Jirawattanasomkul et al. 2021
Production of biofuels	Bhattacharya and Kumar 2010, Sukarni et al. 2019
Biofertiliser production	Gupta et al. 2007
Biogas production	Unpaprom et al. 2021
Use of water hyacinth as duck feed	Jianbo et al. 2008
Wastewater treatment	Gupta et al. 2012
Use as raw material for greaseproof paper making	Goswami and Saikia 1994
Use of water hyacinth as animal feedstuff	Akankali and Elenwo 2019, de Vasconcelos et al. 2016
Fish food	Mahmood et al. 2018
Vermicompost	Goswami et al. 2017, Singh and Kalamdhad 2013
Bioethanol	Madian et al. 2019

5.1.1 Bibliometric Study

In the bibliometric study, four sets of keywords were used in a search on Web of Science; the keywords used were "Heavy metal contaminated water bodies", "Role of Water Hyacinth in Phytoremediation of Heavy Metal(loid)s", "Water Hyacinth: An alternative source of bioenergy", and "Rhizospheric Microorganisms of Water Hyacinth". The search was filtered to include literature from 2000–2021, which returned 7321 articles. Analysis was performed using VOSviewer with the lowest series of keyword occurrences in the outcomes outlined to be 35. A network map of the keywords most utilized was produced; the outcomes are represented in Fig. 5.1. The color represents the cluster category, the number of occurrences is depicted by the size of the cluster, and the lines represent times where keywords are used together. The clusters that were formed are HM toxicity and contaminated aquatic systems (red cluster); Phytoremediation of HM (green cluster); Wastewater treatment of HM contaminated water (blue cluster); Effects of HM on aquatic systems (yellow cluster). The keyword analysis demonstrates a positive outlook on the phytoremediation of HM utilizing water hyacinth with microbial representatives, which is an unexplored area for HM remediation.

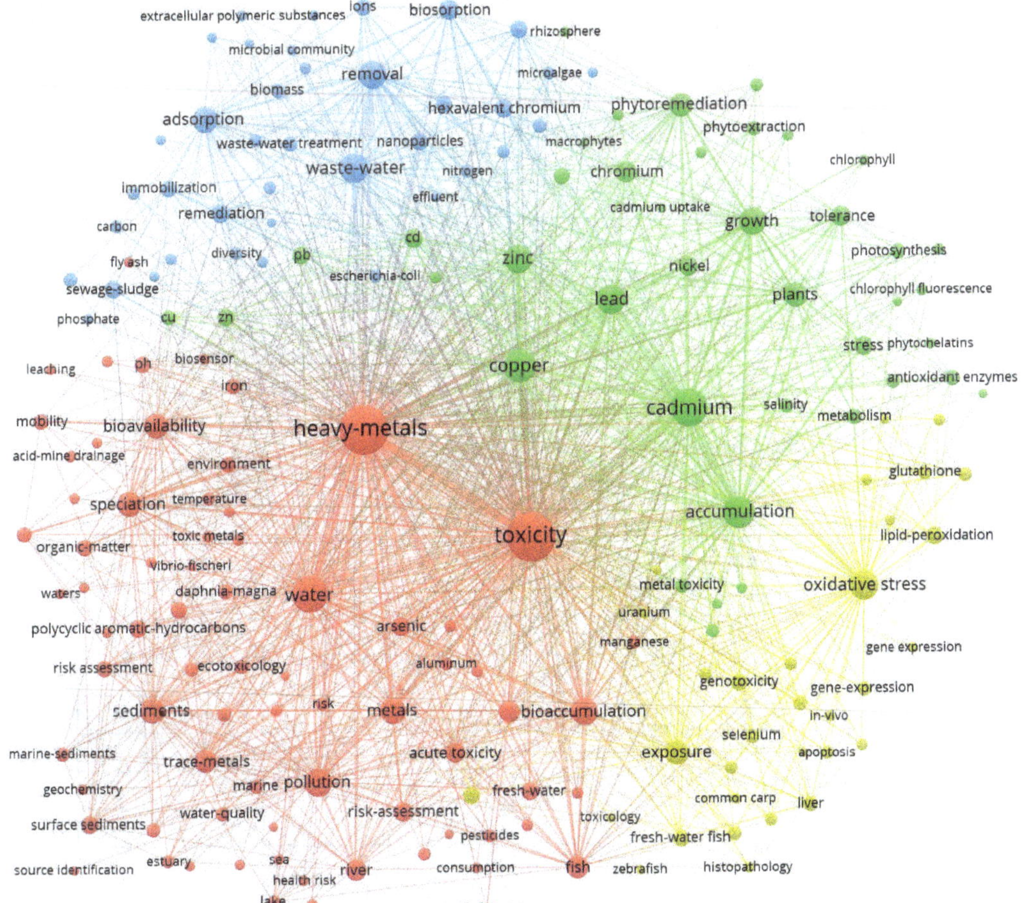

Fig. 5.1. Keywords Co-Occurrence Network Analysis Using VOSviewer Software.

5.2 Heavy Metals in the Environment

HM occur naturally in the Earth's crust and have a highly persistent nature (Fergusson 1990, Length 2007). HM can be toxic at low concentrations, and therefore, the contamination of aquatic systems is one of the significant environmental problems and may cause severe health hazards (Khan et al. 2005).

HM have a huge variation in chemical properties, leading to various applications; Table 5.2 shows some major industries that utilize HM. Historically, the use of HM generates large amounts of waste that have been poorly treated or released untreated, resulting in a high number of case studies that demonstrate large scale poisoning. One of the first cases of HM poising is Minamata Bay, Japan; wastewater containing a high concentration of Hg was discharged into the bay and bioaccumulated in the biota (Harada 1995). The seafood caught from the bay contained high levels of Hg and resulted in serious issues occurring in the population. Methyl mercury now has some of the strictest regulations of any pollution and must not be present in drinking water in any concentration (US EPA 2015).

HM have a background level that is almost indistinguishable across the globe; however, the impact of anthropogenic activities, predominantly industrial, agricultural and domestic, has led to significant increases in HM contamination in aquatic systems (Förstner and Wittmann 1981, Bradl 2005). The main sources of HM pollution are described in Table 5.3, the most significant of which come from mining, combustion, and waste disposal; these are particularly prevalent in developing countries due to the high cost of disposal and use of low-quality fuels (GEA 2012). This increased pollution of aquatic systems is of high concern as water is the most fundamental requirement of life (Vanloon and Duffy 2005).

Table 5.2. Use of Heavy Metals in Industry (Adapted from Shailaja and Gautam 2017, Sharma 2014, Bradl 2005, Förstner and Wittmann 1981).

Industry	*Heavy Metal*							
	As	**Cd**	**Cr**	**Cu**	**Ni**	**Hg**	**Pb**	**Zn**
Battery		✓			✓		✓	
Catalyst					✓	✓		
Chemical and Pharmaceutical		✓	✓	✓		✓	✓	✓
Electronic	✓		✓		✓	✓		
Electroplating			✓	✓	✓		✓	✓
Fertilizer and Animal Feed	✓	✓						✓
Glass, Ceramic, Textile, and Tanning	✓		✓		✓		✓	
Medical	✓				✓	✓		✓
Metallurgy	✓	✓	✓	✓	✓		✓	✓
Pesticide, Insecticide, etc.	✓					✓		
Pigment, Dye, and Paint	✓	✓	✓	✓	✓		✓	✓
Steam Generation Power Plant	✓	✓	✓					✓
Wood and Paper	✓		✓	✓	✓	✓	✓	✓

Table 5.3. Sources of heavy metals in water bodies and their toxicity.

Element	Toxic Forms	Probable Sources into Water Bodies	Toxicity	References
Zinc (Zn)	Zn II	Mining, purifying of zinc, lead, and cadmium ores, steel production, coal burning, and burning of wastes**	Metal fume fever, stomach cramps, Nausea, vomiting**	**ATSDR 2005
Arsenic (As)	Arsines, As III, As V*	Industrial source: Processing of glass, pigments, textiles, paper, metal adhesives, wood preservatives, and ammunition; Dietary Sources: Fish, shellfish, meat, poultry, dairy products, and cereals**	Nausea, vomiting, damage of blood vessels, the lower production rate of blood cells are held on low-level exposure. A high level of ingestion of these metals can cause the death of humans. Skin is dark, 'corns' or 'warts' are found in soles, palms, etc., are held on long-term, low exposure***	*Whitacre and Pears 1974 **WHO 2018 ***Baker and Chesnin 1975
Lead (Pb)	Organic Lead	Lead-acid batteries, solder, alloys, cable sheathing, pigments, rust inhibitors, ammunition, glazes, and plastic stabilizers. Mining and smelting of ore, industrial effluents, pesticides, herbicides and fertilizers, municipal waste, paints, water pipes*	Anaemia, encephalopathy, hepatitis, nephritic syndrome, and also causes kidney, nervous system, brain damage**	*WHO 1989 **David et al. 2003, Singh et al. 2012
Copper (Cu)	$CuSO_4$	Natural weathering of soil and rocks, disturbances of soil, effluent from sewage treatment plants, copper smelters, and refineries**	Gastrointestinal side effects such as abdominal pain, hematemesis, melena, jaundice, anorexia, severe thirst, diarrhoea, and vomiting associated with erosive gastropathy, headache, coma***	**Dorsey et al. 2004 ***Royer and Sharman 2020
Cadmium (Cd)	$CdCl_2$, $CdSO_4$, and cadmium oxide, CdO*	Fertilizers produced from phosphate ores, anticorrosive, electroplated onto steel, pigments in plastics, electric batteries, electronic components, and nuclear reactors**	Itai-Itai disease**	*NRC 1997 **Garrison and Ader 1966
Nikkel (Ni)	Nickel sulfide NiS and nickel sulfate $NiSO_4$*	Inexpensive jewellery, cosmetics, keys, cell phones, eyeglass frames, paper clips, orthodontic braces, stainless steel articles, nickel-plated articles, clothing fasteners zippers, snap buttons, belt buckles, electrical equipment, armaments, alloy, metallurgical and food processing industries, pigments, catalysts**	Contact dermatitis, headaches, gastrointestinal manifestations, respiratory manifestations, lung fibrosis, cardiovascular diseases, lung cancer; nasal cancer, epigenetic effects**	*Costa et al. 2005 **Genchi et al. 2020
Chromium (Cr)	Cr III, and Cr VI*	Effluents discharged from metal finishing, electroplating, different industries like textile, leather, ceramic and dyeing, etc.**	Dermatitis, allergic and eczematous skin reactions, skin and mucous membrane ulceration, perforation of the nasal septum, allergic asthmatic reactions, bronchial carcinomas, gastro-enteritis***	*Authman 2015 **Farag et al. 2006, Arunkumar et al. 2000 ***Baruthio1992
Mercury (Hg)	Methylmercury [CH3Hg]*	Heart disease, irritation of the eye, skin problems**	Brain dysfunction, thyroid dysfunction, renal tubular necrosis, or autoimmune glomerulonephritis***	*Liu et al. 2012 **Martin and Griswold 2009 ***Bernhoft 2012

5.3 Role of Water Hyacinth in Phytoremediation of Heavy Metals

5.3.1 Water Hyacinth

Water hyacinth, a plant native to tropical South America, belongs to the pickerelweed family, Pontederiaceae (Bailey 1949, Chillers 1991), and is one of the most productive yet destructive aquatic plants in the world (Zhang et al. 2010, Charudattan 1986). Water hyacinth has a high reproduction rate through asexual budding and seeds, though sexual reproduction has been described as "unimportant", likely due to lack of suitable pollinators (Gopal 1987). It has been shown that water hyacinth biomass can double within 14 days, and approximately ten plants of water hyacinth can cover the water surface of one and a half hectares within 8 months (Gunnarsson and Petersen 2007).

Water hyacinth demonstrates four different morphological variations (Center et al. 1999, Penfound and Earle 1948); however, two of these occur more frequently. In open water, and often lower nutrient concentrations, the plant demonstrates short bulbous petioles and leaves; this morphological form favors lateral movement (Center and Spencer 1981). If the growth rate increases, likely due to increased nutrients, and dense mats are formed, the crowded scenario allows the plant to develop into its secondary morphological form. This results in an elongation of the petioles and sphering of the leaves, as well as a significant development of the rhizome. This form favors vertical growth due to the crowded conditions and can reach a height of up to 1 meter in favorable conditions. In low nutrient conditions, the roots can be up to 3 meters long, whereas the petioles are likely to be < 30 cm, whereas in high nutrient conditions, the petioles can be over three times the length of the roots. Both morphological forms show evidence of asexual and sexual reproduction; however, the larger form is significantly less likely to develop flowers (Center et al. 1999, Penfound and Earle 1948). Despite the absence of known pollinators (Barrett 1980), water hyacinth can produce a large inflorescence with up to 25 violet flowers (Gopal 1987). Each flower can contain upwards of 500 seeds that sink to the bottom of the water and can remain dormant for over 20 years in the sediments (Matthews 1967, Gopal 1987, Malik 2007).

Water hyacinth has been known to grow in a large variety of water bodies, including lakes, streams, ponds, waterways, ditches, and backwater areas in temperate and tropical zones (Gopal 1987, McKeown and Bugyi 2015). Water hyacinth mats are able to stabilize pH levels and temperature in still water; this prevents stratification and the mixing of gases into and within the water column. The stabilization of physical factors by the mats produces a water body with higher carbon dioxide tension, lower oxygen, a more uniform temperature, lower light penetration, and higher acidity (Penfound and Earle 1948). The ideal conditions for water hyacinth are slightly acidic, high light levels, nutrient rich water, and mesophilic temperatures, approximately 28–30°C (Methy et al. 1990, Gopal 1987). However, water hyacinth can survive under a huge variety of conditions, including slight increases of salinity, nutrient deficient water, short-term drought, and a temperature range of approximately 10–37°C, though it can survive short periods outside this range (Gopal 1987, Penfound and Earle 1948, Jones et al. 2018).

The major threats posed by water hyacinth are to human health, economic development and biodiversity. The dense mats formed by water hyacinths are breeding grounds for vectors of various diseases, such as malaria, encephalitis, and filariasis, which leads to increased spread where the water is used for bathing, drinking, or recreation. However, this varies depending on the site. In Kenya, malaria causing mosquitoes showed a low percentage due to water turbulence and predatory fish, whereas there was an increased spread of schistosomiasis from snails in the vicinity of the mats (Ofulla et al. 2010). These mats also interfere with water transport, fishing, recreation, irrigation, and power generation, causing significant impacts on economic development (Malik 2007, Ofulla et al. 2010, Patel 2012). The reduction in fishing is due to the impact of water hyacinth mats on biodiversity: the reduction in light penetration results in decreased phytoplankton growth and,

Table 5.4. Accumulation of different heavy metals by Water Hyacinth; NM- not mentioned.

Heavy Metal/ Metalloids	Source of Water	Initial Amount Concentration mg/L	Remediated Amount %	Treatment Period Days	References
As	Arsenic contaminated water	4	20.00	10	Jasrotia et al. 2015
As	Solution of $NaAsO_2$	0.5	83	10	de Souza et al. 2018
As III	Diluting stock solution in drinking water	1	84.00	0	Brima and Haris 2014, Rezania et al. 2015
As V	Diluting stock solution in drinking water	0.25	80.00	0	Brima and Haris 2014, Rezania et al. 2015
Cd	NM	5	100.00	21	Das et al. 2016
	Artificial wastewater	50	100.00	0	Zhang et al. 2015, Rezania et al. 2015
	Wastewater Treatment	0.068	95.59	35	Ogunbayo et al. 2012
	Wastewater Treatment	0.078	94.87	35	Ogunbayo et al. 2012
	Wastewater Treatment	0.062	93.55	35	Ogunbayo et al. 2012
	NM	20	91.89	21	Das et al. 2016
	Pulp and Paper mill Effluent	2.45	88.98	15	Kumar et al. 2016
	Solution of $CdSO_4$ 100 $\mu g\ L^{-1}$	15	79.00	8	Romanova et al. 2021
	Industrial Water	1	76.00	15	Rai 2019, Mustafa and Hayder 2021
	Industrial wastewater	5.40	72.2	63	Ayaz et al. 2020
	Aqueous solution	8	80	7	Peng et al. 2020
	Textile Process Effluent	0.1	60	18	Kulkarni et al. 2007
	Textile Process Effluent	47.4	89.74	7	Tabinda et al. 2019
	Textile Waste water	0.19	36.8	14	Wickramasinghe et al. 2018
	Glass industry effluent	0.23	91	40	Singh et al. 2021
	Fresh Water Body	0.05	50.00	8	Smolyakov 2012
	Solution of $CdNO_{32}$ 100 $\mu g\ L^{-1}$	23	50.00	8	Romanova et al. 2021
	Aqueous Solution	1	48	14	Aisien et al. 2010
	Leachate	0.021	31.60	11	Putra and Hastika 2018
	Fresh Water Body	0.05	24.00	8	Smolyakov 2012
	Wastewater	0.21774	0.00	15	Victor et al. 2016
Cr	Wastewater	10	84	11	Mishra and Tripathi 2009
	Textile Wastewater	5	94.78	4	Mahmood et al. 2005a
	Industrial mines wastewater	2.0	0.01	99.5	Saha et al. 2017
	Dye industry	1.07	46	7	Panneerselvam and Priya 2021
	Textile Process Effluent	0.41	73.17	18	Kulkarni et al. 2007
	Industrial Water	1	66.00	15	Rai 2019, Mustafa and Hayder 2021

Table 5.4 contd. ...

...Table 5.4 contd.

Heavy Metal/ Metalloids	Source of Water	Initial Amount Concentration mg/L	Remediated Amount %	Treatment Period Days	References
Cu	Synthetic Wastewater	1.5	97.30	21	Mokhtar et al. 2011, Rezania et al. 2015
	Synthetic Wastewater	2.5	95.60	21	Mokhtar et al. 2011, Rezania et al. 2015
	Textile Wastewater	6	94.44	4	Mahmood et al. 2005a
	Pulp and Paper mill Effluent	5.64	94.33	15	Kumar et al. 2016
	NM	101	80.00	7	Putra et al. 2016
	Industrial wastewater	14.5	82.3	63	Ayaz et al. 2020
	Aqueous solution	8	80	7	Peng et al. 2020
	Landfill leachate	0.62	87.09	15	Abbas et al. 2019
	Glass industry effluent	0.62	93.5	40	Singh et al. 2021
	Textile Process Effluent	24.6	98.64	7	Tabinda et al. 2019
	Papermill effluent	0.23	63	20	Mishra et al. 2013
	Industrial Water	1	63.00	15	Rai 2019, Mustafa and Hayder 2021
	Synthetic Wastewater	5.5	61.60	21	Mokhtar et al. 2011, Rezania et al. 2015
	Textile Process Effluent	0.48	52.08	18	Kulkarni et al. 2007
	Fresh Water Body	0.25	24.00	8	Smolyakov 2012
	Leachate	0.507	22.00	11	Putra and Hastika 2018
	Fresh Water Body	0.25	8.00	8	Smolyakov 2012
	Wastewater Treatment	0.015	6.67	35	Ogunbayo et al. 2012
Fe	Pulp and Paper mill Effluent	8.95	91.84	15	Kumar et al. 2016
	Wastewater Treatment	0.044	90.91	35	Ogunbayo et al. 2012
	Landfill leachate	0.77	87.56	15	Abbas et al. 2019
	Glass industry effluent	4.59	92.8	40	Singh et al. 2021
	Textile Process Effluent	110.8	89.16	7	Tabinda et al. 2019
	Industrial Water	1	83.00	15	Rai 2019, Mustafa and Hayder 2021
	Leachate	0.463	77.80	11	Putra and Hastika 2018
	Textile Process Effluent	1.62	55.56	18	Kulkarni et al. 2007
Ni	Textile Process Effluent	0.42	76.19	18	Kulkarni et al. 2007
	Landfill leachate	1.41	81.56	15	Abbas et al. 2019
	Industrial Water	1	67.00	15	Rai 2019, Mustafa and Hayder 2021

Table 5.4 contd. ...

...Table 5.4 contd.

Heavy Metal/ Metalloids	Source of Water	Initial Amount Concentration mg/L	Remediated Amount %	Treatment Period Days	References
Pb	Pulp and Paper mill Effluent	1.02	90.20	15	Kumar et al. 2016
	Aqueous Solution	5	77.8	14	Aisien et al. 2010
	NM	100	60.00	7	Putra et al. 2016
	Industrial wastewater	7.29	78.1	63	Ayaz et al. 2020
	Aqueous solution	8	80	7	Peng et al. 2020
	Landfill leachate	0.86	84.41	15	Abbas et al. 2019
	Glass industry effluent	0.29	89.6	40	Singh et al. 2021
	Dye industry	0.037	43	7	Panneerselvam and Priya 2021
	Textile Process Effluent	7.8	98.1	7	Tabinda et al. 2019
	Textile Process Effluent	0.32	50	18	Kulkarni et al. 2007
	Leachate	0.042	30.00	11	Putra and Hastika 2018
	Fresh Water Body	0.25	26.00	8	Smolyakov 2012
	Fresh Water Body	0.25	11.00	8	Smolyakov 2012
Zn	Fresh Water Body	0.5	18.00	8	Smolyakov 2012
	Fresh Water Body	0.5	18.00	8	Smolyakov 2012
	Textile Wastewater	6	96.88	4	Mahmood et al. 2005a
	Wastewater	10	94.00	11	Mishra and Tripathi 2009
	Pulp and Paper mill Effluent	6.9	90.14	15	Kumar et al. 2016
	Industrial Water	1	79.00	15	Rai 2019, Mustafa and Hayder 2021
	Aqueous Solution	5	61.8	14	Aisien et al. 2010
	Textile Process Effluent	0.31	43.87	18	Kulkarni et al. 2007
	Chemistry Laboratory water	NA	0.10	0	Kelly et al. 2000
	Textile Wastewater	0.5	100	14	Wickramasinghe et al. 2018
	Glass industry effluent	0.72	99.4	40	Singh et al. 2021
	Landfill leachate	1.42	90.18	15	Abbas et al. 2019

therefore, a disruption in the food chain. This is further exemplified by reducing dissolved oxygen, which can be < 2 mg L^{-1}, a level that is intolerable to most angling fish (Shillinglaw 1981).

Water hyacinth has been shown to remove large quantities of HM from a variety of wastewaters, see Table 5.4; despite this, it shows a low translocation factor, < 1, suggesting that metals are concentrated in the roots (Preussler et al. 2015). This allows the plant to tolerate high levels of HM; it is also able to convert HM to non-toxic forms (Emerhi 2011, Rezania et al. 2015). WH is described as a hyperaccumulator, meaning it can exhibit a bioconcentration factor (BCF) of > 100 (Hakeem et al. 2014) and can reach a BCF of > 1000 for certain metals, including Cd, Cu, Cr, Fe, Mn, Pb and Zn (Victor et al. 2016, Eid et al. 2021). The results shown in Table 5.4 demonstrate that the removal of HM from wastewater is highly site, concentration, and type dependent; however, it is considered a hyperaccumulator across a range of conditions, for a number of metals, with most studies showing a significant concentration of HM in the roots.

In addition to phytoremediation, water hyacinth has been utilized for the production of various energy carriers, fertilizer, animal feed, and bio-materials, as shown in Table 5.1. By combining the two processes, it is possible to increase the value of water hyacinth biomass, therefore increasing the potential profits of harvesting the biomass.

5.3.2 Phytoremediation: A Green Technology

Phytoremediation is defined as the ability of selected organisms to remove, detoxify or incapacitate pollutants; it is an environmentally friendly and cost-effective technique for the removal of metals from sediment and water (Hakeem et al. 2014, Saha et al. 2017). Research into phytoremediation has yielded a large variety of organisms, including mollusks (Lionetto et al. 2001), macrophytes (Mayes and McIntosh 1975, Fawzy et al. 2012), alligators (Tellez and Merchant 2015), and fish (Aissaoui et al. 2017). Phytoremediation can be achieved through phytoextraction/accumulation, rhizofiltration, phytostabilisation, phytotransformation/degradation, and phytovolatilization (Hakeem et al. 2014, Saha et al. 2017). It can be applied to the removal of organic compounds, radionuclides, and metals in freshwater systems. There are many advantages of phytoremediation, including its use both *in* and *ex situ*; however, *in situ* is more common due to the low start-up costs (Ali et al. 2020). Another example is that phytoremediation is not selective and can remediate multiple pollutants simultaneously whilst reducing metal leaching.

5.3.3 Role of Water Hyacinth in Phytoremediation of Heavy Metals

Water hyacinth plays a pivotal role in HM removal from the contaminated or wastewater bodies and acts as "Nature's unorthodox medicine". This chapter critically analyses HM remediation by the water hyacinth.

Water hyacinth has been shown to remove 80–84% of As in contaminated drinking water, whereas in a natural system, it was 20%, refer to Fig. 5.2a (Brima and Haris 2014, Rezania et al. 2015, Jasrotia et al. 2015). Similarly, Pb was removed by up to 26, 50, 31, 90, and 78% from freshwater, textile process effluent, leachate, pulp and paper mill effluent, and an aqueous solution, respectively, depicted in Fig. 5.2b (Kulkarni et al. 2007, Aisien et al. 2010, Smolyakov 2012, Kumar et al. 2016, Putra et al. 2016, Putra and Hastika 2018).

Cd removal by water hyacinth has been reported from a range of wastewaters: 84, 95, 66, and 73% were removed from wastewater, textile wastewater, industrial water, and textile process effluent, respectively, refer Fig. 5.2c (Mahmood et al. 2005a, Kulkarni et al. 2007, Mishra and Tripathi 2009, Rai 2019, Mustafa and Hayder 2021). Similarly, the removal of Ni from textile process effluent and industrial water was 76 and 67%, respectively, shown in Fig. 5.2d (Kulkarni et al. 2007, Rai 2019, Mustafa and Hayder 2021). These levels are similar to that shown for Pb and Cu removal. In comparison, the removal of Cd from similar water systems was up to 100%: the studies utilized freshwater bodies, artificial wastewater, industrial water, wastewater, leachate, pulp, and paper mill effluent, and textile process effluent, removing 100, 100, 76, 95, 32, 89 and 60%, respectively, see Fig. 5.2e (Kulkarni et al. 2007, Aisien et al. 2010, Ogunbayo et al. 2012, Kumar et al. 2016, Putra and Hastika 2018, Victor et al. 2016, Romanova et al. 2021).

For Cu contaminated systems, the systems studied include textile wastewater, freshwater, synthetic wastewater, industrial water, wastewater, leachate, pulp, and paper mill effluent, and textile process effluent; the remediation potential was 94, 24, 97, 63, 7, 22, 94 and 52%, respectively, shown in Fig. 5.2f (Mahmood et al. 2005a, Kulkarni et al. 2007, Mokhtar et al. 2011, Smolyakov

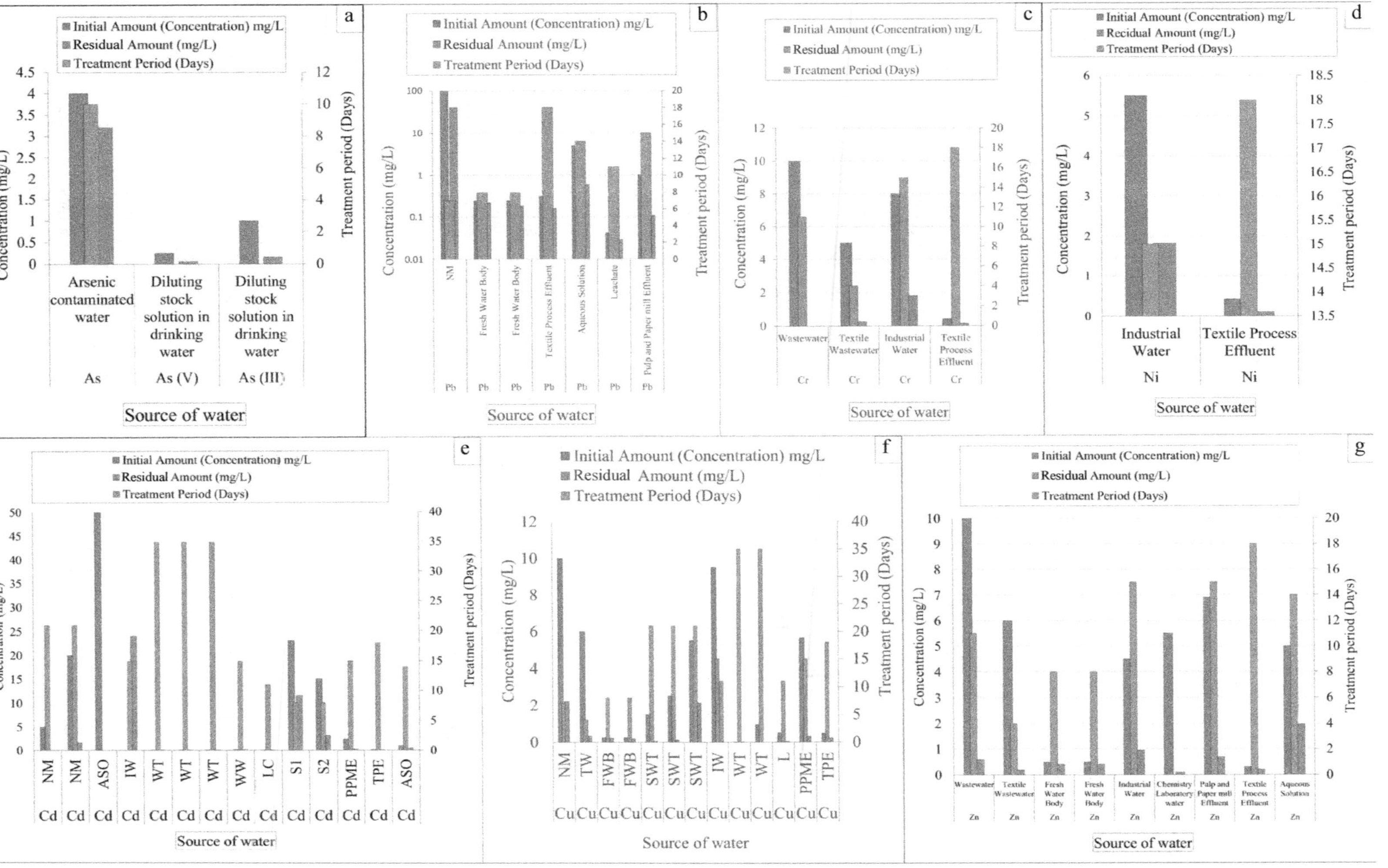

Fig. 5.2 caption contd. ...

2012, Ogunbayo et al. 2012, Rezania et al. 2015, Kumar et al. 2016, Putra et al. 2016, Putra and Hastika 2018, Rai 2019, Mustafa and Hayder 2021). This showed that Pb removal was similar to Cu from freshwater, textile process effluent, leachate and paper, and pulp effluent, but the removal was higher for Cu for spiked solutions.

The removal of Zn from textile process effluent, pulp and paper mill effluent, chemistry laboratory water, industrial water, freshwater bodies, textile wastewater, and wastewater were 44, 90, 79, 18, 97, and 94%, respectively, by water hyacinth, see Fig. 5.2g (Mahmood et al. 2005a, Kulkarni et al. 2007, Mishra and Tripathi 2009, Kelly et al. 2000, Smolyakov 2012, Kumar et al. 2016, Rai 2019, Mustafa and Hayder 2021).

For investigating the remediation characteristics of water hyacinth, a total of 63 data sets have been analyzed, displayed in Fig. 5.3. Using this data, a heatmap has been generated, and the 'k-means clustering' method has been applied to merge sub-clusters. Ten re-runs along with 300 iterations have been considered. The heatmap shows the formation of 3 distinct clusters along with 30 sub-clusters (Fig. 5.3a and b).

Several important inferences can be drawn from the generated heatmap. Firstly, 3 distinct clusterings can be observed in the heatmap. The C1 cluster includes HM found in specific sources with the highest initial and final concentrations and the lowest remediation periods. Cluster C2 includes HM found in specific sources having the lowest initial and final concentrations with moderate to high remediation periods and medium to low remediation rates. Cluster C3 includes HM found in specific sources having moderate to high remediation rates and moderate treatment periods with moderate initial and final concentrations. Figure 5.3a shows that the study by Das et al. (2016), Zhang et al. (2015), and Rezania et al. (2015) demonstrates that Cd has the fastest rate of remediation. Water hyacinth had removed a significant amount (more than 95%) of the element from artificial and synthetic wastewater. In the case of Zn and Cu, they exhibit lower amounts of remediation. The studies by Ogunbayo et al. (2012) and Smolyakov (2012) show 7 and 8% removal of Cu from wastewater and freshwater, whereas the moderate removal efficiency of Zn up to 18% has been demonstrated by the study of Smolyakov (2012). In continuation with these observations, Cu and Pb in freshwater also have low remediation rates and low effectiveness. In contrast to the previous reports, both forms of As in drinking water can be remediated effectively in less time by water hyacinth (Brima and Haris 2014, Rezania et al. 2015).

On the other hand (Fig. 5.3b), indicates different sources of wastewater, the highest amount of HM remediated from the study of Das et al. (2016), Zhang et al. (2015), and Rezania et al. (2015). Apart from this, HM like Cu, Zn, and Cd from synthetic water, wastewater, and textile water has been remediated up to 97.3%, 96.88%, and 95.59% by the study of Mokhtar et al. (2011), Rezania et al. (2015), Mahmood et al. (2005b), and Ogunbayo et al. (2012).

...Fig. 5.2 caption contd.

Fig. 5.2. Water Hyacinth Mediated Heavy Metal(iod)s Removal Potentialities Among Different Waste Water (NM = Not Mentioned) (Textile Wastewater = TW; Fresh Water Body = FWB; Synthetic Waste Water = WW; Industrial Water = IW; Wastewater Treatment = WT; Leachate = L; Pulp and Paper mill Effluent = PPME; Textile Process Effluent = TPE; Aqueous Solution = ASO; Leachate = LC; Solution of $Cd(NO_3)_2$, 100 μg L^{-1} = S1; Solution of $CdSO_4$, 100 μg L^{-1} = S2).

Fig. 5.2a. Arsenic removal efficiency of water hyacinth from different wastewater.

Fig. 5.2b. Lead removal efficiency of water hyacinth from different wastewater.

Fig. 5.2c. Chromium removal efficiency of water hyacinth from different wastewater.

Fig. 5.2d. Nickel removal efficiency of water hyacinth from different wastewater.

Fig. 5.2e. Cadmium removal efficiency of water hyacinth from different wastewater.

Fig. 5.2f. Copper removal efficiency of water hyacinth from different wastewater.

Fig. 5.2g. Zinc removal efficiency of water hyacinth from different wastewater.

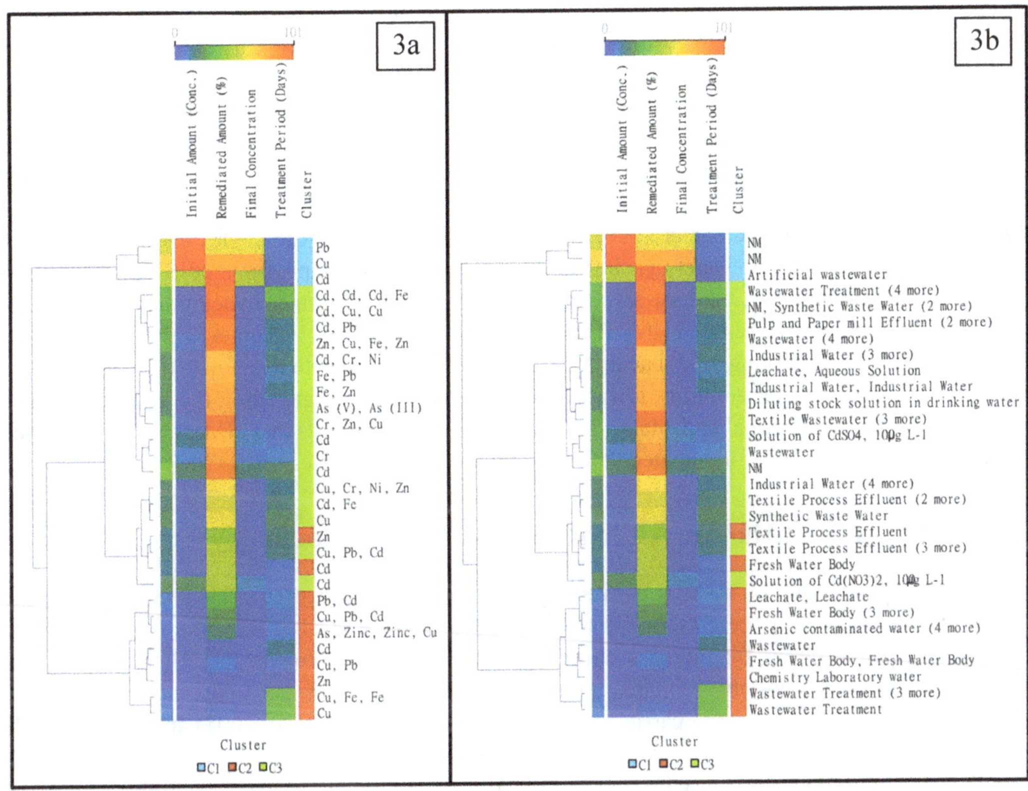

Fig. 5.3. Heat map of accumulation of the Heavy Metals by Water Hyacinth (C1 = HMs found in specific sources having highest initial and final concentrations with lowest remediation periods); (C2 = HMs found in specific sources having lowest initial and final concentrations with moderate to high remediation periods along with medium to low remediation rates); (C3 = found in specific sources having moderate to high remediation rates and moderate treatment period with moderate initial and final concentrations).

Fig. 5.3a. Accumulation of the Heavy Metals by Water Hyacinth on the basis of Heavy Metal.

Fig. 5.3b. Accumulation of the Heavy Metals by Water Hyacinth on the basis of different water sources.

5.3.4 Role of Rhizospheric Microorganisms of Water Hyacinth

To overcome the effects of HM on aquatic systems, phytoremediation, especially water hyacinth and their rhizospheric microorganisms, can be utilized to "restore harmony". The uses of microbes in industrial are extensive and varied due to their low cost, large yield, vast accessibility, chemical steadiness, energy-efficiency, eco-friendliness, and flexibility, etc. (Banerjee et al. 2020, Victor et al. 2016, Inomata et al. 2012). This suggests that not utilizing microbial approaches is a wasted opportunity. Water hyacinth and rhizospheric microbes have undeniable restrictions with their individual potentialities to remediate pollutants like HM. However, when used in synergy, both rhizospheric microorganisms (increases the availability of HM to the macrophytes) and water hyacinth (removal of HM via absorption) conquer many such limitations and therefore deliver a useful basis for augmented remediation of contaminated aquatic environments (Malik 2007).

In continuation with these hypotheses, the morphology and molecular ecology of water hyacinth allow the plant to remove high quantities of heavy metals and make the plant an ideal target as phytoremediation: larger biomass, dense fibrous root, and root-associated microbial consortium (Rai 2019). It has been noted that the natural microbial communities of contaminated aquatic systems are less effective in the remediation of HM than those associated with water hyacinth (Grenni et al. 2019). After four days of treatment, the metal concentration was reduced to 1.2 and 5.8% (from 1 mg L^{-1}). Microbial compositions detected in this study were Alpha-Proteobacteria, Beta-Proteobacteria, Gamma-Proteobacteria, Delta-Proteobacteria, Epsilon-Proteobacteria, Firmicutes, Actinobacteria, Bacteroidetes, Sphingobacteria, and Bacteroidetes, etc. (Grenni et al. 2019). Additionally, rhizospheric bacteria, such as *Pseudomonas diminuta, Brevundimonas diminuta, Nitrobacteria irancium, Ochrobactrum anthropi*, and *Bacillus cereus* of water hyacinth, also demonstrated high Cr and Zn removal potential, of up to 95% in the roots (Abou-Shanab et al. 2007). Another rhizospheric bacterial strain CU-1, recognized with Cu^{2+}, Ni^{2+}, and Zn^{2+} removal capability, was isolated from water hyacinth and is thought to enhance phytoremediation capabilities (So et al. 2003). Moreover, the root-associated microorganisms of water hyacinth can be isolated from compost and used to remove HM, *in vitro* (Pb, Ni, Zn, and Cd) (Vishan et al. 2017).

Plant growth promoting rhizobacteria (PGPR) should be targeted in future studies as they perform several beneficial activities towards the host plant growth, including the increase of HM removal capabilities (Singhal and Mahto 2004). The analysis suggests that HM phytoremediation, utilizing water hyacinth and associated PGPR, has great scope for research.

5.4 Water Hyacinth: An Alternative Source of Bioenergy

Water hyacinth is highly efficient at converting solar energy into biomass (Cheng et al. 2010) and is difficult to eradicate once established (Malik 2007, Coetzee et al. 2012, Nkuna et al. 2019), making it a prime target for biomass cultivation and subsequent conversion. By incorporating phytoremediation, it is possible to increase the value of this cultivation due to the high potential for the production of various energy carriers, such as ethanol (Mishima et al. 2008) and methane rich biogas (Verma et al. 2007, Singhal and Rai 2003, Mathew et al. 2013).

Water hyacinth has been used alone or as a constituent of mixed feed to produce methane by anaerobic digestion (Ghosh et al. 1981). The anaerobic digestion of water hyacinth is highly dependent on substrate composition (Sarto et al. 2019) due to the highly variable nature of water hyacinth (Gunnarsson and Petersen 2007) and the various inoculums that could be utilized; this could solve various issues related to Water hyacinth in different locations. The potential for water hyacinth to produce biogas and the effects of the different substrates to inoculum ratios has been shown in Table 5.5.

Table 5.5. Potentiality of water hyacinth in the production of methane.

Inoculum	Substrate	Ratio	Methane Concentration	References
Cow Dung	Water Hyacinth	0.5:1	406 L kg^{-1} VS	Bhui et al. 2018
Cow Dung and Kitchen Waste mixture 2:1	Water Hyacinth	2:1	540 L kg^{-1} VS	Bhui et al. 2015
Cow Dung Mixture	Water Hyacinth	2:1	552 L kg^{-1} VS	Mathew et al. 2015
Methanogenic sludge mix of primary wastewater sludge and food waste for more than 2 years	Water Hyacinth		292 ± 43 Lkg^{-1} VS	O'Sullivan et al. 2010
Cow Dung	Water hyacinth, pretreated by *Citrobacter werkmanii* VKVVG4 with a dosage of 109 CFU/mL for 4 days	F/M Ratio 1.5	156 ± 19 mL CH_4/g VS	Barua and Kalamdhad 2017

5.5 Future Perspectives

Phytoremediation is a novel, cost-effective, environmentally-friendly technology and is still under development. Various aquatic weeds, including water hyacinth, are detected as hyperaccumulators of HM and can be utilized to restore harmony to contaminated aquatic systems. This technology has little requirement for energy or start-up costs, having a greater potential in developing countries. Using rhizobacteria in tandem with biomass can greatly increase the phytoremediation capabilities by trapping HM or promoting growth. Utilizing post-phytoremediation biomass can increase the value of the biomass and therefore increase the profitability of harvesting the biomass. Water hyacinth has shown potential for the production of biofuels, fertilizer, animal feed, and various bio-materials. To advance and promote green technology using water hyacinth, more research and innovation are required practically.

5.6 Conclusion

This study has analyzed the phytoremediation capabilities of water hyacinth for various HM, including As, Cd, Cr, Cu, Hg, Ni, Pb, and Zn, from a variety of wastewaters. The high growth rate, morphology, persistence, and presence of rhizobacteria have meant that water hyacinth can remove high levels of these metals in various environmental conditions and demonstrate that it could be used in large scale applications. This application is cost-effective, environmentally friendly, and is recommended for broader applications. The utilization of post-phytoremediation biomass has been demonstrated in biofuels, fertilizer, animal feed, and biomaterials.

The utilization of phytoremediation, and subsequent harvesting and conversion, could aid in the removal of contaminants and increase the value of the biomass, therefore reducing the impacts of the weed on local communities and potentially introducing a new market into their economies.

Acknowledgement

Apurba Koley and Douglas Bray are thankful to the BBSRC, United Kingdom, for receiving funding from the BEFWAM project: Bioenergy, Fertiliser and Clean water from Invasive Aquatic macrophytes [Grant Ref: BB/S011439/1] for financial support and research fellowship. Sandipan Banerjee is thankful to the Department of Biotechnology, Govt. of India, for granting DBT Twinning Project and Research Fellowship [No. BT/PR25738/NER/95/1329/2017 dated December 24, 2018].

While conducting the study, the authors are thankful for the support received from Dr. Aishiki Banerjee, Mr. Binoy Kumar Show, and Mr. Anudeb Ghosh.

Declaration of competing interest

The authors declare that they have no known competing financial interests or personal relationships that could have appeared to influence the work reported in this paper.

References

Abbas, Z., F. Arooj, S. Ali, I. E. Zaheer, M. Rizwan and M. A. Riaz. 2019. Phytoremediation of landfill leachate waste contaminants through floating bed technique using water hyacinth and water lettuce. Int. J. Phytoremediat. 2113: 1356–1367.

Abou-Shanab, R. A. I., J. S. Angle and P. Van Berkum. 2007. Chromate-tolerant bacteria for enhanced metal uptake by *Eichhornia crassipes* (Mart.). Int. J. Phytoremediat. 9(2): 91–105.

Aisien, F. A., O. Faleye and E. T. Aisien. 2010. Phytoremediation of heavy metals in aqueous solutions. Leonardo J. Sci. 17: 37–46.

Aissaoui, A., D. Sadoudi-Ali Ahmed, N. Cherchar and A. Gherib. 2017. Assessment and biomonitoring of aquatic pollution by heavy metals (Cd, Cr, Cu, Pb and Zn) in Hammam Grouz Dam of Mila (Algeria). Int. J. Environ. Stud. 74(3): 428–442.

Akankali, J. A. and E. I. Elenwo. 2019. Use of water hyacinth as feed stuff for animals in Niger delta, Nigeria. Int. J. Adv. Sci. Res. Rev. 4: 91–7.

Ali, S., Z. Abbas, M. Rizwan, I. E. Zaheer, I. Yavaş, A. Ünay, M. M. Abdel-Daim, M. Bin-Jumah, M. Hasanuzzaman and D. Kalderis. 2020. Application of floating aquatic plants in phytoremediation of heavy metals polluted water: a review. Sustainability 12(5): 1927.

Ansari, A. A., R. Gill, L. Newman, S. S. Gill and G. R. Lanza. 2016. Phytoremediation: Management of Environmental Contaminants, 3. Springer Nature, Switzerland, pp. 1–576.

Arunkumar, R. I., P. Rajasekaran and R. D. Michael. 2000. Differential effect of chromium compounds on the immune response of the African mouth breeder, *Oreochromis mossambicus* (Peters). Fish Shellfish Immunol. 10: 667–676.

ATSDR. 2005. Zinc. Public Health Statement Zinc, CAS: 7440-66-6: 1–7.

Authman, M. M. 2015. Use of fish as bio-indicator of the effects of heavy metals pollution. J. Aquac. Res. Dev. 06(04). https://doi.org/10.4172/2155-9546.1000328.

Ayaz, T., S. Khan, A. Z. Khan, M. Lei and M. Alam. 2020. Remediation of industrial wastewater using four hydrophyte species: A comparison of individual (pot experiments) and mix plants (constructed wetland). J. Environ. Manage. 255: 109–833.

Bailey, L. H. 1949. Manual of Cultivated Plants. Macmillan, New York.

Baker, D. E. and L. Chesnin. 1975. Chemical monitoring of soils for environmental quality and animal and human health. Advances in Agronomy. 27(C). 305–374. https://doi.org/10.1016/S0065-2113(08)70013-0.

Banerjee, S., T. K. Maiti and R. N. Roy. 2020. Production, purification, and characterization of cellulase from *Acinetobacter junii* GAC 16.2, a novel cellulolytic gut isolate of *Gryllotalpa africana*, and its effects on cotton fiber and sawdust. Ann. Microbiol. 70(1): 1–16.

Barrett, S. C. H. 1980. Sexual reproduction in Eichhornia crassipes (Water Hyacinth) II. Seed production in natural populations. J. Appl. Ecol. 17: 113–124.

Barua, V. B. and A. S. Kalamdhad. 2017. Biochemical methane potential test of untreated and hot air oven pretreated water hyacinth: A comparative study. J. Clean. Prod. 166: 273–284. https://doi.org/10.1016/j.jclepro.2017.07.231.

Baruthio, F. 1992. Toxic effects of chromium and its compounds. Biol. Trace Elem. Res. 32(1): 145–153.

Basu, A., A. K. Hazra, S. Chaudhury, A. B. Ross and S. Balachandran. 2021. State of the art research on sustainable use of water hyacinth: a bibliometric and text mining analysis. Informatics 8(2): 38.

Bernhoft, R. A. 2012. Mercury toxicity and treatment: a review of the literature. J. Environ. Public Health, Article ID 460508, doi:10.1155/2012/460508.

Bhattacharya, A. and P. Kumar. 2010 Water hyacinth as a potential biofuel crop. Electron J. Environ. Agric. Food Chem. 9(1): 112–22.

Bhui, I., S. N. Banerjee, S. Chaudhury and S. Balachandran. 2015. Biogas production by co-digestion of locally available aquatic weeds (*Eichornia crassipes* and *Salvinia cucullata*) with kitchen waste, Proceedings of International Conference on Renewable Energy and Sustainable Environment. Dr. Mahalingam College of Engineering and Technology, Pollachi-642003, India. August 10-13, 2015, ISBN: 978-93-5235-155-8.

Bhui, I., A. K. Mathew, S. Chaudhury and S. Balachandran. 2018. Influence of volatile fatty acids in different inoculum to substrate ratio and enhancement of biogas production using water hyacinth and salvinia. Bioresour. Technol. 270: 409–415.

Bradl, H. 2005. Heavy Metals in the Environment: Origin. Interaction and Remediation Elsevier, San Diego, USA.

Brima, E. I. and P. I. Haris. 2014. Arsenic removal from drinking water using different biomaterials and evaluation of a phytotechnology based filter. Int. Res. J. Environ. Sci. 3: 39–44.

Center, T. D. and N. R. Spencer. 1981. The phenology and growth of water hyacinth (*Eichhornia crassipes* (Mart.) Solms) in a eutrophic Northcentral Florida lake. Aquat. Bot. 10: 1–32.

Center, T. D., F. A. Dray Jr., G. P. Jubinsky and A. J. Leslie. 1999. Water hyacinth weevils (*Neochetina eichhorniae* and *N. bruchi*) inhibit waterhyacinth (*Eichhornia crassipes*) colony development. Biol. Control. 15: 39–50.

Charudattan, R. 1986. Integrated control of water hyacinth (*Eichornia crassipes*) with a pathogen, insects, and herbicides. Weed Sci. 34(1): 26–30.

Cheng, J., B. Xie, J. Zhou, W. Song and K. Cen. 2010. Cogeneration of H_2 and CH4 from water hyacinth by two-step anaerobic fermentation. Int. J. Hydrog. Energy. 35: 3029–3035.

Chillers, C. J. 1991. Biological control of water hyacinth, *Eichornia crassipes* (Pontederiaceae) in South Africa. Agric. Ecosys. Environ. 37: 207–217.

Coetzee, J. A. and M. P. Hill. 2012. The role of eutrophication in the biological control of water hyacinth, *Eichhornia crassipes* in South Africa. BioControl. 57(2): 247–261.

Costa, M., T. L. Davidson., H. Ke. Q. Chen, P. Zhang,Y. Yan and T. Kluz. 2005. Nickel carcinogenesis: epigenetics and hypoxia signaling. MUTAT Res-Fund Mol. M. 592(1-2): 79–88.

Dalvi, A. A. and S. A. Bhalerao. 2013. Response of plants towards heavy metal toxicity: an overview of avoidance, tolerance and uptake mechanism. Ann. Plant. Sci. 2(9): 362–368.

Das, S., S. Goswami and A. Das Talukdar. 2016. Physiological responses of water hyacinth, *Eichhornia crassipes* (Mart.) Solms, to Cadmium and its phytoremediation potential. Turk. J. Biol. 40: 84–94; doi:10.3906/biy-1411-86.

David, S. A., S. S. Babu and J. M. Vitek. 2003. Welding: Solidification and microstructure. Jom. 55(6): 14–20.

de Souza, T. D., A. C. Borges, A. Teixeira de Matos, R. W. Veloso and A. F. Braga. 2018. Optimization of arsenic phytoremediation using *Eichhornia crassipes*. Int. J. Phytoremediat. 2011: 1129–1135.

de Vasconcelos, G. A., R. M. L. Véras, J. de Lima Silva, D. B. Cardoso, P. de Castro Soares, N. N. G. de Morais and A. C. Souza. 2016. Effect of water hyacinth (*Eichhornia crassipes*) hay inclusion in the diets of sheep. Trop. Anim. Health. Prod. 48(3): 539–544.

Dhote, S. and S. Dixit 2009. Water quality improvement through macrophytes - A review. Environ. Monit. Assess. 152(1–4): 149–153.

Dorsey, A., L. Ingerman and S. Swarts. 2004. Relevance to public health background and environmental exposures to copper in the United States. Agency for Toxic Substances and Disease Registry. 11–19. http://www.atsdr.cdc.gov/toxprofiles/tp132.pdf.

Eid, E. M., K. H. Shaltout, S. A. Alamri, A. Alrumman, A. A. Hussain, N. Sewelam and D. Barcelo. 2021. Prediction models based on soil properties for evaluating the uptake of eight heavy metals by tomato plant (*Lycopersicon esculentum* Mill.) grown in agricultural soils amended with sewage sludge. J. Environ. Chem. Eng. 9(5): 105977.

Emerhi, E. A. 2011. Physical and combustion properties of briquettes produced from sawdust of three hardwood species and different organic binders. Adv. Appl. Sci. Res. 2: 236–246.

Farag, A. M., T. May, G. D. Marty, M. Easton and D. D. Harper. 2006. The effect of chronic chromium exposure on the health of Chinook salmon (*Oncorhynchus tshawytscha*). Aquat. Toxicol. 76: 246–257.

Fawzy, M. A., N. E. S. Badr, A. El-Khatib and A. Abo-El-Kassem. 2012. Heavy metal biomonitoring and phytoremediation potentialities of aquatic macrophytes in River Nile. Environ. Monit. Assess. 184(3): 1753–1771.

Fergusson, J. E. 1990. The heavy elements: chemistry, environmental impact and health effects. No. 628.53 F4.

Förstner, U. and G. T. W. Wittmann. 1981. Metal Pollution in the Aquatic Environment, 2nd ed. Springer, Berlin Heidelberg.

Garrison, G. E. and O. L. Ader. 1966. Sodium in drinking water. Arch. Environ. Health. 13(5): 551–553.

Genchi, G., A. Carocci, G. Lauria, M. S. Sinicropi and A. Catalano. 2020. Nickel: Human health and environmental toxicology. Int. J. Environ. Res. Public Health. 17(3).

GEA. 2012. Global Energy Assessment - Toward a Sustainable Future [Internet]. Cambridge University Press, Cambridge, UK and New York, NY, USA. The International Institute for Applied Systems Analysis, Laxenburg, Austria.

Ghosh, S., D. L. Klass and D. P. Chynoweth. 1981. Bioconversion of Macrocystis pyrifera to methane. J. Chem. Technol. Biotechnol. 31(1): 791–807.

Gopal, B. 1987. Aquatic Plant Studies. Water Hyacinth. Elsevier Publishing, New York, New York, USA.

Goswami, L., A. Nath, S. Sutradhar, S. S. Bhattacharya, A. Kalamdhad, K. Vellingiri and K. H. Kim. 2017. Application of drum compost and vermicompost to improve soil health, growth, and yield parameters for tomato and cabbage plants. J. Environ. Manage. 200: 243–252.

Goswami, T. and C. N. Saikia. 1994. Water hyacinth—a potential source of raw material for greaseproof paper. Bioresour. Technol. 50(3): 235–8.

Grenni, P., A. B. Caracciolo, L. Mariani, M. Cardoni, C. Riccucci, H. Elhaes and M. A. Ibrahim. 2019. Effectiveness of a new green technology for metal removal from contaminated water. Microchem. J. 147: 1010–1020.

Gunnarsson, C. C. and C. M. Petersen. 2007. Water hyacinths as a resource in agriculture and energy production: A literature review. Waste Manag. 27: 117–129.

Gupta, P., S. Roy and A. B. Mahindrakar. 2012. Treatment of water using water hyacinth, water lettuce and vetiver grass–a review. Resourc. Environ. 49: 50.

Gupta, R., P. K. Mutiyar, N. K. Rawat, M. S. Saini and V. K. Garg. 2007. Development of a water hyacinth based vermireactor using an epigeic earthworm *Eisenia foetida*. Bioresour. Technol. 98(13): 2605–10.

Hakeem, K., M. Sabir, M. Ozturk and A. R. Mermut. 2014. Soil Remediation and Plants: Prospects and Challenges. Academic Press, London, pp. 1–739.

Harada, M. 1995. Minamata disease: methylmercury poisoning in japan caused by environmental pollution. Crit. Rev. Toxicol. 25(1): 1–24. 10.3109/10408449509089885.

Ilo, O. P., M. D. Simatele, N. M. Mkhize and N. G. Prabhu. 2020. The benefits of water hyacinth (*Eichhornia crassipes*) for Southern Africa: A review. Sustainability 12(21): 9222.

Inomata, Y., M. Kajino, K. Sato, T. Ohara, J. I. Kurokawa, H. Ueda, N. Tang, K. Hayakawa, T. Ohizumi and H. Akimoto. 2012. Emission and atmospheric transport of particulate PAHs in Northeast Asia. Environ. Sci. Technol. 46(9): 4941–4949.

Jasrotia, S., A. Kansal and A. Mehra. 2015. Performance of aquatic plant species for phytoremediation of arsenic-contaminated water. Appl. Water Sci. 7: 889–896.

Jianbo, L. U., F. U. Zhihui and Y. I. Zhaozheng. 2008. Performance of a water hyacinth (*Eichhornia crassipes*) system in the treatment of wastewater from a duck farm and the effects of using water hyacinth as duck feed. Res. J. Environ. Sci. 20(5): 513–9.

Jirawattanasomkul, T., H. Minakawa, S. Likitlersuang, T. Ueda, J. G. Dai, N. Wuttiwannasak and N. Kongwang. 2021. Use of water hyacinth waste to produce fibre reinforced polymer composites for concrete confinement: Mechanical performance and environmental assessment. J. Clean. Prod. 292: 126041.

Jones, J. L., R. O. Jenkins and P. I. Haris. 2018. Extending the geographic reach of the water hyacinth plant in removal of heavy metals from a temperate Northern Hemisphere river. Sci. Rep. 8(1): 1–15.

Kelly, C., K. K. Gaither, A. B. Spry and B. J. Cruickshank. 2000. Incorporation of phytoremediation strategies into the introductory chemistry laboratory. Chem. Educator. 5: 140–143.

Khan, R., S. H. Israili, H. Ahmad and A. Mohan. 2005. Heavy metal pollution assessment in surface water bodies and its suitability for irrigation around the Neyevli lignite mines and associated industrial complex, Tamil Nadu, India. Mine Water Environ. 24(3): 155–161.

Kulkarni, B. V., S. V. Ranade and A. I. Wasif. 2007. Phytoremediation of textile process effluent by using water hyacinth- A polishing treatment. Jr. of Ind. Pollut. Control. 23(1): 97–101.

Kumar, V., A. K. Chopra, J. Singh, R. K. Thakur, S. Srivastava and R. K. Chauhan. 2016. Comparative assessment of phytoremediation feasibility of water caltrop (*Trapa natans* L.) and water hyacinth (*Eichhornia crassipes Solms.*) using pulp and paper mill effluent. Arch. Agr. Environ. Sci. 1(1): 13–21.

Length, F. 2007. Heavy metal pollution and human biotoxic effects. Phys. Sci. Int. J. 2(5): 112–118.

Li, F., X. He, A. Srishti, S. Song, H. T. W. Tan, D. J. Sweeney, S. Ghosh and C. H. Wang. 2021. Water hyacinth for energy and environmental applications: A review. Bioresour. Technol. 124809.

Lionetto, M. G., M. E. Giordano, R. Caricato, M. F. Pascariello, L. Marinosci and T. Schettino. 2001. Biomonitoring of heavy metal contamination along the Salento coast (Italy) by metallothionein evaluation in Mytilus galloprovincialis and Mullus barbatus. Aquat. Conserv. 11(4): 305–310.

Liu, G., Y. Cai, N. O'Driscoll, X. Feng and G. Jiang. 2012. Overview of mercury in the environment. Environmental chemistry and toxicology of mercury. 1–12.

Luu, K. T. and K. D. Getsinger. 1990. Seasonal biomass and carbohydrate allocation in water hyacinth. J. Aquat. Plant Manag. 28(1): 3–10.

Madian, H. R., N. M. Sidkey, M. M. Abo Elsoud, H. I. Hamouda and A. M. Elazzazy. 2019. Bioethanol production from water hyacinth hydrolysate by *Candida tropicalis* Y-26. Arab. J. Sci. Eng. 44(1).

Madikizela, L. M. 2021. Removal of organic pollutants in water using water hyacinth (*Eichhornia crassipes*). J. Environ. Manage. 295: 113–153.

Mahmood, Q., P. Zheng, E. Islam, Y. Hayat, M. J. Hassan, G. Jilani and R. C. Jin. 2005a. Lab scale studies on water hyacinth (*Eichhornia crassipes* Marts Solms) for biotreatment of textile wastewater. Caspian J. Env. Sci. 3(2): 83–88.

Mahmood, S., A. Hussain, Z. Saeed, and M. Athar. 2005b. Germination and seedling growth of corn (*Zea mays* L.) under varying levels of copper and zinc. Int. J. Environ. Sci. Technol. 2(3): 269–274.

Mahmood, S., N. Khan, K. J. Iqbal, M. Ashraf and A. Khalique. 2018. Evaluation of water hyacinth (*Eichhornia crassipes*) supplemented diets on the growth, digestibility and histology of grass carp (*Ctenopharyngodon idella*) fingerlings. J. Appl. Anim. Res. 46(1): 24–28.

Majumdar, A., A. Barla, M. K. Upadhyay, D. Ghosh, P. Chaudhuri, S. Srivastava and S. Bose. 2018. Vermiremediation of metal (loid) s via *Eichornia crassipes* phytomass extraction: a sustainable technique for plant amelioration. J. Environ. Manage. 220: 118–125.

Malik, A. 2007. Environmental challenge vis a vis opportunity: The case of water hyacinth. Environ. Int. 33(1): 122–138.

Martin, S. and W. Griswold. 2009. Human Health Effects of Heavy Metals. Environmental Science and Technology Briefs for Citizens. 15.

Mathew, A. K., R. Goswami, S. N. Banerjee, A. K. Chakraborty, A. Shome, S. Balachandran and S. Chaudhury. 2013. Comparison of Biogas Production from Water Hyacinth (*Eichhornia crassipes*) and Salvinia (*Salvinia cucullata*), IVth International Conference on Advances in Energy Research Indian Institute of Technology Bombay, Mumbai.

Mathew, A. K., I. Bhui, S. N. Banerjee, R. Goswami, A. K. Chakraborty, A. Shome, S. Balachandran and S. Chaudhury. 2015. Biogas production from locally available aquatic weeds of Santiniketan through anaerobic digestion. Clean Technol. Environ. Policy. 17(6): 1681–1688.

Matthews, L. J. 1967. Seedling establishment of water hyacinth. PANS(C) 13: 7–8.

Mayes, R. and A. McIntosh. 1975. Use of aquatic macrophytes as indicators of trace metal contamination in fresh water lakes. In Trace Substances in Environmental Health; Proceedings of University of Missouri's Annual Conference.

McKeown, A. E. and G. Bugyi. 2015. Impact of water pollution on human health and environmental sustainability. IGI Global, pp. 1–361.

Methy, M., P. Alpert and J. Roy. 1990. Effects of light quality and quantity on growth of the clonal plant *Eichhornia crassipes*. Oecologia. 84(2): 265–271.

Mishima, D., M. Kuniki, K. Sei, S. Soda, M. Ike and M. Fujita. 2008. Ethanol production from candidate energy crops: water hyacinth (*Eichhornia crassipes*) and water lettuce (*Pistia stratiotes* L.). Bioresour. Technol. 99(7): 2495–500.

Mishra, S., M. Mohanty, C. Pradhan, H. K. Patra, R. Das and S. Sahoo. 2013. Physico-chemical assessment of paper mill effluent and its heavy metal remediation using aquatic macrophytes—a case study at JK Paper mill, Rayagada, Environ. Monit. Assess. 1855: 4347–4359.

Mishra, V. K. and B. D. Tripathi. 2009. Accumulation of chromium and zinc from aqueous solutions using water hyacinth (*Eichhornia crassipes*). J. Hazard. Mater. 164: 1059–1063. doi:10.1016/j.jhazmat.2008.09.020.

Mokhtar, H., N. Morad and F. F. A. Fizri. 2011. Phytoremediation of copper from aqeous solutions using *Eichhornia crassipes* and *Centella asiatica*. Int. J. Environ. Sci. Dev. 2: 205–210.

Mustafa, H. M. and G. Hayder. 2021. Recent studies on applications of aquatic weed plants in phytoremediation of wastewater: A review article. Ain Shams Eng. J. 12: 355–365.

Nkuna, R., R. Adeleke and A. Roopnarain. 2019. Effects of organic loading rates on microbial communities and biogas production from water hyacinth: A case of mono- and co-digestion. J. Chem. Technol. Biotechnol. 94(4): 1294–1304.

NRC. 1997. National Research Council. Toxicologic assessment of the army's zinc cadmium sulfide dispersion tests.

O'Sullivan, C., B. Rounsefell, A. Grinham, W. Clarke and J. Udy. 2010. Anaerobic digestion of harvested aquatic weeds: Water hyacinth (*Eichhornia crassipes*), cabomba (*Cabomba caroliniana*) and salvinia (*Salvinia molesta*). Ecol. Eng. 36(10): 1459–1468. https://doi.org/10.1016/j.ecoleng.2010.06.027.

Ofulla, A. V .O., D. Karanja, R. Omondi, T. Okurut, A. Matano, T. Jembe, R. Abila, P. Boera and J. Gichuki. 2010. Relative abundance of mosquitoes and snails associated with water hyacinth and hippo grass in the Nyanza gulf of Lake Victoria. Lakes Reserv. Res. Manag. 15(3): 255–271.

Ogunbayo, A. O. and T. O. Ajayi. 2012. Achieving environmental sustainability in wastewater treatment by phytoremediation with water hyacinth (*Eichhornia crassipes*). J. Sustain. Dev. 5(7): 1913–9071.

Panneerselvam, B. and K. S. Priya. 2021. Phytoremediation potential of water hyacinth in heavy metal removal in chromium and lead contaminated water. Int. J. Environ. Anal. Chem. 1–16.

Patel, S. 2012. Threats, management and envisaged utilizations of aquatic weed *Eichhornia crassipes*: an overview. Rev. Environ. Sci. Biotechnol. 11(3): 249–259.

Patnaik, P., T. Abbasi and S. A. Abbasi. 2022. Salvinia (*Salvinia molesta*) and water hyacinth (*Eichhornia crassipes*): Two pernicious aquatic weeds with high potential in phytoremediation. pp. 243–260. *In*: Siddiqui, N. A., S. M. Tauseef, S. A. Abbasi, R. Dobhal and A. Kansal (eds.). Advances in Sustainable Development. Springer, Singapore.

Penfound, W. T. and T. T. Earle. 1948. The biology of the water hyacinth. Ecol. Monogr. 447–472.

Peng, H., Y. Wang, T. L. Tan and Z. Chen. 2020. Exploring the phytoremediation potential of water hyacinth by FTIR Spectroscopy and ICP-OES for treatment of heavy metal contaminated water. Int. J. Phytoremediat. 229: 939–951.

Prasad, M. N. V. 2004. Phytoremediation of metals and radionuclides in the environment: the case for natural hyperaccumulators, metal transporters, soil-amending chelators and transgenic plants. pp. 345–391. Second Edition Heavy Metal Stress in Plant, Springer, Berlin, Heidelberg.

Preussler, K. H., C. F. Mahler and L. T. Maranho. 2015. Performance of a system of natural wetlands in leachate of a posttreatment landfill. Int. J. Environ. Sci. Technol. 12(8): 2623–2638.

Putra, R. S., D. Novarita and F. Cahyana. 2016. Remediation of Lead (Pb) and Copper (Cu) Using Water Hyacinth [*Eichornia crassipes* (Mart.) Solms] with Electro-Assisted Phytoremediation (EAPR). Towards the sustainable use of biodiversity in changing environment: From basic to applied research AIP Conf. Proc. 1744. 020052-1-020052-6; doi: 10.1063/1.4953526.

Putra, R. S. and F. Y. Hastika. 2018. Removal of heavy metals from leachate using electro-assisted phytoremediation (EAPR) and up-take by water hyacinth (*Eichornia crassipes*). Indones. J. Chem. 18(2): 306–312.

Rai, P. K. and B. D. Tripathi. 2007. Heavy metals removal using nuisance blue green alga Microcystis in continuous culture experiment. Environ. Sci. 4(1): 53–59.

Rai, P. K. 2010. Phytoremediation of heavy metals in a tropical impoundment of industrial region. Environ. Monit. Assess. 165(1–4): 529–537.

Rai, P. K. 2019. Heavy metals/metalloids remediation from wastewater using free floating macrophytes of a natural wetland. Environ. Technol. Innov. 15: 100393.

Rezania, S., M. F. M Din, M. Ponraj, F. M. Sairan and S. F. Binti Kamaruddin. 2013. Nutrient uptake and wastewater purification with Water Hyacinth and its effect on plant growth in batch system. J. Environ. Treat. Tech. 1(2): 81–85.

Rezania S., M. Ponraj, A. Talaiekhozani, S. E. Mohamad, M. F. Md Din, S. M. Taib, F. Sabbagh and F. Md Sairan. 2015. Perspectives of phytoremediation using water hyacinth for removal of heavy metals, organic and inorganic pollutants in wastewater. J. Environ. Manag. 163: 125–133.

Romanova, T. E., L. A. Belchenko, O. V. Shuvaeva and M. V. Kurbatova. 2021. The study of cadmium uptake by Water Hyacinth (*Eichhornia crassipes*) using a natural modelling approach. Chem. J. Mold. 7(1): 133–139. https://doi.org/10.19261/cjm.2012.07(1).25.

Royer, A. and T. Sharman. 2020. Copper toxicity. https://europepmc.org/article/nbk/nbk557456.

Saha, P., O. Shinde and S. Sarkar. 2017. Phytoremediation of industrial mines wastewater using water hyacinth. Int. J. Phytoremediat. 191: 87–96.

Sarto, S., R. Hildayati and I. Syaichurrozi. 2019. Effect of chemical pre-treatment using sulfuric acid on biogas production from water hyacinth and kinetics. Renew. Energ. 132: 335–350.

Shailaja, G. and G. Gautam. 2017. Risk assessment of trace elements distribution in soils of basaltic aquifers, southern Maharashtra, India. Int. Res. J. Earth Sci. 5(8): 22–31.

Sharma, R., H. Saini, D. R. Paul, S. Chaudhary and S. P. Nehra. 2021. Removal of organic dyes from wastewater using *Eichhornia crassipes*: a potential phytoremediation option. Environ. Sci. Pollut. Res. 28(6): 7116–7122.

Sharma, S. 2014. Heavy Metals In Water: Presence, Removal and Safety, Royal Society of Chemistry.

Shillinglaw, S. N. 1981. Dissolved oxygen depletion and nutrient uptake in an impoundment infested with *Eichhornia crassipes* (Mart.) Solms. J. Limnol. Soc. South Africa. 7(2): 63–66.

Singh, A. P., P. C. Srivastava and P. Srivastava. 2008. Relationships of heavy metals in natural lake waters with physico-chemical characteristics of waters and different chemical fractions of metals in sediments. Water Air Soil Pollut. 188(1–4): 181–193. https://doi.org/10.1007/s11270-007-9534-6.

Singh, D., A. Tiwari and R. Gupta. 2012. Phytoremediation of lead from wastewater using aquatic plants. J. Agric. Sci. Technol. 8(1): 1–11.

Singh, J. and A. S. Kalamdhad. 2013. Reduction of bioavailability and leachability of heavy metals during vermicomposting of water hyacinth. Environ. Sci. Pollut. Res. 20(12): 8974–8985.

Singh, J., V. Kumarand and P. Kumar. 2021. Kinetics and prediction modeling of heavy metal phytoremediation from glass industry effluent by water hyacinth *Eichhornia crassipes*. Int. J. Environ. Sci. Technol. 1–12.

Singhal, P. K. and S. Mahto. 2004. Role of water hyacinth in the health of a tropical urban lake. J. Environ. Biol. 25(3): 269–277.

Singhal, V. and J. P. N. Rai. 2003. Biogas production from water hyacinth and channel grass used for phytoremediation of industrial effluents. Bioresour. Technol. 86(3): 221–5.

Smolyakov, B. S. 2012. Uptake of Zn, Cu, Pb, and Cd by water hyacinth in the initial stage of water system remediation. J. Appl. Geochem. 27(6): 1214–1219.

So, L. M., L. M. Chu and P. K. Wong. 2003. Microbial enhancement of Cu^{2+} removal capacity of *Eichhornia crassipes (Mart.)*. Chemosphere. 52(9): 1499–1503.

Sucharit, S., C. Harinasuta, T. Deesin and S. Vutikes. 1981. Studies of aquatic plants and grasses as breeding hosts for mosquitoes. Southeast Asian J. Trop. Med. 1: 464–465.

Sukarni, S., Y. Zakaria, S. Sumarli, R. Wulandari, A. A. Permanasari and M. Suhermanto. 2019. Physical and chemical properties of water hyacinth (*Eichhornia crassipes*) as a sustainable biofuel feedstock. In IOP Conference Series: Mater. Sci. Eng. 515(1): 012070.

Tabinda, A. B., R. A. Arif, A. Yasar, M. Baqir, R. Rasheed, A. Mahmood and A. Iqbal. 2019. Treatment of textile effluents with *Pistia stratiotes, Eichhornia crassipes* and *Oedogonium* sp. Int. J. Phytoremediat. 2110: 939–943.

Tellez, M. and M. Merchant. 2015. Biomonitoring heavy metal pollution using an aquatic apex predator. The American alligator, and its parasites. PLoS One. 10(11): 0142522.

Unpaprom, Y., T. Pimpimol, K. Whangchai and R. Ramaraj. 2021. Sustainability assessment of water hyacinth with swine dung for biogas production, methane enhancement, and biofertilizer. Biomass Convers. Biorefin. 11(3): 849–60.

US EPA. 2015. National Recommended Water Quality Criteria - Human Health Criteria Table.

Vanloon, G. W. and S. J. Duffy. 2005. Environmental Chemistry: A Global Perspective. (2nd edn). Oxford University Press, New York. USA.

Verma, V. K., Y. P. Singh and J. P. N. Rai. 2007. Biogas production from plant biomass used for phytoremediation of industrial wastes. Bioresour. Technol. 98(8): 1664–9.

Victor, K. K., Y. Seka, K. K. Norbet, T. A. Sanogo and A. B. Celestin. 2016. Phytoremediation of wastewater toxicity using water hyacinth (*Eichhornia crassipes*) and water lettuce (*Pistia stratiotes*). Int. J. Phytoremediat. 18(10): 949–955.

Vishan, I., S. Sivaprakasam and A. Kalamdhad. 2017. Isolation and identification of bacteria from rotary drum compost of water hyacinth. Int. J. Recycl. Org. Waste Agric. 6(3): 245–253.

Whitacre, R. W. and C. S. Pearse. 1974. Arsenic and the Environment. Miner. Ind. Bull. 17: 1–19.

WHO. 1989. World Health Organization. 1989. Lead—environmental aspects. Geneva, (Environmental Health Criteria, No. 85).

WHO. 2018. World Health Organization. Arsenic. Retrieved from https://www.who.int/news-room/fact-sheets/detail/arsenic on 30th October 2021.

Wickramasinghe, S. and C. K. Jayawardana. 2018. Potential of aquatic macrophytes *Eichhornia crassipes*, *Pistia stratiotes* and *Salvinia molesta* in phytoremediation of textile wastewater. Water Secur. 4: 1–8.

Yan, A., Y. Wang, S. N. Tan, M. L. Mohd Yusof, S. Ghosh, and Z. Chen. 2020. Phytoremediation: a promising approach for revegetation of heavy metal-polluted land. Front. Plant Sci. 11: 359.

Zhang, F., X. Wang D. Yin, B. Peng, C. Tan, Y. Liu, X. Tan, and S. Wu. 2015. Efficiency and mechanisms of Cd removal from aqueous solution by biochar derived from water hyacinth (*Eichhornia crassipes*). J. Environ. Manag. 153: 68–73.

Zhang, Y., D. Zhang and S. Barrett. 2010. Genetic uniformity characterizes the invasive spread of water hyacinth (*Eichhornia crassipes*), a clonal aquatic plant. Mol. Ecol. 19: 1774–1786.

CHAPTER 6

Bacterial Mechanisms for Metal(loid)s Remediation

Insha Sultan and *Qazi. Mohd Rizwanul Haq.**

6.1 Introduction

The elements present on Earth have been categorized as metals, metalloids, and nonmetals. Some are crucial, but others can be harmful. As stated by World Health Organization (WHO), 13 metals such as copper, mercury, manganese, nickel, lead, tin, titanium, arsenic, cadmium, cobalt, thallium, aluminum, and chromium are considered hazardous heavy metals (Index 2018); amongst these, lead, cadmium, mercury, and arsenic show significant toxicity.

The metals and metalloids comprise a class of heavy metals having an atomic density of approximately 4000 kg/m^3. Heavy metals such as Cu, Zn, Sn, Sr, Ni, Mo, Pb, Ti, Cr, Co, V, and Hg (Satyanarayana et al. 2019, Shah et al. 2018) can be toxic at lower, moderate as well as at higher concentrations. Meanwhile, metals like Cu, Zn, Co, Ni, and Mo play a role in cellular growth but can prove fatal above the permissible limit. There are metals like Pb, As, and Cd, which are not required for growth but can interfere with cellular mechanisms if exposed to them (Rudakiya et al. 2018). The cause of water and soil pollution by metal(loid)s can be attributed to direct sewage discharge from industries, untreated wastewater, hospital sewage, and domestic sources. Heavy metals are not only responsible for polluting soil and water but also have a direct influence on the food chain, as at every tropic stage, there is an accumulation of metals, thus creating disturbances in the cell functions (Vardhan et al. 2019, Rudakiya and Pawar 2013).

Due to the higher persistence of heavy metals, non-biodegradability, high toxicity, and potential to disturb the food chains, these elements have become a serious threat to water and land ecosystems (Chowdhury et al. 2016). Metallic compounds lead to the production of reactive oxygen species (ROS) like superoxide and hydroxyl, and hydrogen peroxide (H_2O_2) (Chibuike and Obiora 2014), causing severe illnesses like cancer and inflammation, etc. (Ghanem et al. 2020, Dixit et al. 2015). The heavy metal accumulation by aquatic species above the allowed concentrations causes severe toxicity among aquatic animals decreasing their survival rates; simultaneously, heavy metals can give rise to genetic mutations, thus creating biological variations and health problems (Hong et al. 2020).

Department of Biosciences, Jamia Millia Islamia, New Delhi 110025, India.
* Corresponding author: qhaque@jmi.ac.in; haqqmr@gmail.com

The rapid escalation in the pollution load of the terrestrial and aquatic environment and harmful effects caused due to metal(loid)s have induced an alarming situation globally (Sharma et al. 2021a,b, Nkwunonwo et al. 2020, Zhou et al. 2020, Deng et al. 2020), which needs to be considered and addressed at a faster pace (Sharma and Pandey 2014).

The term bioremediation is an amalgam of two words, *bios* which means life, and remediation deals with fixation of an issue or a problem. Therefore, it is explained as the utilization of organic matter to fix the pollution in the environment (USEPA 2016). Bioremediation is a green approach to dealing with metal pollution. Bioremediation follows the principle that microorganisms, mostly bacteria used to remediate the pollutants, should be proficient at transforming them into less toxic species. The factors that control the process of bioremediation involve identification of microbial inhabitants proficient in degrading the contaminants, availability of pollutants to microbes, the chemical composition of pollutants, water activity, surfactant, pH, moisture, carbon to nitrogen ratio, minerals, nitrogen source, the interaction between microbes and the pollutants, and environmental factors viz. temperature, humidity, luminosity, air composition, CO_2 and O_2 concentrations (AMGA 2004).

In order to treat the polluted sites, the support material used must be cost-effective, pollution free, non-biodegradable, safe and should diffuse easily, must show strong mechanical and chemical stability, and must possess the least attachment with the other organisms (Martins et al. 2013).

The process of bioremediation is sustainable and economical in comparison to the conventionally used methods. It provides a fixed solution to manage the contaminants either by their degradation or transformation into less lethal substances (Strong and Burgess 2008). The usefulness of bioremediation to treat the polluted environment has been extensively studied through the interaction of microbes available naturally or by the inoculation of bacteria artificially (Vidali 2001) from other sources in the polluted sites (Tiedje 1993).

There are many other approaches in progress for bioremediation strategies, including applying immobilized enzymes at the pollution site, inoculation of microorganisms with specific biotransformation capabilities, and use of plant sources, i.e. phytoremediation. The presence of microbes in nature presents an optimum remedy for the scrapping of garbage, dead and decomposing litter that pollute the planet Earth.

The recognition of microbes by Antonie Van Leeuwenhoek in the 1670s (Lane 2015) demonstrated that microscopic entity could be categorized as unicellular or multicellular organisms (Madigan and Martinko 2006). The diverse groups of bacteria show immense potential for bioremediation strategies. The available literature provides evidence that many organisms have a natural tendency to biosorb harmful metalloids (Singh et al. 2014). Generally, bacteria and fungi are used for bioremediation; however, recently, archaea have been found to show excellence towards bioremediation. The known microorganisms like *Phanerochaete, Mycococcus, Arthrobacter, Trametes, Xanthofacter, Methylosinus, Acaligenes, Actinobacteria, Flavobacterium, Nitrosomonas, Rhizoctomia, Mycobacterium, Nocardia, Bacillins, Berijerinckia, Acinetobacter, Penicillium, Pseudomonas, Serratia,* and *Methylosinus* have been recognized as active members of microbial groups to be vigorously involved in bioremediation among diverse environments (Singh et al. 2014). Besides, some bacteria which have been used to treat heavy metal pollution are *Sporosarcina ginsengisoli* (Achal et al. 2012), *Arthrobacter* sp. (Roane et al. 2001), *Burkholderia* sp. (Jiang et al. 2008), *Bacillus cereus* (Kanmani et al. 2012), *Kocuria flava* (Achal et al. 2011), and *Pseudomonas veronii* (Vullo et al. 2008).

Exposure to heavy metals has proven hazardous for all life forms. Their tendency to penetrate through the food chain via waste disposal into water channels leads to further accumulation at each tropic level. Heavy metals have the tendency to generate reactive oxygen species (ROS). These ROS interact with the proteins, lipids, and nucleic acids, thus causing lipid per-oxidation, membrane damage, and enzyme inactivation, disturbing cellular growth (Foyer 1997). The secretions produced by various microorganisms like algae and bacteria attract toxic metals. Information by Bang et al.

(2000) outlined that the thiosulfate reductase gene of *Salmonella typhimurium* in *Escherichia coli* increased the effectiveness of heavy metal elimination and buildup of cadmium up to 150 mM among 98% cells. Exploration of biochemical and genetic competence of bacteria for the amendment of pollution due to heavy metals were examined by Valls and Lorenzo (2002). They interpreted the significance of using biological strategies that comprise improved specificity, sustainability, and the possibility for genetic up-gradation. To remediate heavy metals, the bioremediation approaches used include biosorption, bioaccumulation, biopolymers, biotransformation, biodegradation, and flocculation, etc.

6.2 Strategies and Mechanisms for Bioremediation using Bacteria

6.2.1 Biosorption

The process of biosorption is described as the capacity of microbes to gather heavy metals from polluted sites utilizing their physiological and metabolic pathways (Fourest and Roux 1992). This process comprises two phases, one solid and the other liquid; solid phase is called as sorbent which is made up of organic substances where the liquid phase a solvent usually water contains species which needs to be removed like metallic ions also known as sorbate. Because of the affinity between the sorbent and sorbate (solid-liquid) phase the metal ions are removed. The process of biosorption continues till equilibrium is achieved between sorbent and sorbate. The affinity by which sorbent attaches to sorbate explains its diffusion among both the liquid and solid phases.

The powerful biosorbent response of microorganisms towards metal ions is attributed to the chemical constituents of bacterial cells. The biosorption capacity varies; some biological materials show a broad spectrum of binding where they bind most heavy metals, whilst some are very specific. In the quest to find the best biosorbent, various research groups have used naturally accessible biomass, while some obtained distinct strains and tried to improve the efficiency of biosorption. Research is currently focused on using industrial by-products or waste materials as biosorbents, such as mycelia produced by fermentation, a byproduct from olive oil processing (Pagnanelli et al. 2002), and residues obtained from wastewater purification facilities (Hammaini et al. 2003) have been used.

The mechanisms involved in the process of biosorption are complicated, involving various procedures such as adsorption, ion exchange, chelation, entrapment by diffusion, and the creation of concentration gradient between cell wall and membrane.

The focus on biosorption research was established in the 1980s (Volesky 2001, Volesky and Holan 1995). Initially, the attention was given to bioremediation involving microorganisms only to break down the organic compounds; however, later, the search to explore different biomass for their binding capacity for a number of heavy metals has intensified (Chen and Wang 2008).

Generally, biosorption involves the physical and chemical interactions of metallic ions with the cell exterior of microbes that can facilitate or hamper the intracellular trafficking and can also impact the methods of extracellular precipitation, transformation, or biomineralization (Gavrilescu 2004, Davis et al. 2003). The process of biosorption can be widely classified into two types: (i) metabolism independent, which is a passive process; and (ii) metabolism dependent, which is also known as active biosorption (Wang and Chen 2006).

6.2.1.1 Metabolism dependent biosorption

The movement of metals within the cellular membrane leads to their buildup intracellularly, which relies on the cellular metabolism indicating that this kind of sorption mechanism can happen only in the live cells. This active biosorption mechanism is generally found to be linked with the defense strategies/mechanisms of microorganisms that get activated by the existence of lethal metal ions.

6.2.1.2 Metabolism independent biosorption

In this biosorption, the metal assimilation takes place using physical and chemical association of the effective groups available at the cell exterior of bacteria and the metal ions. The processes involved in this comprise adsorption, ion exchange, and chemical interactions that are independent of the cellular metabolic processes. The cell wall of bacteria is constituted by proteins, lipids, and carbohydrates, which contain functional groups like –COOH, $-NH_2$, PO_4^{3-} and SO_4^{2-} which aid in binding of metallic ions with microbial biomass. The metabolism independent biosorption is reversible and quick (Kuyucak and Volesky 1988). Some of the commonly used strategies of biosorption used by microorganisms in response to heavy metal tolerance are discussed below:

6.2.2 Transportation Inside Cellular Membranes

The transport of heavy metals inside bacterial cell membrane uses the same pathway as followed by ions like Na^{2+}, K^+, and Mg^{2+}, which are vital for cellular metabolism. The ionic diffusion taking place gets confused with the metallic ions having identical charges and diameter as those of vital ions. This kind of transport is metabolism dependent, where metallic ions are diffused inside the cellular membrane (Nourbakhsh et al. 1994, Huang et al. 1990, Gadd et al. 1988).

6.2.3 Physical Adsorption

In this process, metal ions get physically adsorbed using van der Waal's forces of attraction. Kuyucak and Volesky (1988) explained that biosorption of cadmium, copper, zinc, cobalt, and uranium on microbial biomass happens through electrostatic interchange among metallic ions present in the solution and the microorganism cell wall. Biosorption of copper in *Zoogloea ramigera* bacterium and the *Chiarella vulgaris* alga (Asku 1992) and chromium in *Ganoderma lucidum* and *Aspergillus niger* was found to happen using electrostatic interactions.

6.2.4 Ion Exchange

The ionic exchange has proven to be used for biosorption procedures as the cell wall of bacteria is made up of sugars containing charged ionic groups. The bivalent ions of heavy metals interchange with the ions of available sugars. The marine algae alginate appears as salts of cations (Na^+, K^+, Mg^{2+}, and Ca^{2+}). These cations can get exchanged against ions like Co^{2+}, Cu^{2+} Cd^{2+}, and Zn^{2+} following the biosorption of heavy metal ions (Kuyucak and Volesky 1988). The uptake of Cu in *Ganoderma lucidium* (Muraleedharan and Venkobachr 1990) and *Aspergillus niger* was possible through the ionic exchange process.

6.2.5 Precipitation and Adsorption

The precipitation process can work either reliant or free from metabolism procedures. The cellular-dependent procedures work as defensive mechanisms of microbes that get activated on exposure to harmful metallic compounds that indulge in the process of precipitation. Metabolism independent precipitation works using chemical exchange among the metals and the cell walls of microbes.

The adsorption happens when the metal ions get attached to electrolytes of bacterial cell walls involving electrostatic, redox, covalent associations, and Van der Waal's forces to become electroneutral. The process of adsorption is independent of metabolism, so it is a reversible process with many advantages. It is used to treat large volumes of wastewater containing low concentrations of pollutants (Nishitani et al. 2010, Kuroda and Ueda 2010, Ahluwalia and Goyal 2007).

The adsorption process occurs as a result of exchange among the negatively charged cell wall of bacteria and the metallic ions in a pH-dependent manner. The role of pH in the accumulation of metallic ions has been found to be alike in bacteria, yeast, and algae. At low pH (< 2), the collection of metal ions is negligible in yeast, as at the lower pH, the functional sites of the cell wall are surrounded by protons, obstructing the sorption of metal cations by causing repulsion. As the pH increases, the active sites are positioned with a negative charge favoring the attraction of metallic cations, thus helping in the adsorption at the surface of the microbe, which eventually causes depletion in the solubility of metal ions in the solution, thereby causing the precipitation of metal ions, and hence, decreasing their bioavailability (Kuroda and Ueda 2011, Nishitani et al. 2010, Chen and Wang 2008, Esposito et al. 2002). Generally, the pH in the range of 4-8 is considered optimum for the biosorption of heavy metals using different biomass (Machado et al. 2010, Wang and Chen 2006).

6.2.6 Redox Reactions

The microbes have the potential to mobilize and immobilize metalloids, ions, and organometallic compounds, leading to oxidation-reduction processes. The prokaryotic organisms have the tendency to oxidize metal ions, viz. Cu, AsO_2^-, $Mn^{2+,}$ Fe^{2+}, Co^{2+}, Se^0 or SeO_3^{2-} and reduce ions like Co^{3+}, AsO_4^{2-}, Mn^{4+}, Fe^{3+}, SeO_4^{2} or SeO_3^{2}, while harnessing energy for their growth via redox reactions (Gavrilescu 2004). These redox reactions assist in the immobilization of heavy metals by reducing them to their lower oxidation states, creating metallic elements that are biologically least active (Gadd 2004, Valls and Lorenzo 2002).

6.3 Biomethylation

The process of biomethylation involves adding methyl groups to metals and metalloids, thus converting the toxic ions to a less harmful state. The ions such as Se, Sn, Te, Hg, As, Cd, and Pb could be methylated aerobically and anaerobically by various microorganisms such as bacteria, yeast, and fungi, causing an increase in the mobility of ions and improving their fitness and involvement in the procedures which lead to decreased toxicity of metal ions. Biomethylation is an enzyme-dependent pathway that includes the transfer of CH_3(methyl) group to metal(loid)s and the methylated compounds thus generated show diverse solubility, toxicity, and volatility pattern (Gadd 2004, Roane and Pepper 2001). For instance, methyl and dimethyl forms of mercury (Hg) are relatively more toxic than Hg ions, but it supports the processing of Hg^0. Likewise, inorganic forms of arsenic are relatively more toxic than their counter methylated groups (Roane and Pepper 2001, Tabak et al. 2005).

6.4 Generation of Cysteine Rich Peptides (Metallothioneins, Glutathione, and Phytochelatin)

The exposure to heavy metals above permissible concentrations induces the formation of peptides abundant in cysteines such as metallothioneins (MTs), glutathione (GSH), or phytochelatin (PCs). These peptides have low molecular weight and are non-enzyme molecules that resist precipitation and heat coagulation. The remarkable trait of these molecules is the complex formation with divalent metals alone and with the thiol group that forms essential intermediates required to circumvent the effects of reactive oxygen species (Bae et al. 2000, 2001).

6.4.1 MTs

These are low molecular weight proteins with an abundance of cysteine and are present among all the living organisms functioning as antioxidants. The available thiol group (-SH) in the chemical structure of cysteine helps in the uptake of ions like Zn^{2+}, Fe^{2+}, Hg^{2+}, Cu^{2+}, and Cd^{2+}. Structurally, MTs consist of 2 domains, α domain at C terminal and β domain present at N terminal. The number of cysteine residues vary in both the domains; β possesses 9 Cys residues binding three divalent ions, α has 11 Cys residues with the uptake of four ions; altogether seven ions are taken up by individual MTs (Thirumoorthy et al. 2007, Cobbett and Goldsbrough 2002).

MTs have various advantages, and they work against the generation of ROS, minimize oxidative stress, aid in the removal of heavy metals, and the maintenance of cell homeostasis (Smith et al. 2007, Cobbett and Goldsbrough 2002).

6.4.2 GSH

GSH is an L-Glutamyl-L-cysteinyl-glycine as it is formed of 3 amino acids, Glu-Cys-Gly, and the presence of -SH group at the active site is important for the biochemical functions (Mendoza-Cozatl et al. 2005, Penninckx 2000, 2002, Bae and Mehra 1997). GSH is unique as it can regulate its synthesis and is part of various biologically vital processes such as maintaining redox balance, inhibiting ROS, and transferring GSH conjugated enzymes in various detoxification reactions. GSH is found in abundance in the mammalian liver; the organelles with major reservoirs of GSH are nucleus and mitochondria, which is beneficial for the defense of these organelles from ROS (Mendoza-Cozatl et al. 2005, Inouhe 2005, Penninckx 2002). The intercellular toxicity due to Cd^{2+} in yeast is dealt with by the creation of GSH-Cd^{2+} complex, which minimizes the chance of lipid peroxidation of the cellular membrane and transfers the conjugate of GSH to vacuoles which ensue the decline of hazardous metallic ions in the cytosol and therefore favor the reduction in oxidative stress (Preveral et al. 2006, Adamis et al. 2004, Penninckx 2000).

6.4.3 Formation of Natural Phytochelatins and Synthetic Phytochelatin

Polychelatins are cysteine abundant molecules with a framework of (Glu-Cys) nGly (n = 2–11). PCs are made from GSH with the help of the enzyme PC synthase (Cobbett 2000, Grill et al. 1985).

PCs have the ability to tether heavy metal ions by utilizing their SH and COOH residues (Inouhe 2005, Kobayashi et al. 2006). Although PCs are categorized as MT-III, they have a higher tendency to get attached with ions of heavy metals. The PCs were articulated in recombinant *Escherichia coli*, but the hindrance in the expression was caused by γ type bond between Glu-Cys residues instead of α type usually found in chains of amino acids of every protein; γ type bond requires the involvement of multiple enzymes (Bae et al. 2001, Mendoza-Cozatl 2005, Wu et al. 2006, Inouhe 2005, Penninckx 2000). To solve this problem, synthetic substitute ECs were created *in vitro* having a structure of (Glu-Cys) nGly, where all the amino acids were linked through α linkage and were able to synthesize proteins similar to natural PCs. ECs are comprised of 20 repeated units of Glu-Cys (EC20). This synthetic counterpart binds efficiently with metal ions than natural phytochelatins.

The recombinant bacterial strains have been prepared (Wu et al. 2006, Bae et al. 2000, 2001). Research is done by Schmitt et al. (2006) where EC20 was expressed having an objective to discover microbes for bioremediation having improved binding efficiency for heavy metal ions.

To create such recombinant bacterial strains, genetic engineering approach has been used (Kim et al. 2005, Nishitani et al. 2010). PCs, ECs, and MTs were investigated to make them get attached outside the cell surface of microbes to improve the adsorption process compared to non-recombinant microbes (Kuroda and Ueda 2003, Bae et al. 2000, 2001, Jiang et al. 2007).

Biosorption holds a range of advantages in the process of bioremediation, which comprises the removal of heavy metals from water, land, soil, and landfill sites (Tran et al. 2015, Fomina and Gadd 2014). Microorganisms are proven as a prospective source of biosorbents that involve bacterial species like *Magnetospirillum gryphiswaldense* and *Bacillus subtilis*, fungi *Rhizopus arrhizus*, yeast *Saccharomyces cerevisiae*, algae *Chaetomorpha linum*, and seaweed (Zhou et al. 2012, Wang and Chen 2008). Additionally, biomass has been considered a prospective affordable and economical way of dealing with heavy metal toxicity. Industrial wastes and products such as waste from fermentation and food industries (*Saccharomyces cerevisiae*), agricultural waste products like corn core and various polysaccharides have been explored and examined (Wang and Chen 2008). Compared to other microorganisms, bacteria are regarded as excellent biosorbents due to increased surface-to-volume ratios and the availability of effective substances like teichoic acid (Beveridge 1989) on their cell surface, which helps in efficient biosorption. From the literature survey, it is recommended that dead microbial strains provide improved biosorbent capacity compared to their viable strains. The chromium biosorption by dead *Bacillus sphaericus* showed an increase of 13–20% biosorption compared to the viable cells of similar bacterial strains (Velásquez and Dussan 2009).

6.5 Bioaccumulation

It is explained as a procedure that happens when the rate of absorption for contaminants is higher than its evasion, which causes the buildup or accumulation of the contaminant inside microbes (Chojnacka 2010). It is a toxicokinetic method that impacts the sensitivity pattern of microorganisms towards chemical entities. Living organisms can tolerate exposure to chemicals until a particular concentration above which these chemicals lead to toxicity and death of the organisms. The resistance pattern of organisms differ variably that depends on the type of chemical and organism included (Mishra and Malik 2013). The potential organisms for bioaccumulation need to have an increased tolerance for one or more chemicals up to higher concentrations and additionally, should possess biotransformational qualities having the capacity of transforming a toxic form into a less toxic species, which help the microbes in the reduction of toxicity level when it is contained inside it (Mishra and Malik 2013, Ashauer et al. 2012). A diversified group of organisms has been investigated to study bioaccumulation and use them as pollution indicators, which includes bacteria (Diepens et al. 2015), fungi (Almeida et al. 2015), plants (Somdee et al. 2015), fish (Ding et al. 2016), and mussels (Diop et al. 2016). As discussed earlier, bacteria have the tendency to manufacture MTs in response to increased metal toxicity, which helps them to bind efficiently to the metal ions (Samuelson et al. 2000, Wernerus et al. 2001). For instance, a recombinant *Escherichia coli* strain fused with LamB protein consisting of a higher number of histidine sequences showed high tolerance towards higher concentration of Cd^{2+} and its counterions Cu^{2+} or Zn^{2+} (Kotrba et al. 1999). To find the appropriate microorganisms for the process of bioaccumulation, evaluation of the genes and mechanisms accountable for the bioaccumulation process is needed. To achieve this objective, investigation techniques involving molecular biology studies have been used (Mishra and Malik 2013). GeoChip is an extensively studied microarray approach (He et al. 2007), with an exceptional capacity to cover 4,24,000 genes having different functional groups (~ 4000) vital to carry out various biological functions. This technique was utilized to study other microbes for the uranium bioaccumulation (Van Nostrand et al. 2009). Microarray technique like DNA microarray has been investigated to study the genes expressed under exposure to higher heavy metal concentrations (Gorfer et al. 2009). The proteomics technique such as mass spectrometry has been explored to learn about the stress caused by the heavy metals at the protein level, which causes the change/mutation in the protein framework and expression due to the buildup of metallic contamination in the interior of cells (Italiano et al. 2009). Sequencing approaches such as whole genomic sequences of microbes with the bioaccumulation tendency is a promising strategy to target microbes for enhanced

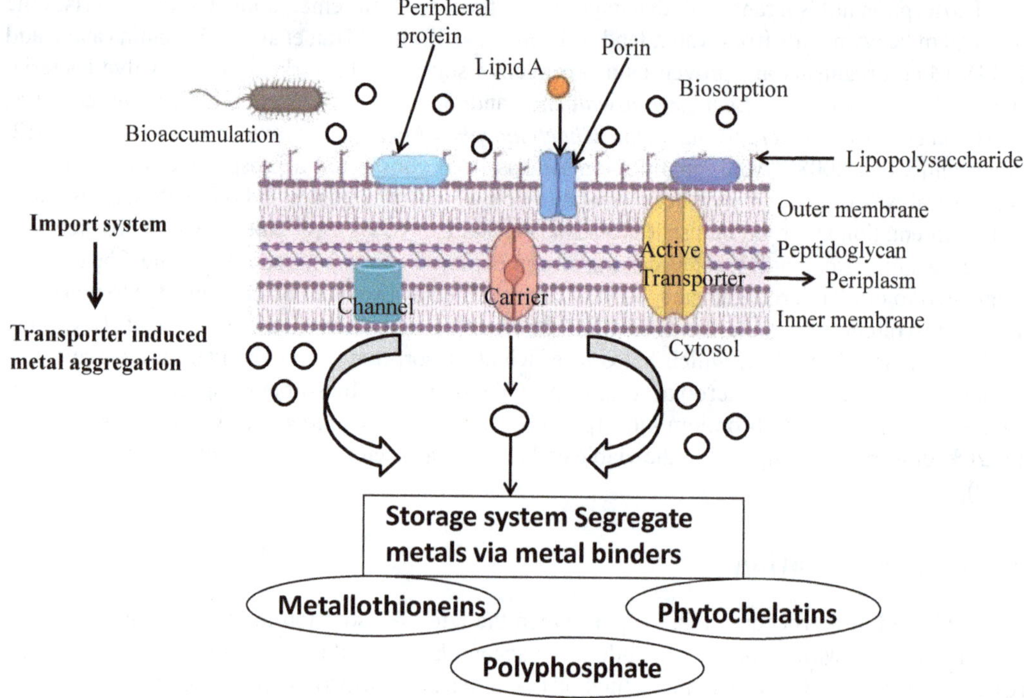

Fig. 6.1. Various bioremediation strategies offered by bacteria for the deletion of metallic ions from the surroundings.

bioaccumulation (Choi et al. 2015, Tan et al. 2015). Investigations to the extent of transcriptomics have been utilized to depict the function of vital genes required for bioaccumulation. Moreover, the response of different organs of the same organisms towards heavy metals was also studied using transcriptomics (Shi et al. 2015, Leung et al. 2014). The implication of bioinformatics tools and mathematical models hold an advantage to evaluate the efficiency of potential microbes for the estimation of chemical concentrations which they can tolerate (Stadnicka et al. 2012) (Fig. 6.1).

6.6 Formation of Siderophores

A siderophore is a small molecular weight complex that selectively binds with iron. These iron chelators are generated from the microorganisms such as fungi and bacteria. Siderophores show higher attraction towards trivalent metal ions Fe^{3+}, predominant in the oxygenated environment (Chu et al. 2010). Siderophores have been divided into three groups: carboxylate, catecholate (phenolates), and hydroxamate siderophores. The catecholates are well produced by bacteria like *Salmonella typhimurium* and *Escherichia coli* (Searle et al. 2015). These siderophores bind to Fe by complex formation with octahedral hexadentate. The carboxylate type bind to Fe via hydroxyl and carboxyl functional groups and are predominantly synthesized by bacteria like *Rhizobium* and many other types of bacteria. A study by Chu et al. (2010) explained that metal ions like Cr, Al, and Ga, sharing similar chemistry with Fe, can lead to the production of siderophores. Therefore, it was deduced that the application of using siderophores in bioremediation processes is not only restricted to Fe but is also applicable for the removal of other toxic ions as well. The Hydroxamate group comprises C (=O) N-(OH) R; the R represents an amino acid. This kind of siderophore is most commonly synthesized by bacteria and fungi (Fig. 6.2). The production of siderophores by the exposure to heavy metals in combination with high Fe concentrations was reported in bacteria such as *Pseudomonas aeruginosa* and *Alcaligenes eutrophus* (Koedam et al. 1994). Moreover,

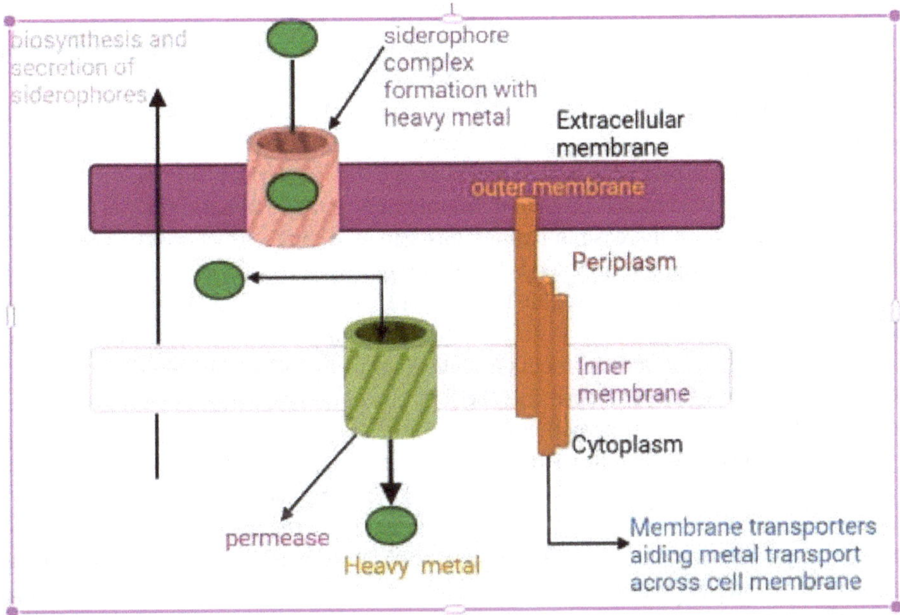

Fig. 6.2. Mechanism of bioremediation of heavy metals by bacteria by the formation of siderophores.

the reduced toxicity of copper was seen in cyanobacteria using siderophore stimulated complex formation (Stone and Timmer 1975). The available literature has mentioned that the proteins used for the uptake of siderophores with their detailed sequences about the outer membrane and periplasmic siderophore binding proteins have been discovered (Chu et al. 2010). Various advanced biological techniques comprising of systems biology approaches have been explored to investigate the extraordinary production of siderophores.

Advancement in biological technologies such as the mass spectrometry-based proteomic method is very helpful for identifying novel siderophores and related proteins along with their functional properties. The platform named Proteomic Investigation of Secondary Metabolism (PrISM) has been explored for the understanding of 2 novel siderophores (gobichelin A and B) from the species of *Streptomyces* (Chen et al. 2013).

The evaluation of genomic sequencing and data using high throughput proteomics (proteogenomics) of *S. lilacinus* has hastened the recognition of fresh DHB (dihydroxybenzoic acid) and threonine siderophores (Albright et al. 2014). Systems biology holds an encouraging potential for gene identification, their processing, and signaling routes for studying siderophores. Techniques like bioinformatics along with high throughput methodology are effective in dealing with the problems of spotting the genes responsible for secondary metabolites synthesis in bacteria (Park et al. 2013) and can be used for the creation of an online database for the siderophore studies (Flissi et al. 2016).

6.7 Formation of Polymers/Biosurfactants

The use of polymers produced by microbes for the remediation of metal(loid)s holds a promising future. The use of biopolymers is a cost-effective and ecofriendly approach that minimizes the toxic production of various contaminants.

These biopolymers are versatile and can withstand a broader range of pH and temperature that have a direct effect on interfaces. As pH can have a direct influence on the functional moieties used for the binding of metal(loid)s, these polymers provide different binding sites for varying pH values at the microbial surface. For instance, at an elevated pH, there are few protons (H^+) available to interfere with the binding of metal(loid)s on the polymers, while at neutral pH, negatively charged functional groups are available, which have the potential of forming stable metallo organic complexes using different chemical associations (Li and Yu 2014). The rise in temperature enhances the adsorption of metal(loid)s on biopolymers (Kiran and Kaushik 2008). It is well known that at the elevated temperatures, compatibility for binding sites is enhanced, which approves the formation of direct contact among the functional groups of biopolymers and the metal(loid)s present (Kiran and Kaushik 2008). But, a sharp increase in temperature can lead to the structural disintegration of the biopolymers and active groups that impedes the effectiveness of biopolymer (Li and Yu 2014). The bonding affinity of biopolymers is based on temperature, pH, specificity, nature of biomass, and the available active moieties (Guibaud et al. 2009, Li and Yu 2014). The characteristic of a biopolymer depends upon its descent, extraction process involved, and adsorption capacity (Li and Yu 2014).

The formation of biopolymers represents the first line of defense strategies against metal(loid)s. It takes part in the protection of the cell environment of microbes (Li and Yu 2014). For the easy removal of metal(loid)s and the prevention of toxicity, the isolation of biopolymers becomes important (Fig. 6.3).

Biosurfactant production is an example of microbial polymers, which has been a useful means of metalloid complex formation. Biosurfactants are amphiphilic surface active substances consisting of both hydrophilic and hydrophobic groups. Their amphiphilic nature helps them to accumulate and form a micellar structure at the surface that changes the existing surface state by adsorption, thereby reducing the surface tension among liquid and solid or between liquids (Reznik et al. 2010, Ron and Rosenberg 2001). The hydrophilic moieties of biosurfactants consist of carbohydrates (saccharides), amino acids, ions (cations and anions), and proteins (peptides), whilst the hydrophobic moieties are formed of fatty acids (branched, hydroxylated, saturated and unsaturated) (Kitamoto et al. 2009, Singh and Cameotra 2004). Generally, biopolymers are utilized for the depletion of organic pollutants, yet various studies have explained that surfactants can produce a complex and remove heavy metals like Zn, Cd, and Pb (Mulligan 2005, Maier and Soberon-Chavez 2000).

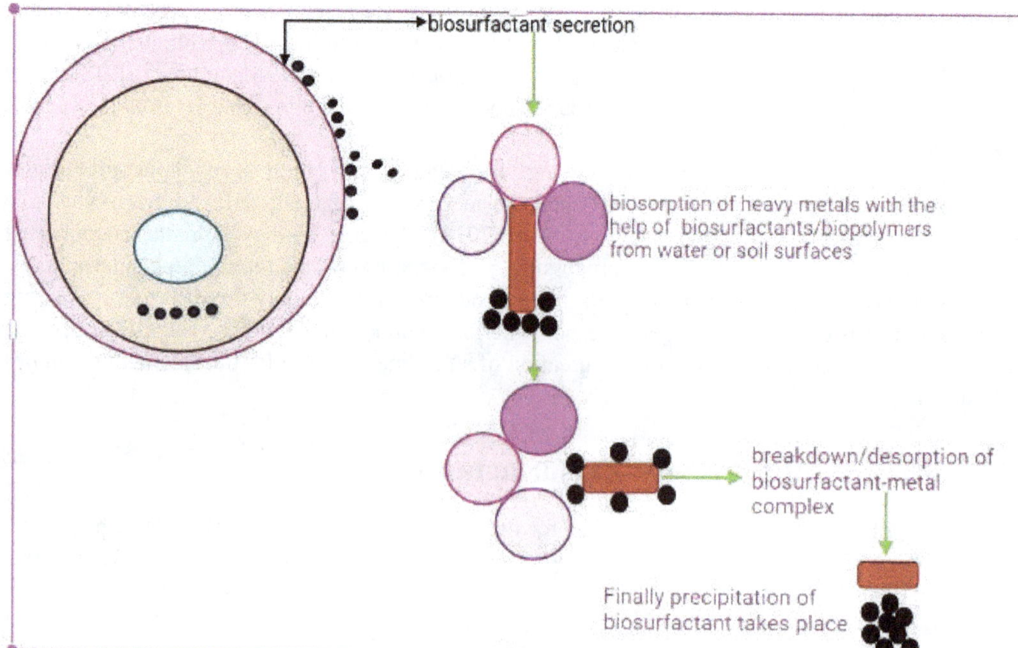

Fig. 6.3. Bioremediation mechanism of biosurfactants takes place through biosorption, desorption, and precipitation of heavy metals.

The class of surfactants called Rhamnolipids is a major type of biosurfactant obtained from *Pseudomonas aeruginosa* and other microbes (Maier and Soberon-Chavez 2000). Rhamnolipid is a class of glycolipids accompanied by a rhamnose group consisting of a glycosyl head and a tail of HAA (3-hydroxy alkanoyl oxy alkanoic acid) (Desai and Banat 1997). They are widely used in industries and are a strategy for eliminating pollution (Muller et al. 2012). Rhamnolipids can take up the metallic ions by involving complexation and electrostatic methods; complexation (complex formation) can improve the solubilization of ions (Rufino et al. 2012), thus reducing the bioavailability of metal ions usually present as byproducts of metabolism, which leads to the formation of low soluble metals salts along with the sulfide and phosphate precipitates (Maier et al. 2000). Various other surfactants are synthesized by microbes (bacteria, fungi) like exocellular polymeric surfactants, which appear as proteins, saccharides, lipoproteins, lipopolysaccharides, or amalgam of all these compounds. For example, among the various species of *Acinetobacter*, the synthesis of high molecular weight emulsifiers was confirmed (Ron and Rosenberg 2001).

The increasing demand for biopolymers has various uses as compared to their chemical counterparts owing to the presence of more than one chiral and functional groups, diversity in structure, ability to withstand extensive scale of pH, temperature, and salinity, the exceptional capacity of aggregation and formation of liquid crystals, versatility, adsorption potential, higher surface activity with reduced surface and micelle tension, improved biodegradation and reduced toxicity and their renewability (Sriram et al. 2011, Kitamoto et al. 2009, Henkel et al. 2017, Bezza and Chirwa 2016, Dadrasnia and Ismail 2015).

The biosurfactants improve the mass transfer of hydrophobic groups by acting as mediators, thus making these groups easily accessible to microorganisms. Apart from this, biosurfactants also alter the properties of the cell membrane leading to efficient microbial binding. This process speeds up the uptake if two immiscible liquids are present at the surface (Dadrasnia and Ismail 2015). Furthermore, they also support overcoming the difficulties caused during diffusion linked mass transport (Bezza and Chirwa 2016).

The cationic, anionic, and neutral charged biosurfactants are used as laundry detergents. They remove, solubilize, break down and desorb metalloids from dug out sediments in a washing element (Dahrazma and Mulligan 2007). A number of bacterial species such as *Arthrobacter*, *Bacillus*, and *Pseudomonas* are being recognized to synthesize surfactants with metal(loid)s attachment properties.

6.7.1 Bacterial Mechanisms to Remove Metalloids by Biosurfactants

Generally, microbes use the reduction pathway compared to the oxidation mechanism for biotransformation of metal(loid)s. This is attributed to the lower toxic characteristics of the reduced state of metalloids. The methods employed to carry this process involve a bacterial efflux system to remove or reduce metalloid exposure, bioaccumulation, and complexation. Plasmid genes clustered in bacteria are used as a resistance mechanism on exposure to metalloids (Gnanamani et al. 2010). The surfactants work as a connecting link among liquid interfaces because of their amphiphilic characteristics, decreasing the surface tension. When the surface tension of water is reduced, it causes the increase in the mobility of metalloids from unsaturated soils, thus making exclusion of metal(loid)s feasible.

It has been evaluated that biosurfactants use two basic principles for the elimination of metal(loid)s from solutions. The biosorption of metals aided by biosurfactants is made possible by the complex formation of free metallic ions. This method follows Le Chatelier's principle, where metalloid activity is declined from the solution phase, thus allowing desorption of metalloids. The second mechanism involves direct contact with the metalloids in a reduced interfacial stress in a solid-solution boundary. This helps in the accumulation of biosurfactants at the solid-solution boundary, which improves the binding of metalloids (Chakraborty and Das 2014, Singh and Cameotra 2004). The contamination of soil is treated following two simple methods. The first process involves immobilizing metalloids on a strong solid matrix to minimize movement, but it does not provide a permanent solution as reuse of soil is partial, and it demands continuous supervision. Therefore, it is usually restricted to radioactive waste and highly toxic compounds. Another mechanism allows movement and transfer of metalloids into liquid state by the solubilization and desorption process; this technique is promising as it provides a long-lasting strategy for pollution control.

The processes used by biosurfactants for the extraction of metalloids comprise ionic exchange, counter binding, and the precipitation of metallic ions. In the process of ionic exchange, anionic (negative charge) biosurfactants are used to bind with cationic (positive) metal to create a bond between the metal ion and surfactant. The bond formed is stronger than the usual bond between the soil and metallic ions. The cellular surface becomes negatively charged due to biosurfactants; therefore, the electrostatic interactions with the metal ion are enhanced. The polar head group of micelles created by surfactants binds with the ions, thus increasing their solubility. After this, the micelles push the ions into the solution, helping in the easy metal revival by simple washing, pumping, and flushing (Asci et al. 2010, Mulligan 2005). The complex formed between metalloid-biosurfactant creates a high strength bond so that removal with water uproots the complex from the soil environment (Mulligan 2005). The positively charged biosurfactants swap the similarly charged ions by fighting for the negatively charged surface (Sarubbo et al. 2015, Franzetti et al. 2014). The biosurfactants can be recovered/recycled and, therefore, employed for various bioremediation approaches. The biosurfactant rhamnolipid produced by *Pseudomonas aeruginosa* consists of a carboxyl group that has an affinity toward divalent cations such as Zn^{2+}, Pb^{2+}, and Cd^{2+} (Franzetti et al. 2014, Juwarkar et al. 2007). The complex formation that takes place among the metal and a biosurfactant is harmless for bacteria. During the process of precipitation, surfactants are added to the soil at the time of washing, and due to the foaming properties of surfactants, the complex formed between the metal and the surfactant is recovered from the soil by adding air to create foams. The surfactant is also retrieved and reused by the process of precipitation at low pH environment (Mulligan 2005).

The concentration of a biosurfactant at which the accumulation of ions (micellar development) begins is perceived as critical micelle concentration (CMC), a concentration where the surfactant accumulates in an ordered molecular system (Chakraborty and Das 2014). At a concentration above CMC, assembly of surfactant molecules leads to aggregation, formation of micelles, and vesicular configurations that depend upon the pH of the solution (Asci et al. 2010). At levels lower than CMC, surfactants are present as monomers in the solution. CMC is determined by the structure and framework of biosurfactants, obtainability of organic groups in the solution, temperature, and ionic composition (Bustamante et al. 2012).

Surfactants are spread on the contaminated soils in small patches utilizing a cement mixer to extrude metals. After this, the complex created among the surfactant and the metal ion is flushed, and soil is placed back, following the precipitation of surfactant from the complex, leaving the metal ions behind. The bond strength is high in surfactant-metal ions that remove the complex from the soil surface if flushed out with water. This process can be replicated for deeply polluted surfaces (*in situ*); however, the flushing activity has to be increased. The surfactants present in the broth without cells can be put in as a whole or in a diluted state at the contamination area; they are balanced and work efficiently in the manufacturing of medium. The biopolymers improve the disintegration of organic pollutants in the presence of metalloids, therefore minimizing the toxicity of metalloids and organic contaminants (Pacwa-Płociniczak et al. 2011).

Decontamination tests have been carried out at metal polluted soils and sludge using various surfactants. The results displayed the efficiencies of rhamnolipids, sophorolipids, lipopeptides, and surfactin for the elimination of metals. A study by Dahrazma and Mulligan (2007) demonstrated the capacity of rhamnolipids for the removal of metal ions from contaminated soils. It was seen that the rhamnolipid was capable of removing 37% Cu, 33% Ni, and 75% Zn when put on in an incessant flow pattern. Similarly, rhamnolipids of *Pseudomonas* were able to remove 19% Fe and 52% Zn using a cyclic pattern.

The other type of surfactants that are efficient in removing metal(loid)s from the soil are lipopeptides obtained from *Bacillus subtilis*. This surfactant is capable of removing metals such as Co, Pb, Ni, Cd, Zn, and Cu from the hydrocarbon and metal contaminated soil (Singh and Cameotra 2013). The removal efficiency varied from 26% for Cu to 44% for Cd. Another biosurfactant, sophorolipids from *Torulopsis bombicola*, has shown promising results in depleting metal ions (Mulligan et al. 2001); this surfactant has displayed the tendency to remove 60% of Zn and 25% of Cu in an experiment of soil sediment washing. In an investigation conducted by Wang and Mulligan (2009), a commercialized rhamnolipid JBR425 obtained from *P. aeruginosa* was used in mine tailings for the removal of arsenic and other metals such as copper, lead, and zinc. It was experienced that 0.1% of rhamnolipid JBR425 could eliminate 148, 74, 2379, and 259 mg/kg of As, Cu, Pb, and Zn, respectively, in an alkaline environment. Therefore, they put forward that under alkaline conditions and in the presence of other metals, the mobility of arsenic improved, thus promoting the metal bridging leading to biosurfactant-metal complex formation.

In present times, applications of biosurfactants have been improved significantly by utilizing advanced high throughput and system biology methods. The metagenomic strategy has been involved in identifying new and improved surfactants for the marine ecosystem (Jackson et al. 2015). Mass spectrometry based proteomics approach was utilized to decipher the localization process of sophorolipids biosurfactant obtained from *Starmerella bombicola.* The results achieved led to the identification of regulators, which helps in the localization process (Ciesielska et al. 2014). Increasingly available understanding of biosurfactants due to the efforts of bioinformaticians and computational biologists has made it possible to develop computational methods and online resources customized for biosurfactant discovery and collection of data in a centralized database. For example, the BioSurf database gives carefully chosen and well organized details about biosurfactants along with their source organisms, genes, metabolic pathways, and related bioinformatics tools and algorithms helpful for discovering biosurfactants (Oliveira et al. 2015).

6.8 Removal of Metalloids using Microbial Flocculation

Flocculants are the medium used for the detachment of solid liquid suspensions (Okaiyeto et al. 2016). They participate in colloid accumulation to form floc, which is supported by a chemical entity known as flocculant. Generally, flocculants have been categorized into three main types: (i) inorganic flocculants, comprising ferrous sulfate, ferric chloride, poly-aluminum chloride, aluminum chloride and sulfate, and alum; (ii) natural flocculants, consisting of gum and mucilage, starch, tannin, microbial flocculant, cellulose, chitosan; and (iii) organic polymeric flocculants, made up of polyethylene amine and polyacrylamide (Sajayan et al. 2017, Lee et al. 2014, Salehizadeh and Yan 2014).

The synthetic flocculants have shortcomings of being hazardous, leading to health problems, production of sludge at large scale, and are non-biodegradable (Lee et al. 2014, Zhai et al. 2012, Farag et al. 2014). Hence, the demand for using bioflocculants has increased as they are ecofriendly, non-hazardous, do not produce pollutants, remove metal(loid)s and solids suspended in wastewater (Guo 2015).

Flocculants obtained from microbes are called bioflocculants. These are polymeric substances manufactured during bacterial growth and consist of a variety of macromolecular polyelectrolytes formed by diverse organisms. Bioflocculants consist of organic groups like glutamic and aspartic acid present in their protein part, glucuronic and galacturonic acid in polysaccharide component, and uronic acids in the carboxylic acid and carbonyl segment. All these groups help in the attachment of metals (Lin and Harichund 2011). The available carboxyl moiety in the polymeric chain cause stretching of the chain due to repulsive forces (electrostatic repulsion); the stretched chains then present an effective area for the purpose of attachment. The metals bind in the form of a complex with the functional groups such as amino and carboxyl groups present in the bioflocculant, thus neutralizing and stabilizing the remaining charge as the length of attachment is reduced (Pathak et al. 2014). The effectiveness of flocculation is determined by the area of contact among flocculants and the adsorbate, particle size, structure, and concentration of adsorbate, ionic composition, temperature, pH, and the organism required for producing flocculent (Okaiyeto et al. 2016). The selection of flocculent contributes significantly to the flocculation action, as it controls the accumulation and aggregation of particles and decides the strength and number of bonds developed in the course of flocculation (Zhang et al. 2014).

6.8.1 Flocculation Mechanism

Flocculation works by the formation of an interface between metal(loid)s ions and the bioflocculants. However, the interface formation is dependent on ionic groups viz. carboxyl, hydroxyl, and amino of the bioflocculant, which help in the interaction between a metal and flocculant (Deng et al. 2005, Sathiyanarayanan et al. 2016). The process of flocculation is independent of metabolism, where linking/bridging amongst metal ions and the active groups of flocculants is achieved by utilizing phenomena of physical adsorption and precipitation, complexation, immobilization, and ion exchange (Sathiyanarayanan et al. 2016).

Other mechanisms utilized by microbes to achieve flocculation comprise polymer bridging, electrostatic patch, and the neutralization of charges. Usually, metal ions are present as cations, and bioflocculants have negatively charged groups, thus making it easier to achieve neutralization of charges. The cationic exchange creates electrostatic force among the flocculants and the metal ions (Lee et al. 2014). When the neutralization occurs, the charge density of the surface particles is decreased by the adsorption on bioflocculant; therefore, the particles get accumulated effectively, and thus the force of attraction for particles is also increased (Salehizadeh and Shojaosadati 2001).

During the process of bridging, the bioflocculant acts as a connecting link for the complex formation which occurs between a particle and a biopolymer as particle-polymer-particle. During the contact between the metal ion and the bioflocculant, some of its functional groups adsorb the metal ions, while some metal ions remain unadsorbed, which get pulled out into the solution. Another biopolymer having vacant adsorption sites makes contact with the metal ions present in the solution, hence favoring attachment. Thus, a metal-polymer-metal complex is created where the biopolymer is helping as a connecting link (Lee et al. 2014). In order to make flocculation possible, the biopolymer should withstand the electrostatic repulsive forces occurring between particles, and it should be able to make an intense contact (binding) with the particles (Okaiyeto et al. 2016, Wang et al. 2011).

To perform bridging properly, many characteristics are to be taken into consideration, which involve the weight and net charge of bioflocculant, suspension ionic strength, type of suspended particles, and the type of mixing. The metalloids also exhibit different effects on varying biopolymers, which rely on the valence and concentration of metal ions (Okaiyeto et al. 2016, Wu and Ye 2007) (Fig. 6.4).

6.9 Bacterial Genetic Modification (Engineering) for Bioremediation

The remediation of pollutants using bacteria or microbes is a significant approach to clean up the environment polluted with metal(loid)s (Li et al. 2021). The growing number of hazardous contaminants in the surroundings is a matter of consideration to safeguard the environment. Methods such as biomolecular engineering have been explored for the genetic modification of bacteria and the enzymes required to break down toxins/contaminants.

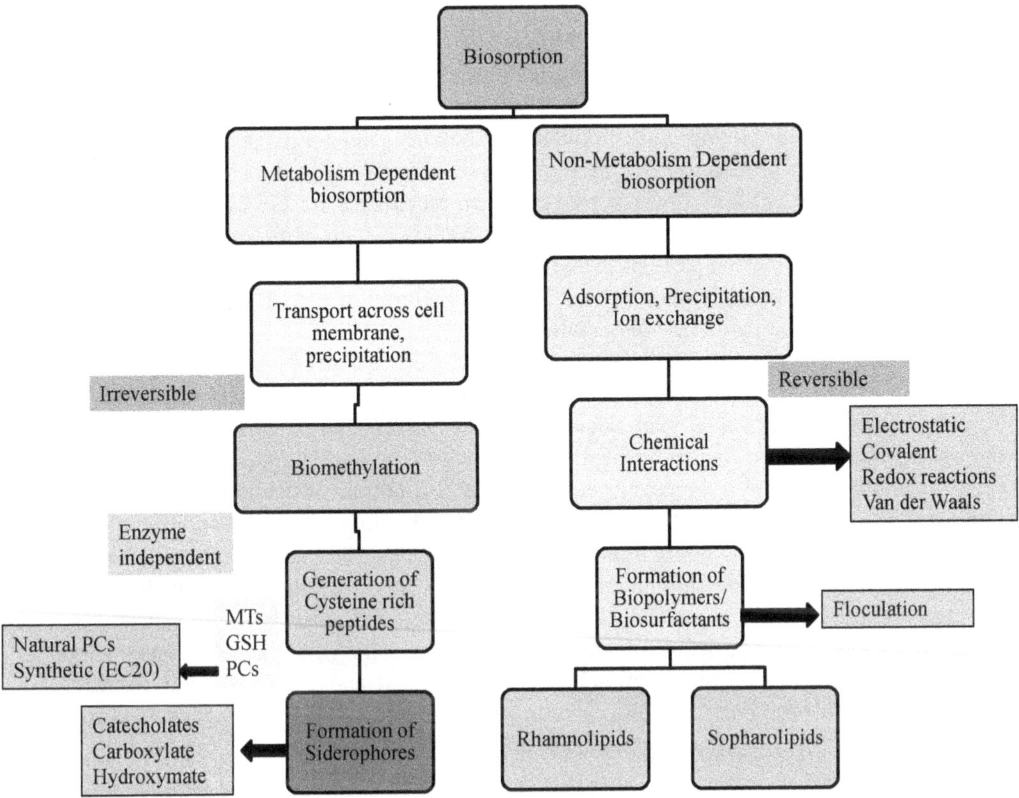

Fig. 6.4. A flow chart summarizing different bioremediation approaches utilized by the microbes for attenuation of heavy metals from surroundings.

Bacteria hold a capacity to carry out the oxidation of pollutants present in the environment efficiently. Therefore, making use of bacterial metabolic activities, it has been found that bacteria can deteriorate distinct toxic compounds (Parrilli et al. 2010, Parales and Haddock 2004). The limitations posed in bacterial bioremediation led to the establishment of technologically advanced metabolic processes that possess Genetically Engineered Microorganisms (GEM)s; they were also given the name of designer biocatalysts (Paul et al. 2005). It is a bacterial tendency or defense mechanism to form resistance and virulence genes when exposed to higher amounts of toxic heavy metals such as Hg. The mercury resistance genes (*mer*)genes appear as an operon that helps bacteria in the detoxification of Hg^{2+} into an unstable Hg using the enzyme mercury reductase. In a three-step process, the slight Hg resistance for mercury was observed to degrade sulfur in a manner similar to that followed for Hg^{2+} transfer inside the cell, the enzymatic NADPH-dependent (nicotinamide adenine dinucleotide phosphate) conversion of ionic Hg to a less toxic elemental form (Hg^0) and finally regulation of mercury transport and conversion of Hg genes into its functional form (Singh et al. 2011). The mercury operon comprises a different set of functional genes responsible for the reduction of mercury, genes such as *mer* A and B, while *mer* T and P work as transport regulators, and *mer* E and H function as membrane transporters (Ruiz and Daniell 2009). The engineered *E. coli* with MTs and mercury genes (*mer T* and *P*) has proven useful in extracting mercury from wastewaters. The metal binding capacity of bacteria is enhanced using metal binding peptides (PCs, MTs) discussed earlier. The selective isolation of metal ions from multi-component toxins using engineered bacteria that have enhanced affinity for metal ions and improved bioaccumulation capability is found to be useful for decontaminating hazardous metal ions. A transgenic *E. coli*

JM109 has been successfully employed to bioaccumulate cadmium from heavy metal polluted sites (Deng et al. 2007). Furthermore, other genetically redesigned bacteria such as *Caulobacter crescentus*, *Pseudomonas putida*, and *Mesorhizobium huakuii* are found to amass cadmium using PCs engineered proteins EC20 and RsaA-6His (Patel et al. 2010). It has been evaluated that the engineered microbes have higher bioaccumulation efficiency than normal strains. The expression of ArsR metal regulatory protein in *E. coli* removed arsenic from polluted locations. In various other microbes, an enhanced affinity towards Ni^{2+} has been recognized for GST-MT fusion protein and *nix* A protein carrying out transport of membrane proteins (Singh et al. 2011). It is believed that overexpression of these fusion proteins in engineered bacteria helps in the efficient bioaccumulation of Ni^{2+}. Taking metal contamination as a worrisome situation for the environment and public health, GEMs hold a significant part in the remediation strategies.

The genetically modified microbes have added new perspectives in creating engineered microorganisms with better and targeted metal attachment capacity, which can be used as biosorbents. The surface display system among bacteria like *Staphylococcus xylosus* and *Staphylococcus carnosus* expressed two polyhistidyl peptides (His_3-Glu-His_3 and His_6). These chimeric proteins helped in improved metal binding and easy surface accessibility (Samuelson et al. 2000). In a similar manner, *Escherichia coli* and *Pseudomonas putida* exhibited better biosorption of phosphate by immobilizing phosphate-binding protein (PBP) on the cell surface (Li et al. 2009). Another study involving *Escherichia coli* recombination with mice MT1 reported efficient metal absorption, thus explaining the possibilities of genetic engineering in developing microbes with efficient potential for the biosorption (Almaguer-Cantu et al. 2011). The inclusion of advanced mechanization as a high throughput genome editing approach (Bao et al. 2016) and whole genome sequences (El-Metwally et al. 2014) for genetic studies can allow exploring many other organisms that can be used in the future application processes of bioremediation.

6.10 Conclusion

Bioremediation seems to be a promising approach for dealing with increasing pollution. At present, the harmful effects imposed by the heavy metals in soil and aquatic environment create a consortium of health hazards for human health. Bioremediation is a green perspective to amend the harmful effects of heavy metals and toxic pollutants. A thorough analysis of microbial metabolism and genetic makeup is recommended to understand the role of bacteria in the binding and decontamination of metals and metal(loid)s. This can be accomplished by altering the metabolic pathways, and insertion of modified genes with high efficiency in the bacteria by utilizing aspects of genetic engineering. The detailed understanding of bacterial genes, proteins, and enzymes accountable for the binding mechanisms of metals and metal(loid)s holds significance to introduce required alterations under different environmental circumstances. It is also important to design and develop novel procedures to improve the elimination and recovery of metal(loid)s using multidisciplinary approaches, including molecular, cellular, genomics, proteomics, bioinformatics, and biotechnological.

References

Achal, V., X. Pan and D. Zhang. 2011. Remediation of copper-contaminated soil by Kocuria flava CR1, based on microbially induced calcite precipitation. Ecol. Eng. 37(10): 1601–1605.

Achal, V., X. Pan, Q. Fu and D. Zhang. 2012. Biomineralization based remediation of As (III) contaminated soil by *Sporosarcina ginsengisoli*. J. Hazard. Mater. 201-202: 178–184.

Adamis, P. D., D. S. Gomes, M. L. Pinto, A. D. Panek and E. C. Eleutherio. 2004. The role of glutathione transferases in cadmium stress. Toxicol. Lett. 154(1): 281–288.

Ahluwalia, S. S. and D. Goyal. 2007. Microbial and plant derived biomass for removal of heavy metals from wastewater. Bioresour. Technol. 98(12): 2243–2257.

Albright, J. C., A. W. Goering, J. R. Doroghazi, W. W.Metcalf and N. L. Kelleher. 2014. Strain-specific proteogenomics accelerates the discovery of natural products via their biosynthetic pathways. J. Ind. Microbiol. Biotechnol. 41: 451–459.

Almaguer-Cantú, V., L. H. Morales-Ramos and I. Balderas-Rentería. 2011. Biosorption of lead (II) and cadmium (II) using *Escherichia coli* genetically engineered with mice metallothionein I. Water Sci. Technol. 63: 1607–1613.

Almeida, S. M., S. H. Umeo, R. C. Marcante, M. E. Yokota, J. S. Valle, D. C. Dragunski et al. 2015. Iron bioaccumulation in mycelium of *Pleurotus ostreatus*. Braz. J. Microbiol. 46: 195–200.

AMGA. 2004. The Australian Mushroom Growers Association (AMGA), Locked Bag 3, 2 Forbes St., Windsor, NSW, 2756, Australia.

Asci, Y., M. Nurbas and Y. S. Acikel. 2010. Investigation of sorption/desorption equilibria of heavy metal ions on/from quartz using rhamnolipid biosurfactant. J. Environ. Manag. 91: 724–731.

Ashauer, R., A. Hintermeister, I. O'Connor, M. Elumelu, J. Hollender and B. I. Escher. 2012. Significance of xenobiotic metabolism for bioaccumulation kinetics of organic chemicals in *Gammarus pulex*. Environ. Sci. Technol. 46: 3498–3508.

Asku, Z. 1992. The biosorption of Cu (II) by *C. vulgaris* and *Z. ramigera*. Environ. Tech. 13(1): 579–586.

Bae, W. and R. K. Mehra. 1997. Metal-Binding characteristics of a phytochelatin analog (Glu-Cys)2Gly. Elsevier Sci. 201–210.

Bae, W., W. Chen, A. Mulchandani and R. K. Mehra. 2000. Enhanced bioaccumulation of heavy metals by bacterial cells displaying synthetic phytochelatins. Biotechnol. Bioeng. 70: 518–24.

Bae, W., W. Chen, A. Mulchandani and R. K. Mehra. 2001. Genetic engineering of *Escherichia coli* for enhance uptake and bioaccumulation of mercury. Appl. Environ Microbiol. 67: 5335–5338.

Bang, S. W., D. S. Clark and J. D. Keasling. 2000. Engineering hydrogen sulfide production and cadmium removal by expression of the thiosulfate reductase gene (phsABC) from *Salmonella enterica Serovar Typhimurium* in *Escherichia coli*. Appl. Environ. Microbiol. 66: 3939–3944.

Bao, Z., R. E. Cobb and H. Zhao. 2016. Accelerated genome engineering through multiplexing. Wiley Interdiscip. Rev. Syst. Biol. Med. 8: 5–21.

Beveridge, T. J. 1989. Role of cellular design in bacterial metal accumulation and mineralization. Annu. Rev. Microbiol. 43: 147–171.

Bezza, F. A. and E. M. N. Chirwa. 2016. Biosurfactant-enhanced bioremediation of aged polycyclic aromatic hydrocarbons (pahs) in creosote contaminated soil. Chemosphere. 144: 635–644.

Bustamante, M., N. Duran and M. Diez. 2012. Biosurfactants are useful tools for the bioremediation of contaminated soil: A review. J. Soil Sci. Plant Nutr. 12: 667–687.

Chakraborty, J. and S. Das. 2014. Biosurfactant-based bioremediation of toxic metals. Microb. Biodegrad. Bioremediat. 167–201.

Chen, C. and J. Wang. 2008. Removal of Pb^{2+}, Ag^{+}, Cs^{+}, and Sr^{2+} from aqueous solution by brewery's waste biomass. J. Hazard. Mater. 151: 65–70.

Chen, Y., M. Unger, I. Ntai, R. A. McClure, J. C. Albright, R. J. Thomson et al. 2013. Gobichelin A and B: mixed-ligand siderophores discovered using proteomics. Med Chem. Comm. 4: 233–238.

Chibuike, G. U. and S. C. Obiora. 2014. Heavy metal polluted soils: effect on plants and bioremediation methods. Appl. Environ. Soil Sci. 2014: 1–12.

Choi, D. H., Y. M. Kwon, K. K. Kwon and S. J. Kim. 2015. Complete genome sequence of *Novosphingobium pentaromativorans* US6-1(T). Stand. Genomic Sci. 10: 107.

Chojnacka, K. 2010. Biosorption and bioaccumulation-the prospectsfor practical applications. Environ. Int. 36: 299–307.

Chowdhury, S., M. J. Mazumder, O. Al-Attas and T. Husain. 2016. Heavy metals in drinking water: occurrences, implications, and future needs in developing countries. Sci. Total Environ. 569: 476–488.

Chu, B. C., A. Garcia-Herrero, T. H. Johanson, K. D. Krewulak, C. K. Lau, R. S. Peacock et al. 2010. Siderophore uptake in bacteria and the battle for iron with the host; a bird's eye view. Biometals. 23: 601–611.

Ciesielska, K., I. N. Van Bogaert, S. Chevineau, B. Li, S. Groeneboer, W. Soetaert et al. 2014. Exoproteome analysis of *Starmerella bombicola* results in the discovery of an esterase required for lactonization of sophorolipids. J. Proteomics. 98: 159–174.

Cobbett, C. and P. Goldsbrough. 2002. Phytochelatins and Metallothioneins: Roles in heavy metal detoxification and homeostasis. Annu. ReVol. Plant Biol. 53: 159–82.

Cobbett, C. S. 2000. Phytochelatins and their roles in heavy metal detoxification. Plant Physiol. 123: 825–832.

Dadrasnia, A. and S. Ismail. 2015. Biosurfactant production by *Bacillus salmalaya* for lubricating oil solubilization and biodegradation. Int. J. Environ. Res. Public Health. 12: 9848–9863.

Dahrazma, B. and C. N. Mulligan. 2007. Investigation of the removal of heavy metals from sediments using rhamnolipid in a continuous flow configuration. Chemosphere. 69: 705–711.

Davis, T. A., B. Volesky and A. A. Mucci. 2003. Review of the biochemistry of heavy metal biosorption by brown algae. Water Res. 37(18): 4311–4330.

Deng, M., X. Yang, X. Dai, Q. Zhang, A. Malik and A. Sadeghpour. 2020. Heavy metal pollution risk assessments and their transportation in sediment and overlay water for the typical Chinese reservoirs. Ecol. Indic. 112: 106166.

Deng, S., G. Yu and Y.P. Ting. 2005. Production of a bioflocculant by *Aspergillus parasiticus* and its application in dye removal. Colloids Surf. B Biointerfaces. 44: 179–186.

Deng, X., X. E. Yi and G. Liu. 2007. Cadmium removal from aqueous solution by gene modified *Escherichia coli* JM109. J. Hazard. Mater. 139(2): 340–344.

Desai, J. D. and I. M. Banat. 1997. Microbial production of surfactants and their commercial potential. Microbiol. Mol. Biol. Rev. 61: 47–64.

Diepens, N. J., M. R. Dimitrov, A. A. Koelmans and H. Smidt. 2015. Molecular assessment of bacterial community dynamics and functional end points during sediment bioaccumulation tests. Environ. Sci. Technol. 49: 13586–13595.

Ding, J., G. Lu and Y. Li. 2016. Interactive effects of selected pharmaceutical mixtures on bioaccumulation and biochemical status in crucian carp (*Carassius auratus*). Chemosphere 148: 21–31.

Diop, M., M. Howsam, C. Diop, J. F. Goossens, A. Diouf and R. Amara. 2016.Assessment of trace element contamination and bioaccumulation in algae (*Ulva lactuca*), mussels (*Perna perna*), shrimp (*Penaeus kerathurus*), and fish (*Mugil cephalus*, *Sarotherondon melanotheron*) along the Senegalese coast. Mar. Pollut. Bull. 103: 339–343.

Dixit, R., D. Malaviya, K. Pandiyan, U. B. Singh, A. Sahu and R. Shukla et al. 2015. Bioremediation of heavy metals from soil and aquatic environment: an overview of principles and criteria of fundamental processes. Sustainability 7(2): 2189–2212.

El-Metwally, S., O. M. Ouda and M. Helmy. 2014. Next Generation Sequencing Technologies and Challenges in Sequence Assembly. 1st Edn. New York, NY: Springer. doi: 10.1007/978-1-4939-0715-1.

Esposito, A., F. E. PAgnanelli and F. Veglio. 2002. pH-related equilibria models for biosorption in single metal systems. Chem. Eng. Sci. 57(3): 307–313.

Farag, S., S. Zaki, M. Elkady and D. Abd-El-Haleem. 2014. Production and characteristics of a bioflocculant produced by *pseudomonas* sp. Strain 38a. J. Adv. Biolo. 4: 286–295.

Flissi, A., Y. Dufresne, J. Michalik, L. Tonon, S. Janot, L. Noe et al. 2016. Norine, the knowledge base dedicated to non-ribosomal peptides, is now open to crowd sourcing. Nucleic Acids Res. 44: D1113–D1118.

Fomina, M. and G. M. Gadd. 2014. Biosorption: current perspectives on concept, definition and application. Bioresour. Technol. 160: 3–14.

Fourest, E. and J. C. Roux. 1992. Heavy metal adsorption by fungal mycelial by-products: influence of pH. Appl. Microbiol. Biotechnol. 37: 399–403.

Foyer, C. H. 1997. Oxygen metabolism and electron transport in photosynthesis. pp. 587–621. *In*: Scandalios, J. (ed.). The Molecular Biology of Free Radical Scavenging Systems. CSHL, New York.

Franzetti, A., I. Gandolfi, L. Fracchia, J. Van Hamme, P. Gkorezis and R. Marchant et al. 2014. Biosurfactant use in heavy metal removal from industrial effluents and contaminated sites. In Biosurfactants: Production and Utilization-Processes, Technologies, and Economics; Kosaric, N.F.V.S., Ed.; CRC Press: Boca Raton, FL, USA. 159: 361–369.

Gadd, G. M. and L. De Rome. 1988. Biosorption of copper by fungal melanine. Appl. Microbiol. Biotech. 29(6): 610–617.

Gadd, G. M. 2004. Microbial influence on metal mobility and application for bioremediation. Geoderma. 122(2-4): 109–119.

Gavrilescu, M. 2004. Removal of heavy metals from the environment by biosorption. Eng. Life Sci. 4(3): 219–232.

Ghanem, K. Z., M. Z. Mahran, M. M. Ramadan, H. Z. Ghanem, M. Fadel and M. H. Mahmoud. 2020. A comparative study on flavour components and therapeutic properties of unfermented and fermented defatted soybean meal extract. Sci. Rep. 10(1): 1–7.

Gnanamani, A., V. Kavitha, N. Radhakrishnan, G. S. Rajakumar, G. Sekaran and A. Mandal. 2010. Microbial products (biosurfactant and extracellular chromate reductase) of marine microorganism are the potential agents reduce the oxidative stress induced by toxic heavy metals. Colloids Surf. B Biointerfaces. 79: 334–339.

Gorfer, M., H. Persak, H. Berger, S. Brynda, D. Bandian and J. Strauss. 2009. Identification of heavy metal regulated genes from the root associated ascomycete *Cadophora finlandica* using a genomic microarray. Mycol. Res. 113: 1377–1388.

Grill, E., E. L. Winnacker and M. H. Zenk. 1985. Phytochelatins: the principal heavy-metal complexing peptides of higher plants. Science 230: 674–676.

Guibaud, G., E. van Hullebusch, F. Bordas, P. d'Abzac and E. Joussein. 2009. Sorption of cd (ii) and pb (ii) by exopolymeric substances (eps) extracted from activated sludges and pure bacterial strains: Modeling of the metal/ligand ratio effect and role of the mineral fraction. Bioresour. Technol. 100: 2959–2968.

Guo, J. 2015. Characteristics and mechanisms of cu (ii) sorption from aqueous solution by using bioflocculant mbfr10543. Appl. Microbiol. Biotechnol. 99: 229–240.

Hammaini, A. 2003. Simultaneous uptake of metals by activated sludge. Miner. Eng. 16: 723–729.

He, Z., T. J. Gentry, C. W. Schadt, L. Wu, J. Liebich, S. C. Chong et al. 2007. GeoChip: a comprehensive microarray for investigating biogeochemical, ecological and environmental processes. ISME J. 1: 67–77.

Henkel, M., M. Geissler, F. Weggenmann and R. Hausmann. 2017. Production of microbial biosurfactants: Status quo of rhamnolipid and surfactin towards large-scale production. Biotechnol. J. 12: 1600561.

Hong, Y. J., W. Liao, Z. F. Yan, Y. C. Bai, C. L. Feng, Z. X. Xu et al. 2020. Progress in the research of the toxicity effect mechanisms of heavy metals on freshwater organisms and their water quality criteria in China. J. Chem. 2020: 1–12.

Huang, C., C. Huang and A. L. Morehart. 1990. The removal of copper from dilute aqueous solutions by *Saccharomyces cerevisiae*. Water Res. 24: 433–439.

Index, E. P. 2018. EPI report. Yale University. https://epi.envirocenter.yale.edu/2018-epi-report/ introduction. Accessed 17 Oct 2018.

Inouhe, M. 2005. Phytochelatins. Braz. J. Plant Physiol. 17(1): 650–678.

Italiano, F., A. Buccolieri, L. Giotta, A. Agostiano, L. Valli, F. Milano et al. 2009. Response of the carotenoid less mutant *Rhodobacter sphaeroides* growing cells to cobalt and nickel exposure. Int. Biodeterior. Biodegradation. 63: 948–957.

Jackson, S. A., E. Borchert, F. O'Gara and A. D. W. Dobson. 2015. Metagenomics for the discovery of novel biosurfactants of environmental interest from marine ecosystems. Curr. Opin. Biotechnol. 33: 176–182.

Jiang, C. Y., X. F. Sheng, M. Qian and Q. Wang. 2008. Isolation and characterization of heavy metal resistant Burkholderia species from heavy metal contaminated paddy field soil and its potential in promoting plant growth and heavy metal accumulation in metal polluted soil. Chemosphere. 72: 157–164.

Jiang, Z. B., H. T. Song, N. Gupta, L. X. Ma and Z. B. Wu. 2007. Cell surface display of functionally active lipases from *Yarrowia lipolytica* in *Pichia pastoris*. Protein Expr. Purif. 56: 35–39.

Juwarkar, A. A., A. Nair, K. V. Dubey, S. Singh and S. Devotta. 2007. Biosurfactant technology for remediation of cadmium and lead contaminated soils. Chemosphere. 68: 1996–2002.

Kanmani, P., J. Aravind and D. Preston. 2012. Remediation of chromium contaminants using bacteria. Int. J. Environ Sci. and Tech. 9: 183–193.

Kim, S. K., B. S. Lee, D. B. Wilson and E. K. Kim. 2005. Selective cadmium accumulation using recombinant *E. coli*. J. Biosci. Bioeng. 99(2): 109–114.

Kiran, B. and A. Kaushik. 2008. Chromium binding capacity of *Lyngbya putealis* exopolysaccharides. Biochem. Eng. J. 38: 47–54.

Kitamoto, D., T. Morita, T. Fukuoka, M. A. Konishi and T. Imura. 2009. Self-assembling properties of glycolipid biosurfactants and their potential applications. Curr. Opin. Colloid Interface Sci. 14: 315–328.

Kobayashi, I., S. Fujiwara, H. Saegusa, M. Inohe, H. Metsumoto and M. Tsuzuki. 2006. Relief of arsenate toxicity by Cd-stimulated phytochelatin synthesis in the green alga *Chlamydomonas reinhardtii*. Mar. Biotechnol. 8: 94–101.

Koedam, N., E. Wittouck, A. Gaballa, A. Gillis, M. Hofte and P. Cornelis. 1994. Detection and differentiation of microbial siderophores by isoelectric focusing and chrome azurol S overlay. Biometals. 7: 287–291.

Kotrba, P., L. Doleckova, V. de Lorenzo and T. Ruml. 1999. Enhanced bioaccumulation of heavy metal ions by bacterial cells due to surface display of short metal binding peptides. Appl. Environ. Microbiol. 65: 1092–1098.

Kuroda, K. and M. Ueda. 2003. Bioadsorption of cadmium ion by cell surface-engineered yeasts displaying metallothionein and hexa-His. Appl. Microbiol. Biotechnol. 63: 182–186.

Kuroda, K. and M. Ueda. 2010. Engineering of microorganisms towards recovery of rare metal ions: Mini Review. Appl. Microbiol. Biotechnol. 87: 53–60.

Kuroda, K. and M. Ueda. 2011. Yeast biosorption and recycling of metal ions by cell surface engineering. Microbial Biosorption of Metals, Springer. 10: 235–247.

Kuyucak, N. and B. Volesky. 1988. Desorption of cobalt-laden algal biosorbent. Biotechnol. Bioeng. 33: 815–822.

Lane, N. 2015. The Unseen World: Reflections on Leeuwenhoek. 1677. Concerning Little Animal. Philos. Trans. R. Soc. of London. 370(1666): 1–10.

Lee, C. S., M. F. Chong, J. Robinson and E. Binner. 2014. A review on development and application of plant-based bioflocculants and grafted bioflocculants. Ind. Eng. Chem. Res. 53: 18357–18369.

Leung, P. T. Y., J. C. H. Ip, S. S. T. Mak, J. W. Qiu, P. K. S. Lam, C. K. C. Wong et al. 2014. *De novo* transcriptome analysis of *Perna viridis* highlights tissue-specific patterns for environmental studies. BMC Genomics. 15: 804.

Li, Q., Z. Yu, X. Shao, J. He and L. Li. 2009. Improved phosphate biosorption by bacterial surface display of phosphate-binding protein utilizing ice nucleation protein. FEMS Microbiol. Lett. 299: 44–52.

Li, Q., J. Li, L. Jiang, Y. Sun, C. Luo and G. Zhang. 2021. Diversity and structure of phenanthrene degrading bacterial communities associated with fungal bioremediation in petroleum contaminated soil. J. Hazard. Mater. 403: 123895.

Li, W. W. and H. Q. Yu. 2014. Insight into the roles of microbial extracellular polymer substances in metal biosorption. Bioresour. Technol. 160: 15–23.

Lin, J. and C. Harichund. 2011. Isolation and characterization of heavy metal removing bacterial bioflocculants. Afr. J.Microbiol. Res. 5: 599–607.

Machado, M. D., E. V. Soares, M. V. M. Helena and H. M. V. M. Soares. 2010. Removal of heavy metals using a brewer's yeast strain of *Saccharomyces cerevisiae*: Chemical speciation as a tool in the prediction and improving of treatment efficiency of real electroplating effluents. J. Hazard. Mat. 180: 347–353.

Madigan, M. and J. Martinko (eds.). 2006. Brock Biology of Microorganisms (13th ed.). Pearson Education. p. 1096. ISBN 0-321-73551-X.

Maier, R. M. and G. Soberon-Chavez. 2000. *Pseudomonas aeruginosa* rhamnolipids: biosynthesis and potential applications. Appl. Microbiol. Biotechnol. 54: 625–633.

Maier, R. M., I. L. Pepper and C. P. Gerba. 2000. Environmental Microbiology. Houston, TX: Gulf Professional Publishing.

Martins, S. C. S., C. M. Martins and S. T. Santaella. 2013. Immobilization of microbial cells: a promising tool for treatment of toxic pollutants in industrial wastewater. Afr. J. Biotechnol. 12: 4412–4418.

Mendoza-Cozatl, D., H. L. Tavera, A. H. Navarro and R. M. Sanchez. 2005. Sulfur assimilation and glutathione metabolism under cadmium stress in yeast, protists and plants. FEMS Microbiol. 29(4): 653–671.

Mishra, A. and A. Malik. 2013. Recent advances in microbial metal bioaccumulation. Crit. Rev. Environ. Sci. Technol. 43: 1162–1222.

Muller, M. M., J. H. Kugler, M. Henkel, M. Gerlitzki, B. Hörmann, M. Pohnlein et al. 2012. Rhamnolipids–next generation surfactants? J. Biotechnol. 162: 366–380.

Mulligan, C. N., R. N. Yong and B. F. Gibbs. 2001. Heavy metal removal from sediments by biosurfactants. J. Hazard. Mater. 85: 111–125.

Mulligan, C. N. 2005. Environmental applications for biosurfactants. Environ. Pollut. 133: 183–198.

Muraleedharan, T. R. and C. Venkobachar. 1990. Mechanism of biosorption of copper (II) by *Ganoderma lucidum*. Biotechnol. Bioeng. 35: 320–325.

Nishitani, T., M. Shimada, K. Kuroda and M. Ueda. 2010. Molecular design of yeast cell surface for adsorption and recovery of molybdenum, one of rare metals. Appl. Microbiol. Biotechnol. 86: 641–648.

Nkwunonwo, U. C., P. O. Odika and N. I. Onyia. 2020. A review of the health implications of heavy metals in food chain in Nigeria. Sci. World J. 2020: 1–11.

Nourbakhsh, M., Y. Sag, D. Ozer, Z. Aksu, T. Kustal and A. Caglar. 1994. A comparative study of various biosorbents for removal of chromium (VI) ions from industrial waste waters. Process Biochem. 29: 1–5.

Okaiyeto, K., U. U. Nwodo, S. A. Okoli, L. V. Mabinya and A. I. Okoh. 2016. Implications for public health demands alternatives to inorganic and synthetic flocculants: Bioflocculants as important candidates. Microbiology Open. 5: 177–211.

Oliveira, J. S., W. Araujo, A. I. Lopes Sales, A. de Brito Guerra, S. C. da Silva Araujo, A. T. R. de Vasconcelos et al. 2015. BioSurfDB: knowledge and algorithms to support biosurfactants and biodegradation studies. Database 2015: bav033.

Pacwa-Plociniczak, M., G. A. Plaza, Z. Piotrowska-Seget and S. S. Cameotra. 2011. Environmental applications of biosurfactants: Recent advances. Int. J. Mol. Sci. 12: 633–654.

Pagnanelli, F., L. Toro and F. Veglio. 2002. Olive mild solid residues as heavy metal sorbent material: a preliminary study. Waste Management. 22: 901–907.

Parales, R. E. and J. D. Haddock. 2004. Biocatalytic degradation of pollutants. Curr. Opin. Biotechol. 15(4): 374–379.

Park, H. M., B. G. Kim, D. Chang, S. Malla, H. S. Joo, E. J. Kim et al. 2013. Genome-based cryptic gene discovery and functional identification of NRPS siderophore peptide in *Streptomyces peucetius*. Appl. Microbiol. Biotechnol. 97: 1213–1222.

Parrilli, E., R. Papa, M. L. Tutino and G. Sannia. 2010. Engineering of a psychrophilic bacterium for the bioremediation of aromatic compounds. Bioeng. Bugs. 1(3): 213–216.

Patel, J., Q. Zhang, R. M. L. McKay, R. Vincent and Z. Xu. 2010. Genetic engineering of *Caulobacter crescentus* for removal of cadmium from water. Appl. Biochem. Biotechnol. 160(1): 232–243.

Pathak, M., A. Devi, H. K. Sarma and B. Lal. 2014. Application of bioflocculating property of *Pseudomonas aeruginosa* strain iasst201 in treatment of oil-field formation water. J. Basic Microbiol. 54: 658–669.

Paul, D., G. Pandey, J. Pandey and R. K. Jain. 2005. Accessing microbial diversity for bioremediation and environmental restoration. Trends Biotechnol. 23(3): 135–142.

Penninckx, M. A. 2000. A short review on the role of glutathione in the response of yeasts to nutritional, environment and oxidative stresses. Enzyme Microbiol. Technol. 26: 737–742.

Penninckx, M. J. 2002. An overview on glutathione in Saccharomyces versus nonconventional yeasts. FEMS Yeast Res. 2: 295–305.

Preveral, S., E. Ansoborlo, S. Mari, A. Vavasseur and C. Forestier. 2006. Metal(loid)s and radionuclides cytotoxicity in *Saccharomyces cerevisiae*. Role of YCF1, glutathione and effect of buthionine sulfoximine. Biochimie. 88(11): 1651–1663.

Reznik, G. O., P. Vishwanath, M. A. Pynn, J. M. Sitnik, J. J. Todd, J. Wu et al. 2010. Use of sustainable chemistry to produce an acyl amino acid surfactant. Appl. Microbiol. Biotechnol. 86: 1387–1397.

Roane, T. M. and I. L. Pepper. 2001. Environmental microbiolology. *In*: Roane, T. M. and Pepper, I. L. (ed.). Microorganisms and Metal Pollutants. Academic Press. 17: 403–423.

Roane, T. M., K. L. Josephson and I. L. Pepper. 2001. Dual-bioaugmentation strategy to enhance remediation of co-contaminated soil. Appl. Environ. Microbiol. 67(7): 3208–3215.

Ron, E. Z. and E. Rosenberg. 2001. Natural roles of biosurfactants. Environ. Microbiol. 3: 229–236.

Rudakiya, D. M. and K. S. Pawar. 2013. Evaluation of remediation in heavy metal tolerance and removal by *Comamonas acidovorans* MTCC 3364. IOSR J. Environ. Sci. Toxicol. Food Technol. 5: 26–32.

Rudakiya, D. M., V. Iyer, D. Shah, A. Gupte and K. Nath. 2018. Biosorption potential of Phanerochaete chrysosporium for arsenic, cadmium, and chromium removal from aqueous solutions. Global Chall. 2(12): 1800064.

Rufino, R. D., J. M. Luna, G. M. Campos-Takaki, S. R. M. Ferreira and L. A. Sarubbo. 2012. Application of the biosurfactant produced by *Candida lipolytica* in the remediation of heavy metals. Chem. Eng. Trans. 27: 61–66.

Ruiz, O. N. and H. Daniell. 2009. Genetic engineering to enhance mercury phytoremediation. Curr. Opin. Biotechnol. 20(2): 213–219.

Sajayan, A., G. Seghal Kiran, S. Priyadharshini, N. Poulose and J. Selvin. 2017. Revealing the ability of a novel polysaccharide bioflocculant in bioremediation of heavy metals sensed in a vibrio bioluminescence reporter assay. Environ. Pollut. 228: 118–127.

Salehizadeh, H. and S. A. Shojaosadati. 2001. Extracellular biopolymeric flocculants: Recent trends and biotechnological importance. Biotechnol. Adv. 19: 371–385.

Salehizadeh, H. and N. Yan. 2014. Recent advances in extracellular biopolymer flocculants. Biotechnol. Adv. 32: 1506–1522.

Samuelson, H., M. Wernerus and S. Svedberg Stahl. 2000. *Staphylococcal* surface display of metal-binding polyhistidyl peptides, Appl. Environ. Microbiol. 66: 1243–1248.

Sarubbo, L., R. Jr. Rocha, J. Luna, R. Rufino, V. Santos and I. Banat. 2015. Some aspects of heavy metals contamination remediation and role of biosurfactants. Chem. Ecol. 31: 707–723.

Sathiyanarayanan, G., S. K. Bhatia, H. J. Kim, J. M. Jeon, Y. G. Kim, S. H. Park et al. 2016. Metal removal and reduction potential of an exopolysaccharide produced by arctic psychrotrophic bacterium *Pseudomonas* sp. Pamc 28620. RSC Adv. 6: 96870–96881.

Satyanarayana, T., S. K. Deshmukh and M. V. Deshpande (eds.). 2019. Advancing frontiers in mycology & mycotechnology: basic and applied aspects of fungi. Springer Nature, Singapore, pp. 1–433.

Schmitt, M., P. Schwanewilm, J. Ludwig and L. H. Fraté. 2006. Use of PMA1 as a housekeeping Biomarker for Assessment of Toxicant-induced Stress in *Saccharomyces cerevisiae*. Appl. Environ. Microbiol. 72(2): 1515–1522.

Searle, L. J., G. Meric, I. Porcelli, S. K. Sheppard and S. Lucchini. 2015. Variation in siderophore biosynthetic gene distribution and production across environmental and faecal populations of *Escherichia coli*. PLoS ONE. 10: e0117906.

Shah. D., D. M. Rudakiya, V. Iyer and A. Gupte. 2018. Simultaneous removal of hazardous contaminants using polyvinyl alcohol coated *Phanerochaete chrysosporium*. Int. J. Agri. Environ. Biotechnol. 11(2): 235–241.

Sharma, P. and S. Pandey. 2014. Status of phytoremediation in world scenario. Int. J. Environ. Bioremediat. Biodegrad. 2: 178–191.

Sharma, P., A. K. Pandey, A. Udayan and S. Kumar. 2021a. Role of microbial community and metal-binding proteins in phytoremediation of heavy metals from industrial wastewater. Bioresour. Technol. 326: 124750.

Sharma, P., D. Purchase and R. Chandra. 2021b. Residual pollutants in treated pulp paper mill wastewater and their phytotoxicity and cytotoxicity in *Allium cepa*. Environ. Geochem. Health. 1–22.

Shi, B., Z. Huang, X. Xiang, M. Huang, W. X. Wang and C. Ke. 2015. Transcriptome analysis of the key role of GAT2 gene in the hyper-accumulation of copper in the oyster *Crassostrea angulata*. Sci. Rep. 5: 17751.

Singh, A. K. and S. S. Cameotra. 2013. Efficiency of lipopeptide biosurfactants in removal of petroleum hydrocarbons and heavy metals from contaminated soil. Environ. Sci. Pollut. Res. 20: 7367–7376.

Singh, J. S., P. C. Abhilash, H. B. Singh, R. P. Singh and D. P. Singh. 2011. Genetically engineered bacteria: an emerging tool for environmental remediation and future research perspectives. Gene. 480(1-2): 1–9.

Singh, P. and S. S. Cameotra. 2004. Enhancement of metal bioremediation by use of microbial surfactants. Biochem. Biophys. Res. Commun. 319: 291–297.

Singh, R., P. Singh and R. Sharma. 2014. Microorganism as a tool of bioremediation technology for cleaning environment: a review. Proc. Int. Acad. Ecol. Environ. Sci. 4(1): 1–6.

Smith, M. C., E. R. Sumner and S. V. Avery. 2007. Gluthatione and Gts1p drive beneficial variability in the cadmium resistances of individual yeast cells. Mol. Microbiol. 66(3): 699–712.

Somdee, T., B. Thathong and A. Somdee. 2015. The removal of cyanobacterial hepatotoxin [Dha(7)] microcystin-LR via bioaccumulation in water lettuce (*Pistia stratiotes* L.). Bull. Environ. Contam. Toxicol. doi: 10.1007/s00128-015-1715-1.

Sriram, M. I., K. Kalishwaralal, V. Deepak, R. Gracerosepat, K. Srisakthi and S. Gurunathan. 2011. Biofilm inhibition and antimicrobial action of lipopeptide biosurfactant produced by heavy metal tolerant strain bacillus cereus nk1. Colloids Surf. B Biointerfaces. 85: 174–181.

Stadnicka, J., K. Schirmer and R. Ashauer. 2012. Predicting concentrations of organic chemicals in fish by using toxicokinetic models. Environ. Sci. Technol. 46: 3273–3280.

Stone, E. L. and V. R. Timmer. 1975. On the copper content of some northern conifers. Can. J. Bot. 53: 1453–1456.

Strong, P. J. and J. E. Burgess. 2008. Treatment methods for wine related ad distillery wastewaters: a review. Bioremediat. J. 12(2): 70–87.

Tabak, H. H., P. Lens, E. D. V. Hullebusch and W. Dejonghe. 2005. Developments in bioremediation of soil and sediments polluted with metals and radionuclides-1. Microbiolal processes and mechanisms affecting bioremediation of metal contamination and influencing meal toxicity. Rev. Environ. Sci. Biotechnol. 4: 115–156.

Tan, B., C. Ng, J. P. Nshimyimana, L. L. Loh, K. Y. H. Gin and J. R. Thompson. 2015. Next-generation sequencing (NGS) for assessment of microbial water quality: current progress, challenges, and future opportunities. Front. Microbiol. 6: 1027.

Thirumoorthy, N., K. T. M. Kumar, A. S. Sundar, L. Panayappan and M. Chatterjee. 2007. Metallothionein: An overview. World J. Gastroenterol. 13(7): 993–996.

Tiedje, J. M. 1993. Bioremediation from an ecological perspective. *In situ* bioremediation: When does it work. 110–120.

Tran, V. S., H. H. Ngo, W. Guo, J. Zhang, S. Liang, C. Ton-That et al. 2015. Typical low cost biosorbents for adsorptive removal of specific organic pollutants from water. Bioresour. Technol. 182: 353–363.

USEPA. 2016. United States Environmental Protection Agency. https://www3.epa.gov/(Accessed May 2016).

Valls, M. and V. Lorenzo. 2002. Explointing the genetic and biochemical capacities of bacteria for the remediation of heavy metal pollution. FEMS Microbiol. Rev. 26(4): 327–338.

Van Nostrand, J. D., W. M. Wu, L. Wu, Y. Deng, J. Carley, S. Carroll et al. 2009. GeoChip-based analysis of functional microbial communities during the reoxidation of a bioreduced uranium-contaminated aquifer. Environ. Microbiol. 11: 2611–2626.

Vardhan, K. H., P. S. Kumar and R. C. Panda. 2019. A review on heavy metal pollution, toxicity and remedial measures: current trends and future perspectives. J. Mol. Liq. 111: 197.

Velásquez, L. and J. Dussan. 2009. Biosorption and bioaccumulation of heavy metals on dead and living biomass of *Bacillus sphaericus*. J. Hazard. Mater. 167: 713–716.

Vidali, M. 2001. Bioremediation: an overview. Pure and Applied Chemistry 73(7): 1163–1172.

Volesky, B. 2001. Detoxification of metal-bearing effluents: biosorption for the next century. Hydrometallurgy 59(2-3): 203–216.

Volesky, B. and Z. R. Holan. 1995. Biosorption of heavy metal. Biotechnol. Prog. 11: 235–250.

Vullo, D. L., H. M. Ceretti, M. A. Daniel, S. A. M. Ramirez and A. Zalts. 2008. Cadmium, zinc and copper biosorption mediated by *Pseudomonas veronii* 2E. Bioresour. Technol. 99(13): 5574–5581.

Wang, J. and C. Chen. 2006. Biosorption of heavy metal by *Saccharomyces cerevisiae*. Biotechnol. Adv. 24: 427–451.

Wang, J. and C. Chen. 2008. Biosorbents for heavy metals removal and their future. Biotechnol. Adv. 27: 195–226.

Wang, L., F. Ma, Y. Qu, D. Sun, A. Li, J. Guo and B. Yu. 2011. Characterization of a compound bioflocculant produced by mixed culture of *Rhizobium radiobacter* f2 and *Bacillus sphaeicus* f6. World J. Microbiol. Biotechnol. 27: 2559–2565.

Wang, S. and C. N. Mulligan. 2009. Rhamnolipid biosurfactant-enhanced soil flushing for the removal of arsenic and heavy metals from mine tailings. Process Biochem. 44: 296–301.

Wernerus, H., J. Lehtio, T. Teeri, P. A. Nygren and S. Stahl. 2001. Generation of metal-binding staphylococci through surface display of combinatorially engineered cellulose-binding domains. Appl. Environ. Microbiol. 67: 4678–4684.

Wu, C. H., T. K. Wood, A. Mulchandani and W. Chen. 2006. Engineering Plant-microbe symbiosis for rhizo remediation of heavy metal. Appl. Environ. Microbiol. 72(2): 1129–1134.

Wu, J. Y. and H. F. Ye. 2007. Characterization and flocculating properties of an extracellular biopolymer produced from a *Bacillus subtilis* dyu1 isolate. Process Biochem. 42: 1114–1123.

Zhai, L. F., M. Sun, W. Song and G. Wang. 2012. An integrated approach to optimize the conditioning chemicals for enhanced sludge conditioning in a pilot-scale sludge dewatering process. Bioresour. Technol. 121: 161–168.

Zhang, Z., C. Wu, Y. Wu and C. Hu. 2014. Comparison of coagulation performance and floc properties of a novel zirconium-glycine complex coagulant with traditional coagulants. Environ. Sci. Pollut. Res. 21: 6632–6639.

Zhou, Q., N. Yang, Y. Li, B. Ren, X. Ding, H. Bian et al. 2020. Total concentrations and sources of heavy metal pollution in global river and lake water bodies from 1972 to 2017. Glob. Ecol. Conserv. 22: e00925.

Zhou, W., Y. Zhang, X. Ding, Y. Liu, F. Shen, X. Zhang et al. 2012. Magnetotactic bacteria: promising biosorbents for heavy metals. Appl. Microbiol. Biotechnol. 95: 1097–1104.

CHAPTER 7

Bioremediation Potential of *Trichoderma* species for Metal(loid)s

Mohd. Kashif Kidwai,[1] *Anju Malik,*[1,*] *Sanju Bala Dhull,*[2] *Pawan Kumar Rose*[1] and *Vinod Kumar Garg*[3]

7.1 Introduction

Environmental pollution is a global issue which demands interventions at national as well as international platforms. Human-driven activities are the root cause of different types of pollution affecting the overall ecological setup in the ecosystems. More than a billion people are deprived of access to potable water and more than 150 million human population is bound to consume polluted water for their daily activities (Rahman and Singh 2020, Paria et al. 2021). It has been reported that ~ 22 million hectares land is polluted due to various anthropogenic activities such as urbanization, industrialization, intensive agriculture, etc. Various types of environmental pollution are responsible of causing economical and ecological losses in both developed and developing nations. Pollution is reported as the prime cause of human mortality at the global level accounting for approximately sixteen percent of the total deaths (Akhtar and Mannan 2020). Various environmental issues that emerged in the recent past due to pollution include drinking water quality deterioration, poor ambient air quality, contamination in food commodities, habitat degradation, excessive consumption of natural resources, extinction of different organisms, municipal and industrial waste generation, excessive use of agrochemicals, altering soil properties, yield and nutritive quality of the agricultural produce, discharge of heavy metals causing toxicological diseases thereby affecting human health, alteration in ecosystem services, human-induced global warming, climate change, etc. All these issues are a challenge for the survival of present and future generations of humans (Anand et al. 2006, Bishnoi et al. 2007, Chauhan and Rai 2009, Dixit et al. 2011, Bano et al. 2018, Sall et al. 2020, Tarekegn et al. 2020, Aibeche et al. 2022).

Heavy metals have been of vital importance for humans since ages. Distinct intricate processes were discovered several thousand years ago for the elimination of metals. Significant metabolic role is played by nickel, zinc, copper, iron, etc., in the different organisms and are classified as essential trace elements (Siddiquee et al. 2013, Kumar and Dwivedi 2021). Heavy metals play a very important role in different industrial processes and contribute to economic development all

[1] Department of Energy and Environmental Sciences, Chaudhary Devilal University, Sirsa, Haryana, India.
[2] Department of Food Science and Technology, Chaudhary Devilal University, Sirsa, Haryana, India.
[3] Department of Environmental Science and Technology, Central University of Punjab, Bathinda, Punjab, India.
* Corresponding author: anjumalik27@yahoo.com, anjumalik@cdlu.ac.in

over the world. However, some environmental issues and challenges are also associated with the exposure of heavy metals all over the world. Industrialization, urbanization, intensive agricultural practices, etc., are some of the human-induced processes held responsible for the indiscriminate release of heavy metals. Various health-related threats posed by heavy metals to humans are given in Fig. 7.1 (White et al. 1997, Arriagada et al. 2009, Ali et al. 2019, Dusengemungu et al. 2020, Rahman and Singh 2020, Sall et al. 2020, Zhang et al. 2020, Haider et al. 2021). According to Rahman and Singh (2020), more than 60 million people all over the globe are experiencing various toxicities and hazards due to exposure to heavy metals such as cadmium, chromium, mercury, lead, etc. The persistent nature of heavy metals is one of the major reasons for their accumulation in the environment (Zhang et al. 2020, Priyadarshini et al. 2021). Each of the metal(loid)s possesses a natural capacity to bind with biomolecules such as proteins and inhibit enzymatic activities affecting cellular biochemical processes and functional ability of cell membrane in different organisms (Siddiquee et al. 2013). Leaching is one of the prime processes for the mobilization of different heavy metals in different ecosystems. Untreated industrial discharge of electroplating, electronics, metallurgical, pharmaceutical, agrochemical, industries, acid mine drainage, untreated sewage discharge, fly ash from thermal power plants, etc., are the anthropogenic activities which contribute heavy metals into the environment (Briffa et al. 2020). Various national and international organizations have given standards for safe disposal of heavy metals to regulate the release of heavy metals in wastewaters entering in receiving water bodies. Heavy metals are responsible for inducing different toxicities in humans and other organisms even at low quantity and possess non-biodegradable properties due to persistence in nature (Anand et al. 2006, Anju 2017, Sall et al. 2020, Aibeche et al. 2022, Kidwai and Dhull 2021).

Metals are of two types, i.e., cationic metals (positively charged metals) and anionic metals (negatively charged metallic elements). Metals such as nickel, mercury, chromium, cadmium, lead, copper, zinc, etc., are cationic metals and arsenic is among the common anionic metalloids (Hrynkiewicz and Baum 2014, Rahman and Singh 2020). The efficient removal and management of heavy metals from different ecosystems is a challenge globally. The ecofriendly management of heavy metals induced pollution by applying processes like bioremediation provides a cost-effective strategy as other conventional methods have been reported for their high economic cost (Say et al. 2001, Paria et al. 2021).

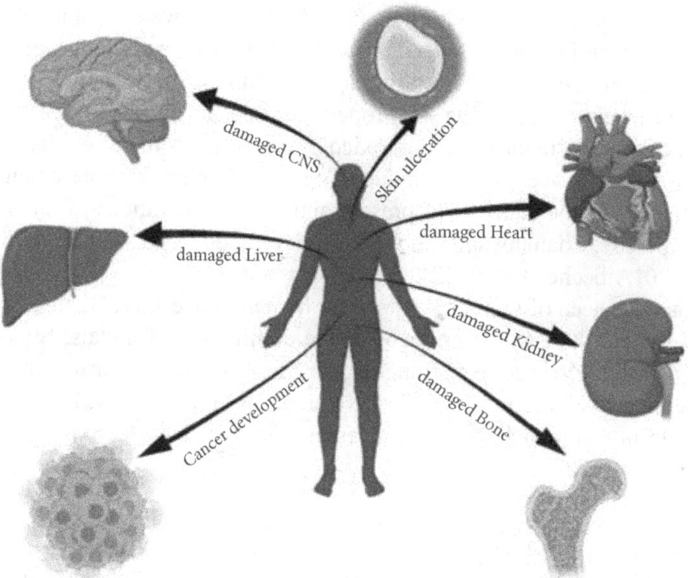

Fig. 7.1. Human health issues associated with heavy metals.

According to the International Agency for Research on Cancer (IARC), chromium, arsenic and cadmium have been classified as group I carcinogens, causing cancer in humans, whereas lead, cobalt, copper, zinc, etc., are non-carcinogenic to humans but cause disruption of enzymatic activities due to oxidative damage in human body (Zakaria et al. 2021). Various physic-chemical remediation methods such as photocatalysis method, membrane filtration method, magnetic separation, ion exchange method, flotation, coagulation, flocculation, hydrodynamic separation, etc., are applied for the remediation of these metal(loid)s, but due to constraints such as high costs and poor efficiency, scientists and researchers are exploring the bioremediation strategies for the sustainable remediation of toxic metal(loid)s.

7.2 Bioremediation

Microorganisms are omnipresent and provide various ecological services including remediation of different pollutants, i.e., metal(loid)s, pesticides, dyes, organic chemicals, etc. Therefore, microorganisms have been applied in various biotechnological strategies for detoxification of polluted environment (Bano et al. 2018, Rahman and Singh 2020). Microorganisms are applied in various *in situ* and *ex situ* bioremediation strategies such as bioleaching, biosorption, biotransformation, bioaccumulation, etc. Microbe-assisted phytoremediation, technically known as phytobial remediation, is a successful bioremediation strategy involving association of both plants and microorganisms. Both rhizospheric and endophytic microorganisms support the process of phytobial remediation in soils, polluted due to metal(loid)s, by positively supporting the plant growth alongwith the induction of enzymes and secondary metabolites for the management of metal(loid)s induced oxidative stress in plants (Hrynkiewicz and Baum 2014).

Microorganisms possess the capacity to accumulate metal ions and thus play an important role in bioremediation. Microorganism induced bioremediation process obtains required nutrients and energy from metal(loid)s by converting them into soluble forms, which further gets removed from the ecosystem. Microbially induced bioremediation for the treatment of polluted soils is a cost-effective alternative in comparison to conventional remediation processes such as adsorption, ion exchange, flocculation, etc. (Briffa et al. 2020). Microorganisms employ various mechanisms such as oxidation, reduction, bio-precipitation, bioleaching, bio-volatilization, bioaccumulation, and biosorption for bioremediation (Rahman and Singh 2020, Kumar et al. 2021).

7.3 Mycoremediation

Bioremediation of toxic pollutants by fungi is known as mycoremediation. Mycoremediation is an eco-friendly, natural, reusable, less time consuming and cost-effective strategy to intervene the challenges posed by various pollutants in terrestrial and aquatic ecosystems. Fungi are an important class of microorganisms involved in providing various ecosystem services. They play significant roles in various ecological processes such as nutrient recycling, decomposition of organic matter, degradation and solubulization of diverse pollutants, etc. (Malik 2004, Zafar et al. 2007, Siddiquee et al. 2015, Sen et al. 2016, Oladipo et al. 2018, Hu et al. 2019, Lin et al. 2020, Rose 2022). Various fungi are identified and applied in the bioremediation of various toxic pollutants including metal(loid)s (Bisht et al. 2014) as exhibited in Table 7.1. Fungi are special microorganisms that are reported to survive in harsh conditions viz. very high or very low temperatures, alteration of pH, scarcity of desired nutrients, presence of heavy metals in excess quantities, etc. Fungi have innate metal biosorption ability known as mycosorption and its biomass is known as mycosorbent. Mycosorption has fascinated researchers all over the world due to its high efficiency. Mycological adsorption strategies using living and non-living fungal biomass are applied for the removal of various pollutants including metal(loid)s. Fungi produce various extracellular enzymes which biometabolize the complex biomolecules into less toxic or eco-friendly products enabling their

Table 7.1. Fungal species used in bioremediation of diverse pollutants.

Sr. No.	Pollutant	Fungi	References
1.	Copper, Mercury, Arsenic, Nickle Cadmium, Selenium Lead	*Saccharomyces cerevisiae* *Mucor rouxii* *Aspergillus niger* *Trichoderma harzianum* *Aspergillus* species *Aspergillus penicillioides* *Fusarium* species *Penicillium* species *Phanerochaetechrysosporium* *Rhodotorulamucilaginosa* *Clavisporalusitaniae* *Trichoderma harzianum* *Trametes versicolor* *Pleurotussajur-caju* *Wickerhamomycesanomalus*	Price et al. 2001, Park et al. 2003, Yan and Viraraghavam 2003, Arica et al. 2004, Park et al. 2005, Arriagada et al. 2009, Puglisi et al. 2012, Siddiquee et al. 2015, Zhang et al. 2015, Lin et al. 2020, Aibeche et al. 2022, Paria et al. 2021
2.	Dyes	*Clitocybuladusenii* *Aspergillus flavus* *Aspergillus niger* *Penicillium.oxalicum* *Coriolus versicolor* *Trametes versicolor* *Phanerochaetechrysosporium* *Irpexlacteus* *Trichoderma viridi* *Trichoderma asperellum* *Trichoderma harzianum* *Trichoderma virens*	Wesenberg et al. 2002, Argumedo-delira et al. 2012, Chew and Ting 2015, Saroj et al. 2015, Sen et al. 2016, Argumedo-delira et al. 2021
3.	Pesticides	*Aspergillus terreus* *Penicillium citrinum* *Trichoderma harzianu* *Cunninghamella elegans* *Pleurotusostreatus*	Katayama and Matsumura 1993, Pothuluri et al. 1996, Purnomo et al. 2013, Oliveira et al. 2015
4.	Polycyclic Aromatical Hydrocarbon	*Trematophoma* species (UTMC-5003) *Aspergillus* species *Trichoderma viridi* *Trichoderma pseudokoningii* *Trichoderma longibrachiatum* *Pleurotusostreatus* *Ganoderma lucidum* *Irpexlacteus*	Baldrian et al. 2000, Cjathaml et al. 2008, Ye et al. 2011, Argumedo-delira et al. 2012, Bhattacharya et al. 2014, Moghimi et al. 2017, Agrawalet al. 2018
5.	Antibiotics	*Trametes versicolor* *Phanerochaetechrysosporium* *Pleurotusostreatus*	Marco-Urrea et al. 2010, Singh et al. 2017

viability for detoxification of variety of pollutants. Due to the process of evolution, diverse fungal species have developed natural defence mechanisms such as bioaccumulation, biosorption, chelation of metals, association with plants, etc., to survive and develop heavy metal resistance (Gupta et al. 2017, Priyadarshini et al. 2021). Basidiomycetes, Ascomycetes, Hyphomycetes classes of fungi are widely investigated for the remediation of different pollutants including metal(loids)s as shown in Table 7.1 (Karn et al. 2021). Among all types of fungi, filamentous fungi are more efficient biosorbent than others fungi. The mycelial surface of fungi possesses various functional groups viz. hydroxyl, thiol, carboxyl, phosphate, etc., which enable the negatively charged mycelial surface

Fig. 7.2. *Trichoderma* species reported for the resistance and bioremediation of metal(loid)s.

for biosorption (Arica et al. 2004, Chew and Ting 2015). Fungal agents are equipped with diverse mechanisms and properties such as enzymatic diversity, production of metallothioneins, resistance to different abiotic factors such as pH, salt, metal(loid)s, etc., that enable them to be applied in bioremediation of metal(loid)s (Gharieb and Gadd 2004, Deshmukh et al. 2016, Garcia-Hernandez et al. 2017, Yaghoubian et al. 2019, Wang et al. 2021).

The bioremediation of metal(loid)s by fungi is conducted by employing various strategies such as (i) absorption of metallic ions on the surface of living and dead fungal biomass, (ii) uptake of metal ions by living fungal cells, and (iii) the transformation of the chemical state of metal ions within a living fungal cell.

Mycosorption include metabolism independent and metabolism dependent processes. Metabolism independent process uses live or dead fungal biomass, whereas metabolism dependent involves the production of extracellular metabolites for the chemical transformation of metal(loid) s. Fungi are reported to remove both soluble and insoluble types of metal(loid)s. For solubilization of metal(loid)s, fungi such as *Trichoderma* species, *Aspergillus* species, etc., produce diverse metabolites, organic acids, phosphatases, etc. White et al. (1997) discussed the application of heterotrophic fungal agents as potential bioremediators due to their tolerance for wide range of pH along with the ability to produce organic acids for solubilization of different metal(loid)s.

7.4 Trichoderma

Fungal genus *Trichoderma*, a filamentous fungi mainly found in soil, belongs to class sordariomycetes, Order Hypocreales, phylum Ascomycota and family Hypocreaceac, and is widely distributed in various climatic zones enabling them to be omnipresent in nature (Samuels 1996, Kredics et al. 2001, Kubicek et al. 2003, Harman et al. 2004, Sadfi-Zouaoui et al. 2009, Argumedo-delira et al. 2012, Kredics et al. 2012, Błaszczyk et al. 2014, Nongmaithem et al. 2016, Kidwai and Nehra 2017, Ayad et al. 2018, Copete-Pertuz et al. 2019, Rahman and Singh 2020, Ram et al. 2020, Argumedo-delira et al. 2021, Umar 2021). *Trichoderma* species and its distribution is influenced by abiotic and biotic factors which include microclimate, substrates, competition, interaction with other organisms, coexistence, etc. (Hoyos-Carvajal et al. 2009). However, due to its adaptability to survive in different environmental conditions, it is reported to occur in various ecosystems including aquatic and terrestrial ecosystems. Diverse species of genus *Trichoderma* have been isolated from marine ecosystems, as endophytes, rhizhospheric zone, etc. It is pertinent to mention that different species of genus *Trichoderma* like *T. viride*, *T. polysporum*, etc., are reported to thrive in low temperature environment as they are found in temperate regions, whereas species like *T. harzianum* is commonly reported to occur in warmer environment near tropical regions. *T. pseudokoningii* occurs in environmental conditions having high humid environment (Kubicek et al. 2003, Kredics et al. 2012). *Trichoderma* species are significant fungi widely investigated and reported for providing the beneficial impact on humans all over the globe. These unique filamentous fungi are reported for various applications. They are popularly used as biological fungicides against various plant pathogens; they have also been reported for production of industrially important enzymes, production of biofuels, secondary metabolites of agricultural and clinical significance, production of specific proteins, etc. In the soil, *Trichoderma* species are used in the bioremediation of organic and inorganic pollutants including metal(loids)s (Kredics et al. 2001, Rigot and Matsumura 2002, Errasquin and Vazquez 2003, Kredics et al. 2012, Mastouri et al. 2012, Tripathi et al. 2013, Bisht et al. 2014, Malgorzata et al. 2014, Petrovic et al. 2014, Sen et al. 2016, Kidwai and Nehra 2017, Mohammadian et al. 2017, Oladipo et al. 2018, Maurya et al. 2019, Bisht et al. 2019, Dusengemungu et al. 2020, Tandon et al. 2020, Ram et al. 2020).

Trichoderma species derive the nutrients either as saprotrophs including necrotroph and mycotroph or as biotroph. *Trichoderma* species are efficient decomposer and are reported to generate compost of high quality. *Trichoderma* species are mainly isolated from rhizospheric region of plants (Tandon et al. 2020, Sarangi et al. 2021). Various innate activites of *Trichoderma* species have established them as potential biocontrol agent and plant growth promoter, along with the ability to produce specific enzymes utilized in various industries (Yedidia et al. 1999, Howell 2003, Benítez et al. 2004, Harman et al. 2004, Mastouri et al. 2012, Nongmaithem et al. 2016, Kidwai and Nehra 2017, Copete-Pertuz et al. 2019, Ram et al. 2020, Umar 2021). Even mere occurrence of some *Trichoderma* species in soil is considered as a positive ecological indicator for soil health (Harman et al. 2004).

7.5 Bioremediation Strategies Employed by *Trichoderma* species

Diverse species of *Trichoderma* as exhibited in Table 7.2 are reported to bioremediate different metal(loid)s present in different ecosystems by employing various strategies as exhibited in Fig. 7.3. Some of the *Trichoderma* species have resistance for different metal(loid)s which helps them in bioremediation of heavy metals and metalloids. For example, *Trichoderma viride*, *Trichoderma harzianum*, *Trichoderma virens*, *richoderma atroviridi*, *Trichoderma asperellum*, *Trichoderma reesei*, *Trichoderma inhamatum*, *Trichoderma gamsii*, *Trichoderma ghanense*, *Trichoderma koningii*, *Trichoderma koningiopsis*, *Trichoderma saturnisporum*, *Trichoderma lixii*, *Trichoderma aureoviride*, *Trichoderma brevicompactum*, etc., are reported to have the potential to

Table 7.2. *Trichoderma* species reported for bioremediation of different metal(loid)s.

Sr. No.	*Trichoderma* species	Metal(loid)	References
1.	*Trichoderma viride*	Chromium (IV)	Bishnoi et al. 2007, Sugasini and Rajagopal 2015, Migahed et al. 2017, Priyadarshini et al. 2021
		Copper	Anand et al. 2006
		Aluminum	White et al. 1997
		Cadmium	Ali et al. 2007, Sahu et al. 2012
		Lead	Ali et al. 2007, Sahu et al. 2012
2.	*Trichoderma harzianum*	Cadmium	Kredics et al. 2001, Adams et al. 2007, Freitas et al. 2011, Faedda et al. 2012, Fiorentino et al. 2013, Hoseinzadeh et al. 2017, Ayad et al. 2018, Haider et al. 2021
		Lead	Kredics et al. 2001, Adams et al. 2007, Siddiquee et al. 2013, Hoseinzadeh et al. 2017, Tansengco et al. 2018
		Nickle	Kredics et al. 2001, Sarkar et al. 2010, Siddiquee et al. 2013, Hoseinzadeh et al. 2017, Tansengco et al. 2018, Padua et al. 2021
		Iron	Kredics et al. 2001
		Copper	Kredics et al. 2001, Fiorentino et al. 2013, Siddiquee et al. 2013, Mohammadian et al. 2017, Tansengco et al. 2018
		Cobalt	Kredics et al. 2001, Mohammadian et al. 2017
		Chromium	Transengco et al. 2018
		Mercury	Puglisi et al. 2012
		Arsenic	Lynch and Moffat 2005, Arriagada et al. 2009
		Uranium	Akhtar et al. 2007, Akhtar et al. 2009
3.	*Trichoderma virens*	Cadmium	Dixit et al. 2011
		Nickle	Padua et al. 2021
		Copper	Siddiquee et al. 2013, Dusengemungu et al. 2020, Tansengco et al. 2018
		Chromium	Tansengco et al. 2018
		Nickle	Siddiquee et al. 2013, Tansengco et al. 2018
		Lead	Siddiquee et al. 2013
		Cobalt	Babu et al. 2014, Dusengemungu et al. 2020
4.	*Trichoderma atroviridi*	Copper,	Lopez and Vazquez 2003, Yazdani et al. 2009
		Cadmium	Lopez and Vazquez 2003, Cao et al. 2008
		Nickle	Cao et al. 2008
		Zinc	Kacprzak and Malina 2005
		Barium	Kacprzak and Malina 2005
		Iron	Kacprzak and Malina 2005
5.	*Trichoderma asperellum*	Arsenic	Zeng et al. 2010, Su et al. 2017
		Lead	Hoseinzadeh et al. 2017, Maurya et al. 2019, Kumar and Dwivedi 2021
		Copper	Tan and Ting 2012
		Cadmium	Hoseinzadeh et al. 2017, Maurya et al. 2019, Kumar and Dwivedi 2021
		Nickle	Hoseinzadeh et al. 2017
		Chromium	Kumar and Dwivedi 2021
		Aluminum	Kumar and Dwivedi 2021

Table 7.2 contd. ...

...Table 7.2 contd.

Sr. No.	***Trichoderma* species**	**Metal(loid)**	**References**
6.	*Trichoderma reesei*	Chromium	Ng et al. 2013
		Copper	Kim et al. 2003
		Cadmium	Kim et al. 2003
7.	*Trichoderma inhamatum*	Nickle	Padua et al. 2021
		Chromium	Morales and Cristiani 2008
8.	*Trichoderma gamsii*	Chromium	Kavita et al. 2012, Transengco et al. 2018
		Copper	Tansengco et al. 2018
		Nickle	Tansengco et al. 2018
		Lead	Tansengco et al. 2018
9.	*Trichoderma ghanense*	Cadmium	Oladipo et al. 2018
		Copper	Oladipo et al. 2018
		Lead	Oladipo et al. 2018
		Arsenic	Oladipo et al. 2018
10.	*Trichoderma koningii*	Cadmium	Wang et al. 2009
11.	*Trichoderma koningiopsis*	Chromium	Cheng et al. 2012
		Copper	Salvadori et al. 2014
12.	*Trichoderma saturnisporum*	Copper	Transengco et al. 2018
		Chromium	Tansengco et al. 2018
		Nickle	Tansengco et al. 2018
		Lead	Tansengco et al. 2018
13.	*Trichoderma lixii*	Copper	Kumar and Divedi 2021
14.	*Trichoderma aureoviride*	Lead	Siddiquee et al. 2013
		Nickle	Siddiquee et al. 2013
		Copper	Siddiquee et al. 2013
15.	*Trichoderma brevicompactum*	Chromium	Zhang et al. 2020, Priyadarshini et al. 2021
		Lead	
		Cadmium	
		Copper	

detoxify various metal(loid)s using different innate strategies such as biotransformation, bosorption, bBioaccumulation and phytobial remediation. Some of the *Trichoderma* species can bioremediate more than one metal(loid)s at a time (Cao et al. 2008).

7.5.1 Bioaccumulation

Bioaccumulation is a process that involves living cells for the removal of metal(loid)s by initiating metal reflux through energy-based processes. Diverse *Trichoderma* species are widely reported for their potential to tolerate and accumulate metal(loid)s viz. cadmium, silver, copper, nickel, mercury, etc. (Errasquin and Vazquez 2003, Ting and Choong 2009, Zeng et al. 2010, Anand et al. 2006, Tripathi et al. 2013, Hoseinzadeh et al. 2017). The use of *Trichoderma viride* for the bioaccumulation and removal of copper through an energy-based process was confirmed by electro micrographs of the mycelia. The majority of the copper was reported to be accumulated on the cell wall followed by cytoplasm (Anand et al. 2006). Hoseinzadeh et al. (2017) reported the bioaccumulation potential

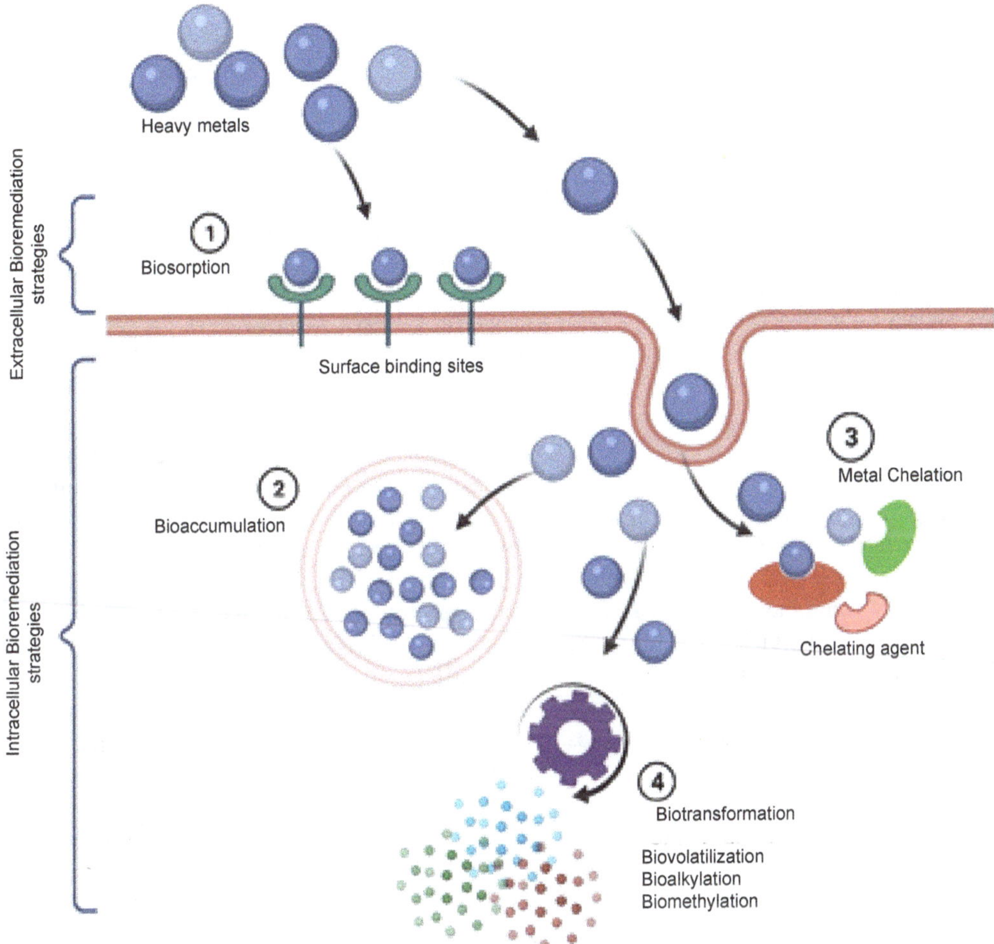

Fig. 7.3. Bioremediation strategies employed by *Trichoderma* species.

of *T. harzianum* TS103 and *Trichoderma asperellum* TS141. Yazdani et al. (2009) reported the bioaccumulation of copper from polluted water in *in vitro* condition by *Trichoderma atroviride*. Intracellular sequestration by producing secondary metabolites such as metal transport proteins is reported in *Trichoderma virens* for bioremediation of chromium and copper (Tansengco et al. 2018).

7.5.2 Biotranformation

Biomethylation is a type of biotransformation process that involves the change of organic and inorganic methylated compounds of metals into volatile and non-methylated volatile forms by the enzymatic action at the cellular level, whereas in the case when an alkyl group is associated with metal(loid)s, it is known as bioalkylation (Bentley and Chasteen 2002). *Trichoderma* species such as *T. asperellum*, *T. viridi*, etc., are reported for the transformation of As(V) to less toxic state As (III) by employing the process of biovolatilization (Zeng et al. 2010, Tripathi et al. 2013). Tripathi et al. (2017) discussed the methylation of toxic arsenic As(V) present in soil by tolerant *Trichoderma* species, which induced the biotransformation of the arsenic present in soil into dimethyl arsenic acid (DMAV), Monomethyl arsenic acid (MMA), and Trimethyl arsine (TMA) in comparison to uninoculated soil. The biomethylation of As(V) by *Trichoderma* species also limited the uptake of different species of arsenic in *Trichoderma* inoculated chickpea plants. Similarly, Su et al. (2017) discussed the potential of chlamydospore of *Trichoderma asperellum* in inducing biomethylation of arsenic in arsenic polluted soils.

7.5.3 Biosorption

Biosorption is a process employed by the material of biological origin including cells of microorganisms along with cells of other life forms to accumulate heavy metals from waste in the aqueous state by employing various biochemical processes involving the binding of free groups as negatively charged ions with several biopolymers (Akhtar et al. 2007, Zafar et al. 2007, Vankar and Bajpai 2008, Ting and Choong 2009, Iskandar et al. 2011, Tripathi et al. 2013, Shukla and Vankar 2014, Tripathi et al. 2017, Nongmaithem et al. 2016). Nongmaithem et al. (2016) reported the potential of *Trichoderma harzianum* in biosorption of various metal(loid)s by an eco-friendly strategy such as using continuous packed bed columns for removal of metal(loid)s from polluted wastewater. Similarly, Bishnoi et al. (2007) reported the use of *Trichoderma viride* for the removal of Cr(VI) and discussed various mechanisms such as ion exchange, complexation, bioprecipitation, etc., employed by *Trichoderma* species. The presence of hydroxyl (-OH) and amide (-NH) functional groups in biomass of *Trichoderma viride* played a significant role in the biosorption of Cr(VI). Similarly, Yazdani et al. (2009) reported the biosorption of copper from polluted water in *in vitro* condition by using the *Trichoderma atroviride* isolated from polluted environment. *Trichoderma* species are potential biosorbent for the removal of metal(loid)s in agricultural soil polluted with Cd and Ni. Metal(loid)s posed non-significant fungistatic effects on various *Trichoderma* isolates (Nongmaithem et al. 2016). The presence of polysaccharides in the cells of *Trichoderma* species enables their ability for high rate of biosorption of metal(loid)s that results in detoxification of polluted soil and industrial wastewater (Vankar and Bajpai 2008).

7.5.4 Phytobial remediation

Trichoderma species are reported to associate with different plant species, thereby helping the plants to remediate the pollutants and positively influencing growth parameters of plants. Microorganisms are reported to enhance the metal extraction potential of plants known by the process of phytobial remediation (Khalid et al. 2021). Detoxification of metal(loid)s by the integrated use of microorganisms such as fungi along with plants is a unique strategy for bioremediation of heavy metal-induced pollution. Association of *Trichoderma* and plants is applicable for remediation of soils with multiple pollutants as *Trichoderma* species are reported to have various mechanisms for metal detoxification along with other qualities including the potential to degrade organic contaminants enabling the association of plants and *Trichoderma* for efficient bioremediation of metal(loid)s. *Trichoderma atroviridae* is investigated for having potential for the translocation of various metal(loid)s in *Brassica juncea* L. (Cao et al. 2008), *Trichoderma* species like *Trichoderma harzianum* enhanced the bioaccumulation coefficients of Cr, Cu, Ni , Pb and Cd in *P. arundinacea, Salix, M. giganteus, P. virgatum* plants (Kacprzak et al. 2014); similarly, *Trichoderma koningii* is also reported to enhance the phytoextraction potential of *Brassica napa* L. plants in cadmium contaminated soils (Wang et al. 2009). The detoxification of potassium cyanide by *T. harzianum* enhanced the phytoremediation potential of *Pteris vittata* L. used for bioaccumulation of arsenic (Lynch and Moffat 2005). The combined application of mycorrhizal fungi and *T. harzianum* magnified the tolerance level of *Eucalyptus globulus* for high concentrations of Al and As in soils. Application of *T. harzianum* along with *P. fluorescens* assisted and enhanced the heavy metal uptake capacity of *B. juncea*. Devi et al. (2017) reported the protective effect of *T. logibrachiatum* on sunflower (*Helianthus annus* L.) plants by elevating the quantity of potential antioxidant enzymes like superoxide dimutases (SOD), peroxidase (POD) and catalase (CAT) enzymes, enabling the plants to tolerate lead-induced toxicity. *Trichoderma harzianum* reduced the adverse impact of Cd induced toxicity and improved the photosynthetic process along with the growth of soybean (*Glycine max* L.) plants (Lynch and Moffat 2005, Adams et al. 2007, Cao et al. 2008, Arriagada et al. 2009, Chauhan and Rai 2009, Dixit et al. 2011, Tripathi et al. 2013, Devi et al. 2017, Ram et al. 2020, Haider et al. 2021). The transgenic plants such as transgenic tobacco (*Nicotiana tabacum* L.) having Glutathione transferase (GST) gene from *Trichoderma virens* enabled the transgenic tobacco plants to induce the production of antioxidant enzymes having high tolerance against Cd-induced toxicity. *Trichoderma* species are reported to express metal(loid)s metabolism associated genes such as thmea1 gene, copper amine oxidase gene, copper chaperone gene for the production of specific metallothioneins (MT), which are low molecular weight cysteine rich metal binding proteins, copper chaperone proteins in case of copper induced pollution, and enzymes such as glutathione (GSH), glutathione peroxidase in state of the toxic environment as a strategy for successful bioremediation of metal(loid)s (Faedda et al. 2012, Mei et al. 2018).

7.5 Conclusion

Anthropogenic processes are solely blamed for the release of metal(loid)s in different ecosystems causing various toxicities and affecting different organisms. All available conventional physico-chemical remediation strategies are having high cost and low efficiency issues. The application of plants and various types of microorganisms for remediation of environmental pollutants is an eco-friendly approach known as bioremediation. Application of fungal organisms in mycoremediation is an established strategy for the environmental management of various hazardous pollutants including heavy metals. Diverse fungal agents have been identified and applied in mycoremediation against various heavy metals. Among all fungal organisms, various non-pathogenic *Trichoderma* species have proved to be effective organisms for bioremediation of various metal(loid)s such as nickel, copper, arsenic, lead, mercury, chromium, etc. More investigations are required to identify the bioremediation potential of novel unreported *Trichoderma* species as the distribution and efficiency of *Trichoderma* species are influenced by various abiotic factors. Modern molecular tools must be applied to develop mutant and genetically modified *Trichoderma* species having enhanced potential for biosorption, bioaccumulation and biotransformation strategies. The development of genetically modified plants having genes associated with tolerance and resistance metabolism from different *Trichoderma* species is a potential strategy for enhancing the phytoremediation potential of various plants against metal(loid)s mainly from the soil. Several studies discussed the biosorption potential of various non-pathogenic *Trichoderma* species that may be applied for safe and economically feasible removal of metal(loid)s from water bodies, thereby playing a key role in the water purification process.

References

Adams, P., F. M. M. De-leij and J. M. Lynch. 2007. *Trichoderma harzianum* Rifai 1295-22 mediates growth promotion of crack willow (*Salix fragilis*) saplings in both clean and metal-contaminated soil. Microb. Ecol. 54(2): 306–313.

Agrawal, N., P. Verma and S. K. Shahi. 2018. Degradation of polycyclic aromatic hydrocarbons (phenanthrene and pyrene) by the ligninolytic fungi *Ganoderma lucidum* isolated from the hardwood stump. Bioresour. Bioprocess. 5(11): 1–9.

Aibeche, C., N. Selami, F. ZitouniHaouar, K. Oeunzar. A. Addou, M. Harche and A. Djabeur. 2022. Bioremediation potential and lead removal capacity of heavy metaltolerant yeasts isolated from DayetOumGhellaz Lake water (northwest of Algeria). Int. Microbiol. 25(1): 61–73.

Akhtar, K., M. W. Akhtar and A. M. Khalid. 2007. Removal and recovery of uranium from aqueous solutions by *Trichoderma harzianum*. Water Res. 41: 1366–1378.

Akhtar, K., K. M. Khalid, M. W. Akhtar and M. A. Gauri. 2009. Removal and recovery of uranium from aqueous solutions by Ca-alginate immobilized *Trichoderma harzianum*. Bioresour. Technol. 100: 4551–4558.

Akhtar, N. and M. A. Mannan. 2020. Mycoremediation: Expunging environmental pollutants. Biotechnol. Rep. 26: e00452.

Ali, E. H. and M. Hashem. 2007. Removal efficiency of the heavy metals Zn(II), Pb(II) and Cd(II) by *Saprolegnia delica*and *Trichoderma viride*at different pH values and temperature degrees. Mycobiol. 35(3): 135–144.

Ali, H., E. Khan and I. Ilahi. 2019. Environmental chemistry and ecotoxicology of hazardous heavy metals: environmental persistence, toxicity, and bioaccumulation. J. Chem. 1–14.

Anand, P., J. Isar, S. Saran and R. K. Saxena. 2006. Bioaccumulation of copper by *Trichoderma viride*. Bioresour. Technol. 97: 1018–1025.

Anju, M. 2017. Biotechnological strategies for remediation of toxic Metal(loid)s from environment. pp. 315–359. *In*: Gahlawat, S., R. Salar, P. Siwach, J. Duhan, S. Kumar and P. Kaur (eds.). Plant Biotechnology: Recent Advancements and Developments. Springer, Singapore.

Arica, M. Y., C. Arpa, B. Kaya, S. Bektas, A. Denizli and O. Genc. 2004. Comparative biosorption of mercuric ions from aquatic systems by immobilized live and heat inactivated *Trametes versicolor* and *Pleurotussajur-caju*. Bioresour. Technol. 89: 145–154.

Argumedo-delira, R., A. Alorcon, R. Ferrara and J. J. Almaraz. 2012. Tolerance and growth of 11 *Trichoderma* strains to crude oil, naphthalene, phenanthrene and benzo[a]pyrene. J. Environ. Manag. 95: S291–S299.

Argumedo-delira, R., M. Gomez Martinez and R. Kaffure. 2021. *Trichoderma* biomass as an alternative for removal of congo red and malachite green industrial dyes. Appl. Sci. 11: 448.

Arica, M. Y., C. Arpa, B. Kaya, S. Bektas, A. Denizli and O. Genc. 2004. Comparative biosorption of mercuric ions from aquatic systems by immobilized live and heat inactivated *Trametes versicolor* and *Pleurotussajur-caju*. Bioresour. Technol. 89: 145–154.

Arriagada, C., E. Aranda, I. Sampedro, I. Garcia-Romera and J. A. Ocampo. 2009. Contribution of the saprobic fungi *Trametes versicolor* and *Trichoderma harzianum* and the arbuscular mycorrhizal fungi *Glomus deserticola* and *G. claroideum* to arsenic tolerance of Eucalyptus globules. Bioresour. Technol. 100: 6250–6257.

Ayad, F., A. Matallah-Boutiba, O. Rouane-Hacene, M. Bouderbala and Z. Boutiba. 2018. Tolerance of *Trichoderma* sp. to heavy metals and its antifungal activity in Algerian marine environment. J. Pure. Appl. Microbiol. 12(2): 855–870.

Babu, A. G., J. Shim K. S. Bang, P. Shea and B. T. Oh. 2014. *Trichoderma virens* PDR-28: a heavy metal-tolerant and plant growth-promoting fungus for remediation and bioenergy crop production on mine tailing soil. J. Environ. Manag. 132: 129–134.

Baldrian, C. in Der Wiesche, J. Gabriel, F. Nerud and F. Zadrazil. 2000. Influence of cadmium and mercury on activities of ligninolytic enzymes and degradation of polycyclic aromatic hydrocarbons by *Pleurotusostreatus* in soil. Appl. Environ. Microbiol. 66: 2471–2478.

Bano, A., J. Hussain, A. Akbar, K. Mehmood, M. Anwar, M. S. Hasni, S. Ullah, S. Sajid and I. Ali. 2018. Biosorption of heavy metals by obligate halophilic fungi. Chemosphere. 199: 218–222.

Benítez, T., A. M. Rincon, M. C. Limon, and A. C. Codon. 2004. Biocontrol mechanisms of *Trichoderma* strains. Int. Microbiol. 7: 249–260.

Bentley, R. and T. G. Chasteen. 2002. Microbial methylation of mettaloids: arsenic, antimony and bismuth. Microbiol. Mol. Biol. Rev. 66(2): 250–271.

Bhattacharya, S., A. Das, K. Prashanthi, M. Palaniswamy and J. Angayarkanni. 2014. Mycoremediation of Benzo[a]pyrene by *Pleurotusostreatus* in the presence of heavy metals and mediators. Biotechnol. 4: 205–211.

Bisht, J., N. S. Harsh, L. M. S. Palni and V. Pande. 2014. Effect of repeated application of endosulfan on fungal population of pine forest soil. Biotechnol. Int. 7(1): 11–20.

Bisht, J., N. S. Harsh, L. M. S. Palni, V. Agnihotri and A. Kumar. 2019. Biodegradation of chlorinated organic pesticides endosulfan and chlorpyrifos in soil extract broth using fungi. Remediat. 29: 63–77.

Bishnoi, N. R., R. Kumar and K. Bishnoi. 2007. Biosorption of Cr (VI) with *Trichoderma viride* immobilized fungal biomass and cell free Ca alginate beads. Indian J. Exp. Biol. 45: 657–664.

Błaszczyk, L., S. Marek, S. Krzysztof, L. Jolanta and J. Małgorzata. 2014. *Trichoderma* spp. application and prospects for use in organic farming and industry. J. Plant Prot. Res. 54: 4.

Briffa, J., E. Sinagra and R. Blundell. 2020. Heavy metal pollution in the environment and their toxicological effects on humans. Heliyon. 6: e04691.

Cao, L., M. Jiang, Z. Zeng, A. Du, H. Tan and Y. Liu. 2008. *Trichoderma atroviride* F6 improves phytoextraction efficiency of mustard (*Brassica juncea* (L.) Coss. var. foliosa Bailey) in Cd, Ni contaminated soils, Chemosphere. 71(9): 1769–1773.

Chauhan, J. S. and J. P. N. Rai. 2009. Phytoextraction of soil cadmium and zinc by microbes inoculated Indian mustard (*Brassica juncea*). J. Plant Interact. 4(4): 279–287.

Cheng, L., J. Lin, J. Lin and J. Du. 2012. Influence factors and adsorption mechanism of Cr (VI) by the powder of *Trichoderma koningiopsis*. J. Quanzhou. Normal. Univ. 30: 6.

Chew, S. Y. and A. D. Y. Ting. 2015. Common filamentous *Trichoderma asperellum* for effective removal of triphenylmethane dyes. Desalin. Water Treat. 1–6.

Cjathaml, T., P. Erbanova, A. Kolman, C. Novotny, V. Sasek and C. Mougin. 2008. Degradation of PAHs by Ligninolytic Enzymes of *Irpexlacteus*. Folia Microbiol. 53(4): 289–294.

Copete-Pertuz, L. S., F. Alandete-Novoa, P. Jersson, G. A. Correa-Londoño and A. L. Mora-Martínez. 2019. Enhancement of ligninolytic enzymes production and decolourisingactivity in *Leptosphaerulina* sp. by co–cultivation with *Trichoderma viride* and *Aspergillus terreus*. Sci. Total Environ. 646: 1536–1545.

Deshmukh, R., A. A. Khardenavis and H. J. Purohit. 2016. Diverse metabolic capacities of fungi for bioremediation. Indian J. Microbiol. 56(3): 247–264.

Devi, S. S., S. Srinivasulu and K. V. B. Rao. 2017. Protective role of *Trichoderma logibrachiatum* (WT2) on lead induced oxidative stress in *Helianthus annus* L. Indian J. Exp. Biol. 55: 235–241.

Dixit, P., P. K. Mukherjee, V. Ramachandran and S. Eapen. 2011. Glutathione transferase from *Trichoderma virens* enhances cadmium tolerance without enhancing its accumulation in transgenic *Nicotiana tabacum*. PLoS ONE. 6(1): e16360.

Dusengemungu, L., K. George, C. Gwanama and K. O. Ouma. 2020. Recent advances in biosorption of copper and cobalt by filamentous fungi. Front. Microbiol. 11: 582016.

Errasquin, E. L. and C. Vazquez. 2003. Tolerance and uptake of heavy metals by *Trichoderma atroviride* isolated from sludge. Chemosphere. 50: 137–143.

Faedda, R., I. Puglisi, V. Sanzaro, G. Petrone and S. O. Cacciola. 2012. Expression of genes of *Trichoderma harzianum* in response to the presence of cadmium in the substrate. Nat. Prod. Res. 1–8.

Fiorentino, N., M. Fagnano, P. Adamo, A. Impagliazzo, M. Mori, O. Pepe et al. 2013. Assisted phytoextraction of heavy metals: compost and *Trichoderma* effects on giant reed (*Arundo donax* L.) uptake and soil N-cycle microflora. Italian J. Agron. 8: e29.

Freitas, A. D. L., G. F. De Moura and M. A. B. De Lima. 2011. Role of the morphology and polyphosphate in *Trichoderma harzianum*related to cadmium removal. Mol. 16(3): 2486–2500.

Garcia-Hernandez, M. A., J. F. Villarreal-Chiu and M. T. Garza-Gonzalez. 2017. Metallophilic fungi research: an alternative for its use in the bioremediation of hexavalent chromium. Int. J. Environ. Sci. Technol. 14: 2023–2038.

Gharieb, N. M. and G. M. Gadd. 2004. Role of gluthatione in detoxification of metal (loid) by *Saccharomycescervisiae*. Biometals. 17: 183–188.

Gupta, S., A. Wali, M. Gupta and S. K. Annepu. 2017. Fungi: An Effective Tool for Bioremediation. pp. 593–606. *In*: Singh, D.P. et al. (eds.). Plant-Microbe Interactions in Agro-Ecological Perspectives.

Haider, F. U., A. J. Coulter, S. A. Cheema, M. Farooq, J. Renzhi, Z. Guo and S. C. Liqun. 2021. Co-application of biochar and microorganisms improves soybean performance and remediate cadmium-contaminated soil. Ecotoxicol. Environ. Saf. 214: 112112.

Harman, G. E., C. R. Howell, A. Viterbo, I. Chet and M. Lorito. 2004. *Trichoderma* species-opportunistic, avirulent plant symbionts. Nat. Rev. Microbiol. 2: 43–56.

Hoseinzadeh, S., S. Shahabivand and A. A. Aliloo. 2017. Toxic Metals accumulation in *Trichoderma asperellum* and *T. harzianum*. Microbiol. 86(6): 728–736.

Howell, C. R. 2003. Mechanisms employed by *Trichoderma* species in the biological control of plant diseases; the history and evolution of current concept. Plant Dis. 87: 4–10.

Hoyos-Carvajal, L., S. Orduz and J. Bissett. 2009. Genetic and metabolic biodiversity of *Trichoderma* from Colombia and adjacent neotropic regions. Fungal Genet. Biol. 46: 615–631.

Hrynkiewicz, K. and C. Baum. 2014. Application of microorganisms in bioremediation of environment from heavy metals. pp. 215–228. *In*: Malik, A. et al. (eds.). Environmental Deterioration and Human Health.

Hu, Y., S. D. Veresoglou, L. Tedersoo, T. Xu, T. Ge, L. Liu, B. Chen, Y. Chen, Z. Hao, Y. Su and M. C. Rillig. 2019. Contrasting latitudinal diversity and co-occurrence patterns of soil fungi and plants in forest ecosystems. Soil Biol. Biochem. 131: 100–110.

Iskandar, N. L., N. A. Zainudin and S. G. Tan. 2011. Tolerance and biosorption of copper (Cu) and lead (Pb) by filamentous fungi isolated from a freshwater ecosystem. J. Environ. Sci. 23: 824–830.

Kacprzak, M. and G. Malina. 2005. The tolerance and Zn^{2+}, $Ba^{2+}$and Fe^{3+}accumulation by *Trichoderma atroviride* and *Mortierella exigua* isolated from contaminated soil. Can. J. Soil Sci. 85(2): 283–290.

Kacprzak, M. J., K. Rosikon, K. Fijalkowski and A. Grobelak. 2014. The effect of *Trichoderma* on heavy metal mobility and uptake by *Miscanthus giganteus*, *Salix* sp., *Phalaris arundinacea*, and *Panicum virgatum*. Appl. Environ. Soil Sci. 506142. doi.org/10.1155/2014/506142.

Katayama, A. and F. Matsumura. 1993. Degradation of organochlorine pesticides, particularly endosulfan, by *Trichoderma harzianum*. Environ. Toxicol. Chem. 12: 1059–1065.

Kavita, B. and H. Keharia. 2012. Biosorption potential of *Trichoderma gamsii* biomass for removal of Cr (VI) from electroplating industrial effluent. Int. J. Chem. Eng. 1–7.

Karn, R., N. Ojha, S. Abbas and S. Bhugra. 2021. A review on heavy metal contamination at mining sites and remedial techniques. Earth Environ. Sci. 796: 1–30.

Khalid, M., S. Ur-Rehman, D. Hassani, K. Hayat, P. Zhou and N. Hui. 2021. Advances in fungal-assisted phytoremediation of heavy metals: A review. Pedosphere. 31(3): 475–495.

Kidwai, M. K. and M. Nehra. 2017. Biotechnological applications of *Trichoderma* species for environmental and food security. *In*: Gahlawat, S., R. Salar, P. Siwach, J. Duhan, S. Kumar and P. Kaur (eds.). Plant Biotechnology: Recent Advancements and Developments. Singapore. Springer Publication.

Kidwai, M. K. and S. B. Dhull. 2021. Heavy metals induced stress and metabolic responses in fenugreek (*Trigonella foenum-graecum* L.) plants. *In*: Naeem, M. et al. (eds.). Fenugreek: Biology and Applications. Singapore. Springer Nature, Singapore Pte.

Kim, S. K., B. P. Chun, M. K. Yoon and S. Y. Hyun. 2003. Biosorption of cadmium and copper by *Trichodermareesei* RUT C30. J. Ind. Eng. Chem. 9: 403–406.

Kredics, L., L. Antal, L. Manczinger and E. Nagy. 2001. Breeding of mycoparasitic *Trichoderma* strains for heavy metal resistance. Lett. Appl. Microbiol. 33: 112–116.

Kredics, L., M. Laday, P. Kormoczi, L. Manczinger, G. Rakhely, C. Vagvolgyi and A. Szekeres. 2012. Genetic and biochemical diversity among *Trichoderma* isolates in soil samples from winter wheat fields of the Pannonian Plain. Acta Biol. Szeg. 56: 141–149.

Kubicek, C., J. Bissett, I. Druzhinina, C. Kulling-Grandiger and G. Szakacs. 2003. Genetic and metabolic diversity of *Trichoderma*: a case study on South East Asian isolates. Fungal Genet. Biol. 38: 310–319.

Kumar, A., A. N. Yadav, R. Mondal, D. Kour, G. Subrahmanyam, A. A. Shabnam et al. 2021. Myco-remediation: A mechanistic understanding of contaminants alleviation from natural environment and future prospect. Chemosphere. 284: 131325.

Kumar, V. and S. K. Dwivedi. 2021. Mycoremediation of heavy metals: processes, mechanisms, and affecting factors. Environ. Sci. Pollut. Res. 28: 10375–10412.

Lin, Y., W. Xiao, Y. Ye, C. Wu, Y. Hu and H. Shi. 2020. Adaptation of soil fungi to heavy metal contamination in paddy fields-a case study in eastern China. Environ Sci. Pollut. Res. 27: 27819–27830.

Lopez, E. E. and C. Vazquez. 2003. Tolerance of uptake of heavy metals by *Trichoderma atroviride* isolated from sludge. Chemosphere. 50: 137–143.

Lynch, J. M. and A. Moffat. 2005. Bioremediation—prospects for the future application of innovative applied biological research. Ann. Appl. Biol. 146: 217–221.

Malgorzata, J. Kacprzak, K. Rosikon, K. Fijalkowski and A. Grobelak. 2014. The Effect of *Trichoderma* on heavy metal mobility and uptake by *Miscanthus giganteus*, *Salix* sp., *Phalaris arundinacea*, and *Panicum virgatum*. Appl. Environ. Soil Sci. 506142.

Malik, A. 2004. Metal bioremediation through growing cells. Environ. Int. 30: 261–278.

Marco-Urrea, E., M. Pérez-Trujillo, P. Blánquez, T. Vicent and G. Caminal. 2010. Biodegradation of the analgesic naproxen by *Trametes versicolor* and identification of intermediates using HPLC-DAD-MS and NMR. Bioresour. Technol. 101: 2159–2166.

Mastouri, F., T. Björkman and G. E. Harman. 2012. *Trichoderma harzianum* enhances antioxidant defense of tomato seedlings and resistance to water defcit. Mol. Plant-Microb. Int. 25: 1264–1271.

Maurya, M., Rashk-E-Eram, S. K. Naik, J. S. Choudhary and S. Kumar. 2019. Heavy metals scavenging potential of *Trichoderma asperellum* and *Hypocrea nigricans* isolated from acid soil of Jharkhand. Indian J. Microbiol. 59(1): 27–38. https://doi.org/10.1007/s12088-018-0756-7.

Mei, J., L. Wang, X. Jiang, B. Wu and L. Mei. 2018. Functions of the C_2H_2 transcription factor gene thmea1 in *Trichoderma harzianum* under copper stress based on transcriptome analysis. BioMed Res. Int. 8149682. doi: 10.1155/2018/8149682.

Migahed, F., A. Abdelrazak and G. Fawzy. 2017. Batch and continuous removal of heavy metals from industrial effluents using microbial consortia. Int. J. Environ. Sci. Technol. 14: 1169–1180.

Moghimi, H. T., R. Heidary and J. Hamedi. 2017. Assessing the biodegradation of polycyclic aromatic hydrocarbons and laccase production by new fungus *Trematophoma* sp. UTMC 5003. World J. Microbiol. Biotechnol. 33(7): 136.

Mohammadian, E., A. B. Ahari, M. Arzanlou, S. Oustan and S. H. Khazaei. 2017. Tolerance to heavy metals in filamentous fungi isolated from contaminated mining soils in the Zanjan Province, Iran. Chemosphere. 185: 290–296.

Morales, B. L. and U. E. Cristiani. 2008. Hexavalent chromium removal by a *Trichoderma inhamatum* fungal strain isolated from tannery effluent. Water Air Soil Pollut. 187: 327–336.

Ng, I. S., X. Wu, X. Yang, Y. Xie, Y. Lu and C. Chen. 2013. Synergistic effect of *Trichoderma reesei* cellulases on agricultural tea waste for adsorption of heavy metals Cr(VI). Bioresour. Technol. 145: 297–301.

Nongmaithem, N., A. Roy and P. Bhattacharya. 2016. Screening of *Trichoderma* isolates for their potential of biosorption of nickel and cadmium. Braz. J. Microbiol. 47(2): 305–313.

Oladipo, O., O. O. Awotoye, A. Olayinka, C. C. Bezuidenhout and M. S. Maboeta. 2018. Heavy metal tolerance traits of filamentous fungi isolated from gold and gemstone mining sites. Braz. J. Microbiol. 49: 29–37.

Oliveira, B. R., A. Penetra, V. V. Cardoso, M. J. Benoliel, M. T. Barreto Crespo, A. Samson and V. J. Pereira. 2015. Biodegradation of pesticides using fungi species found in the aquatic environment. Environ. Sci. Pollut. Res. 22: 11781–11791.

Padua, J. C. and D. E. Cruz. 2021. Isolation and characterization of Nickel-Tolerant *Trichoderma* strains from marine and terrestrial environments. J. Fungi. 7: 591.

Park, J. K., J. W. Lee and J. Y. Jung. 2003. Cadmium uptake capacity of two strains of *Saccharomyces cerevisiae* cells. Enzyme Microb. Technol. 33: 371–378.

Park, D., Y. S. Yun, J. H. Jo and J. M. Park. 2005. Mechanism of hexavalent chromium removal by dead fungal biomass of *Aspergillus niger*. Water Res. 39: 533–540.

Paria, K., S. Pyne and S. K. Chakraborty. 2021. Optimization of heavy metal (lead) remedial activities of fungi *Aspergillus penicillioides* (F12) through extra cellular polymeric substances. Chemosphere. 286: 131874.

Petrovic, J. J., G. Danilovic, N. Curcic, M. Milinkovic, N. Stosic, D. Pankovic and R. Vera. 2014. Copper tolerance of *Trichoderma* species. Arch. Biol. Sci. Belgrade. 66: 137–142.

Pothuluri, J. V., F. E. Evans, D. R. Doerge, M. I. Churchwell and C. E. Cerniglia. 1996. Metabolism of metolachlor by the fungus *Cunninghamella elegans*. Arch. Environ. Contam. Toxicol. 32: 117–125.

Price, M. S., J. J. Classen and G. A. Payne. 2001. *Aspergillus niger* absorbs copper and zinc from swine wastewater. Bioresour. Technol. 77: 41–49.

Priyadarshini, E., S. S. Priyadarshini, B. G. Cousins and N. Pradhan. 2021. Metal-Fungus interaction: Review on cellular processes underlying heavy metal detoxification and synthesis of metal nanoparticles. Chemosphere. 274: 129976.

Puglisi, I., R. Faedda, V. Sanzaro, A. R. L. Piero, G. Petrone and S. O. Cacciola. 2012. Identification of di_erentially expressed genes in response to mercury I and II stress in *Trichoderma harzianum*. Gene. 506: 325–330.

Purnomo, A., T. Mori, S. R. Putra and R. Kondo. 2013. Biotransformation of heptachlor and heptachlor epoxide by white-rot fungus *Pleurotusostreatus*. Int. Biodeterior. Biodegrad. 82: 40–44.

Rahman, Z. and V. P. Singh. 2020. Bioremediation of toxic heavy metals (THMs) contaminated sites: concepts, applications and challenges. Environ. Sci. Pollut. Res. 27(22): 27563–27581.

Ram, R.M., A. Vaishnav and H. B. Singh. 2020. *Trichoderma* spp.: Expanding potential beyond agriculture. pp. 351–367. *In*: Manoharachary, C., H. B. Singh and A. Varma (eds.). *Trichoderma*: Agricultural Applications and Beyond. Soil Biology. Springer.

Rigot, J. and F. Matsumura. 2002. Assessment of the rhizosphere competency and Pentachlorophenol-metabolising activity of a pesticide degrading strain of *Trichoderma harzianum* introduced into the root zone of corn seedlings. J. Environ. Sci. Health. 37: 201–210.

Rose, P. K. 2022. Bioconversion of agricultural residue into biofuel and high-value biochemicals: recent advancement. pp. 233–268. *In*: Nandabalan, Y. K, V. K. Garg, N. K. Labhsetwar, A. Singh (eds.). *Zero* Waste Biorefinery. Energy, Environment, and Sustainability. Singapore: Springer Publisher.

Sadfi-Zouaoui, N., I. Hannachi, M. Rouaissi, M. Hajlaoui, M. Rubio, E. Monte, A. Boudabous and M. Hermosa. 2009. Biodiversity of *Trichoderma* strains in Tunisia. Can. J. Microbiol. 55: 154–162.

Sahu, A., A. Mandal, J. Thakur, M. C. Manna and A. S. Rao. 2012. Exploring bioaccumulation efficacy of *Trichodermaviride*: an alternative bioremediation of cadmium and lead. Nat. Acad. Sci. Lett. 35: 299–302.

Sall, M. L., A. K. D. Diaw, D. G. Sall, S. E. Aaron and J. J. Aaron. 2020. Toxic heavy metals: impact on the environment and human health and treatment with conducting organic polymers, a review. Environ Sci. Pollut. Res. 27: 29927–29942.

Salvadori, M. R., R. A. Ando, C. Nascimento and B. Correa. 2014. Bioremediation from wastewater and extracellular synthesis of copper nanoparticles by the fungus *Trichoderma koningiopsis*. J. Environ Sci. Health Part A. 49(11): 1286–1295.

Samuels, G. J. 1996. *Trichoderma*: a review of biology and systematics of the genus. Mycol. Res. 100: 923–935.

Sarangi, S., H. Swain, T. Adak, P. Bhattacharyya, A. K. Mukherjee, G. Kumar and S. T. Mehetre. 2021. *Trichoderma*-mediated rice straw compost promotes plant growth and imparts stress tolerance. Environ. Sci. Pollut. Res. 28: 44014–44027.

Sarkar, S., A. Satheshkumar, R. Jayanthi and R. Premkumar. 2010. Biosorption of Nickel by live biomass of *Trichoderma harzianum*. Res. J. Agric. Sci. 1(2): 69–74.

Saroj, S., S. Dubey, P. Agarwal, R. Prasad and R. P. Singh. 2015. Evaluation of the efficacy of a fungal consortium for degradation of azo dyes and simulated textile dye effluent. Sustain. Water Resour. Manag. 1: 233–243.

Say, R., A. Denizil and Y. Arica. 2001. Biosorption of cadmium (II), Lead (II) and copper (II) with filamentous fungus *Phenarochaetechrysosporium*. Bioresourc. Technol. 76: 67–70.

Sen, S., S. Raut, P. Bandopadhaya and S. Raut. 2016. Fungal decolouration and degradation of azo dyes: A review. Fungal Biol. Rev. 30(3): 112–133.

Shukla, D. and P. S. Vankar. 2014. Role of *Trichoderma* species in bioremediation process: biosorption studies on hexavalent chromium. pp. 405–414. *In*: Gupta, V. K., M. Schmoll, A. Herrera-Estrella, R. S. Upadhyay, I. Druzhinina, M. G. Tuohy (eds.). Biotechnology and Biology of Trichoderma. Vol 30. USA: Elsevier Publishers

Siddiquee, S., S. N. Aishah, S. A. Azad, S. N. Shafawati and L. Naher. 2013. Tolerance and biosorption capacity of Zn^{2+}, Pb^{2+}, Ni^{3+} and Cu^{2+} by filamentous fungi (*Trichodermaharzianum*, *T. aureoviride* and *T. virens*). Adv. Biosci. Biotechnol. 4: 570–583.

Siddiquee S., K. Rovina, S. Azad, L. Naher, S. Suryani and P. Chaikaew. 2015. Heavy metal contaminants removal from wastewater using the potential filamentous fungi biomass: a review. J. Microb. Biochem. Technol. 7: 6. DOI: 10.4172/1948-5948.1000243.

Singh, S. K., R. Khajuria and L. Kaur. 2017. Biodegradation of ciprofloxacin by white rot fungus *Pleurotusostreatus*. Biotechnol. 7: 69.

Su, Shiming, X. Zeng, L. Bai, P. N. Williams, Y. Wang, L. Zhang and C. Wu. 2017. Inoculating chlamydospores of *Trichoderma asperellum* SM-12F1 changes arsenic availability and enzyme activity in soils and improveswater spinach growth. Chemosphere. 175: 497–504.

Sugasini, A. and K. Rajagopal. 2015. Hexavalent chromium removal from aqueous solution using *Trichoderma viride*. Int. J. Pharm. Bio. Sci. 6(1): 485–495.

Tan,W. S. and A. S. Y. Ting. 2012. Efficacy and reusability of alginate-immobilized live and heat-inactivated *Trichoderma asperellum* cells for Cu (II) removal from aqueous solution. Bioresour. Technol. 123: 290–295.

Tandon, A., T. Fatima, Anshu, D. Shukla, P. Tripathi, S. Srivastava and P. C. Singh. 2020. Phosphate solubilization by *Trichoderma koningiopsis* (NBRI-PR5) under abiotic stress conditions. J. King Saud Univ. - Sci. 32(1): 791–798.

Tansengco, M., J. Tejano, F. Coronado, C. Gacho and J. Barcelo. 2018. Heavy metal tolerance and removal capacity of *trichoderma* species isolated from mine tailings in Itogon, Benguet. Environ. Nat. Resour. J. 16(1): 39–57.

Tarekegn, M. M., F. Z. Salillih and A. I. Ishetu. 2020. Microbes used as a tool for bioremediation of heavy metal from the environment. Cogent Food Agric. 6: 1783174.

Ting, A. S. and C. C. Choong. 2009. Bioaccumulation and biosorption efficacy of *Trichoderma* isolate SP2F1 in removing copper (Cu(II)) from aqueous solutions. World J. Microbiol. Biotechnol. 25: 1431–1437.

Tripathi, P., P. C. Singh, A. Mishra, P. S. Chauhan, S. Dwivedi, R. T. Bais and R. D. Tripathi. 2013. *Trichoderma*: a potential bioremediator for environmental clean up. Clean Technol. Environ. Policy. 4. https://doi.org/10.1007/s10098-012-0553-7.

Tripathi, P., P. Singh, A. Mishra, S. Srivastava, R. Chauhan, S. Awasthi, S. Mishra et al. 2017. Arsenic tolerant *Trichoderma* sp. reduces arsenic induced stress in chickpea (*Cicer arietinum*). Environ. Pollut. 223: 137–145.

Umar, A. 2021. Screening and evaluation of laccase produced by different *Trichoderma* species along with their phylogenetic relationship. Arch. Microbiol. 203: 4319–4327.

Vankar, P. S. and D. Bajpai. 2008. Phyto-remediation of chrome-VI of tannery effluent by *Trichoderma* species. Desalin. 222: 255–262.

Wang , B., L. Liu, Y. Gao and J. Chen. 2009. Improved phytoremediation of oilseed rape (*Brassica napus*) by *Trichoderma* mutant constructed by restriction enzyme-mediated integration (REMI) in cadmium polluted soil. Chemosphere. 74: 1400–1403.

Wang, S., T. Liu, X. Xiao and S. Luo. 2021. Advances in microbial remediation for heavy metal treatment: a mini review. J. Leather Sci. Eng. 3: 1.

Wesenberg, D. Fred′ eric Buchon, Spiros and N. Agathos. 2002. Degradation of dye-containing textile effluent by the agaric white-rot fungus *Clitocybuladusenii*. Biotechnol. Lett. 24: 989–993.

White, C., J. A. Sayer and G. M. Gadd. 1997. Microbial solubulization and immobilization of toxic metals: key biogeochemical processes for treatment of contamination. FEMS Microbiol. Rev. 20: 503–516.

Yaghoubian, Y., S. A. Siadat, M. R. Telavat, H. Pirdashti and I. Yaghoubia . 2019. Bio-removal of cadmium from aqueous solutions by filamentous fungi: *Trichoderma* spp. and *Piriformospora indica*. Environ. Sci. Pollut. Res. 26: 7863–7872.

Yan, G. and T. Viraraghavam. 2003. Heavy metal removal from aqueous solution by fugus *Mucor rouxii*. Water Res. 37: 4486–4496.

Yazdani, M., C. K. Yap, F. Abdullah and S. G. Tan. 2009. *Trichoderma atroviride*as a bioremediator of Cu pollution: an in vitro study. Toxicol. Environ. Chem. 91(7): 1305–1314.

Ye, J.-S., H. Yin, J. Qiang, H. Peng, H.-M. Qin, N. Zhang and B.-Y. He. 2011. Biodegradation of anthracene by *Aspergillus fumigatus*. J. Hazard. Mater. 185: 174–181.

Yedidia, I. I., N. Benhamou and I. Chet. 1999. Induction of defense responses in cucumber plants (*Cucumis sativus* L.) by the biocontrol agent *Trichoderma harzianum*. Appl. Environ. Microbiol. 65: 1061–1070.

Zafar, S., F. Aqil and I. Ahmad. 2007. Metal tolerance and biosorption potential of filamentous fungi isolated from metal contaminated agricultural soil. Bioresour. Technol. 98: 2557–2561.

Zakaria, Z., N. S. Zulkafflee, N. M. Redzuan, J. Selamat, M. R. Ismail, S. M. Praveena, G. Tóth and A. F. Razis. 2021. Understanding potential heavy metal contamination, absorption, translocation and accumulation in rice and human health risks. Plants. 10: 1070.

Zhang, D., C. Yin, N. Abbas, Z. Mao and Y. Zhang. 2020. Multiple heavy metal tolerance and removal by an earthworm gut fungus *Trichoderma brevicompactum* QYCD-6. Sci. Rep. 10: 6940.

Zhang, Q., G. Zeng, G. Chen et al. 2015. The effect of heavy metal-induced oxidative stress on the enzymes in white rot fungus *Phanerochaetechrysosporium*. Appl. Biochem. Biotechnol. 175: 1281–1293.

Zeng, X., S. Su, X. Jiang, L. Li, L. Bai and Y. Zhang. 2010. Capability of pentavalent arsenic bioaccumulation and biovolatilization of three fungal strains under laboratory conditions. Clean: Soil Air Water. 38: 238–241.

CHAPTER 8

Trends in Waste Water Treatment using Phycoremediation for Biofuel Production

Anuchaya Devi,[1] *Anita Singh,*[2] *Monika Mahajan,*[1] *Sinha Sahab,*[1] *Vaibhav Srivastava,*[1] *Pooja Singh*[3] and *Rajeev Pratap Singh*[1,*]

8.1 Introduction

The aquatic environment is getting polluted every day with anthropogenic activities, and it is a serious environmental concern. Waste water (WW) runoff from the municipality is one of the major sources of contribution (Tripathi et al. 2019). With no waste water treatment, they bring inorganic, organic, xenobiotic, and other toxic compounds to the river system (Paxéus 1996). In various developing countries, only conventional methods of pollutant treatment are adapted where pollutants like esters, surfactants, benzene, phenol, phthalates, and heavy metals (Zn, Cd, Cr, Fe, Ni, Pb, etc.) are not effectively separated by such conventional waste water treatment (WWT) steps; therefore it is very timely to develop advanced methods for waste water treatment to stop the hazards of toxic pollutants (Gupta et al. 2017). In this context, WWT method using biological sources, such as the phycoremediation system, emerges to overcome the constraints mentioned above. Phycoremediation is the term coined for the use of micro or macroalgae in waste water to bio-transform the concentrations of the contaminant into its cell. This route was chosen to find solutions to the bottlenecks of conventional methods in industrial waste water treatment. The main bottleneck is the elimination of phosphorus and nitrogen, which can be removed in tertiary treatment using algae (followed by the release of waste water), and then the WW can be released to the environment (Abinandan et al. 2013, Olguin 2003). But, the pace of pollutant accumulation load by microalgae exhibits variation among different microalgal species (Dominic et al. 2009). Advantages of using phycoremediation techniques are: (i) phycoremediation is a less expensive, eco-friendly, and hazardless process; (ii) phycoremediation keeps balance in bacterial population; (iii) algae uses WW as well as removes CO_2 from the air, which thereby leads to carbon sequestration and contributes to reducing greenhouse gas effect; (iv) they are photosynthetic non-pathogenic microorganisms, and

[1] Institute of Environment and Sustainable Development, Banaras Hindu University, Varanasi-221005.
[2] Department of Botany, Faculty of Science, Banaras Hindu University, Varanasi-221005.
[3] Department of Science, Society for Higher Education & Practical Applications (SHEPA), Varanasi, India.
* Corresponding author: rps.iesd@bhu.ac.in, rajeevprataps@gmail.com

do not emit toxic gases to the environment; (v) the algal biomass has potential values and can be used for the production of value-added products like fuels and chemicals (Gani et al. 2015). Because of the several benefits of algae in many countries, oxidation ponds have been designed for sewage effluent treatment where the photosynthetic bacteria and algae are cultivated to oxidize the WW with the elimination of O_2 and absorption of CO_2. Extensive cultivations of algae for waste removal from waste water and industrial establishment can produce an increased amount of biomass which has high calorific value, has immense potential to be used as a substitute for coal, and can be utilized for other applications such as biofuel, biochar, and biofertilizers. Among the various biofuels, biodiesel, bioethanol, and biohydrogen are manufactured by making use of algal biomass feedstocks. With the technique hydrothermal liquefaction (HTL), bio-oil from crude microalgae can be extracted and directly applied as an effective biofuel. The cellular structure of algae has the capacity to produce macromolecules like lipids and active metabolites, which can further be employed in nutraceuticals, medicines, preservatives, and colorants. Moreover, through the genetic engineering process, the cellular factories of algae can be managed to enhance the production of highly valued compounds for achieving greater technological aspects (Bansal et al. 2018). In modern times, many researchers have utilized phycoremediation as a sustainable tool in treating various types of waste water with varying micro and macro algae species. There are several WW that have been investigated till now, and they are WW from dairy, greywater, municipal and domestic waste water, leather effluent, palm oil mill waste water, food processing waste water, etc., using different species of microalgae such as *Botryococcusbraunii, Chlorella vulgaris, Oscillatorialimosa, Rhizocloniumhieroglyphicum, Nostoc* species, *Spirulina* species (Gani et al. 2016). Here in this chapter, we are focusing on the use of phycoremediation for dual purposes, i.e., waste water treatment as well as biofuel production through the biomass cultivated in the waste water. Few real instances where the technique is used will be discussed here; the technological constraints, the bottlenecks, and the scope of future studies will also be addressed. This is a step to help and promote the readers about the prospects of application-based research in phycoremediation technology toward the welfare of humans and our environment.

8.2 Physico-Chemical-Biological Parameters of Waste water

Characteristics of waste water may be sorted into physical, chemical, and biological parameters, and their sources are reproduced in Fig. 8.1 (Rawat et al. 2011).

8.2.1 Physical Characteristics

Among all the physical characteristics, temperature is considered to be a very determining factor for aquatic life because it governs chemical and biological reactions of the aqueous phase. Extreme high temperatures may enhance unwanted planktonic and fungal growth. It is also imperative in determining other characteristics such as electrical conductivity, various forms of alkalinity, pH, the saturation level of gases, etc. The color of the waste water also tells many facts about it; the color of domestic waste water usually gives an indication of its duration of existence. Household septic tanks become black in color due to reactions of organic and inorganic materials in the biological environment. The color of industrial waste water depends upon the nature of the secondary or tertiary product formed due to the reactions of materials in it. The odor present in waste water is mainly due to impurities dissolved; it happens due to the organic processes such as decaying of the materials. In most cases, the odor in the septic water is due to the generation of hydrogen gas. Total solid content is defined by different types of dissolved and suspended materials which stay there in the water as residue (Metcalf et al. 1991, Rawat et al. 2011).

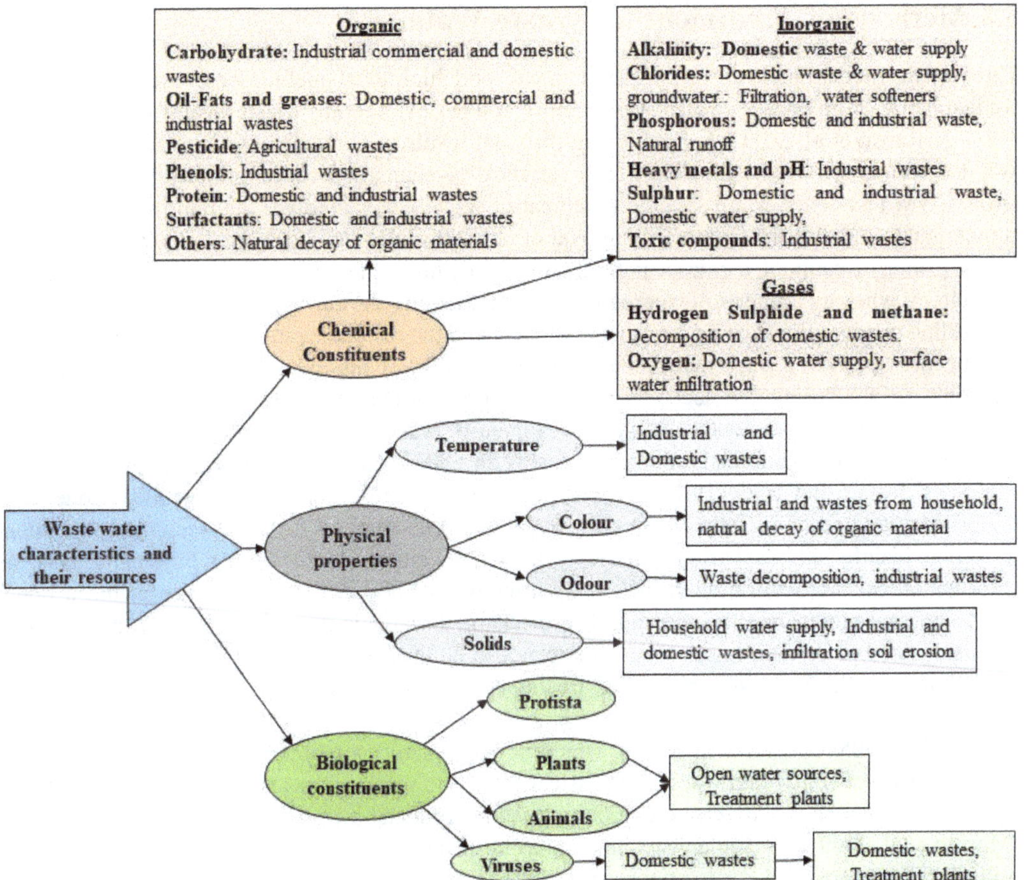

Fig. 8.1. The physico-chemical and biological parameters of waste water and their resources (reproduced from Rawat et al. 2011).

8.2.2 Chemical Characteristics

In organic waste water, the typically found combination is C, H, O, and other essential elements such as phosphorous, sulfur, ammonia, and iron. Ammonia found in waste water can be considered as it is the chemical proof of organic contamination (Fig. 8.1).

The prime units of waste water are carbohydrates, lipids, oils, proteins, and urea. Minute quantities of numerous synthetic organic chemicals can also be found. The other common inorganic constituents in waste water are nitrogen, chloride, hydrogen, iron, phosphorus, sulfur, and of heavy metals in trace amounts (Jorgensen and Weatherley 2003, Rawat et al. 2011).

8.2.3 Biological Characteristics

The role of the biological parameter can be considered one of the most significant to WWT because naturally, waste water contains huge amounts of macro and microscopic living entities (Fig. 8.1). The treatment effectiveness can be determined by a number of both entities of micro and macro-organisms as well as aquatic animals in a collecting body of waste water (Rawat et al. 2011, Renuka et al. 2016).

8.3 Methods of Treatment for Waste Water

Three basic modes such as physical, chemical, and biological methods are used to eliminate contaminants from waste water. Common methods of treating waste water are shown in Fig. 8.2 (Rawat et al. 2011). For achieving different levels of contaminant elimination, individual WWT methods are used collectively in a variety of systems and called primary, secondary, and tertiary WWT systems. More specific treatment of waste water involves the removal of particular contaminants and also the controlled removal of nutrient molecules (Yang et al. 2007).

In primary treatment, a quiescent basin temporarily holds the sewage where solids get settled in the bottom, whereas lighter solids, grease, and oil floats into the surface area (Fig. 8.2). Following the settling process, the floating items are removed, and the residual water may either be discharged or sent to a secondary treatment facility. The elimination of dissolved and suspended biological materials occurs during the secondary treatment process (Mantzavinos and Kalogerakis 2005). Water-borne microbes conduct secondary treatment of waste water. Prior to discharge or tertiary treatment, secondary treatment sometimes needs an additional separation procedure to eliminate bacteria (Metcalf and Eddy 1991). Tertiary treatment is considered as any treatment done after the primary and secondary treatment (Fig. 8.2). Water discharged from the secondary treatment is further disinfected via various means of treatment, either chemically or physically, such as chlorination or micro-filtration, which is an effective method before the discharge into a pond, river, stream, lagoon, bay, or wetland. The discharged water is sometimes used in irrigation as well, subjected to its cleanliness. The water can also be utilized for groundwater rejuvenation and agricultural applications. The biological treatment of waste water is considered to be the most environmentally benign and cost-effective among the other treatment methods (Mantzavinos and Kalogerakis 2005, Gasperi et al. 2008). This process uses microorganisms playing a crucial role in the breakdown of compounds present in waste materials to valorize the residues by producing value-added products such as a diverse range of high-value products, including biofuels and biopolymers (Morillo et al. 2009). For some years now, it has been spurred that the microalgae can be intensely utilized for low cost and for treating waste water in an environmentally friendly way, in contrast to other more generally available treatment protocols (Thangam et al. 2021). The correct strain of microalgae for WWT, as well as for generating alternative fuel sources, requires sustainable growth media such as domestic waste water load. Most waste waters contain an incredibly high amount of nutrients, primarily total N and P concentration along with toxic metals; hence, there is no need for expensive chemical-based treatment media (Morillo et al. 2009). According to (Thangam et al. 2021), microalgal species *Scenedesmus* has preferentially takes 97.7% sulfate, 89.6% phosphorus,

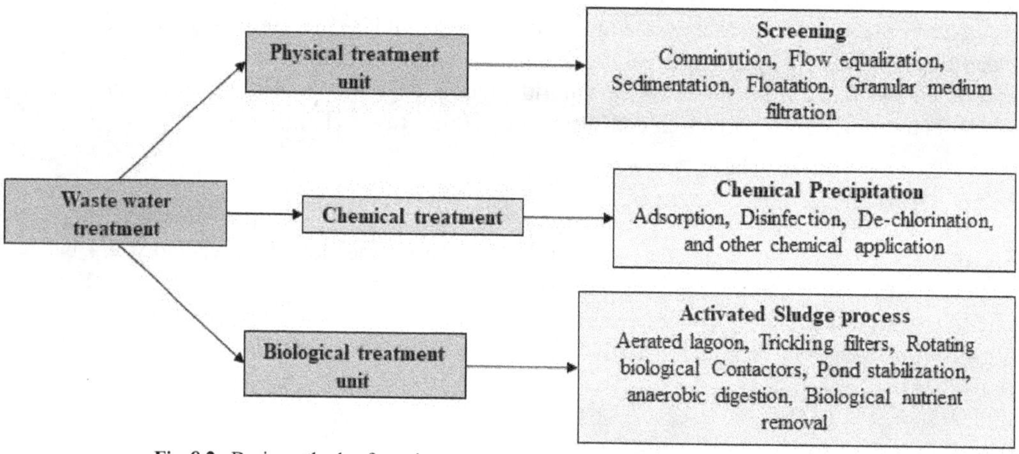

Fig. 8.2. Basic methods of treating waste water (reproduced from Rawat et al. 2011).

71.2% nitrate, 30% ammonia, 23.5% calcium, 15.2% alkalinity, 10.9% chloride, and 2% magnesium from the waste water, which contributes to the waste water treatment (Thangam et al. 2021).

The sustainability of treating waste water has been firmly established by applying microalgae for a long time (De-Bashan and Bashan 2010). The dual role of microalgae on WWT as well as energy production has been proposed and discussed for a very long time span (Oswald and Golueke 1960, Benemann et al. 1977, Hoffmann 1998, Mallick 2002, De-Bashan and Bashan 2010). However, in recent years the popularity of microalgae seems to be high among researchers for this purpose because of their non-complicated cultivation and their utmost possibility of use as alternative biomass for biofuel generation. The high alarming issue of global warming, along with the depletion of fossil fuel and the necessity for mitigation of greenhouse gas (GHS) emissions, has made microalgal application for biological waste water treatment and biofuel production a need.

8.4 Phycoremediation

Phycoremediation is understood in a broad sense as the use of macro or microalgae for the cleansing or biotransformation of pollutants which may include nutrients and xenobiotics from degraded water or waste water. The name of the process of waste water treatment, "Phycoremediation", was introduced by John in the year 2000 (John 2000). Biological treatment is a successful process for the removal of nutrients, pathogenic organisms, and heavy metals. They also contribute oxygen to heterotrophic aerobic bacteria for mineralizing organic pollutants, consuming, in turn, the carbon dioxide liberated from bacterial respiration (Munoz and Guieysse 2006). Because of its advantages, photosynthetic aeration facilitates minimizing operation costs and sets a benchmark for volatilizing pollutants under mechanical aeration. Studies have demonstrated that microalgae can help the aerobic degradation of (WWT) hazardous contaminants (Safonova et al. 2004, Munoz and Guieysse 2006). Examples are shown in Table 8.1, which depicts the function of algae in helping the elimination of environmental contaminants, either directly or indirectly, via the microbial population in the system. The biomass generated as a result of waste water treatment may be collected using various flocculation methods and turned into a variety of value-added products as well as biofuel. The application of the produced biomass also depends on the species of microalgae used for the purpose.

8.4.1 Uses of Biomass after Phycoremediation Process

8.4.1.1 As biofertilizers

Microalgae has excellent application in agriculture as a soil corrector. It has a distinctive property of fixing atmospheric nitrogen in the soil, and because of this purpose, it is used for increasing fertility and acts as a biofertilizer. When added to the field, it gives results in improving physico-chemical properties such as electrical conductivity, pH, the existence of significant elemental components in the soil, for example, nitrogen, phosphorus, and potassium, etc. (Renuka et al. 2016). *Spirulina, Nostoc, Gloeotrichia, AnabaenaandAulosira,* which are members of blue-green algae, are employed as biofertilizers to capture atmospheric nitrogen and fix it in the soil (Priyadarshani and Rath 2012). Microalgae are extremely essential in rice agriculture owing to their capacity to fix nitrogen. It has a high yield, cheap cost, and is eco-friendly (Dineshkumar et al. 2018).

8.4.1.2 Production of biochar

Slow pyrolysis of algae biomass may be used to create biochar, which is a carbon-dense substance rich in carbon. This kind of biochar has a much higher pH than other types of biochar, which may help to increase the pH of acidic soils overall. Moreover, it may enhance the nutritional content of (N) nitrogen, (P) phosphorus, and other inorganic elements in the soil, further improving the soil's

Table 8.1. Contribution of microalgae to the elimination of environmental pollutants (Rawat et al. 2011).

Sr. No.	Microalgae	Aquatic microalgae	Waste water types	References
1.	*Prototheca zopfii*	Fresh water	Hydrocarbons from degraded petroleum present in Louisiana crude and motor oils waste	Walker et al. 1975
2.	*Chlamydomonas* species	Fresh water	Waste water from meta cleavage	Jacobson and Alexander 1981
3.	*Chlorella pyrenoidosa*	Fresh and Brackish water	Role in azo dyes degradation in waste water	Jinqi and Houtian1992
4.	*Chlorella* species	Fresh and marine water	Anaerobic digested dairy waste	Wang et al.2010
5.	*Ankistrodesmus* and *Scenedesmus*	Fresh water	Olive oil mill wastewaters and paper industry waste waters	Tran et al. 2010
6.	*Spirulina platensis*	Brackish and Freshwater	Domestic wastewater treatment	Lalibert et al. 1997
7.	*Chlorella sorokiniana*	Freshwater	Wastewater treatment in aerobic dark heterotrophic condition	Ogbonna et al. 2000
8.	*Botryococcus braunii*	Freshwater	In secondarily treated sewage for continuous and batch cultures	Sawayama et al. 1994
9.	*Scenedesmus*	Freshwater	Elimination of ammonia from anaerobic digestion discharge effluent containing excess levels of ammonia and alkalinity	Jin et al. 2005

fertility for agricultural uses (Chaiwong et al. 2013). The existence of several functional moieties and inorganic components in algal biochar plays a crucial part as an accumulation surface in the remediation process for waste water purification (Yu et al. 2017).

8.4.1.3 Conversion to biofuels

Due to their high photosynthetic efficiency and capacity to retain large quantities of lipids in their intracellular structure, the biomass produced by phycoremediation is capable of generating biodiesel (Fathi et al. 2013). Microalgae are widely used for their bioremediation properties, which enable them to clean waste water by eliminating the majority of inorganic components at the source. Microalgal species such as *Chlorella vulgaris* have been utilized to produce biofuels in the past (Das et al. 2019). Microalgae have been accepted for utilization to produce biodiesel because of their high ability to metabolize CO_2, abundantly apply waste water, and yield a higher amount of intracellular lipids (Sharma et al. 2020).

8.4.1.4 Bio−hydrogen from microalgae

Hydrogen gas is preferred as eco-friendly and non-contributor to global warming fuel. The major issue in using hydrogen as an energy source is the sustainable production of hydrogen gas (Jones et al. 2012). Several groups of photosynthetic organisms were identified to generate hydrogen in the presence of water and sunlight. *Chlamydomonas reinhardtii*, which is a eukaryotic organism, acts as an important hydrogen producer. In the same way, microalgae have immense potential for the production of bio-hydrogen by appropriately designing photobioreactors for large-scale production of hydrogen gas (Kumaran et al. 2016).

8.5 Detailed Information of Algal fuel Production Treating Waste Water

8.5.1 Harvesting

Successful treatment of waste water and the production of algal biomass for biofuels require harvesting of biomass from water. The selection of an appropriate harvesting method is an important aspect as it affects the economics of biofuels. The cost of the harvesting process can be 20–30% of the net charge of production. Therefore, adequate selection of harvesting method is crucial, and it depends on the characteristics of the microalgal culture grown (Brennan and Owende 2010). Algae, which are appropriate for the remediation of waste water and production of biofuels, are unicellular and have low density. Biomass recovery methods involve centrifugation, flocculation, floatation, filtration, and sedimentation. The continuous centrifugation process is the preferred method for biomass separation owing to its rapid, efficient, and universal nature (Kong et al. 2010). Sedimentation due to gravity is also a common method of harvesting algal biomass. The process is age-old but works for different types of algae and is significantly energy effective. Another common method used for biomass collection is filtration. The flocculation method works on the dispersion of the charge mechanism. Microalgae are generally negatively charged, and the addition of metal salts plays a role in displacing the charge, stimulating aggregation of the algal biomass, and creating efficient sedimentation or filtration. Potash alum is generally used in conventional WWT, where it suitably flocculates algal biomass (Grima et al. 2003). Flocculation may also be done through the addition of alkali substances to increase the pH or by using cationic polymers. The cost of using cationic polymers is high and may adversely disrupt the techno-economy (Antoni et al. 2007). Auto-flocculation is a process where spontaneous aggregation of algal particles takes place, resulting in sedimentation of the microalgae. Filtration is a common method of solid-liquid separation of culture medium. Vacuum filtration can also be adopted for larger algae (greater than 70 μm) in combination with a filtering tool. Smaller algae generally require membrane micro-filtration or ultrafiltration for an effective harvesting process (Danquah et al. 2009). Immobilization is also used as a technique to separate biomass from culture media, and immobilized biomass after harvesting can be used for biofuel conversion by thermal or fermentation process. The floatation process in algal biomass has shown promising results in producing tiny and unicellular algae in laboratory-scale experiments (Cheng et al. 2011).

8.5.2 Methods of Lipid Extraction from Algal Biomass

Microalgal lipid extraction is a major step for the production of biodiesel. Lipid extraction is performed by physical methods such as oil pressing technique and chemical method which involve solvent extraction. The extraction method employed should be fast, effective, and non-damaging to lipid molecules (Cheng et al. 2011). The modified method of lipid extraction is the most commonly used technique (Bligh and Dyer 1959). The use of a particular solvent for lipid extraction depends on the type of microalgae used. In a sustainable process, solvents must be non-polar, cheap, non-toxic, volatile, and poor extractors of other cellular components. Depending on the biomass, pre-treatment of samples may be needed. Solvents usually break cells; thus, this isn't required for wet biomass extraction. Cell disruption may be achieved via homogenization, sonication, bead beating, grinding, or freezing (Singh and Singh 2010). Cell disruption also includes the methods of osmotic shock, autoclaving, freeze-drying, and microwaving (Bligh and Dyer 1959). The selected protocols of cell disruption will be performed by the state of biomass, the type of biomass, and the scale that requires to be utilized. Cell disruption methods will be chosen based on the condition of the biomass, the kind of biomass, and the size required. The direct esterification and transesterification of microalgal fatty acids may be carried out on both wet as well as dry biomass, making it a potential technique for biofuel production. The method consists of many stages and involves solvent extraction, ultrasonication, high-pressure heating (3.5 atm), filtering, liquid density separation, and solvent and oil recovery through evaporation for drying (Singh and Singh 2010).

8.5.3 Characterization of Extracted Lipid

It is critical to detect lipids since the lipid fraction will have a direct effect on the produced biodiesel. Lipids can be qualified and quantified using a variety of techniques, including Nile red spectrofluorometry, Nile red fluorescence microscopy, thin-layer chromatography (TLC), Fourier transform infrared microspectroscopy (FTIR), high-pressure liquid chromatography (HPLC), gas chromatography (GC), or any chromatography combined with another mass spectrometry technique (Singh and Singh 2010). Nile red microscopy has mainly been used to detect the existence of lipid fractions inside cells as a preliminary screen for lipid accumulation and as a semi-quantitative method for lipid storage monitoring. Another method known as Nile red spectrofluorometry may also be used to determine the lipid content in a semi-quantitative manner. However, these techniques do not provide information on the kind of lipid present or the lipid percentage. FTIR may be used to determine the identity, amount, and composition of lipids and carbohydrates and is an effective technique for monitoring lipid accumulation (Singh and Singh 2010).

8.5.4 Transesterification

Raw microalgal oil has a high level of viscosity. Therefore, it requires conversion to lower molecular weight constituents in the form of fatty acid (FA) alkyl esters to meet the specifications of biodiesel (Singh and Singh 2010). Transesterification is the method wherein the microalgal lipid in raw form (free fatty acids (FFA)/triacylglycerols) gets converted into harmless, renewable biodiesel fuel involving alcohol and catalyst, which is biodegradable (Fig. 8.3). The nature of the reaction is reversible and hence requires the supply of surplus amount alcohol to maintain equilibrium shift in the product side with a fast reaction rate (Demirbas 2011). The transesterification reaction is performed with oil/fat substrate mixing with short-chain alcohol; usually, it is methanol in the presence of a catalyst. Outcomes of the reaction are FA methyl esters and glycerol (Devi et al. 2015). Acids, bases, or enzymes are commonly used as catalysts for transesterification reactions. Base catalysis is a faster reaction, but it can convert all free fatty acids (FFA) into methyl ester, and the FFA contents come in the range of 20–50%. It is responsible for saponification during base-catalyzed transesterification. Acid catalysis is suitable for the transesterification of oils containing high levels of free fatty acids (Devi et al. 2017). Several renewable heterogeneous catalysts have been investigated to carry out transesterification reactions and have shown significant results (Eq. 8.1). More research is still required and vital to establishing the efficiency of this technology for the generation of biodiesel (Gohain et al. 2017).

Eq. 8.1. Broad equation of Transesterification reaction.

$$R{-}C({=}O){-}OR' + R''{-}OH \rightleftharpoons R{-}C({=}O){-}OR'' + R'OH$$

8.6 Pros of Phycoremediation Process

- Algae have the biological ability to combat different problems associated with waste water treatment which is not doable by usual chemical treatment. For example, algae can perform

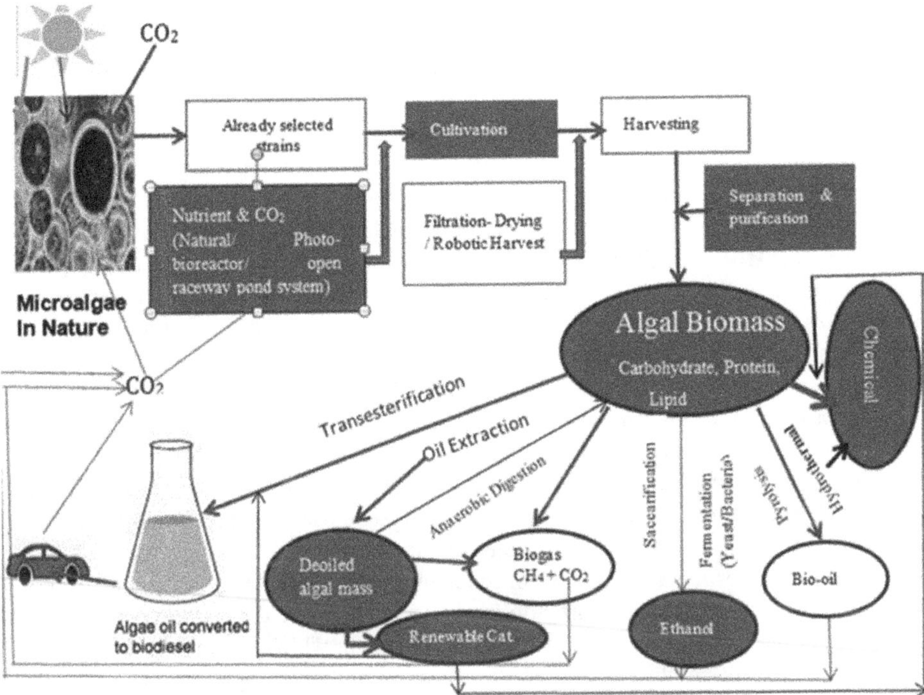

Fig. 8.3. Bio-refinery concept involving microalgae.

bioremediation and produce biomass that can be used for biofuel production and value-added products to meet the energy crisis.

- Algal photobioreactors can be activated according to the need. The photobioreactor system is possible to use in a case-specific mode in batches, semi-continuous mode, as well as in continuous mode for treating waste water.
- Algae have commercial value because they can produce valuable products, and can be acquired from algae as active metabolites such as eicosa hexaenoic acid, astaxanthin, and β-carotene.
- Algal technology does not create disturbance to industrial processes. The primary requirements of algal growth incorporate water, proper nutrient, and land.
- Phycoremediation is considered a cost-effective technique and has no involvement of toxic chemicals in its large scale operation.
- Algae play a crucial role in the environment by means of CO_2 sequestration, and it is a potential remedy for the risk of global warming. Oxygen is liberated and, in turn, can be supportive in cleansing the surrounding environment.
- Phycoremediation can thrive in fluctuating growth conditions; they can sustain irrespective of the quality as well as quantity of the nutrient fed to it such as the waste water with variable amounts of pollutants that exist in it.
- Phyco-technological techniques are tech-friendly, which is relatively easy because skilled personnel are not required for its operation and maintenance.
- This technique has an explicit role in eliminating pollutants from the contaminated site without stressing the environment.
- Lastly, this technology has tremendous scope to be explored further to unravel the advantages of this method so that it can be exhibited at a large scale or industrial scale (Bansal et al. 2018).

8.7 Cons of Phycoremediation Technology

The core disadvantages of the phycoremediation process are (i) the space needed for the luxuriant growth of algal culture (ii) the time taken in the process involved is also a matter of concern. Though numerous algal species are preferred for producing abundant bioactive compounds that have great pharmaceutical application, the purpose is not yet accomplished due to the minimal production. In this process, the main drawback lies in the less production of required compounds produced in the algal cells; therefore, the primary key belongs to the algal cell itself, where it has to behave as cell factories (Fu et al. 2016). Genetically modified (GM) algae can resolve the issue of low production, but these are not allowed to develop for the purpose of bioremediation in an open system. The algal species modifies itself with respect to the environmental variation and carry on growing efficiently in an open system. In an open system, the probability of contamination is there; however, this situation can also be resolved by using mild concentrated antibiotics. Thorough sterilization of the photobioreactor where the algal strains are being produced may be maintained in order to eliminate the threat of contamination. A considerable amount of input, such as a constant supply of CO_2, is required for the development of algae in an effective photobioreactor, and it requires a significant amount of work and is thus costly. For the growth of algae, closed and functional photobioreactors with sufficient light, temperature, and CO_2 conditions are needed. Moreover, downstream processing factors such as the harvesting and recovery of secondary metabolites produced in the algae cells are expensive way. When algal cultures are applied in large areas for the purpose of cleaning waste water or algal raceway ponds, the pace is relatively sluggish, which allows it to be tolerated for a long time and, as a result, inhibits economic growth in the particular region (Bansal et al. 2018).

8.8 Adopted Means to Solve the Bottlenecks of Phycoremediation

8.8.1 Genetic Engineering

In this technique, genetic information of the algal cells can be modified, and it is performed through mutagenesis or genetic recombination. Selective and specific breeding is performed via physical and chemical mutagens. It can be attained through the arbitrary and specific introduction of the external DNA molecule. The genetically altered algae cells are also referred to as genetically modified organisms (GM) since they have the ability to accelerate the pace at which the desired products are formed. The most crucial objective is to isolate or cultivate a strain of algae with advantageous characteristics distinct from the parent algal species. Biotechnology provides the opportunity for the development of genetically modified strains in which the target gene encoding the desired characteristic can be incorporated into algal cells, which will prompt further to manifest the required attributes in the target algal cells in future cultures or applications (Hlavova et al. 2015, Guihéneuf et al. 2016).

8.8.2 Function in Biorefinery

Algae are a significant source to build and functionalize a biorefinery. It is an essential step to establish connections between algae and several industrial establishments. Many inputs as well as output goods will be necessary to monitor a well-settled biorefinery system. Additionally, algal biomass provides numerous energy and non-energy resources as feedstock material and is depicted in Fig. 8.3 (Trivedi et al. 2015). The proteinaceous portion of biomass can be utilized as food and feed; the lipid portion can be employed to give biodiesel, syngas, methane, and kerosene; the carbohydrate fragment helped in the synthesis of alcohols and acids. Likewise, in addition to producing useful chemicals; substances such as pigments or carotenoids, antioxidants, polyunsaturated fatty acids (PUFA), vitamins and others medicinal goods are also synthesized from algae. Algal biomass may also be utilized to produce biofuels. Algae are a powerful source of energy that may help to reduce our reliance on petroleum-based fuels and the release of greenhouse gases into the environment. Economic, as well as environmental sustainability, can only be accomplished via value-added products production in combination with the production of biofuels (Foley et al. 2011).

Petroleum-based industries are posing tremendous adverse impacts on the environment. Due to their numerous advantages, algae have been anticipated as an important biofuel source. The advantages of algae over other biofuel sources outweigh the disadvantages. Life cycle assessments are important for determining the system's economic, environmental, and social performance (Kumar et al. 2017). After realizing the broad utilization of algae in several domains, it is significant to achieve its life-cycle assessment studies.

8.8.3 Contamination Problem in Algae Culture

The production of algal biomass in massive amounts is hindered by biological contaminants, which prevent the industrial procedure to establish. In most cases, algal culture is polluted by zooplankton and bacteria. Therefore, further research is necessary to recognize the cohabitation of zooplanktons, microalgae, and bacteria (Wang et al. 2013). Mainly, algal contamination is triggered by zooplankton, bacteria, other algae, and virus. Amongst the zooplanktons, mostly ciliates, rotifers, and cladophora feed on microalgae in extensive cultivation systems and hinder the purpose. *Flavobacterium* sp., *Alteromonas* sp., *Pseudomonas* sp., and *Bacillus* sp. are bacterial species that result in breakage of the microalgal cells. Microalgae in bioreactors can be spoiled by the presence of other microalgal species because of direct cell contact or rivalry for the nutrient requirement. Viral species are also found to affect cyanobacterial species and eukaryotic algae; the viral species are the LPP virus and CCV virus, respectively (Wang et al. 2013). These contaminations by biological factors can be eradicated by tactics like filtration, changing culture conditions, using chemical agents, etc. (Wang et al. 2013). Furthermore, the development of basic knowledge of the interaction of microalgal species with pollutants may pave the way for the development of adequate controls for biological contaminants in the future. Solutions such as the cultivation of genetically resistant algae species, the modification of the contamination survival particular inhibitor, and constant monitoring of contamination may result in some progress. Additionally, in recent times it has been proposed that contamination due to microorganisms can be prohibited by introducing the cationic biopolymer α-Poly-l-lysine (α-PLL) in culture systems. The polymeric chain of this compound helps in achieving harvest quickly. Apart from quick growth, the biological impurities of the algal biomass can be restricted owing to its intrinsic anti-microbial activity of α-PLL, and therefore, it can be applied to comfort the harvesting and avoid contamination of the algal biomass (Noh et al. 2018).

8.9 Prospects of Future Research

The treatment of industrial waste water is urgently required to rehabilitate polluted water bodies such as ponds, lakes, and rivers to provide safe drinking water. Since algae can clean contaminated water to a certain degree, it may be dumped without harming open water bodies. Also, understanding initial metal ion concentration, biosorption capability, and pre-treatment are required to create engineered strains with higher ion absorption and biomass concentration. This knowledge will help create inexpensive algal strains with higher bioremediation capability (Zeraatkar et al. 2016). The study of growth optimization of *Spirulina* sp. has been performed under diverse temperature and pH conditions as well as several media compositions for finding out its best growth conditions and the yield (Thirumala 2012). Algae can be used as a suitable expression media for recombinant protein expression in comparison to higher plants and animals. There are some limitations that need to be addressed (Hempel et al. 2011). There are protocols for genetic manipulation of the typical organisms such as *Phaeodactylumtricornutum* (diatom) and the *Chlamydomonasreinhardtii* (green algae). Genome editing, algal transgenic expression, and the discovery of genetic components in algal chromosomes have all seen significant advancements in recent years. These advances have the potential scope for the development of algae biotechnology in the future (Scaife and Smith 2016). Algae have been studied for their ability to synthesize silver nanoparticles as well as their anti-fungal capabilities, among other things. It is possible to increase the biocompatibility of silver nanoparticles for human use by using algae-mediated production and evaluating their anti-fungal efficacy against clinical infections, which will help to broaden the scope of the study (Rajeshkumar et al. 2014). When it comes to the production of metal oxide nanoparticles, microalgae have also been used, and this algae-based method has the potential to significantly reduce environmental contamination while also avoiding health risks associated with the use of toxic chemicals in the production of metal oxide nanoparticles, which is in the current scenario (Fawcett et al. 2017). It is necessary to take an innovative approach in order to fight the eutrophication problem by putting in place a system in which novel notions, mechanisms, and protocols are discussed and implemented on a continuous basis in order to resolve the difficulties of eutrophication and water management. Numerous researches should be conducted on the destruction of coral reefs, whose destruction poses risk to the environment, and the solutions to save this delicate ecosystem by purifying the water system, among other things (Van Ginkel 2012). There is an increasing need for assessing the purity of water and introduce some appropriate methods for improving the quality of water utilizing algae for eutrophication management. Ocean acidification assessments should be performed more often since marine ecosystems have a direct impact on human existence. These environments promote algae growth and reduce acidity levels, rejuvenating the marine ecology and allowing the re-colonization of live creatures. Due to the economic significance of this field of study, it must be evaluated for advanced research in the near future (Tu et al. 2017). A viable alternative feedstock for algae is challenging to discover. But one of its difficulties is industrial or large-scale manufacturing economic feasibility (slow and less biomass from algae, harvesting of biomass spending high energy increases cost of production). Humanity now needs more biomass with more energy-efficient harvesting methods. So, sophisticated and viable downstream processes are required now. Considering the current situation, advanced harvesting techniques, photobioreactors for algae growth, and biorefining concepts will decrease the overall cost of algal biofuel production (Behera et al. 2015).

8.10 Conclusions

There are enormous challenges that are forced on phycoremediation technology. However, one needs to be optimistic to employ the technology for both phycoremediation and industrial applications of microalgae. Until this time, higher lipid producing microalgae strain as a source for phycoremediation as well for biofuel generation has gained much attention. This chapter has included the latest advancements in experimenting with microalgae for the treatment of waste water and the biodiesel generation process. Microalgae have contributed widely for tertiary treatment of prevailing conventional WWT systems and more closely for BOD and nutrient accumulation in engineered systems such as high rate algal pond culture systems. Phycoremediation has proved to be a very suitable method surpassing the problems associated with the treatment of waste water sources with the use of expensive, toxic, environment degrading chemical compounds. Phycoremediation process has challenges such as land space need, nutrient demand, supply, the suitable algal strain of specific species requirement, and mixing of pollutants from unwanted interfering species in the environment. Further, effective and advanced research is needed to produce sufficient algal biomass for future large-scale or industrial uses. Researchers, industrialists, and students must constantly assess and evaluate algal-based systems and technology in industrial and academic contexts so that this method can be implemented most appropriately for humankind's holistic advancement and development without disturbing environmental health.

Acknowledgment

The authors are highly thankful to the Director, Dean, and Head, Institute of Environment and Sustainable Development, for providing necessary help. RPS is grateful to the authorities of Banaras Hindu University, Varanasi, for providing support under the IOE (Institute of Excellence) scheme under Dev scheme No. 6031. AD is indebted to the Department of Biotechnology, Ministry of Science and Technology, Govt. of India for providing a DBT-RA fellowship (Award Letter No. DBT-RA/2021/January/NE/633) to her.

References

Abinandan, S., M. Premkumar, K. Praveen and S. Shanthakumar. 2013. Nutrient removal from sewage-An experimental study at laboratory scale using microalgae. Int. J. Chem. Tech. Res. 5: 2090–2095.

Antoni, D., V. V. Zverlov and W. H. Schwarz. 2007. Biofuels from microbes. Appl. Microbiol. Biotechnol. 77: 23–35.

Bansal, A., O. Shinde and S. Sarkar. 2018. Industrial waste water treatment using phycoremediation technologies and co-production of value-added products. J Bioremediat. Biodegrad. 9: 2.

Benemann, J. R., J. C. Weissman, B. L. Koopman and W. J. Oswald. 1977. Energy production by microbial photosynthesis. Nature. 268: 19–23.

Behera, S., R. Singh, R. Arora, N. K. Sharma, M. Shukla and S. Kumar. 2015. Scope of algae as third generation biofuels. Front. Bioeng. Biotechnol. 2: 90.

Bligh, E. G. and W. J. Dyer. 1959. A rapid method of total lipid extraction and purification. Can. J. Biochem. Physiol. 37: 911–917.

Brennan, L. and P. Owende. 2010. Biofuels from microalgae-a review of technologies for production, processing, and extractions of biofuels and co-products. Renew. Sustain. Energy Rev. 14: 557–577.

Chaiwong, K., T. Kiatsiriroat, N. Vorayos and C. Thararax. 2013. Study of bio-oil and bio-char production from algae by slow pyrolysis. Biomass Bioenergy 56: 600–606.

Cheng, Y. L., Y. C. Juang, G. Y. Liao, P. W. Tsai, S. H. Ho, K. L. Yeh et al. 2011. Harvesting of *Scenedesmus obliquus* FSP-3 using dispersed ozone flotation. Bioresour. Technol. 102: 82–87.

Danquah, M. K., B. Gladman, N. Moheimani and G. M. Forde. 2009. Microalgal growth characteristics and subsequent influence on dewatering efficiency. Chem. Eng. J. 151: 73–78.

Das, V., A. Devi, R. Das, M. C. Kalita and D. Deka. 2019. Microalgal lipid augmentation of *Chlorella* sp. and algal biodiesel production using CaO as catalyst-A green outlook. Algal Res. 10: 43–53.

De, K., K. Venkataraman and B. Ingole. 2017. Current status and scope of coral reef research in India: A bio-ecological perspective. Indian J. Geo-Marine Sci. 46(4): 647–662.

De-Bashan, L. E. and Y. Bashan. 2010. Immobilized microalgae for removing pollutants: review of practical aspects. Bioresour. Technol. 101: 1611–1627.

Demirbas, A. 2011. Competitive liquid biofuels from biomass. Appl. Energy 88: 17–28.

Devi, A., V. K. Das and D. Deka. 2015. Designer biodiesel: Preparation of biodiesel blends by mixing several vegetable oils at different volumetric ratios and their corresponding fuel quality enhancement. Res. J. Chem. Sci. 60–65.

Devi, A., V. K. Das and D. Deka. 2017. Ginger extract as a nature based robust additive and its influence on the oxidation stability of biodiesel synthesized from non-edible oil. Fuel. 187: 306–314.

Dineshkumar, R., R. Kumaravel, J. Gopalsamy, M. N. A. Sikder and P. Sampathkumar. 2018. Microalgae as bio-fertilizers for rice growth and seed yield productivity. Waste Biomass Valoriz. 9: 793–800.

Dominic, V. J., S. Murali and M. C. Nisha. 2009. Phycoremediation efficiency of three micro algae chlorella vulgaris, Synechocystis Salina and Gloeocapsagelatinosa. SB Acad. Rev. 16: 138–146.

Fathi, A. A., M. M. Azooz and M. A. Al-Fredan. 2013. Phycoremediation and the potential of sustainable algal biofuel production using waste water. Am. J. Appl. Sci. 10: 189.

Fawcett, D., J. J. Verduin, M. Shah, S. B. Sharma and G. E. J. Poinern. 2017. A review of current research into the biogenic synthesis of metal and metal oxide nanoparticles via marine algae and seagrasses. J. Nanosci. 1–15.

Foley, P. M., E. S. Beach and J. B. Zimmerman. 2011. Algae as a source of renewable chemicals: opportunities and challenges. Green Chem. 13: 1399–1405.

Fu, W., A. Chaiboonchoe, B. Khraiwesh, D. R. Nelson, D. Al-Khairy, A. Mystikou et al. 2016. Algal cell factories: approaches, applications, and potentials. Mar. Drugs. 14: 225.

Gani, P., N. M. Sunar, H. Matias-Peralta, A. A. A. Latiff, U. K. Parjo and A. R. A. Razak. 2015. Phycoremediation of waste waters and potential hydrocarbon from microalgae: a rev. Adv. Environ. Biol. 9: 1–8.

Gani, P., N. M. Sunar, H. Matias-Peralta and A. A. A. Latiff. 2016. Application of phycoremediation technology in the treatment of food processing waste water by freshwater microalgae *Botryococcus* sp. J. Eng. Appl. Sci. 11: 7288–7292.

Gasperi, J., S. Garnaud, V. Rocher and R. Moilleron. 2008. Priority pollutants in waste water and combined sewer overflow. Sci. Total Environ. 07: 263–272.

Gohain, M., A. Devi and D. Deka. 2017. Musa *balbisiana Colla* peel as highly effective renewable heterogeneous base catalyst for biodiesel production. Ind. Crops Prod. 109: 8–18.

Grima, E. M., E. H. Belarbi, F. A. Fernández, A. R. Medina and Y. Chisti. 2003. Recovery of microalgal biomass and metabolites: process options and economics. Biotechnol. Adv. 20: 491–515.

Guihéneuf, F., A. Khan and L. S. P. Tran. 2016. Genetic engineering: a promising tool to engender physiological, biochemical, and molecular stress resilience in green microalgae. Front. Plant Sci. 7: 400.

Gupta, A., M. Kumar and I. S. Thakur. 2017. Analysis and optimization of process parameters for production of polyhydroxyalkanoates along with waste water treatment by *Serratia* sp. ISTVKR1. Bioresour. Technol. 242: 55–59.

Hempel, F., J. Lau, A. Klingl and U. G. Maier. 2011. Algae as protein factories: expression of a human antibody and the respective antigen in the diatom *Phaeodactylumtricornutum*. PloS one. 6: 28424.

Hlavova, M., Z. Turoczy and K. Bisova. 2015. Improving microalgae for biotechnology-From genetics to synthetic biology. Biotechnol. Adv. 33: 1194–1203.

Hoffmann, J. P. 1998. Waste water treatment with suspended and nonsuspended algae. J. Phycol. 34: 757–763.

Jacobson, S. N. and M. Alexander. 1981. Enhancement of the microbial dehalogenation of a model chlorinated compound. Appl. Environ. Microbiol. 42: 1062–1066.

Jin, Y., M. C. Veiga and C. Kennes. 2005. Bioprocesses for the removal of nitrogen oxides from polluted air. J. Chem. Technol. Biotechnol.: Int. Res. Process, Environ. Clean Technol. 80: 483–494.

Jinqi, L. and L. Houtian. 1992. Degradation of azo dyes by algae. Environ. Pollut. 75: 273–278.

John, J. 2000. A self-sustainable remediation system for acidic mine voids. In 4th International conference of diffuse pollution. 1: 506–511.

Jones, C. S. and S. P. Mayfield. 2012. Algae biofuels: versatility for the future of bioenergy. Curr. Opin. Biotechnol. 23: 346–351.

Jorgensen, T. C. and L. R. Weatherley. 2003. Ammonia removal from waste water by ion exchange in the presence of organic contaminants. Water Res. 37: 1723–1728.

Kong, Q. X., L. Li, B. Martinez, P. Chen and R. Ruan. 2010. Culture of microalgae Chlamydomonasreinhardtii in waste water for biomass feedstock production. Appl. Biochem. Biotechnol. 160: 9–18.

Kumar, V., R. P. Karela, J. Korstad, S. Kumar, R. Srivastava and K. Bauddh. 2017. Ecological, economical and life cycle assessment of algae and its biofuel. In Algal biofuels. Springer. Cham: 451–466.

Kumaran, M., A. Khalid, H. Salleh, A. Razali, A. Sapit, N. Jaat and N. Sunar. 2016. Effect of algae-derived biodiesel on ignition delay, combustion process and emission. Conference Series: Mater. Sci. Eng. 160: 012031.

Lalibert, C. G., E. J. Olguin and J. de la Noiie. 1997. Mass cultivation and waste water treatment using Spirulina. In Spirulina Platensis Arthrospira. CRC Press: 177–192.

Mallick, N. 2002. Biotechnological potential of immobilized algae for waste water N, P and metal removal: a review. Biometals. 15: 377–390.

Mantzavinos, D. and N. Kalogerakis. 2005. Treatment of olive mill effluents: Part I. Organic matter degradation by chemical and biological processes-an overview. Environ. Int. 31: 289–295.

Metcalf, L., H. P. Eddy and G. Tchobanoglous. 1991. Waste water engineering: treatment, disposal and reuse (Vol. 4). New York: McGraw-Hill.

Morillo, J. A., B. Antizar-Ladislao, M. Monteoliva-Sánchez, A. Ramos-Cormenzana and N. J. Russell. 2009. Bioremediation and biovalorisation of olive-mill wastes. Appl. Microbiol. Biotechnol. 82: 25–39.

Munoz, R. and B. Guieysse. 2006. Algal-bacterial processes for the treatment of hazardous contaminants: a review. Water Res. 40: 2799–2815.

Noh, W., J. Kim, S. J. Lee, B. G. Ryu and C. M. Kang. 2018. Harvesting and contamination control of microalgae *Chlorella ellipsoidea* using the bio-polymeric flocculant α-poly-l-lysine. Bioresour. Technol. 249: 206–211.

Ogbonna, J. C., H. Yoshizawa and H. Tanaka. 2000. Treatment of high strength organic waste water by a mixed culture of photosynthetic microorganisms. J. Appl. Phycol. 12: 277–284.

Olguin, E. J. 2003. Phycoremediation: key issues for cost-effective nutrient removal processes. Biotechnol. Adv. 22: 81–91.

Oswald, W. J. and C. G. Golueke. 1960. Biological transformation of solar energy. Adv. Appl. Microbiol. 2: 223–262.

Paxéus, N. 1996. Organic pollutants in the effluents of large waste water treatment plants in Sweden. Water Res. 30: 1115–1122.

Priyadarshani, I. and B. Rath. 2012. Commercial and industrial applications of micro algae-A review. J. Algal Biomass Util. 3: 89–100.

Rajeshkumar, S., C. Malarkodi, K. Paulkumar, M. Vanaja, G. Gnanajobitha and G. Annadurai. 2014. Algae mediated green fabrication of silver nanoparticles and examination of its antifungal activity against clinical pathogens. Int. J. Metals. 8.

Rawat, I., R. R. Kumar, T. Mutanda and F. Bux. 2011. Dual role of microalgae: phycoremediation of domestic waste water and biomass production for sustainable biofuels production. Appl. Energy. 88: 3411–3424.

Renuka, N., R. Prasanna, A. Sood, A. S. Ahluwalia, R. Bansal, S. Babu, R. Singh, Y. S. Shivay and L. Nain. 2016. Exploring the efficacy of waste water-grown microalgal biomass as a biofertilizer for wheat. Environ. Sci. Pollut. Res. 23: 6608–6620.

Safonova, E., K. V. Kvitko, M. I. Iankevitch, L. F. Surgko, I. A. Afti and W. Reisser. 2004. Biotreatment of industrial waste water by selected algal-bacterial consortia. Eng. Life Sci. 4: 347–353.

Sawayama, S., S. Inoue and S. Yokoyama. 1994. Continuous culture of hydrocarbon-rich microalga Botryococcusbraunii in secondarily treated sewage. Appl. Microbiol. Biotechnol. 41: 729–731.

Scaife, M. A. and A. G. Smith. 2016. Towards developing algal synthetic biology. Biochem. Soc. Trans. 44: 716–722.

Sharma, B., P. Paul, A. Devi, M. C. Kalita, D. Deka and V. V. Goud. 2020. Cost effective biomass production of Chlorella homosphaera and Scenedesmus Obliquus-Two biofuel potent microalgae from northeast India. Asian J. Microbiol. Biotechnol. Environ. Sci. 22: 486–490.

Singh, S. P. and D. Singh. 2010. Biodiesel production through the use of different sources and characterization of oils and their esters as the substitute of diesel: a review. Renew. Sustain. Energy Rev. 14: 200–216.

Thangam, K. R., A. Santhiya, S. R. Abinaya Sri, D. MubarakAli, S. Karthikumar, R. Shyam Kumar et al. 2021. Bio-refinery approaches based concomitant microalgal biofuel production and waste water treatment. Sci. Total Environ. 785: 147267.

Thirumala, M. 2012. Optimization of growth of Spirulina platensis LN1 for production of carotenoids. Int. J. Life Sci. Biotechnol. Pharm. Res. 1: 152–157.

Tran, N., H. Bartlett, J. R. Kannangara, G. S. K. Milev, A. S. Volk and M. A. Wilson. 2010. Catalytic upgrading of biorefinery oil from micro-algae. Fuel. 89-265-274.

Trivedi, J., M. Aila, D. P. Bangwal, S. Kaul and M. O. Garg. 2015. Algae based biorefinery-how to make sense? Renew. Sustain. Energy Rev. 47: 295–307.

Tripathi, R., A. Gupta and I. S. Thakur. 2019. An integrated approach for phycoremediation of waste water and sustainable biodiesel production by green microalgae, *Scenedesmus* sp. ISTGA1. Renew. Energy 135: 617–625.

Tu, Q., M. Eckelman and J. Zimmerman. 2017. Meta-analysis and harmonization of life cycle assessment studies for algae biofuels. Environ. Sci. Technol. 51: 9419–9432.

Van Ginkel, C. E. 2012. Algae, phytoplankton and eutrophication research and management in South Africa: past, present and future. Afr. J. Aquat. Sci. 37: 17–25.

Walker, J. D., R. R. Colwell, Z. Vaituzis and S. A. Meyer. 1975. Petroleum-degrading achlorophyllous alga Protothecazopfii. Nature. 254: 423–424.

Wang, H., W. Zhang, L. Chen, J. Wang and T. Liu. 2013. The contamination and control of biological pollutants in mass cultivation of microalgae. Bioresour. Technol. 128: 745–750.

Wang, L., Y. Wang, P. Chen and R. Ruan. 2010. Semi-continuous cultivation of *Chlorella vulgaris* for treating undigested and digested dairy manures. Appl. Biochem. Biotechnol. 162: 2324–2332.

Yang, Q., Z. H. Chen, J. G. Zhao and B. H. Gu. 2007. Contaminant removal of domestic waste water by constructed wetlands: effects of plant species. J. Integr. Plant Biol. 49: 437–446.

Yu, K. L., B. F. Lau, P. L. Show, H. C. Ong, T. C. Ling, W. H. Chen et al. 2017. Recent developments on algal biochar production and characterization. Bioresour. Technol. 246: 2–11.

Zeraatkar, A. K., H. Ahmadzadeh, A. F. Talebi, N. R. Moheimani and M. P. McHenry. 2016. Potential use of algae for heavy metal bioremediation a critical review. J. Environ. Manag. 181: 817–831.

CHAPTER 9

Lignocellulosic Waste as Adsorbent for Water Pollutants

A Step towards Sustainability and Circular Economy

Andrea Beatriz Saralegui, Maria Natalia Piol, Victoria Willson, Nestor Caracciolo, Cristina Vázquez* and *Susana Patricia Boeykens*

9.1 Introduction

Waste production and disposal by different industries or processes are a direct consequence of the historically used linear economy. The new millennium brought change to this economic-environmental paradigm and to sustainable development and circular economy concepts. The notion of process sustainability was initially introduced in 1987 (United Nations 1987) as the development that meets current needs without compromising the future generations. On the other hand, a circular economy describes an enclosed system of interactions between economy and environment. This concept appeared for the first time in 1980 in a book written with an economic environmental vision. These authors demonstrated how recycling and substitution processes could modify the finite natural resources exploitation, leading to the development of new technologies (Pearce and Turner 1990). Currently, the circular economy is considered a strategy that aims to reduce the entry of new raw materials into production systems and waste generation, thus closing the "loops" of economic and ecological resource flows. The Pact of Amsterdam defines it as the re-use, repair, refurbishment, and recycling of existing materials and products to promote new growth and job opportunities. This Pact focuses on waste management, turning it into a resource, on a collaborative economy, and improving process efficiency (European Union 2016). It differs from the linear economy, in which every raw material enters the production cycle and becomes a waste to be disposed of, hence generating an infinite resource need (Picone and Seraffini 2019). Therefore, the close link between sustainable development, the environment, and the economy must be recognized (Panchal et al. 2021, Macedo 2005). Rethinking water decontamination treatments as a cycle linked to other

Universidad de Buenos Aires, Facultad de Ingeniería, Instituto de Química Aplicada a la Ingeniería (IQAI), Laboratorio de Química de Sistemas Heterogéneos (LaQuíSiHe). Av. Paseo Colón 850, C1063ACV. Buenos Aires, Argentina.
* Corresponding author: laquisihe@fi.uba.ar

production cycles offers a new vision and a more economically and environmentally efficient result (European Commission 2018).

Wastewaters have a diverse composition and its pollutants, both organic and inorganic substances, coexist causing different effects on biota and the environment (Li et al. 2019, Liu et al. 2008, Di Giulio and Newman 2008, EPA 2000). The sources of these substances are also varied. Tanneries, electroplating, metallurgic, among other industries, generate large amounts of highly toxic and bioaccumulative inorganic pollutants.

Metals' presence in concentrations higher than the permitted limits happens to be quite risky as they are not biodegradable and are indeed persistent and bioaccumulative. This last characteristic gives them the ability to biomagnify due to the amount of metal increase in each upper link of the food chain. By these means, chronic or acute pollutant exposure can occur in individuals not directly exposed to the water source. Some metals such as copper and zinc are essential and are required to maintain homeostasis in individuals in a certain range of concentrations. Others, such as lead, nickel, and metalloids such as arsenic, are considered carcinogenic in humans by the International Agency for Research on Cancer (IARC). Chromium, on the other hand, as Cr(III), is considered essential; however, in the oxidation state of Cr(VI), it proves to be highly toxic and also deemed carcinogenic by the IARC (IARC 2012, Liu et al. 2008, Goyer and Clarkson 2001). The case of limiting nutrients is contrasting since when in excess, they accelerate the eutrophication of water bodies and, in the case of nitrate, it produces toxicity (Ansari and Singh Gill 2014, Di Giulio and Newman 2008, Manahan 2003).

Chromium, lead, copper, and nickel metals, as well as limiting nutrients such as phosphates and nitrates, are some examples of pollutants found in body fluids. The simultaneous contaminants' presence could produce synergistic or antagonistic effects which do not take place in individual systems (Wang and Peng 2010).

Since 2007, Australia and New Zealand have included the effect of pollutant mixture in their Water Quality Guidelines (Norwood et al. 2007, 2003, Playle 2004). In order to globalize these regulations, the development of toxic agent mixture studies and the evaluation of the effects from joint exposure to various chemical substances is being encouraged (Shuhaimi-Othman and Pascoe 2007, Feron and Groten 2002).

The search for low-cost alternatives for pollutants removal is an objective to be attained in order to improve the application feasibility to industrial processes and for water treatment in low-income communities. The application decision of a water treatment system would imply a direct decrease in industries' environmental effects. To achieve this goal, a system that is also efficient, easy to implement, and has low maintenance should be assembled.

The biosorption process is considered an adequate alternative treatment due to its easy operation and simple design. The addition of low-cost quality requires the use of new materials, among which waste represents a substantial resource.

Lignocellulosic residue refers to the residual plant biomass from different processes. These wastes are attractive alternative materials in emerging countries due to their low cost and abundance. As examples of available sources for this type of waste, the following can be mentioned:

- Natural vegetation which, due to its rapid growth, must be removed from its habitat since its proliferation can affect the ecosystem's harmonious functioning, as in the case of some floating aquatic macrophytes that are considered pests by the Food and Agriculture Organization (Saralegui et al. 2021, Montaño Heredia 2018, Saralegui 2018, Redding and Midlen 1992).
- Food industry waste is generated in large volumes near the area where the effluent treatment will be established. As examples: peanut, avocado, walnut, and banana shells; sugarcane bagasse; wheat bran; avocado and peach pit, etc. (Piol et al. 2021, Ighalo and Adeniyi 2020, Joseph et al. 2019, Piol et al. 2019, Boeykens et al. 2018, Montaño Heredia 2018, Boeykens et al. 2017, De Gisi et al. 2016, Dehghani et al. 2016, Bhatnagar et al. 2015, Gautam et al. 2014).

In most cases, lignocellulosic adsorbents have been used as raw materials without activation, and the adsorption depends on the functional groups on their surface. In order to understand and predict adsorption mechanisms that would take place, surface characterization of the adsorbent is needed. For this purpose, Fourier Transform Infrared Spectra, Nitrogen Adsorption curve, and Scanning Electron Microscopy analysis could be useful.

In addition to physical adsorption, chemical adsorption in the forms of micro-precipitation, complexation, ion exchange, and redox reactions should be considered. Knowledge about the influence on these processes of variables such as temperature, concentration, pH, ionic strength, and interaction between ions, is required for their optimization (Joseph et al. 2019, Al-Qahtani 2016, Lalley et al. 2016, Miretzky and Cirelli 2010, Miretzky et al. 2006, 2004, Naja et al. 2005, Volesky and Naja 2005, Volesky and Holan 1995). Fewer references have been found in the literature regarding the physicochemical mechanisms involved in the adsorption process using lignocellulosic adsorbents.

When choosing the adsorbent, the evaluation of its retention capacity, its regeneration and reuse possibility, as well as its final disposal options must be considered (Simón et al. 2021, European Commission 2018, Piol et al. 2019, Saralegui 2018, Bhatnagar et al. 2015).

In this chapter, the biosorption of metals on lignocellulosic residues for water decontamination is discussed.

9.2 Methodology

The procedure to obtain a correct study methodology and a subsequent effective discussion on this subject should have a strong experimental base that should also be as similar as possible. In all the cases discussed in this work, an attempt was made to choose results from similar experiments to enable its comparison.

The absorbents mentioned were first washed and dried, then ground and sieved until each case reached the specified particle size.

Tests and results described in this work carried out with batch (discontinuous) adsorption reactors were performed with continuous agitation and enough time so as to reach equilibrium.

The conditions of each test were monitored as follows: the pH with a pH meter (such as Orion™ 4-star Plus pH-ISE), the ionic strength was tracked with a conductivity meter (such as Hanna™ HI8733) or by specifying a salt concentration, and lastly, the temperature was surveilled with thermostatic equipment (such as ICSA™ HH-S4).

Tests carried out in fixed-bed reactors were executed at slow flows, the adsorbent masses were previously measured, and the particle size used in different types of reactors was preserved.

9.3 Biosorption Processes and Influencing Factors

Biosorbents can adsorb heavy metals through ion exchange, micro-precipitation, complexation, and redox reaction mechanisms, as schematized in Fig. 9.1. Conditions such as pH, ionic strength, and temperature affect the adsorption capacity (Zeraatkar et al. 2016).

The metal removal from the solution can take place through a coordination complex formation on the biosorbent surface by the interaction between the metal and one or more functional groups (-SH, $-NH_2$, -COOH, -OH, OCH_3) of its surface (Sag and Kursal 2001).

The pH has a strong influence on the speciation of heavy metals in aqueous solutions. At low pH, the majority of heavy metals are present as cationic forms, and these forms are more mobile and soluble in water. As pH increases, the formation of hydroxides and other anions begins (Joseph et al. 2019). As an example, various chromium ions (CrO_4^{-2}, $HCrO_4^-$, $Cr_2O_7^{-2}$, and H_2CrO_4) are present at different pH values (Sarı and Tuzen 2008).

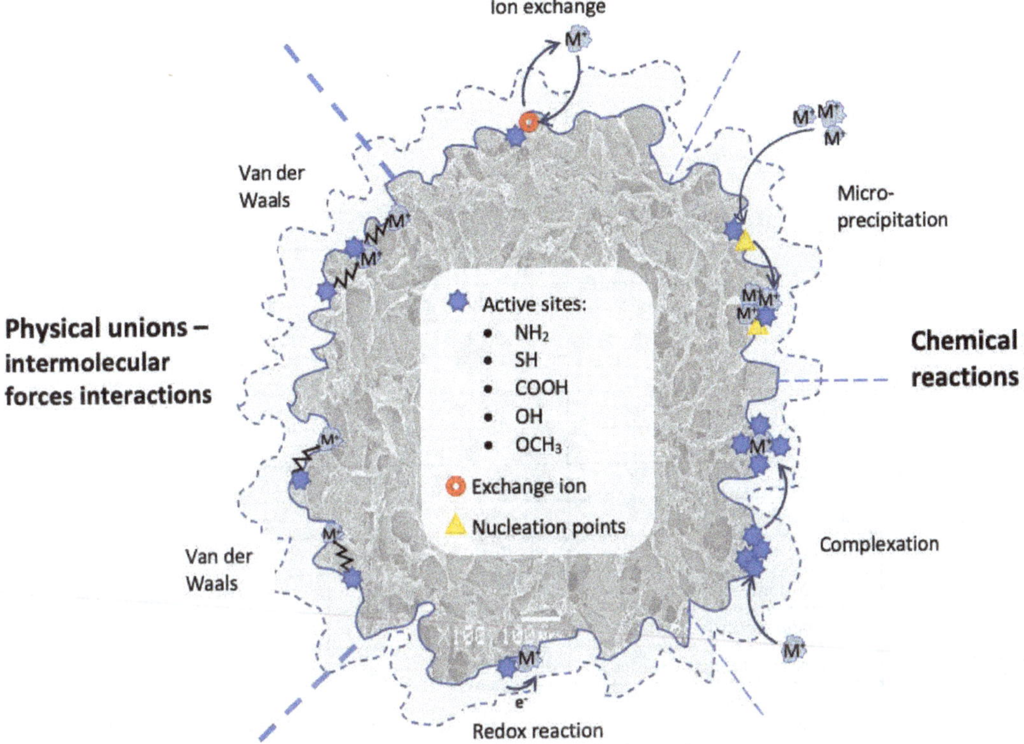

Fig. 9.1. Biosorption mechanisms involved in metal removal from a solution.

In addition, pH also has an influence on the biosorbent behavior as a result of chemical changes of the different functional groups present on the surface. A pH increase tends to improve adsorption since the adsorbent surface becomes more negatively charged and more readily interacts with heavy metal cations (Krishnani et al. 2008).

Micro-precipitation is the chemical reaction between metal ions and groups on the surface of the adsorbent, and it is generally preceded by binding to specific sites that provide nucleation points. Local deviations in physical conditions such as pH are necessary to create these nucleation points. This process contributes to the efficiency of metal removal (Ahalya et al. 2006). The interaction during the biosorption process depends on the type and quantity of active sites present on the surface, whether they are ionized, occupied by protons, or other ions. This depends on the pH and pKa of the respective functional group. The mechanisms can range from London-Van der Waals interactions or physical adsorption to chemical reactions (Uluozlu et al. 2008, Won et al. 2008). An increase in the pH of the metal ion solution during biosorption causes protons to be released from the surface, thus creating ion exchange sites (Svecova et al. 2006).

On the other hand, ionic strength may affect heavy metals biosorption. The presence of other ions could result in the formation of heavy metal complexes that are highly soluble in water (Alvim Ferraz and Lourenço 2000). Surface chemistry theories indicate a decrease in the electric double layer as the ionic strength and electrostatic interactions increase, resulting in decreased adsorption of heavy metals (Fontecha-Cámara et al. 2007).

While biosorption processes are not entirely affected by temperature, as the enthalpy involved is negligible, some surface complexation and ion exchange mechanisms of heavy metals removal are improved at higher temperatures (Chen et al. 2010). Some studies focused on biosorbents, and the experimental conditions for heavy metal removal in batch reactors are shown in Table 9.1.

Table 9.1. Studies on metals removal using lignocellulosic waste performed in batch reactors (q, adsorption capacity; T, temperature).

Metal	Biosorbent	q (mg/g)	T (°C)	pH	References
Cd(II)	Pine sawdust	1.39	NI*	3–4	Simón et al. 2021
	Moringa stenopetala	23.26	30	5	Kebede et al. 2018
	Raw sawdust (papullus)	0.74	25	5	Asadi et al. 2008
		0.20	25	2	
		0.60	25	7	
	Raw rice hull	0.80	25	5	
		0.19	25	2	
		0.63	25	7	
Cr(VI)	Wheat bran	0.50	25	5.5	Boeykens et al. 2018
	Avocado peel	0.30	25	5.5	
	Pecan shell	1.00	25	5.5	
	Peanut shell	0.80	25	5.5	
	Sugarcane bagasse	0.20	25	5.5	
	Avocado stone	0.70	25	5.5	Boeykens et al. 2019
	Azolla pinnata (macrophyte)	0.32	25	6	Saralegui et al. 2021
	Lemna minor (macrophyte)	0.08	25	6	
	Limnobium laevigatum (macrophyte)	0.47	25	6	
	Pistia stratiotes (macrophyte)	0.67	25	6	
	Salvinia molesta (macrophyte)	0.20	25	6	
	Banana peel	1.16	25	6	Piol et al. 2021
Cu(II)	Cashew leaf	1.73	40	6	Pereira et al. 2021
	Carnauba straw	9.51	50	6	
	Moringa stenopetala	10.20	30	5	Kebede et al. 2018
	Azolla pinnata (macrophyte)	18.62	25	6	Saralegui et al. 2021
	Lemna minor (macrophyte)	14.20	25	6	
	Limnobium laevigatum (macrophyte)	12.72	25	6	
	Pistia stratiotes (macrophyte)	14.37	25	6	
	Salvinia molesta (macrophyte)	11.25	25	6	
	Raw sawdust (papullus)	0.69	25	5	Asadi et al. 2008
		0.25	25	2	
		0.80	25	7	
	Raw rice hull	0.68	25	5	
		0.27	25	2	
		0.82	25	7	
Hg(II)	Banana peel	46.8	25	6	Lavanya et al. 2021

Table 9.1 contd. ...

...Table 9.1 contd.

Metal	Biosorbent	q (mg/g)	T (ºC)	pH	References
Ni(II)	Pine sawdust	5.56	NI*	3-4	Simón et al. 2021
	Raw sawdust (papullus)	0.49	25	5	Asadi et al. 2008
		0.27	25	2	
		0.65	25	7	
	Raw rice hull	0.57	25	5	
		0.30	25	2	
		0.67	25	7	
	Holly oak stem	0.14	22	4	Prasad and Freitas 2000
		0.20	22	5	
		0.22	22	6	
		0.21	22	7	
Pb(II)	Wheat bran	24.70	25	5.5	Boeykens et al. 2018
	Peanut shell	23.60	25	5.5	
	Banana peel	18.90	25	5.5	
	Avocado peel	23.00	25	5.5	
	Pecan shell	17.60	25	5.5	
	Sugarcane bagasse	8.10	25	5.5	
	Avocado stone	12.10	25	5.5	Boeykens et al. 2019
	Moringa stenopetala	16.13	30	5	Kebede et al. 2018
	Azolla pinnata (macrophyte)	43.14	25	6	Saralegui et al. 2021
	Lemna minor (macrophyte)	41.09	25	6	
	Limnobium laevigatum (macrophyte)	42.02	25	6	
	Pistia stratiotes (macrophyte)	42.22	25	6	
	Salvinia molesta (macrophyte)	39.51	25	6	
	Raw sawdust (papullus)	0.82	25	5	Asadi et al. 2008
		0.15	25	2	
		0.70	25	7	
	Raw rice hull	0.84	25	5	
		0.15	25	2	
		0.70	25	7	
	Sarsaparrilla root	25	30	2	Chandra Sekhar et al. 2003
		23	30	5	
		20	30	7	
	Algae gelidium	19.2	20	3	Vilar et al. 2005
		39	20	5	
		36.8	35	5	
		37.4	45	5	

Table 9.1 contd. ...

...Table 9.1 contd.

Metal	Biosorbent	q (mg/g)	T (°C)	pH	References
Zn(II)	Pine sawdust	0.80	NI*	3-4	Simón et al. 2021
	Raw sawdust (papullus)	0.62	25	5	Asadi et al. 2008
		0.20	25	2	
		0.80	25	7	
	Raw rice hull	0.70	25	5	
		0.23	25	2	
		0.82	25	7	
	Sarsaparrilla root	12	30	2	Chandra Sekhar et al. 2003
		16	30	5	
		15	30	7	

* Not informed

According to Table 9.1, it can be observed that at working pHs, the cations tend to be more adsorbed than anions for all biosorbents. In the case of Cr(VI), as it is more voluminous and has a negative charge, it can be observed that the adsorption capacity is lower on avocado stone, *Lemna minor*, wheat bran, avocado peel, and almost for all adsorbents tested.

Normally, most of the tests are carried out at a pH between 5 and 6 (Prasad and Freitas 2000, Asadi et al. 2008). In most cases, the influence of ionic strength has not been studied, making it impossible to assess the real situation of an effluent with high salt content. In all studied cases, a decrease in adsorption capacity is observed when increasing ionic strength (Vilar et al. 2005).

To improve chemical and physical stability, researchers have implemented biomass in blends with inorganic materials as clay particles, zeolites, or dolomite with cation exchange properties and excellent chemical resistance (Emami Moghaddam et al. 2020, Rashid et al. 2016). Some examples of biosorbents and their mixtures with other materials used for heavy metal removal are provided in Table 9.2.

The use of adsorbent mixtures is justified when considering a real effluent since its composition may contain several ions that can be adsorbed on the same material, generating a competition against the adsorption of the metal or pollutant of interest. In some cases, in order to optimize treatment, knowledge about the specificity of adsorbents for a certain contaminant permits the comparison between a reactor filled with an adsorbents mixture and sequenced reactors, each containing one adsorbent (Piol et al. 2021).

Table 9.2. Adsorbent mixtures used for different contaminants' removal.

Biosorbent	Other adsorbents	Adsorbate 1	Adsorbate 2	References
Banana peel	Dolomite	Cr(VI)	P(V)	Piol et al. 2021
Algal-alginate beads (ZABs)	Zeolite 13X	Cu(II)	---	Emami Moghaddam et al. 2020
Algal biorefinery waste	Office paper waste	Fe(II)	Congo red	Fawzy and Gomaa 2020
Coffee waste*	Limestone*	Orange II	Methylene blue	Iakovleva et al. 2017
Tea waste	Dolomite	Cu(II)	Methylene blue	Albadarin et al. 2014

*co-granulated

9.4 Fixed-bed Column Reactor Design

An adsorption reactor is a device that brings an adsorbent into intimate contact with the adsorbates present in a usually aqueous process effluent. Figure 9.2 describes three types of treatments by adsorption: the discontinuous tank (batch reactor), the continuous tank, and the tubular reactor.

The most elementary and economical devices are tubular reactors, which could use waste material from another industrial process as its filling. The adsorbent granulometry is crucial since if it is exceedingly fine, the reactor would suffer a substantial pressure drop and require additional pumping expense, and if it is overly coarse, the surface area available for mass transfer would be reduced, and the equipment would lose removal efficiency. That is why an intermediate compromise solution is adopted for adsorbent granulometry in order to avoid these unwished effects.

For the purpose of designing the reactor for certain contaminant retention, the adsorption mechanism is indistinct, whether it is physical or chemical, since the adsorption heats involved are not significant. The initial tests leading to the prototype design are performed in a normal laboratory environment, and the process is considered isothermal. The initial metal concentration (C_0) and the effluent volumetric flow (F_v) to be treated is considered available data, as well as the desired final exit concentration (C_e) at the reactor outlet, which is generally determined according to the applicable legislation in the place where the equipment is located (Treybal 1981).

If the process must operate continuously, two parallel reactors and a valve system should be installed to allow the exhausted-fill reactor to be replaced without any service interruption.

In this work, to estimate the approximate volume (V) of the reactor required for the adsorbate removal, the following considerations are proposed.

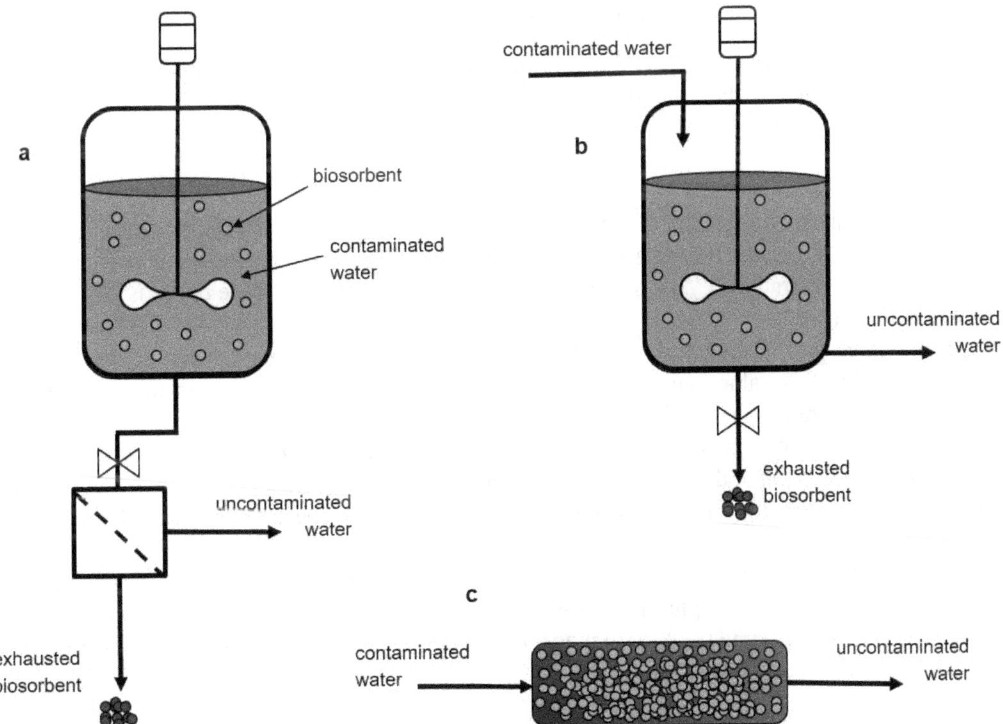

Fig. 9.2. Representation of three treatment systems for aqueous effluents decontamination by adsorption: a. Discontinuous tank (batch reactor), b. Continuous tank and c. Tubular reactor.

The mass balance for a continuous reactor could be expressed in terms of the adsorption rate (r) and the effluent molar flow to be treated ($Fm = \frac{mol\ of\ adsorbate}{t}$). Then, assuming stationary condition (approximation) the following expression is obtained:

$$r\ dV = dF_m \qquad \text{(Eq. 9.1)}$$

The rate can be written as an expression of the reaction rate using the experimental reaction order obtained from the kinetics studies in batch reactors, resulting in:

$$r = -kC^n \qquad \text{(Eq. 9.2)}$$

Where k is the rate constant, n is the reaction order, and C is moles of adsorbate per volume unit.

Therefore, as the volumetric flow, (F_v), represents the volume of solution that enters the reactor as a function of time; the relationship between both flows could be written as:

$$dF_m = dF_v\ dC \qquad \text{(Eq. 9.3)}$$

As F_v does not vary with time or the reactor length, it is more convenient to work with this parameter. From Eq. 1 and 3, the following is obtained:

$$V = F_v \int_{C0}^{Ce} \frac{dC}{r} \qquad \text{(Eq. 9.4)}$$

In order to use the reactor for a proposed treatment, the reactor residence time ($\tau = \frac{V}{F_v}$) is defined as the time to reach the maximum exit concentration (C_e) (Thirunavukkarasu et al. 2021) and could be obtained by the following expression:

$$\tau = \int_{C0}^{Ce} \frac{dC}{r} \qquad \text{(Eq. 9.5)}$$

Figure 9.3 shows the graphical representation of what happens inside the reactor in operation, while the graph illustrates the relationship between the outlet reactor solution concentration (C) and the inlet concentration (C_0) as a function of the time; this sigmoid-type curve is known as the "breakthrough curve".

In the upper part of the graph, the filling exhaustion is observed until reaching the saturation time (t_s). Beyond this value, any variable changes over time. This is the maximum time that the reactor should operate since, for longer periods, its filling is already exhausted.

To estimate a prototype design volume, firstly, the most suitable filler for batch experiments is sought, and the equation of the system's adsorption rate (kinetics) is obtained. Then, experimentally, a laboratory-scale reactor filled with the selected adsorbent is operated with a volumetric flow during the required time.

Experimentally, for a defined reactor, its total volume without filling ($V_{reactor} = \pi r^2 h$), which is equal to the free volume plus the volume occupied by the adsorbent, is measured, and the experimental residence time (τ_{exp}) for that adsorbent/adsorbate pair is calculated according to:

$$\tau_{exp} = \frac{V_{reactor}}{F_v} \qquad \text{(Eq. 9.6)}$$

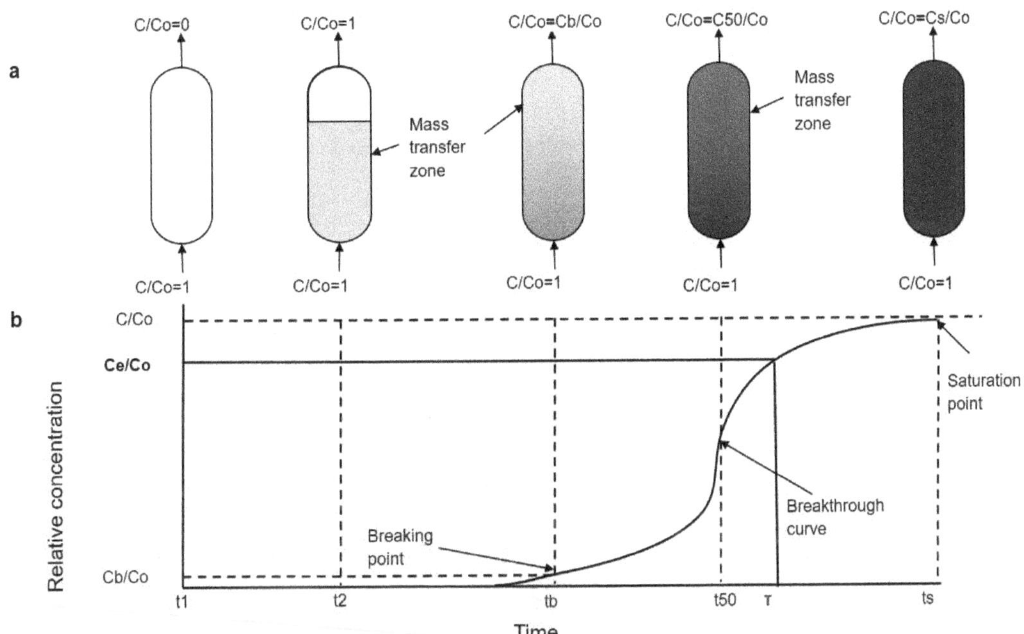

Fig. 9.3a. Representation of different reactor states during its operation. b. Breakthrough curve representing the relationship between concentrations as a function of time.

If the reactor meets the plug flow assumptions,[1] the residence time (τ) should be calculated as the suitable volume for flow inside the reactor, which is the empty volume (V_{empty}) (considering the free space inside the reactor), divided by the F_v flow (Eq. 9.7), and must also coincide with the experimental residence time (τ_{exp}) until reaching C_e.

$$\tau = \frac{V_{empty}}{F_v} \qquad \text{(Eq. 9.7)}$$

The porosity of the bed, which depends on the adsorbent particle size and the packing conditions, is calculated by Eq. 9.8:

$$\varepsilon_L = \frac{V_{empty}}{V_{reactor}} \qquad \text{(Eq. 9.8)}$$

From Eq. 9.6 and 9.8, the for a reactor of a given volume can be calculated with the bed porosity as in Eq. 9.9:

$$\tau = \varepsilon_L \; \tau_{exp} \qquad \text{(Eq. 9.9)}$$

The criteria to be used when scaling to a larger volume design is the conservation of the experimental residence time. Several scales can be made in order to design the final prototype by maintaining the same packing conditions and particle size of the bed (the porosity could also be kept constant).

[1] An accepted way of verifying that the reactor works in plug flow is to carry out a fluid dynamic test with some inert material such as glass beads of the same granulometry as the actual filling. The dispersion module D/(u*L) is thus evaluated (Levenspiel 1997).

Table 9.3. Obtained results from experimental laboratory-scale fixed-bed reactors.

Metal	Biosorbent	Column parameters	F_v (mL min^{-1})	t_b (min) (breaking point)	t_s (min) (saturation point)	References
Cd(II)	Sugarcane bagasse	$W_{adsorbent} = 7.15$ g $V_{reactor} = 49.5$ cm^3	1.60	70	128	Vera et al. 2018
Pb(II)		$W_{adsorbent} = 7.15$ g $V_{reactor} = 49.5$ cm^3	1.60	74	168	
Cr(VI)	Corn stalk	$W_{adsorbent} = 3.0$ g $V_{reactor} = 8.36$ cm^3	10.0	45	220	Chen et al. 2012
	Pistia stratiotes (macrophyte)	$W_{adsorbent} = 2.8$ g $V_{reactor} = 30.54$ cm^3	0.15	213	366	Saralegui 2018
Cu(II)	*Azolla pinnata* (macrophyte)	$W_{adsorbent} = 4.3$ g $V_{reactor} = 30.54$ cm^3	0.50	116	200	

Finally, knowing the real flow to be treated (F_v^{real}), the volume of the prototype to fulfill the service can be estimated as:

$$V_{prototype} = \tau_{reactor}\ F_v^{real} \qquad \text{(Eq. 9.10)}$$

Another widely used method to estimate the reactor design parameters, even with lignocellulosic adsorbents, is the application of different models for the breakthrough curve, such as the model used by Thomas, Yoon-Nelson, and Adam-Bohart (Pereira de Sá Costa et al. 2021, Piol et al. 2021, Thirunavukkarasu et al. 2021, Chu 2020, Saralegui 2018, Jara Enriquez 2017, Chen et al. 2012, Ahmad and Hameed 2010, Aksu and Gönen 2004).

Table 9.3 shows some results found using fixed-bed reactors. Each metal/biosorbent pair presents different operating times. As can be seen, although the systems are composed of metal adsorption on lignocellulosic waste, no generalizations can be found, and it is concluded that for each system, it is necessary to carry out the experimental process that allows us to design its industrial application.

In order to design plants working with adsorption operations, the main factors to be established are the contact time required between phases, the allowable flow rate, and the energy required to carry out the industrial operation.

9.5 Biosorbent Reuse and Final Disposal

When the adsorbent is saturated, its regeneration should be evaluated using the appropriate desorbent to allow its reuse in a new cycle. It will always be sought to achieve as many adsorption-desorption reuse cycles as possible. In the case of Copper adsorption by *Azolla pinnata* macrophyte, the ethylenediaminetetra acetic acid (EDTA) showed to be the most convenient desorbent (Saralegui et al. 2021). For aluminum adsorbed on brown algae, the best desorbent showed to be nitric acid (Pereira de Sá Costa et al. 2021), and in the case of chromate and banana peel mixture, no satisfactory results were found with the studied desorbents (Piol et al. 2021). Lead, nickel, cadmium, copper, and chromium were desorbed from holly oak biomass stems with EDTA 10 mM obtaining removal percentages between 70 and 82% (Prasad and Freitas 2000).

In the case of micronutrients adsorption such as Mn(II), Zn(II), Cu(II), Co(II), and Cr(III) ions on algal biomass, once saturated and mixed in adequate proportions with other feeds, this biomass can be used as a biological feed additive to meet the nutritional needs of farm animals (Michalak et al. 2013).

Composting and its use in soils as a nutrient source (Lei et al. 2022) are viable alternatives for final disposal as long as the final concentration of the toxic metals is taken into consideration.

Other final biomass disposal alternatives include metals recovery options (Liu and Tran 2021) as extraction and recuperation by calcination, among others. In the case that this biomass is previously used for an effluent treatment process, the application of circular economy principles would be reinforced.

9.6 Conclusion

In this chapter, a discussion about different metal biosorption processes using lignocellulosic waste and its influencing factors has been presented. Availability, accessibility, and biodegradability are important reasons for choosing this kind of biosorbents. The adsorption capacity of the same biomass collected from different parts of the world will be different due to its phytochemical constituents depending largely on its maturity degree, region of growth, and adaptive capacities. On the other hand, the effluent pH and the ionic strength level can alter the complex chemical structure of the adsorbent. In addition to the previously mentioned advantages, biosorbents are low-cost and environmentally friendly solutions for wastewater treatment. Different adsorbents have been tested, noting that the metal to be removed and the characteristics of the effluent to be treated will determine the most convenient adsorbent as well as the optimum working conditions.

The basic information for reactors' industrial-scale outline presented in this work allows the design of customized reactors, which should be conducted to solve specific application problems. This development involves previous experimental studies in order to save costs regarding materials and time.

Acknowledgments

Authors acknowledge the financial support from Universidad de Buenos Aires (UBACyT N°20020190100323BA, N°20020190200302BA, PDE 032/2020).

References

Ahalya, N., R. Kanamadi and T. Ramachandra. 2006. Biosorption of iron (III) from aqueous solutions using the husk of Cicer arietinum. Indian J. Chem. Technol. 13: 122–127.

Ahmad, A. A. and B. H. Hameed. 2010. Fixed-bed adsorption of reactive azo dye onto granular activated carbon prepared from waste. J. Hazard. Mater. 175: 298–303.

Aksu, Z. and F. Gönen. 2004. Biosorption of phenol by immobilized activated sludge in a continuous packed bed: Prediction of breakthrough curves. Process Biochem. 39: 599–613.

Albadarin, A. B., J. Mo, Y. Glocheux, S. Allen, G. Walker and C. Mangwandi. 2014. Preliminary investigation of mixed adsorbents for the removal of copper and methylene blue from aqueous solutions. Chem. Eng. J. 255: 525–534.

Al-Qahtani, K. M. 2016. Water purification using different waste fruit cortexes for the removal of heavy metals. J. Taibah Univ. Sci. 10: 700–708.

Alvim Ferraz, M. C. M. and J. C. N. Lourenco. 2000. Influence of organic matter content of contaminated soils on the leaching rate of heavy metals. Environ. Prog. 19: 53–58.

Ansari, A. and S. Singh Gill. 2014. Eutrophication: Causes, Consequences and Control. Netherlands: Springer.

Asadi, F., H. Shariatmadari and N. Mirghaffari. 2008. Modification of rice hull and sawdust sorptive characteristics for remove heavy metals from synthetic solutions and wastewater. J. Hazard. Mater. 154: 451–458.

Bhatnagar, A., M. Sillanpää and A. Witek-Krowiak. 2015. Agricultural waste peels as versatile biomass for water purification—A review. Chem. Eng. J. 270: 244–271.

Boeykens, S. P., M. N. Piol, L. Samudio Legal, A. B. Saralegui and C. Vázquez. 2017. Eutrophication decrease: Phosphate adsorption processes in presence of nitrates. J. Environ. Manage. 203: 888–895.

Boeykens, S., A. Saralegui, N. Caracciolo and M. N. Piol. 2018. Agroindustrial waste for lead and chromium biosorption J. Sustain. Dev. Energy Water Environ. Syst. 6: 341–350.

Boeykens, S. P., N. Redondo, R. A. Obeso, N. Caracciolo and C. Vázquez. 2019. Chromium and Lead adsorption by avocado seed biomass study through the use of Total Reflection X-Ray Fluorescence analysis. Appl. Radiat. Isot. 153: 108809.

Chandra Sekhar, K., C. T. Kamala, N. S. Chary and Y. Anjaneyulu. 2003. Removal of heavy metals using a plant biomass with reference to environmental control. Int. J. Miner. Process. 68: 37–45.

Chen, H., J. Zhao, G. Dai, J. Wu and H. Yan. 2010. Adsorption characteristics of Pb(II) from aqueous solution onto a natural biosorbent, fallen *Cinnamomum camphora* leaves. Desalination 262: 174–182.

Chen, S., Q. Yue, B. Gao, Q. Li, X. Xu and K. Fu. 2012. Adsorption of hexavalent chromium from aqueous solution by modified corn stalk: A fixed-bed column study. Bioresource Technol. 113: 114–120.

Chu, K. H. 2020. Breakthrough curve analysis by simplistic models of fixed bed adsorption: In defense of the century-old Bohart-Adams model. Chem. Eng. J. 380: 122513.

De Gisi, S., G. Lofrano, M. Grassi and M. Notarnicola. 2016. Characteristics and adsorption capacities of low-cost sorbents for wastewater treatment: A review. Sust. Mat. Technol. 9: 10–40.

Dehghani, M. H., D. Sanaei, I. Ali and A. Bhatnagar. 2016. Removal of chromium (VI) from aqueous solution using treated waste newspaper as a low-cost adsorbent: Kinetic modeling and isotherm studies. J. Mol. Liq. 215: 671–679.

Di Giulio, R. T. and M. C. Newman. 2008. Ecotoxicology. pp. 1157–1187. *In*: Klaasen, C. D. (ed.). Casarett and Doull's. Toxicology. The basic science of poisons. Kansas: McGraw-Hill.

Emami Moghaddam, S. A., R. Harun, M. N. Mokhtar and R. Zakaria. 2020. Kinetic and equilibrium modeling for the biosorption of metal ion by Zeolite 13X-Algal-Alginate Beads (ZABs). J. Water Process Eng. 33: 101057.

EPA, Environmental Protection Agency. 2000. *In situ* treatment of soil and groundwater contaminated with chromium-technical resource. 625/R-00/004. https://cfpub.epa.gov/si/si_public_record_report.cfm?Lab=NRMRL&dirEntryId=64150.

European Commission. 2018. Report on Critical Raw Materials and the Circular Economy. https://op.europa.eu/en/publication-detail/-/publication/d1be1b43-e18f-11e8-b690-01aa75ed71a1/language-en/format-PDF/source-80004733.

European Union. 2016. URBAN AGENDA FOR THE EU. Pact of Amsterdam. Netherlands. https://ec.europa.eu/regional_policy/sources/policy/themes/urban-development/agenda/pact-of-amsterdam.pdf.

Fawzy, M. A. and M. Gomaa. 2020. Use of algal biorefinery waste and waste office paper in the development of xerogels: A low cost and eco-friendly biosorbent for the effective removal of congo red and Fe (II) from aqueous solutions. J. Environ. Manage. 262: 110380.

Feron, V. J. and J. P. Groten. 2002. Toxicological evaluation of chemical mixtures. Food Chem. Toxicol. 40: 825–839.

Fontecha-Camara, M. A, M. V. Lopez-Ramon, M. A. Alvarez-Merino and C. Moreno-Castilla. 2007. Effect of surface chemistry, solution pH, and ionic strength on the removal of herbicides diuron and amitrole from water by an activated carbon fiber. Langmuir. 23: 1242–7.

Gautam, R. K., A. Mudhoo, G. Lofrano and M. C. Chattopadhyaya. 2014. Biomass-derived biosorbents for metal ions sequestration: Adsorbent modification and activation methods and adsorbent regeneration. J. Environ. Chem. Eng. 2: 239–259.

Goyer, R. A. and T. W. Clarkson. 2001. Toxic effects of metals. pp. 861–867. *In*: Klaasen, C. D. (ed.). Casarett and Doull's. Toxicology. The basic science of poisons. New York: McGraw-Hill.

Iakovleva, E., M. Sillanpää, P. Maydannik, J. T. Liu, S. Allen, A. B. Albadarin and C. Mangwandi. 2017. Manufacturing of novel low-cost adsorbent: Co-granulation of limestone and coffee waste. J. Environ. Manage. 203: 853–860.

IARC, International Agency for Research on Cancer. 2012. A review of human carcinogens. Monographs on the evaluation of the carcinogenic risk of chemicals in humans, 100C. Lyon, France.

Ighalo, J. O. and A. G. Adeniyi. 2020. Adsorption of pollutants by plant bark derived adsorbents: An empirical review. J. Water Process Eng. 35: 101228.

Jara Enriquez, I. 2017. Adsorption-desorption of copper in dolomite studied in a fixed-bed reactor (in Spanish). MS Thesis. Universidad de Buenos Aires, Facultad de Ingeniería, Argentina.

Joseph, L., B.-M. Jun, J. R. V. Flora, C. M. Park and Y. Yoon. 2019. Removal of heavy metals from water sources in the developing world using low-cost materials: A review. Chemosphere. 229: 142–159.

Kebede, T. G., A. A. Mengistie, S. Dube, T. T. I. Nkambule and M. M. Nindi. 2018. Study on adsorption of some common metal ions present in industrial effluents by *Moringa stenopetala* seed powder. J. Environ. Chem. Eng. 6: 1378–1389.

Krishnani, K. K., X. Meng, C. Christodoulatos and V. M. Boddu. 2008. Biosorption mechanism of nine different heavy metals onto biomatrix from rice husk. J. Hazard. Mat. 153: 1222–1234.

Lalley, J., C. Han, X. Li, D. D. Dionysiou and M. N. Nadagouda. 2016. Phosphate adsorption using modified iron oxide-based sorbents in lake water: Kinetics, equilibrium, and column tests. Chem. Eng. J. 284: 1386–1396.

Lavanya, K. M., J. A. K. Florence, B. Vivekanandan and R. Lakshmipathy. 2021. Comparative investigations of raw and alkali metal free banana peel as adsorbent for the removal of Hg^{2+} ions. Materials Today: Proceedings (in press).

Lei, L., X. Cui, C. Li, M. Dong, R. Huang, Y. Li, Y. Li, Z. Li and J. Wu. 2022. The cadmium decontamination and disposal of the harvested cadmium accumulator *Amaranthus hypochondriacus* L. Chemosphere. 286: 131684.

Levenspiel, O. 1997. Chemical Reaction Engineering (in Spanish), 6th Ed., Spain: Reverté.

Li, Y., M. A. Taggart, C. McKenzie, Z. Zhang, Y. Lu, S. Pap and S. Gibb. 2019. Utilizing low-cost natural waste for the removal of pharmaceuticals from water: Mechanisms, isotherms and kinetics at low concentrations. J. Cleaner Prod. 227: 88–97.

Liu, J., R. A. Goyer and M. P. Waalkes. 2008. Toxic effects of metals. pp. 931–979. *In*: Klaasen, C. D. (ed.). Casarett and Doull's.Toxicology. The Basic Science of Poisons. New York, United State: McGraw-Hill.

Liu, Z. and K.-Q. Tran. 2021. A review on disposal and utilization of phytoremediation plants containing heavy metals. Ecotoxicol. Environ. Saf. 226: 112821.

Macedo, B. 2005. The concept of sustainability (in Spanish). UNESCO OREALC/2005/PI/H/12. https://unesdoc.unesco.org/ark:/48223/pf0000162177.

Manahan, S. E. 2003. Toxicological chemistry and biochemistry. New York: CRC Press.

Michalak, I., K. Chojnacka and A. Witek-Krowiak. 2013. State of the art for the biosorption process—a review. Appl. Biochem Biotechnol. 170: 1389.

Miretzky, P., A. Saralegui and A. F. Cirelli. 2004. Aquatic macrophytes potential for the simultaneous removal of heavy metals (Buenos Aires, Argentina). Chemosphere. 57: 997–1005.

Miretzky, P., A. Saralegui and A. Fernández Cirelli. 2006. Simultaneous heavy metal removal mechanism by dead macrophytes. Chemosphere. 62: 247–254.

Miretzky, P. and A. F. Cirelli. 2010. Cr(VI) and Cr(III) removal from aqueous solution by raw and modified lignocellulosic materials: a review. J. Hazard. Mater. 180: 1–19.

Montaño Heredia, G. 2018. Use of weeds as biomass for the removal of arsenic and nickel (in Spanish). Graduate Thesis. Universidad de Morón. Argentina.

Naja, G., V. Diniz and B. Volesky. 2005. Predicting metal biosorption performance. Proc.16th Internat. Biotechnol. Symposium, S. Harrison, D. Rawlings and J. Petersen (eds.). IBS – Compress Co., South Africa.

Norwood, W. P., U. Borgmann, D. G. Dixon and A. Wallace. 2003. Effects of metal mixtures on aquatic biota: a review of observations and methods. Human Ecol. Risk Assess. 9: 795–811.

Norwood, W. P., U. Borgmann and D. G. Dixon. 2007. Interactive effects of metals in mixtures on bioaccumulation in the amphipod *Hyalella azteca*. Aquat. Toxicol. 84: 255–267.

Panchal, R., A. Singh and H. Diwan. 2021. Does circular economy performance lead to sustainable development? – A systematic literature review. J. Environ. Manage. 293: 112811.

Pearce, D. W. and R. K. Turner. 1990. Economics of natural resources and the environment. United States: International Monetary Fund Joint Library, pp. 35–42.

Pereira de Sá Costa, H., M. G. C. da Silva and M. G. A. Vieira. 2021. Fixed bed biosorption and ionic exchange of aluminum by brown algae residual biomass. J. Water Process Eng. 42: 102117.

Pereira, J. E. S., R. L. S. Ferreira, P. F. P. Nascimento, A. J. F. Silva, C. E. A. Padilha and E. L. Barros Neto. 2021. Valorization of carnauba straw and cashew leaf as bioadsorbents to remove copper (II) ions from aqueous solution. Environ. Technol. Innov. 23: 101706.

Picone, J. L. and G. Seraffini. 2019. Plastics recycling and circular economy (in Spanish). In the waste we generate: its sustainable management, a great challenge, ed. T. Pérez. Argentina. National Academy of Exact, Physical and Natural Sciences.

Piol, M. N., M. Paricoto, A. B. Saralegui, S. Basack, D. Vullo and S. P. Boeykens. 2019. Dolomite used in phosphate water treatment: Desorption processes, recovery, reuse and final disposition. J. Environ. Manage. 237: 359–364.

Piol, M. N., C. Dickerman, M. P. Ardanza, A. Saralegui and S. P. Boeykens. 2021. Simultaneous removal of chromate and phosphate using different operational combinations for their adsorption on dolomite and banana peel. J. Environ. Manage. 288: 112463.

Playle, R. C. 2004. Using multiple metal-gill binding models and the toxic unit concept to help reconcile multiple-metal toxicity results. Aquat. Toxicol. 67: 359–370.

Prasad, M. N. V. and H. Freitas. 2000. Removal of toxic metals from solution by leaf, stem and root phytomass of *Quercus ilex* L. (holy oak). Environ. Pol. 110: 277–283.

Rashid, A., H. N. Bhatti, M. Iqbal and S. Noreen. 2016. Fungal biomass composite with bentonite efficiency for nickel and zinc adsorption: A mechanistic study. Ecol. Eng. 91: 459–471.

Redding, T. A. and A. B Midlen. 1992. Aquatic vegetation and the problems it poses in irrigation systems. In Study of fish production in irrigation canals (in Spanish). FAO Technical Document on Fisheries. No. 317. Rome, FAO. 114p. http://www.fao.org/docrep/003/T0401s/T0401S04.htm.

Sag, Y. and T. Kursal. 2001. Recent trends in the biosorption of heavy metals: a review. Biotechnol. Bioprocess Eng. 6: 376–385.

Saralegui, A. 2018. Use of macrophyte biomass in reactors for the removal of heavy metals from aqueous effluents (in Spanish). PhD Thesis, Universidad de Buenos Aires, Facultad de Ingeniería, Argentina.

Saralegui, A. B., V. Willson, N. Caracciolo, M. N. Piol and S. P. Boeykens. 2021. Macrophyte biomass productivity for heavy metal adsorption. J. Environ. Manage. 289: 112398.

Sarı, A. and M. Tuzen. 2008. Biosorption of total chromium from aqueous solution by red algae (*Ceramium virgatum*): Equilibrium, kinetic and thermodynamic studies J. Hazard. Mater. 160: 349–355.

Shuhaimi-Othman, M. and D. Pascoe. 2007. Bioconcentration and depuration of copper, cadmium, and zinc mixtures by the freshwater amphipod *Hyalella azteca*. Ecotoxicol. Environ. Safety. 66: 29–35.

Simón, D., S. Gass, C. Palet and A. Cristóbal. 2021. Disposal of wooden wastes used as heavy metal adsorbents as components of building bricks. J. Building Eng. 40: 102371.

Svecova, L., M. Spanelova, M. Kubal and E. Guibal. 2006. Cadmium, lead and mercury biosorption on waste fungal biomass issued from fermentation industry. I. Equilibrium studies. Sep. Purif. Technol. 52: 142–153.

Thirunavukkarasu, A., R. Nithya and R. Sivashankar. 2021. Continuous fixed-bed biosorption process: A review. Chem. Eng. J. Adv. 8: 100188.

Treybal, R. E. 1981. Mass-transference operation. Singapore: McGraw - Hill.

Uluozlu, O., A. Sari, M. Tuzen and M. Soylak. 2008. Biosorption of Pb(II) and Cr(III) from aqueous solution by lichen (*Parmelina tiliacea*) biomass. Bioresour. Technol. 99: 2972–2980.

United Nations 1987. Report of the World Commission on Environment and Development, "Our Common Future". https://undocs.org/en/A/42/427.

Vera, L. M., D. Bermejo, M. F. Uguña, N. Garcia, M. Flores and E. González. 2018. Fixed bed column modeling of lead(II) and cadmium(II) ions biosorption on sugarcane bagasse. Environ. Eng. Res. 24: 31–37.

Vilar, V. J. P., C. M. S. Botelho and A. R. Boaventura. 2005. Influence of pH, ionic strength and temperature on lead biosorption by *Gelidium* and agar extraction algal waste. Process Biochem. 40: 3267–3275.

Volesky, B. and Z. R. Holan. 1995. Biosorption of heavy metals. Biotechnol. Progress.11: 235–1150.

Volesky, B. and G. Naja. 2005. Biosorption: Application Strategies. Proc. 16th Internat. Biotechnol. Symposium, S. Harrison, D. Rawlings and J. Petersen (eds.). IBS – Compress Co., South Africa.

Wang, S. and Y. Peng. 2010. Natural zeolites as effective adsorbents in water and wastewater treatment. Chem. Eng. J. 156: 11–24.

Won, S., M. Han and Y. Yun. 2008. Different blinding mechanisms in biosorption of reactive dyes according to their reactivity. Water Research 42: 4847–4855.

Zeraatkar, A. K., H. Ahmadzadeh, A. F. Talebi, N. R. Moheimani and M. P. McHenry. 2016. Potential use of algae for heavy metal bioremediation, a critical review. J. Environ. Manage. 181: 817–831.

Section II
Bioremediation of Specific Toxic Metal(loid)s

CHAPTER 10

Bioremediation of Arsenic
A Sustainable Approach in Managing Arsenic Contamination

Loveleena Khanikar and *Md. Ahmaruzzaman**

10.1 Introduction

The use of heavy metals has increased many-fold since industrialization, due to which large quantities of heavy metal containing wastes are released into the environment (Azubuike et al. 2016). Due to precipitation and water run-off, the heavy metals available in air and soil are released into water bodies. Their non-biodegradable nature results in adverse health effects on humans and animals living in that area (Jiang et al. 2019). One such heavy metal is arsenic, which is a metalloid. Arsenic forms approximately 1.5 ppm of the Earth's crust and is released through various natural and anthropogenic processes. There are four oxidation states of Arsenic (+5, +3, –3, 0), out of which arsenite and arsenate are the most frequent forms, and the noxiousness of As(III) is higher than As(V) (Satyapal et al. 2018). When food and drinking water are concerned, arsenic in both organic and inorganic forms exists depending on the type of food, whereas drinking water consists of inorganic arsenic, i.e., As(III) and As(V) (Tantry et al. 2015). Due to its germicidal power and decay-resistant nature, arsenic is mainly used in insecticides, herbicides, or wood preservatives. Some other uses of arsenic are in industries, electronics, and medicine (Chung et al. 2014). Various arsenic compounds can disperse in water bodies (in the form of rain, snow, or discarded industrial wastes), thus contaminating lakes, rivers, and ground water. Arsenic contamination in the ground, as well as surface water, is becoming a matter of serious concern due to the undesirable impact of arsenic on human health, thus affecting a huge population in several countries. Reports establish that groundwater contamination of arsenic is a major worldwide issue, since drinking arsenic contaminated water has affected around 100 million individuals all over the world (Bahar et al. 2013). The countries most affected by arsenic complications in groundwater are in certain areas of Argentina, Hungary, Bangladesh, Chile, India (West Bengal), Romania, Northern China, Mexico, Taiwan (China), and numerous parts of the USA (Pal and Paknikar 2011). Groundwater arsenic concentrations in various countries of the world are presented in Table 10.1. In the 1970's, the UNICEF teamed up with the Department of Public Health Engineering to install tubewells in Bangladesh, while the testing for arsenic contamination during the installation process was not

Department of Chemistry, National Institute of Technology Silchar-788010, Assam, India.
* Corresponding author: mda2002@gmail.com

Table 10.1. Arsenic concentration in groundwater reported in various countries (Shankar et al. 2014*, Adeloju et al. 2021).

Country	Region	Groundwater As level (μgL^{-1})	Permissible Limit (μgL^{-1})
Bangladesh*	Noakhali	< 1–4730	50(WHO)
India*	West Bengal Uttar Pradesh	10–3200	50(WHO)
China*	-------	50–4440	50(WHO)
Nepal*	Rupandehi	Up to 2620	50
Pakistan*	Muzaffargarh	Up to 906	50
Argentina	-----	10–1000	50(WHO)
Chile	------	900–1040	50(WHO)
China Inner Mongolia	-------	1–2400	50(WHO)
Lao PDR	Laos	277.8	50(WHO)
Peru	-------	500	50(WHO)
Mexico*	Lagunera	8–620	25
Afghanistan*	Ghazni	10–500	10(WHO)
Brazil*	Minas Gerais (Sotheastern Brazil)	0.4–350	10(WHO)
Canada*	Nova Scotia (Halifax country)	1.5–738.8	10(WHO)
Finland*	Southwest Finland	17–980	10(WHO)
Germany	-------	10–150	10(WHO)
New Zealand	Taupo Volcanic zone	21	10(WHO)
Spain	------	< 1–100	10(WHO)
Romania	------	10–176	10(WHO)
Taiwan*		10–1820	10(WHO)

included in the standard water testing procedures, and this resulted in mass poisoning from the intake of arsenic contaminated water (Choe and Sheppard 2016).

Humans continuously exposed to arsenic contaminated drinking water for longer durations have increased jeopardy of developing skin related symptoms like skin cancer, hyperkeratosis, and hyperpigmentation, which can induce cancer of other organs like the liver, lungs, kidney, and bladder, damage of mucous membranes, digestive, respiratory, circulatory and nervous systems (Murugesan et al. 2006, Upadhyay et al. 2018). Animals and animal products such as poultry, fish, cow milk, etc., also contain traces of arsenic contamination. In developing countries, the growth of plants is also affected by longterm arsenic exposure since arsenic contaminated water is used for soil irrigation, thus accumulating arsenic in the plants, which either degrades the quality of crops or leads to death of plants (Choe and Sheppard 2016). With the aim of minimizing adverse health effects, an upper concentration limit (MCL) of 10 μg/L for drinking water containing arsenic was set up by the World Health Organisation (WHO). Still, many developing countries are unsuccessful in maintaining this limit, and the concentration of arsenic is higher than 10 μg/L (Bahar et al. 2013, Pal and Paknikar 2011).

Removal of arsenic from water has become very necessary, considering the lethal outcomes of arsenic on both human and animal health (Bahar et al. 2013). There are several conventionally applied techniques available for the elimination of arsenic from polluted water. Some of them are chemical precipitation, filtration, oxidation-reduction, ion exchange processes, etc. (Satyapal et al. 2018). But there exist certain drawbacks of these technologies, including their high cost and release of toxic by-products. Most of the chemical methods that are used causes an increase in certain factors like pH and conductivity, along with an increase in the overall burden of dissolved matter present in wastewater (Renuka et al. 2014). Thus, increased interest to acquire a sustainable, cost-effective, and environment-friendly technique is crucial, involving minimal infrastructure and inputs for the management of arsenic infected water (Renuka et al. 2014).

It is well known that bioremediation technology is a swiftly emerging area of environmental restoration that considers the microbial community for removal of noxious components from the environment and reduce the resulting noxiousness of the pollutants (Choe and Sheppard 2016). Several categories of the biological processes exist that depends on the type of organisms used, viz. microbes, plants, or both (Shukla and Srivastava 2017). This chapter emphasizes the available bioremediation techniques for the treatment of arsenic contamination and discusses the future aspects of the microorganisms in reducing the environmental contamination.

10.2 Origins of Arsenic Release into the Environment

Arsenic is a highly toxic element that is mainly initiated into nature through innumerable natural and man-made processes (Adeloju et al. 2021). The vital causes of arsenic discharge from natural sources are volcanic eruptions, atmospheric emissions, wind deposition, biological activity, and weathering of rocks (Bahar et al. 2013).

On the other hand, some important human-caused sources that release arsenic are mining of coal, burning fossil fuels, manufacturing sulfide ores, and wood preservatives. Industries manufacturing products that contain arsenic are likely to generate solid and liquid wastes loaded with arsenic. It is used in fertilizers such as insecticides and pesticides due to its germicidal power. Arsenic is also used in manufacturing alloy, pharmaceuticals and glass industries, semiconductor industries, pulp and paper production, cement manufacture, leather preservatives, antifouling paints, livestock and industrial manufacturing, and manufacture of glassware, timber preservation, etc. Due to the lack of proper disposal of wastes, they contribute to groundwater contamination (Pal and Paknikar 2011, Hue 2015). The use of certain compounds as herbicides in orchards, for managing aquatic weeds in ponds, which includes $PbHAsO_4$, $Ca_3(AsO_4)_2$, $NaAsO_2$, $Mg_3(AsO_4)_2$, $Zn_3(AsO_4)_2$, $Zn_3(AsO_3)_2$, and Paris green also play a role in soil and groundwater impurity (Pal and Paknikar 2011).

Out of the several possible sources of environmental arsenic, groundwater used for drinking and agricultural utilities is considered to have a high concentration of inorganic arsenic (Smedley and Kinniburgh 2002). Groundwater contamination is considered as a severe risk to human health. Above 200 minerals are found to contain arsenic, amidst which Arsenopyrite (FeAsS), Realgar(As_4S_4), Orpiment, Arsenolite, and Anargite(Cu_3AsS_4) serve as the major contributors for groundwater contamination. Several geochemical processes result in the release of arsenic into groundwater. Iron and aluminum oxides found in sediments play a notable role in the contamination of groundwater (Shankar et al. 2014). The most copious arsenic source is the mineral arsenopyrite (Shankar et al. 2014). There have also been pieces of evidence that suggest that the alluvial regions of Bangladesh are found to be rich in arsenopyrite, and the oxidation of arsenopyrite is a major source of arsenic release into groundwater (Adeloju et al. 2021). Besides this, certain anthropogenic sources lead to groundwater contamination. Mining activities carried out at the Toroku mine (located in Takachilo, Nishiusuki District, Miyazaki, Japan) resulted in air and water pollution and many people suffered from arsenicosis after inhaling polluted air and drinking polluted water (Masuda 2018). However, compared to the natural sources, the extent of contamination of groundwater by anthropogenic activities is considerably less.

10.3 Arsenic Toxicity and its Impact on Health

Arsenic has been categorized as a group I human carcinogen by the International Agency for Research on Cancer (IARC) (Choe and Sheppard 2016). It is a matter of serious concern when the water used for consumption and cooking purposes exceeds the maximum concentration limit that is set for arsenic in drinking water. Consuming water contaminated with arsenic has many adverse effects on child development, and child growth and gives rise to a plethora of diseases in adults (Smith and Steinmaus 2009). It is associated with skin and internal cancers and also other noncancerous diseases like peripheral neuropathy, diabetes, and cardiovascular diseases. Compared to organic arsenic, the toxicity of inorganic arsenic is more. It overpowers the activities of various enzymes engaged in cellular respiration, glutathione metabolism, DNA synthesis and affects the progress of the fetal nervous system by passing through the placenta (Hong et al. 2014). A well-known fact is that inorganic As(III) is believed to be 60 times more lethal than As(V). The enzyme methyl transferase methylates inorganic arsenic in the presence of S-adenosylmethionine (methyl donor) and glutathione (GSH, essential co-factor) to organic arsenic metabolites where the final arsenic metabolites Mono-methylarsonic acid (MMA V) and Di-methylarsinic acid (DMAV) have lesser toxicity than inorganic arsenic, but the intermediate metabolites Mono-methylarsonous (MMA III) acid and Di-methyl arsinous acid (DMA III) have a much higher degree of toxicity than inorganic arsenic (Shankar et al. 2014). One of the most useful indicators of arsenic exposure is urinary arsenic, since it reflects the digestive and respiratory exposures (Hong et al. 2014). The presence of MMA(III) and DMA(III) was noticed in the urine of the people that were subjected to repeated exposure of arsenic contaminated water (Hughes 2002). The various arsenic species and their related toxicity is in the order (Murugesan et al. 2006): As V < MMA V < DMA V < As III < MMA III ≈ DMA III.

The prominent features of acute exposure to arsenic poisoning are nausea, seizures, vomiting, anorexia, fever, severe abdominal pain (especially in babies), diarrhea, excessive salivation, melanosis, acute psychosis, hematological abnormalities, coma, convulsions, renal failure, cardiac arrhythmia and eventually cardiovascular failure (Ratnaike 2003).

Continuous exposure to arsenic for a long term (5–10 years) can result in 'arsenicosis', a term used for numerous health disorders of arsenic, of which skin cancer, lung cancer, high blood pressure, reproductive disorders, neurological disorders, liver disorders, diabetes, etc., are a few (Sharma et al. 2014). The health consequences associated with chronic arsenic exposure can be carcinogenic or noncarcinogenic.

The carcinogenic effects of prolonged exposure to arsenic are:

(a) Skin cancer: Skin is a primary target organ that displays the effects of long-term arsenic exposure (Dwivedi et al. 2015). It can be identified as the earliest human carcinogen (Rossman et al. 2004). Several studies suggested arsenic polluted drinking water as a source of skin cancers. Oral arsenic exposure gives rise to skin lesions that occur before skin cancers, which include hypopigmentation, hyperpigmentation (found on eyelids, neck, nipple and extends to abdomen, chest, and back), hyperkeratosis (occur mostly on palms and soles), parakeratosis, and Bowen's disease (squamous cell carcinoma *in situ*) (Rossman et al. 2004, Shukla 2017). Skin cancers associated with arsenic are mostly due to squamous cell carcinoma and basal cell carcinoma caused by keratinization (Hong et al. 2014).

(b) Lung cancer: One of the established causes of lung cancer is drinking arsenic contaminated water, and preliminary evidence also suggest that arsenic ingestion during *in utero* or early childhood has noticeable pulmonary effects, which increases the death rate in young adults due to the incidence of malignant and nonmalignant lung diseases (Hughes 2002, Smith et al. 2006). Reports also justify that the combined effect of exposure to cigarette smoking and ingested arsenic increases the possibility of lung cancer (Chen et al. 2004). Chances of lung cancer are also observed in the pesticide and smelter workers (Shukla and Srivastava 2017).

(c) Other cancers due to arsenic: Besides skin and lung cancers, exposure to high concentrations of arsenic may also cause other cancers, although they have no association with exposure to low levels of arsenic (Hong et al. 2014, Chiang et al. 1993, Lin et al. 2013, Heck et al. 2014). The epidemiologic studies in Taiwan with endemic black foot disease reported a high mortality rate in both men and women due to lung, bladder, kidney, nasal cancers (Centeno et al. 2002).

Some of the noncancerous diseases associated with arsenic include diabetes mellitus, neurological effects, cardiovascular diseases, and *in utero* diseases during pregnancy. Sufficient human data were available to report the linkage of high arsenic exposures (150 µg/L in drinking water) to diabetes (Young et al. 2018). High arsenic exposure and its connection with type 2 diabetes was reported by Grau-Perez et al. (2018). Epidemiological studies suggested a strong correlation of arsenic exposure with enhanced risks of mortality due to cardiovascular ailments (Tantry et al. 2015). Chronic arsenic exposure leads to myocardial injury, cardiac arrhythmias, and cardiomyopathy (Ratnaike 2003). A peripheral vascular disease known as the Blackfoot disease has been observed in individuals of China that were persistently subjected to arsenic in drinking water which eventually led to gangrene of the foot (Ratnaike 2003). Effects of long-term exposure of arsenic contaminated groundwater in pregnant women were first reported in Bangladesh in 2001 (Milton et al. 2005). The outcomes of pregnancy are related to serious consequences like unplanned abortion (up to 28 weeks of pregnancy), miscarriages, reduced birth weight, and infant mortality (Smith and Steinmaus 2009). Milton et al. (2005) reported that childbirth and abortion chances were witnessed in people exposed to long-term arsenic concentrations of greater than 50% g/L. Reduction of newborn birth weight due to maternal arsenic exposure in the early stage of pregnancy was reported in Bangladesh (Huyck et al. 2007). The effect of arsenic exposure on cognitive function was reported in Bangladesh, which showed reduced intellectual function in children when arsenic levels in water exceeded 50 µg/L, with remarkably poor performances and full-scale scores compared to those with lesser arsenic levels than 5.5 µg/L (Wasserman et al. 2004). Decrements in intellectual functional scores with increasing concentration of arsenic in urine was reported by von Ehrenstein et al. (2007). Utero arsenic exposures are related to the occurrence of a grownup disease during childhood, and increases the risk of childhood cardiovascular diseases, cancer, and death (Young et al. 2018). All the findings point towards the fact that arsenic exposure is connected with an enhanced risk of developing various major ailments.

10.4 Conventional Treatment Methods

The conventionally used technologies mainly utilized for the treatment of arsenic contaminated groundwater are described below:

(i) Oxidation and reduction: This process usually involves As(III) oxidation as a pre-required step, which is catalyzed by chemicals including inorganic iron, manganese oxides, hydrogen peroxide, gaseous chlorine, etc. (Pal and Paknikar 2011, Adeloju et al. 2021).

(ii) Precipitation: Calcium, magnesium, Mn(II), or Fe(II) salts, when added to dissolved arsenic, gives rise to a low-solubility solid mineral which can be eradicated through the process of sedimentation or filtration. (Pal and Paknikar 2011).

(iii) Coagulation-filtration: The oxidation of As(III) is followed by adsorption upon the coagulated flocs (produced by the adding of flocculants) and can be eliminated by the process of filtration (Pal and Paknikar 2011).

(iv) Adsorption: This involves the adsorption of arsenic onto adsorbents like iron oxide or hydroxides, activated alumina, titanium oxide, cerium oxide, and zero-valent iron (Pal and Paknikar 2011, Adeloju et al. 2021).

(v) Ion exchange: Linkage of synthetic ion exchange resins to functional groups such as quarternary amine groups $N + (CH_3)_3$ can be helpful in the arsenic elimination (Pal and Paknikar 2011).

(vi) Membrane process: This technique uses the processes like nanofiltration, electrodialysis, and reverse osmosis for arsenic removal (Adeloju et al. 2021).

Despite the efficiency of these methods, there are certain disadvantages due to which there is a requirement to build sustainable and eco-friendly technologies. The conventional arsenic removal technologies and the factors accountable for reducing their efficiencies are summarized in Table 10.2.

Table 10.2. Effectiveness of the conventional technologies for removal of arsenic (Duarte et al. 2009).

Process	Chemical reagent	As(III) removal efficiency (%)	As(V) removal efficiency (%)	Ideal condition
Oxidation reduction				
Precipitation (including lime softening)	Sulphates (aluminium, copper, ammonia)	< 30	80–90	pH 6–6.5
Coagulation-filtration	Ferric chloride	< 30	90–95	pH 6–8
Adsorption	Activated alumina or carbon	30–60	> 95	pH 5.5–6
	Iron hydroxide (granular)	30–60	> 95	pH near 8
Ion exchange	Anionic resins	< 30	80–95	$[SO_4^{2-}] < 20$ mg/L $[SDT] < 500$ mg/L
Membrane filtration (nanofiltration and reverse osmosis)		60–90	> 95	Presence of dissolved As
		80–95	> 95	

10.5 Bioremediation Technologies

The term bioremediation initially appeared in a scientific literature during 1987, and presently it is one of the swiftly emerging fields of environmental restoration. It is the usage of natural microbes to eliminate the toxic environmental pollutants released due to several natural and man-made causes, thus leading to the contamination of soil and water (Dua et al. 2002). Its environmental compatibility, potential cost-effectiveness (high input to output ratio), and high-efficiency property are the reasons why it is often preferred over the conventional technologies for the removal of hazardous wastes from heavily contaminated sites. Several surveys were conducted on the removal, recovery, or detoxification of organic and inorganic metals by means of metal-microbe interactions (Gadd 2004). Certain categories of bioremediation were classified based on the different biological processes, which include mycoremediation (fungi), phytoremediation (plants), phytobial remediation (endophytic organisms), phycoremediation (algae), and microbial remediation (microorganisms) (Shukla and Srivastava 2017). The microbes used in bioremediation either degrade, reduce, eliminate or transform the contaminants and utilize them as a means of food and energy, which in turn stimulates the growth of the microbes. Several factors such as site conditions, quantity and toxicity of contaminant present, and the microorganism population present naturally decide which bioremediation technology will be suitable for a particular site. Maintaining the required conditions is very necessary for bioremediation to be effective, as it allows the right microbes to grow and multiply and eat more contaminants. Numerous strategies like exclusion, extrusion, accommodation, biotransformation, and methylation and demethylation have been developed by the organisms in order to survive in a metal stressed environment (Choe and Sheppard 2016). Based on this, three basic principles of bioremediation: bioactivity, biochemistry, and bioavailability, are considered with the aim of selecting the most appropriate strategy for the treatment of a particular site (Dua et al. 2002).

Arsenic removal from drinking water is necessary to prevent the masses from numerous health hazards of arsenic contamination. The treatment of groundwater containing arsenic using bioremediation techniques have now been widely practiced in many places since it doesn't require the use of chemical reagents and depends on the activity of the microbes to reduce, mobilize or immobilize arsenic through various processes either solitarily or by combining with the conventional methods (Zouboulis and Katsoyiannis 2005, Wang and Zhao 2009). The biological oxidation-reduction reactions catalyzed by microorganisms is a very well-known technique for the treatment of arsenic contamination. Several studies have been conducted based on different biomass types like bacteria, algae, and fungi to identify highly efficient biological procedures aimed at heavy metal removal from groundwater (Kermani et al. 2012). The microorganisms that are naturally present in the contaminated area interact with the different forms of arsenic (Pal and Paknikar 2011). The vast range of microbial mechanisms coupled with the conventional technologies can be considered an inexpensive and efficient way for arsenic bioremediation in toxic environments (Pal and Paknikar 2011).

In the following discussion, the focus will be on the different bioremediation techniques that are available, and their potential as remediation techniques in arsenic elimination, mainly from polluted groundwater and also other environmental sources.

10.5.1 Arsenic-Bacterial Interaction

10.5.1.1 Microbial redox reactions

The first report of arsenite metabolism was identified by Green in 1918, which was a bacterium of the genus *Achromobacter* (Lièvremont et al. 2009). Since then, several investigations have illustrated the ability of microorganisms in the oxidation of As(III). As(III) oxidation to As(V) is required to facilitate the process of arsenic removal from contaminated aqueous systems. But the conventional methods for As(III) oxidation are often performed using chemical reagents, leading to high expenses and the release of undesirable by-products. Although oxidation of arsenic by physicochemical methods is thermodynamically favourable, these oxidation reactions are considered extremely slow and kinetically hindered without a biological or chemical catalyst (Garcia-Dominguez et al. 2008). So, bacterial oxidation of arsenite to arsenate is considered an alternative technology for the purification of waters contaminated with arsenic (Katsoyiannis and Zouboulis 2004, Ike et al. 2008). There are reports of over 30 strains that represent at least nine genera of arsenite oxidizing prokaryotes, which are categorized under α-, β- and γ-Proteobacteria, Deinocci (i.e., *Thermus*), and Crenarchaeota (Stolz et al. 2002). The physiologically diverse bacteria that oxidize arsenite include heterotrophic species and the more recently described chemoautotrophic species (Stolz et al. 2002). Heterotrophic oxidation is generally considered as a detoxification mechanism for the conversion of As(III) to the less toxic form As(V), rather than one that can support growth (Garcia-Dominguez et al. 2008). Studies have shown that with an initial concentration of 1500 mg L^{-1} As(III), a mixed culture of bacteria (*Haemophilus*, *Micrococcus*, and *Bacillus*) displayed high As(III)-oxidizing activity in the temperature range 25–35°C and pH value 7–10, which enhanced the removal of arsenic due to efficient chemisorption of As(V) onto activated alumina (Ike et al. 2008). In contrast, the chemolithoautotrophic As(III) oxidizer uses As(III) to donate electrons, which are accepted by either oxygen or nitrate, and CO_2 is used as the source of carbon (Bahar et al. 2013, Wang and Zhao 2009). The energy produced from the oxidation of As(III) is used for the fixation of CO_2 into organic cellular material, which offers bacteria the carbon required for their growth (Lièvremont et al. 2009). Researches revealed that the As(III) bacterial oxidation was catalyzed by a periplasmic arsenite oxidase, *aox* operon (Cai et al. 2009). The periplasmic enzyme consists of two subunits, and the genes that encode them are *aoxA*/*aroB*/*asoB* (small Fe-S rieske subunit) and *aoxB*/*aroA*/*asoA* (large Mo-pterin subunit), respectively (Cai et al. 2009). *AoxR* regulates the expression of the *aox* operon, after which the complex AoxAB involved in the As(III) oxidation is synthesized and exported to the periplasm (Satyapal et al. 2018). As(III) oxidizing bacteria such as β-proteobacterial strain ULPAs1 (possessing *aoxA* and *aoxB* genes) and

Agrobacterium tumefaciens (possessing a signal transduction system; *aoxS-aoxR-aoxA-aoxB-cytc2*) isolated from arsenic-contaminated soil and water have been reported (Chang et al. 2009, Muller et al. 2003). A bacterium designated as NT-26, was isolated from a gold mine rock by Santini et al. (2000), which was from the *Agrobacterium/Rhizobium* branch of the *α-Proteobacteria.* The bacterium was found to develop chemoautotrophically with As(III), oxygen, along with bicarbonate or carbon dioxide for carbon supply, and the doubling time was 7.6 h. The uniqueness of NT-26 lies in its capacity to grow very quickly with arsenite chemoautotrophically, thus oxidizing As(III) with an enzyme (Aro) existing in the periplasmic space. Liao et al. (2011) stated that the aerobic arsenite oxidizing bacterium strain AR-11 efficiently oxidized As(III) to As(V) at concentrations relevant to environmental groundwater samples. There are also several reports showing the growth of chemolithotrophic arsenite oxidizers with several organic and inorganic sources of energy (Garcia-Dominguez et al. 2008, Duquesne et al. 2007). Removal of about 98.77% arsenic from groundwater was possible when a facultative chemolithotrophic arsenite oxidizing strain of *Delftia* spp. BAs29 was combined with a natural biosorbent (Biswas et al. 2019).

The type of species used determines the range of pH in As(III) microbial oxidation (Bahar et al. 2013). Dastidar and Wang (2009) stated that for the oxidation of As(III) (about 98.8%), the species *Thiomonas arsenivorans* required an optimum pH of 6, although a significant quantity of As(III) was oxidized (around 90.5%) at pH value 4. Many species oxidize arsenite at a pH that is near to neutral. The As(III) oxidation by *Alcaligenes faecalis* Strain O1201 was investigated at pH 4, 5, 6, 7, 8, and 9, and the optimal pH was found to be 7 (Suttigarn and Wang 2005). For bioremediation to be effective, modification of pH for different cultures is necessary as As(III) oxidation is maximum at optimum pH (Bahar et al. 2013).

A further significant procedure of arsenic decontamination is the microbial reduction of arsenate to arsenite. Arsenate reduction can be explained by two mechanisms. The detoxification and resistance mechanism which is most thoroughly analyzed is based on arsenical resistance (*ars*) operon. Bacterial ars operon comprises of a minimum of 3 genes (*arsR*, *-B*, *-C*), while an extended one comprising of 5 genes (*arsR*, *-D*, *-A*, *-B*, *-C*) was described in certain bacteria (Lièvremont et al. 2009). The regulatory genes are *arsR* and *arsD*, *arsC* gene reduces As(V) to As(III) as it encodes As(V) reductase enzyme, while energy was provided to *arsB* by the gene *arsA* to form a transmembrane efflux pump in order to export As(III) from the cytoplasm (Liao et al. 2011). Because of structural resemblances with the phosphate ions, the arsenate enters the cell via phosphate transporters (Lièvremont et al. 2009). After reaching the cytoplasm of the cell, a process mediated by the As(V) reductase enzyme *arsC* reduces As(V) to As(III), while arsB, the As(III)- specific transporter exports the As(III) from the cell (Oremland and Stolz 2005).

On the other hand, dissimilatory arsenate-reducing prokaryotes (DARPs) give rise to another process, identified as dissimilatory reduction (Lièvremont et al. 2009). The DARPs are isolated from sediments of freshwater, soda lakes, estuaries, gold mines, and hot springs (Pal and Paknikar 2011). During anaerobic respiration, the bacterial cells accept electrons using As(V) (Oremland and Stolz 2005). The subunits in arsenate reductase, ArrA, and ArrB are recognized as membrane-bound heterodimer proteins (Lièvremont et al. 2009). Although significant similarities are found to exist in the arsenite oxidases of chemoautotrophs and heterotrophs, but the arsenate reductases of arsenic resistant microbes (ARMs) and dissimilatory arsenate-respiring prokaryotes (DARPs) are found to have notable differences (Oremland and Stolz 2005). The anaerobic bacterium, *Chrysiogenes arsenatis* (strain BAL-1^T), was isolated from a gold mine by Macy et al. (1996), whose growth was achieved by reducing arsenate to arsenite. Zobrist et al. (2000) investigated that reduction of arsenate was achieved by cell suspensions of the anaerobic bacterium *Sulfurospirillium barnesii* when dissolved in solution or adsorbed onto ferrihydrite surface, along with Fe(III) reduction to soluble Fe(II) in ferrihydrite.

Thus, microbial redox reactions are one of the most efficient and extensively used systems for the treatment of arsenic contaminated water.

10.5.1.2 Methylation

Another mechanism of heavy metal remediation is microbial methylation. Inorganic arsenic can be methylated by certain bacteria and fungi that lead to the formation of methylarsenicals (Pal and Paknikar 2011). Fungi and eukaryotes played a major role in arsenic methylation, while very little was known regarding the bacterial systems (Bahar et al. 2013). Challenger (1945) and his co-workers confirmed the arsenic biomethylation to trimethylarsine, based on the study of the fungus *Scopulariopsis brevicaulis*, and it involved repeated sequencing of As(V) to As(III) and subsequent addition of a methyl group by oxidative addition. But due to a shortage of analytical methods, he could not signify the bacterial role in arsenic methylation.

At present, frequent studies on arsenic methylation by bacteria have been reported (Cullen 2014). They are found chiefly in anaerobic systems like sewage sludge, compost formed by the decomposition of organic matter, and freshwater sediments. The involvement of anaerobic bacteria (found in sewage disasters) in the methylation of arsenic was reported by Michalke (2000) and his coworkers. Several microorganisms (i.e., methanogenic archaea, involving *Methanobacterium formicicum, Methanosarcina barkeri, Methanobacterium thermoautotrophicum*; sulfate-reducing bacteria composed of *Desulfovibrio vulgaris* and *D. gigas*; *Clostridium collagenovorans* which was a peptolytic bacterium) examined in pure cultures with regard to their capability in synthesizing volatile metals showed *M. formicicum* as the most efficient.

The As(V) reduction and subsequent methyl group addition gives rise to methyl arsenicals which are discharged as gaseous products from the microbes (Pal and Paknikar 2011). The process is also known as biovolatilization. The methylated arsenic species consist of monomethyl arsonate (+5), monomethylarsonite (+3), dimethylarsinate (+5), dimethylarsenite (+3), and trimethylarsine oxide, while mono-, di-, and trimethylarsines are considered volatile (Bentley and Chasteen 2002). The stepwise transformation of As(III) into mono-, di, and trimethylated products is presented in the scheme below (Thomas et al. 2004):

$$As(III)O_3^{3-} + CH_3^+ \rightarrow CH_3As(V)O_3^{2-} + 2e^- \rightarrow CH_3As(III)O_2^{2-} + CH_3^+ \rightarrow (CH_3)_2As(V)O_2^- + 2e^- \rightarrow (CH_3)_2As(III)O^- + CH_3^+ \rightarrow (CH_3)_3As(V)O^- + 2e^- \rightarrow (CH_3)_3As(III)$$

Here, As(V) reduction to As(III) occurs, and the As(III) is then methylated. The methylation step oxidizes As(III), and the As(III) formed is again reduced and then methylated. In these reactions, the source of methyl groups is a form of methionine (Hue 2015). The *Penicillium* sp. strain sequestered from evaporation pond water by Huysmans and Frankenberger (1991) established that it was able to methylate and subsequently volatilize inorganic arsenic. The optimum conditions required for the production of trimethylarsine were monitored, and the data found could be beneficial for implementing the technique of bioremediation for arsenic removal from water or soil. Anaerobic enhancement cultures consisting of iron, manganese, sulfate-reducing heterotrophs, and broad-spectrum anaerobic heterotrophs were isolated from a lake sediment contaminated with arsenic, all of which had the ability to produce mono-, di- and trimethyl arsenicals, with the sulfate-reducing bacteria producing the highest concentration of methylarsenicals (Bright et al. 1994).

The demethylation mechanisms are much less when compared to methylation mechanisms (Stolz et al. 2006). There are reports which suggest the microbial dealkylation of cacodylic acid and sodium arsonate, and the species *Alcaligenes*, *Pseudomonas*, and *Mycobacterium* demethylated compounds of monomethyl and dimethyl arsenic (Bentley and Chasteen 2002, Stolz et al. 2006).

A significant role may be played by methylation and demethylation mechanisms that will have an effect on the toxicity and arsenic mobility in groundwater and soils, in which the volatile arsenic escapes from the surfaces of water or soil by volatilization. The methylation process does not necessarily contribute to the detoxification mechanism since current data suggests the formation of MMA(III) and DMA(III) by methylation to be toxic (Qin et al. 2006). The trivalent form of methylated arsenic may also take part in DNA damage (Stolz et al. 2006). Increasing order for strength of DNA follows the sequence (Wang and Chen 2006): trimethylarsine oxide [TMAO(V)] < DMAA(V) < MMAA(V) < [As(III), As(V)] < MMAA(III) < DMAA(III).

10.5.1.3 Arsenic bioremediation by sulfate–reducing bacteria

Many researchers have successfully demonstrated the ability of sulfate-reducing bacteria in arsenic removal from waters and wastewaters. Arsenic present in groundwater may be efficiently removed by the application of biological sulfate reduction technologies. Under strong reducing conditions, sulfate is microbially reduced to sulfide, and the sulfide formed is precipitated from the solution with iron (Rittle et al. 1995). Arsenic and other metals can be eliminated from water by coprecipitating with iron sulfides either as orpiment (As_2S_3), realgar (AsS), arsenopyrite (FeAsS)-like phase, or by adsorption in the presence of iron upon biogenic mackinawite (FeS), greigite (Fe_3S_4) or pyrite (FeS_2)-like phase (Alam and McPhedran 2019). Major arsenic tolerant SRBs that take part in the bioremediation of arsenic are *Desulfovibrio, Deltaproteobacteria, Desulfomicrobium*, and *Desulfosporosinus*. The ability of SRB in reducing sulfate to sulfide followed by diminution of As(III) concentration in water was first studied by Rittle et al. (1995). The microbes control the heavy metal mobility in reducing environments, since sulfidogenic bacteria is responsible for the reduction of sulfate to sulfide at surface temperatures and pressure (Rittle et al. 1995). Based on the mechanism, the SRB uses sulfate to accept electrons during the process of respiration, whereas the agro-industrial and the discarded hydrocarbons are utilized to donate electrons or as sources of carbon, thus generating hydrogen sulfide (Hussain et al. 2016). Metal sulfides are formed due to vigorous reaction between H_2S and the dissolved metals, which being insoluble, settle down as precipitates (Hussain et al. 2016). Arsenic immobilization by sulfide occurs through a complex pathway and is strongly dependent on pH, as the reactions controlling the solubility of arsenic in the presence of sulfide are dependent on pH (Keimowitz et al. 2007). The formation of thioarsenite species at high pH limits the elimination of soluble arsenic by biomineralization (Alam and McPhedran 2019). Due to this, either neutral (Teclu et al. 2008) or mildly acidic pH ranges are mostly preferred. The SRB produced adsorbent is considered to have a higher specific capacity of uptake of different metal ions in solution when compared to the other adsorbents (Teclu et al. 2008).

Kirk et al. (2004) conducted studies on SRB and suggested that by stimulating the sulfate-reducing bacteria in the subsurface, remediation of naturally occurring arsenic in groundwater may be effective. Thereafter, several other investigations continued to determine the ability of SRB in arsenic removal from groundwater. Studies conducted by Teclu et al. (2008) on arsenic bioremoval from groundwater using a mixed culture of SRB showed that when the precipitate produced by SRB was bought into contact with 1 mg/L of arsenite or arsenate, about 77% arsenate and 55% arsenite removal was observed. The data obtained for adsorption followed the Langmuir isotherm and Freundlich isotherm. Saalfield and Bostick (2009) investigated the sulfur, iron, and arsenic speciation that was affected by dissimilatory sulfate and secondary iron reduction, suggesting that As(V) bacterial reduction was required for As sequestration in sulfides. Another analysis by Upadhyaya et al. (2012) showed that key factor for the removal of arsenic by almost 90% was the existence of SRB in high quantities and colocation of sulfate and arsenate-reducing activities in the presence of iron(II). Experiments conducted by Sun et al. (2016) using sediments in a local groundwater showed that the remediation of arsenic by using SRB in order to produce arsenic-bearing sulfides might not always be fruitful. Apart from the laboratory conducted studies, few studies have also been conducted in the contaminated sites. For example, Ludwig et al. (2009) designed a permeable reactive barrier (PRB) to analyze its efficiency in eliminating arsenic and also other heavy metals (Pb, Cd, Zn, Ni) from groundwater by promoting sulfate reduction by SRB and sulfide mineral precipitation and sorption of arsenic. Efficient As(III) and As(V) removal was observed.

This draws towards the conclusion that SRB can be used as a method for handling arsenic contamination in groundwater.

10.5.2 Mycoremediation (Fungal Bioremediation)

Mycoremediation is considered to be an environment-friendly, economical and successful treatment strategy in soil and water pollution. Fungi are eukaryotic organisms and are known for their role as active agents of microorganism-mediated bioremediation. The fungi groups with chief practical importance are yeasts, mushrooms, and molds (Ayele et al. 2021). Based on the mode of sexual reproduction, the classification of kingdom Fungi includes phyla Zygomycota (conjugated fungi), Chytridiomycota (chytrids), Deuteromycota (imperfect fungi), Basidiomycota (club fungi), Ascomycota (sac fungi), and Glomeromycota (Silva et al. 2019). The sorption capacity of fungi is superior to the rest of the microorganisms. The several characteristics that mark fungi as an attractive candidate for the remediation of several pollutants are: its robust growth, high surface area to volume ratio, heavy metal resistant property, production of versatile extracellular ligninolytic enzymes, presence of metal-binding proteins, and its adaptable nature to the fluctuating pH and temperature (Akhtar and Mannan 2020). The ability of fungi to endure sudden pH or humidity fluctuations is because of its physiological and colonization strategy (Silva et al. 2019). Compared to the other biosorption agents, fungal biomass consists of a high proportion of cell wall material, which displays outstanding properties of metal-binding (Ayele et al. 2021). The vast hyphal network and longer lifespan provides an advantage to fungi over bacteria for exclusion of the toxic pollutants from the environment, and more specifically, in soil bioremediation (Singh et al. 2015). The high adaptability of filamentous fungi can be determined from their ability to grow in extreme areas consisting of highly toxic chemical compounds (Choe and Sheppard 2016). Three processes are mainly involved in mycoremediation: biosorption, intracellular accumulation, and biovolatilization (Bahar et al. 2013). Biosorption is considered a biological practice that utilizes microbes as biosorbent, that are renewable, and consists of materials of bacterial, fungal, plant, or animal origin (Chojnacka 2010). On the other hand, the intracellular accumulation of living organisms by sorbate is defined as bioaccumulation (Chojnacka 2010). Different scholars carried out investigations on active sorption processes performed using live biomass and also passive sorption processes performed using dead fungal biomass. The dead biomass has greater tolerance to toxicity, high environmental resistance, comparatively fast regeneration and reuse absorbance, and a high recovery rate of sorbed metals (Ayele et al. 2021). Moreover, there is no requirement to maintain a specific culture medium for the growth and metabolism of lifeless fungal biomas, which serves as a further advantage (Singh et al. 2016). Fungal biosorption is widely used for the remediation of industrial wastes coming from food, pharmaceuticals, or wastewater treatment (Loukidou et al. 2003).

Pokhrel and Viraraghavan (2006) isolated the nonliving fungal biomass of *Aspergillus niger*, covered with iron oxide, which was examined for its ability in removing aqueous arsenic. It was found that at a pH of 6, iron oxide coated biomass of *A. niger* displayed maximum removal of As(III) (around 75%) and As(V) (about 95%). Murugesan et al. (2006) examined the ability of the tea fungus (produced as waste during fermentation of black tea) to isolate metal ions present in groundwater. Autoclaved tea fungal mat and $FeCl_3$ pre-treated tea fungal mat were used for As(III), As(V) and Fe(II) removal from samples of groundwater and found that the efficiency of $FeCl_3$ pretreated fungal mats was greater due to affinity of iron towards arsenic to form arsenic-iron oxides. The adsorption procedure followed the model of Freundlich isotherm. An additional advantage of the biosorption process over the chemical adsorbents is that, once the adsorption process is complete, desorption of metals from the mat surface is possible, and the mat can be degraded easily (Murugesan et al. 2006). Another novel biosorbent for the removal of aqueous arsenic was prepared when polyethylenimine was chemically grafted on *Penicillium chrysogenum* fungal biomass by Deng and Ting (2007). The reaction consisted of two steps, and at a pH of 10.4, the zeta potential of modified biomass was zero.

The biosorbent was beneficial in eliminating arsenic from water due to the protonation of amine groups on the biosorbent surface.

The above results indicate the efficiency of fungal biosorption as an effective treatment for the reduction of arsenic in groundwater.

10.5.3 Phycoremediation (Algal Bioremediation)

Using macro or micro algae for abatement or biotransformation of contaminants like toxic chemicals or nutrients from wastewater is known as phycoremediation (Kumar et al. 2018). The use of microalgae is desirable for bioremediation mainly because the nutrients like nitrates and phosphates, along with certain heavy metals, act as food sources for algae and support their fast cell cycles leading to their growth (Dwivedi 2012). Wastewater treatment by algae is considered to be an effective approach without any secondary pollution if the produced biomass is recycled and efficient cycling of nutrients is permitted (Rawat et al. 2011). Along with the removal of toxic pollutants, microalgae can also generate biomass for biofuel production with concomitant carbon dioxide sequestration (Rawat et al. 2011). Li et al. (2011) investigated the feasibility of growing *Chlorella* sp. in municipal wastewater centrate and biomass production. The results showed that the algae were capable of removing total phosphorus (80.9%), ammonia (93.9%), total nitrogen (89.1%), and chemical oxygen demand (90.8%) from the raw centrate.

The numerous advantages of bioremediation by algal species over the other processes are: (a) synthesis of algal biomass is not required; (b) the biomass produced can be renewed and reutilized in further processes of adsorption or desorption; (c) toxic by-products are not released; (e) high uptake capacity and efficient removal of heavy metal; (d) the biomass produced by algae can be applied in continuous or discontinuous regimes; (e) Suitable for anaerobic and aerobic sewage treatment units; and (f) cost-effective (Salama et al. 2019).

The uptake of metals by algae occurs through the process of adsorption. The first step is the physical adsorption of metal ions above the cell surface in very less time, and the ions are slowly transported into the cytoplasm through the process of chemisorption (Dwivedi 2012). Algae release a protein called metallothionein, which binds to the metal chemically as a defense mechanism for metal elimination from its usual cellular activity. It is thus responsible for metal uptake by algae in aqueous media (Pal and Paknikar 2011). The algal species *Chlorella*, *Spirulina* and *Scenedesmus* are most widely used for metal uptake and nutrient removal (Dwivedi 2012). Suhendrayatna et al. (1999) exposed a freshwater alga, *Chlorella vulgaris*, to arsenite at intensities from 0 to 100 mg As cm^{-3} to examine the tolerance, accumulation, modification, and excretion properties of arsenic species. Up to a concentration of 50 mg As cm^{-3}, the cell growth remained unaffected. Suppression of algal growth was observed at concentrations of arsenite higher than 50 mg As cm^{-3}. The predominant arsenic metabolite in the algal cell was arsenate. Wang et al. (2013) exposed *Microcystis aerugonisa* to different concentrations of arsenic for a period of 15 days in culture media of BG11. The algae *M. aerugonisa* showed that in both cells and their growth media, the arsenite uptake was more predominant than arsenate after 15 days of exposure to arsenate or arsenite. Algal experiments performed by Jahan et al. (2006) specified that the common green algae, *Scenedesmus abundans*, was capable of removing arsenic. The biotransformation of arsenic by *Cyanidioschyzon* sp. was investigated by Qin et al. (2009). On incubating the algal culture with 20 μM of arsenic, As(V) formed by the oxidation of As(III) is further reduced to arsenite, which is succeeded by arsenite methylation to form trimethylarsine oxide.

Bioaccumulation of heavy metals in the cell interior hampers the activity of photosynthesis which decreases algal growth (Salama et al. 2019). The use of dead immobilized algal cells is another mechanism of wastewater treatment, and this process eliminates the most difficult harvesting step (Dwivedi 2012). Higher cell density increases the reaction rates of the immobilized cells, and thus they are preferred over their free-living counterparts (Dwivedi 2012).

Algal bioremediation can thus be considered as a sustainable technology since several species of algae have been established as favourable candidates for the treatment of waters rich in heavy metals along with massive biofuel production.

10.5.4 Phytoremediation

Among the various biological and eco-friendly bioremediation techniques, the use of green plants in cleaning up soil and water contaminated with arsenic is considered to be an innovative technique that requires a comparatively lower level of monetary and technical input (Mahmud et al. 2008). Phytoremediation is an *in situ* technology that is low-cost and non-destructive and whose main goal is to utilize plants for lessening the amount of metals and metalloids from polluted soil and water (Rahman and Hasegawa 2011). Remediation of soil and water contamination is possible using both terrestrial and aquatic plants (Rahman and Hasegawa 2011). The three main biotechnological engineering methods for improving plant tolerance to toxic metal accumulation are: (i) the transporter genes and uptake system of metal and metalloid are manipulated; (ii) enhancement of the metal and metalloid ligand production; (iii) metal and metalloids conversion to forms that are less toxic (Mosa et al. 2016). Several processes are included in phytoremediation, while the most widely used strategies include phytoextraction, phytostabilization, rhizofiltration, phytotransformation, and phytovolatilization (Rahman and Hasegawa 2011, Anju 2017). Phytoextraction is known as the most important technique that is primarily utilized in contaminated soil treatment (Yan et al. 2020). Here, the toxic metals are concentrated and precipitated from contaminated soils by the hyperaccumulating plants, into the above-ground biomass. This method of heavy metal elimination is considered permanent (Yan et al. 2020). Also, it decreases the amount of waste material to be disposed off, and in many cases, it is possible to recycle the contaminants from the contaminated biomass (Etim 2012). Phytostabilization is another important mechanism that uses certain species of plants which can tolerate metals to immobilize and prevent contaminant migration into soil and groundwater, thus reducing its bioavailability through heavy metal precipitation or metal valence reduction in the rhizosphere, absorption and confiscation inside the root tissues or adsorption onto the root cells (Yan et al. 2020, Etim 2012). At heavily contaminated sites, vegetation cover can be reformed through this technique by underground stabilization of the heavy metals, reducing their leaching to groundwater, and decreasing the migration of pollutants by wind (Etim 2012). Another phytoremediation strategy is phytovolatilization, in which the plants uptake pollutants from the soil, volatilize it, and then release the volatile degradation product via leaves or foliage system (Yan et al. 2020, Etim 2012). The added advantage of this technique is the gaseous dispersion of contaminants without further requirement of plant harvesting and disposal (Yan et al. 2020). In phytotransformation, a definite acceptable range of solubility and hydrophobicity is required for plants to uptake contaminants, and after uptake, the complex organic molecules are broken into simpler ones, or the molecules are incorporated into plant tissues (Etim 2012). The use of plant roots for the elimination of pollutants from water is known as rhizofiltration (Yan et al. 2020). The rhizosphere soils have higher microbial counts since the plants discharge amino acids, sugars, enzymes, and other composites, which stimulate microbial growth (Etim 2012). The several categories of phytoremediation processes are concised in Table 10.3.

Pteris vittate, a Chinese brake fern, was the first known hyperaccumulator plant which is considered an excellent arsenic hyperaccumulating plant and is mainly used for remediation of soils that are contaminated (Fazi et al. 2015). Ma et al. (2001) reported that around 12-64 mg arsenic/kg was accumulated in the fronds of the hyperaccumulator *P. vitatta* from uncontaminated soils containing 0.5–7.5 mg arsenic/kg and up to 22,630 mg arsenic/kg from soils with 1500 mg arsenic/kg.

Phytoremediation of water contaminated with arsenic is a comparatively new approach. Tu et al. (2004) examined the effectiveness of *Pteris vittate* L. in a hydroponic growing setup where a plant cultivated in 600 ml of groundwater was effective in reducing the concentration of arsenic

Table 10.3. Different types of phytoremediation processes (Vamerali et al. 2009).

Phytoextraction	This process is associated with the hyperaccumulating plants that uptake pollutants from the environment by concentrating them in harvestable plant biomass.
Phytostabilization	This process uses plants to decrease the movement and bioavailability of environmental pollutants.
Phytovolatilization	The plants uptake contaminants available in soil and water with their release into the atmosphere.
Phytotransformation	The pollutants are chemically modified by the hyperaccumulating plants resulting in the inactivation, degradation (phytodegradation), or immobilization (phytostabilization).
Rhizofiltration	This process is usually performed by plant of aquatic origin that are involved in pollution sorption from the aquatic environment.

from 46 to less than 10 μg/L in a period of 3 days while the rate of arsenic uptake by re-used plants was slower. Also, the younger fern plants were found to be more sufficient than the older ones of the same size for arsenic removal.

Phytoremediation of contaminated aquatic environment can be achieved through biosorption or bioaccumulation of the dissolved contaminants present in water by the aquatic macrophytes (the roots take part in adsorption or accumulation of pollutants) or by other floating plants (the entire body of the plant is used for accumulation) (Rahman and Hasegawa 2011). The most studied species of aquatic macrophyte are *Lemna gibba* L. and *Lemna. minor* L. of *Lemnaceae* family (Mkandawire et al. 2004, Mkandawire and Dudel 2005). Mishra et al. (2008) reported that the aquatic macrophytes known as *Eichhornia crassipes*, *Lemna minor*, and *Spirodela polyrrhiza*, were efficient in isolating a large quantity of Hg and As in open coalmine sewage. The order of these macrophytes in their arsenic removal tendencies was *E. crassipes* > *L. minor* > *S. polyrrhiza*. The accumulation of metals took place mainly in the plant roots, than its leaves.

Thus, numerous advantages like utilization of natural processes to clean up pollutants, easy maintenance, no-requirement of expensive equipment, and eco-friendly nature make phytoremediation fruitful.

10.6 Future Perspectives of Bioremediation

This chapter has validated that the presence of arsenic in groundwater can have dire consequences on human health as it impacts the availability of safe and good quality water for domestic use. Also, the boundaries associated with conventional technologies for arsenic remediation increased the requirement of a sustainable and green remediation strategy.

One of the finest alternative and economical techniques for managing arsenic contamination is bioremediation, which encompasses the interaction between the micro-organisms found naturally in the contaminated areas and the different chemical species of arsenic.

This chapter consists of certain biologically interceded conversions of arsenic along with their mechanisms. Of them, the microbial arsenic oxidation-reduction reactions are the most attractive one, as it possesses a great potential in replacing the chemical oxidation process. Several bacteria mediate the arsenite oxidation to form the less noxious arsenate, which can be eliminated easily from water. Reports showing the involvement of both autotrophic and heterotrophic bacteria have been presented in the chapter. Microbial oxidation, in combination with the conventional technologies, provide affordable and very effectual techniques aimed at arsenic removal in the target zones.

Biovolatilization is a process in which the volatile arsenic forms existing in water and soil sample are released into the environment through methylation by microbes. But rather than the treatment of arsenic contamination, the microbial methylation and demethylation processes may give rise to negative effects in terms of atmospheric pollution, unless it is practiced under controlled conditions.

A number of investigators also studied the function of SRB in effective arsenic bioremediation.

Certain algae and fungi possess high bioaccumulation ability. Another promising technology is arsenic biosorption, as the arsenic that is sorbed can be renewed, thus reducing chances of unsuitable disposal and pollution. Biosorbent surfaces can be further modified in order to increase the surface charge or expose more functional groups. But the chemicals used should also be taken into consideration in order to minimize its secondary adverse effects.

Arsenic remediation by the use of plant species is also an emerging and successful technique. However, there are certain areas in phytoremediation that need further modification for the process to be efficient.

So the bioremediation technology needs further improvement and research in all the fields discussed above.

10.7 Conclusion

In conclusion, we can say that the bioremediation techniques of arsenic display very convincing, attractive, and incredible opportunities to remove arsenic from groundwater. Arsenic affected people are mostly found in the poor regions since they are completely dependent on groundwater for drinking and household use, as they fail to manage to pay for the treatment techniques used in water purification. Bioremediation techniques and their further advances, with their simplicity, profitability, eco-friendliness and economic nature, have a long way to go, and become the sustainable remediation technology required for solving the problems of arsenic contamination in groundwater.

References

Adeloju, S. B., S. Khan and A. F. Patti. 2021. Arsenic contamination of groundwater and its implications for drinking water quality and human health in under-developed countries and Remote communities—a review. Appl. Sci. 11(4): 1926.

Akhtar, N. and M. A.-ul. Mannan. 2020. Mycoremediation: Expunging environmental pollutants. Biotechnol. Reports. 26: e00452.

Alam, R. and K. Mc. Phedran. 2019. Applications of biological sulfate reduction for remediation of arsenic—a review. Chemosphere 222: 932–944.

Anju, M. 2017. Biotechnological strategies for remediation of toxic metal(loid)s from environment. pp. 315–360. *In*: Gahlawat, S. K. et al. (eds.). Plant Biotechnology: Recent Advancements and Developments, Springer, ISBN: 978-981-10-4731-2.

Ayele, A., S. Haile, D. Alemu and M. Kamaraj. 2021. Comparative utilization of dead and live fungal biomass for the removal of heavy metal: A concise review. Sci. World J. 1–10.

Azubuike, C. C., C. B. Chikere and G. C. Okpokwasili. 2016. Bioremediation techniques–classification based on site of application: Principles, advantages, limitations and prospects. World J. Microbiol. Biotechnol. 32(11).

Bahar, M. M., M. Megharaj and R. Naidu. 2013. Bioremediation of arsenic-contaminated water: Recent advances and future prospects. Water Air Soil Pollut. 224: 1722.

Bentley, R. and T. G. Chasteen. 2002. Microbial methylation of metalloids: Arsenic, antimony, and bismuth. Microbiol. Mol. Biol. Rev. 66(2): 250–271.

Biswas, R., V. Vivekanand, A. Saha, A. Ghosh and A. Sarkar. 2019. Arsenite oxidation by a facultative chemolithotrophic *Delftia* spp. BAS29 for its potential application in groundwater arsenic bioremediation. Int. Biodeterior. Biodegrad. 136: 55–62.

Bright, D. A., S. Brock, K. J. Reimer, W. R. Cullen, G. M. Hewitt and J. Jafaar. 1994. Methylation of arsenic by anaerobic microbial consortia isolated from lake sediment. Appl. Organometal. Chem. 8(4): 415–422

Cai, L., G. Liu, C. Rensing and G. Wang. 2009. Genes involved in arsenic transformation and resistance associated with different levels of arsenic-contaminated soils. BMC Microbiol. 9(1): 4.

Centeno, J. A., F. G. Mullick, L. Martinez, N. P. Page, H. Gibb, D. Longfellow et al. 2002. Pathology related to chronic arsenic exposure. Environ. Health Perspect. 110: 883–886.

Challenger, F. 1945. Biological methylation. Chem. Rev. 36(3): 315–361.

Chang, J.-S., I.-H. Yoon, J.-H. Lee, K.-R. Kim, J. An and K.-W. Kim. 2009. Arsenic detoxification potential of AOX genes in arsenite-oxidizing bacteria isolated from natural and constructed wetlands in the Republic of Korea. Environ. Geochem. Health 32(2): 95–105.

Chen, C.-L., L.-I. Hsu, H.-Y. Chiou, Y.-M. Hsueh, S.-Y. Chen, M.-M. Wu et al. 2004. Ingested arsenic, cigarette smoking, and Lung Cancer Risk. JAMA 292(24): 2984.

Chiang, H. S., H. R. Guo, C. L. Hong, S. M. Lin and E. F. Lee. 1993. The incidence of bladder cancer in the black foot disease endemic area in Taiwan. Br. J. Urol. 71(3): 274–278.

Choe, S.-I. and D. C. Sheppard. 2016. Bioremediation of arsenic using an aspergillus system. pp. 267–274. *In*: Gupta, V. K. (ed.). New and Future Developments in Microbial Biotechnology and Bioengineering: Aspergillus System Properties and Applications, Elsevier.

Chojnacka, K. 2010. Biosorption and bioaccumulation—the prospects for practical applications. Environ. Int. 36(3): 299–307.

Chung, J.-Y., S.-D. Yu and Y.-S. Hong. 2014. Environmental source of arsenic exposure. J. Prev. Med. Public Health 47(5): 253–257.

Cullen, W. R. 2014. Chemical mechanism of arsenic Biomethylation. Chem. Res. Toxicol. 27(4): 457–461.

Dastidar, A. and Y.-T. Wang. 2009. Arsenite oxidation by batch cultures of Thiomonas Arsenivorans strain B6. J. Environ. Eng. 135(8): 708–715.

Deng, S. and Y. P. Ting. 2007. Removal of AS(V) and as(iii) from water with a Pei-modified Fungal biomass. Water Sci. Technol. 55(1-2): 177–185.

Dua, M., A. Singh, N. Sethunathan and A. K. Johri. 2002. Biotechnology and bioremediation: Successes and limitations. Appl. Microbiol. Biotechnol. 59: 143–152.

Duarte, A. A., S. J. Cardoso and A. Alçada. 2009. Emerging and innovative techniques for arsenic removal applied to a small water supply system. Sustainability 1(4): 1288–1304.

Duquesne, K., A. Lieutaud, J. Ratouchniak, D. Muller, M.-C. Lett and V. Bonnefoy. 2007. Arsenite oxidation by a chemoautotrophic moderately acidophilic *Thiomonas* sp.: From the strain isolation to the gene study. Environ. Microbiol. 1(10): 228–237.

Dwivedi, A. K., S. Srivastava, S. Dwivedi and V. Tripathi. 2015. Natural bio-remediation of arsenic contamination: A short review. Hydrol. Current Res. 6: 186.

Dwivedi, S. 2012. Bioremediation of heavy metal by algae: Current and future perspective. J. Adv. Lab. Res. Biol. 3: 195–199.

Etim, E. 2012. Phytoremediation and its mechanisms: A review. Int. J. Environ. Bioenergy. 2: 120–136.

Fazi, S., S. Amalfitano, B. Casentini, D. Davolos, B. Pietrangeli, S. Crognale et al. 2015. Arsenic removal from naturally contaminated waters: A review of methods combining chemical and biological treatments. Rendiconti Lincei. 27(1): 51–58.

Gadd, G. 2004. Microbial influence on metal mobility and application for bioremediation. Geoderma. 122(2-4): 109–119.

Garcia-Dominguez, E., A. Mumford, E. D. Rhine, A. Paschal and L. Y. Young. 2008. Novel autotrophic arsenite-oxidizing bacteria isolated from soil and sediments. FEMS Microbiol. Ecol. 66(2): 401–410.

Grau-Perez, M., A. Navas-Acien, I. Galan-Chilet, L. S. Briongos-Figuero, D. Morchon-Simon, J. D. Bermudez et al. 2018. Arsenic exposure, diabetes-related genes and diabetes prevalence in a general population from Spain. Environ. Pollut. 235: 948–955.

Heck, J. E., A. S. Park, J. Qiu, M. Cockburn and B. Ritz. 2014. Risk of leukemia in relation to exposure to Ambient Air Toxics in pregnancy and early childhood. Int. J. Hyg. Environ. Health 217(6): 662–668.

Hong, Young-Seoub et al. 2014. Health effects of chronic arsenic exposure. J. Prev. Med. Public Health Yebang Uihakhoe chi 47(5): 245–52.

Hue N. 2015. Bioremediation of arsenic toxicity. pp. 155–163. *In*: Chakrabarty, N. (ed.). Arsenic Toxicity: Prevention and Treatment, CRC Press.

Hughes, M. F. 2002. Arsenic toxicity and potential mechanisms of action. Toxicol. Lett. 133(1): 1–16.

Hussain, A., A. Hasan, A. Javid and J. I. Qazi. 2016. Exploited application of sulfate-reducing bacteria for concomitant treatment of metallic and non-metallic wastes: A mini review. Biotech. 6(2).

Huyck, K. L., M. L. Kile, G. Mahiuddin, Q. Quamruzzaman, M. Rahman, C. V. Breton et al. 2007. Maternal arsenic exposure associated with low birth weight in Bangladesh. J. Occup. Environ. Med. 49(10): 1097–1104.

Huysmans, K. and W. Frankenberger. 1991. Evolution of trimethylarsine by a *Penicillium* sp. isolated from agricultural evaporation pond water. Science Total Environ. 105: 13–28.

Ike, M., T. Miyazaki, N. Yamamoto, K. Sei and S. Soda. 2008. Removal of arsenic from groundwater by arsenite-oxidizing bacteria. Water Sci. Technol. 58(5): 1095–1100.

Jahan, K., P. Mosto, C. Mattson, E. Frey and L. Derchak. 2006. Microbial removal of arsenic. Water, Air, and Soil Pollut. 6(1-2): 71–82.

Jiang, Y., B. Xi, R. Li, M. Li, Z. Xu, Y. Yang et al. 2019. Advances in Fe(III) bioreduction and its application prospect for groundwater remediation: A review. Front. Environ. Sci. Eng. 13(6).

Katsoyiannis, I. A. and A. I. Zouboulis. 2004. Application of biological processes for the removal of arsenic from Groundwaters. Water Res. 38(1): 17–26.

Keimowitz, A. R., B. J. Mailloux, P. Cole, M. Stute, H. J. Simpson and S. N. Chillrud. 2007. Laboratory investigations of Enhanced sulfate reduction as a groundwater Arsenic remediation strategy. Environ. Sci. Technol. 41(19): 6718–6724.

Kermani, A. J. N., M. F. Ghasemi, A. Khosravan, A. Farahmand and M. R. Shakibaie. 2012. Cadmium bioremediation by metal-resistant mutated bacteria isolated from active sludge of industrial effluent. Iran. J. Environ. Health. Sci. Eng. 7: 279–286.

Kirk, M. F., T. R. Holm, J. Park, Q. Jin, R. A. Sanford, B. W. Fouke et al. 2004. Bacterial sulfate reduction limits natural arsenic contamination in groundwater. Geology 32(11): 953.

Kumar, P. K., S. Vijaya Krishna, K. Verma, K. Pooja, D. Bhagawan and V. Himabindu. 2018. Phycoremediation of sewage wastewater and industrial flue gases for biomass generation from microalgae. South African J. Chem. Eng. 25: 133–146.

Li, Y., Y.-F. Chen, P. Chen, M. Min, W. Zhou, B. Martinez et al. 2011. Characterization of a microalga *Chlorella* sp. well adapted to highly concentrated municipal wastewater for nutrient removal and biodiesel production. Bioresource Technol. 102(8): 5138–5144.

Liao, V. H.-C., Y. -J. Chu, Y.-C Su, S.-Y. Hsiao, C.-C Wei, C.-W. Liu et al. 2011. Arsenite-oxidizing and arsenate-reducing bacteria associated with arsenic-rich groundwater in Taiwan. J. Contam. Hydrol. 123(1-2): 20–29.

Lièvremont, D., P. N. Bertin and M.-C. Lett. 2009. Arsenic in contaminated waters: Biogeochemical cycle, microbial metabolism and Biotreatment Processes. Biochimie 91(10): 1229–1237.

Lin, H.-J., T.-I. Sung, C.-Y. Chen and H.-R. Guo. 2013. Arsenic levels in drinking water and mortality of liver cancer in Taiwan. J. Hazard. Mater. 262: 1132–1138.

Loukidou, M. X., K. A. Matis, A. I. Zouboulis and M. Liakopoulou-Kyriakidou. 2003. Removal of As(V) from wastewaters by chemically modified Fungal biomass. Water Res. 37(18): 4544–4552.

Ludwig, R. D., D. J. Smyth, D. W. Blowes, L. E. Spink, R. T. Wilkin, D. G. Jewett et al. 2009. Treatment of Arsenic, heavy metals, and Acidity using a Mixed ZVI-COMPOST PRB. Environ. Sci. Technol. 43(6): 1970–1976.

Ma, L. Q., K. M. Komar, C. Tu, W. Zhang, Y. Cai and E. D. Kennelley. 2001. A fern that hyperaccumulates arsenic. Nature. 409(6820): 579–579.

Macy, J. M., K. Nunan, K. D. Hagen, D. R. Dixon, P. J. Harbour, M. Cahill et al. 1996. *Chrysiogenes arsenatis* gen. nov., sp. nov., a new arsenate-respiring bacterium isolated from gold mine wastewater. Int. J. Syst. Bacteriol. 46(4): 1153–1157.

Mahmud, R., N. Inoue, S.-Ya Kasajima and R. Shaheen. 2008. Assessment of potential indigenous plant species for the phytoremediation of Arsenic-Contaminated areas of Bangladesh. Int. J. Phytoremediat. 10(2): 119–132.

Masuda, H. 2018. Arsenic cycling in THE Earth's crust AND hydrosphere: Interaction between naturally occurring arsenic and human activities. Progr. Earth and Planet. Sci. 5(1).

Michalke, K., E. B. Wickenheiser, M. Mehring, A. V. Hirner and R. Hensel. 2000. Production of volatile derivatives of metal(loid)s by microflora involved in anaerobic digestion of sewage sludge. Appl. and Environ. Microbiol. 66(7): 2791–2796.

Milton, A. H., W. Smith, B. Rahman, Z. Hasan, U. Kulsum, K. Dear et al. 2005. Chronic arsenic exposure and adverse pregnancy outcomes in Bangladesh. Epidemiol. 16(1): 82–86.

Mishra, V. K., A. R. Upadhyay, V. Pathak and B. D. Tripathi. 2008. Phytoremediation of mercury and arsenic from tropical opencast coalmine effluent through naturally occurring Aquatic Macrophytes. Water, Air, and Soil Pollut. 192(1-4): 303–314.

Mkandawire, M., Y. V. Lyubun, P. V. Kosterin and E. G. Dudel. 2004. Toxicity of arsenic species *Tolemna gibba* L. and the influence of Phosphate on arsenic bioavailability. Environ. Toxicol. 19(1): 26–34.

Mkandawire, M. and E. G. Dudel. 2005. Accumulation of arsenic In *Lemna gibba* L. (duckweed) in TAILING waters of two Abandoned uranium mining sites in saxony, Germany. Sci. Total Environ. 336(1-3): 81–89.

Mosa, K. A., I. Saadoun, K. Kumar, M. Helmy and O. P. Dhankher. 2016. Potential biotechnological strategies for the cleanup of heavy metals and metalloids. Front. Plant Sci. 7: 303.

Muller, D., D. Lièvremont, D. D. Simeonova, J.-C. Hubert and M.-C. Lett. 2003. Arsenite oxidase AOX genes from a metal-resistant β-proteobacterium. J. Bacteriol. 185(1): 135–141.

Murugesan, G. S., M. Sathishkumar and K. Swaminathan. 2006. Arsenic removal from groundwater by pretreated waste tea Fungal biomass. Bioresource Technol. 97(3): 483–487.

Oremland, R. S. and J. F. Stolz. 2005. Arsenic, microbes and contaminated aquifers. Trends Microbiol. 13(2): 45–49.

Pal, A. and K. M. Paknikar. 2011. Bioremediation of arsenic from contaminated water. pp. 477–523. *In*: Satyanarayana, T., B. N. Johri and A. Prakash (eds.). Microorganisms in Environmental Management: Microbes and Environment, Springer, Dordrecht.

Pokhrel, D. and T. Viraraghavan. 2006. Arsenic removal from an aqueous solution by a MODIFIED Fungal biomass. Water Res. 40(3): 549–552.

Qin, J., B. P. Rosen, Y. Zhang, G. Wang, S. Franke and C. Rensing. 2006. Arsenic detoxification and evolution of trimethylarsine gas by a microbial arsenites-adenosylmethionine methyltransferase. Pro. Natl. Acad. Sci. USA. 103(7): 2075–2080.

Qin, J., C. R. Lehr, C. Yuan, X. C. Le, T. R. McDermott and B. P. Rosen. 2009. Biotransformation of arsenic by a Yellowstone thermoacidophilic eukaryotic alga. Pro. Natl. Acad. Sci. 106(13): 5213–5217.

Rahman, M. A. and H. Hasegawa. 2011. Aquatic arsenic: Phytoremediation using floating macrophytes. Chemosphere 83(5): 633–646.

Ratnaike, R. N. 2003. Acute and chronic arsenic toxicity. Postgrad. Med. J. 79(933): 391–396.

Rawat, I., R. Ranjith Kumar, T. Mutanda and F. Bux. 2011. Dual role of microalgae: Phycoremediation of domestic wastewater and biomass production for sustainable biofuels production. Appl. Energy 88(10): 3411–3424.

Renuka, N., A. Sood, R. Prasanna and A. S. Ahluwalia. 2014. Phycoremediation of wastewaters: A synergistic approach using microalgae for bioremediation and biomass generation. Int. J. Environ. Sci. Technol. 12(4): 1443–1460.

Rittle, K. A., J. I. Drever and P. J. S. Colberg. 1995. Precipitation of arsenic during bacterial sulfate reduction. Geomicrobiol. J. 13(1): 1–11.

Rossman, T., A. N. Uddin and F. J. Burns. 2004. Evidence that arsenite acts as a cocarcinogen in skin cancer. Toxicol. Appl. Pharmacol. 198(3): 394–404.

Saalfield, S. L. and B. C. Bostick. 2009. Changes in Iron, sulfur, and arsenic speciation associated with Bacterial sulfate reduction in Ferrihydrite-rich systems. Environ. Sci. Technol. 43(23): 8787–8793.

Salama, E.-S., H.-S. Roh, S. Dev, M. A. Khan, R. A. Abou-Shanab, S. W. Chang et al. 2019. Algae as a green technology for heavy metals removal from various wastewater. World J. Microbiol. Biotechnol. 35(5).

Santini, J. M., L. I. Sly, R. D. Schnagl and J. M. Macy. 2000. A new chemolithoautotrophic arsenite-oxidizing bacterium isolated from a gold mine: Phylogenetic, physiological, and preliminary biochemical studies. Appl. Environ. Microbiol. 66(1): 92–97.

Satyapal, G. K., S. K. Mishra, A. Srivastava, R. K. Ranjan, K. Prakash, R. Haque et al. 2018. Possible bioremediation of arsenic toxicity by isolating indigenous bacteria from the middle Gangetic plain of BIHAR, INDIA. Biotechnol. Reports. 17: 117–125.

Shankar, S., U. Shanker and Shikha. 2014. Arsenic contamination of Groundwater: A review of Sources, Prevalence, health risks, and strategies for mitigation. Sci. World J. 1–18.

Sharma, A. K., J. C. Tjell, J. J. Sloth and P. E. Holm. 2014. Review of arsenic contamination, exposure through water and food and low cost mitigation options for rural areas. Appl. Geochem. 41: 11–33.

Shukla, A. and S. Srivastava. 2017. Emerging aspects of bioremediation of arsenic. Green Technol. Environ. Sustain.: 395–407.

Shukla, S. 2017. Arsenic-exposure, mechanism of action and toxicity analysis: A prenatal view. J. Med. Sci. Clinical Res. 5(7): 25108–121.

Silva, A., C. Delerue-Matos, S. Figueiredo and O. Freitas. 2019. The use of algae and fungi for removal of pharmaceuticals by bioremediation and biosorption processes: A review. Water 11(8): 1555.

Singh, M., P. K. Srivastava, P. C. Verma, R. N. Kharwar, N. Singh and R. D. Tripathi. 2015. Soil fungi for mycoremediation of arsenic pollution in agriculture soils. J. Appl. Microbiol. 119(5): 1278–1290.

Singh, N. K., A. S. Raghubanshi, A. K. Upadhyay and U. N. Rai. 2016. Arsenic and other heavy metal accumulation in plants and algae growing naturally in contaminated area of West BENGAL, INDIA. Ecotoxicol. Environ. Saf. 130: 224–233.

Smedley, P. L. and D. G. Kinniburgh. 2002. A review of the source, behaviour and distribution of arsenic in natural waters. Appl. Geochem. 17(5): 517–568.

Smith, A. H., G. Marshall, Y. Yuan, C. Ferreccio, J. Liaw, O. von Ehrenstein et al. 2006. Increased mortality from lung cancer and bronchiectasis in young adults after exposure to arsenic *in utero* and in early childhood. Environ. Health Perspect. 114(8): 1293–1296.

Smith, A. H. and C. M. Steinmaus. 2009. Health effects of arsenic and Chromium in drinking WATER: Recent human Findings. Annu. Rev. Public Health 30(1): 107–122.

Stolz, J. F., P. Basu and R. S. Oremland. 2002. Microbial transformation of elements: The case of arsenic and selenium. Int. Microbiol. 5(4): 201–207.

Stolz, J. F., P. Basu, J. M. Santini and R. S. Oremland. 2006. Arsenic and selenium in microbial metabolism. Annu. Rev. Microbiol. 60(1): 107–130.

Suhendrayatna, A. Ohki, T. Kuroiwa and S. Maeda. 1999. Arsenic compounds in the freshwater Green microalga *Chlorella vulgaris* after exposure to arsenite. Appl. Organometal. Chem. 13(2): 127–133.

Sun, J., A. N. Quicksall, S. N. Chillrud, B. J. Mailloux and B. C. Bostick. 2016. Arsenic mobilization from sediments In Microcosms under sulfate reduction. Chemosphere. 153: 254–261.

Suttigarn, A. and Y.-T. Wang. 2005. Arsenite oxidation byalcaligenes faecalisstrain O1201. J. Environ. Eng. 131(9): 1293–1301.

Tantry, B. A., D. Shrivastava, I. Taher and M. Nabi Tantry. 2015. Arsenic exposure: Mechanisms of action and related health effects. J. Environ. Anal. Toxicol. 05(06).

Teclu, D., G. Tivchev, M. Laing and M. Wallis. 2008. Bioremoval of arsenic species from contaminated waters by sulphate-reducing bacteria. Water Res. 42(19): 4885–4893.

Thomas, D. J., S. B. Waters and M. Styblo. 2004. Elucidating the pathway for arsenic methylation. Toxicol. Appl. Pharmacol. 198(3): 319–326.

Tu, S., L. Q. Ma, A. O. Fayiga and E. J. Zillioux. 2004. Phytoremediation of Arsenic-Contaminated groundwater by the arsenic Hyperaccumulating Fernpteris vittatal. Int. J. Phytoremediat. 6(1): 35–47.

Upadhyaya, G., T. M. Clancy, J. Brown, K. F. Hayes and L. Raskin. 2012. Optimization of arsenic removal water treatment system through characterization of terminal electron accepting processes. Environ. Sci. Technol. 46(21): 11702–11709.

Upadhyay, M. K., P. Yadav, A. Shukla and S. Srivastava. 2018. Utilizing the potential of microorganisms for managing arsenic contamination: A feasible and sustainable approach. Front. Environ. Sci. 6: 24.

Vamerali, T., M. Bandiera and G. Mosca. 2009. Field crops for phytoremediation of metal-contaminated land. A review. Environ. Chem. Lett. 8(1): 1–17.

von Ehrenstein, O. S., S. Poddar, Y. Yuan, D. G. Mazumder, B. Eskenazi, A. Basu et al. 2007. Children??s intellectual function in relation to arsenic exposure. Epidemiol. 18(1): 44–51.
Wang, J. and C. Chen. 2006. Biosorption of heavy metals by saccharomyces cerevisiae: A Review. Biotechnol. Adv. 24(5): 427–451.
Wang, S. and X. Zhao. 2009. On the potential of biological treatment for arsenic contaminated soils and groundwater. J. Environ. Management. 90(8): 2367–2376.
Wang, Z., Z. Luo and C. Yan. 2013. Accumulation, transformation, and release of inorganic arsenic by the Freshwater cyanobacterium Microcystis Aeruginosa. Environ. Sci. Pollut. Res. Int. 20(10): 7286–7295.
Wasserman, G. A., X. Liu, F. Parvez, H. Ahsan, P. Factor-Litvak, A. van Geen et al. 2004. Water arsenic exposure and children's intellectual function in Araihazar, Bangladesh. Environ. Health Perspect. 112(13): 1329–1333.
Yan, A., Y. Wang, S. N. Tan, M. L. Mohd Yusof, S. Ghosh and Z. Chen. 2020. Phytoremediation: A promising approach for revegetation of heavy metal-polluted land. Front. Plant Sci. 11: 359.
Young, J. L., L. Cai and J. C. States. 2018. Impact of prenatal arsenic exposure on chronic adult diseases. Syst. Biol. Reprod. Med. 64(6): 469–483.
Zobrist, J., P. R. Dowdle, J. A. Davis and R. S. Oremland. 2000. Mobilization of arsenite by dissimilatory reduction of adsorbed arsenate. Environ. Sci. Technol. 34(22): 4747–4753.
Zouboulis, A. I. and I. A. Katsoyiannis. 2005. Recent advances in the bioremediation of arsenic-contaminated groundwaters. Environ. Int. 31(2): 213–219.

CHAPTER 11

Phytoremediation of Uranium and Other Radionuclides in Soil and Water and Effects of Biogeochemical Conditions

Naira Ibrahim[1,*] and *Fengxiang Han*[2]

11.1 Introduction

Due to the growth of the world population and economy, energy demand has continuously increased. Since conventional oil, gas, and other nonrenewable fossil fuels are mainly used for energy consumption, the world may experience an energy crisis. Nuclear energy, on the other hand, may provide a partial solution to the energy demand. However, the nuclear energy industry may generate nuclear wastes, and if not managed properly, they may heavily contaminate soil and water with U and other radionuclides (Bleise et al. 2003, Chang et al. 2005, Gavrilescu et al. 2009). Radionuclides emit α, β, and γ radiations at particular rates with various half-lives. In nature, uranite (UO_2^{+2}) and pitchblende ($U_2O_8^{+2}$) are major oxidized uranite, while complex oxides, silicates, and phosphates are examples of secondary minerals. The percentage of U in soil ranges between 80 to 90% in the form of U (VI) oxidation state as the uranyl (UO_2^{+2}) cation, which is considered the mobile form (Ebbs et al. 1998). Also, the stable form for U in water is (UO_2^{+2}) ions and as soluble carbonate complex as $(UO_2)^2 CO_3(OH)^{-3}$, $UO_2 CO_3$, $UO_2 (CO_3)^{-2}$, $UO_2 (CO_3)_3^{-4}$ and $(UO_2)^3(CO_3)^6$. Moreover, there are other complex forms for U in water and soil with sulfate and phosphate as well as carbonate and hydroxide where these forms increase the solubility of U (Shahandeh and Hossner 2002b, Ebbs et al. 1998). The most prevalent forms, U_3O_8, UO_2, and UO_3, all dissolve in bodily fluid (weeks for UO_3 to years for UO_2 and U_3O_8) (Lin et al. 1993).

Furthermore, all soluble chemical forms are absorbed in the human body within days, according to Bleise et al. (2003), whereas insoluble chemical forms take months to years. As a result, the soluble form of U was dominated by its harmful chemical effects, while the insoluble forms have typical radiation effects as their particles settle in the lungs and lymph nodes. According to Priest (2001),

[1] Department of Biology, Jackson State University, 1400 Lynch St, Building John Peoples, Jackson, MS, USA.
[2] Department of Chemistry and Biochemistry, Jackson State University, 1400 Lynch St, Building John Peoples, Jackson, MS, USA. Email: fengxiang.han@jsums.edu
* Corresponding author: naira.a.ibrahim@jsums.edu

most U in the human body comes from foods such as fruits, cereals, and table salt. De Boulois et al. (2008) discovered that U 238 has the highest chemical toxicity with the least radioactive. Chang et al. (2005) found that consuming a significant volume of U damaged the kidneys and increased the risk of cancer because of its alpha radioactivity. The chemical toxicity of U has posed public health concerns, according to Bednar et al. (2007). Some of the most popular remediation procedures for U in soil and water include the soil excavation or removal of a topsoil layer, transfer to designated reservoirs, electrokinetic, and ion exchange (Dushenkov 2003). Furthermore, these approaches have the potential to deteriorate soil quality and destroy other critical soil components for plant growth (Choudhary and Sar 2010). Phytoremediation uses plants to remove contaminants from contaminated soil or water (Malaviya and Singh 2012). Vandenhove and Hees (2004) discovered that it is one of the most promising strategies for long-term U-contamination cleaning up in soils and water. Additionally, the presence of plants over the treatment process reduces erosion from wind and water. Plants can absorb, accumulate, and degrade contaminants from soils and water.

11.2 Phytoremediation of Radionuclides

Negri and Hinchman (2000) discovered that the mechanisms for phytoremediation include (a) phytoextraction—high biomass plants take up radionuclides and accumulate in the lower part of the shoots. These plants are cultivated using conventional farming methods; (b) rizofilteration—aquatic plants take up and accumulate radionuclides from a contaminated environment, e.g., water by the roots of plants; (c) phytovolatilization—plants remove contaminants by uptaking from the soil and then volatilize them from the foliage; and (d) stabilization—the plants have the capacity to stabilize radionuclide in the rhizosphere.

11.2.1 Phytoextraction

Plants remove radionuclides from the soil by uptaking and bioaccumulating them in their shoots, with minimum disruption to the soil's structure and impact on its fertility (Erakhrumen and Agbontalor 2007). Cline and Rickard (1972), Adriano et al. (1980), and Ebbs et al. (1998) found that the bioaccumulation coefficient is used to measure phytoextraction efficiency (BC). Furthermore, the transfer coefficient (Soil–Plant Transfer Factor) is determined as ratios of concentrations in plants over soils. In addition, gross radionuclide removal (mg/L) is determined by multiplying plant biomass by radionuclide concentration. Also, Huang et al. (1998) and Dushenkov et al. (1999) found that radiostrontium is distinguished by its mobility in the soil and availability for absorption by plant roots. The application of phytoextraction enhanced the availability of radiostrontium in the soil (Ebbs et al. 1998, Echevarria et al. 2001). Moreover, Tc had high mobility in soil (Echevarria et al. 1997). Also, Bunzl et al. (1999) noticed that the soil polluted with 137Cs might also be contaminated with other heavy metals deposited from the smelter processes. According to Fesenko et al. (1997), radionuclide bioavailability is impacted significantly by radionuclide deposition and aging. Furthermore, the uptake of 137Cs by plants is ascertained by the exchangeable and mobile sources of radioactivity in the soil (Fesenko et al. 1997).

11.2.2 Rhizofiltration

Timofeeva-Ressovskaia et al. (1962) in Russia witnessed the first use of water plants to clean up radionuclide pollution in the early 1950s. Also, it's found that the water plants had the ability to absorb large quantities of radionuclides, with Bioaccumulation Concentration (BC) values for 1910 for 90Sr, while 1230 for 137Cs for *Cladophora glomerata* (L.) (Timofeeva-Ressovskaia et al. 1962). In addition, many engineering approaches have recently been investigated to extract contaminants from aqueous streams by using terrestrial plant roots (Dushenkov et al. 1995) or seedlings as previously described (Salt et al. 1997). Moreover, a removal kinetic for Cs and Sr from the water was discovered by Sorochinsky et al. (1998) and Dunshenkov et al. (1997). As a result, rhizofiltration is the most effective method for removing radionuclides from aqueous stream.

11.2.3 Phytostabilization

It is one of the processes ideal for areas polluted by radionuclides because it allows to retain the radionuclides in the soil, preventing other pollutants from being exposed. Berti and Cunningham (2000) observed that the development of vegetated roots prevents humans from encountering radionuclides in polluted sites by avoiding windblown dust.

11.2.4 Phytovolatilization

This process allows plants to take up a large amount of water and then evaporate into the air, which is well-known in ^{3}H remediation. It is worth noting that the best way to reduce the risk of ^{3}H is to adjust the direction in the atmosphere rather than to remove ^{3}H or isolate it from the water (Fulbright et al. 1996). Murphy (2001) created a basic uptake model for ^{3}H, in which the simulations showed the decreased concentration of ^{3}H in water as the rate of ^{3}H decayed. Moreover, the installation performance of phytoremediation should include monitoring the concentration of ^{3}H in the air (Negri and Hinchman 2000).

11.3 Remediation Technologies of Uranium Contaminated Soils and Water

There are three methods for remediating uranium from soil and water:

11.3.1 Physical Methods

Coagulation, evaporation, extraction, and membrane separation technologies are some of the most common methods for removing uranium contamination from water and soil, especially in small areas (Tang et al. 2003).

11.3.2 Chemical Methods

Certain chemicals, such as FeS_2 and MnO_2, are used to strip U from water or soil in the same way as zero-valent iron (ZVI) is (Chicgoua et al. (2005). Although these chemical methods have high efficiency in extracting uranium and are low in cost, they are still considered experimental.

11.3.3 Biological Methods

According to Dushenkov (2003) and Kalin et al. (2004), it may be possible to minimize or fix uranium-polluted environments using microorganisms (e.g., bacteria and fungi) or plants because of their ability to alter extracellular binding sites and pH in soil and water by altering the bioavailability of U.

11.3.3.1 Application of phytoremediation in remediating uranium pollution

Phytoremediation is based on the plants' ability to accumulate heavy metals, chemical or radioactive contaminants competently. Plants have been found to clean up contaminants in soil and water through rhizosphere filtration, absorption, stability, degradation, volatilization, etc. The roots of some plants, such as sunflower, Indian mustard, and others, can absorb a large amount of U and transfer it to their shoots.

11.3.3.1.1 Soil

This section will go through some common plants that are used to remove U from the soil. Each of these plants can remove U from the soil in a unique way. For example, Brooks (1983) discovered that the Leguminosae family's Spring Vetch (*Vicia sativa*) and Hairy or Winter Vetch (*Vicia villosa*) can grow in soil rich in U, which is used as an indicator of U. Furthermore, the Giant Sunflower (*Helianthus giganteus*) was discovered to be one of the plants that can accumulate U through rhizofiltration (Dushenkov et al. 1997a). Due to its ability to efficiently absorb a variety of heavy metals such as chromium, lead, cadmium, zinc, copper, and nickel (Baker 1994, Chaney 1998), the Indian Mustard (*Brassica juncea*) of the Cruciferae is considered a uranium accumulator. One-Seed Juniper (*Juniperus monosperma*), a perennial species, has been used as a phytoremediation species due to its ability to remove uranium from soil (Brooks 1983, Cannon 1957). Furthermore, due to its resistance to heavy metal toxicity, the willow (*Salix smithiana*) has the capacity to absorb U from soil (Mihalik et al. 2010). Vyslouzilova (2006) studied willow plants for U phytoextraction. The pH of the soil increased in response to plant types and U sources (Meng et al. 2018). Furthermore, UO_3 and $UO_2(NO_3)_2$ have better solubility and bioavailability than UO_2, which is thought to be a key factor in phytoremediation efficacy. Furthermore, several statistical studies revealed that planting and initial aging significantly affected uranyl and UO_3 treatments but did not affect UO_2 treatments. As a result, both aging and planting dramatically reduced bioavailable and labile U, owing to the weak acid soluble U in both the UO_3 and Uranyl treatments, when compared to U in soils after one month with no plants. Besides, there are significant differences among various U species. Yang and Pan (2013) found that sunflower root exudates had considerable effects on the bioavailability, toxicity, and phytoavailability of U in soil, owing to its proclivity for metal complexation. Also, Meng et al. (2018) confirmed the same results as shown by previous reports that the redistribution of U among the soil fractions was governed by both U sources and plant species (Lu et al. 2005, Jalali and Khanlari 2008, Kim et al. 2010).

11.3.3.1.2 Water

Rhizofiltration is characterized by the low cost and its high efficiency in the purification of water. By using the technology of rhizofiltration, Salt et al. (1995) suggested the use of terrestrial plants as these are able to remediate water contaminated with U. Dittmer (1937) observed that terrestrial plants had a broad root system and a sophisticated uptake mechanism, with a total length of roots including root hairs of about 620 km and a total surface area of over 3000 m^2. Furthermore, Kabata-Pendias and Pendias (1989) found that these plants are unable to differentiate between isotopes of the same compound, leading to the use of radioactive isotopes such as C_{14}, O_{18}, P_{32}, S_{35}, and Fe_{52} as a tracer in plant physiology and biochemistry. Dushenkov et al. (1995) discovered that the roots of terrestrial plants, such as sunflower (*Helianthus annuus* L.), Indian mustard (*Brassica juncea* L.), and various grasses, can eliminate radionuclide and toxic metals such as U, Cd^{+2}, Cu^{+2}, and others from the water.

11.3.3.1.3 Other applications of phytoremediation of U in soil and water

11.3.3.1.3.1 Coupled electro-kinetic remediation and phytoremediation of uranium contaminated soil and water

Li et al. (2019) carried out a study on the sunflower and Indian mustard to examine the efficiency of electrokinetic field (EKF) enhanced phytoremediation in soil contaminated with U. Also, it has been discovered that EKF treatments have a considerable impact on soil pH in the anode region while raising it in the cathode zone (Li et al. 2019). Therefore, the redistribution of U inside various solid-phase components is different in both anode and cathode regions. Furthermore, with UO_3, uranyl, and UO_2, the electrokinetic field (EFK) treatments boosted the removal of U efficiency by 35–50%, 35–47%, and 26–62%, respectively, from the soil.

11.3.3.1.3.2 Coupled electrokinetic and phytoremediation was used to remove other radionuclides such as Cs

Mao et al. (2016a) carried out a study to examine the distribution and solubility of Cesium (Cs) by using the application of electrokinetic field and phytoremediation coupled with electrokinetic. It has been noticed that Cs is one of the radionuclides with high solubility and bioavailability in soil compared to lead (Pb). On the other hand, the electrokinetic field (EKF) was found to greatly improve the solubility and bioavailability of additional Pb and Cs. Also, the EKF treatment enhanced the lowering of pH near the anode, which is essential for the dissolution of metal(loids). However, the acidification increased the solubility (SOL) and exchange (EXC) of Pb and Cs in soils. As a result of the high OH-concentration, there was element mobilization at the anode and the likelihood of precipitation near the cathode. Also, it was noticed that the bioaccumulation of Pb and Cs was increased significantly due to the enhancement of the electro-kinetic field. Besides, the increase in the SOL and exchange (EXC) forms increased the plants' ability to accumulate metals in the roots and shoots. In addition, it's observed that EKF plays a significant role in enhancing consecutive soil washing as it improves the performance of removing the Pb and Cs from soil (Mao et al. 2016b). Also, the acidification near the anode was essential for the release of Pb and Cs from the solid-phase fractions. On the other hand, alkalization near the cathode may have resulted in metal precipitation. Furthermore, it was discovered that utilizing $CaCl_2$ and citric acid instead of Na_2 EDTA to increase the remediation of Pb in soil was more efficient in the nodal area of the soil. As a result, as the electrokinetic field (EKF) intensities grew, the performance of washing soil also improved. On the other hand, Mao et al. (2018) conducted research on the use of non-uniform electric intensity (EI) and polarity reversal (PR) to improve EKF in the cleanup of radioactive contamination of soil with 137Cs. As can be shown, EI and PR both increased the effectiveness of eliminating 137Cs with about 5% with the enhancement of EKF in 168 hrs as they have the ability to focus on the phenomena of the alkalization pH of the soil.

11.3.3.2 Application of microbes in mitigating uranium pollution

Recently, microorganisms such as pseudomonas and others were used to accumulate uranium with high efficiency (Malekzadeh et al. 2002).

Vegetables and some selected plants, such as mustard, have aerial sections that can accumulate U higher than their roots, making them hyperaccumulators for U (Tang et al. 2009). On the other hand, several plants such as spinach have demonstrated their capacity to take up U by their roots (Xu et al. 2009). Sunflower may remove U from water because of its enormous bio-volume (Dushenkov et al. 1997b). Furthermore, Chen et al. (2011) discovered that raising the soil conditioner led to increased bioavailability of U from the soil. The soil microorganism forms symbiotic associations with the plant's roots. As a result, a combination of microbial remediation and phytoremediation technology has been observed to extract the U from a contaminated area. Merroun and Selenska-Pobell (2008) found that microbial interactions with U and other radionuclides include some applications, which are:

11.3.3.2.1 Bioreduction

It is one of the uranium bioremediation strategies that use an electron donor to induce the enzymatic system to convert the uranium U(VI) reduction to an insoluble U(IV) (Wilkins et al. 2007, Begg et al. 2011, Law et al. 2011). Recently, Kelly et al. (2008), Bernier-Latmani et al. (2010), and Alessi et al. (2012) found that the bioreduced forms of uranium are uraninite (UO_2) and U(IV).

11.3.3.2.2 Biomineralization

It is a microbiological process in which metals are precipitated as sulfide, phosphates, carbonates, and hydroxides at the cell surface (Macaskie et al. 1992). The process of biomineralization for uranium in the soil takes place by using species of microbes which are *Citrobacter*, i.e., *Serratia* species and a glycerol phosphate (Pattanapipitpaisal et al. 2002). As the glycerol phosphate plays vital role in activating the phosphatase enzyme in the cell of the bacteria by converting the organic form of phosphate to an inorganic form (hydrogen uranyl phosphate ($HUO_2\,PO_4$)).

11.3.3.2.3 Bioaccumulation

Suzuki and Banfield (1999) suggested that uranium penetrated easily inside cells due to the increased permeability of a membrane resulting in uranium toxicity. Furthermore, many investigations have found that *Pseudomonas* species can successfully bioremediate uranium, particularly in the form of uranyl phosphate (Kazy et al. 2009, VanEngelen et al. 2010, Choudhary and Sar 2011). Therefore, bioaccumulation of uranium was considered as a suitable technique for bioremediating U either in land or water.

11.3.3.2.4 Biosorption

It is one of the mechanisms for uranium's passive uptake by living and nonliving microbiological cells (Lloyd and Macaskie 2000). Schiewer and Volesky (2000) noticed that alkaline groups such as carboxyl, hydroxyl, amine, sulfhydryl, and phosphate bound metals through chemical sorption. Therefore, gram-positive and gram-negative bacterial cells hold an electronegative charge which could attract the cations of metals, which are transferred to the surface. In addition, the binding of metals with the cell walls is more rapid than the uptake into the cell. Thus, it is easier to remove the metals from the surface of the cell to restore the bio sorbent (Schiewer and Volesky 2000).

11.4 Geochemical Modeling on Evaluation of Uptake of Uranium from Soil and Water

11.4.1 Soil

The model describes the uptake of uranium species from soil and water through some statistical equations. Nriagu (1991) and Abreu et al. (2008) calculated the Soil-Bioconcentration Coefficient (S-BC) by using soil chemical extraction to measure the available U from the soil that was absorbed by the plants as mentioned in the Eq. 11.1 and 11.2.

$$S-BC_{plant/soil}=\frac{[U\ concentration\ in\ (root+leaf)]}{available\ in\ soil\ fraction,\ extracted\ by\ NH_4-acetate} \quad \text{(Eq. 11.1)}$$

$$S-BC_{leaf/soil}=\frac{[U\ concentration\ in\ leaf]}{available\ in\ soil\ fraction,\ extracted\ by\ NH_4-acetate} \quad \text{(Eq. 11.2)}$$

11.4.2 Water

In the case of water, Water-Bioconcentration Coefficient (W-BC) was calculated by knowing the absorption of U from water using plants as in the Equations 11.3 and 11.4.

$$W-BC_{plant/water}=\frac{[U\ concentration\ in\ (root+leaf)]}{dissolved\ in\ water} \quad \text{(Eq. 11.3)}$$

$$W - BC_{leaf/water} = \frac{[U\ concentration\ in\ leaf]}{dissolved\ in\ water} \qquad \text{(Eq. 11.4)}$$

The Translocation Coefficient (TC) was used to represent the element translocation from soil to the plant and is defined by the following Eq. 11.5.

$$TC = [leaf]/[root + leaf] \qquad \text{(Eq. 11.5}$$

11.5 Effects of Biogeochemical Conditions on Phytoremediation of Uranium

Biogeochemical interactions regulate the mobility of U and other radionuclides (such as Tc, Np, and Pu), as well as metabolic processes like microbial respiration (Laura et al. 2014). Therefore, interactions between microbes-radionuclide, especially uranium, in the environment control their fate mobility and availability (Parrish et al. 2008, Handley-Sidhu et al. 2010). Thus, adsorption-desorption, precipitation-dissolution, chelation-dissociation, mineralization-assimilation, and protonation-deprotonation are all biogeochemical processes (He et al. 1998). In addition, the involvement of each process is influenced by the soil type and rhizosphere influences (Gobran et al. 1999).

11.5.1 pH

Soil pH is considered one of the most essential factors in the availability of metals in the soil (Barancíkova et al. 2004). Moreover, Tudoreanu and Phillips (2004) observed an indirect linear relationship between the uptake of metals in soil and the pH of the soil. For example, if the pH of the soil decreases, the availability of the metals increases and vice versa (Kirkham 2006). Kabat-Pendias (2000) reported that the mobility increased over the pH range between 4.5–5.5; when pH was remarkably high, it converts Cd to insoluble form as carbonate and phosphate.

11.5.2 Soil organic matter (SOM)

It is noticed that the majority of organic matter, in the form of bulk organic components or coated particle matter, collects on the soil's surface (Thompson and Goyne 2012). Also, it is a reactive component in the soil that has the capability to capture the cations of the metals. Kirkham (2006) found that the quality of SOM plays a vital role in pollutant accumulation and binding. Organic soils, such as biosolid, sewage sludge, pig manure, and other organic soils, have a high Cd binding (30 times higher than other organic soils) and were utilized as soil additions to reduce Cd availability (Jadia and Fulekar 2009, Sarwar et al. 2010, Hao et al. 2012). Zhang et al. (2004) discovered that spreading muck on the surface of sandy soil reduced Cd leaching. On the other hand, the application of lime and base fertilizers made the pH close to neutral; thus, the uptake of heavy metals was limited (Menon et al. 2005).

11.5.3 Clay Minerals and Cation Exchange Capacity

Both clay minerals and CEC affect the mobility and availability of metals in the soil (Gothberg et al. 2004). Also, Gothberg et al. (2004) found that the presence of other ions may influence the availability of Cd due to the ionic strength (Degryse et al. 2004) and the sites of exchange on the surface of the roots (Tlustos et al. 2006). As a result, if the ionic strength of the growth medium is low, the uptake of metals by plants increases (Gothberg et al. 2004). Furthermore, increased surface charge caused by phosphate adsorption, liming, and organic matter application promotes Cd adsorption (Bolan and Duraisamy 2003).

11.5.4 Fertilizers

Mohamed et al. (2015) found that fertilizers modified metal speciation via complex creation and subsequent transport and absorption of metals by roots. Also, the effect of nitrogen on metal uptake from soil was dependent on plant species, nitrogen source, rate, and time of application (Zaccheo et al. 2006). For example, in the field experiments increasing the dosage of N led to an increase in the concentration of the available Cd in the soil (Mitchell et al. 2000). In addition, fertilizers caused the distribution of nutrients in the rhizosphere, the growth of the roots and the plants, thus modifying the availability of metals and their absorption and accumulation in plants (Chaney et al. 2008). On the other hand, fertilization may not be an appropriate approach for removing Cd from the soil in agriculture (Ji et al. 2011). Adding an appropriate quantity of NaCl to the halophyte *Spartina alterniflora* resulted in a considerable rise in Cd content in roots and shoots (Chai et al. 2013), as well as increased Cd absorption and translocation in *Sesuvium portulaca* (Ghnaya et al. 2007).

11.5.5 Redox Potential

The standard hydrogen electrode (SHE) is used to measure the zero point of Eh. Eh is extensively utilized in fields involving live organisms, such as microbial ecology (Alexander 1964). It is noticed that the reduced forms of trace metals lead to a decrease in mobility. Reduction of soluble Cr (VI) to other Cr (III) decreased mobility (Lovely 1993). Weber et al. (2009) reported that the speciation of trace metals was affected by microbial respiration through some changes in sorption and precipitation. It is noticed that the reductive dissolution of metal oxides may cause a loss in the capacity of sorption (Zachara et al. 2001). On the other hand, Weber (2009) noticed that releasing Fe^{+2} may participate with the cations of trace elements for sorption on mineral and organic sorbents. Zachara et al. (2001) observed that the effect of mobilization of metals might be disturbed by the adsorption or precipitation in the presence of Fe (II) and its precipitates as siderite, vivianite, and green rust (Parmar et al. 2001). In addition, Kirk (2004) and Tufano and Fendorf (2008) observed that microbial respiration increased the formation of these precipitates due to an increase in the concentration of Fe^{2+}, carbonate, or phosphate.

11.6 Conclusion

This chapter clearly showed different methods of remediation of U and other radionuclides in soil and water. Remediation of U includes three types of processes, which are physical, chemical, and biological. Phytoremediation is one of those biological processes that are environment friendly and cost-effective, using plants to collect and clean up U and other radionuclides from soils and water. Moreover, the capacity of the electrokinetic field (EKF) to boost the bioavailability and solubility of certain radionuclides in soil and water has been demonstrated. Furthermore, utilizing plants, biochemical modeling is used to calculate the bioconcentration and translocation coefficient of these radionuclides, particularly U, in soil and water. This chapter also discussed biogeochemical factors that affect metal(loid)'s and radionuclide's solubility and availability in soil and water, such as pH, organic molecules, fertilizers, and redox potential. As a result, this chapter provides information regarding radionuclide remediation to researchers and graduate students who are beginning to write proposals or projects.

References

Abreu, M. M., M. T. Tavares and M. J. Batista. 2008. Potential use of *Erica andevalensis* and *Erica australis* in phytoremediation of sulphide mine environments: São Domingo's, Portugal. J. Geochem. Explore. 96: 210–222.

Adriano, D. C, A. Wallace and E. M. Romney. 1980. Uptake of transuranic nuclides from soil by plants grown under controlled environmental conditions. pp. 336–360. *In*: Hanson, W. (ed.). Transuranic Elements in the Environment. Technical Information Center, USDOE.

Alessi, D. S., B. Uster, H. Veeramani, E. I. Suvorova, J. S. Lezama-Pacheco, J. E. Stubbs et al. 2012. Quantitative separation of monomeric U(IV) from UO_2 in products of U(VI) reduction. Environ. Sci. Technol. 46: 6150–6157.

Alexander, M. 1964. Biochemical ecology of soil microorganisms. Annu. Rev. Microbial. 18: 217–250.

Baker, A. J. M. 1994. The possibility of *in-situ* heavy metal decontamination of polluted soils using crops of metal-accumulating plants. Res. Conserve. Recycle. 11: 41–49.

Barancíkova, G., M. Madaras and O. Rybar. 2004. Crop contamination by selected trace elements. J. Soils Sediment. 4: 37e–42.

Bednar, A. J., V. F. Medina, D. S. Ulmer-Scholle, B. A. Frey, B. L. Johnson, W. N. Brostoff and S. L. Larson 2007. Effects of organic matter on the distribution of uranium in soil and plant matrices. Chemosphere. 70: 237–247.

Begg, J. D. C., I. T. Burke, J. R. Lloyd, C. Boothman, S. Shaw, J. M. Charnock and K. Morris. 2011. Bioreduction behavior of U(VI) sorted to sediments. Geomicbio. J. 28: 160–171.

Bernier-Latmani, R., H. Veeramani, E. D. Vecchia, P. Junier, J. S. Lezama-Pacheco, W. R. Berti and S. D. Cunningham. 2000. Phytostabilization of metals. pp. 71–88. *In*: Raskin, I. (ed.). Phytoremediation of Toxic Metals: Using Plants to Clean Up the Environment. Wiley-Interscience, John Wiley and Sons, Inc. New York, NY.

Bleise, A., P. R. Danesi and W. Burkart. 2003. Properties, use and health effects of depleted uranium (DU): a general overview. J. Environ. Radioact. 64: 93–112.

Bolan, N. S. and D. Duraisamy. 2003. Role of inorganic and organic soil amendments on immobilization and phytoavailability of heavy metals: a review involving specific case studies. Aust. J. Soil Res. 41: 533e–555.

Brooks, R. R. 1983. Biological Methods of Prospecting for Minerals, John Wiley & Sons, NY.

Bunzl, K., B. P. Albers, W. Shimmack, K. Rissanen, M. Suomela, M. Puhakainen, T. Rahola and E. Steinnes. 1999. Soil to plant uptake of fallout 137 Cs by plants from boreal areas polluted by industrial emissions from smelters. Sci. Total Environ. 234: 213–221.

Cannon, H. L. 1957. US Geological Survey Bull. 1030M., 399.

Chai, M. W., F. C. Shi, R. L. Li, F. C. Liu, G. Y. Qiu and L. M. Liu. 2013. Effect of NaCl on growth and Cd accumulation of halophyte Spartina alterniflora under CdCl2 stress. S. Afr. J. Bot. 85: 63e–69.

Chaney, R. L. 1998. Land Treatment of Hazardous Wastes. Noyes Data Corporation, Park Ridge, NJ, 50–76.

Chaney, R. L., K. Y. Chen, Y. M. Li, J. S. Angle and A. J. M. Baker. 2008. Effects of calcium on nickel tolerance and accumulation in *Alyssum* species and cabbage grown in nutrient solution. Plant Soil 311: 131e–140.

Chang, P., K. W. Kim, S. Yoshida and S. Y. Kim. 2005. Uranium accumulation of crop plants enhanced by citric acid. Environ. Geochem. Health 27: 529–538.

Chen, B. D., M. M. Chen and R. Bai. 2011. The potential role of *Arbuscular mycorrhizal* in controlling uranium contamination in the environment. Environ. Sci. 32: 809–815 (in Chinese).

Chicgoua, N., M. Günther and J. M. Broder. 2005. Investigating the mechanism of uranium removal by zerovalent iron. Environ. Chem. 2: 235–242.

Choudhary, S. and P. Sar. 2010. Identification and characterization of uranium accumulation potential of a uranium mine isolated *Pseudomonas* strain. World J. Microbial. Biotech. 27: 1795–1801.

Choudhary, S. and P. Sar. 2011. Uranium biomineralization by a metal resistant *Pseudomonas aeruginosa* strain isolated from contaminated waste. J. Hazard. Mater. 186: 336–343.

Cline J. F. and W. H. Rickard. 1972. Radioactive strontium and cesium in cultivated and abandoned field plots. Health Phys. 23: 317–324.

De Boulois, H. D., E. J. Joner, C. Leyval, I. Jakobsen, B. D. Chen, P. Roos, Y. Thiry, G. Rufyikiri, B. Delvaux and S. Declerck. 2008. Impact of *Arbuscular mycorrhizal* fungi on uranium accumulation by plants. J. Environ. Radioact. 99: 775–784.

Degryse, F., J. Buekers and E. Smolders. 2004. Radio-labile cadmium and zinc in soils as affected by pH and source of contamination. Eur. J. Soil Sci. 55: 113e–121.

Dittmer, H. J. 1937. A Quantitative study of the roots and root hairs of a winter rye plant (*Secale Cereale*). Am. J. Bot. 24: 417–420.

Dushenkov, S. 2003. Trends in phytoremediation of radionuclides. Plant Soil. 249: 167–175.

Dushenkov, S., Y. Kapulnik, M. Blaylock, B. Sorochinsky, I. Raskin and B. Ensley. 1997a. Phytoremediation: a novel approach to an old problem. pp. 563–572. *In*: Wise, D. L. (ed.). Global Environmental Biotechnology. Elsevier Science, Amsterdam.

Dushenkov, S., D. Vasudev, Y. Kapulnik, D. Gleba, D. Fleisher, K. C. Ting and B. Ensley. 1997b. Removal of uranium from water using terrestrial plants. Environ. Sci. Technol. 31: 3468–3474.

Dushenkov, S., A. Mikheev, A. Prokhnevsky, M. Ruchko and B. Sorochinsky. 1999. Phytoremediation of radiocesium-contaminated soil in the vicinity of Chernobyl, Ukraine. Environ. Sci. Technol. 33: 469–475.
Dushenkov, V., N. P. B. A. Kumar, H. Motto and I. Raskin. 1995. Rhizofiltration: the use of plants to remove heavy metals from aqueous streams. Environ. Sci. Technol. 29: 1239–1245.
Ebbs, S. D., D. J. Brady and L. V. Kochian. 1998a. Role of uranium speciation in the uptake and the translocation of uranium by plants. J. Exp. Bot. 49: 1183–1190.
Echevarria, G., P. C. Vong and J. L. Morel. 1997. Bioavailability of Technetium-99 as affected by plant species and growth, application form, and soil incubation. J. Environ. Qual. 26: 947–956.
Echevarria, G., N. I. Sheppard and J. Morel. 2001. Effect of pH on sorption of uranium in soils. J. Environ. Radioact. 53: 257–264.
Fesenko, S. V., S. I. Spiridono, N. I. Sanzharova and R. M. Alexakhin. 1997. Dynamics of 137Cs bioavailability in soil-plant system in areas of the Chernobyl Nuclear Power Plant accident zone with a different physico-chemical composition of radioactive fallout. J. Environ. Radioact. 34: 287–313.
Fulbright, H. H., A. L. Schwirian-Spann, K. M. Jerome, B. B. Looney and V. V. Brunt. 1996. Status and practicality of detritiation and tritium reduction strategies for environmental remediation. Aiken, SC, Westinghouse Savannah River Company. 1–10.
Gavrilescu, M., L. V. Pavel and I. Cretescu. 2009. Characterization and remediation of soils contaminated with uranium. J. Hazard. Mater. 163: 475–510.
Ghnaya T., I. Slama, D. Messedi, C. Grignon, M. H. Ghorbel and C. Abdelly. 2007. Cadmium effects on growth and mineral nutrition of two halophytes: *Sesuvium portulacastrum* and *Mesembryanthemum crystallinum*. J. Plant Physiol. 162: 1133–1140. 10.1016/j.jplph.2004.11.011.
Gobran, G. R., S. Clegg and F. Courchesne. 1999. The rhizosphere and trace element acquisition in soils. *In*: Selim, H. M. and I. K. Iskandar (eds.). Fate and Transport of Heavy Metals in the Vadose Zone. Lewis Publ., Boca Raton, FL. 225e–250.
Gothberg, A., M. Greger, K. Holm and B. E. Bengtsson. 2004. Influence of nutrient levels on uptake and effects of mercury, cadmium, and lead in water spinach. J. Environ. Qual. 33: 1247–e1255.
Handley-Sidhu, S., M. J. Keith-Roach, J. R. Lloyd and D. J. Vaughan. 2010. A review of the environmental corrosion, fate, and bioavailability of munitions grade depleted uranium. Sci. Total Environ. 408: 5690–5700.
Hao, X. Z., D. M. Zhou, D. D. Li and P. Jiang. 2012. Growth, cadmium, and zinc accumulation of ornamental sunflower (*Helianthus annuus* L.) in contaminated soil with different amendments. Pedosphere. 22(5): 631e–639.
He, Z. L., A. K. Alva, D. V. Calvert, Y. C. Li and D. J. Banks. 1999. Effects of nitrogen fertilization of grapefruit trees on soil acidification and nutrient availability in a Riviera fine sand. Plant Soil. 206: 11e–19.
Huang, J. W., M. J. Blaylock, Y. Kapulnik and B. D. Ensley. 1998. Phytoremediation of uranium-contaminated soils: role of organic acids in triggering uranium hyperaccumulation in plants. Environ. Sci. Technol. 32: 2004–2008.
Jadia, C. and H. M. Fulekar. 2009. Phytoremediation of heavy metals: recent techniques. African J. Biotechnol. 8(6): 921–928.
Jalali, M. and Z. V. Khanlari. 2008. Effect of aging process on the fractionation of heavy metals in some calcareous soils of Iran. Geoderma. 143(1): 26–40.
Ji, P. H., T. H. Sun, Y. F. Song, M. L. Ackland and Y. Liu. 2011. Strategies for enhancing the phytoremediation of cadmium-contaminated agricultural soils by *Solanum nigrum* L. Environ. Pollut. 159: 762–768.
Kabata-Pendias, A. and H. Pendias. 1989. Trace elements in soils and plants; CRC Press: Boca Raton, FL.
Kabata-Pendias, A. 2000. Trace elements in soils and plants. 3rd Edition, CRC Press, Boca Raton. http://dx.doi.org/10.1201/9781420039900.
Kalin, M., W. N. Wheeler and G. Meinrath. 2004. The removal of uranium from mining wastewater using algal/microbial biomass. J. Environ. Radioact. 78: 151–177.
Kazy, S. K., S. F. D'Souza and P. Sar. 2009. Uranium and thorium sequestration by a *Pseudomonas* sp.: mechanism and chemical characterization. J. Hazard. Mater. 163: 65–72.
Kelly, S. D., K. M. Kemner, J. Carley, C. Criddle, P. M. Jardine, T. L. Marsh, D. Phillips, D. Watson and W. M. Wu. 2008. Speciation of uranium in sediments before and after *in situ* bio stimulation. Environ. Sci. Technol. 42: 1558–1564.
Kim, K. R., G. Owens and R. Naidu. 2010. Heavy metal distribution, bioaccessibility and phytoavailability in long-term contaminated soils from Lake Macquarie. Aust. J. Soil Res. 47: 166e–176.
Kirk, G. 2004. The Biogeochemistry of Submerged Soils. Ann. Bot. 96(1): 165. doi: 10.1093/aob/mci162.
Kirkham, M. B. 2006. Cadmium in plants on polluted soils: effect of soil factors, hyperaccumulation and amendments. Geoderma. 137: 19e–32.
Law, G. T .W., A. Geissler, I. T. Burke, F. R. Livens, J. R. Lloyd, J. M. McBeth and K. Morris. 2011. Uranium redox cycling in sediment and biomineral systems. Geomicrobio. J. 28: 497–506.
Li, J., J. Zhang, S. L. Larson, J. H. Ballard, K. Guo, Z. Arslan, Y. Ma, C. A. Waggoner, J. R. White and F. X. Han. 2019. Electrokinetic-enhanced phytoremediation of uranium-contaminated soil using Sunflower and Indian mustard. Inter. J. Phytoremediat. 21(12): 1197–1204. doi.org/10.1080/15226514.2019.1612847.
Lin, R. H., L. J. Wu, C. H. Lee and S. Y. Lin-Shiau. 1993. Cytogenetic toxicity of uranyl nitrate in Chinese hamster ovary cells. Mutate. Res. 319: 197–203.

Lloyd, J. R and L. E. Macaskie. 2000. Bioremediation of radionuclide-containing wastewaters. *In*: Lovely, D. R. (ed.). Environmental Microbe–Metal Interactions. ASM Press, Washington D. C. 277–327.

Lovely, D. R. 1993. Dissimilatory metal reduction. Annu. Rev. Microbiol. 47: 231–261.

Lu, A., S. Zhang and X. Q. Shan. 2005. Time effect on the fractionation of heavy metals in soils. Geoderma. 125(3): 225–234.

Macaskie, L. E., R. M. Empson, A. K. Cheetham, C. P. Grey and A. J. Skarnulis. 1992. Uranium bioaccumulation by a *Citrobacter* sp. because of enzymically mediated growth of polycrystalline HUO_2 PO_4. Sci. 257: 782–784.

Malaviya, P. and A. Singh. 2012. Constructed wetlands for management of urban stormwater runoff. Crit. Rev. Environ. Sci. Technol. 20: 2153–2214.

Malekzadeh, F., A. Farazmand and H. Ghafourian. 2002. Uranium accumulation by a bacterium isolated from electroplating effluent. World J. Microbiol. Biotech. 18: 295–300.

Mao, X., F. X. Han, X. Shao, K. Guo, J. McComb, S. Njemanze, Z. Arslan and Z. Zhang. 2016a. The distribution and elevated solubility of lead, arsenic and cesium in contaminated paddy soil enhanced with the electro-kinetic field. Int. J. Environ. Sci. Technol. 13: 1641–1652. DOI 10.1007/s13762-016-1007-2.

Mao, X., Z. Shao, F. X. Han, Z. Arslan, R. Zhang, J. McComb and K. Gao. 2016b. Remediation of lead, arsenic, and cesium contaminated soil using consecutive washing enhanced with electro-kinetic field. J. Soil. Sediment. 10: 2344–2353. DOI: 10.1007/s11368-016-1435-0.

Mao, X., X. Shao, Z. Zhang and F. X. Han. 2018. Mechanism and optimization of enhanced electro-kinetic remediation on 137Cs contaminated kaolin soils: A semi-pilot study based on experimental and modeling methodology. Electrochimica. Acta. 284: 38–51. https://doi.org/10.1016/j.electacta.2018.07.136.

Meng, F., D. Jin, K. Guo, S. L. Larson, J. H. Ballard, L. Chen, Z. Arslan, G. Yuan, C. A. Waggoner, J. White, Y. Ma and F. X. Han. 2018. Influences of U sources and forms on its bioaccumulation in Indian mustard and Sunflower. Water Air Soil Pollut. 229–369. https://doi.org/10.1007/s11270-018-4023-7.

Menon, M., S. Hermle, K. C. Abbaspour, S. E. Oswald and R. Schulin. 2005. Water regime of metal-contaminated soil under juvenile forest vegetation. Plant Soil. 271: 227e–241.

Merroun, M. L. and S. Selenska-Pobell. 2008. Bacterial interactions with uranium: an environmental perspective. J. Contam. Hydrol. 102: 285–295.

Merten, D., E. Kothe and G. Buchell. 2004. Studies on microbial heavy metal retention from uranium mine drainage water with special emphasis on rare earth elements. Mine Water Environ. 23: 34–43.

Mohamed, F. J., L. Wenshu, E. B. Rebecca, A. E. Kathleen, L. Ann, M. H. Kathy, S. Paul, B. J. Mohammad, M. Oliver, H. Sara, J. M. Barbara, D. Foday and T. R. John. 2015. Impact of Ebola experiences and risk perceptions on mental health in Sierra Leone. BMJ Global Health. doi: 10.1136/ bmjgh-2017-000471.

Mihalík, J., P. Tlustoš and J. Szaková. 2010. Comparison of willow and sunflower for uranium phytoextraction induced by citric acid. J. Radioanal. Nucl. Chem. 279–285.

Mitchell, L. G., C. A. Grant and G. J. Racz. 2000. Effect of nitrogen application on concentration of cadmium and nutrient ions in soil solution and in durum wheat. Can. J. Soil Sci. 80: 107e–115.

Murphy, C. E. J. 2001. An estimate of the history of tritium inventory in wood following irrigation with tritiated water. Aiken, SC, Westinghouse Savannah River Company: 1–10.

Negri, C. M. and R. R. Hinchman. 2000. The use of plants for the treatment of radionuclides. pp. 107–132. *In*: Raskin, I. (ed.). Phytoremediation of Toxic Metals: Using Plants to Clean Up the Environment. Wiley-Interscience, John Wiley and Sons, Inc. New York, NY.

Nriagu, J. O. 1991. Human influence on the global cycling of trace metals. pp. 1–15. *In*: Farmer, J. D. (ed.). Heavy metals in the environment, Vol. 1. Edinburgh: CEP Consultants.

Parrish, R. R., M. Horstwood, J. G. Arnason, S. Chenery, T. Brewer, N. S. Lloyd and D. O. Carpenter. 2008. Depleted uranium contamination by inhalation exposure and its detection after approximately 20 years: implications for human health assessment Sci. Total Environ. 390: 58–68.

Parmar, N., A. Y. Gorby, J. T. Beveridge and F. Grant Ferris. 2001. Formation of green rust immobilization of nickel in response to bacterial reduction of hydrous ferric oxide. Geomicrobio. 18(4). DOI: 10.1080/014904501753210549.

Pattanapipitpaisal, P., A. N. Mabbett, J. A. Finlay, A. J. Beswick, M. Paterson-Beedle, A. Essa, J. Wright, M. R. Tolley, U. Badar, N. Ahmed, J. L. Hobman, N. L. Brown and L. E. Macaskie. 2002. Reduction of Cr (VI) and bioaccumulation of chromium by Gram-positive and Gram-negative microorganisms not previously exposed to Cr-stress. Environ. Technol. 23: 731–745.

Priest, N. D. 2001. Toxicity of depleted uranium. Lancet 357: 244–246.

Raskin, I. and B. Ensley. 2000. Phytoremediation of Toxic Metals: Using Plants to Clean Up. Open J. Ecology 5(8).

Salt, D. E., M. Blaylock, N. P. B. A. Kumar, V. Dushenkov, B. D. Ensley, I. Chet and I. Raskin. 1995. Phytoremediation: A novel strategy for the removal of toxic metals from the environment using plants. Biotechnol. 13: 468–474.

Sarwar, N., S. S. Saifullah-Malhi, H. M. Zi, A. Naeem, S. Bibia and G. Farida. 2010. Role of mineral nutrition in minimizing cadmium accumulation by plants. J. Sci. Food Agric. 90: 925e–937.

Schiewer, S. and B. Volesky. 2000. Biosorption processes for heavy metal removal. pp. 329–362. *In*: Lovley, D. R. (ed.). Environmental Microbe–Metal Interactions. ASM Press, Washington D. C.

Shahandeh, H. and L. R. Hossner. 2002b. Role of soil properties in phytoaccumulation of uranium. Water Air Soil Pollut. 141: 165–180.

Sorochinsky, B. V., A. N. Mikheev, M. V Kuchko and A. T. Prokhrevsky. 1998. Decontamination of small water reservoirs of the 10-km zone of chernobyl npp by rhizofiltration. In Problems of Chernobyl Exclusion Zone. 97–102. Naukova Dumka. Kiev.

Suzuki, Y. and J. F. Banfield. 1999. Geomicrobiology of uranium. Uranium, mineralogy, geochemistry and the environment. pp. 393–424. *In*: Burns, P., Finch, R. (eds). Reviews in Minerology, Vol. 38. Washington, DC, USA: Mineralogy Society of America.

Tang, L., Y. Bai and C. Deng. 2009. Restoration of uranium-contaminated soil hyperaccumulator and accumulation characteristics of the screening. Nuclear Technol. 32: 136–140 (in Chinese).

Tang, Z. J., P. Zhang and S. Q. Zuo. 2003. Technology research of treatment for uranium wastewater of low concentration. Ind. Water Waste. 4: 9–12 (in Chinese).

Timofeeva-Ressovskaia, E. A., B. M. Agafonov and N. V. Timofeev- Ressovsky. 1962. On radioisotopes fate in water bodies. Proc. Inst. Biol. 22: 49–67.

Tlustos, P., J. Szakova, K. Korinek, D. Pavlikova, A. Hanc and J. Balik. 2006. The effect of liming on cadmium, lead and zinc uptake reduction by spring wheat grown in contaminated soil. Plant Soil Environ. 52: 16–e24.

Tudoreanu, L. and C. J. C. Phillips. 2004. Modelling cadmium uptake and accumulation in uptake and effects of mercury, cadmium, and lead in water spinach. J. Environ. Qual. 33.

Tufano, K. T. and S. Fendorf. 2008. Contrasting impacts of iron reduction on arsenic retention. Environ. Sci. Technol. 42: 4777–4783.

Vandenhove, H. and M. V. Hees. 2004. Phytoextraction for clean-up of low-level uranium contaminated soil evaluated. J. Environ. Radioact. 72: 41–45.

VanEngelen, M. R., E. K. Field, R. Gerlach, B. D. Lee, W. A. Apel and B. M. Peyton. 2010. UO_2^{+2} speciation determines uranium toxicity and bioaccumulation in an environmental *Pseudomonas* sp. isolate. Environ. Toxicology. Chem. 29: 763–769.

Vyslouzilova, M. 2006. Rhizosphere characteristics, heavy metal accumulation and growth performance of two willow (Salix 9 Ruben's) clones. Plant Soil Environ 52: 353–361.

Weber, F. A., A. Voegelin and R. Kretzschmar. 2009. Multi-metal contaminant dynamics in temporarily flooded soil under sulfate limitation. Geochim. Cosmochim. Acta. 73: 5513–5527.

Wilkins, M. J., F. R. Livens., D. J. Vaughan, I. Beadle and J. R. Lloyd. 2007. The influence of microbial redox cycling on radionuclide mobility in the subsurface at a low-level radioactive waste storage site. Geobiol. 5: 293–301.

Xu, J., Y. B. Gong and Q. C. Zhang. 2009. Differences of patience and uranium uptake and accumulation of uranium in soil for three plants. Chem. Res. Appl. 21: 322–326 (in Chinese).

Yang, J. and X. Pan. 2013. Root exudates from sunflower (*Helianthus annuus* L.) show a strong adsorption ability toward Cd (II). J. Plant Interact. 8(3): 263–270.

Yao, J., Y. Ma and H. He. 2010. The potential applications for phytoremediation in controlling radioactive uranium pollution in the uranium mining. Environ Sichuan. 29: 48–51 (in Chinese).

Zaccheo, P., C. Laura and D. M. P. Valeria. 2006. Ammonium nutrition as a strategy for cadmium metabolization in the rhizosphere of sunflower. Plant Soil. 283: 43–e56.

Zachara, J. M., C. T. Resch and S. C. Smith. 1994. Influence of humic substances on Co2+ sorption by a subsurface mineral separate and its mineralogic components. Geochim. Cosmochim. Acta 58: 553–566.

Zachara, J. M., C. S. Smith and J. K. Fredrickson. 2001. The effect of biogenic Fe(II) on the stability and sorption of Co(II) EDTA (2-) to goethite and subsurface sediment. Geochimi. Et. Cosmochimi. Acta. 64(8): 1345–1362.

Zhang, M. K., Z. L. He., D. V. Calvert and P. J. Stoffella. 2004. Leaching of minerals and heavy metals from muck-amended sandy soil columns, Soil Sci. 169: 528e–540.

CHAPTER 12

Bioremediation Strategies for Removal of Chromium from Polluted Environment

Preksha Palsania, Mohd. Ashraf Dar and *Garima Kaushik**

12.1 Introduction

Rapid advancement of society and globalization comes up with the increment of man-made and industrial activities leading to environmental pollution which has a contrary impact on the ecosystem. Metals and metalloids are well-known contaminants found in the air, water, sediments, and soils. They are naturally occurring constituents of rocks and minerals, and break into surroundings in high concentrations, which is an outcome of exhaustive processing and human-induced, including manufacturing, mining, coloring agents, smelters, and refinery (Dixon and Weed 1977, Raskin and Ensley 2000). Chromium, lead, mercury, cadmium, and arsenic are some of the heavy metals that have been reported for posing significant environmental repercussions when present in high concentrations in surroundings. Amid significant metals, Cr, Pb, Hg, Cd, Cu, Zn, Co, Mn, and Mo are given special consideration in ecotoxicology; also, these metals are essential elements for natural life forms but can be harmful to health if intake exceeds the limit. Toxic metals are exposed to soil, water, and air by both natural-activity, i.e., soil and rock fragmentation, biogenic sources, forest fires, and volcanic eruption, as well as anthropogenic-activity, i.e., metal handling, mining, metallurgy, farming, concoction industry, and trash dumping (Grace Pavithra et al. 2019). Even at their lowest concentrations, these heavy metals (HMs) are reported to be toxic, mutagenic, and carcinogenic. When HMs, particularly non-essential elements, are introduced into soil/air/water, they have a devastating impact on ecosystem biodiversity (Ali et al. 2019). The high water solubility of HMs makes them easily accessible to diverse biological systems. HMs are persistent in the environment for a long time because of their physic-chemical properties of not further fragmentation into simpler forms, so they remain in the surroundings and gravitate towards accumulation and magnification in the trophic levels of the food chain (Appenroth 2009, Li et al. 2015). The waste load from industries and domestic carries a considerable number of HMs such as Cd, Cu, Pb, and Hg with them in excretory wastes, detergents, gadgets, water supply corrosion and distribution pipelines, etc.

Department of Environmental Science, School of Earth Sciences, Central University of Rajasthan, BandarSindri, Ajmer, 305817, Rajasthan, India.
* Corresponding author: garimakaushik@curaj.ac.in

(Sahni 2011). Effluent treatment does not aid the entire elimination of contaminants from waste load and eliminates around 50 percent of the metal content (Akpor et al. 2014). As a result, wastewater with substantial metal loadings discharged into the open even after the treatment process and by-product sludge is also abundant in heavy metal concentration (Arjoon et al. 2013, Sharma and Bhattacharya 2017). Detergents used for domestic purposes are also sources of many HMs like Ar, Fe, Cr, Co, Zn, Ni, and Mn (Jenkins 1998). Likewise, anthropogenic activities including platting, mining, vehicular pollution, fertilizers and pesticides, pharma and cosmetic sectors, paints, computers, communication-based technological sectors, and more add to an uncontrolled increase in HM concentrations into the ecosystem (Sahni 2011). Humans and other life forms are vulnerable to HMs for a long duration of time by epidermal contact, inhalation, or food containing HMs; they develop various illnesses (Alissa and Ferns 2011, Jaishankar et al. 2014, Sharma et al. 2014, Anyanwu et al. 2018).

Surface water bodies and soil get contaminated when wastewaters from the industrial, residential, and agricultural areas having HMs are released in ecosystems. According to data issued by public health organizations, heavy metal-related disorders affect millions of individuals worldwide. Continuing heavy metal susceptibility can cause degenerative physical/muscular and neurological conditions that imitate Parkinson's, muscular dystrophy, multiple sclerosis, and Alzheimer's like diseases and is also a reason for cancer (Anyanwu et al. 2018). If HMs exist in the environs exceeding the permitted limit, they are reported to cause chronic and acute toxicities, which as a result, affect the functioning of the mental and central nervous system by impairing or altering, and mutilate vital organs, i.e., kidney, lung, liver, etc., along with alteration in blood composition (Singh and Kalamdhad 2011). The symptoms of heavy metal poisoning are ambiguous and difficult to diagnose in the early stages (Harada 1995, Järup 2003, Huff et al. 2007, Sahni 2011, Tchounwou et al. 2012). Physicochemical, biological, or their combination technologies are employed to eliminate Cr(VI), although many of these technologies are not ecologically sound or cost feasible. There isn't a single approach that claims to completely degrade Cr(VI) (Emenike et al. 2018). However, coagulation-flocculation, sedimentation, electro flotation, electrochemical treatment, filtration, and ion exchange are cost-feasible but non-ecologically sound physicochemical treatment procedures (Song et al. 2003). To address these issues, biotic approaches like bioremediation procedures, which are both economically and environmentally sustainable, perform a crucial part in removing Cr(VI). The application of microorganisms (like bacteria, algae, and fungi) and other microbial-integrated processes and plants (phytoremediation) has no harmful impact on our surroundings. Different strategies are identified using microbes for Cr remediation, which comprise biotransformation (i.e., reduction and oxidation), biosorption, and/or bioaccumulation. They are appraised as alternatives to reduce Cr toxicity and become more common in Cr(VI) remediation in recent years (Jeyasingh and Philip 2005). The focus of this chapter is to provide an overview of different strategies for bioremediation of the chromium polluted environment.

12.1.1 Chromium

In 1797, chromium was discovered by Louis Vauquelin, the French chemist. The origin of Cr comes from the Greek word "chroma" because of the various colors detected in Cr comprising compounds (Barnhart 1997, Guertin et al. 2016). Cr element in the periodic table lies at 24th with 52 g/mol atomic weight and is placed between vanadium (V) and manganese (Mn). It is a transition metal with a high melting point, steely grey look, lustrous, and hard that has high gloss. Chromium is a versatile metal element that is very utilitarian in many sectors, including metallurgic, refractory, leather tanning, chrome plating, chemical, and pigment production. The metallurgical industry consumes 90 percent of total Cr production. With about 100 mg/L, Cr is reported as 21st among the most abundant element in the Earth's crust, having the highest reservoirs in countries South Africa and Kazakhstan (Barnhart 1997). Cr exists in different oxidation states, like –2, –1, 0, +1, +2, +3,

+4, +5, and +6, of which it predominantly occurs in +2, +3, and +6 states, but Cr^{2+} is an unstable form that further oxidizes to Cr^{3+} (von Burg and Liu 1993, Mohan and Pittman 2006, Malaviya and Singh 2016). The modification of Cr to acidic, antacid, or amphoteric oxides depends on oxidation expressed. Highly stable oxidation states are Cr(III) and Cr(VI), and in the presence of oxygen, it turns into green chromic oxide. When exposed to oxygen, Cr is found to be unstable since it forms a thin oxide layer that is impervious to oxygen and protects metal beneath. Cr has metallic radiance, steely-dark shading, hard, delicate, resists staining, and has melting and breaking limits of 1907°C and 2671°C (Haynes 2014). It exists in oxidation state zero in metallic form, as Cr(III)-chromite salt, and as Cr(VI)-dichromate salt. Cr(VI) rapidly dissolves in water and drifts, but Cr(III) is known for its stable form (Elangovan et al. 2006). Figure 12.1 shows that MnO_2, found in soil/sediments, is in partial equilibrium with ambient oxygen and spontaneously oxidizes Cr(III) to Cr(VI). Meantime, certain carbon compounds will reduce Cr(VI) to Cr(III) (Jobby et al. 2018).

Cr(III) exist in diverse forms such as $Cr(OH)_3$, $Cr(OH)\ ^{2+}$, $Cr(OH)\ ^{2-}$, and $Cr(OH)\ ^{4-}$ while Cr(VI) occurs in CrO_4^{2-}, Cr_2O^{2-}, and $HCrO^{4-}$ forms (Xia et al. 2019).

Figure 12.2 represents that Cr(VI) is influenced by the solution pH, total Cr concentration, presence of oxidizing and reducing agents, and kinetics of redox reactions (Zhang and Tian 2020). CrO_4^{2-} is the existing prime form of Cr(VI) above pH 7, while $HCrO^{4-}$ is the existing key form when the pH lies within 1 to 6 (Jobby et al. 2018). Cr(VI) is still considered to be a human cancer-causing agent, according to various certified and non-managerial corporations. The USEPA has proclaimed Cr(VI) as one of seventeen toxins associated with risk to human health. Trivalent chromium is non-hazardous, having stability, but Cr(VI) is lethal, unstable, and water dissolving. The dissolvability of Cr(III) in water mixtures is less distinct than that of Cr(VI). Cr(III) forms hydrated mixtures that are less dissolvable at pH levels below 5.

Generally, Cr is a toxic metal that enters the body via inhalation, ingestion, or dermal proximity (Ertani et al. 2017). As Cr(III), an essential component of human anatomy, it also plays a crucial part in maintaining normal physiological functions. Excessive consumption, on the other hand, is inimical to humans (Khattar et al. 2014). Growth delay, diabetes increases atherogenesis,

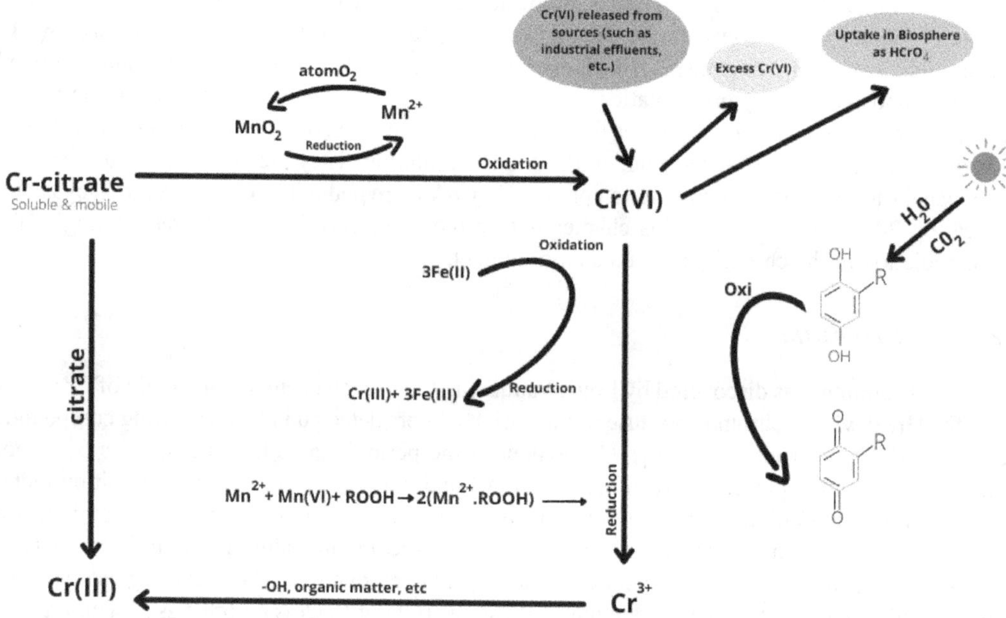

Fig. 12.1. Chromium Speciation in the environment (Adapted from Chen et al. 2021).

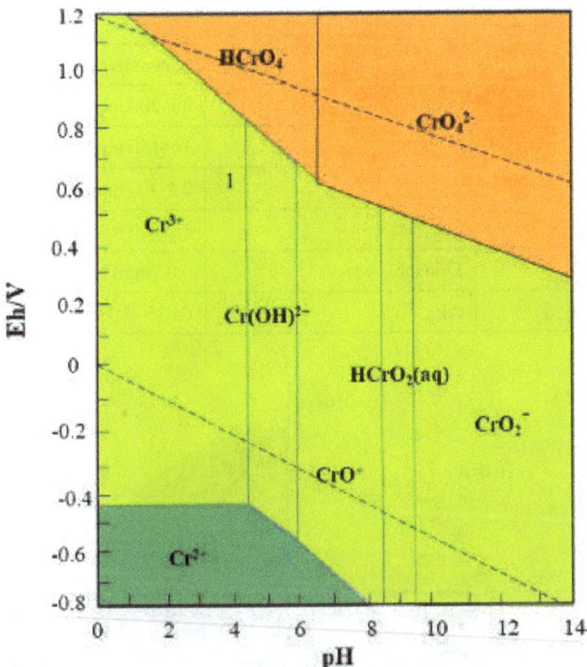

Fig. 12.2. In aqueous solution, a Pourbaix illustration for Cr species (Adapted from Pradhan et al. 2017).

hypercholesterolemia, and hyperinsulinemia are caused by a deficit of Cr(III) in the diet (Coetzee et al. 2018, Lapenna and Ciofani 2020). Cr(III) is found in various sections of human anatomy, with the highest concentration in hair follicles (Pradhan et al. 2017). Various studies show Cr(VI) is 100–1000 times extra toxic than element Cr(III) (Qian et al. 2017). Abscess on skin, hypersensitivity, and malignancy are a result of it (Elangovan et al. 2006). Shanker and Venkateswarlu's study shows that Cr(VI) can also attach to DNA, interfering with typical physiological functions like repair and replication (Shanker and Venkateswarlu 2011). Cr(VI)'s toxicity in cells is due to its simple migration through cell walls, where free radicals created by the process of reduction destroy cellular DNA, whereas Cr(III) is less toxic due to its lower potential to enter cells (Thatoi et al. 2014). Cr(VI) is transformed into Cr(III) after being absorbed by the human body and is eliminated in the urine; nevertheless, produced complexes may persist in the body for a long duration (Coetzee et al. 2018). Chromium compounds are absorbed and stored by plant roots. The build-up of Cr in various plant parts is highest in roots, followed by leaves and fruits. Cr in high concentration adversely affects plants by slowing their growth, causing chlorosis and necrosis (Singh et al. 2013). Table 12.1 shows different concentrations of chromium in various components of the environment.

Cr(VI) is broadly employed in a variety of industrial processes, including leather processing, electroplating, metallurgy, pigment synthesis, and wood preservation, among others. Among industries, leather/tanning industry is one of the most valuable industrial sectors in India. The tanning industry's wastewater comprises a large quantity of organic and inorganic waste, having high quantities of chromate and sulfate (Durai and Rajasimman 2011). Over and above 17,000 tonnes of Cr wastes are released into the environment by tannery effluent from these productions (Kamaludeen et al. 2003). Because of insoluble hydroxides production in water bodies, it is reported that Cr(VI) is highly toxic than Cr(III), which gets precipitated when the pH is over 5.5. WHO prescribed levels for total Cr in drinkable and surface inland water bodies are 0.050 and 0.10 mg L^{-1}, respectively, but in the case of industrial wastewater, the dispense ranges from 0.1–400 mg L^{-1} (WHO 2004). The necessity for the ferrochrome to be engaged in a wide variety of applications in making various materials drives the worldwide requirements for chromite (Al, Fe, Mg) Cr_2O_4 (Cramer et al.

Table 12.1. Concentrations of chromium in various environmental components (Kamaludeen et al. 2003).

Sr. No.	Environmental Component	Concentrations
1.	Crust of continent	80–200 mg kg^{-1}
2.	Soil	10–150 mg kg^{-1}
3.	Freshwater	0.1–6.0 mg L^{-1}
4.	Seawater	0.2–5.0 mg L^{-1}
5.	Drinkable water	0.05 mg L^{-1}
6.	Air	0.015–0.03 mg m^{-3}

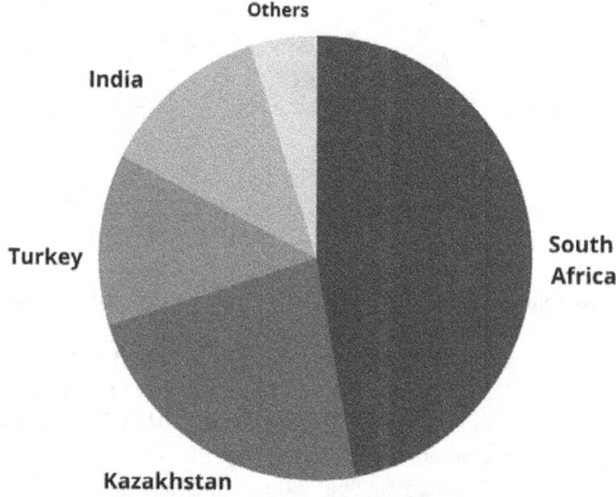

Fig. 12.3. Production of chromite across the world in 2016 (Adapted from Coetzee et al. 2020).

2004). South Africa has around 70–80 percent of the world's chromite mineral deposits and is also the single largest manufacturer. According to the U.S. Geological Survey, South Africa is known to be one of the largest chromite mine manufacturers in 2016 as the average production was around 14,000 MT of chromite Fig. 12.3 (U.S. Department of the Interior and U.S. Geological Survey, January 2017). Hence, mining for chromite and ferrochrome pollutes not just ground and water, but they also contribute considerably to polluting air.

12.1.2 Chromium Toxicity

Cr(III) is a dietary supplement that aids in the digestion of glucose, lipids, and amino acids. According to studies, chromium was found in the tissues of newborn babies of humans and hatchlings. The greatest concentration of Cr is found in hair, which in lower focus causes negative effect on cell structure. Extreme Cr(VI) exposure causes mental as well as immunological alterations. Because of inward inhalation of materials comprising Cr(VI) causes irritations to the larynx and liver, crevices in the nasal septum, bronchitis, and asthma. Contact with Cr(VI) mixtures causes skin hypersensitivity, dermal putrefaction, dermatitis, and dermal ingestion. Cr(VI) can bind to two-stranded DNA, causing gene duplication, replication, and repair to be altered. Workers in the chromium industry were not the only ones affected (Shankar and Venkateswarlu 2011). A substantial amount of Cr is released into the air, water bodies, and soil by the production industries. The form of Cr determines its adaptability and bioavailability. In contrast to Cr(III), which is very meagrely soluble in water, Cr(VI) is immensely movable and bio-available. In contrast, Cr(III) complexes are impervious to cell layers that are 100 times less harmful than Cr(VI). Cr with +3 oxidation state is intracellular, which leads to a reduction in Cr(VI), builds amino acids nucleotides structures and is a possible mutagen (Swaroop et al. 2019, Costello et al. 2019).

Cr(VI) is derived via both human-induced and natural outsources and has a great resilience for changing environmental conditions (Zhitkovich 2011). Cr(VI) is known to be a malignant growth-promoting agent that is let out from industries or drained into soil and then into groundwater. It is stated to be dissoluble and toxic at low concentrations. Cr(III), a key component for fat, carbohydrate, and protein digestion support, is mildly cytotoxic in high doses. Because of its toxic qualities and mutagenic nature, Cr(VI) is well recognized for its natural causing impact and adverse wellbeing (Pavithra et al. 2019). The human body holds roughly 0.03 mg/L Cr; thus, administrating 20–200 g/day is safe and adequate. The fatal portion of Cr(VI) is about 1–3 g, and in several states, LD_{50} noxiousness is in the range of 40–100 mg/kg for Cr(VI) and 1900–3100 mg/kg for Cr(III). Because Cr(VI) is more malignant than the other oxidation forms, it is classified as a distinct species by WHO, which means it has no explicit limit and is difficult to assess.

Cr(III)'s beneficial effects include lowering triglyceride levels, cholesterol, muscle-to-fat ratio, and bulk gain by restricting insulin to cell surface receptors (Pechova and Pavlata 2007). Browning found out that Cr(VI) particulates can result in lung carcinoma through the system of carcinogenesis (Browning and Wise Sr 2017). Cr^{+6} interferes with the route of homologous recombination fixation of dsDNA fragmented with alterations in RAD51 subcellular restriction, indicating that Cr(VI) and Cr(III) have the potential to harm lymphocyte DNA. A persistent, supplementary chromosomal alteration that causes lung cancer is observed (Medeiros et al. 2003). Cr(VI) is linked to a decrement in take-up affecting photosynthesis in plants, resulting in a poor growth rate. It has a variety of effects on the responsive oxygen species in plant cells, including physiological, morphological, and biochemical effects. In plants, the noxiousness of Cr is cited for the reason of chlorosis and deception (Shahid et al. 2017). Cr is attracted to water and is retained in the form of residue or aggregate by oceanic plants. Metal uptake in plants is dependent on the synthetic metal sort and life type of macrophytes. Free-floating plants like water hyacinth and water lettuce contribute to high levels of chromium build-up (Tabinda et al. 2018). Figure 12.4 depicts Cr pollution and its effects on soil and water bodies.

12.1.3 Chromium Hazard

Human life is at risk due to chromium hazard - both anthropogenic and natural sources are contributors to it; however, the contribution of anthropogenic activities is around 50–70 percent of chromium in total air emission (ATSDR 2015). Key industries where laborers are exposed to chromium are nearly doubled; the general public includes ferrochrome alloy producers, stainless steel manufacturers, tannery industries, pigment production, and chrome plating (Pellerin and Booker 2000, ATSDR 2015). The passages commonly followed for chromium hazards is through ingestion, inhalation, and skin contact.

- Chromium toxicity by ingestion

The primary sources of Cr exposure in the human body are through intake of food and drinking water containing Cr, while it has been observed that exposure in children is through ingestion of soil. One of the studies reported a higher level of Cr (0.05–155 g/L) via breastfeeding, implying that exposure through breastfeeding is also possible (Casey and Hambidge 1984, ATSDR 2015). Also, the plant products and animal products contain high concentrations (typically < 10–1300 μg/kg) of Cr. It is also found that there is also the addition of Cr through the utensils (ATSDR 2015). The chromium concentration in drinking water is > 25 μg/L which is also a significant contributor to

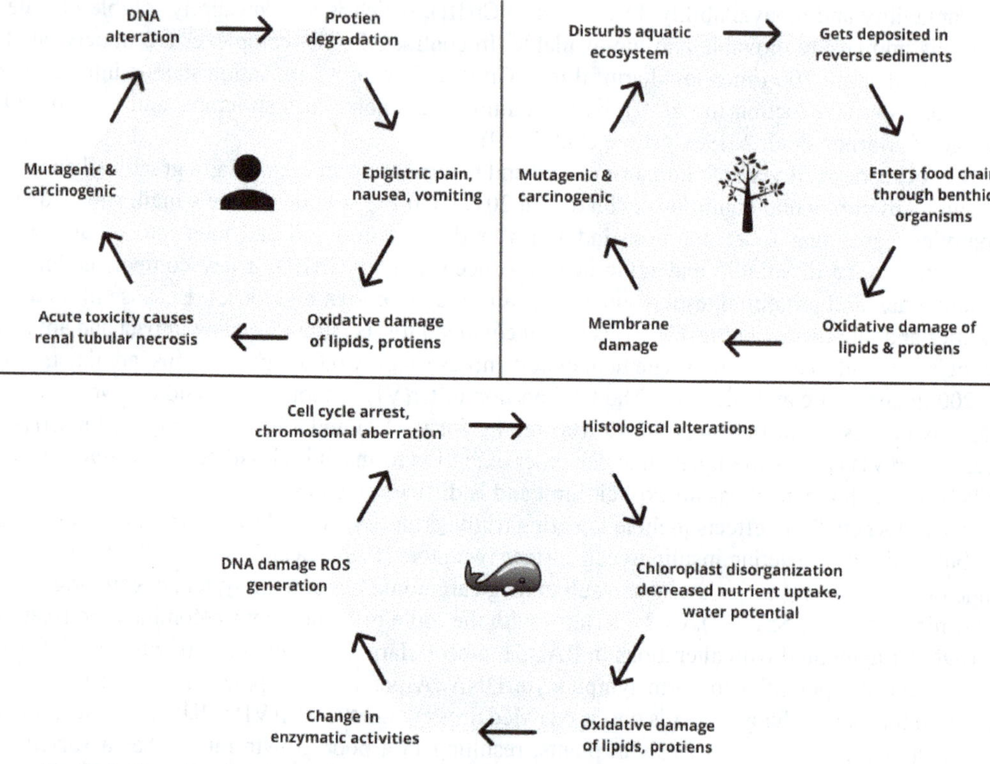

Fig. 12.4. Chromium impact on soil and water bodies (Adapted and modified from Grace Pavithra et al. 2019).

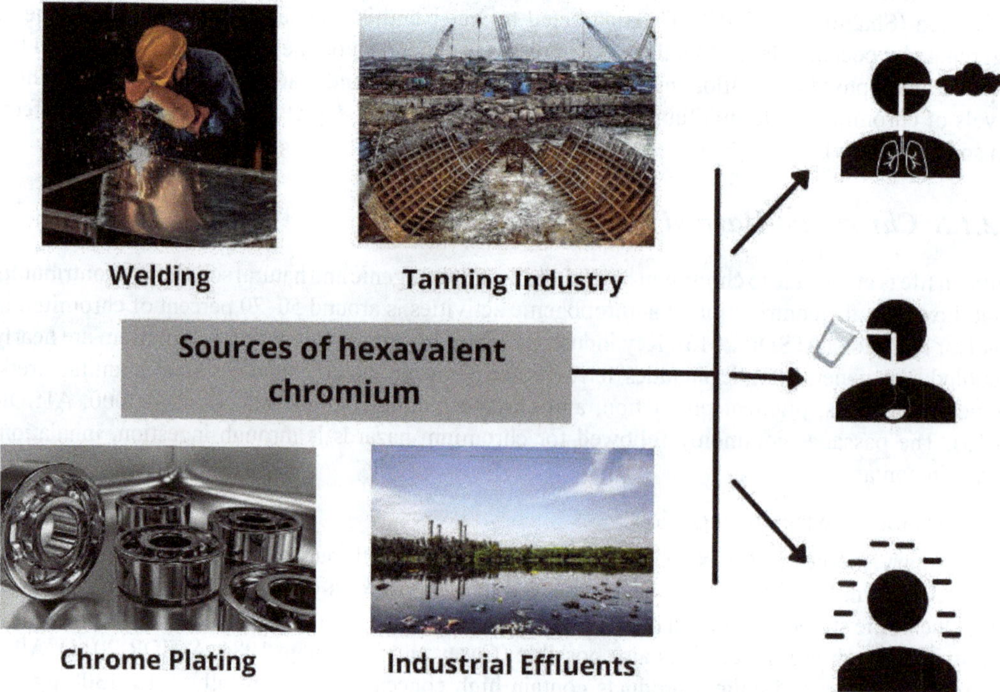

Fig. 12.5. Routes of exposure of chromium in humans.

chromium toxicity. Gastric fluids convert Cr(VI) to Cr(III) once it enters the body by ingestion, and only 2–3 percent of Cr penetrates the gastrointestinal tract (Guertin 2004). There is not sufficient evidence that Cr is carcinogenic when consumed; nonetheless, some studies demonstrate an increase in gastrointestinal problems owing to Cr(VI) (about 20 mg/L)—contaminated areas (Sharma et al. 2012). However, Cr(VI) is reported to cause ulcers in the stomach and small intestine (ATSDR 2015).

- Chromium toxicity by inhalation

Chromium's equilibrium vapor pressure is around $16 \times 10^{-}$ mmHg, making it extremely non-volatile. In the air, Cr can be found in the forms of fumes, dust, and aerosols. Through inhalation, daily intake of chromium is approximately < 0.2–0.6 μg, as reported by (WHO 2003). Nasal damage, breathing issues, runny nose, asthma, and allergies are all caused by inhaling Cr(VI). About 53–85 percent of the Cr(VI) that enters the lungs is absorbed into blood or mucus, leaving the rest of the Cr in the lungs to induce lung cancer over the passage of time (Guertin 2004, ATSDR 2015). Workers in the chrome industry and chromite plating have been demonstrated to be more susceptible to cancer by inhaling Cr(VI).

- Chromium toxicity by skin contact

Although dermal Cr exposure is not ought to be carcinogenic, certain levels of Cr do enter the skin when it comes into contact with Cr-containing products or soil. As Cr(VI) is more highly soluble in water than Cr(III), the former is readily absorbed through the skin (Guertin 2004).

- Other effects of chromium on humans

Essential micronutrients are important for metabolism in plants and animals. Chromium is also one of the essential micronutrients but when it accumulates in higher concentrations, it may cause serious health hazards. The U.S. EPA listed it as Class A carcinogenic as chromium is very toxic in nature (Richard and Bourg 1991, Gibb et al. 2000). Cr(VI) enters in sulfate-utilizing organisms via cell walls. Under normal circumstances, intercellular reductants spontaneously interact and produce temporary intermediates Cr(V) and Cr(VI). The oxidation result releases radicals that can easily bind with DNA-protein. Cr(VI) also combines with cellular components and prevents them from performing their anatomical roles (Cheung and Gu 2007, Dhal et al. 2013).

In humans, Cr(VI) causes eczema, indigestion, diarrhea, internal bleeding, liver and kidney damage, and respiratory issues after acute exposure. Nasal septum irritation and ulceration, respiratory hypersensitivity, and acute toxicity result from inhalation (Mohan and Pittman 2006). Cr(VI) builds up in the placenta in mammals, limiting fetus growth (Cheung and Gu 2007). Erstwhile Cr(VI) enters into the body, gets transformed into Cr(III) and passes through the urinary system within a week, but a slight residue remains in the body for years. Exposures of more than 0.11 mg g^{-1} body weight deemed fatal (Richard and Bourg 1991). In mammalian metabolism, trivalent chromium is an essential microelement. It is liable for lowering levels of glucose in the blood and is used to control some types of diabetes, in addition to insulin. By lowering the concentration of low-density lipids in the blood, it reduces cholesterol levels also (Mohan and Pittman 2006). Low Cr in the diet has been linked to sluggish growth and fertility, diabetic hyperinsulinemia, intense atherogenesis, hypercholesterolemia, etc. (Katz and Salem 1993). Table 12.2 summarizes concentration of chromium in human tissues and fluids.

12.2 Microbial Mechanism of Cr(VI)

Several indigenous microbes have the ability to colonize and adapt to the contaminated environmental conditions, which are unlivable for both plants and animals. As we know, bioremediation is an effective method for cleaning the resistant compounds; so for conducting future bioremediation

Table 12.2. Chromium concentration in different tissues and fluids in human body (Pradhan et al. 2017).

Organ/tissue/fluid total	Chromium concentration
Serum	0.01–0.38 mg L^{-1}
Blood	0.12–0.67 mg L^{-1}
Urine	0.05–1.80 mg L^{-1}
Breast milk	0.06–1.56 mg L^{-1}
Saliva	0.55–0.70 mg L^{-1}
Skeleton	5–15 mg kg^{-1}
Hair	0.234–3.80 mg kg^{-1}
Teeth	7.20–35.00 mg kg^{-1}
Nail	0.52–172.92 mg kg^{-1}
Liver	5–15 mg kg^{-1}
Spleen	7–29 mg kg^{-1}
Lung	130–1375 mg kg^{-1}
Skin	50–200 mg kg^{-1}
Muscle	5–10 mg kg^{-1}
Average amount per human body	0.4–6 mg

processes, it is important to undergo purification of isolated Cr-tolerant bacteria that is naturally sustained in an environment, whether contaminated or not, for further steps.

The study of the interaction between microbes and HMs is ecologically and economically feasible and gaining popularity. Bioremediation is a process that uses mainly microorganisms to degrade and reduce or detoxify harmful chemical compounds and pollutants to a safe level in the environment (Asha and Sandeep 2013). Bacteria/fungi/yeast/algae and plants have proven their remediation ability, with bacteria and fungi shown to be the most effective. Metal uptake biotransformation (i.e., reduction) and bioaccumulation/biosorption, which changes the oxidative state of the metal, are two detoxifying methods evolved by these microbes (Konczyk et al. 2010). Bioreduction of water-soluble, mobile, and more toxic Cr (VI) to less toxic, insoluble, and immobile Cr(III) is an intriguing possibility among them (Guillen-Jiménez et al. 2009). Due to low conducting cost, reusability, and lack of related pollution, biosorption is an optimistic and efficient approach for removing HMs (Kumar et al. 2008, Uluozlu et al. 2008, Wu et al. 2008, Sepeher et al. 2012). Cr^{+6} microbial detoxification processes are next discussed in further depth.

12.2.1 Biosorption of Cr(VI)

Biosorption is the ability of biological materials to interact with the materials of biological cell surfaces when soluble substances are present (Kaduková and Virčíková 2005). It is a passive, fast and remedial physical and chemical interaction among metal parts (sorbate) and the biological substance (biosorbent) (Ahluwalia and Goyal 2007). Hence, the process is passively carried out in the absence of active or inactive microorganisms. Microorganisms are important biosorbents as they have a good ratio of high surface-to-volume and also the inexpensive cost of production. Algae, microalgae, and as well as cyanobacteria (Khoubestani et al. 2015, Kwak et al. 2015), yeasts (Martorell et al. 2012, Mahmoud et al. 2015), fungi (Khani et al. 2012), and bacteria (Wu et al.

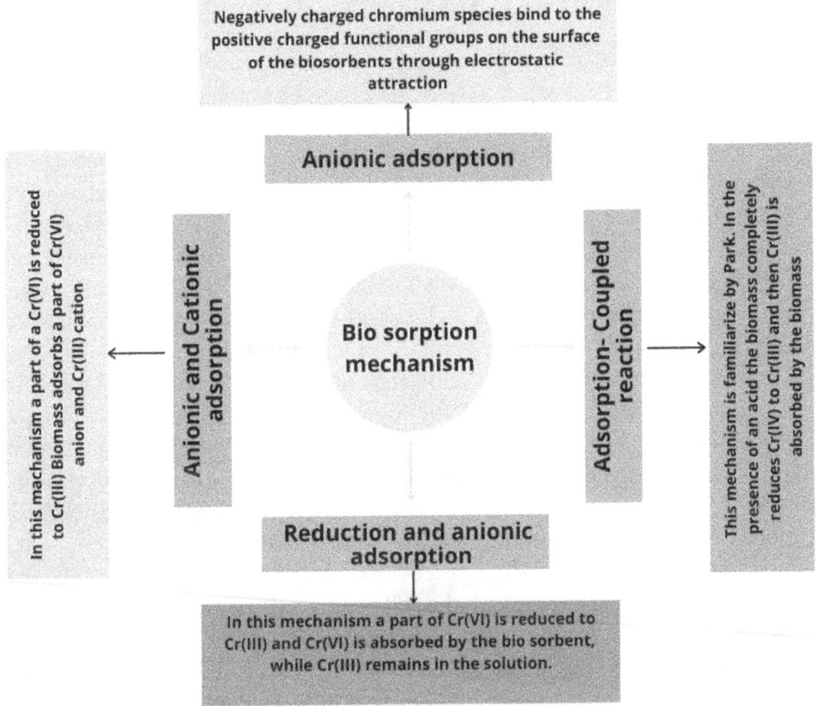

Fig. 12.6. Biosorption mechanisms of Cr(VI) (Adapted and modified from Deepa and Mishra 2020).

2015, Huang et al. 2016) are some microorganisms used to remove HMs and showed the capability to remove Cr(VI) in biosorption. Sludge, aquatic macrophytes, plant, fruit, and vegetable wastes, as well as inorganic chemicals, should all be considered (Lima et al. 2011, Chen et al. 2015, Khelaifia et al. 2016, Wang et al. 2016). The cell wall of microorganisms maintains its firmness and has several functional groups that can attach HMs ions (such as carboxylate, hydroxyl, phosphate, and amino acid) (Scott and Karanjkar 1992, Fernandez et al. 2013), and the energy will be demanded by the cell after active sorption. The second step is linked to the existence of citoplasmatica metal-binding proteins (Doble and Kumar 2005, Nemr et al. 2015). Through an oxidation-reduction mechanism, which is carried out by microbes as a biosorbent, it first absorbs metal from tannery wastewater and changes it from high toxic form to low toxic (Gupta et al. 2019). Several studies claimed that with the aid of Cr resistant bacterium isolates like *Pseudomonas* sp., *Desulfovibrio* sp., *Microbacterium* sp., and barn bagasse by *Enterobacter* sp., *Shewanella algae*, *Escherichiacoli*, *Aspergillus terreus* or *Rhizopus sexualis*, etc., is capable of removing Cr from tannery loads (Saranraj and Sujitha 2013). Cr binding in fungi is mediated via amine, amide, carboxyl, phosphate, and alkane groups (Ahluwalia and Goyal 2010). *Rhizopus arrhizus* cell walls include chitin and chitosan, which are involved in chromium absorption (Ismael et al. 2004). A study conducted by Iram et al. (2013) recorded that *Aspergillus fumigatus*, isolated from a polluted source, possesses high biosorption capacity for certain HMs. Srinath et al. (2002) studied that the biosorption of Cr by *Bacillus coagulans* (live cells), *Bacillus coagulans* (dead cells), *Bacillus megaterium*, and *Bacillus circulans* can remove Cr (VI) efficiently by 47.6 percent, 79.8 percent, 69 percent, and 64 percent, respectively, withholding of 24 h, while *Brevibacterium* sp. CrT-12 exhibited 100 percent removal efficiency in a time period of 72 hours (Faisal and Hasnain 2004). Other studies also outlined that *Cellulosimicrobium funkei* AR6 strain and *Acinetobacter* sp. are efficient in removing Cr(VI) to 80.43 and 100 percent with a duration of 120 h and 72 h, respectively (Srivastava et al. 2007, Karthik et al. 2017).

12.2.2 Bioaccumulation of Cr(VI)

CrO_4^{2-} is considered to be toxic as well as carcinogenic that enters cells through the same transport routes as the structurally identical anion SO_4^{2-}. Cr(VI) gets absorbed by the cell via nonselective anion channels and sulfate transporters that lead passage (Joutey et al. 2015). In the inner part of the cell, oxidized CrO_4^{2-} leads to the destruction of DNA by one-electron reduction, resulting in transitory Cr(V) and Cr(VI) species and the ongoing production of free radicals such as RS^- and OH^-, which cause changes in DNA by alternating the bond cleavage as gene expression changes as shown in Fig. 12.7. This is how Cr(VI) is swiftly reduced to Cr(III). Thus, on both side of the plasma membrane, there will be no equilibrium of Cr(VI) concentration; Cr(VI) is bioaccumulated by the cell's reduction capacity, which will be the major source of energy (Pattanapipitpaisal et al. 2002, Joutey et al. 2015).

Cr(III) is known to be a stable, non-carcinogenic compound that is not permeable to biological tissues; Cr(III) quickly forms complexes with biologically relevant ligand molecules and is picked up by cells (Ksheminska et al. 2005). Many studies revealed that the results obtained by utilizing bacteria, fungus, algae, and plant-derived biomasses have the capacity to actively incorporate metals from the aqueous solutions into their cells (Gavrilescu 2004, Chojnacka 2010). Microorganisms' metal absorption is influenced by the initial concentration of metal and contact duration (Cheung and Gu 2007). So, it's a metabolism-dependent process that only happens in viable cells and demands the use of energy to transport Cr(VI) through the cell membranes. However, when the concentration of metal is high, the development of the cell is inhibited, which is a severe restriction (Dönmez and Aksu 1999). The microbiological accumulation of Cr(VI) is critical for developing techniques for concentrating, removing, and recovering Cr(VI) from polluted solutions. Ksheminska et al. (2008) proposed in their study that Cr(VI) molecules (anions SO_4^{2-} and PO_4^{3-}) enter cells via enhanced diffusion via non-selective and oxidative state-sensitive anion channels. Cr(III) is rarely permeable to biological tissues, perhaps because they create complexes with limited solubility (Wang and Chen 2009).

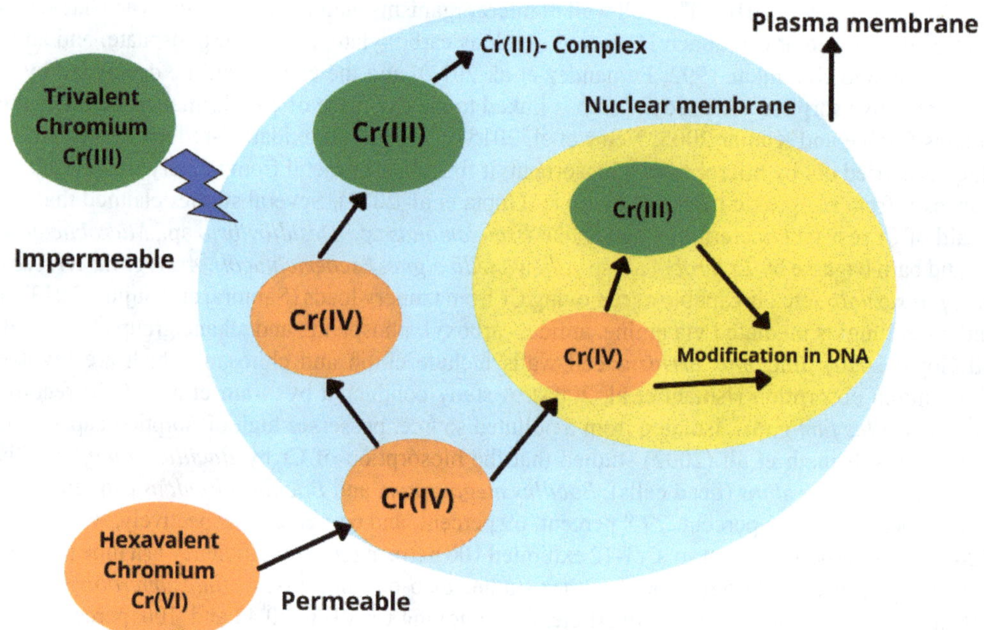

Fig. 12.7. Mechanism of Hexavalent Cr toxicity and mutagenicity of a microbial cell. (Adapted from Narayani and Shetty 2013).

12.2.3 Biotransformation of Cr(VI)

Cr (III) is stable and less toxic than Cr(VI); biotransformation of Cr (VI) is identified as a detoxifying process. In recent exploration, the extracellular reductase activity of chromate has been found in supernatants of microbial cultures (Rath et al. 2014, Dey and Paul 2016). Enzymatic reduction of Cr(VI) in bacteria (including *Arthrobacter*, *Pseudomonas*, and *Bacillus*) might be linked to dissoluble cytosolic proteins and/or impenetrable plasma membrane enzymes (Viti et al. 2014). Indeed, under aerobic or anaerobic (or occasionally both) circumstances, various chromate reductase (ChrR, LpDH, NemA, and YieF) have been identified in the cytoplasmic fraction (which is considered to be soluble) or attached to the membrane catalyzing the process (Baldiris et al. 2018, Thatoi et al. 2014).

The study conducted by Ksheminska et al. (2008) reported that for the bioremediation process, several species of yeast could be applied. Some species like *Pichia guilliermondii*, *Rhodotorula mucilaginosa*, and *Cyberlindnera jadinii* and *Wickerhamomyces anomalus* (Fernández et al. 2013, 2016, respectively) ought to have biologically transforming Cr(VI) to Cr(III) capabilities. *Candida maltosa* shows chromate-reducing functions which are dependent on NAD in its soluble protein potion in the study of (Ramírez-Ramírez et al. 2004, Chatterjee et al. 2012). Cr(VI) decontamination can take place through indirect releasing of chemicals (i.e., sulfate and riboflavin) by yeast cells into extracellular media (Ksheminska et al. 2006, Fedorovych et al. 2009). Although the biotransformation of Cr(VI) to Cr(III) is the prime reason for the tolerance to Cr in yeasts is due to their restricted

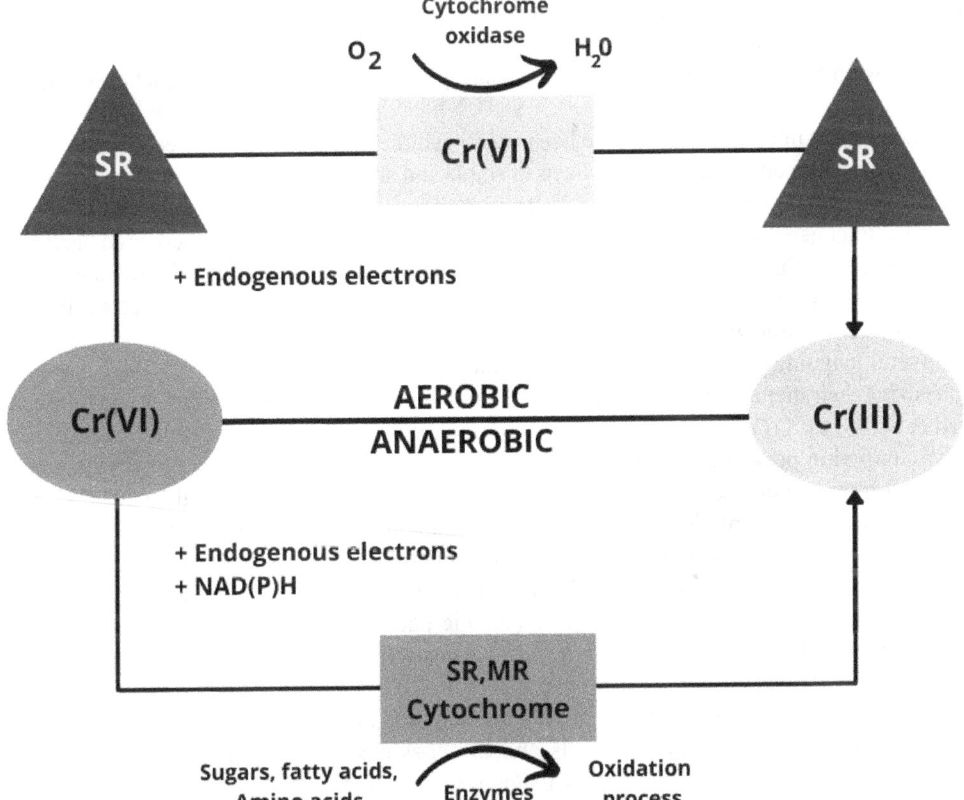

Fig. 12.8. Reduction of Cr(VI) by using microbiologically induced aerobic and anaerobic processes. (Adapted and modified from Joutey et al. 2015).

ion absorption (Ksheminska et al. 2005). Gu et al.'s (2015) study found a chromate reductase in a soluble portion of *Aspergillus niger*, while Singh and Bishnoi (2015) discovered a reduction of chromium through enzymes of *Aspergillus flavus* through the process of biosorption. Metabolic processes like the anaerobic processes and their end products Fe and SO_4^{2-} reducing bacteria (Fe(II) and H_2S) amino acids, organic acids, sugars, vitamins, nucleotides, and glutathione are the results of non-enzymatic reduction (Somasundaram et al. 2009, Dhal et al. 2013). The fungal-originated citrate acid and oxalates also reduce Cr(VI) via Fe^{+3} photocatalysis or Mn in the absence of sunlight (Barrera-Diaz et al. 2012).

Table 12.3 summarises various biological processes of chromate resistance in diverse bacteria, fungus, and algal species, as reported in previous studies.

12.3 Future thoughts on Chromium Removal with Possible uses

Heavy metal exposure is a stressful state for diverse organisms, including microorganisms. The capability of specific organisms for metal resistance may be related to their capability to live in harsh environmental conditions as the expression of particular genes may be engaged in processes to combat stress (Fernández et al. 2017). In recent transcriptomic and proteomic research in response to Cr(IV), diverse bacterial and fungal species have been utilizing it. Remediation of heavy metal pollution in the environment through the development of effective biological processes (together with a worldwide macromolecule analysis) reveals diverse possibilities. Exposure to chromium altered around 22 percent of the genome in yeasts. After exposure, double arrays of the metal-response genome profiles were created, where one array is linked with metal exposure (which gives information on the basis of transcriptional alterations), and the other one provides information on the relationship between expression of unnecessary gene and susceptibility to metal exposing (Jin et al. 2008).

Recombinant microorganisms have been used in studies involving genetic engineering. Robins et al. (2013) developed a method on the basis of stable and active reductase of chromate immobilized on granules of polyhydroxyalkanoate by fusing *E. coli* (nemA gene) with the polyhydroxyalkanoate synthase extract gene phaC from *Ralstonia eutropha*. Similarly, researchers discovered bacteria that overexpresses a mutant enzyme called ChrR6, which has reduced the activity of chromate 200 times in wild-type strains. It is essential to continue improving this technology since it represents a viable and cost-effective strategy for *ex situ* Cr (VI) remediation.

Several ongoing researches are concentrating on the utilization of microbial association, emphasizing their metabolic ascendance for eliminating metals and adaptability for field application (Qian et al. 2016). Cr(VI) at 100 milligrams concentration, consortium of *B. subtilis*, *A. junii*, and *E. coli* cultured in packed bed reactor (continuous) enhanced the elimination of Cr^{+6} via a process called absorption, demonstrating the efficiency of 50–70 percent (Samuel et al. 2013). Taking into consideration, bioaugmentation is ought to be one of the most beneficial bioremediation strategies as it refers to the act of introducing suitably adapted microorganisms strain or microbial consortium to the target site having contamination (Herrero and Stuckey 2015). The bioaugmentation process via microbial consortium furnishes varied metabolic pathways and resilience necessary for wide applications (He et al. 2014); furthermore, it is a cost-effective and environmentally acceptable bioremediation option.

The Cr (VI) level can be reduced, and the stability of enzymes may be improved by the application of immobilized microbe cells and the enzymes in combination with the technique named nanotechnology (i.e., like carbon nanotubes infused into Ca^{+2} alginate beads). Applying metal-reducing bacteria in combination with nano-materials (siderite) that act as electron donors in the enzymatic Cr(VI) reduction and immobilisation (Cr(III)-containing precipitates) is another approach connected to the remediation of places contaminated with Cr at the nanoscale (Seo and Roh 2015, Gutiérrez-Corona et al. 2016).

Table 12.3. Microorganisms reported being applied for remediation of Cr(VI) (Adopted from Fernández et al. 2018).

BIOSORPTION	
Type of Microorganisms	**References**
Algae	
Spirulina platensis, Spirulina sp.	Magro et al. 2013, Rezaei 2013, Kwak et al. 2015
Sargassum filipendula	Bertagnolli and Silva 2013
Pelvetia canaliculata	Hackbarth et al. 2016
Bacteria	
E. coli	Liu et al. 2015
Acinetobacter haemolyticus, A. junii	Yahya et al. 2012, Mrudula et al. 2012
Pseudomonas putida V1	Cabral et al. 2014
Corynebacterium paurometabolum	Prabhakaran et al. 2016
Pantoea sp.	Ontañon et al. 2014
Bacillus amyloliquefaciens	Rath et al. 2014
Bacillus subtilis SS-1	Sukumar et al. 2014
Bacillus coagulans (live cells), Bacillus coagulans (dead cells)	Srinath et al. 2002
Enterococcus casseliflavus	Saranraj et al. 2010
Arthrobacter ps-5	Shuhong et al. 2014
Yeast	
Pichia anomala, Wickerhamomyces anomalus, and *Cyberlindnera fabianii*	Joutey et al. 2015
Candida tropicalis	Bahafid et al. 2013
Filamentous fungi	
Aspergillus spp., *A. niger*	Samuel et al. 2015, Sivakumar 2016
A. flavus	Singh et al. 2016
Fusarium oxysporum NCBT-156	Amatussalam et al. 2011
A. fumigatus	Balaji and David 2016
*P. griseofulvum*MSR1	Abigail et al. 2015
Saccharomyces cerevisiae (immobilized beads)	Mahmoud and Mohamed 2017
Phaneroch-aete chrysosporium (viable cells)	Pal and Vimala 2011
Hypocrea tawa, Trichoderma inhamatum	Morales-Barrera et al. 2008
BIOTRANSFORMATION	
Type of Microorganisms	**References**
Bacteria	
E. coli	Robins et al. 2013
Acinetobacter haemolyticus	Ahmad et al. 2013
Leucobacter sp. G161	Ge et al. 2014

Table 12.3 contd. ...

...*Table 12.3 contd.*

BIOTRANSFORMATION	
Type of Microorganisms	**References**
Halomonas spp.	Lara et al. 2016
Pannonibacter phragmitetus	Xu et al. 2011
Stenorophomonas maltophilia	Bhange et al. 2016
Ochrobactrum intermedium	Rida et al. 2012
Bacillus subtilis	Zheng et al. 2013
Bacillus cereus	Zhao et al. 2012
Nesterenkonia sp.	Amoozegar et al. 2007
Pediococcus acidilactici	Lytras et al. 2017
Yeast	
Rhodotorula mucilaginosa	Chatterjee et al. 2012
Cyberlindnera jadinii, Wickerhamomyces anomalus	Fernández et al. 2013
Pichia jadinii, Pichia anomala	Martorell et al. 2012

12.4 Conclusion

Our environment is getting contaminated daily due to various industrial and human activities. Cr(VI) and Cr(III) exhibit different characteristics; Cr(VI) having high mobility and solubility enhances the possibilities of diffusing past through cellular linings, which is the reason for its carcinogenic, mutagenic, and teratogenic nature, whereas the Cr(III) is considered negligibly toxic and relatively not soluble in aqueous system. To extract this heavy metal from water, soil, and sediment, an environmentally safe, flexible, and low-cost technique is needed. Bioremediation is a technique based on microorganisms' heavy metal removal processes such as biotransformation, bioaccumulation, and biosorption. Among these, the biosorption method has proved to be successful in terms of technological advancement. The biological detoxification of Cr(VI) has been extensively studied. Enough that, by expanding genetics knowledge and dedicating experimental sites for transfer bioremediation technology, these prospects offer the potential for significant advancements; however, extensive research in this field is still required to solve the problems associated with the large-scale remediation process.

References

Abigail, E. A. M., M. S. Samuel and R. Chidambaram. 2015. Hexavalent chromium biosorption studies using *Penicillium griseofulvum* MSR1 a novel isolate from tannery effluent site: Box Behnken optimization, equilibrium, kinetics and thermodynamic studies. J. Taiwan Inst. Chem. Eng. 49: 156–164. https://doi.org/10.1016/j.jtice.2014.11.026.

ATSDR. 2015. Toxicological profile for chromium. Agency for Toxic Substance and Disease Registry U.S Department of Health and Human Services, Public Health Services, ATSDR, Atlanta.

Ahluwalia, S. S. and D. Goyal. 2007. Microbial and plant derived biomass for removal of heavy metals from wastewater. Bioresour. Technol. 98: 2243–2257.

Ahluwalia, S. S. and D. Goyal. 2010. Removal of Cr(VI) from aqueous solution by fungal biomass. Eng. Life Sci. 10: 480–485.

Ahmad, W. A., W. H. W. Ahmad, N. A. Karim, A. S. S. Raj and Z. A. Zakaria. 2013. Cr (VI) reduction in naturally rich growth medium and sugarcane bagasse by *Acinetobacter haemolyticus*. Int. Biodeterior. Biodegrad. 85: 571–576. https://doi.org/10.1016/j.ibiod.2013.01.008.

Akpor, O. B., G. O. Ohiobor and T. D. Olaolu. 2014. Heavy metal pollutants in wastewater effluents: sources, effects and remediation. Adv. Biosci. Bioeng. 2(4): 37.

Ali, H., E. Khan and I. Ilahi. 2019. Environmental chemistry and ecotoxicology of hazardous heavy metals: environmental persistence, toxicity, and bioaccumulation. J. Chem. 2019: 1–14.

Alissa, E. M. and G. A. Fern. 2011. Heavy metal poisoning and cardiovascular disease. J. Toxicol. 2011: 870125.

Amatussalam, A., M. N. Abubacker and R. B. Rajendran. 2011. *In situ* Carica papaya stem matrix and *Fusarium oxysporum* (NCBT-156) mediated bioremediation of chromium. Indian J. Exp. Biol. 49: 925–931.

Amoozegar, M. A., A. Ghasemi, M. R. Razavi and S. Naddaf. 2007. Evaluation of hexavalent chromium reduction by chromate-resistant moderately halophile, *Nesterenkonia* sp. strain MF2. Process Biochem. 42(10): 1475–1479.

Anyanwu, B., A. Ezejiofor, Z. Igweze, O. Orisakwe, B. O. Anyanwu, A. N. Ezejiofor and Z. N. Igweze. 2018. Heavy metal mixture exposure and effects in developing nations: an update. Toxics 6(4): 65.

Appenroth, K. J. 2009. Definition of "heavy metals" and their role in biological systems. pp. 19–29. *In*: Soil heavy metals. Soil biology, vol 19. Springer, Berlin, Heidelberg.

Arjoon, A., A. O. Olaniran and B. Pillay. 2013. Co-contamination of water with chlorinated hydrocarbons and heavy metals: challenges and current bioremediation strategies. Int. J. Environ. Sci. Technol. 10(2): 395–412.

Asha, L. P. and R. S. Sandeep. 2013. Review on bioremediation-potential tool for removing environmental pollution. Int. J. Basic Appl. Chem. Sci. 3(3): 21–33.

Bahafid, W., N. T. Joute, H. Sayel, I. Boulara and N. Ghachtouli. 2013. Bioaugmentation of chromium-polluted soil microcosms with *Candida tropicalis* diminishes phytoavailable chromium. J. Appl. Microbiol. 115: 727–734. http://dx.doi.org/10.1111/jam.12282.

Balaji, R. and E. David. 2016. Isolation and characterization of Cr(VI) reducing bacteria and fungi their potential use in bioremediation of chromium containing tannery effluent (Ambur and Ranipet, Vellore dist, Tamilnadu). Adv. Res. J. Life Sci. 2: 1–4.

Baldiris, R., N. Acosta-Tapia, A. Montes, J. Hernández and R. Vivas-Reye. 2018. Reduction of hexavalent chromium and detection of chromate reductase (ChrR) in *Stenotrophomonas maltophilia*. Molecules 23(2): 406.

Barnhart, J. 1997. Occurrences, uses, and properties of chromium. Regul. Toxicol. Pharmacol. 26: S3–S7.

Barrera-Díaz, C. E., V. Lugo-Lugo and B. Bilyeu. 2012. A review of chemical, electrochemical and biological methods for aqueous Cr(VI) reduction. J. Hazard. Mater. 223: 1–12.

Bertagnolli, C. and M. G. C. Silva. 2013. Bioadsorção de cromo na alga Sargassum filipendula e em seus derivados. PhD thesis. State University of Campinas, Campinas, Brazil.

Bhange, K., V. Chaturvedi and R. Bhatt. 2016. Feather degradation potential of *Stenotrophomonas maltophilia* KB13 and feather protein hydrolysate (FPH) mediated reduction of hexavalent chromium. 3 Biotech. 6(1): 42.

Browning, C. L. and Sr. J. P. Wise. 2017. Prolonged exposure to particulate chromate inhibits RAD51 nuclear import mediator proteins. Toxicology and Applied Pharmacology. 331: 101–107.

Cabral, L., P. Giovanella, A. Kellerman, C. Gianello, F. M. Bento and F. A. O. Camargo. 2014. Impact of selected anions and metals on the growth and *in vitro* removal of methylmercury by *Pseudomonas putida* V1. Int. Biodeterior. Biodegrad. 91: 29–36. https://doi.org/10.1016/j.ibiod.2014.01.021.

Casey, C. E. and K. M. Hambidge. 1984. Chromium in human milk from American mothers. Br. J. Nutr. 52: 73–77.

Chatterjee, S., N. C. Chatterjee and S. Dutta. 2012. Bioreduction of chromium (VI) to chromium (III) by a novel yeast strain *Rhodotorula mucilaginosa* (MTCC 9315). Afr. J. Biotechnol. 11: 14920–14929.

Chen, J. and Y. Tian. 2021. Hexavalent chromium reducing bacteria: mechanism of reduction and characteristics. Environmental Science and Pollution Research, 1–17.

Chen, T., Z. Zhou, S. Xu, H. Wang and W. Lu. 2015. Adsorption behavior comparison of trivalent and hexavalent chromium on biochar derived from municipal sludge. Bioresour. Technol. 190: 388–394.

Chen, Z., L. Zou, H. Zhang, Y. Chen, P. Liu and X. Li. 2014. Thioredoxin is involved in hexavalent chromium reduction in *Streptomyces violaceoruber* strain LZ-26-1 isolated from the Lanzhou reaches of the Yellow River. Int. Biodeterior. Biodegrad. 94: 146–151. https://doi.org/10.1016/j.ibiod.2014.07.013.

Cheung, K. H. and J. D. Gu. 2007. Mechanism of hexavalent chromium detoxification by microorganisms and bioremediation application potential: a review. Int. Biodeterior. Biodegrad. 59: 8–15.

Chojnacka, K. 2010. Biosorption and bioaccumulation-the prospects for practical applications. Environ. Int. 36: 299–307.

Coetzee, J. J., N. Bansal and E. M. Chirwa. 2018. Chromium in the environment, its toxic effect from chromite-mining and ferrochrome industries, and its possible bioremediation. Exposure and Health 12: 51–62.

Coetzee, J. J., N. Bansal and E. M. Chirwa. 2020. Chromium in the environment, its toxic effect from chromite-mining and ferrochrome industries, and its possible bioremediation. Exposure and Health 12(1): 51–62.

Coreño-Alonso, A., A. Solé, E. Diestra, I. Esteve, J. F. Gutiérrez-Corona, G. E. López, F. J. Fernández and A. Tomasini. 2014. Mechanisms of interaction of chromium with *Aspergillus niger var tubingensis* strain Ed8. Bioresour. Technol. 158: 188–192. https://doi.org/10.1016/j.biortech.2014.02.036.

Costello, R. B., J. T. Dwyer and J. M. Merkel. 2019. Chromium supplements in health and disease. The Nutritional Biochemistry of Chromium (III). Elsevier, pp. 219–249.

Cramer, L., J. Basson and L. Nelson. 2004. The impact of platinum production from UG2 ore on ferrochrome production in South Africa. J. S. Afr. Inst. Min. Metall. 104: 517–527.

Deepa, A. and B. K. Mishra. 2020. Microbial biotransformation of hexavalent chromium [Cr (VI)] in tannery wastewater. Microbial Bioremediation & Biodegradation. Springer, Singapore, 143–152.

Dey, S. and A. K. Paul. 2016. Evaluation of chromate reductase activity in the cell-free culture filtrate of *Arthrobacter* sp. SUK 1201 isolated from chromite mine overburden. Chemosphere. 571, 156: 69–75.

Dhal, B., H. N. Thatoi, N. N. Das and B. D. Pandey. 2013. Chemical and microbial remediation of hexavalent chromium from contaminated soil and mining/metallurgical solid waste: a review. J. Hazard. Mater. 250-251: 272–291.

Dixon, J. B. and S. B. Weed. 1977. Minerals in soil environments. Soil Science Society of America, Madison, WI. 948 pp.

Doble, M. and A. Kumar. 2005. Biotreatment of industrial effluents. Elsevier Inc., Burlington.

Dönmez, G. and Z. Aksu. 1999. The effect of copper (II) ions on the growth and bioaccumulation properties of some yeasts. Process. Biochem. 35: 135–142.

Durai. G. and M. Rajasimman. 2011. Biological treatment of tannery wastewater—a review. J. Environ. Sci. Technol. 4(1): 1–17.

Elangovan, R., S. Abhipsa, B. Rohit, P. Ligy and K. Chandraraj. 2006. Reduction of Cr(VI) by a *Bacillus* sp. Biotechnol. Lett. 28: 247–252.

Emenike, C. U., B. Jayanthi, P. Agamuthu and S. H. Fauziah. 2018. Biotransformation and removal of heavy metals: a review of phytoremediation and microbial remediation assessment on contaminated soil. Environ. Rev. 26(2): 156–168.

Ertani, A., A. Mietto, M. Borin and S. Nardi. 2017. Chromium in agricultural soils and crops: a review. Water Air Soil Pollut: 228.

Faisal, M. and S. Hasnain. 2004. Microbial conversion of Cr(VI) into Cr(III) in industrial effluent. Afr. J. Biotechnol. 3(11): 610–617.

Fedorovych, D. V., M. V. Gonchar, H. P. Ksheminska, T. M. Prokopiv, H. I. Nechay, P. Kaszycki, H. Koloczek and A. A. Sibirny. 2009. Mechanisms of chromate detoxification in yeasts. Microbiol. Biotechnol. 3: 15–21.

Fernández, P. M., M. E. Cabral, O. D. Delgado, J. I. Fariña and L. I. C. Figueroa. 2013. Textile dye polluted waters as an unusual source for selecting chromate-reducing yeasts through Cr(VI) enriched microcosms. Int. Biodeter. Biodegr. 79: 28–35.

Fernández, P. M., E. L. Cruz, S. C. Viñarta and L. I. C. Figueroa. 2016. Optimization of culture conditions for growth associated with Cr (VI) removal by *Wickerhamomyces anomalus* M10. Bull. Environ. Contam. Toxicol. 98(3): 400–4006. https://doi.org/10.1007/s00128-016-1958-5.

Fernández, P. M., M. M. Martorell, M. G. Blaser, L. A. M Ruberto, L. I. C. Figueroa and W. P. Mac Cormack. 2017. Phenol degradation and heavy metals tolerance of Antarctic yeasts. Extremophiles 21(3): 445–457.

Fernández, P. M., S. C. Viñarta, A. R. Bernal, E. L. Cruz and L. I. C. Figueroa. 2018. Bioremediation strategies for chromium removal: current research, scale-up approach and future perspectives, Chemosphere. https://doi:10.1016/j.chemosphere.2018.05.166.

Gavrilescu, M. 2004. Removal of heavy metals from the environment by biosorption. Eng. Life Sci. 4: 219–232.

Ge, S., W. Zheng, X. Dong, M. Zhou, J. Zhou and S. Ge. 2014. Distributions of soluble hexavalent chromate reductase from *Leucobacter* sp. G161 with high reducing ability and thermostability. J. Pure. Appl. Microbiol. 8: 1893–1900.

Gibb, H. J., P. S. Lees, P. F. Pinsky and B. C. Rooney. 2000. Clinical findings of irritation among chromium chemical production workers. Am. J. Ind. Med. 38: 127–131.

Grace Pavithra, K., V. Jaikumar, P. S. Kumar and P. Sundar Rajan. 2019. A review on cleaner strategies for chromium industrial wastewater: present research and future perspective. Journal of Cleaner Production. 228: 580–593.

Gu, Y., W. Xu, Y. Liu, G. Zeng, J. Huang, X. Tan, H. Jian, X. Hu, F. Li and D. Wang. 2015. Mechanism of Cr(VI) reduction by *Aspergillus niger*: enzymatic characteristic, oxidative stress response, and reduction product. Environ. Sci. Pollut. R. 22: 6271–6279. https://doi.org/10.1007/s11356-014-3856-x.

Guertin, J. 2004. Toxicity and health effects of chromium (all oxidation states). Chromium (VI) Handbook. CRC Press, Boca Raton.

Guertin, J., J. A. Jacobs and C. P. Avakian. 2016. Chromium (VI) handbook. CRC Press, Boca Raton.

Guillen-Jiménez, F., A. R. Netzahuatl-Muñoz, L. Morales-Barrera and E. Cristiani-Urbina. 2009. Hexavalent chromium removal by *Candida* sp. in a concentric draft-tube airlift bioreactor. Water Air Soil Pollut. 204: 43–51. https://doi.org/10.1007/s11270-009-0024-x.

Gupta, P., R. Rani, A. Chandra, S. Varjani and V. Kumar. 2019. The role of microbes in chromium bioremediation of tannery effluent. In. Water and wastewater treatment technologies. Springer, Singapore, pp. 369–377.

Gutiérrez-Corona, J. F., P. Romo-Rodríguez and F. Santos-Escobar. 2016. Microbial interactions with chromium: basic biological processes and applications in environmental biotechnology. World J. Microbiol. Biotechnol. 32: 191. https://doi.org/10.1007/s11274- 016-2150-0.

Hackbarth, F. V., D. Maass, A. Au Souza, V. J. P. Vilar and S. M. A. U. G. Souza. 2016. Removal of hexavalent chromium from electroplating wastewaters using marine macroalga *Pelvetia canaliculata* as natural electron donor. Chem. Eng. J. 290: 477–489. https://doi.org/10.1016/j.cej.2016.01.070.

Harada, M. 1995. Minamata disease: methylmercury poisoning in Japan caused by environmental pollution. Crit. Rev. Toxicol. 25(1): 1–24.

Haynes, W. M. 2014. CRC handbook of chemistry and physics. CRC press.

He, Z., Y. Yao, Z. Lu and Y. Ye. 2014. Dynamic metabolic and transcriptional profiling of *Rhodococcus* sp. strain YYL during the degradation of tetrahydrofuran. Appl. Environ. Microbiol. 80: 2656–2664. doi: 10.1128/AEM.04131-13.

Herrero, M. and D. C. Stuckey. 2015. Bioaugmentation and its application in wastewater treatment: A review. Chemosphere 140: 119–128.

Huff, J., R. M. Lunn, M. P. Waalkes, L. Tomatis and P. F. Infante. 2007. Cadmium-induced cancers in animals and in humans. Int. J. Occup. Environ. Health 13(2): 202–212.

Iram, S., G. Uzma, S. Rukh and T. Ara. 2013. Bioremediation of heavy metals using isolates of filamentous fungus collected from polluted soil of Kasur, Pakistan. Int. Res. J. Biol. Sci. 66–73.

Ismael, A. R., X. Rodriguez, C. Gutierrez and M. G. Moctezuma. 2004. Biosorption of Chromium(VI) from aqueous solution onto fungal biomasss. Bioinorg. Chem. Appl. 2: 1–7.

Jaishankar, M., T. Tseten, N. Anbalagan, B. B. Mathew and K. N. Beeregowda. 2014. Toxicity, mechanism and health effects of some heavy metals. Interdiscip. Toxicol. 7(2): 60–72.

Järup, L. 2003. Hazards of heavy metal contamination. Br. Med. Bull. 68(1): 167–182.

Jenkins, D. 1998. The effect of reformulation of household powder laundry detergents on their contribution to heavy metals levels in wastewater. Water Environ. Res. 70(5): 980–983.

Jeyasingh, J. and L. Philip. 2005. Bioremediation of chromium contaminated soil: optimization of operating parameters under laboratory conditions. J. Hazard Mater. 118(1-3): 113–120.

Jin, Y. H., P. E. Dunlap, S. J. McBride, H. Al-Refai, P. R. Bushel and J. H. Freedman. 2008. Global transcriptome and deletome profiles of yeast exposed to transition metals. Plos. Genet. 25: 4(4): 1000053.

Jobby, R., P. Jha, A. K. Yadav and N. Desai. 2018. Biosorption and biotransformation of hexavalent chromium [Cr(VI)]: A comprehensive review. Chemosphere. 207: 255–266.

Joutey, N. T., H. Sayel, W. Bahafid and N. El Ghachtouli. 2015. Mechanisms of hexavalent chromium resistance and removal by microorganisms. pp. 45–69. *In*: Reviews of environmental contamination and toxicology, vol 233. Springer, Cham.

Kaduková, J. and E. Virčíková. 2005. Comparison of differences between copper bioaccumulation and biosorption. Environ. Int. 31(2): 227–232.

Kamaludeen, S. P. B., K. R. Arunkumar and K. Ramasamy. 2003. Bioremediation of chromium contaminated environments.

Karthik, C., V. S. Ramkumar, A. Pugazhendhi, K. Gopalakrishnan and P. I. Arulselvi. 2017. Biosorption and biotransformation of Cr (VI) by novel *Cellulosimicrobium funkei* strain AR6. J Taiwan Inst. Chem. Eng. 70: 282–290.

Katz, S. A. and H. Salem. 1993. The toxicology of chromium with respect to its chemical speciation: a review. J. Appl. Toxicol. 13: 217–224.

Khani, M. H., H. Pahlavanzadeh and K. Alizadeh. 2012. Biosorption of strontium from aqueous solution by fungus *Aspergillus terreus*. Environ. Sci. Pollut. Res. Int. 19: 2408–2418.

Khattar, J. I. S., S. Parveen, Y. Singh, D. P. Singh and A. Gulati. 2014. Intracellular uptake and reduction of hexavalent chromium by the cyanobacterium *Synechocystis* sp. PUPCCC 62. J. Appl. Phycol. 27: 827–837.

Khelaifia, F. Z., S. Hazourli, S. Nouacer, H. Rahima and M. Ziati. 2016. Valorisation of raw biomaterial waste-date stones-for Cr (VI) adsorption in aqueous solution: thermodynamics, kinetics and regeneration studies. Int. Biodeterior. Biodegrad. 114: 76–86.

Khoubestani, R. S., N. Mirghaffari and O. Farhadian. 2015. Removal of three and hexavalent chromium from aqueous solutions using a microalgae biomass-derived biosorbent. Environ. Prog. Sustain. Energy. 34: 949–956.

Kirubanandam Grace, P., V. Jaikumar, P. Senthil Kumar and P. Sundarrajan. 2019. A review on cleaner strategies for chromium industrial wastewater: present research and future perspective. Journal of Cleaner Production. doi: 10.1016/j.jclepro.2019.04.117.

Konczyk, J., C. Kozlowski and W. Walkowiak. 2010. Removal of chromium(III) from acidic aqueous solution by polymer inclusion membranes with D2EHPA and Aliquat 336. Desalination 263: 211–216.

Ksheminska, H., D. Fedorovych, L. Babyak, D. Yanovych, P. Kaszycki and H. Koloczek. 2005. Chromium (III) and (VI) tolerance and bioaccumulation in yeast: a survey of cellular chromium content in selected strains of representative genera. Process Biochem. 40(5): 1565–1572.

Ksheminska, H. P., T. M. Honchar, G. Z. Gayda and M. V. Gonchar. 2006. Extracellular chromate—reducing activity of the yeast cultures. Central Eur. J. Biol. 1: 137–149.

Ksheminska, H., D. Fedorovych, T. Honchar, M. Ivash and M. Gonchar. 2008. Yeast tolerance to chromium depends on extracellular chromate reduction and Cr(III) chelation. Food Technol. Biotechnol. 46: 419–426.

Kumar, P. A., S. Chakraborty and M. Ray. 2008. Removal and recovery of chromium from wastewater using short chain polyaniline synthesized on jute fiber. Chem. Eng. J. 141: 130–140.

Kwak, H. W., M. K. Kim, J. Y. Lee, H. Yun, M. H. Kim, Y. H. Park and K.H. Lee. 2015. Preparation of bead-type biosorbent from water-soluble *Spirulina platensis* extracts for chromium (VI) removal. Algal Res. 7: 92–99. https://doi.org/10.1016/j.algal.2014.12.006.

Lapenna, D. and G. Ciofani. 2020. Chromium and human low-density lipoprotein oxidation. J. Trace Elem. Med. Biol. 59: 126411.

Lara, P., E. Morett and K. Juárez. 2017. Acetate biostimulation as an effective treatment for cleaning up alkaline soil highly contaminated with Cr(VI). Environ. Sci. Pollut. Res. 24: 25513–2552. https://doi.org/10.1007/s11356-016-7191-2.

Li, Q., H. Liu, M. Alattar, S. Jiang, J. Han, Y. Ma and C. Jiang. 2015. The preferential accumulation of heavy metals in different tissues following frequent respiratory exposure to PM 2. 5 in rats. Sci. Rep. 5(5): 16936.

Lima, L. K. S., S. J. Kleinübing, E. A. Silva and M. G. C. Silva. 2011. Removal of chromium from wastewater using macrophytes Lemna Minoras biosorbent. Chem. Eng. Trans. 25: 303–308.

Liu, X., G. Wu, Y. Zhang, D. Wu, X. Li and P. Liu. 2015. Chromate reductase YieF from *Escherichia coli* enhances hexavalent chromium resistance of human HepG2 cells. Int. J. Mol. Sci. 16, 11892-11902.

Lytras, G., C. Lytras, D. Argyropoulou, N. Dimopoulos, G. Malavetas and G. Lyberatos. 2017. A novel two-phase bioreactor for microbial hexavalent chromium removal from wastewater. J. Hazard Mater. 336: 41–51. https://doi.org/10.1016/j.jhazmat.2017.04.049.

Magro, C. D., M. C. Deon, A. Thome, J. S. Piccin and L. M. Colla. 2013. Biossorção passiva de crômio (VI) através da microalga *Spirulina platensis*. Rev. Quím. Nova 36: 1104–1110. http://dx.doi.org/10.1590/S0100-40422013000800011.

Mahmoud, M.S. and S.A. Mohamed. 2017. Calcium alginate as an eco-friendly supportingmaterial for baker's yeast strain in chromium bioremediation. Housing Build. Nat. Res. Center J. 13: 245e254.

Malaviya, P. and A. Singh. 2016. Bioremediation of chromium solutions and chromium containing wastewaters. Critical Reviews in Microbiology 42(4): 607–633.

Martorell, M. M., P. M. Fernández, J. I. Fariña and L. I. C. Figueroa. 2012. Cr (VI) reduction by cell-free extracts of *Pichia jadinii* and *Pichia anomala* isolated from textile-dye factory effluents. Int. Biodeterior. Biodegradation 71: 80–85.

Medeiros, M., A. Rodrigues, M. Batoreu, A. Laires, J. Rueff and A. Zhitkovich. 2003. Elevated levels of DNA-protein crosslinks and micronuclei in peripheral lymphocytes of tannery workers exposed to trivalent chromium. Mutagenesis 18(1): 19–24.

Mohan, D. and C. U. Pittman. 2006. Activated carbons and low cost adsorbents for remediation of tri- and hexavalent chromium from water. J. Hazard. Mater. 137: 762–811.

Morales-Barrera, L., F. M. Guillen-Jiṁenez, O. Moreno, Alicia, Villegas- Garrido, Thelma Lilia, Sandoval-Cabrera, Antonio, Hernandez-Rodrıguez, Cesar Hugo, Cristiani-Urbina, Eliseo. 2008. Isolation, identification and characterization of a Hypocrea awa strain with high Cr (VI) reduction potential. J. Biochem. Eng. 40: 284e292.

Morales-Barrera, L. and E. Cristiani-Urbina. 2015. Bioreduction of hexavalent chromium by *Hypocrea tawa* in a concentric draft-tube airlift bioreactor. J. Environ. Biotechnol. Res. 1: 37–44.

Mrudula, P., S. Jamwal, J. Samuel, N. Chandrasekaran and A. Mukherjee. 2012.Enhancing the hexavalent chromium bioremediation potential of Acinetobacter junii VITSUKMW2 using statistical design experiments. J. Microbiol. Biotechnol. 22: 1767e1775.

Narayani, M. and K. V. Shetty. 2013. Chromium-resistant bacteria and their environmental condition for hexavalent chromium removal: a review. Crit. Rev. Environ. Sci. Technol. 43(9): 955–1009.

Nemr, A. E., A. El-Sikaily, A. Khaled and O. Abdelwahab. 2015. Removal of toxic chromium from aqueous solution, wastewater and saline water by marine red alga *Pterocladia capillacea* and its activated carbon. Arab. J. Chem. 8: 105–117.

Ontañon, O. M., P. S. González, L. F. Ambrosio, C. E. Paisio and E. Agostini. 2014. Rhizoremediation of phenol and chromium by the synergistic combination of a native bacterial strain and *Brassica napus* hairy roots. Int Biodeterior. Biodegrad. 88: 192–198. https://doi.org/10.1016/j.ibiod.2013.10.017.

Pal, S. and Y. Vimala. 2011. Bioremediation of chromium from fortified solutions byphanerochaete chrysosporium (MTCC 787). J. Bioremed. Biodegrad. 2: 127. https://doi.org/10.4172/2155-6199.1000127.

Pattanapipitpaisal, P., A. N. Mabbett, J. A. Finlay, A. J. Beswick, M. Paterson-Beedl and A. Essa. 2002. Reduction of Cr (VI) and bioaccumulation of chromium by gram positive and gram negative microorganisms not previously exposed to Cr-stress. Environ. Technol. 23(7): 731–745.

Pechova, A. and L. Pavlata. 2007. Chromium as an essential nutrient: a review. VETERINARNI MEDICINA-PRAHA-52(1): 1.

Pellerin, C. and S. M. Booker. 2000. Reflections on hexavalent chromium: health hazards of an industrial heavyweight. Environ. Health Perspect. 108: A402–A407.

Prabhakaran, C. Divyasree and S. Subramanian. 2016. Studies on the bioremediation of chromium from aqueous solutions using *Corynebacterium paurometabolum*. Trans. Indian Inst. Met. 70: 497–509.

Pradhan, D., L. B. Sukla, M. Sawyer and P. K. S. M. Rahman. 2017. Recent bioreduction of hexavalent chromium in wastewater treatment: a review. J. Ind. Eng. Chem. 55: 1–20.

Qian, J., L. Wei and R. Liu. 2016. An exploratory study on the pathways of Cr (VI) reduction in sulfate-reducing up-flow anaerobic sludge bed (UASB) reactor. Sci. Rep. 6: 23694.

Qian, J., J. Zhou, L. Wang, L. Wei, Q. Li, D. Wang and Q. Wang. 2017. Direct Cr (VI) bio-reduction with organics as electron donor by anaerobic sludge. Chem. Eng. J. 309: 330–338.

Ramírez-Ramírez, R., C. Calvo-Méndez, M. Ávila-Rodriguez, P. Lappe, M. Ulloa, R. Vázquez-Juárez and J. F. Gutiérrez-Corona. 2004. Cr (VI) reduction in a chromate-resistant strain of *Candida maltosa* isolated from the leather industry. A. Van. Leeuw. 85: 63–68.

Raskin, I. and B. D. Ensley. 2000. Phytoremediation of toxic metals. John Wiley and Sons, New York.

Rath, B. P., S. Das, P. K. D. Mohapatra and H. Thatoi. 2014. Optimization of extracellular chromate reductase production by *Bacillus amyloliquefaciens* (CSB 9) isolated from chromite mine environment. Biocatal. Agric. Biotechnol. 3: 35–41. https://doi.org/10.1016/j.bcab.2014.01.004.

Rezaei, H. 2013. Biosorption of chromium by using *Spirulina* sp. Arab. J. Chem. 9: 846–853. https://doi.org/10.1016/j.arabjc.2013.11.008.

Richard, F. C. and A. C. M. Bourg. 1991. Aqueous geochemistry of chromium: a review. Water Res. 25: 807–816.

Rida, B., K. Yrjälä and S. Hasnain. 2012. J. Microbiol. Biotechnol. 22–547.

Robins, K. J., D. O. Hooks, B. H. A. Rehm and D. F. Ackerley. 2013. *Escherichia coli* NemA is an efficient chromate reductase that can be biologically immobilized to provide a cell free system for remediation of hexavalent chromium. PLoS One 8: 1–8. https://doi.org/10.1371/journal.pone.0059200.

Sahni, S. K. 2011. Hazardous metals and minerals pollution in India: sources, toxicity and management. pp. 1–29. Indian National Science Academy. Angkor Publishers (P) Ltd.

Samuel, J., M. Pulimi and M. L. Paul. 2013. Batch and continuous flow studies of adsorptive removal of Cr(VI) by adapted bacterial consortia immobilized in alginate beads. Bioresour. Technol. 128: 423–430.

Samuel, M. S., M. E. A. Abigail and C. Ramalingam. 2015. Isotherm modelling, kinetic study and optimization of batch parameters using response surface methodology for effective removal of Cr(VI) using fungal biomass. PLoS One 10(3): 0116884. https://doi.org/10.1371/journal.pone.0116884.

Saranraj, P., D. Stella, D. Reetha and K. Mythili. 2010. Bioadsorption of chromium resistant *Enterococcus casseliflavus* isolated from tannery effluent. J. Ecobiotechnol. 2: 17–22.

Saranraj, P. and D. Sujitha. 2013. Microbial bioremediation of chromium in tannery effluent: a review. Int. J. Microbiol. Res. 4(3): 305–306.

Scott, J. A. and A. M. Karanjkar. 1992. Repeated cadmium biosorption by regenerated *Enterobacter aerogenes* biofilm attached to activated carbon. Biotechnol. Lett. 14: 737–740.

Seo, H. and Y. Roh. 2015. Biotransformation and its application: biogenic nano-catalyst and metal-reducing-bacteria for remediation of Cr (VI)-contaminated water. J. Nanosci. Nanotechnol. 15: 5649–5652.

Sepehr, M. N., S. Nasseri, M. Zarrabi, M. R. Samarghandi and A. Amrane. 2012. Removal of Cr (III) from tanning effluent by *Aspergillus niger* in airlift bioreactor. Sep. Purif. Technol. 96: 256–262. https://doi.org/10.1016/j.seppur.2012.06.013.

Shahid, M., S. Shamshad, M. Rafiq, S. Khalid, I. Bibi, N. K. Niazi, C. Dumat and M. I. Rashid. 2017. Chromium speciation, bioavailability, uptake, toxicity and detoxification in soil-plant system: A review. Chemosphere 178: 513–533.

Shanker, A. and B. Venkateswarlu. 2011. Chromium: environmental pollution, health effects and mode of action. pp. 650–659. *In*: Jerome O.N. (ed.). Encyclopedia of Environmental Health, vol 65. Elsevier, Burlington.

Sharma, B., S. Singh and N. J. Siddiqi. 2014. Biomedical implications of heavy metals induced imbalances in redox systems. Biomed. Res. Int. 2014: 1–26.

Sharma, P., V. Bihari, S. K. Agarwal, V. Verma, C. N. Kesavachandran, B. S. Pangtey, N. Mathur, K. P. Singh, M. Srivastava and S. K. Goel. 2012. Groundwater contaminated with hexavalent chromium [Cr(VI)]: a health survey and clinical examination of community inhabitants (Kanpur, India). PLoS ONE 7: 47877.

Sharma, S. and A. Bhattacharya. 2017. Drinking water contamination and treatment techniques. Appl. Water. Sci. 7(3): 1043–1067.

Shuhong, Y., Z. Meiping, Y. Hong, W. Han, X. Shan, L. Yan and W. Jihui. 2014. Biosorption of Cu(2+), Pb(2+) and Cr(6+) by a novel exopolysaccharide from *Arthrobacter* ps-5. Carbohydr. Polym. 101: 50–56. https://doi.org/10.1016/j.carbpol.2013.09.021.

Singh, H. P., P. Mahajan, S. Kaur, D. R. Batish and R. K. Kohli. 2013. Chromium toxicity and tolerance in plants. Environ. Chem. Lett. 11: 229–254.

Singh, J. and A. S. Kalamdhad. 2011. Effects of heavy metals on soil, plants, human health and aquatic life. Int. J. Res. Chem. Environ. 1(2): 15–21.

Singh, R. and N. R. Bishnoi. 2015. Biotransformation dynamics of chromium (VI) detoxification using *Aspergillus flavus* system. Ecol. Eng. 75: 103–109.

Sivakumar, D. 2016. Biosorption of hexavalent chromium in a tannery industry wastewater using fungi species. Glob. J. Environ. Sci. Manag. 2: 105–124. https://doi.org/10.7508/gjesm.2016.02.002.

Somasundaram, V., L. Philip and S. M. Bhallamudi. 2009. Experimental and mathematical modeling studies on Cr (VI) reduction by CRB, SRB and IRB, individually and in combination. J. Hazard. Mater. 172: 606–617.

Song, Z., C. J. Williams and R. G. J. Edyvean. 2003. Tannery wastewater treatment using an upflow anaerobic fixed biofilm reactor (UAFBR). Environ. Eng. Sci. 20(6): 587–599.

Srinath, T. Verma, P. W. Ramteke and S. K. Garg. 2002. Chromium (VI) biosorption and bioaccumulation by chromate resistant bacteria. Tannery. Technol. 48: 427–435.

Srivastava, S., A. H. Ahmad and I. S. Thakur. 2007. Removal of chromium and pentachlorophenol from tannery effluents. Bioresour. Technol. 98(5): 1128–1132.

Sukumar, C., V. Janaki, K. Kamala-Kannan and K. Shanthi. 2014. Biosorption of chromium (VI) using *Bacillus subtilis* SS-1 isolated from soil samples of electroplating industry. Clean Technol. Environ. Policy 16: 405–413. https://doi.org/10.1007/s10098-013-0636-0.

Swaroop, A., M. Bagchi, H. Preuss, S. Zafra-Stone, T. Ahmad and D. Bagchi. 2019. Benefits of chromium (III) complexes in animal and human health, The Nutritional Biochemistry of Chromium (III). Elsevier, pp. 251–278.

Tabinda, A. B., R. Irfan, A. Yasar, A. Iqbal and A. Mahmood. 2018. Phytoremediation potential of *Pistia stratiotes* and *Eichhornia crassipes* to remove chromium and copper. Environmental Technology, 1–6.

Tchounwou, P. B., C. G. Yedjou, A. K. Patlolla and D. J. Sutton. 2012. Heavy metal toxicity and the environment. Exp. Suppl. 101: 133–164.

Thatoi, H., S. Das, J. Mishra, B. P. Rath and N. Das. 2014. Bacterial chromate reductase, a potential enzyme for bioremediation of hexavalent chromium: a review. J. Environ. Manag. 146: 383–399.

U.S. Department of the Interior & U.S. Geological Survey January 2017. Mineral Commodity Summaries 2017. U.S.

Viti, C., E. Marchi, F. Decorosi and L. Giovannetti. 2014. Molecular mechanisms of Cr (VI) resistance in bacteria and fungi. FEMS Microbiol. Rev. 38: 633–659.

Von Burg, R. and D. Liu. 1993. Chromium and hexavalent chromium. J. Appl. Toxicol. 13: 225–230.

Wang, J. and C. Chen. 2009. Biosorbents for heavy metals removal and their future. Biotechnol. Adv. 27: 195–226.

Wang, C., L. Gu, X. Liu, X. Zhang, L. Cao and X. Hu. 2016. Sorption behavior of Cr(VI) on pineapple-peel-derived biochar and the influence of coexisting pyrene. Int. Biodeterior. Biodegrad. 111: 78–84.

World Health Organization. 2004. Guidelines for drinking-water quality, vol 1. WHO, Geneva.

World Health Organization (WHO). 2003. Chromium in drinking water. Background document for development of WHO Guidelines for Drinking-water Quality, WHO/SDE/WSH/03.04/04.

Wu, D., Y. Sui, S. He, X. Wang, C. Li and H. Kong. 2008. Removal of trivalent chromium from aqueous solution by zeolite synthesized from coal fly ash. J. Hazard. Mater. 155: 415–423.

Wu, X., X. Zhu, T. Song, L. Zhang, H. Jia and P. Wei. 2015. Effect of acclimatization on hexavalent chromium reduction in a biocathode microbial fuel cell. Bioresour. Technol. 180: 185–191.

Xia, S., Z. Song, P. Jeyakumar, S. M. Shaheen, J. Rinklebe, Y. S. Ok, N. Bolan and H. Wang. 2019. A critical review on bioremediation technologies for Cr (VI)-contaminated soils and wastewater. Crit. Rev. Environ. Sci. Technol. 49: 1027–1078.

Xu, L., M. Luo, W. Li, X. Wei, K. Xie, L. Liu, C. Jiang and H. Liu. 2011. Reduction of hexavalent chromium by *Pannonibacter phragmitetus* LSSE-09 stimulated with external electron donors under alkaline conditions. J. Hazard Mater. 185: 1169–1176.

Yahya, S. K., Z. A. Zakaria, J. Samin, A. S. Raj and W. A. Ahmad. 2012. Isotherm kinetics of Cr (III) removal by non-viable cells of *Acinetobacter haemolyticus*. Colloids Surf. B. Biointerfaces. 1: 362–368. https://doi.org/10.1016/j.colsurfb.2012.02.016.

Zhang, R. and Y. Tian. 2020. Characteristics of natural biopolymers and their derivative as sorbents for chromium adsorption: a review. J. Leather Sci. Eng. 2.

Zhao, C., Q. Yang, W. Chen and B. Teng. 2012. Removal of hexavalent chromium in tannery wastewater by *Bacillus cereus*. Can. J. Microbiol. 58: 23–28.

Zheng, Z., Y. Li, X. Zhang, P. Liu, J. Ren, G. Wu, Y. Zhang, Y. Chen and X. Li. 2015. A *Bacillus subtilis* strain can reduce hexavalent chromium to trivalent and an nfr A gene is involved. Int. Biodeterior. Biodegrad. 97: 90–96. https://doi.org/10.1016/j.ibiod.2014.10.017.

Zhitkovich, A. 2011. Chromium in drinking water: sources, metabolism, and cancer risks. Chemical research in toxicology 24(10): 1617–1629.

CHAPTER 13

Environmental Evidence and Behaviour of Mercury Emissions, Biogeochemical Cycle, and Remediation in Earth Systems

Ramamoorthy Ayyamperumal,[1,*] *Xiaozhong Huang,*[1] *Mohamed Khalith S.B.,*[2] *Natchimuthu Karmegam,*[3] *Kantha Deivi Arunachalam,*[2] *Diksha Sharma,*[4,5] *Manikanda Bharath Karuppasamy,*[6] *Gnachandrasamy Gopala Krishnan*[7] and *Balasubramani Ravindran*[8,*]

13.1 Introduction

For millions of years, mercury (Hg) has been accessible to human beings. The present knowledge about mercury is that it cannot be produced or removed, and the same quantity has existed on the planet Earth since the creation of the universe. It is a molten material at normal temperatures that appears in different forms, such as solid in mercury sulfide known as cinnabar, as a vapor in its elemental form of mercury, a water-soluble medium of organic mercury, etc. Hg, one of the planet's most commonly known toxic elements, has induced environmental toxicity, long-term transport, persistence, and bioaccumulation (UNEP 2013). It doesn't undergo degradation in the environment and the persistence of Hg in the atmosphere makes it an environmental issue in the present world

[1] Key Laboratory of Western China's Environmental system, College of Earth and Environmental Sciences, Lanzhou University, Lanzhou-730000, P.R. China.
[2] Center for Environmental Nuclear Research, Directorate of Research, SRM Institute of Science and Technology, Kattankulathur, Chennai, Tamil Nadu, 603203, India.
[3] Department of Botany, Government Arts College Autonomous, Salem, 636 007, Tamil Nadu, India.
[4] Ministry of Education Key Laboratory of Cell Activities and Stress Adaptations, School of Life sciences, Lanzhou University, Lanzhou-730000, P.R. China.
[5] Gansu Key Laboratory of Biomonitoring and Bioremediation for Environment Pollution, School of Life Science, Lanzhou University, Lanzhou-730000, P.R. China.
[6] Institute for Ocean Management, Anna University, Chennai, 600025, Tamil Nadu, India.
[7] Department of Environmental Energy and Engineering, Kyonggi University Youngtong-Gu, Suwon, Gyeonggi-Do, 16227, Republic of Korea.
[8] School of Geography and Planning, Sun Yat-Sen University, Guangzhou, 510275, P.R. China.
* Corresponding authors: ramamoorthy@lzu.edu.cn; kalamravi@gmail.com

(Bishop et al. 2020, Gworek et al. 2020). Mercury is one of the most studied environmental toxins among heavy metals. A substantial body of research suggests that a complex combination of transport and transformation (Goss and Schwarzenbach 1998) allows for reinvestment of this attribute into air temperatures, soil, and water-based ecosystems. Indeed, recent progress has been made in providing important information sets covering areas in the world where there was a lack of previous Hg environmental data. This is because of its unstable state, especially in its methyl form that creates high ecological mobility. Mercury can be transferred to the atmosphere thousands of kilometers, making it a global pollutant (Selin and Jacob 2008). In the past few decades, substantial scientific awareness has been gained on the sources and emissions of mercury, its pathways and cycling through the environment, exposure to humans, and its impacts on human and environmental health (Pacyna 2020, UN Environment 2019).

Since the industrial revolution of the 1850s, the usage of mercury has increased. Combined with knowledge from ancient times on the human Hg additive background, the developed model has shown much higher impacts of human activities on the global Hg cycle (Streets et al. 2017, 2011). An increase in the mining of mercury ores and conversion into various chemical compounds has resulted in excessive contamination. As a result, mercury concentrations in the earth's surface, atmosphere, and marine areas have increased significantly. The element has no role to play in the anabolism or catabolism of humans and other organisms (Selin et al. 2018). Global Hg-cycling has also been used to estimate the incorporation of these data into international and regional models; modern methods, such as stable Hg-isotope analysis, provide reasonable source limitations and equipment. Current research and updated models indicate that the intensive convection of these HgII reservoirs is strong in tropical and semi-tropic regions. Nevertheless, other microbes turn the product into organic compounds known as methylmercury (MeHg). It is absorbed and biomagnified in animals and human beings. An increase in the concentration of MeHg causes toxic effects. Its toxic effect was understood when it was recognized as the cause of the Minamata disease in 1956. The Minamata Convention was signed in 2013 by the United Nations and came into force in August 2017 (Selin et al. 2018). Changes in anthropogenic pollution are continuing and will persist for the long term, as well as significant changes in foreign source areas related to existing emission levels (Giang et al. 2015, Gray et al. 2015). For a better understanding of the mercury cycle, scientific uncertainties and variability of climate limits capacity to link varying pollution conditions to environmental and pollutant concentrations and obscures capacity for adjustments as a result of the Convention's implementation. (Kwon and Selin 2016, Selin 2014) and (Hsu-Kim et al. 2018) offer a more comprehensive description of the issues affecting the processing of marine Hg, Hg methylation and demethylation mechanisms and human and human sensitivity towards Hg in the sense of environmental change and disruption, while critical theoretical perspectives for foreign policies are presented.

Mercury gets circulated in the atmosphere, lithosphere, biosphere, etc., until it gets translocated to deep-sea sediments. Techniques are urgently required to forecast the effect of anthropogenic disruptions and consider the dynamics of local Hg cycling. The objective of this chapter is to summarize the available scientific information related to mercury and its cycling in the earth systems; furthermore, this chapter presents the global understanding of the biogeochemical cycle associated with mercury, including the processes and characteristics of the environmental cycling of mercury including different pathways of emissions and its ecological impact including human exposure. The present chapter will provide in-depth information regarding the understanding of the long-term anthropogenic impact of Hg and present environmental policy acts. Ultimately, it outlines the mercury processes in earth systems with a comprehensive overview of the atmospheric processes related to mercury.

13.2 Physicochemical Characteristics of Mercury

Mercury is a sparkling, metallic silver element that exists as a liquid with a density of 13.6 g cm^{-3} at room temperature (melting point 234 K). It is denoted by the symbol "Hg", which has an atomic number of 80 and an atomic mass of 200.59. Mercury is present in numerous consumer goods, and its salts are used as purgatives, antisyphilia, disinfectants, astringents, and for other medicinal purposes. It is the only substance that occurs in a liquid state at ambient temperature. It has relatively high vapor pressure, which vaporizes into colorless vapors, with the highest volatility of any product. This is highly reactive with an ionizing potential greater than any other electro positively manufactured materials, except hydrogen. It is a good conductor of electricity but a bad conductor of fire. Mercury exists in three potential states of electrical charge or valence. Elemental mercury (Hg^0) does not have any electrical effect. Mercury is also present in two positively charged or cationic states, i.e., (i) mercurous mercury Hg(I). (Hg^+) is not widespread but is present in stable molecules like calome (Hg_2Cl_2) (UNEP 2013) and (ii) mercuric mercury Hg(II) (Hg^{2+}). Mercury's cation is more water-soluble than most other inorganic compounds like phosphorus, ammonia and calcium. It is also more commonly associated with hydroxyl ions (mercuric chloride). Hg^{2+} is also present in organic tissue (Government of Canada 2013). Dimethylmercury, for example, is more harmful than inorganic mercury. Since mercury can easily be assimilated into small soil minerals particles, certain scientists use the Hg(p) notation to represent metal ions attached to or swallowed within a particle. Mercury is naturally present in the environment with different stable isotopes (max. ^{202} Hg 29.9%, lowest ^{196} Hg 0.15%).

13.3 Mercury Cycle

Mercury may follow a variety of pathways. The global mercury cycle includes the earth, soil, liquid systems, seas, atmosphere, and human activity. Each domain communicates directly or implicitly with other domains. Mercury circulation in global domains is unique because of its emission sources, external disturbance, and deposition. Hg is transported and recycled between these domains until it is finally eliminated from the system by burying in the sediments of coastal and deep oceans, lakes, and subsurface soils (AMAP/UN Environment 2019, UN Environment 2019). Figure 13.1 shows a simplified representation of the biogeochemical cycle of mercury (Stein et al. 2009). This section addresses important processes for Hg cycling between broad environmental areas (atmosphere, terrestrial, and aquatic [freshwater and ocean]) as exhibited in (Fig. 13.2) (Environment and Climate Change Canada - Canada.ca 2021). The Gaseous Elemental Mercury (GEM), Reactive Gas Mercury (RGM), and Total Particulate Mercury (Hgp) are known to be found in the atmosphere as airborne mercury species with a global average concentration of about 1.6 ng m^{-3}. Due to their solubility, Hg(II) and Hg(p) mercury are found in water sediments at a concentration ranging between 1 and 100 pg m^{-3}. Mercury may be removed and dissolved in water, soil, and sediment. Methyl mercury in water bodies can accumulate and be re-emitted in the atmosphere through the food chain.

13.3.1 Earth's Crust

Earth's crust is the prime resource for mercury availability. Mercury is emitted through volcanoes, degassing from Hg bearing minerals, geothermal processes, erosion of Hg bearing minerals, and re-emission through various processes (Gworek et al. 2020, 2016). Human beings mined Hg from the crust of the earth and used it for different applications. Mine waste and other chemical forms are discarded on the open ground adjacent to the mines. Various analyses suggest a varying degree of contamination of the earth by mercury with an optimum contamination length of approximately

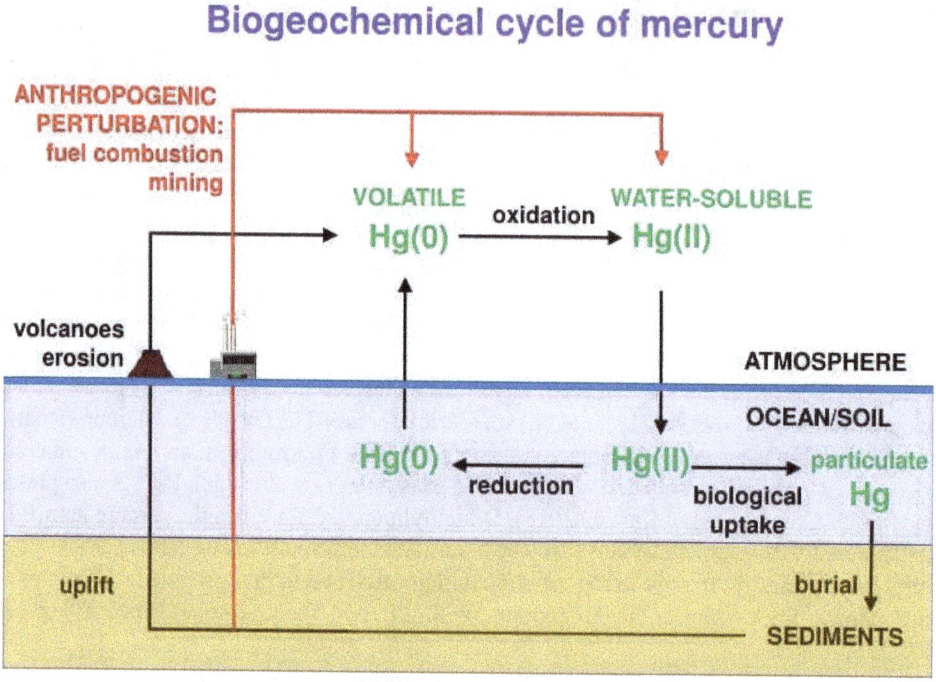

Fig. 13.1. Biogeochemical cycling sources and pathways of mercury in the environment.

600 m (Beal et al. 2014, Gray et al. 2015). Terrestrial materials absorb Hg through moisture deposition, which is only about 3000 mg/y (Holmes et al. 2010). Once mercury reaches the earth's surface, it escapes to the atmosphere or reacts with biomass, microbes, and inorganic oxidants or dissolves in water bodies. In crustal samples, the abundance of Hg varies between 0.9 and 8 ppb (Canil et al. 2015). The global distribution of mercury deposits in 26 mineral belts is widely divided into three categories (Rytuba 2003): (i) silica-carbonate, (ii) hot spring, and (iii) Almaden mines.

Some of the most commonly available mercury-bearing minerals are cinnabar (HgS), cordierite ($Hg_3S_2Cl_2$), schwartzite [$(HgCuFe)_{12}Sb4S_{13}$], livingstonite ($HgSb_4S_7$), etc. It also occurs in other mercury-bearing sulfides such as pyrite (FeS), sphalerite (ZnS), and others (Rytuba 2003). Wet and dry land, ocean, volcanic eruptions, and biomass firing are included in the generally determined emission (Caldwell et al. 2006). Major terrestrial ecosystem emissions include the burning of biomass and HgO volatilization from enriched soils (Obrist et al. 2018). Often Hg from biomass litters and fertile soils is mobilized by wildfires (Burke et al. 2010). International biomass burning emissions contribute 210 and 680 mg of mercury per annum (Amos et al. 2014). Figure 13.3 shows the global mercury emission levels and the concentration of the mercury cycling process in earth systems.

Mercury is dispersed on the Earth's surface by air and its circulation. Nitrogen oxide is involved in the application of dry Hg0 and wet Hg(II) on plants and soils (Choi and Holsen 2009). Biomass is one of the leading sinks for both wet and dry deposits of Hg. Forests, livestock, and fields are covered by biomass (Pirrone et al. 2010). Though mercury is transported from biomass to the environment, more Hg is retained in plant seeds, roots, radicals, and soil. The most important Hg pools are located in soils within the terrestrial environment (Grigal 2003, Obrist et al. 2012). According to World Soil Carbon Resource extraction, Hg pools are estimated to be approximately 300 Gg (Hararuk et al. 2013). Other than the possible dissolution of mercury compounds on the ocean floor, the world's oceans have no clear pollution to the earth's surface. Transportation in freshwater bodies undergoes

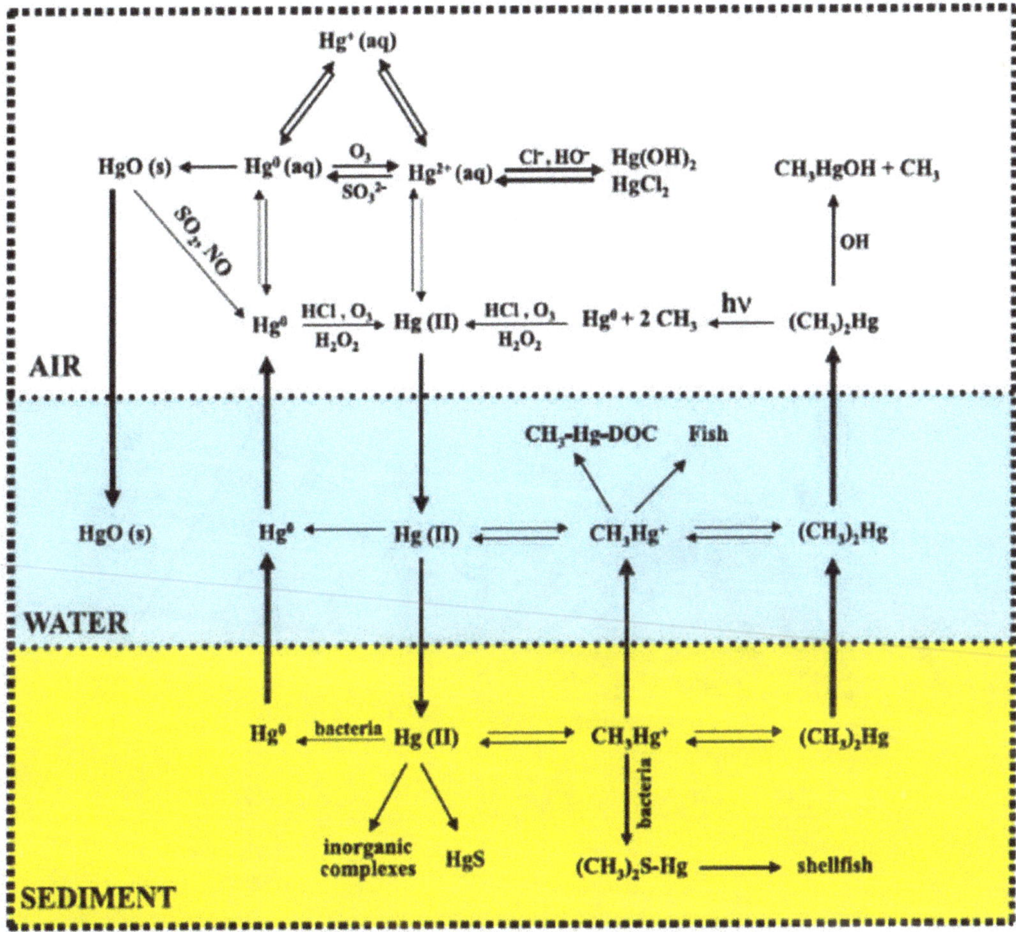

Fig. 13.2. Shows a schematic diagram of mercury cycling in an Earth system.

corrosion, and Hg gets released and gets mixed in canals and rivers. Mercury is also deposited on the bottom of the lakes and ponds (Araujo et al. 2017).

13.3.2 Freshwater

Lakes, rivers, canals, ponds, glaciers, ice caps, groundwater, etc., are categorized as freshwater bodies. The primary sources of Hg for glaciers, icecaps, and icebergs are the atmosphere and discharge in rivers, canals, oceans, etc., whereas Hg contamination in groundwater is very limited as scarce literature is available on the presence of Hg in groundwater. Mercury in the freshwater bodies is contributed by geogenic processes, biological activities, vegetation cover, accumulation of organic sediments, climatic variations, etc. (Nasr and Arp 2017). The prevalent Hg origins include clear Hg effluent hardening, stream drainage comprising Hg deposits accreted in terrestrial habitats, and Hg accumulation in the atmosphere. Later lakes are particularly important with large surface-to-volume ratios and low catchment-to-lake surface areas (Obrist et al. 2018). Water is commonly used in various industrial processes which contaminate lakes and rivers with Hg. Ultimately, wastewater from such factories is released into water sources or on open surfaces. Most anthropogenic mercury waste is concentrated in freshwater bodies. In freshwater bodies, pharmacological property converts HgO and other mercury into different forms of methyl mercury (MeHg). Artisanal and small-scale

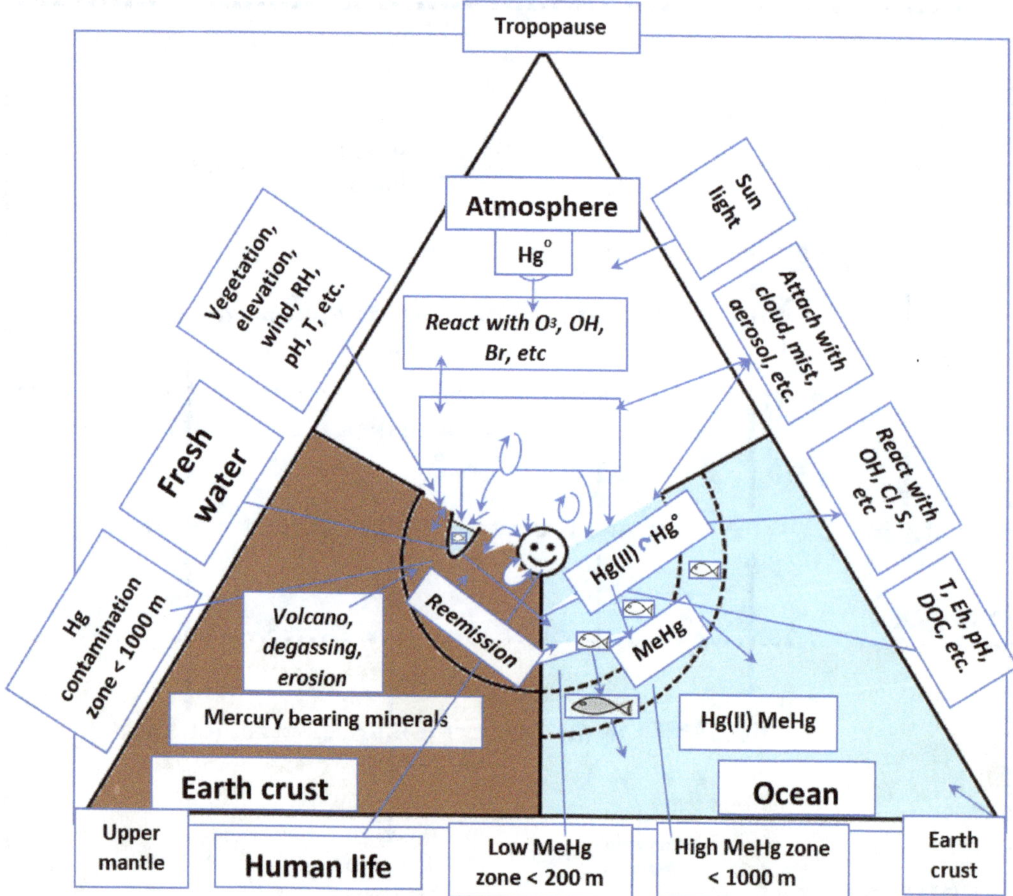

Fig. 13.3. A comprehensive view of the mercury cycle.

mining are also reasons for exposure to mercury in the aquatic ecosystem. The Hg water intake for marine environments is calculated to be between 800 and 2200 mg/year (Kobal et al. 2017, Kocman et al. 2017). The Indian and Chinese industrial Hg outlets (Chlor-Alkali, Hg, large-scale gold mining, non-ferrous metalworking, Hg waste) release about 86 Mg/year into lakes and waterways and account for its presence in water reserves (Kocman et al. 2017). Some Hg moves with runoff water from the terrestrial landscape into aquatic ecosystems. Even when the annual runoff flux is combined with the yearly atmospheric exchange flux, the total amount is much smaller than the store of Hg in the soils of the landscape (UN Environment 2019, Wu et al. 2016).

13.3.3 Ocean

The mercury emissions from anthropogenic activities are released to air, rivers, lakes, oceans, open ground, etc. Similarly, natural processes such as volcanic emissions result in a release to air or water bodies. The ocean does not contaminate the earth's crust or freshwater bodies directly. Some other path to mercury persistence and pollution is human consumption of seafood from the oceans. As the largest sink of natural and anthropogenic Hg emissions, the oceans directly or indirectly play a major role in the mercury cycle. Oceans also release most of the atmospheric mercury. Hg is found on the ocean's surface as Hg0, Hg (II), and PHg. Hg0 is less prone and erratic. Hg concentration in the air-water partition is extremely low at 0.117 (Gworek et al. 2016). Hg quickly spreads from the surface water to the atmosphere, or undergoes conversion into other chemical processes, or is absorbed as methyl mercury (MeHg) in diverse microbes. In the oxidizing environment, Hg (II) reacts with the ions OH and Cl. However, in the reduction reaction, Hg (II) reacts with S (Gworek et al. 2016). Mercury from the ocean does not directly contaminate the Earth's crust or freshwater bodies. The intake of seafood by humans is another pathway for mercury pollution. Based on the concentration of MeHg, there are three zones in the ocean. The epipelagic zone extending up to 200 m has a lower level of MeHg. The microbes which convert mercury to MeHg are consumed by organisms at different trophic levels leading to the bioaccumulation of Hg in tissues of other organisms. MeHg concentration is comparatively high in the mesopelagic zone as the fish's prey is already accumulated with MeHg. This is because of mercury bioaccumulation in the food web. The Hg (II) reaction is confined to S in the deep ocean with a decrease in the atmosphere and fewer cells. Any mercury in the water exists either as inorganic Hg (II) or as organic forms such as methyl mercury, where the latter forms due to microbial activity (Horowitz et al. 2017). The ocean's gross mercury is 350 g mercury with a trust level of 270–450 g (UN Environment 2017). Maximum oceanic mercury deposition contributes to 5500 ± 2700 Mg/year (Amos et al. 2014).

On the other hand, the ocean is the largest supplier of atmospheric Hg in the world, when elemental mercury is removed from the ocean by around 2900 mg/year (range 1900–4200 mg/year). Stern et al. (2012) revised various changes that will take place and are expected to take place in the future in the Arctic Ocean. However, they also underlined current restrictions on achieving several simultaneous effects on the Hg cycle and the process of bioaccumulation (Stern et al. 2012).

13.3.4 Atmosphere

Atmosphere hosts almost all emissions from the earth's surface, freshwater bodies, oceanic surface, and anthropogenic emissions. As mentioned earlier, Hg occurs in the air in three main forms: gaseous elemental mercury (GEM), divalent mercury (RGM), and particulate mercury (PHg), respectively (Fu et al. 2012, Karthik et al. 2017). Atmospheric Hg is also classified as an entirely different material, along with total gaseous mercury (TGM) and particulate-based mercury (PBM or PHg). The atmospheric Maximum Gaseous Mercury (MGM) concentration levels were measured in the 1970s, with reliable observations from ~ 1990 (Lindberg et al. 2007, Temme et al. 2007). Further, TGM will be split into Hg (GEM or Hg0) and gaseous Hg (RGM or Hg^{2+}) (Ci et al. 2012, Schroeder and Munthe 1998). The elemental mercury is essentially harmless in its vapor form. Therefore, it flies long stretches before being settled on the surface until it gets released into the atmosphere. HgOs react in a way that is transformed into Hg(II), Ozone, OH radicals, Br radicals, etc. Sunlight converts Hg(II) to HgO. Mercury is present in the form of particulate mercury (PHg) in clouds, dust, dung, humidity, aerosols, etc. The bulk of anthropogenic mercury Hg(II) is released. Since Hg's adsorption coefficient in soot type is large in crude oil, Hg can easily be set in soot (Pirrone et al. 2000). The release of this mercury or its interaction with the particulate matter in the atmosphere and the earth's deposits is uncertain. Mercury is most certainly extracted through the quantization of suspended particles from two primary outlets. Mercury is the first outlet for particulate matter immediately generated from factories and power plants. The second cause of Hg correlated with particulate matter is gas or aqueous (Goss and Schwarzenbach 1998) adsorption of HgO or Hg (II). Hg (II) and PHg easily get deposited (within two weeks) to the ocean or land. The removal of HgO from the atmosphere is through oxidation with the formation of Hg (II). While the oxidized deposition of Hg (II) in the oceans is dominant, HgO deposition is prevalent in several terrestrial ecosystems (Selin and Jacob 2008, Sunderland and Mason 2007). Naturally, over several years, the Hg flow has increased by atmospheric deposition through geochemical reserves and also because of human activity, including mining and fossil fuel combustion. Mean Total Gaseous Mercury (TGM) concentrations in the background, severe in the northern hemisphere, tropics, and southern hemisphere, are 1.3–1.6, 1.1–1.3, and 0.8–1,1 Ng m^{-3} (Slemr et al. 2011, Sprovieri et al. 2016).

Dissolution of HgO in rainwater or adsorption with atmospheric aerosol results in precipitation and deposition in soil or vegetation or dissolves in the water body. Elemental mercury is additionally present within the atmospheric water, like fog or mist. However, the elemental mercury is less reactive with water. According to Henry's theorem, aqueous mercury remains in equilibrium with the gas process or reacts to compounds present in water. PHg and Hg(II) species are deposited on the land or sea, as they are less volatile. Besides, mercury forms complexes with atmospheric water SO_3^{2-}, OH^-, Cl^- ions (Jiang et al. 2013, Ma et al. 2014). Temperature-increasing sulfides significantly promote divalent mercury reduction (Chang et al. 2017). Hgo responses are triggered by the development of species Hg(II) with oxidants such as Br, O_3, and OH (Bieser et al. 2017, Jiao and Dibble 2017). The Br reduction process first appeared to be a two-stroke procedure, by which Hgo reacts to HgBr with Br radical and HgBr reacts to developing entirely distinct Hg (II) with different specific oxidants (Br, I, OH, BRO, IO, NO_2, etc.) (Gratz et al. 2015, Hsu-Kim et al. 2018). The global reserve of Hg is approximately 4,400 and 5,300 mg (Streets et al. 2017). The approximate magnitude of ambient mercury has risen by 3 to 5-fold from the baseline levels compared to 1850 (Amos et al. 2013, Engstrom et al. 2014, Goss and Schwarzenbach 1998). The gross estimated input (i.e., total anthropogenic, natural, and re-emission) of direct phylogeny Hg emissions in the environment amounts to approximately 13% (Pirrone et al. 2010, Streets et al. 2017).

13.3.5 Human life

Human life is involved in various anthropogenic activities related to mercury, including food consumption of MeHg contaminated seafood. Several anthropogenic processes such as mercury production, coal, oil and gas burning, intentional biomass burning, cement industries, chlor-alkali industries, biomedical industries, dentistry, batteries, LED, etc., release mercury into the atmosphere. Mercury is used for various reasons. Anthropogenic activities play an important role in the emission, transport, and deposition of mercury. Although mercury has been reported to be used by humans for more than 4000 years, its use impacted the industrial revolution of 1850. Numerous anthropogenic activities, such as fossil fuel combustion, and a considerable amount of mercury are released into the environmental pollutants (Kwon and Selin 2016). Liquid and solid wastes contaminated with mercury are discharged into freshwater bodies or open surfaces. Humans are consuming MeHg-contaminated seafood, including fishes all over the globe. At present, mercury is mined in China, Mexico, Indonesia, and the Republic of Kyrgyzstan. Peru and other countries are extracting too little mercury. Mercury mining between 1850 and 2010 is estimated to be 720 Gg (Streets et al. 2011). It is calculated that the last 500 years of mercury extraction was 923 Gt (Hylander and Meili 2003). According to the study (Pirrone and Mason 2009, UN Environment 2017), the gross global mercury production in 2015 is reported to be between 1630 and 2150 mg. Mercury can be a residual contaminant in nonferrous ores, especially in zinc, gold, and copper ores, and in some cases in silver and antimony. Mercury by-products, in particular, are calomel (Hg_2Cl_2) or metallic mercury may be captured during other phases of the mining process on granular activated carbon. The size will range from 765–1455 mg (UNEP 2013). This produces at least 30 to 100 mg of mercury from natural gas. Mercury collected from the oil and gas field in 2015 is estimated to be approximately 20–30 mg (UN Environment 2017). In 2016, when a global gas supply of 3.59 trillion cubic meters is considered, for a 359 kg mercury, 0.1 μg Nm^{-3} will be its density, whereas 50 μg/Nm^3 will represent total mercury of 180 mg Nm^{-2}.

Mercury recovered as well as marketed by the mercury electronic versions is estimated to be 370 to 450 mg in 2015. Mercury in chlor-alkali cells and associated warehouses is estimated to be 10,000 Mg (UN Environment 2017). To effectively monitor the mercury level in susceptible populations as well as the general public, it is crucial to choose adequate indicators. Biomarkers besides human mercury exposure are blood, hair, and urine, which are economic indicators for the population in general screening (United Nations Environment Programme and World Health

Organization 2008) and guide to identifying individuals at risk from mercury exposure. United Nations Environment Plan (UNEP) and Global Mercury Assessment 2013 reported that dental mercury accounted globally for 270–341 metric tons in 2010 (AMAP/UN Environment 2019), which represents 10 percent of total mercury use (AMAP Assessment 2015, Veiga and Baker 2004). Mercury extraction is one of the major sources of mercury in China, UK, United States, Australia, etc. The industry has gained attention after the introduction of mercury restrictions in Europe and North America (Den Hond et al. 2015). The quantities of mercury recycled from various plants are estimated to vary between 1,040 and 1,410 mg (UN Environment 2017). The highest contribution was estimated to be 1540 Gt (Streets et al. 2017) from the cumulative manufacturing of Hg in air, ground, and water from 1850 to 2010. The natural process is the single largest source of emission of mercury per annum. Anthropogenic activities are one-third of the total emissions of mercury (Kwon and Selin 2016). There are several factors related to the management of mercury emissions, mobility, transport, and accumulation. A large model is provided for the awareness of the mercury cycle (Fig. 13.3). The information is to assess the long-term impact of human activities, together with those affected by NOx-specific and different climate change policies, on global Hg cycling. Hg-induced toxicological symptoms include tremors, nervousness, neuromuscular changes, headaches, and at very high concentrations causes kidney, lungs, and thyroid dysfunction, changes in vision, and deafness. Inorganic mercury compounds adversely affect the internal organs, for example, kidneys (United Nations Environment Programme and World Health Organization 2008). Biogeochemical circulations of the natural and anthropogenic Hg differ in principle, but at least interfere with the progress of natural processes and alter them and make them similar to the processes taking place in ecosystems contaminated with anthropogenic Hg. For the year 2015, the global anthropogenic emission of Hg was estimated to have risen 205 times more than 2010 (UN Environment 2019). Anthropogenic activities have augmented the concentrations of total atmospheric mercury to about 450 percent beyond natural values. This rise is mirrored in mercury masses of some aquatic food-webs exhibiting levels of concern for environmental and public health. It is a universal contaminant that travels vast distances and accumulates in the environment, thus polluting diverse ecosystems and posing significant risks to human health. It adversely affects the populations that are exposed to various sources of mercury emissions, such as urban and industrial sites. To take adequate prevention measures, it is of critical importance to widen the mercury monitoring networks and to enhance the collaborative initiatives globally.

13.4 Hg Remediation

Hg pollution in soil and water is primarily caused by various point sources of mercury contamination. Hg can be found at different levels of the food chain due to its bioaccumulation in significant amounts (Selin 2009). WHO advises a weekly dose of 1.6 mg g^{-1} of CH_3Hg; according to the Environmental Protection Agency (EPA), adults may consume 0.1 mg g^{-1} of body weight per day. Because mercury and other heavy metals cannot be degraded in the environment, bioremediation techniques should be applied to the removal/immobilization. Adsorption, desorption, oxidation, reduction, stability, and confinement are all mechanisms used in removal technologies. The primary goal of these technologies is to extract mercury from the polluted area or to convert harmful mercury species into less dangerous mercury species (Hinton and Veiga 2001). Stabilization and confinement are the most extensively used immobilization strategies, which inhibit the mobility of mercury through physical trapping or chemical complexation, respectively (Bengtsson and Picado 2008).

13.4.1 Hg Remediation of Soil

Because of its unique physical and chemical features, mercury is both persistent and volatile. It can enter the soil by a number of routes and exists in various forms in the soil, resulting in substantial soil pollution. As a result, soil Hg pollution treatment and remediation have got a lot of attention. The physical and chemical features of soil, as well as the occurrence of Hg, all have an impact on remediation efficiency. As a result, a suitable remediation strategy should be applied based on the soil quality and morphology of Hg in the soil. Remediation techniques of Hg polluted soils can be classified into chemical, physical, and biological treatment.

13.4.1.1 Physical treatment

Physical treatment is based on a number of factors, the most important of which are:

I. Physical separation
II. Soil vapor extraction
III. Soil replacement
IV. Fixed/stabilized soil
V. Vitrification
VI. Thermal desorption
VII. Electrokinetic remediation technology

Physical treatment requires injecting chemical reagents into the soil to improve the cleanup efficiency. Thermal desorption cleans polluted soil by heating the pollutants based on the soil's low boiling point directly or indirectly, further separating them. Thermal desorption is a two-step technology.

I. Tail gas treatment
II. Heating contaminated soil to volatilize pollutants

The volatility of contaminants is the fundamental premise for thermal treatment remediation. Mercury is the only heavy metal that exists as a liquid at ambient temperature. HgO has melting and boiling temperatures of 38.8°C and 356.7°C, respectively, as well as a vapor tension of 0.18 Pa, indicating that it is a volatile metal. With increasing temperature, the rate of vaporization increases. In contaminated soil, thermal desorption may transform Hg into gas and be collected by heating. Thermal desorption can be applied to remove Hg from polluted soil (He et al. 2015).

Heavy metal contaminated soils can be treated effectively with electrokinetic remediation. The basic technique is to insert two inert electrodes on both ends of polluted soil and supply voltage to create an appropriate electric field gradient, which allows the heavy metals present in the soil to be fixed by electromigration, electrophoresis, or electro-osmosis (Teng et al. 2020).

13.4.1.2 Chemical treatment

Chemical remediation is a method for removing contaminants that uses chemical reagents, reactions, and principles. These treatments normally induce contaminants to degrade, eliminate or reduce soil toxicity. The most prevalent chemical cleanup treatments are currently:

I. Soil washing,
II. Chemical stabilization,
III. Oxidation
IV. Reduction
V. Reduction dechlorination
VI. Solvent extraction
VII. Soil performance improvement remediation technology

Soil washing is a remediation technique that uses a variety of chemical reagents to remove contaminants from the soil (Teng et al. 2020). Soil washing reagents help pollutants to dissolve or migrate to the soil. While in use, the removal process can be separated into two types:

- *In situ* leaching—in this process, leaching solution is poured into damaged soil before recovering the heavy metals from the filtrate.
- *Ex situ* leaching—before undergoing cleaning by liquid treatment and other procedures, the polluted soil should be mined and placed in the appropriate treatment facility (Khalid et al. 2017).

The method of Hg remediation in the soil through soil washing, alkali, acids, and chelating agents are widely used to remove water-soluble Hg. The dissolution of Hg-containing components adsorbed by soil components determines the use of chelating agents or alkali, although acidic reagents can eliminate Hg by complexation (Wang et al. 2016, Song et al. 2017).

Commonly used eluents are:

- Organic chelating reagents (citric acid, humic acid, oxalic acid, malonic acid, malic acid, and fulvic acid).
- Salt leaching reagents (NaCl, KI, and $Na_2S_2O_3$).
- Acidic leaching reagents (HCl, H_2SO_4, HNO_3, and H_3PO_4).
- Artificial chelating reagents (Methyl Glycine-N, N-diacetic acid (MGDA), Nitrilotriacetic acid (NTA), Diethylene Triamine Penta Acetic acid (DTPA), Ethylene diamine tetra acetic acid (EDTA), Ethylene Diamine Disuccinate (EDDS).

Chemical stabilization includes adding chemical reagents or materials to the soil to immobilize heavy metals by precipitation, complexation, and adsorption processes, which decrease the mobility and bioavailability of heavy metals (Teng et al. 2020). Chemical stabilization cannot eliminate or remove contaminants from the soil but is able to considerably decrease their concentration in pore water and solubility/mobility, lowering the rate at which heavy metals are transferred to microbes, plants, and water (Shen et al. 2009).

HgS is a sulfide that is easily formed and maintains its stability under reducing conditions. Since then, some specialists have conducted additional research on the viability of CMC–FeS (carboxymethylcellulose) nanoparticles for mercury removal, discovering that CMC–FeS nano compounds exhibit greater stability and a greater affinity to Hg. Other materials that have been investigated as fixatives include basic materials, activated carbon, biochar, organic matter, phosphate, layered silicate minerals, and ferromanganese oxide compounds.

13.4.1.3 Biological treatment

Bioremediation is a method that uses biological elements to decrease, eliminate, or immobilize hazardous chemicals in the soil and purify it (Xu et al. 2015). Bioremediation is broadly divided into two classes.

- Microbial treatment
- Plant treatment (phytoremediation)

Plants and their associated rhizospheric bacteria are used in phytoremediation to remove or immobilize the contaminants in soils, sediments, subsurface, and surface water (Dermont et al. 2008). Thus, Hg absorption and accumulation by plant roots, as well as interactions between Hg and rhizosphere components, can reduce the bioavailability of Hg and its mobility in soil. Precipitation, metal valence reduction, complexation, and adsorption are the major methods of phytostabilization for polluted soil remediation (Ghosh and Singh 2005). Phytostabilization technology, in general, does not eliminate Hg from the soil but transforms the Hg into different states so that it gets accumulated and precipitated in the roots, reducing the mobility of Hg in the soil.

Microorganisms are ubiquitous and have been reported to possess the ability to survive in heavy metal polluted areas; hence, microbes are crucial in the conversion of hazardous heavy metals to harmless ones (Tajudin et al. 2016). Microorganisms that are Hg resistant have a *mer operon* and can efficiently manage Hg contamination in different environments (Teng et al. 2020). *Bacillus thuringiensis* PW-05, having *mer-operon,* was isolated from the coast of Orissa. Simultaneously, AAS investigations and real-time PCR monitoring evidenced that the bacteria have the ability to bioremediate Hg (Marques et al. 2009). Chemical extraction was followed by a microbial reduction in a two-stage method. The findings of the experiments indicate that $(NH_4)_2S_2O_3$ is an effective extracting agent that does not alter bacterial growth. Therefore, combining chemical extraction with bioremediation appears to be a feasible remediation strategy (Sharma et al. 2021).

13.4.2 Hg Removal from Water

13.4.2.1 Hg removal by biochar

Biochar-based mercury removal is a cutting-edge method due to its high removal efficacy, environmental friendliness, and cost-effectiveness (Boszke et al. 2007). The fundamental process of adsorption is the chemical bonding of Hg(II) to active sites of the adsorbent surface (eSH, eNH_2, eOH, eCOOH). According to X-ray absorption data, Hg binds to S in sulfur-rich biochar and Cl and O in sulfur-poor biochar.

13.4.2.2 Hg removal by algae based biomasses

The use of marine macroalgae to remove mercury from saline waters is a cost-effective method (Henriques et al. 2015). Green algae (Phaeophyta), Brown algae (Phaeophyta), and red algae (Rhodophyta) are the three types of marine algae (Chlorophyta). Henriques et al. (2015) compared the Hg (II) removal effectiveness of three varieties of algae and observed that *Ulva lactuca*, a green macroalgae with numerous functional groups such as -OH, $-NH_3$, S, and > CO, performed the best.

13.4.3 Hg Removal from Air

13.4.3.1 Treatment by adsorbents

The use of solid adsorbents in the flue duct is one of the viable solutions, as it is reported as a direct technique for the removal of mercury. Researchers have studied an extensive range of materials having potential properties for the removal of Hg over the last few decades. The two excellent adsorbent have been discovered for good adsorption ability (Zhao et al. 2017a); they are metal oxides and activated carbons (Sjostrom et al. 2010).

Treatment by Activated carbons (AC): Activated carbon is a versatile and widely used sorbent. Several AC manufacturing and functionalization methods are available for a variety of processes. It's surface and textural qualities, as well as other properties, are linked to the removal of mercury. 100 mg of the prepared adsorbent was employed in the mercury removal studies (Fan et al. 2010). A combination of O_2, N_2, NO, CO_2, and SO_2 were used as the carrier gas, with a flow speed of 1000ml min^{-1}. The experiment was carried out at 150 °C with a 200 min test time. The Hg concentration was initially fixed at 20 g/m^3. The AC has a Hg removal effectiveness ranging from 60% to 90%.

Treatment by Zeolites: Zeolites are recognized research materials. Adsorption, ion exchange, and catalytic characteristics are some of the significant properties they possess (Bachu 2008). The crystal structure of zeolite is formed by the spatial bonding of SiO_4 and AlO_4 tetrahedra, also known as primary building units (PBU). According to the results of the HgO removal trials (Wdowin et al. 2014), the incorporation of Ag to the zeolites increased the Hg sorption.

Treatment by Metal-organic frameworks (MOFs): MOFs are structurally varied materials prepared by metal ions and bridging ligands (organic linkers) that are coordinatively bonded and highly porous in nature. In pen metal sites of UiO-66 and MOF Mil-101(Cr) materials were used to remove Hg^0 from flue gas. The sample UiO-66 had an efficiency of around 50 percent, whereas the sample MIL-101(Cr) had an extremely high mercury removal efficiency of over 85 percent (Zhao et al. 2017b).

13.5 Conclusion

This chapter is an effort to review the global cycling of Hg which has progressed in the recent past, concentrating on processes that occur within diverse environmental reservoirs. New constraints on the size of reservoirs and fluxes are possible as a result of the emergence of large global data sets and the improvement of models and analytical techniques. Hg is a natural element found in the environment and has originated from the Earth's crust layer. Hg can neither be formed nor destroyed but natural and anthropogenic activities can restructure mercury with possible dangerous health effects. Hg is a potent neurotoxin of global concern due to its chemical properties. It is a universal contaminant that travels vast distances and accumulates in the environment, thus polluting ecosystems and posing significant risks to human health. It also affects the populations exposed to the source of mercury emissions, such as urban and industrial sites. To take adequate prevention measures, it is of critical importance to upscale mercury monitoring networks and enhance global collaborations. The chapter thoroughly discussed the biogeochemical cycles in context of mercury in different ecological systems.

13.5.1 Future Perspectives and Research Needs

In the future, researchers investigating the mercury cycle from around the world to monitor the presence of mercury in ecological setup, including food chain activities, should continue to boost the efficiency of pollution control initiatives in the future. Researchers have filled a large surveillance vacuum and offered the scientific community a genuinely global view of current mercury emission rates. Several countries had no provisions on mercury monitoring before the Global Mercury Observation System (GMO), etc. To analyze future emissions reduction policies and their effectiveness, continuous monitoring and measurement of total mercury data will be carried out. A regional surveillance system for mercury accumulation in surface air and deposition at different altitudes should be established on websites. The global community has acknowledged mercury's environmental and safety issues. The worldwide monitoring and disposal of mercury depend on specific knowledge on mercury production, leaks, absorption, distribution, and quantities in organisms and the atmosphere. This also seeks to recognize weaknesses in assessing wireless reach and scale, especially in terms of regional and institutional network organization. United Nations Environment Programme and World Health Organization recognizes the importance of mercury monitoring efforts in other media platforms such as water and soil and highlights the importance of compiling such information in the future. To provide sustainability and a long-term source of monitoring capacity in the regions, the focus should be on the increasing number of national initiatives. In both hotspot and remote regions, developing new national human biomonitoring networks and/or expanding global networks is required. A clear picture of how the stimulation and intensification of mercury in living creatures are accomplished from the mercury emissions measurement device would be helpful. On the other hand, the extension and development of baselines will simplify preventive actions to minimize and exterminate global mercury emissions. Meanwhile, the existence of global or regional networks (e.g., GMOs, AMAP, or DEMO COPHES) and several local scientific studies (e.g., the GBMS database) attests to the ongoing efforts to upscale monitoring networks at the global level as well as to enhance the collaborative networks. Support from the global community and governments is welcome and should be encouraged during the development of a complete and more exhaustive overview in the future. Moreover, the quality assessment of the presented networks that track mercury levels in water and soil is critical and should be performed as a follow-up action to develop a better understanding.

Acknowledgments

The first author is grateful to the Key Laboratory of western China's Environmental System, College of Earth, and Environmental Sciences for a post-doctoral researcher (Award No: 252813, dated 10/04/2020). National Key Research Project and Development Program of China and the State Administration of Foreign Experts Affairs (SAFEA) provided financial support for Ramamoorthy Ayyamperumal's academic visit to Lanzhou University.

References

AMAP. 2015. AMAP Assessment 2015: Human Health in the Arctic, Arctic Monitoring and Assessment Programme (AMAP), Oslo, Norway.

AMAP/UN Environment. 2019. Technical Background Report for the Global Mercury Assessment 2018. Arctic Monitoring and Assessment Programme, Oslo, Norway.

Amos, H. M., D. J. Jacob, D. G. Streets and E. M. Sunderland. 2013. Legacy impacts of all-time anthropogenic emissions on the global mercury cycle. Global Biogeochem. Cycles 27: 410–421.

Amos, H. M., D. J. Jacob, D. Kocman, H. M. Horowitz, Y. Zhang and S. Dutkiewicz et al. 2014. Global biogeochemical implications of mercury discharges from rivers and sediment burial. Environ. Sci. Technol. 48: 9514–9522.

Araujo, B. F., H. Hintelmann, B. Dimock, M. G. Almeida and C. E. Rezende. 2017. Concentrations and isotope ratios of mercury in sediments from shelf and continental slope at Campos Basin near Rio de Janeiro, Brazil. Chemosphere. 178: 42–50.

Bachu, S. 2008. CO_2 storage in geological media: Role, means, status and barriers to deployment. Prog. Energy Combust. Sci. 34: 254–273.

Beal, S. A., M. A. Kelly, J. S. Stroup, B. P. Jackson, T. V. Lowell and P. M. Tapia. 2014. Natural and anthropogenic variations in atmospheric mercury deposition during the Holocene near Quelccaya Ice Cap, Peru. Global Biogeochem. Cycles. 28: 437–450.

Bengtsson, G. and F. Picado. 2008. Mercury sorption to sediments: Dependence on grain size, dissolved organic carbon, and suspended bacteria. Chemosphere. 73: 526–531.

Bieser, J., F. Slemr, J. Ambrose, C. Brenninkmeijer, S. Brooks, A. Dastoor et al. 2017. Multi-model study of mercury dispersion in the atmosphere: Vertical and interhemispheric distribution of mercury species. Atmos. Chem. Phys. 17: 6925–6955.

Bishop, K., J. B. Shanley, A. Riscassi, H. A. de Wit, K. Eklöf, B. Meng et al. 2020. Recent advances in understanding and measurement of mercury in the environment: Terrestrial Hg cycling. Sci. Total Environ. 72: 137647.

Boszke, L., A. Kowalski, A. Astel, A. Barański, B. Gworek and J. Siepak. 2007. Mercury mobility and bioavailability in soil from contaminated area. Environ. Geol. 55: 1075–1087.

Burke, M. P., T. S. Hogue, M. Ferreira, C. B. Mendez, B. Navarro, S. Lopez et al. 2010. The effect of wildfire on soil mercury concentrations in southern california watersheds. Water Air Soil Pollut. 212: 369–385.

Caldwell, C. A., P. S. And and E. Prestbo. 2006. Concentration and dry deposition of mercury species in arid south central New Mexico (2001–2002). Environ. Sci. Technol. 40: 7535–7540.

Canil, D., P. W. Crockford, R. Rossin and K. Telmer. 2015. Mercury in some arc crustal rocks and mantle peridotites and relevance to the moderately volatile element budget of the Earth. Chem. Geol. 396: 134–142.

Chang, L., Y. Zhao, H. Li, C. Tian, Y. Zhang, X. Yu, et al. 2017. Effect of sulfite on divalent mercury reduction and re-emission in a simulated desulfurization aqueous solution. Fuel Process. Technol. 165: 138–144.

Choi, H.D. and T.M. Holsen. 2009. Gaseous mercury emissions from unsterilized and sterilized soils: The effect of temperature and UV radiation. Environ. Pollut. 157: 1673–1678.

Ci, Z., X. Zhang and Z. Wang. 2012. Enhancing Atmospheric Mercury Research in China to Improve the Current Understanding of the Global Mercury Cycle: The Need for Urgent and Closely Coordinated Efforts. Environ. Sci. Technol. 46: 5636–5642.

Den Hond, E., E. Govarts, H. Willems, R. Smolders, L. Casteleyn, M. Kolossa-Gehring, et al. 2015. First steps toward harmonized human biomonitoring in Europe: Demonstration project to perform human biomonitoring on a European scale. Environ. Health Perspect. 123: 255–263.

Dermont, G., M. Bergeron, G. Mercier and M. Richer-Laflèche. 2008. Soil washing for metal removal: A review of physical/chemical technologies and field applications. J. Hazard. Mater. 152: 1–31.

Engstrom, D. R., W. F. Fitzgerald, C. A. Cooke, C. H. Lamborg, P. E. Drevnick, E. B. Swain et al. 2014. Atmospheric Hg emissions from preindustrial gold and silver extraction in the americas: a reevaluation from lake-sediment archives. Environ. Sci. Technol. 48: 6533–6543.

Environment and Climate Change Canada - Canada.ca, URL https://www.canada.ca/en/environment-climate-change.html (accessed 9.5.21).

Fan, X., C. Li, G. Zeng, Z. Gao, L. Chen, W. Zhang et al. 2010. Removal of gas-phase element mercury by activated carbon fiber impregnated with CeO_2. Energy Fuels. 24: 4250–4254.

Fu, X., X. Feng, J. Sommar and S. Wang. 2012. A review of studies on atmospheric mercury in China. Sci. Total Environ. 421: 73–81.

Giang, A., L. C. Stokes, D. G. Streets, E. S. Corbitt and N. E. Selin. 2015. Impacts of the minamata convention on mercury emissions and global deposition from coal-fired power generation in Asia. Environ. Sci. Technol. 49: 5326–5335.

Ghosh, M. and S. P. Singh. 2005. A review on phytoremediation of heavy metals and utilization of it's by products. Asian J. Energy Environ. 6: 18.

Goss, K. U. and R. P. Schwarzenbach. 1998. Gas/solid and gas/liquid partitioning of organic compounds: critical evaluation of the interpretation of equilibrium constants. Environ. Sci. Technol. 32: 2025–2032.

Government of Canada, Mercury: chemical properties. URL https://www.canada.ca/en/environment-climate-change/services/pollutants/mercury-environment/about/chemical-properties.html (accessed 9.5.21).

Gratz, L. E., J. L. Ambrose, D. A. Jaffe, V. Shah, L. Jaeglé, J. Stutz et al. 2015. Oxidation of mercury by bromine in the subtropical Pacific free troposphere. Geophys. Res. Lett. 42: 10494–10502.

Gray, J. E., P. M. Theodorakos, D. L. Fey and D. P. Krabbenhoft. 2015. Mercury concentrations and distribution in soil, water, mine waste leachates, and air in and around mercury mines in the Big Bend region, Texas, USA. Environ. Geochem. Health. 37: 35–48.

Grigal, D. F. 2003. Mercury sequestration in forests and peatlands. J. Environ. Qual. 32: 393–405.

Gworek, B., O. Bemowska-Kałabun, M. Kijeńska and J. Wrzosek-Jakubowska. 2016. Mercury in marine and oceanic waters—a review. Water Air Soil Pollut. 227: 1–19.

Gworek, B., W. Dmuchowski and A. H. Baczewska-Dąbrowska. 2020. Mercury in the terrestrial environment: a review. Environ. Sci. Eur. 32: 1–19.

Hararuk, O., D. Obrist and Y. Luo. 2013. Modelling the sensitivity of soil mercury storage to climate-induced changes in soil carbon pools. Biogeosci. 10: 2393–2407.

He, F., J. Gao, E. Pierce, P. J. Strong, H. Wang and L. Liang. 2015. *In situ* remediation technologies for mercury-contaminated soil. Environ. Sci. Pollut. Res. 22: 8124–8147.

Henriques, B., L. S. Rocha, C. B. Lopes, P. Figueira, R. J. R. Monteiro, A. C. Duarte et al. 2015. Study on bioaccumulation and biosorption of mercury by living marine macroalgae: Prospecting for a new remediation biotechnology applied to saline waters. Chem. Eng. J. 281: 759–770.

Hinton, J. and M. Veiga. 2001. Mercury contaminated sites: A review of remedial solutions. Proc. NIMD (National Institute for Minamata Disease) Forum 2001. Mar. 19–20, 2001, Minamata, Japan. http://www.mcilvainecompany.com/Decision_Tree/subscriber/Tree/DescriptionTextLinks/Minamata_Forum_2001.pdf..

Holmes, C. D., D. J. Jacob, E. S. Corbitt, J. Mao, X. Yang, R. Talbot et al. 2010. Global atmospheric model for mercury including oxidation by bromine atoms. Atmos. Chem. Phys. 10: 12037–12057.

Horowitz, H. M., D. J. Jacob, Y. Zhang, T. S. DIbble, F. Slemr, H. M. Amos et al. 2017. A new mechanism for atmospheric mercury redox chemistry: Implications for the global mercury budget. Atmos. Chem. Phys. 17: 6353–6371.

Hsu-Kim, H., C. S. Eckley, D. Achá, X. Feng, C. C. Gilmour, S. Jonsson et al. 2018. Challenges and opportunities for managing aquatic mercury pollution in altered landscapes. Ambio 47: 141–169.

Hylander, L. D. and M. Meili. 2003. 500 years of mercury production: global annual inventory by region until 2000 and associated emissions. Sci. Total Environ. 304: 13–27.

Jiang, Y. Z., C. M. Chen, L. X. Jiang, S. T. Liu and B. Wang. 2013. Study of mercury re-emission from simulated wet flue gas desulfurization liquors. Adv. Mater. Res. 610: 2033–2037.

Jiao, Y. and T. S. Dibble. 2017. First kinetic study of the atmospherically important reactions $BrHg^{\cdot}$ + NO2 and $BrHg^{\cdot}$ + HOO. Phys. Chem. Chem. Phys. 19: 1826–1838.

Karthik, R., A. Paneerselvam, D. Ganguly, G. Hariharan, S. Srinivasalu, R. Purvaja et al. 2017. Temporal variability of atmospheric Total Gaseous Mercury and its correlation with meteorological parameters at a high-altitude station of the South India. Atmos. Pollut. Res. 8: 164–173.

Khalid, S., M. Shahid, N. K. Niazi, B. Murtaza, I. Bibi and C. Dumat. 2017. A comparison of technologies for remediation of heavy metal contaminated soils. J. Geochem. Explor. 182: 247–268.

Kobal, A. B., J. Snoj Tratnik, D. Mazej, V. Fajon, D. Gibičar, A. Miklavčič et al. 2017. Exposure to mercury in susceptible population groups living in the former mercury mining town of Idrija, Slovenia. Environ. Res. 152: 434–445.

Kocman, D., S. J. Wilson, H. M. Amos, K. H. Telmer, F. Steenhuisen, E. M. Sunderland et al. 2017. Toward an assessment of the global inventory of present-day mercury releases to freshwater environments. Int. J. Environ. Res. Public Health. 14: 138.

Kwon, S. Y. and N. E. Selin. 2016. Uncertainties in atmospheric mercury modeling for policy evaluation. Curr. Pollut. Rep. 22: 103–114.

Lindberg, S., R. Bullock, R. Ebinghaus, D. Engstrom, X. Feng, W. Fitzgerald et al. 2007. A synthesis of progress and uncertainties in attributing the sources of mercury in deposition. Ambio. 36: 19–32.

Ma, Y., H. Xu, Z. Qu, N. Yan and W. Wang. 2014. Absorption characteristics of elemental mercury in mercury chloride solutions. J. Environ. Sci. 26: 2257–2265.

Marques, A. P. G. C., A. O. S. S. Rangel and P. M. L. Castro. 2009. Remediation of heavy metal contaminated soils: phytoremediation as a potentially promising clean-up technology. Crit. Rev. Environ. Sci. Technol. 39: 622–654.

Nasr, M. and P. A. Arp. 2017. Mercury and organic matter concentrations in lake and stream sediments in relation to one another and to atmospheric mercury deposition and climate variations across Canada. J. Chem. 2017.

Obrist, D., D. W. Johnson and R. L. Edmonds. 2012. Effects of vegetation type on mercury concentrations and pools in two adjacent coniferous and deciduous forests. J. Plant Nutr. Soil Sci. 175: 68–77.

Obrist, D., J. L. Kirk, L. Zhang, E. M. Sunderland, M. Jiskra and N. E. Selin. 2018. A review of global environmental mercury processes in response to human and natural perturbations: Changes of emissions, climate, and land use. Ambio 47: 116–140

Pacyna, J. M. 2020. Recent advances in mercury research. Sci. Total Environ. 738: 139955.

Pirrone, N., I. M. Hedgecock and L. Forlano. 2000. Role of the ambient aerosol in the atmospheric processing of semivolatile contaminants: A parameterized numerical model (Gas-Particle Partitioning (GASPAR)). J. Geophys. Res. Atmos. 105: 9773–9790.

Pirrone, N. and R. Mason. 2009. Mercury fate and transport in the global atmosphere: Emissions, measurements and models. Springer-Verlag, New York.

Pirrone, N., S. Cinnirella, X. Feng, R. B. Finkelman, H. R. Friedli, J. Leaner et al. 2010. Global mercury emissions to the atmosphere from anthropogenic and natural sources. Atmos. Chem. Phys. 10: 5951–5964.

Rytuba, J. 2003. Mercury from mineral deposits and potential environmental impact James. Environ. Geol. 43: 326–338.

Schroeder, W. H. and J. Munthe. 1998. Atmospheric mercury—An overview. Atmos. Environ. 32: 809–822.

Selin, H., S. E. Keane, S. Wang, N. E. Selin, K. Davis and D. Bally. 2018. Linking science and policy to support the implementation of the Minamata Convention on Mercury. Ambio 47: 198–215.

Selin, N. E. and D. J. Jacob. 2008. Seasonal and spatial patterns of mercury wet deposition in the United States: Constraints on the contribution from North American anthropogenic sources. Atmos. Environ. 42: 5193–5204.

Selin, N. E. 2009. Global biogeochemical cycling of mercury: a review. Annu. Rev. Environ. Resour. 34: 43–63.

Selin, N. E. 2014. Global change and mercury cycling: Challenges for implementing a global mercury treaty. Environ. Toxicol. Chem. 33: 1202–1210.

Sharma, R., T. Jasrotia, S. Sharma, M. Sharma, R. Kumar, R. Vats et al. 2021. Sustainable removal of Ni(II) from waste water by freshly isolated fungal strains. Chemosphere. 282: 130871.

Shen, C., Y. Chen, S. Huang, Z. Wang, C. Yu, M. Qiao et al. 2009. Dioxin-like compounds in agricultural soils near e-waste recycling sites from Taizhou area, China: Chemical and bioanalytical characterization. Environ. Int. 35: 50–55.

Sjostrom, S., M. Durham, C. J. Bustard and C. Martin. 2010. Activated carbon injection for mercury control: Overview. Fuel. 89: 1320–1322.

Slemr, F., E. G. Brunke, R. Ebinghaus and J. Kuss. 2011. Worldwide trend of atmospheric mercury since 1995. Atmos. Chem. Phys. 11: 4779–4787.

Song, B., G. Zeng, J. Gong, J. Liang, P. Xu, Z. Liu et al. 2017. Evaluation methods for assessing effectiveness of *in situ* remediation of soil and sediment contaminated with organic pollutants and heavy metals. Environ. Int. 105: 43–55.

Sprovieri, F., N. Pirrone, M. Bencardino, F. D'Amore, F. Carbone, S. Cinnirella et al. 2016. Atmospheric mercury concentrations observed at ground-based monitoring sites globally distributed in the framework of the GMOS network. Atmos. Chem. Phys. 16: 11915–11935.

Stein, E. D., Y. Cohen and A. M. Winer. 1996. Environmental distribution and transformation of mercury compounds. Crit. Rev. Environ. Sci Technol. 26: 1–43.

Stern, G. A., R. W. Macdonald, P. M. Outridge, S. Wilson, J. Chételat, A. Cole et al. 2012. How does climate change influence arctic mercury? Sci. Total Environ. 414: 22–42.

Streets, D. G., M. K. Devane, Z. Lu, T. C. Bond, E. M. Sunderland and D. J. Jacob. 2011. All-time releases of mercury to the atmosphere from human activities. Environ. Sci. Technol. 45: 10485–10491.

Streets, D. G., H. M. Horowitz, D. J. Jacob, Z. Lu, L. Levin, A. F. H. Schure et al. 2017. Total mercury released to the environment by human activities. Environ. Sci. Technol. 51: 5969–5977.

Sunderland, E. M. and R. P. Mason. 2007. Human impacts on open ocean mercury concentrations. Global Biogeochem. Cycles 21: 4022.

Tajudin, S. A. A., M. A. M. Azmi and A. T. A. Nabila. 2016. Stabilization/solidification remediation method for contaminated soil: a review. IOP Conf. Ser. Mater. Sci. Eng. 136: 012043.

Temme, C., P. Blanchard, A. Steffen, C. Banic, S. Beauchamp, L. Poissant et al. 2007. Trend, seasonal and multivariate analysis study of total gaseous mercury data from the Canadian atmospheric mercury measurement network (CAMNet). Atmos. Environ. 41: 5423–5441.

Teng, D., K. Mao, W. Ali, G. Xu, G. Huang, N.K. Niazi. et al. 2020. Describing the toxicity and sources and the remediation technologies for mercury-contaminated soil. RSC Adv. 10: 23221–23232.

UN Environment. 2017. Toolkit for Identification and Quantification of Mercury Releases Reference Report and Guideline for Inventory Level 2, Geneva, Switzerland.

UN Environment. 2019. Global Mercury Assessment 2018 | UNEP - UN Environment Programme. Geneva, Switzerland.

United Nations Environment Programme. 2013. Global Mercury Assessment 2013: Sources, emissions, releases, and environmental transport. Geneva, Switzerland.

United Nations Environment Programme, World Health Organization. 2008. Guidance For Identifying Populations at Risk from Mercury Exposure. Geneva, Switzerland.

Veiga, M. and R. Baker. 2004. Global mercury project. Protocols for environmental and health assessment of mercury released by artisanal and small-scale gold miners. Vienna (Austria) UNIDO.

Wang, H., H. Song, R. Yu, X. Cao, Z. Fang and X. Li. 2016. New process for copper migration by bioelectricity generation in soil microbial fuel cells. Environ. Sci. Pollut. Res. 23: 13147–13154.

Wdowin, M., M. M. Wiatros-Motyka, R. Panek, L. A. Stevens, W. Franus and C. E. Snape. 2014. Experimental study of mercury removal from exhaust gases. Fuel. 128: 451–457.

Wu, Q., S. Wang, L. Zhang, M. Hui, F. Wang and J. Hao. 2016. Flow analysis of the mercury associated with nonferrous ore concentrates: implications on mercury emissions and recovery in China. Environ. Sci. Technol. 50: 1796–1803.

Xu, J., A. G. Bravo, A. Lagerkvist, S. Bertilsson, R. Sjöblom and J. Kumpiene. 2015. Sources and remediation techniques for mercury contaminated soil. Environ. Int. 74: 42–53.

Zhao, L., Y. Huang, H. Chen, Y. Zhao and T. Xiao. 2017a. Study on the preparation of bimetallic oxide sorbent for mercury removal. Fuel 197: 20–27.

Zhao, S., Y. Duan, T. Yao, M. Liu, J. Lu, H. Tan et al. 2017b. Study on the mercury emission and transformation in an ultra-low emission coal-fired power plant. Fuel 199: 653–661.

Section III

Biotechnological Strategies for Remediation of Toxic Metal(loid)s

CHAPTER 14

Deciphering the Role of Metal Binding Proteins and Metal Transporters for Remediation of Toxic Metals in Plants

Harsimran Kaur,[1] *Sukhmeen Kaur Kohli,*[2] *Kanika Khanna,*[2] *Shalini Dhiman,*[2] *Jaspreet Kour,*[2] *Tamanna Bhardwaj*[2] and *Renu Bhardwaj*[2,*]

14.1 Introduction

Excessive discharge of high density metals and metalloids together referred to as heavy metals (HMs), from anthropogenic activities, industries, and natural phenomena, is a matter of concern worldwide (Hasan et al. 2017). These metals are readily transferred via food chains from one trophic level to another, resulting in their biomagnification (Jain et al. 2018, Caroli et al. 2020). Plants growing at sites contaminated with HMs tend to accumulate these metals in greater amounts, thus contaminating the food chains and acting as a route for HM entry into human tissues, exposing them to the danger of diseases including cancer (Singh et al. 2016, Lu et al. 2021).

Depending upon the physicochemical properties, bioactive-metals are divided into redox and non-redox active metals like manganese (Mn), copper (Cu), iron (Fe), chromium (Cr), nickel (Ni), cadmium (Cd), zinc (Zn), aluminium (Al) and mercury (Hg), respectively. The redox metals induce oxidative injury, leading to reactive oxygen species (ROS) generation in plants, inducing disruption of cellular homeostasis, damaged photosynthetic machinery, DNA breakage, and defragmented proteins/cell membrane, eventually triggering cell death. Non-redox active metals on the other hand inflict indirect oxidative stress by binding to proteins sulfhydryl groups, glutathione depletion, constraining antioxidative enzymes, or persuading ROS-generating enzymes (Emamverdian et al. 2015). HMs in the soil and water hinder optimal plant growth and photosynthetic efficiency (Jain et al. 2018). Plants have evolved multifarious avoidance and tolerance mechanisms to regulate their responses to HM stress. Biosynthesis of cysteine rich metal binding peptides is the core HM detoxification mechanism that immobilizes, sequesters, and detoxifies the metal ions (Viehweger 2014). Despite the significant progress in understanding the role of protein quality regulatory

[1] PG Department of Botany, Khalsa College, Amritsar, Punjab, India.
[2] Department of Botanical and Environmental Sciences, Guru Nanak Dev University, Amritsar, Punjab, India.
* Corresponding author: renubhardwaj82@gmail.com

system in plants systems, the knowledge related to HM stress remains scanty. This book chapter aims at providing improved insight into the regulatory role of proteins and some organic molecules regarding HM tolerance.

14.2 HM Remediation Approaches Employed by Plants

At toxic levels, HMs impede plant functioning and obscure metabolic processes by altering protein structure, hampering the functional groups of major cellular molecules, targeting cytoplasmic membrane integrity, and disturbing the functionality of crucial metal ions in biomolecules, including enzymes and pigments (Emamverdian et al. 2015). These changes result in the suppression of critical cellular events in plants including respiration, photosynthesis, and enzymatic reactions (Hossain et al. 2012). Besides, enhanced production of ROS and cytotoxic molecules like methylglyoxal disturbs the prooxidant-antioxidant homeostasis in plant cells leading to oxidative stress. This further results in DNA, protein, and lipid oxidation, ion leakage, redox imbalance, denatured cellular structure, and membranes, eventually activating programmed cell death (Fryzova et al. 2017). However, plants adopt two defense strategies to overcome HM toxicity, avoidance and tolerance. Employing these strategies, plants can maintain HM concentrations below the toxic threshold levels in the cells (Yan et al. 2020).

The avoidance strategy followed by plants involves limiting heavy metal uptake and restricting its movement in the plant tissues (Emamverdian et al. 2015). At extracellular levels, avoidance is the first line of defense involving an array of mechanisms like metal ion precipitation, root sorption, metal exclusion, and metal immobilization by mycorrhizal association (Dalvi and Bhalerao 2013). Upon HM exposure, plants try to restrain their entry via root sorption or metal ion modification. Root exudates, including organic compounds, organic acids, and amino acids, ligate HM resulting in the formation of stable HM complexes in the rhizosphere (Yan et al. 2020). Certain root exudates can also alter the rhizospheric pH that causes HM to precipitate. Precipitation limits HM bioavailability and thus reduces their toxic effects (Dalvi and Bhalerao 2013). Metal exclusion mechanisms create a barrier between the plant root and shoot systems and restrict the admittance of HMs into the roots, limiting their uptake and transportation to the shoot (Yan et al. 2020).

Furthermore, arbuscular mycorrhizae can also limit HM entry into roots by chelation, adsorption, or absorption of HMs within the rhizosphere (Hossain et al. 2012). Implanting HMs in cell walls is another strategic method limiting HM uptake in plants (Memon and Schröder 2009). The pectins in the cell wall contain negatively charged carboxylic groups capable of binding to HMs. For this reason, the cell wall acts as a cation exchanger and restricts HM ions from gaining entry into the cell (Yan et al. 2020).

If avoidance strategies fail to restrict HMs from gaining entry into plant tissues, in that case, the tolerance mechanisms such as metal chelation with cellular compounds, metal sequestration and compartmentalization, attachment with the cell wall, and accumulation of osmolytes is initiated (Emamverdian et al. 2015). Tolerance mechanisms are basically the second line of defense employed by plants at the intracellular level. To minimize the toxic effects of HM ions amassed inside the cytosol, plants employ detoxification strategies that involve the chelation of these ions with ligands. Chelation, mediated by several organic and inorganic ligands, helps to reduce the content of free metal ions to reasonably low levels (Yan et al. 2020). Following chelation, the HM-ligand complexes are transported into the inactive compartments, such as the vacuole, from the cytosol, wherein these non-reactive complexes are stored (Jutsz and Gnida 2015, Yan et al. 2020). Sequestration and compartmentalization effectively protect the cells from the injurious effects of HMs by removing them from the delicate sites of the cell wherein cell division, and respiration ensue, thus plummeting the connections between HM ions and cellular processes and circumventing injuries to cell functioning (Sheoran et al. 2010).

14.3 Metal Binding Proteins and HM Remediation

Metal binding proteins are detoxification ligands that reduce metal ion toxicity through complexation reactions and consequent transport and accumulation within the vacuoles. It is generally observed that metal ion uptake is facilitated in roots from soils and abridged into the metal ion-ligand complexes, favored by specialized ligands (Baker 1981). The chelators, along with metal transporters, control the sequestration of metal ions within vacuoles of both hyperaccumulators as well as non-hyperaccumulator plants. Metal binding proteins in plants comprise an array of ligands, including phytochelatins (PCs), metallothioneins (MTs), glutathione, ferritins, chaperons, amino acids both proteogenic (histidine and cysteine, etc.) and non-proteogenic (nicotianamine), and numerous other low molecular weight organic acids (Miransari 2011, Singh et al. 2019). All these metal proteins have the potential to decrease metal toxicity within different plant species. Few ligands, namely nicotianamine, MTs, and glutathione are also involved in complex formation with metal ions during mobilization as well as storage and utilized if there is any metal ion requirement. Besides, PCs are only synthesized during metal ion detoxification (Cobbett and Goldsbrough 2002). Generally, in non-hyperaccumulator plants, the metal ions bind strongly to sulfur ligands like PCs and MTs, while in the case of hyperaccumulators, they are bound to oxygen containing ligands with weaker associations. Moreover, non-thiol ligands like histidine and nicotianamine play prominent roles during hyperaccumulation (Leitenmaier and Kupper 2013). Excessive metal ions accumulated within cells also enhance the ROS that alternatively cause cellular and metabolic damages followed by oxidative stress. Alongside, there are several other mechanisms possessed by plants as defense processes in the form of stimulated activities of enzymatic and non-enzymatic antioxidants for the removal of excessive ROS (Mittler et al. 2004).

14.3.1 Phytochelatins

PCs are among the most prominent ligands associated with metal ion detoxification in plants and most suitable for effectively remediating HM pollution. Plants have been observed to synthesize minute cysteine-enriched oligomers known as PCs during the initial phase of HM stress (Pochodylo and Aristilde 2017). It is noteworthy that PC-synthesis is essential in inducing plant resistance towards HMs (Emamverdian et al. 2015). Furthermore, PC-biosynthesis is controlled at both transcriptional as well as post-translational stages in plants subjected to metal toxicity. Although the overexpression of genes encoding phytochelatin synthases, PCS, in plants do not always lead to stimulated tolerance towards metal toxicity. To illustrate, the overexpression of gene *PCS1* in *Arabidopsis thaliana* showed paradoxical hypersensitivity towards some lethal metals (Pb, Zn, and Cd); however, PCs were stimulated at the same time in contrast to wild type plants (Lee et al. 2003). In some cases, PCs in mutant plants enhance the HM-accumulation without having much effect on plant resistance. Therefore, it elucidates another activity of PCs in plants in maintaining metal ion homeostasis, defense responses, antioxidant activities, and nitrogen and sulfur metabolism (Furini 2012). Henceforth, the preclusion of freely moving metal ions within cytosol manifests the most probable mechanism for combating HM-toxicity (Hasan et al. 2017).

The mechanism of action of PCs lies in their binding action followed by their detoxification (Fig. 14.1). It entraps the metal ions by forming stable metal-PC complexes that are further moved to vacuoles for sequestration to maintain the metal homeostasis in plants (Guo et al. 2008). Various metal ions can bind to PCs and broader research is carried in this area to elaborate the same. It has also been known that disparate from MTs, PCs are not involved in regulating HM metabolism during non-toxic conditions. However, plants present hypersensitive responses when PC-synthases are inhibited, or their deficiency is triggered genetically (Singh et al. 2019). The metal ion and PC complexes are confiscated within the vacuoles, and due to acidic pH, the complexes dissociate.

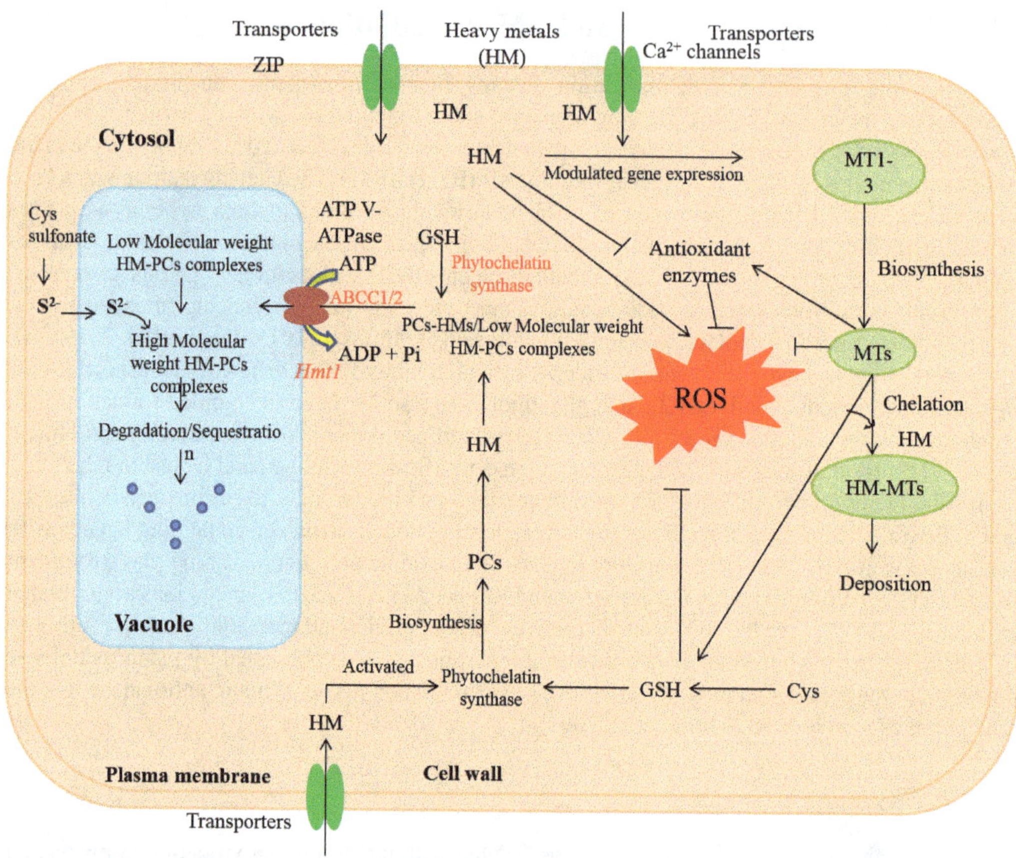

Fig. 14.1. Diagrammatic representation of the functional roles of phytochelatins (PCs) and metallothioneins (MTs) for heavy metal (HM) remediation. HMs result in the activation of phytochelatin synthases (PCS) enzyme and also activate the expression of MTs, followed by the synthesis of low molecular weight HM-PC complexes within the cytosol. Low molecular weight complexes are transferred into vacuoles *via* tonoplast through ATP-binding cassettes and V-type ATPase transporters, mainly ABCC1/2. After the compartmentalization, low molecular weight complexes integrate HM and S^{2-} to generate high molecular weight HMs-PCs complexes. MTs also initiate the detoxification process after the formation of complexes and also regulate redox homeostasis of the cell by modulating the antioxidative defense system and maintaining sufficient concentrations of cellular GSH.

Although, when there is higher complex formation within vacuoles, they may return to cytoplasm followed by degradation (Hartley-Whitaker et al. 2001).

Nevertheless, the PC-mediated HM-detoxification mechanism involves not only chelation but also covers the various aspects of metal accumulation, stabilization, and complexation within the vacuoles by forming PC-metal complexes of higher molecular weight (Furini 2012). Interestingly, metal sequestration is the most effective strategy mainly employed by the organisms for the amelioration of toxic ions. The metal ions complexed are transported towards vacuoles from cytosol with the aid of a range of transporters. Hence, vacuolar sequestration is a crucial mechanism adopted by the plants to regulate HM-homeostasis that is facilitated by ATP-dependent vacuolar pumps such as V-ATPases and VPPases along with abundant tonoplast transporters (Sharma et al. 2016). It is noteworthy that HM-transporter 1 (HMT1) was discovered back in 1995 in *Saccharomyces pombe*, and it is quite an effective PC-transporter that enables accumulation and complexation of Cd metal ions within vacuoles (Mendoza-Cozatl et al. 2011). *HMT1* gene mainly encodes ABC-membranal transporter proteins; therefore, HMT1 along with ATP are necessary elements for the transference of low molecular weight Cd-PC complexes inside the vacuoles. Subsequently, two ABC-transporters, namely, ABCC1 and ABCC2, were also observed as metal ion-PC complex vacuolar transporter in *A. thaliana* and *S. pombe*, respectively (Park et al. 2012). Further, the experiments conducted using double mutants determined that for effective and complete metal detoxification of hazardous metal ions, vacuolar sequestration through ABCC1 and ABCC2 is essential in *Arabidopsis* plants. Moreover, these metal ions also presumably enhance the transport process and trigger PC-synthesis and its transport towards vacuoles as observed in barley. Therefore, it suggests that PCs are both indulged in maintaining metal detoxification in plants and regulating the homeostasis of key metal ions within the cells (Song et al. 2017). Apart from this, some studies provide evidence that phosphorylation induced the regulatory action of ABCC1 critical for vacuolar confiscation of metal ions. Ser^{846}-phosphorylation is a pre-requisite for providing HM tolerance to plants, as exemplified by the studies conducted in *Arabidopsis* in regard to ABCC1 functioning (Zhang et al. 2017).

14.3.2 Metallothioneins and Ferritins

Metallothioneins (MTs), naturally occurring cysteine enriched low molecular weight metal-binding proteins, are synthesized intracellularly within plants (Capdevila and Atrian 2011). Normally, MTs function in plants to maintain metal homeostasis. During stress conditions, they immobilize, sequester, and detoxify metal ions (Fig. 14.1) (Singh et al. 2019). These are mainly synthesized through the translation of mRNA-encoding genes. Moreover, their expression level is also linked to metal ion concentration; therefore, they can be used as effective markers during stress conditions (Morris et al. 1999). Even though MTs were discovered about decades back, their physiological functions are still being elucidated. The prominent role of MTs comprises their active functioning during maintaining metal homeostasis, HM-sequestration, and provide protection from oxidative damage to plants under metal stress (Hossain et al. 2012, Hasan et al. 2017). The role of transitional metals, namely Fe, Zn, Mg, Mn, Cu, Mo, B, etc., is crucial for living organisms as they are critically important for carrying out various physiological attributes. For instance, in plants, Cu is essential for photosynthetic processes, ethylene synthesis, ROS-homeostasis, respiration, and cell wall synthesis (Peñarrubia et al. 2010, Hasan et al. 2017). The role of MTs in regulating metal ion homeostasis, specifically for Cu needed in vegetative growth as well as senescence, has been widely reported. Furthermore, it has also been determined that MT-deficient mutants lead to limited Cu accumulation in plant organs in contrast to wild-type plants, whilst there was no differentiation among the life cycles of wild-type and mutants during varied growth conditions. During the initial vegetative phase, no variation in Cu uptake of both categories of plants (wild type and mutants) was observed. The levels of Cu were doubled in the plant organs of MT-deficient plants compared to wild types. Contrary to this, Cu-levels in seeds of these plants were nearly half the concentration of wild-type seeds (Benatti et al. 2014).

Generally, intracellular HM sequestration in plants also requires the attachment of metal ions onto cysteine-enriched MTs (Sácký et al. 2014). The combinatorial action of elevated thermodynamics along with lower stability and kinetics forms the underlying characteristics of metal ion-MTs complexation. These features lead to firm binding of metals, and a segment of the metal ions is replaced very easily for peptides or proteins for proper binding (Maret 2000). Transgenic plants show overexpression of genes encoding MTs for inducing heavy metal resistance, and they depict a better mechanism of metal accumulation and distribution (Tomas et al. 2015). Vacuoles are the main sites for metal detoxification in plants. Besides, the metal ion chelation by MTs is well established, along with its mechanism for metal-MTs complexes transport from cytoplasm towards vacuoles (Yang et al. 2011). Resultantly, it is advocated that metal ions are not transferred into vacuoles, and *ThMT3* regulates HM amassment in the cytosol. The biological activities of MTs are quite complex. Several studies have reported that along with chelation and metal ion homeostasis, MTs also control the redox homeostasis of cells during unfavorable conditions (Kang 2006). Moreover, HM stress also triggers ROS accumulation in plants that further causes damage towards different biomolecules, specifically protein denaturation, to hinder various metabolic and physiological parameters along with causing many disorders and cell death (Hasan et al. 2017).

Strikingly, MTs have displayed a specific mechanism for protecting plants from stresses and oxidative damage (Ansarypour and Shahpiri 2017). Several studies demonstrate the role of MTs in inducing stress forbearance by acting as scavengers of ROS and regulating ROS homeostasis. During the ROS scavenging process, metal ions are set free from MTs, and ROS moieties are attached to cysteine residues. In addition, many studies have advocated that metal ions may also initiate signaling cascade prior to ROS removal (Hassinen et al. 2011). For instance, under usual conditions, the cell functions through Zn mobilization followed by the transfer from one site to another. The release of Zn from MT during oxidative reaction occurs through a normal pathway where Zn is distributed in cells or restricted during oxidative stressed conditions. Also, Zn is crucial for antioxidative responses, suggesting MTs participate in ROS-homeostasis and protect cells, biomolecules, and macromolecules from ROS (Kang 2006). Furthermore, there are several classes of MTs in plants that possess distinctive functions with tissue-specific gene expression. GUS reporter depicted that *MT1a* and *MT2b* are observed to be expressed in phloem, while *MT2a* and *MT3* within the mesophyll cells as well as root tips, respectively (Hassinen et al. 2011). Alongside, it was documented that *MT2c* gene encoding MT2 is expressed in roots, root tips, leaf sheaths, and leaves of *Oryza sativa*, while downregulation of the expression was observed in seeds (Liu et al. 2015). Taking together both their diverse roles as well as tissue specific expression patterns, it was demonstrated by Irvine et al. (2017) that a lower cost MT-biosensor can be developed, which can induce signal with metal ions. Therefore, simplified sensor technology can be prominently used for monitoring the environment, specifically in regions with metal pollution.

Additionally, ferritins are broadly defined as proteins with the ability to store a large amount of Fe^{2+} and Fe^{3+} in an orderly manner. Apart from metal ion storage, they also induce the detoxification of metal ions. Ferritins belong to the ubiquitous and multimeric family of Fe-storing proteins that enable the sequestration of numerous iron atoms (Harrison and Arosio 1996). In plants, ferritins store only iron, whilst in animals, they can store various metal ions, including Cd, Al, As, Zn, Pb, Cu, etc. (Dedman et al. 1992). These are produced in plants under severe toxic conditions, and genes encoding is regulated by various antioxidants and abscisic acid. The genes encoding these proteins are basically located in plastids. The expression of these genes is regulated and activated by numerous environmental signals, and excessive Fe is one such instance. Depending upon their responses towards Fe-signals, two pathways have been proposed for regulating ferritin related genes in plants. These are controlled by the action of different antioxidants and also serine and threonine phosphatases inhibitors (Savino et al. 1997). Thus, these proteins act as protective agents in plants as they prevent the reactions among free radicals triggered by Fe-ions (Ravet et al. 2009). Moreover, plant ferritins also constitute a significant component during oxidative stress responses and the most important moieties for protecting plastids during oxidative damage through storage of excess of Fe. With the induction in ferritin levels, excess of Fe also induces the concentrations of dehydroascorbate and reduces ascorbate levels. In this way, Fe-induced oxidative stress causes ferritin accrual as well as elevated Fe-uptake by the proteins.

14.3.3 Natural Resistance-Associated Macrophage Proteins (NRAMPs)

NRAMP family constitutes various conserved membrane proteins, including Nramp1 and Nramp2 (also known as DCT1 or DMT1), Smf1p, Smf2p, and Smf3p (Nramp homologous yeast proteins), located in the cells of root apices. These are actively involved in transporting various heavy metals ions like Cd, Co, Pb, and Ni in a wide array of organisms (Yan et al. 2020). Such kinds of evolutionarily conserved membrane proteins are found in diverse kingdoms like plants, animals, fungi, and bacteria. They assist in the transference of lethal metal ions in plants through cellular and vacuolar membranes (Jain et al. 2018). The proton motive force or proton gradient around the cell membranes is responsible for transporting metal-ion through the transporter (Singh et al. 2016). NRAMPs are characterized by 10–12 transmembrane domains with broad metal specificity. Thus, different metals can interact with the same protein (Caroli et al. 2020).

In *Arabidopsis*, NRAMPs controls Fe and Cd transfer, wherein NRAMP1 protein is believed to maintain its metabolic level (Thomine et al. 2000). Previously, NRAMPs were linked with Fe uptake and its transportation in biological systems. Several studies indicate that certain NRAMPs in cells are involved in the inward and outward movement of cations (Singh et al. 2016). NRAMP1 confined to the plasma membrane was also found to restore the capability of Fe-regulated transporter1 mutant for uptake of Co and Fe, specifying that the protein exhibited wide selectivity for HMs (Cailliatte et al. 2010). Besides, NRAMPs are also found to be intricated in the transportation of non-essential HMs. Certain NRAMPs are located on cell membranes, whereas others reside on the vacuolar membranes, emphasizing their probable role in regulating cytoplasmic metal ion concentrations or discharging these ions from the vacuole (Caroli et al. 2020). In rice, it was discovered that OsNRAMP3, localized on the plasma membrane, retorts to environmental alterations in Mn availability. When present in excess, Mn is accumulated in the vesicles and despoiled during post-translational regulation in response to mineral nutrient accessibility in the environment (Socha et al. 2014).

14.3.4 Metal Tolerance Proteins (MTPs)

MTPs are divalent cation transporters, also referred to as cation diffusion facilitators (CDF), involved mainly in HM tolerance, metals homeostasis, and maintenance of mineral nutrients inside plants. MTPs provide metal homeostasis mainly through cation efflux from the cell (Caroli et al. 2020). These are categorized into four discrete groups, with the group I and III being extremely crucial. MTPs are mainly involved in the transportation of HM cations, including Co^{2+}, Ni^{2+}, Zn^{2+}, etc., from the cytosol into the vacuole (Singh et al. 2016). Most of the MTPs are present on the vacuolar membrane and work as Cd^{2+}, Zn^{2+}, and Ni^{2+} antiporters (Ovečka and Takáč 2014) or are positioned in subcellular locations, including endomembrane compartments associated with trans-Golgi or pre-vacuolar organelles (Caroli et al. 2020). AtMTP11, being exceedingly present in leaf hydathodes, is implicated not only in the vesicular transference of Mn but also in exocytosis of the superfluous ions in secretory tissues instead of being stored (Gustin et al. 2011, Socha et al. 2014).

MTP1, localized at plasma and the vacuolar membrane, is a Zn^{2+}/H^{+} antiporter implicated in both Zn accretion and forbearance (Yan et al. 2020). In *Thlaspi goesingense*, MTPs are complexed in the storage of Ni in vacuoles (Persans et al. 2001). High expression of MTPs in non-hyperaccumulators under the influence of HMs suggests their probable role in metal forbearance (Singh et al. 2016). Nearly twelve MTP genes are well characterized in *A. thaliana*, and about ten such genes have also been reported in *O. sativa* (Gustin et al. 2011). The first MTP gene reported and characterized in *A. thaliana* was referred to as Zinc Transporter 1 gene (*ZAT1*) but was retitled as metal tolerance protein 1 (AtMTP1) (Singh et al. 2016). In *A. halleri*, Zn hyperaccumulator, the *AhMTP1* gene is assumed to participate in Zn hyper-tolerance (Shahzad et al. 2010). Another MTP family member, AtMTP11, is reported to transport Mn, besides providing Mn tolerance (Peiter et al. 2007).

14.3.5 CDPKs (Calcium Dependent Protein Kinases)

CDPKs is a group of proteins present at the plasma membrane that regulate key functional roles during HM signaling (Keyster et al. 2020). Intercellular Ca^{2+} activation occurs in response to various stress stimuli, including HM stress, which further leads to activation of various downstream signaling pathways. These sensory proteins in plants belong to the multigene family that directly bind Ca^{2+} before phosphorylating substrates involved during stress signaling responses in plants (Kumar and Trivedi 2016). Activation of CDPKs by transient Ca^{2+} levels regulates various downstream signaling involved in metal sequestration, transportation, metal tolerance, and its metabolism inside plants (Hasanzadeh and Hazrati 2021). CDPKs activation under HM stress results in regulation of downstream proteins, including transcription factors, membrane channels, and NADPH oxidase, all of which individually execute imperative functions. Thus, sustaining appropriate CDPK levels under HM stress becomes vital for plants (Keyster et al. 2020). Exogenous supply of Ca^{2+} improved HM stress tolerance ability and enhanced the performance of antioxidants enzymes (Ahmad et al. 2016). Ca^{2+} is also stated to enhance antioxidant enzymatic activity and provide Cr tolerance in *Setaria italica* (Fang et al. 2014).

During stress, CDPKs overexpression increases plants' stress tolerance (Boudsocq and Sheen 2013). Up-regulated CDPK gene expression has been described in *S. italica* seedlings under the influence of Cr (Fang et al. 2014) and in *Cucurbita pepo* leaves under Ni stress (Valivand et al. 2019). Huang et al. (2014) also suggested that CDPKs enhancement with an increase in Cr stress reflects the role of Ca^{2+} in rice roots exposed to HM stress. Additionally, Yeh et al. (2007) described the importance of CDPKs in the activation of MAPKs during Cd and Cu stress, indicating a relation amongst CDPKs and MAPKs for signal transduction under HM constraint (Keyster et al. 2020).

14.3.6 MAPKs (Mitogen Activated Protein Kinases)

MAPK cascade, comprising three-tier components (MAPKKK, MAPKK, and MAPK), gets activated during the metal stress in the plants that further regulate the signal transduction cascade involved in phytohormone biosynthesis (Jonak et al. 2002, Jalmi et al. 2018). In eukaryotes, the MAPK cascade is implicated in extracellular signal transduction and targets intracellular proteins (Li et al. 2006). Among several abiotic stressors, HMs exhibit an intense effect on the MAPK signaling pathways. ROS generated under HM stress activate MAPKs (Jonak et al. 2004, Jalmi and Sinha 2015). MAPKs can phosphorylate several transcription factors, including bZIP, MYB, ABRE, DREB, WRKY, MYC, and NAC, thereby influencing HM stress response (Tiwari and Lata 2018). Even though plenty of reports depict the role of activated MAPKs in retortion to HMs like Cu, As, Cd, Pb, Fe, and Zn, a detailed MAPK signaling cascade under such conditions remains elusive (Jalmi et al. 2018).

Tobacco plants overexpressing MAPK from *Lycium chinense* exhibited augmented tolerance towards Cd stress (Guan et al. 2016). Additionally, regulation of root growth by MAPKs by further impelling auxin signaling and cell cycle-associated gene expression in rice exposed to Cd stress was observed by (Zhao et al. 2013). Another study revealed that four different types of MAPKs (SAMK, SIMK, MMK3, MMK2) are stimulated by Cd and Cu in Alfalfa (Jonak et al. 2004). In the case of Arabidopsis and rice, Cd induces ATMEKK1 and OsMAPK2, respectively (Shao et al. 2008). Pb stress results in upregulation of four different MAPKs, including MAPK6, MAPK18, MAPK20, and MAPKKK7 in radish (Wang et al. 2013). MAPK can also upregulate the activity of downstream antioxidant enzymes under Cu stress, highlighting the function of MAPK in ROS signal intensification via the antioxidative enzyme system (Liu et al. 2018). Besides, the sulfate assimilation pathway, jasmonic acid, and ethylene signaling pathways are regulated by some of the MAPKs under HM stress (Ahsan et al. 2008, Opdenakker et al. 2012).

14.4 Metal Transporters and HM Remediation

Several genes with specific expression patterns are present in the plant genome and involved in HM uptake and transport in plants. They ensure an appropriate and adequate supply of nutrients and HM in plants for various cellular processes. The members of these gene families encode an array of molecules that include metal ion transporters. Specified transporters are positioned at the root cell plasma membrane and are vital for HM ion uptake. These can transport explicit metals through the cell membranes, thus facilitating metal translocation into the shoots from the roots (DalCorso et al. 2019). Upon accretion of HM ions in the cytosol, these metal ions are then routed into cellular organelles so as to eliminate the HMs from the plant cells. In this course, plants employ proteins as transporters for the allocation of HM ions (Jain et al. 2018). Metal transporters are classified into numerous families, including HMA (Heavy Metal ATPase), MATE (Multidrug and Toxic Compound Extrusion), and ZIP (ZRT and IRT like protein). These transporters are indirectly or directly complexed in the accumulation and transference of HMs (Takahashi et al. 2012). Besides, they also maintain metal homeostasis by dividing these HM ions between several cellular organelles and compartments (Clemens et al. 2002).

14.4.1 HMA Transporters

HMA transporters have been comprehensively analyzed and are reported to be localized in the cellular and vacuolar membrane and the cell endomembrane system. They transport the metal ions across the membranes upon ATP hydrolysis (Caroli et al. 2020). ATP pumps charged substrates across the membranes of the cell. They are distinguished from others due to the formation of intermediate compounds that are charged in the reaction cycle (Lombi et al. 2001). The plant HMAs are categorized into two sets. The first group is intricated in the transportation of monovalent ions, whereas the second group transports the divalent ions (Liu et al. 2017). The P1B-type ATPases of the HMA transporter family are diverse in regard to subcellular localization, tissue distribution, and metal specificity. Based on metal specificity, two subgroups of these transporters are known: the Cu/Ag group and the Zn/Co/Cd/Pb group (Takahashi et al. 2012). These transporters are essential for metal ion homeostasis and forbearance (Williams and Mills 2005, Yan et al. 2020).

In rice, nine HMA genes have been recognized, amongst which *OsHMA1–OsHMA3* encodes for Zn/Co/Cd/Pb subgroup transporters, whereas *OsHMA9* encodes for Cu/Ag subgroup HMA transporters (Takahashi et al. 2012). Vacuolar HMA3 compartmentalizes Zn, Co, Cd, and Pb by sequestering them in the vacuole (Hanikenne and Baurain 2014). Besides, the HMA4 transporter carried out translocation of Cd and Zn from HMA4 overexpression augmented efflux of Cd and Zn in the xylem vessels from root symplasm and encouraged metal tolerance (Verret et al. 2004). SpHMA3, specific Cd transporter localized in the tonoplast and exceedingly expressed in leaves and stems of *Sedum plumbizinicola*, sequesters Cd in the vacuoles resulting in Cd detoxification (Liu et al. 2017).

14.4.2 MATE Transporters

MATE transporters are widely found in archaea, bacteria, and eukaryotes (Dong et al. 2019). MATE family is a group of membrane-restricted efflux transporters elaborated in the excretion of exogenous and endogenous harmful compounds that accumulate in cells (Piddock 2006, Ghori et al. 2019). Most of the characterized MATE transporters export compounds carrying positive charges. Besides, the MATE family is reported in toxic HM efflux and detoxification (Huang et al. 2021). MATE is reported as a secondary carrier as well. It utilizes the energy in the electrochemical gradient to transport the HM against the concentration or electrochemical gradient (Hvorup et al. 2003). AtALF5 in *A. thaliana* was the first MATE transporter to be identified (Diener et al. 2001). In Cloud *Solanum tuberosum StMATE33* is speculated to play a vital role in responses towards Cd^{2+} stress; *StMATE18/60/40/33/5* was up-regulated during Cu^{2+} constraint, whereas *StMATE59* (II) was remarkably persuaded under Ni^{2+} stress (Huang et al. 2021). In Arabidopsis, overexpression of *GsMATE* encoding extrusion transporter conferred enhanced tolerance towards Al toxicity in transgenic plants by plummeting Al accretion in *Arabidopsis* roots (Ma et al. 2018).

14.4.3 ZIP Transporters

ZIP transporters execute a chief function in metal ion movement in plants (Caroli et al. 2020). Some of these proteins are associated with Zn and Fe transport in roots and shoots of *A. thaliana* during HM toxicity. ZIP transporters are involved in the translocation of several ions, including Zn, Fe, Cd, and Mn, contingent with the substrate type and their peculiar identity (Jain et al. 2018). ZIP family transports HMs from organelles or extracellular spaces into the cytoplasm. 15 different genes viz. *ZIP1-12* and *IRT1-3* of the ZIP family are described in Arabidopsis. Amongst these, AtIRT1 is well-studied in plants and known for its contribution in regulating plant Zn and Fe homeostasis (Milner et al. 2013).

Assunção et al. (2001) stated that overexpression of genes belonging to the ZIP family encodes ZTN1 and ZTN2 transporters that stimulated augmented uptake of Zn in *T. caerulescens*. López-Millán et al. (2004) reported six different ZIP family transporters, including MtZIP1, MtZIP3, MtZIP4, MtZIP5, MtZIP6, and MtZIP7 in legume *Medicago truncatula* that function in heavy metals transport. OsZIP3, present in the nodes of rice plants, unloads Zn ions into the vascular bundles via the xylem, whereas translocation of Zn from root to shoot involves OsZIP4, OsZIP5, and OsZIP8 (Sasaki et al. 2015). Mulberry ZIP1 and ZIP5 are intricated in Cd transport (Fan et al. 2018). Amusingly, Fe transporters including IRT1 and IRT2 can uptake Cd in *O. sativa* and *A. thaliana* (Takahashi et al. 2012). Besides, certain ZIP transporters are accountable for metal ion dissemination into the cellular compartments. AtIRT2, located in the endo-membrane vesicles, contributes towards Fe transport into these cellular compartments (Migeon et al. 2010). Similarly, OsZIP1, present both on the plasma membrane as well as endoplasmic reticulum, limits Cu, Cd, and Zn amassing in rice plants (Liu et al. 2019).

14.5 Other Proteins Stimulated under Metal Toxicity

14.5.1 Heat Shock Proteins (HSPs)

Plants synthesize a group of proteins called HSPs by expressing stress genes under unfavorable conditions (Gupta et al. 2010). HSP proteins have low molecular weight ranging between 10–200 KD and are categorized as chaperones, which contribute towards tolerance under stress conditions. HSPs play a crucial role in the proper folding of misfolded proteins or denatured proteins (Park and Seo 2015). These proteins are expressed constitutively in plant cells and help in the folding of newly synthesized polypeptides. HSP70 and co-chaperones like DnaJ ensure proper protein folding before reaching their target location (Park and Seo 2015). HM stress causes disturbed cellular homeostasis by disrupting critical enzymes and destroying protein functioning. Thus, the stimulation of HSPs is thought to be an essential protective, adaptive, and genetically preserved response towards environmental stress in organisms (Hasan et al. 2017). HSP70 family members are widely expressed in response to metal-induced stress in an array of plant genera (Gupta et al. 2010). Although several reports revealed that overexpression of HSP70 genes positively correlates with tolerance acquisition towards HM stress, the cellular functioning of HSP70 functioning during stress is not clear (Hasan et al. 2017). HSP60 chaperons also have a protective role. Rodríguez-Celma et al. (2010) reported that HSP60 inhibits protein denaturation even though HM ions were present in the cytosol.

Differential regulation of HSP26.13p and Cp-sHSPs, chloroplastic HSPs, in *Chenopodium album* provide protection to plants from stress induced by metals like Cu, Ni, and Cd (Haq et al. 2013). Cai et al. (2017) reported a solid association between HSP regulated HM forbearance and melatonin biosynthesis in plants, which is further controlled by heat-shock factor A1a (HsfA1a). Markedly, maximum HSPs are stimulated during the early stages of cellular procedures that follow lethal exposure to HMs much below the fatal dosage and thereby behave as putative biomarkers for environmental contamination (Hasan et al. 2017). As one of the protein quality regulator systems, HSPs maintain optimal cellular machinery functioning under environmental cues.

14.5.2 Glutathione

In plants, glutathione (GSH) is an abundant low molecular weight thiol tripeptide. It is involved in sequestering HMs, providing protection from ROS and several other cellular events. This water-soluble tripeptide with a sulfhydryl group (SH) scavenges metals in plants (Talke et al. 2006). The enzyme glutathione reductase is involved in the conversion of oxidized glutathione (GSSG) to its reduced form, which possesses a disulfide bridge that tends to break under metal stress. This complex acts as an enzyme activator and also has a higher affinity to PCs than metal ions (Jain et al. 2018). GSH aids in scavenging radicals and heavy metal detoxification (Singh et al. 2015). Upon HM exposure, GSH concentration rapidly decreases in the cytoplasm, imposing an abrupt effect on the redox status of the oxidized and reduced GSH forms. Thus, in cells exposed to HM stress, a redox signature is generated (Ghori et al. 2019). Besides being an antioxidant, it is also a precursor for PC synthesis in metal-exposed plants (Fryzova et al. 2017).

GSH complexes with Cd and forms Cd-GS2 complex. Bogs et al. (2003) reported an altered BjGT1 expression in *Brassica juncea* on exposure to Cd in the growth medium. The amount of GSH increased rapidly under Cd stress since GSH acts as an initiator molecule for the synthesis of PCs. Cd stress triggers GSH biosynthesis in plants, which is employed for the synthesis of cysteine (Gupta et al. 2002). Ercal et al. (2001) also observed that plants exposed to the 100 mmol Cd/L resulted in high GSH, cysteine, and γ-glutamyl-cysteine levels. Hence, several reports suggest that GSH plays a crucial role in detoxification and degradation.

14.5.3 Organic Acids

Under HM stress, secretion of organic acids (OAs) increases. OAs possess carboxyl groups that readily chelate with HMs and form non-toxic complexes that may prevent their admittance into the plants (Yu et al. 2019). Kochian et al. (2004) reported that OAs also detoxify toxic metals like Al and Cu in the rhizosphere. OA-metal complexes are reported to be more soluble in wet soils. Several reports indicate the involvement of OAs in HM uptake by plants followed by the detoxification process. HMs bind to OAs, and the resulting complex is immobilized in the vacuoles. Citric acid, malic acid, oxalic acid, tartaric acid, and succinic acid are amongst a few OAs that have been reported to play a crucial role in imparting metal tolerance in plants. Lu et al. (2013) reported that due to the presence of citrate, there is an escalation in the transfer and accretion of Cd in the leaves of Zn/Cd hyperaccumulator *Sedum alfredii.* Another report suggested that citrate chelates Fe^{2+} during the transference of HM ions in the xylem (Durrett et al. 2007). In hyperaccumulator *Thlaspi caerulescens*, malate is responsible for Zn sequestration in the leaf epidermal cell vacuoles (Tolrà et al. 1996). Similarly, in *A. halleri*, Zn is allied with malate in the mesophyll cells (Sarret et al. 2002). In Cd hyperaccumulator, 85% of concentrated Cd is associated with malate, whereas 15% of the HM ions were complexed with citrate (Lu et al. 2013). Ni hyperaccumulation is linked to the concentration of citrate and malate in several plants. A positive correlation was observed between water-soluble Cd content and citrate in *Solanum nigrum* leaves, indicating that citrate is liable for amassment of Cd in the plant foliage (Sun et al. 2008). Anjum et al. (2015) stated that citrate serves as a high affinity molecule for metals including Co, Ni, Zn, and Cd not only in *A. hallerii* but also in other non-hyperaccumulating species. In wheat under Al toxicity, Al–citrate complexes were identified, suggesting its role in ameliorating heavy metal stress (Shen et al. 2004). Chai et al. (2012) described the role of oxalate in sequestration of Cd in root cells vacuoles and Cd transport to the aerial parts of *Spartina alterniflora* grass. The stability of Al-citrate and Al-oxalate complexes was found to be greater than that of Al-ATP, suggesting the role of citrate and oxalate in the core detoxification mechanism, which hinders Al binding to ligands like ATP (Wu et al. 2018).

14.6 Outlook and Challenges

Stress induced by HMs is a critical matter affecting agricultural yields and food healthiness because of their lethal outcomes and hasty accretion in the surrounding environment. The innate mechanisms established in HM sensitive or tolerant plants help them tolerate metal toxicity, particularly via up-regulation of genes intricated in resistance as well as acclimatization under the influence of abiotic stressors by releasing inhibitors, augmenting antioxidants, etc. Remarkable developments and progress accomplished in biological and molecular fields have provided insight into intricate strategies employed at both molecular and cellular levels by plants to overcome HM stress. Molecular versatility and functional diversity of phytochelatins and metallothioneins is gaining importance due to their role in HM detoxification and their ability to maintain cellular ion equilibrium in plants under metal stress. They interact with the antioxidant defense system in plants or translocate and distribute the HM ions amid the root and the shoot system in a tissue specific manner. Protein transporters, including NRAMP and ABC transporters, help translocate the toxic metal ions and discharge them out from the plants. Glutathione maintains cellular redox potential under metal toxicity. Plant chaperones facilitate correct protein folding patterns, which are otherwise disrupted in the presence of HM ions. Thus, analyzing plant proteins complexed in the HM stress forbearance can help develop biotechnological approaches to improve plant tolerance towards HMs. Omics and biochemical studies can provide detailed insights that can help to exploit the potential of stress-responsive proteins. Despite all the advances in protein research, a comprehensive understanding of the underlying mechanisms regulated by signaling cascades that control metal accumulation is still lacking. Thus, developing schemes to minimize HM uptake and translocation by manipulating the protein control system in plant cells appears to be a promising strategy that can possibly guarantee improved agricultural yield and food safety.

References

Ahmad, P., A. A. Abdel Latef, E. F. Abd_Allah, A. Hashem, M. Sarwat, N. A. Anjum and S. Gucel. 2016. Calcium and potassium supplementation enhanced growth, osmolyte secondary metabolite production, and enzymatic antioxidant machinery in cadmium-exposed chickpea (*Cicer arietinum* L.). Front. Plant Sci. 7: 513.

Ahsan, N., D. G. Lee, I. Alam, P. J. Kim, J. J. Lee and Y. O. Ahn. 2008. Comparative proteomic study of arsenic-induced differentially expressed proteins in rice roots reveals glutathione plays a central role during as stress. Proteomics. 8: 3561–3576.

Anjum, N. A., M. Hasanuzzaman, M. A. Hossain, P. Thangavel, A. Roychoudhury, S.S. Gill et al. 2015. Jacks of metal/metalloid chelation trade in plants—an overview. Front. Plant Sci. 6: 192.

Ansarypour, Z. and A. Shahpiri. 2012. Heterologous expression of a rice metallothionein isoform (OsMTI-1b) in *Saccharomyces cerevisiae* enhances cadmium, hydrogen peroxide and ethanol tolerance. Braz. J. Microbiol. 48: 537–543.

Assunção, A. G., P. Da Costa Martins, S. De Folter, R. Vooijs, H. Schat H. and M. G. M. Aarts. 2001. Elevated expression of metal transporter genes in three accessions of the metal hyperaccumulator *Thlaspi caerulescens*. Plant Cell Environ. 24: 217–226.

Baker, A. J. 1981. Accumulators and excluders-strategies in the response of plants to heavy metals. J. Plant Nutr. 3: 643–654.

Benatti, M. R., N. Yookongkaew, M. Meetam, W. J. Guo, N. Punyasuk, S. AbuQamar and P. Goldsbrough. 2014. Metallothionein deficiency impacts copper accumulation and redistribution in leaves and seeds of *Arabidopsis*. New Phytol. 202: 940–951.

Boudsocq, M. and J. Sheen. 2013. CDPKs in immune and stress signaling. Trends Plant Sci. 18: 30–40.

Cai, S. Y., Y. Zhang, Y. P. Xu, Z. Y. Qi, M. Q. Li, G. J. Ahammed et al. 2017. HsfA1a upregulates melatonin biosynthesis to confer cadmium tolerance in tomato plants. J. Pineal. Res. 62: e12387.

Cailliatte, R., A. Schikora, J. F. Briat, S. Mari and C. Curie. 2010. High affinity manganese uptake by the metal transporter NRAMP1 is essential for *Arabidopsis* growth in low manganese conditions. Plant Cell. 22: 904–917.

Capdevila, M. and S. Atrian. 2011. Metallothionein protein evolution: a miniassay. J. Biol. Inorg. Chem. 16: 977–989.

Caroli, M. D., A. Furini, G. DalCorso, M. Rojas and G. P. Di Sansebastiano. 2020. Endomembrane reorganization induced by heavy metals. Plants. 9: 482.

Chai, M. W., R. L. Li, F. C. Shi, F. C. Liu, X. Pan, D. Cao and X. Wen. 2012. Effects of cadmium stress on growth, metal accumulation and organic acids of *Spartina alterniflora* Loisel. Afr. J. Biotechnol. 11: 6091–6099.

Cobbett, C. and P. Goldsbrough. 2002. Phytochelatins and metallothioneins: roles in heavy metal detoxification and homeostasis. Annu. Rev. Plant Biol. 53: 159–182.

DalCorso, G., E. Fasani, A. Manara, G. Visioli and A. Furini. 2019. Heavy metal pollutions: state of the art and innovation in phytoremediation. Int. J. Mol. Sci. 20: 3412.

Dalvi, A. A. and S. A. Bhalerao. 2013. Response of plants towards heavy metal toxicity: an overview of avoidance, tolerance and uptake mechanism. Ann. Plant Sci. 2: 362–368.

Dedman, D. J., A. Treffry and P. Harrison. 1992. Interaction of aluminium citrate with horse spleen ferritin. Biochem. J. 287: 515–520.

Diener, A. C., R. A. Gaxiola and G. R. Fink. 2001. Arabidopsis ALF5, a multidrug efflux transporter gene family member, confers resistance to toxins. Plant Cell. 13: 1625–1638.

Dong, B., L. Niu, D. Meng, Z. Song, L. Wang, Y. Jian et al. 2019. Genome-wide analysis of MATE transporters and response to metal stress in *Cajanus cajan*. J. Plant Interact. 14: 265–275.

Durrett, T. P., W. Gassmann and E. E. Rogers. 2007. The FRD3-mediated efflux of citrate into the root vasculature is necessary for efficient iron translocation. Plant Physiol. 144: 197–205.

Emamverdian, A., Y. Ding, F. Mokhberdoran and Y. Xie. 2015. Heavy metal stress and some mechanisms of plant defense response. Sci. World J. 2015: 756120.

Ercal, N., H. Gurer-Orhan and N. Aykin-Burns. 2001. Toxic metals and oxidative stress part 1: mechanisms involved in metal induced oxidative damage. Curr. Top. Med. Chem. 1: 529–539.

Fan, W., C. Liu, B. Cao, M. Qin, D. Long, Z. Xiang and A. Zhao. 2018. Genome-wide identification and characterization of four gene families putatively involved in cadmium uptake, translocation and sequestration in mulberry. Front. Plant Sci. 9: 879.

Fang, H., T. Jing, Z. Liu, L. Zhang, Z. Jin and Y. Pei. 2014. Hydrogen sulfide interacts with calcium signaling to enhance the chromium tolerance in *Setaria italica*. Cell Calcium 56: 472–481.

Fryzova, R., M. Pohanka, P. Martinkova, H. Cihlarova, M. Brtnicky, J. Hladky and J. Kynicky. 2017. Oxidative stress and heavy metals in plants. Rev. Environ. Contam. Toxicol. 245: 129–156.

Furini, A. 2012. Plants and Heavy Metals. Springer Science & Business Media.

Ghori, N. H., T. Ghori, M. Q. Hayat, S. R. Imadi, A. Gul, V. Altay and M. Ozturk. 2019. Heavy metal stress and responses in plants. Int. J. Environ. Sci. Technol. 16: 1807–1828.

Guan, C., J. Ji, X. Li, C. Jin and G. Wang. 2016. LcMKK, a MAPK kinase from *Lycium chinense*, confers cadmium tolerance in transgenic tobacco by transcriptional upregulation of ethylene responsive transcription factor gene. J. Genet. 95: 875–885.

Guo, Q., J. Zhang, Q. Gao, S. Xing, F. Li and W. Wang. 2008. Drought tolerance through overexpression of monoubiquitin in transgenic tobacco. J. Plant Physiol. 165: 1745–1755.

Gupta, D. K., H. Tohoyama, M. Joho and M. Inouhe. 2002. Possible roles of phytochelatins and glutathione metabolism in cadmium tolerance in chickpea roots, J. Plant Res. 115: 429–437.

Gupta, S. C., A. Sharma, M. Mishra, R. K. Mishra and D. K. Chowdhuri. 2010. Heat shock proteins in toxicology: how close and how far? Life Sci. 86: 377–384.

Gustin, J. L., M. J. Zanis and D. E. Salt. 2011. Structure and evolution of the plant cation diffusion facilitator family of ion transporters. BMC Evol. Biol. 11: 1–13.

Hanikenne, M. and D. Baurain. 2014. Origin and evolution of metal P-type ATPases in Plantae (Archaeplastida). Front. Plant Sci. 4: 544.

Haq, N. U., S. Raza, D. S. Luthe, S. A. Heckathorn and S. N. Shakeel. 2013. A dual role for the chloroplast small heat shock protein of *Chenopodium album* including protection from both heat and metal stress. Plant Mol. Biol. Rep. 31: 398–408.

Harrison, P. M. and P. Arosio. 1996. The ferritins: molecular properties, iron storage function and cellular regulation. BBA Bioenergetics. 1275: 161–203.

Hartley-Whitaker, J., G. Ainsworth, R. Vooijs, W. T. Bookum, H. Schat and A. A. Meharg. 2001. Phytochelatins are involved in differential arsenate tolerance in *Holcus lanatus*. Plant Physiol. 126: 299–306.

Hasan, M., Y. Cheng, M. K. Kanwar, X. Y. Chu, G. J. Ahammed and Z. Y. Qi. 2017. Responses of plant proteins to heavy metal stress—a review. Front. Plant Sci. 8: 1492.

Hasanzadeh, M. and N. Hazrati. 2021. Calcium sensing and signaling in plants during metal/metalloid stress. In Metal and Nutrient Transporters in Abiotic Stress (pp. 169–197). Academic Press.

Hassinen, V. H., A. I. Tervahauta, H. Schat and S. O. Kärenlampi. 2011. Plant metallothioneins–metal chelators with ROS scavenging activity?. Plant Biol. 13: 225–232.

Hossain, M. A., P. Piyatida, J. A. T. da Silva and M. Fujita. 2012. Molecular mechanism of heavy metal toxicity and tolerance in plants: central role of glutathione in detoxification of reactive oxygen species and methylglyoxal and in heavy metal chelation. J. Bot. 2012. doi.org/10.1155/2012/872875.

Huang, T. L., L. Y. Huang, S. F. Fu, N. N. Trinh and H. J. Huang. 2014. Genomic profiling of rice roots with short-and long-term chromium stress. Plant Mol. Biol. 86: 157–170.

Huang, Y., G. He, W. Tian, D. Li, L. Meng, D. Wu and T. He. 2021. Genome-wide identification of MATE gene family in potato (*Solanum tuberosum* L.) and expression analysis in heavy metal stress. Front. Genet. doi.org/10.3389/fgene.2021.650500.

Hvorup, R. N., B. Winnen, A. B. Chang, Y. Jiang, X. F. Zhou and M. H. Saier, Jr. 2003. The multidrug/oligosaccharidyl-lipid/polysaccharide (MOP) exporter superfamily. Eur. J. Biochem. 270: 799–813.

Irvine, G. W., S. N. Tan and M. J. Stillman. 2017. A simple metallothionein-based biosensor for enhanced detection of arsenic and mercury. Biosensors 7: 14.

Jain, S., S. Muneer, G. Guerriero, S. Liu, K. Vishwakarma, D. K. Chauhan et al. 2018. Tracing the role of plant proteins in the response to metal toxicity: a comprehensive review. Plant Signal. Behav. 13: e1507401.

Jalmi, S. K. and A. K. Sinha. 2015. ROS mediated MAPK signaling in abiotic and biotic stress-striking similarities and differences. Front. Plant Sci. 6: 769.

Jalmi, S. K., P. K. Bhagat, D. Verma, S. Noryang, S. Tayyeba, K. Singh et al. 2018. Traversing the links between heavy metal stress and plant signaling. Front. Plant Sci. 9: 12.

Jonak, C., L. Okrész, L. Bögre and H. Hirt. 2002. Complexity, cross talk and integration of plant MAP kinase signalling. Curr. Opin. Plant Biol. 5: 415–424.

Jonak, C., H. Nakagami and H. Hirt. 2004. Heavy metal stress. Activation of distinct mitogen-activated protein kinase pathways by copper and cadmium. Plant Physiol. 136: 3276–3283.

Jutsz, A. M. and A. Gnida. 2015. Mechanisms of stress avoidance and tolerance by plants used in phytoremediation of heavy metals. Arch Environ. Prot. 41: 104–114.

Kang, Y. J. 2006. Metallothionein redox cycle and function. Exp. Biol. Med. 231: 1459–1467.

Keyster, M., L. A. Niekerk, G. Basson, M. Carelse, O. Bakare, N. Ludidi et al. 2020. Decoding heavy metal stress signalling in plants: towards improved food security and safety. Plants. 9: 1781.

Kochian, L. V., O. A. Hoekenga and M. A. Pineros. 2004. How do crop plants tolerate acid soils? Mechanisms of aluminum tolerance and phosphorus efficiency. Annu. Rev. Plant Biol. 55: 459–493.

Kumar, S. and P. K. Trivedi. 2016. Heavy metal stress signaling in plants. pp. 585–603. *In*: Parvaiz Ahmad (ed.). Plant Metal Interaction. Elsevier.

Lee, S., J. S. Moon, T. S. Ko, D. Petros, P. B. Goldsbrough and S. S. Korban. 2003. Overexpression of *Arabidopsis* phytochelatin synthase paradoxically leads to hypersensitivity to cadmium stress. Plant Physiol. 131: 656–663.

Leitenmaier, B. and H. Küpper. 2013. Compartmentation and complexation of metals in hyperaccumulator plants. Front. Plant Sci. 4: 374.

Li, Y., O. P. Dankher, L. Carreira and L. X. Jiang. 2006. The shoot-specific expression of gamma-glutamylcysteine synthetase directs the long-distance transport of thiol-peptides to roots conferring tolerance to mercury and arsenic. Plant Physiol. 141: 288–298.

Liu, H., H. Zhao, L. Wu, A. Liu, F. J. Zhao and W. Xu. 2017. Heavy metal ATPase 3 (HMA3) confers cadmium hypertolerance on the cadmium/zinc hyperaccumulator *Sedum plumbizincicola*. New Phytol. 215: 687–698.

Liu, J., J. Wang, S. Lee and R. Wen. 2018. Copper-caused oxidative stress triggers the activation of antioxidant enzymes via ZmMPK3 in maize leaves. PloS one. 13: e0203612.

Liu, X. S., S. J. Feng, B. Q. Zhang, M. Q. Wang, H. W. Cao, J. K. Rono et al. 2019. OsZIP1 functions as a metal efflux transporter limiting excess zinc, copper and cadmium accumulation in rice. BMC Plant Biol. 19: 1–16.

Liu, Y., C. Zhang, D. Wang, W. Su, L. Liu, M. Wang and J. Li. 2015. EBS7 is a plant-specific component of a highly conserved endoplasmic reticulum-associated degradation system in *Arabidopsis*. Proc. Natl. Acad. Sci. 112: 12205–12210.

Lombi, E., F. J. Zhao, S. J. Dunham and S. P. McGrath. 2001. Phytoremediation of heavy metal-contaminated soils: Natural hyperaccumulation versus chemically enhanced phytoextraction. J. Environ. Qual. 30: 1919–1926.

López-Millán, A. F., D. R. Ellis and M. A. Grusak. 2004. Identification and characterization of several new members of the ZIP family of metal ion transporters in *Medicago truncatula*. Plant Mol. Biol. 54: 583–596.

Lu, H., D. Qiao, Y. Han, Y. Zhao, F. Bai and Y. Wang. 2021. Low molecular weight organic acids increase Cd accumulation in sunflowers through increasing Cd bioavailability and reducing Cd toxicity to plants. Minerals. 11: 243.

Lu, L., S. Tian, X. Yang, H. Peng and T. Li. 2013. Improved cadmium uptake and accumulation in the hyperaccumulator *Sedum alfredii*: the impact of citric acid and tartaric acid. J. Zhejiang Univ. Sci. B 14: 106–114.

Ma, Q., R. Yi, L. Li, Z. Liang, T. Zeng, Y. Zhang et al 2018. GsMATE encoding a multidrug and toxic compound extrusion transporter enhances aluminum tolerance in *Arabidopsis thaliana*. BMC Plant Biol. 18: 1–10.

Maret, W. 2000. The function of zinc metallothionein: a link between cellular zinc and redox state. J. Nutr. 130: 1455S–1458S.

Memon, A. R. and P. Schröder. 2009. Implications of metal accumulation mechanisms to phytoremediation. Environ. Sci. Pollut. Res. 16: 162–175.

Mendoza-Cózatl, D. G., T. O. Jobe, F. Hauser and J. I. Schroeder. 2011. Long-distance transport, vacuolar sequestration, tolerance, and transcriptional responses induced by cadmium and arsenic. Curr. Opin. Plant Biol. 14: 554–562.

Migeon, A., D. Blaudez, O. Wilkins, B. Montanini, M. M. Campbell, P. Richaud, S. Thomine and M. Chalot. 2010. Genome-wide analysis of plant metal transporters, with an emphasis on poplar. Cell. Mol. Life Sci. 67: 3763–3784.

Milner, M. J., J. Seamon, E. Craft and L. V. Kochian. 2013. Transport properties of members of the ZIP family in plants and their role in Zn and Mn homeostasis. J. Exp. Bot. 64: 369–381.

Miransari, M. 2011. Hyperaccumulators, arbuscular mycorrhizal fungi and stress of heavy metals. Biotechnol. Adv. 29: 645–653.

Mittler, R., S. Vanderauwera, M. Gollery and F. Van Breusegem. 2004. Reactive oxygen gene network of plants. Trends Plant Sci. 9: 490–498.

Morris, C. A., B. Nicolaus, V. Sampson, J. L. Harwood and P. Kille. 1999. Identification and characterization of a recombinant metallothionein protein from a marine alga, *Fucus vesiculosus*. Biochem. J. 338: 553–560.

Opdenakker, K., T. Remans, J. Vangronsveld and A. Cuypers. 2012. Mitogen-Activated Protein (MAP) kinases in plant metal stress: regulation and responses in comparison to other biotic and abiotic stresses. Int. J. Mol. Sci. 13: 7828–7853.

Ovečka, M. and T. Takáč. 2014. Managing heavy metal toxicity stress in plants: biological and biotechnological tools. Biotechnol. Adv. 32: 73–86.

Park, C. J. and Y. S. Seo. 2015. Heat Shock Proteins: A review of the molecular chaperones for plant immunity. Plant Pathol. J. 31: 323–333.

Park, J., W. Y. Song, D. Ko, Y. Eom, T. H. Hansen, M. Schiller et al. 2012. The phytochelatin transporters AtABCC1 and AtABCC2 mediate tolerance to cadmium and mercury. Plant J. 69: 278–288.

Peiter, E., B. Montanini, A. Gobert, P. Pedas, S. Husted, F. J. M. Maathuis et al. 2007. A secretory pathway-localized cation diffussion facilitator confers plant manganese tolerance. Proc. Natl. Acad. Sci. 104: 8532–8537.

Peñarrubia, L., N. Andrés-Colás, J. Moreno and S. Puig. 2010. Regulation of copper transport in *Arabidopsis thaliana*: a biochemical oscillator? JBIC J. Biol. Inorg. Chem. 15: 29–36.

Persans, M. W., K. Nieman and D. E. Salt. 2001. Functional activity and role of cation-efflux family members in Ni hyperaccumulation in *Thlaspi goesingense*. Proc. Natl. Acad. Sci. 98: 9995–10000.

Piddock L. J. 2006. Clinically relevant chromosomally encoded multidrug resistance efflux pumps in bacteria. Clin. Microbiol. Rev. 19: 382–402.

Pochodylo, A. L. and L. Aristilde. 2017. Molecular dynamics of stability and structures in phytochelatin complexes with Zn, Cu, Fe, Mg, and Ca: Implications for metal detoxification. Environ. Chem. Lett. 15: 495–500.

Ravet, K., B. Touraine, J. Boucherez, J. F. Briat, F. Gaymard and F. Cellier. 2009. Ferritins control interaction between iron homeostasis and oxidative stress in *Arabidopsis*. Plant J. 57: 400–412.

Rodríguez-Celma, J., R. Rellán-Álvarez, A. Abadía, J. Abadía and A. F. LópezMillán. 2010. Changes induced by two levels of cadmium toxicity in the 2-DE protein profile of tomato roots. J. Proteom. 73: 1694–1706.

Sácký, J., T. Leonhardt, J. Borovička, M. Gryndler, A. Briksí and P. Kotrba. 2014. Intracellular sequestration of zinc, cadmium and silver in *Hebeloma mesophaeum* and characterization of its metallothionein genes. Fungal Genet. Biol. 67: 3–14.

Sarret, G., P. Saumitou-Laprade, V. Bert, O. Proux, J.L. Hazemann, A. Traverse, M.A. Marcus and A. Manceau. 2002. Forms of zinc accumulated in the hyperaccumulator *Arabidopsis halleri*. Plant Physiol. 130: 1815–1826.

Sasaki, A., N. Yamaji, N. Mitani-Ueno, M. Kashino and J. F. Ma. 2015. A nodelocalized transporter OsZIP3 is responsible for the preferential distribution of Zn to developing tissues in rice. Plant J. 84: 374–384.

Savino, G., J. F. Briat and S. Lobréaux. 1997. Inhibition of the iron-induced ZmFer1 maize ferritin gene expression by antioxidants and serine/threonine phosphatase inhibitors. J. Biol. Chem. 272: 33319–33326.

Shahzad, Z., F. Gosti, H. Frerot, E. Lacombe, N. Roosens, P. Saumitou-Laprade et al. 2010. The five AhMTP1 zinc transporters undergo different evolutionary fates towards adaptive evolution to zinc tolerance in *Arabidopsis halleri*. PLoS Genet. 6:e1000911.

Shao, H. B., L. Y. Chu and M. A. Shao. 2008. Calcium as a versatile plant signal transducer under soil water stress. BioEssays. 30: 634–641.

Sharma, S. S., K. J. Dietz and T. Mimura. 2016. Vacuolar compartmentalization as indispensable component of heavy metal detoxification in plants. Plant Cell Environ. 39: 1112–1126.

Shen, R., T. Iwashita and J. F. Ma. 2004. Form of Al changes with Al concentration in leaves of buckwheat. J. Exp. Bot. 55: 131–136.

Sheoran, V., A. S. Sheoran and P. Poonia. 2010. Soil reclamation of abandoned mine land by revegetation: a review. Int. J. Soil Sediment and Water. 3: 13.

Singh, S., P. Parihar, R. Singh, V. P. Singh and S. M. Prasad. 2016. Heavy metal tolerance in plants: role of transcriptomics, proteomics, metabolomics, and ionomics. Front. Plant Sci. 6: 1143.

Singh, V. P., S. Singh, J. Kumar and S. M. Prasad. 2015. Investigating the roles of ascorbate-glutathione cycle and thiol metabolism in arsenate tolerance in ridged Luffa seedlings. Protoplasma. 252: 1217–1229.

Singh, R., A. B. Jha, A. N. Misra and P. Sharma. 2019. Adaption mechanisms in plants under heavy metal stress conditions during phytoremediation. pp. 329–360. *In*: Vimal Chandra Pandey and Kuldeep Bauddh (eds.). Phytomanagement of polluted sites. Elsevier.

Socha, A. L. and M. L. Guerinot. 2014. Mn-euvering manganese: the role of transporter gene family members in manganese uptake and mobilization in plants. Front. Plant Sci. 5: 106.

Song, W. Y., D. G. Mendoza-Cózatl, Y. Lee, J. I. Schroeder, S. N. Ahn, H. S. Lee et al. 2017. Identification of amino acid residues important for the arsenic resistance function of *Arabidopsis* ABCC 1. FEBS Lett. 591: 656–666.

Sun, Y. B., Q. X. Zhou and C. Y. Diao. 2008. Effects of cadmium and arsenic on growth and metal accumulation of Cd-hyperaccumulator *Solanum nigrum* L. Bioresource Technol. 99: 1103–1110.

Takahashi, R., K. Bashir, Y. Ishimaru, N. K. Nishizawa and H. Nakanishi. 2012. The role of heavy-metal ATPases, HMAs, in zinc and cadmium transport in rice. Plant Signal Behav. 7: 1605–1607.

Talke I. N., M. Hanikenne and U. Krämer. 2006. Zinc-dependent global transcriptional control, transcriptional deregulation, and higher gene copy number for genes in metal homeostasis of the hyperaccumulator *Arabidopsis halleri*. Plant Physiol 142: 148–167.

Thomine, S., R. Wang, J. M. Ward, N. M. Crawford and J. I. Schroeder. 2000. Cadmium and iron transport by members of a plant metal transporter family in *Arabidopsis* with homology to Nramp genes. Proc. Natl. Acad. Sci. 97: 4991–4996.

Tiwari, S. and C. Lata. 2018. Heavy metal stress, signaling, and tolerance due to plant-associated microbes: an overview. Front. Plant Sci. 9: 452.

Tolrà, R. P., C. Poschenrieder and J. Barceló. 1996. Zinc hyperaccumulation in *Thlaspi caerulescens*. II. Influence on organic acids. J. Plant Nutr. 19: 1541–1550.

Tomas, M., M. A. Pagani, C. S. Andreo, M. Capdevila, S. Atrian and R. Bofill. 2015. Sunflower metallothionein family characterisation. Study of the Zn (II)-and Cd (II)-binding abilities of the HaMT1 and HaMT2 isoforms. J. Inorg. Biochem. 148: 35–48.

Valivand, M., R. Amooaghaie and A. Ahadi. 2019. Seed priming with H^2S and Ca^{2+} trigger signal memory that induces cross-adaptation against nickel stress in zucchini seedlings. Plant Physiol. Biochem. 143: 286–298.

Verret, F., A. Gravot, P. Auroy, N. Leonhardt, P. David, L. Nussaume et al. 2004. Overexpression of AtHMA4 enhances root–to–shoot translocation of zinc and cadmium and plant metal tolerance. Febs. Lett. 576: 306–312.

Viehweger, K. 2014. How plants cope with heavy metals. Bot. Stud. 55: 1–12.

Wang, Y., L. Xu, Y. Chen, H. Shen, Y. Gong, C. Limera and L. Liu. 2013. Transcriptome profiling of radish (*Raphanus sativus* L.) root and identification of genes involved in response to lead (Pb) stress with next generation sequencing. PLoS One, 8: e66539.

Williams, L. E. and R. F. Mills. 2005. P1B-ATPases—an ancient family of transition metal pumps with diverse functions in plants. Trends Plant Sci. 10: 491–502.

Wu, L., Y. Kobayashi, J. Wasaki and H. Koyama. 2018. Organic acid excretion from roots: a plant mechanism for enhancing phosphorus acquisition, enhancing aluminum tolerance, and recruiting beneficial rhizobacteria. Soil Sci. Plant Nutr. 64: 697–704.

Yan, A., Y. Wang, S. N. Tan, M. L. Mohd Yusof, S. Ghosh and Z. Chen. 2020. Phytoremediation: a promising approach for revegetation of heavy metal-polluted land. Front. Plant Sci. 11: 359.

Yang, J., Y. Wang, G. Liu, C. Yang and C. Li. 2011. *Tamarix hispida* metallothionein-like ThMT3, a reactive oxygen species scavenger, increases tolerance against Cd^{2+}, Zn^{2+}, Cu^{2+}, and NaCl in transgenic yeast. Mol. Biol. Rep. 38: 1567–1574.

Yeh, C. M., P. S. Chien and H. J. Huang. 2007. Distinct signalling pathways for induction of MAP kinase activities by cadmium and copper in rice roots. J. Exp. Bot. 58: 659–671.

Yu, G., J. Ma, P. Jiang, J. Li, J. Gao, S. Qiao and Z. Zhao. 2019. The mechanism of plant resistance to heavy metal. *In* IOP Conference Series: Earth and Environmental Science (Vol. 310, No. 5, p. 052004). IOP Publishing.

Zhang, J., J. U. Hwang, W. Y. Song, E. Martinoia and Y. Lee. 2017. Identification of amino acid residues important for the arsenic resistance function of Arabidopsis ABCC1. FEBS Lett. 591: 656–666.

Zhao, F. Y., F. Hu, S. Y. Zhang, K. Wang, C. R. Zhang and T. Liu. 2013. MAPKs regulate root growth by influencing auxin signaling and cell cycle-related gene expression in cadmium-stressed rice. Environ. Sci. Pollut. Res. 20: 5449–5460.

Chapter 15

Remediation of Toxic Metal(loid)s
Biotechnological Strategies

Manish Singh Rajput,[1] Upasana Jhariya,[2] Kritika Pandey,[1] Shweta Rai,[3] Surbhi Kuril,[1] Pratibha Singh[4] and Sridhar Pilli[5,]*

15.1 Introduction

In today's age of rapid industrialization, the removal of hazardous pollutants from the environment has become the foremost issue. The ubiquitous nature of metallic elements and metalloids results in their accumulation in the environment with time, which causes serious health issues throughout the food chain. Metals and metalloids with a density greater than 5000 kg m^{-3} and an atomic mass greater than 20 are referred to as "heavy metals" in the environment. According to this classification, heavy metals and metalloids include 51 elements from the periodic table.

Heavy metals' ability to enter the food chain, agricultural land, and water resources, pose a major health danger to terrestrial and aquatic ecosystems. Approximately, five million locations across the world are reported to be contaminated with various heavy metals and metalloids (He et al. 2015). The majority of these contaminated locations by heavy metals are in industrialized countries, including 6000 km^2 in the United States, 810,000 km^2 of agricultural land in China, and 250,000 sites in the European countries that have been polluted by heavy metals and metalloids (Khalid et al. 2016). Due to the harmful impacts of these metal(loid)s, there is a greater demand for acceptable techniques and procedures for cleaning up contaminated soils and water. In conventional contaminated site rehabilitation, physical removal and disposal of contaminants are common techniques. The main goal of metal(loid) pollution remediation is to shift metal(loid) pollution from a vast volume to a reduced one. As a result, several attempts have been made for heavy metal bioremediation using microbial systems as an appealing solution because such procedures are cost-effective and environment-friendly (Pandey et al. 2021). Phytoremediation is a type of bioremediation that requires the application of plants to adsorb or decompose contaminants in the soil and water (Awa and Hadibarata 2020).

[1] Department of Biotechnology, Dr. Ambedkar Institute of Technology for Handicapped, Kanpur, U.P. 208024, India.
[2] CSIR-National Environmental Engineering Research Institute, Maharastra, India.
[3] Department of Biotechnology, Motilal Nehru National Institute of Technology, Allahabad, U.P. – 211004, India.
[4] Department of Bioinformatics, University of Hyderabad, Telangana-500046, India.
[5] Department of Civil Engineering, National Institute of Technology, Warangal, 506004, Telangana, India.
* Corresponding author: srenitw@nitw.ac.in

Among the various heavy metallic elements, cadmium, arsenic, chromium, cobalt, copper, manganese, mercury, nickel, tin, lead, and titanium are a few metals classified as hazardous heavy metals by the World Health Organization (WHO). Lead, cadmium, arsenic, and mercury all have toxicity profiles when they are exposed. Metalloids like selenium and arsenic, which have specific gravities of 4.79 g cm^{-3} and 5.73 g cm^{-3}, respectively, are commonly regarded as heavy metals (Wilkin et al. 2018). Additionally, the term "heavy metal(loids)" has been used to represent both metals and metalloids as an expedient shorthand. Several writers have looked at this topic throughout the years because of its global relevance.

Heavy metallic elements can be divided into two major types, i.e., essential and non-essential. Copper (Cu), nickel (Ni), zinc (Zn), iron (Fe), and manganese (Mn) are essential metals that play significant regulatory functions in a variety of biological activities, including electron transport proteins and as cofactors in several enzymes. Non-essential metals like mercury (Hg), cadmium (Cd), and lead (Pb) have no recognized biological roles (Chaffai and Koyama 2011).

15.2 Sources of Metal(loid)s Contamination

Several natural and man-made activities are the major sources of heavy metal contamination in the environment. Various factors contribute to the spread of common and hazardous heavy metals and metalloids, resulting in environmental deposition and contamination.

The parent material from which soil was generated is the most significant and primary source of heavy metals. The earth's crust surface is made up of 95% volcanic rocks and 5% of sedimentary materials (Li et al. 2019). Heavy metals including Cd, Co, Cu, Zn, and Ni are found in higher quantities in ingenious basaltic rocks, whereas shales produced from fine inorganic and organic sediments have higher levels of metal elements such as Cu, Zn, Cd, Mn, and Pb. Carbonates, sulfides, oxides, and salts are the most common forms of heavy metals found in soil (Roane et al. 2015). Increased heavy metal pollution in water supplies and soils can be attributed to anthropogenic sources, including recent advances in industry and agriculture. Smelting and mining operations are substantial point sources of heavy metals in the agro-ecosystem. Environmental contamination is exacerbated by metal-containing chemicals, polluted biosolids, and fertilizers. Anthropogenic metal(loid) forms are more ecologically unstable, therefore, more soluble and accessible than natural forms, providing a greater environmental risk (Czarnecki et al. 2015).

Heavy metallic elements in the soil can be introduced by several anthropogenic activities like the techniques used for the improvement of soil fertility. In addition to fertilizers that can cause increased heavy metal concentrations in the environment, organic substances with higher heavy metal concentrations than normal agricultural soils, such as farm yard manures, composts, and biosolids, are applied to soils for enhancing soil fertility or recovering problem soils. Furthermore, Pb buildup in soils near roadways is caused by emissions from vehicles that use lead-enriched gasoline. Heavy metal contamination of soils has also been observed as a result of mining activity in some places (Singh et al. 2021a, Anju and Banerjee 2011).

15.3 Toxicity of Heavy Metals

Essential as well as non-essential heavy metals, when entering into the ecosystem at larger concentrations, can exert harmful effects. The toxicity levels of these metals vary from organism to organism, toxicity potential of metals, and their bioavailability. The long persistent nature of these elements is mainly responsible for the threat to all living beings. Moreover, some heavy metals are highly toxic even at very low concentrations (Mani and Kumar 2014). Certain heavy metals, such as Zn, Cu, Fe, and Mn, have a dual nature, increasing plant growth or biofortification at low concentrations and benefiting the entire food chain (Imran et al. 2016), whilst a minor rise in concentration can impair not only plant growth but also effect other organisms, such as humans

Table 15.1. Effects of metals on the living beings.

Metals	Health risks associated with metals	References
Cd	It is known to be mutagenic and carcinogenic; by altering calcium (Ca) metabolism in the body, it alters (Ca) metabolism in the body, which can cause hypercalciuria, severe anemia, and renal failure.	Awofolu 2005
Zn	It can produce tiredness and dizziness at high concentration levels.	Schmid et al. 2002
Pb	Renal failure, cardiovascular illness, and lower IQ, part-time memory loss, coordination problems, and impaired learning capacity have all been linked to lead poisoning in children.	Salem and El-Maazawi 2000
Cr	It has the potential to induce hair loss.	Salem and El-Maazawi 2000
Cu	Cu levels over a certain threshold has have been reported to damage the kidney and brain, cause also anemia and intestinal discomfort.	Salem and El-Maazawi 2000
As	Because arsenic in its arsenate form is similar to phosphate, it disturbs cellular functions, including oxidative phosphorylation and ATP production.	Tripathi et al. 2007

(Roy and Mcdonald 2015). Heavy metals have adverse effects on a variety of physiological and biochemical processes in plants. Hence, they have the potential to significantly reduce plant growth and cause cell death to serious level (Popova et al. 2009). When trace metals and metalloids such as, As, Se, Sb reach threshold quantities, they are poisonous to plants and animals (Shukla et al. 2018). The toxicity profile of heavy metals in humans is associated with a variety of health issues, depending on the metal and its oxidation state, among other factors. Table 15.1 contains the summary of health risks caused by metals.

Since metal(loid)s contaminated surroundings regularly pose an unsustainable threat to human and natural wellbeing, heavy metal(loid) pollution cleanup is a major concern for many countries, garnering the attention of many authorities and researchers. The following points highlight the necessity of heavy metal(loid)s contamination remediation. Metal(loid)s cannot be eliminated; they are among the most challenging pollutants to remediate (Li et al. 2019).

Metal(loid)s may only alter their valence and species in nature; they cannot be degraded by any method, including biotreatment. Under specific environmental circumstances, some metals and metalloids can even be transformed from less hazardous to more destructive forms, such as elemental mercury, to more dangerous methylmercury (Jaiswal et al. 2018). Metal(loid)s, unlike organic compounds, are fundamentally nonbiodegradable and, as a result, accumulate in the environment. Metal(loid)s bioaccumulation, and bioaugmentation in the food chain might harm normal physiological functions and put human life at risk (Pratush et al. 2018).

Some of these elements are deadly at low doses and may have carcinogenic, endocrine-disrupting, mutagenic, and teratogenic properties. High levels of these elements in soil cause degradation of agricultural land and hazardous chemical absorption. Contamination directly affects agricultural soil quality, such as phytotoxicity at significant amounts, soil microbial activity, and the introduction of zootoxic chemicals into the human diet. As a result, their remediation methods are required to be further developed (Li et al. 2019).

15.4 Bioavailability of Metals and Metalloids

The fraction of total metals available to plants and other organisms is referred to as bioavailable, which is ready to be incorporated into biota (bioaccumulation). Total metal concentrations do not always correspond with metal bioavailability. Sulfide minerals, for example, may be encased in quartz or another material (Yan et al. 2019). Despite high overall metal concentrations in sediments and soils containing chemically inert minerals challenging for life forms to assimilate, the related environmental consequences may be negligible (Yan et al. 2019, Li et al. 2019). Metals such as

As, Al, Be, Cr, Cu, Cd, Hg, Ni, Si, Sb, and Pb, according to the US Environmental Protection Agency (USEPA), are important in bioavailability analysis. Other metals that the EPA is currently not focusing on include Ag, Na, Co, Mo, Mn, Ba, Tl, V, and Zn. These metals were chosen due to their elevated health risk and potential for human exposure (Yan et al. 2019).

Metal(loids) are abundant in nature. They are recognized to have detrimental consequences not only on plants and animals but also on humans; hence, their treatment is critical (Anju 2017), which presented in Table 15.1. Metal(loid) remediation is a challenge since they are resistant to degradation, persist in the environment, and cannot be destroyed. Metal(loid) pollution remediation is mainly by lowering density, changing their speciation to reduce toxicity, or increasing their volatility. Phytoremediation has evolved as one of the most secure, efficient, environmentally sustainable, solar-powered, and successful techniques for the removal of metal and metalloids from the environment (Muthusaravanan et al. 2018). A range of complementary techniques has been developed to increase the efficacy of plant remediation; chelate-assisted and microbe-mediated phytoremediation are two examples of this type of phytoremediation. Plant growth-promoting bacteria (PGPB) can improve phytoremediation efficiency by modifying metal solubility and bioavailability through biosurfactant activity. The availability of plant absorbing metals has a big impact on the efficiency of phytoextraction for heavy metal cleanup (Anju 2017). Soil metals can be produced in four fractions: free metal ions, soil solution soluble metal complexes, metal ions at sites of exchange of ions and adsorbed into inorganic soil components, organic compounds, and precipitated or unsolvable compounds, in particular carbonates, oxides, and hydroxides. Chemical heterogeneity has an important influence on solubility and potential metals available in soils (Anju 2017). Metal bioavailability often restricts the effectiveness of metal phytoremediation. Changing the soil environment in order to improve the obtainability of metallic elements is crucial for efficient plant uptake (Sumiahadi and Acar 2018).

15.5 Remediation Strategies for Removal of Heavy Metal(loid)s

Physical, chemical, and biological procedures are the three main removal strategies that have evolved over time. Their goal is to either completely remove pollutants or convert them into less hazardous forms. However, the physical and chemical methods have many drawbacks, such as the use of harmful chemicals, the generation of harmful byproducts, etc. To overcome limitations associated with traditional methods, there has been a surge in curiosity in substitute bio-based, environmentally approachable alternatives known as biological remediation. The environment-friendly and cost-effective nature of bioremediation turns it into a most preferable alternative for traditional methods to remove waste from the surrounding. It can be used as *ex situ* and *in situ* approaches according to the requirement (Jin et al. 2018).

Physical techniques including leaching and landfilling, excavation, and calcination provide great elimination efficiency with the treatment of vast soil volume, but they are costly. Physical techniques are simple, highly efficient, and offer rapid application, but they are expensive and can alter some soil characteristics, such as texture or particle size, resulting in a decline in soil fertility (Raffa et al. 2021).

Adsorption, electrochemical remediation, and soil washing are examples of chemical treatments that are exceedingly efficient but can introduce new chemical substituent contaminants into soils. Soil cleaning, for example, might introduce new chemical pollutants into the soil. Physical-chemical methods (precipitation, ion exchange, reverse osmosis, chemical reduction, and evaporation) are simple and convenient for use, but then again, they come at a significant expense. To overcome the side effects of physical and chemical remediation methods, biological approaches are now coming forward. One of the most accepted, environmentally beneficial, and cost-effective approaches is bioremediation (Raffa et al. 2021).

15.6 Bioremediation

Bioremediation of heavy metallic elements utilizing microbial systems has been attempted in numerous ways since such approaches are cost-effective and ecologically friendly. To detoxify or break down pollutants in the environment, bioremediation is a complicated process that relies primarily on microbes, plants, or microbial or plant enzymes (Raj et al. 2021). Biodegradation, biotransformation, and biodeterioration are all included in bioremediation. The process of bioremediation can be accelerated by adding bacteria and fungus. The objective of bioremediation is to decrease pollutant levels to undetectable, non-toxic, or acceptable levels or to fully mineralize organo-pollutants to carbon dioxide. Because bioremediation is a natural process, the public perceives it to have a lower impact on natural ecosystems (Rudakiya and Patel 2021).

Bioleaching, biomineralization, biosorption, bioaccumulation, biodegradation, and biotransformation have all been used in the bioremediation of heavy metals (Sharma et al. 2021a) (Fig. 15.1).

Pollutants are partially transformed or detoxified by bacteria and plants during biodegradation, and even sometimes completely. Bio-mineralization is a more precise word that refers to the process of fungi, bacteria, and plants excreting different organic acids and mineralizing them into insoluble metal forms (Rudakiya et al. 2019). Table 15.2 represents more about the studies related to bioremediation.

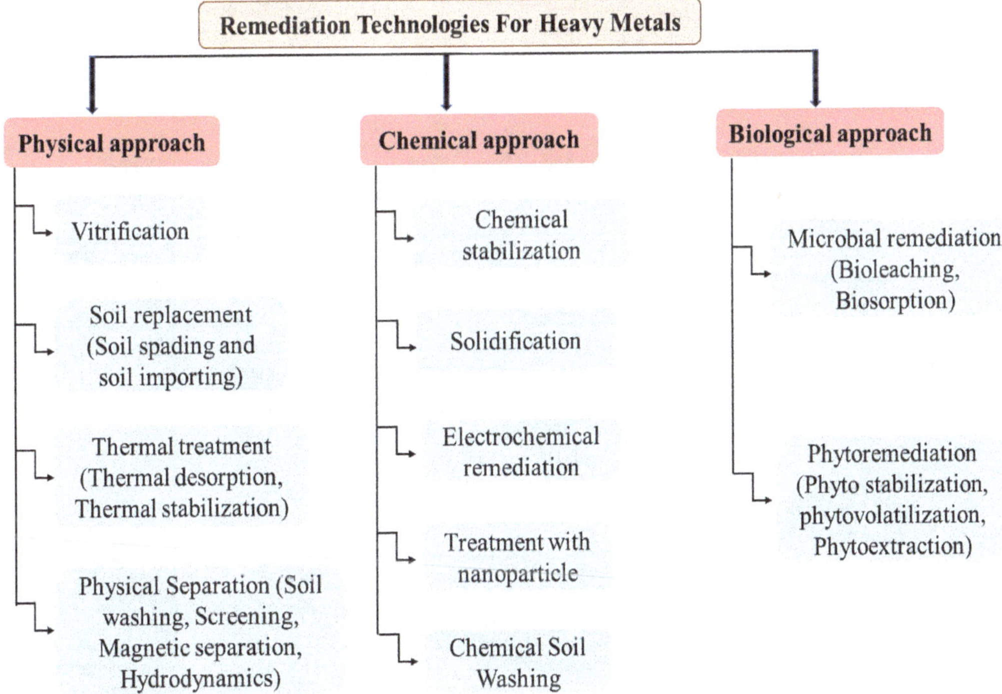

Fig. 15.1. Approaches used for remediation of metals and metalloids.

15.6.1 Important Mechanisms of Bioremediation

15.6.1.1 Biosorption

Biosorption is the process where heavy metals become immobilized on the cellular structure of bacteria, fungi, and algae. This occurs through the extracellular association between the metal ions

Table 15.2. Metal bioremediation using microorganisms and their source of isolation.

Organism	Species	Metal	Source	References
Bacteria	*Bacillus licheniformis, Escherichia coli, Pseudomonas fluorescence, and Salmonella typhi*	Cadmium, Lead, and Zinc	Textile industry effluents	Basha and Rajaganesh 2014
	Arundo donax L., *Stenotrophomonas maltophilia* sp., and *Agrobacterium* sp.	Arsenic	Plant Growth Promoting Bacteria (PGPB) consortium	Guarino et al. 2020
	Enterobacter cloacae KJ-46 and KJ-47	Lead	Soil of abandoned metal mines	Kang et al. 2015
	Sporosarcina pasteurii strains (ATCC11859 and *Terrabacter tumescens* (AS.1.2690)	Nickel, Copper, Lead, Cobalt, Zinc, and Cadmium	Nursery garden soil from Tsinghua University	Li et al. 2013
	Pseudoalteromonas sp.	Cadmium, Lead, Zinc(II)	Soil	Jiang et al. 2017
	Streptococcus equisimilis	Nickel, Cadmium	Spanish Type Culture Collection, University of Valencia	Costa and Tavares 2016
	Bacillus arsenicus	Arsenic(III), Arsenic(V)	Rhizosphere of maize plants	Podder and Majumder 2016
Fungi	*Beauveria bassiana*	Zinc(II), Chromium(VI),Copper(II), Nickel(II), and Cadmium(II)	Entomopathogenic fungi	Gola et al. 2016
	Fomitopsis meliae, Trichoderma ghanense and *Rhizopus microsporus*	Cadmium, Copper, Lead, Arsenic, and Iron	Gold and gemstone mine site soils	Oladipo et al. 2018
	Fomitopsis cf. meliae and *Ganoderma aff. steyaertanum*	Zinc, Copper, Cadmium, and Lead	Thailand soil	Kaewdoung et al. 2016
	Aspergillus niger and *Aspergillus flavus*	Lead and Chromium	Peri-urban agricultural areas	Shazia et al. 2013
	Aspergillus flavus, Aspergillus gracilis, Aspergillus penicillioides, Aspergillus restrictus, and *Sterigmatomyces halophilus*	Cadmium (II), Copper (II), Iron (II), Manganese (II), Lead (II), and Zinc(II)	Solar saltern in Phetchaburi, Thailand	Bano et al. 2018
	Ustilago maydis	Copper (II), Cadmium (II), Chromium Chromium (III), Nickel (II), Zinc(II)	Infected plant	Sargin et al. 2016
Algae	*Enteromorpha* sp.	Chromium (VI)	Coastal area	Rangabhashiyam et al. 2015
	Cystoseira indicia	Copper (II)	Oman sea	Akbari et al. 2015
	Sargassum filipendula	Chromium (II)	Aushadh Agri Science Private Limited, Gujarat	Verma et al. 2017
	Chlorella vulgaris	Chromium (II)	Polluted water	Cheng et al. 2017
	Ulva sp., *Chaetomorpha* sp., *Cystoseira* sp, and \ *Polysiphonia* sp.	Zinc (II)	Coastal area	Deniz and Karabulut 2017

(cations) and cell surface (anions). Binding mechanisms are aided by active functional groups in extracellular materials. Biosorption mechanisms involve complex formation, physical adsorption, reduction, precipitation, and ion exchange. Biosorption efficacy is affected by metal ion process parameters, properties, sorption center density, and immobilization agent types (Velkova et al. 2018). Bacteria, fungi, and algae are the distinctive bio adsorbent owing to the following:

- Functional groups like amino, sulfate, carboxyl, or phosphate can bind and accumulate heavy metal ions on the polysaccharide slime layers of bacteria (Cui et al. 2017).
- Fungi use chitin–chitosan complex structure coordination for ion exchange, phosphate, glucuronic acid, and polysaccharides in their cells to adsorb heavy metals and metalloids from the environment (Hassan et al. 2020).
- Algae produce peptides as a defensive strategy against stress. Heavy metals/metalloids are absorbed on algae. The algal cell wall contains functional groups (amino, carboxyl, sulfonate, and sulfhydryl), and ion exchange increases the adsorption of metallic elements (Taleei et al. 2019).

15.6.1.2 Bioleaching

The bioleaching approach relies on microorganisms' ability to produce secretions, viz. low molecular weight organic acids, having the ability to dissolve heavy metals, and soil particles possess the ability to adsorb heavy metal ions to limit their mobility and stabilize contaminants. Heavy metals are thereby mineralized by microbial metabolism or their metabolites, either directly or indirectly. Biosurfactants, such as lipids, polysaccharides, and lipopeptides, are generated by microorganisms and have a high surface activity that promotes the formation of chelating metals, binding of metal ions, and leaching (Yang et al. 2018).

15.6.1.3 Bioaccumulation

Bioaccumulation occurs when the rate of absorption of the substance is greater than the rate of loss of the substance. As a result, the pollutant is trapped inside the organism and accumulates. Bioaccumulation is a toxicokinetic process that impacts living organisms' chemical sensitivity. Candidate species for bioaccumulation should be able to tolerate a wide range of pollutants, from low levels to high ones. They may also have better bio transformational skills, which allow them to change a hazardous chemical into a non-toxic form, reducing the contaminant's toxicity while keeping it confined. It's vital to study the mechanisms in bioaccumulation, as well as the genes that control sensitivity and tolerance and the maximum concentrations of specific compounds, which could be crucial in choosing suitable bioaccumulation species (Chojnacka 2010).

Changes in redox processes, increasing or reducing solubility by adjusting pH or employing other complex reactions, and adsorption/uptake of a pollutant from the contaminated region are the key ideas in bioremediation. Bioremediation, which involves using microorganisms for detoxification and breakdown of different environmental toxins, has received much attention in recent years. Alterations introduced in the genetic constitution of microorganisms through the genetic modification step are known to increase further biosorption and bioaccumulation (Joutey et al. 2015).

15.7 System Biology Approach in Bioremediation

Enhanced bioremediation approaches require complete knowledge of the complex regulatory system in microorganisms. Pervious scientific studies have developed an advanced platform consisting of various tools, including OMICs approaches for studying the physiological and functional factors of contaminant degraders in a comprehensive manner. System biological methods are basically used for exploring the pathways adopted by plants, bacteria, and fungi, for example, tolerance, accumulation, filtration, and metabolism for effective elimination of metals and metalloids (Chandran et al. 2020). The OMICs approach involves the application of powerful tools including proteomics, transcriptomics, and metabolomics to unwind the types of genes and proteins and their functions involved in the mitigation of contaminants from the environment (Rodríguez et al. 2020).

16S rRNA sequencing helps to identify the chromium remediating bacteria *E. coli* FACU, which is capable to reduce chromate at 100 μg mL^{-1} by converting Cr(VI) into Cr(III) (Mohamed et al. 2020). Moreover, transcriptomics approaches are also used for tracking the heavy metal stress tolerance in microalgae response, including gene expression, protein synthesis, and cellular signaling (Tripathi and Poluri 2021). Furthermore, materials like morpholine-functionalized polycarbonate hydrogels are introduced, and function as metal ions sequesters from the liquid solutions (Kawalec et al. 2013).

15.8 Phytoremediation

Phytoremediation is the technique of reducing the detrimental effects of pollutants in the environment by employing plants and microbes (Gadd 2010). This approach has been found to be a cost-effective and efficient strategy for removing metalloids from contaminated soil compared to engineering procedures such as excavation, solidification, soil burning, washing, and flushing. In 1983, Chaney was the first to suggest phytoremediation as an alternative. Green plant remediation of hazardous metals and metalloids is gaining popularity as a potential alternative to chemical and physical remediation (Sarwar et al. 2017).

Phytoremediation is one of the branches of bioremediation, which involves the unique abilities of plants to intercept, absorb, collect, adsorb, or stabilize pollutants. The phytoremediation method consists in transferring pollutants to less harmful components via the roots of plants or absorbing them into the roots or shoots. Furthermore, the phytoremediation method's having aesthetically pleasing characteristics have made it a popular remediation approach among locals. Phytoremediation is an eco-friendly approach since it decreases the risk of pollutant dispersal while also maintaining the indigenous ecotype by circumventing the excavation of polluted sites. Phytoremediation is also used to reduce risk and recover precious metals, including nickel, thallium, and gold, using phytoextraction (Muthusaravanan et al. 2018). Furthermore, prior studies have discovered the economic advantage of using phytoremediation on durable land since the area may be utilized to produce higher-value crops once the soil quality has been restored by phytoextraction. Phytoremediation techniques include rhizofiltration, phytoextraction, phytovolatilization, photoevaporation, phytostabilization, phytodesalination, rhizodegradation, and phytodegradation (Fig 15.2). *Helianthus annuus*, *B. juncea*, shrub tobacco *Nicotiana glauca*, and yellow poplar *Liriodendron tulipifera* are among the high biomass producing metallophytes with well-established genetic modification processes, making it a viable choice for modified phytoremediation plants (Kotrba 2013).

The plant metal homeostasis procedure includes metal uptake, chelation, trafficking, transport into different cell chambers and organelles, as well as storage or discharge of metal ions under excess situations (Muthusaravanan et al. 2018).

15.8.1 Phytoextraction

When a plant's roots collect pollutants from the soil or water, and transfer them to the aboveground biomass, like the plant's shoots and leaves, phytoextraction or phytoaccumulation occurs. The plant employed in phytoextraction should be capable of generating a large amount of biomass for collecting pollutants. The ratio between the soil and plant metal concentration determines the efficacy of plant species in the phytoextraction of metals. After that, plants that perform phytoextraction are collected

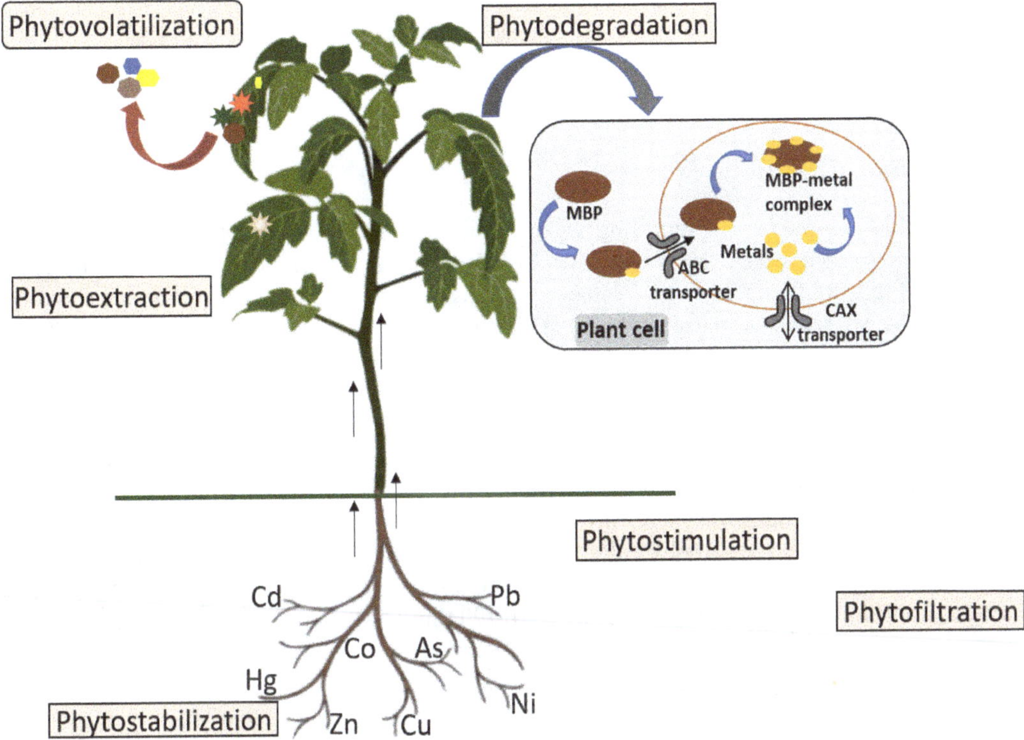

Fig. 15.2. Different techniques used by plants for remediating metals and metalloids contaminated environment.

and burned. The ash created by the burning will be disposed of in a landfill, removing the pollutants from the soil (Cristaldi et al. 2017).

However, phytoextraction efficacy will be limited by the metal's poor bioavailability in soil and its low absorption rate. Furthermore, the efficacy of phytoextraction will be decreased if metals are kept in the roots rather than transported to the leaves and shoots (Cristaldi et al. 2017, Suman et al. 2018).

15.8.2 Phytostabilization

Phytostabilization is a method of restricting pollutant transport in the soil by absorbing heavy metals through the roots or precipitating heavy metals inside the rhizosphere. The plant used to do phytostabilization alters the soil chemistry, allowing heavy metal absorption and precipitation in the soil. Furthermore, during the phytostabilization process, specific redox enzymes secreted by plants transform heavy metals in soil into less hazardous conditions. At the polluted metal mining site, this approach is well-practiced. The phytostabilization method avoids groundwater pollution by preventing additional percolation and mobilization of metallic pollutants. However, because this approach merely deactivates and stabilizes heavy metals rather than eliminating them from the soil or water, it is only a management strategy (Saran et al. 2019).

15.8.3 Phytovolatilization

Phytovolatilization is the procedure by which pollutants are absorbed from the soil, transported through the xylem, converted into a volatile and less harmful form, and released to the environment. Phytovolatilization has been employed to eradicate metals such as selenium and mercury since they are highly volatile. This process is especially advantageous for Hg since the mercuric ion is transformed to a less dangerous elemental state (Cristaldi et al. 2017). This approach provides a temporary solution to the problem since the metal(loid)s transferred into the atmosphere through volatilization can be returned back to the soil by the mean of precipitation (Wang et al. 2012).

15.8.4 Rhizofiltration

Rhizofiltration is a method of removing organic and inorganic contaminants from contaminated groundwater, surface water, and wastewater by adsorbing and precipitating them on the roots of plants. Hypoxia tolerance, metal tolerance, and high surface area for absorption are the most important characteristics to consider when selecting a plant for rhizofiltration (Cristaldi et al. 2017).

15.8.5 Rhizodegradation

Rhizodegradation occurs when soil organic pollutants are degraded by microorganisms in the rhizosphere. Rhizoremediation can be carried out by all types of microorganisms, including fungi, bacteria, and yeasts. The rhizosphere has a higher concentration of microorganisms than the soil. High levels of amino acids, carbohydrates, and flavonoids are present in the exudates produced by the plants in the ground. The nutrient-rich exudates offer nitrogen and carbon supplies to rhizosphere bacteria, boosting their metabolic activity by 10 to 100 times (Ali et al. 2013, Cristaldi et al. 2017).

15.8.6 Potential Techniques for Phytoremediation

The phytoremediation treatment process is sustainable, low-cost, environment-friendly; however, it is a slow process as it depends on plant development, seasonality, and climatic conditions. To overcome these limitations and improve metal removal efficiency, phytoremediation is frequently coupled with other treatment methods, for example, by combining (Awa and Hadibarata 2020):

- Phytoremediation + Bioaugmentation
- Phytoremediation + Chelating agent (such as EDTA) amended soil
- Phytoremediation + Electrokinetic remediation

Plants have been used to remediate hazardous metals and metalloids; however, due to heavy metal phytotoxicity, this process has been sluggish and generally unsuccessful (Mosa et al. 2016). Few biotechnological approaches were employed to speed up the phytoremediation process for better eradication of heavy metal pollutants from the environment. There are three primary biotechnological methods used in combination with phytoremediation (Awa and Hadibarata 2020). Mainly two processes, i.e., (i) Metal/metalloid transporter genes (ii) Uptake system manipulation, enhance the synthesis of metal and metalloid ligands. Thus, metals and metalloids are transformed into less hazardous and volatile forms. Along with these approaches, new OMICs approaches such as genomics, transcriptomics, proteomics, and metabolomics are also being used to unravel the genetic factors and metabolic networks involved in heavy metal and metalloid tolerance in the phytoremediation plants (Mohanty 2021).

15.8.7 Factors Affecting Phytoremediation

Phytoremediation involves three major approaches, viz. phyto-stabilization, phytoextraction, and phytovolatilization for the eradication of heavy metals from contaminated ecosystems. Heavy metals are prevented from entering the food chain through phyto-stabilization. Phyto-stabilization prevents the entry of heavy metals into the food chain, whereas in the phytoextraction process, heavy metals are absorbed from the soil and stored in the plant tissues. Further, phytovolatilization enables the evaporation of heavy metals present in the soil and turns them into less toxic and non-hazardous forms (Umar 2017).

Several variables, including plant species, medium characteristics, metal bioavailability in soil, and the addition of chelating agents, are all key elements that can influence processes in phytoremediation. Phytoremediation generally involves the use of metal hyperaccumulators (Dixit and Dhankher 2011). These plants may collect high concentrations of heavy metals in their biomass while remaining healthy. These metal hyperaccumulators are also a vital factor that can affect the process of phytoremediation (Umar 2017).

15.9 Enhancement in Phytoremediation by using Genetically Engineered Plants

An appropriate plant for phytoremediation must be able to accumulate and tolerate metal(s) in overground tissues, be herbivore-repellent, adapt to native environmental conditions, be easily cultivated with a wide global distribution efficiently grow and produce large biomass (Ozyigit et al. 2020). As wild plants can not satisfy all these conditions, the transgenic technique is being employed to fulfill all these demands. A transgenic approach is a promising future alternative for combining the high deposition capacity for heavy metal(loid)s with these novel properties. The *Brassica juncea*, *Helianthus annuus*, *Liriodendron tulipifera*, and shrub like *Nicotiana glauca* are some of the high bio-metering metallophytes with established technicians for a genetic alteration that appear to be appropriate contenders as engineered plants (Vazquez-Nunez et al. 2018). Changing the solubility of extracellular metals ions by transferring, chelating, or acidifying apoplast, absorbing ion, and/or trafficking within cells, transporting inside cells, and storing metal ion outflows under extreme conditions are all part of the plant metal homeostasis network (Tiwari and Lata 2018). The foremost biotechnological approaches to the development of phyto remedial genetically modified plants include alteration of absorption and conveyance by metal-transporting bio membranes through metal transporters, increasing efficiency in intracellular chelation (glutathione, phytochelatin, metallothioneins) and subcellular compartments (Gunarathne et al. 2019).

15.9.1 Metal Transporters

Transition metal transporters genes play a crucial function in metal(loid)s absorption, cell division, metal distribution to the metal proteins required, and metal sequestration (i.e., vacuoles, chloroplast, mitochondria). They are of two types, viz. transporters for metal retention and transporters for metal efflux (Ozyigit et al. 2020). The multifamily efflux transporter families of ATP-bound transporter families (ABC) of multidrug resistance-associated proteins (MRPs) have been found to be changed in plant genomes and have been nicotianamine synthase (NAS) and SAM families (S-adene-methionine synthesis) (Tomar et al. 2016). Aquaporin transporter family members transport arsenite in plants, according to recent research (Bienert et al. 2008, Mosa et al. 2012, Zhao et al. 2009). Aquaporins from plants and aquaglyceroporins from bacteria, yeast, and mammals are members of the intrinsic protein family. Metal transporter genetic modification has resulted in increased accumulation and metal tolerance in many plant species. For instance, in *Arabidopsis thaliana* yeast YCF1 overexpression has led to an increase in accumulation and tolerance of Cd (II) and Pb (II) (Song et al. 2003). Arsenic, as arsenate (AsO_4^{-3}), is found naturally in soil. Due to chemical similarities between inorganic phosphate Pi (PO_4^{-3}) and arsenate, arsenate (V) is frequently taken up by active phosphate transporters present in the plants. Transgenic *Arabidopsis* has been found to improve arsenic accumulation and tolerance (LeBlanc et al. 2011).

For phytoremediation purposes, genetic manipulation of metal transporters should focus on three levels: soil root metal absorption, root to shoot metal translocation, and cell wall or vacuole metal sequestration. The expression of metal conveyor genes can be enhanced, depending upon the region where the protein is exposed, in terms of metal absorption, translocation, and/or sequestration (Anju 2017).

15.9.2 Metal-Binding Proteins

Metal sequestration and chelation are key processes utilized for the treatment of heavy metal stress by the majority of plants and micro-organisms. The two best-characterized thiol-rich peptides, metallothioneins (MTs) and phytochelatins (PCs), bind together in a range of metals to create stable cytosol compounds and decrease their toxicity in plant cell sequestration to a vacuole (Singh et al. 2021b). In order to cope with heavy metal stress, many plants and microbes use metal sequestration and chelation via specific ligands. Thiol-rich peptides (MTs) and phytochelatin (PCs), two of the most well-studied metal-binding proteins, create stable cytosol complexes and minimize metal toxicity in plant cells through vacuole sequestration. Ligands are largely present in plants, and numerous genetic and physiological studies have shown PCs and MTs to be crucial in metal accumulation and tolerance in plants. PCs and MTs have been frequently utilized to increase the metal accumulating capability of bacteria by producing them at several parts of plants (Yan et al. 2020).

15.9.2.1 Manipulated metal transporters genes

- The NtCBP4 metal transporter gene obtained from the *N. tabacum* source was used in the *N. tabacum* target plant, which produced channel proteins for tobacco calmodulin-binding that describes Ni^{2+} tolerance and Pb^{2+} hypersensitivity in transgenic plants (Daghan 2019).
- *Arabidopsis thaliana* target plant, with AtNRAMP1 metal transporter gene, is the only transgenic plant that survived 600 μm of toxic iron for a long duration (25 days) (Curie et al. 2000).
- The PHT1 or PHT7 with/without YCF1 metal transporter gene acquired from Arabidopsis source was also employed in Arabidopsis as target plant, which generated high-affinity transporters for phosphate that increased phosphate accumulation and tolerance (LeBlanc et al. 2011).
- The yeast source gene AtABCC3 transporter was utilized in *A. Thaliana* as a goal plant to create a transporter type ABC that improved the Cd tolerance (Brunetti et al. 2015).
- The YCF1 metal transporter gene obtained from yeast source is used in *Arabidopsis* and Poplar as target plants. YCF1 is a member of the ATP binding cassette transporter family, which leads to the driving of Cd(II) conjugated to glutathione into vacuoles. Overexpression of YCF1 in *Arabidopsis* showed increased endurance and accumulation of Cd and Pb (Song et al. 2003).
- The AtCAX2 metal transporter genes obtained from *A. thaliana* source were used in the tobacco plant, which produced calcium vacuolar transporter that enhances metal deposition in roots and shoots (Hirschi et al. 2000).
- The ZAT metal transporter gene derived from *A. thaliana* source was utilized to produce a putative Zn^{2+} transporter, which increased Zn^{2+} tolerance and accumulation by two-fold (van der Zaal et al. 1999).
- The metal transporter gene PgIREG1 was obtained from *Psychotria gabriellae*. Mainly PgIREG1 provides Ni tolerance in yeast and genetically modified plant (production of vacuolar transporter) (Merlot et al. 2014).
- HMT1 was utilized as a metal transporter to enhance the tolerance to the accumulated As, Cd, Zn, and Cu obtained from the *S. pombe* source in *A. thaliana* (Huang et al. 2012).

15.9.2.2 Metallothioneins (MTs)

MTs are sulfate-rich proteins of low molecular weight (5–10 kDa) that tie metal ions in metallothioneins classes. MTs are encrypted within the gene family of cyanobacteria, eukaryotic, higher plants, microorganisms, and some prokaryotes, fungi, and mammals (Singh et al. 2021b). Based on the cysteine concentration and structure, three varieties of MTs are classified, i.e.

i. Cys-Cys
ii. Cys-X-Cys
iii. Cys-X-X-Cys

Where "X" is any amino acid

As they form mercaptide connections with metal ions, MTs play a major role in homeostasis, detoxification, metal tolerance, and plant dispersion. In bacteria and plants, including mice, hamsters, yeast, humans, plants, and cyanobacteria, a number of genes have been overexpressed (Singh et al. 2021b).

15.9.2.2.1 Transgenic bacteria and plants—the role of Metallothionein genes

- The mt-1 is an MT gene that is obtained from Mouse *(Mus musculus)* and used in *Lycopersicon esculentum* and *Nicotiana tabacum* (genetically modified plant species). The plants with the MT gene have accumulated 1.6 times more zinc content than in WT (wild type) and can tolerate 200 mM $CdCl_2$ at the seedling level (Pan et al. 1994, Sheng et al. 2007). The *cup1* is an MT gene which is obtained from yeast (*S. cerevisiae*) and used in *N. tabacum* (genetically modified plant species), has 2–3 times higher Cu concentration than the control, but no Cd tolerance (Thomas et al. 2003).
- ScMT II is an MT gene derived from *S. cerevisiae* and used in *N. tabacum* (genetically modified plant species). The transgenic tobacco plant was able to deposit 3.5–4.5 times more Cd than the critical level (100 mg Cd/kg) (Daghan et al. 2013).
- MTL4, an MT gene acquired from humans (*Homo sapiens*) and employed in *Mesorhizobium huakuii* PGP subsp. rengei coupled with the plant *A. sinicus*, would improve Cd^{2+} uptake by two-fold (Sriprang et al. 2002).
- The *smtA* is another MT gene that is obtained from cyanobacteria (*Synechococcus* sp.) and used in *A. thaliana*. Compared to wild plants, the application of smtA to *A. thaliana* plant showed tolerance to higher zinc concentration (Xu et al. 2010).

15.9.2.3 Phytochelatins (PCs)

Some fungi and photosynthetic organisms (angiosperms, gymnosperms, and bryophytes) create enzymatically low-molecular-weight cysteine-rich metal-binding peptides termed as phytochelatins in response to high metal stress. These peptides are crucial in the metal detoxifying ability of plants because they sequestrate many metal(loid) ions in vacuoles in the form of complexes of the metal(loid)-thiolate (Singh et al. 2021b). In the context of its method for metal-intolerant plants, mass manufacturing of PCs appears to be a practical rather than fundamental process. The PCs mainly have thiol-rich peptides, consisting of a non-translational (non-ribosomal) GSH enzyme phytochelatin synthase for producing (-EC)nG molecules (n 2-11) (Singh et al. 2021b).

Metals such as Cd, Hg, Ag, Cu, Ni, Au, Pb, and Zn have demonstrated that they encourage the synthesis of PCs in many plant species, but with the greatest strength, Cd activates. Transgenic plants with high phytochelatin synthase were more tolerant to heavy metals and had greater phytochelatin levels (Kim et al. 2019). Also, related arsenic detoxification is provided by phytochelatins. Heavy metal stress in some engineered bacteria may trigger phytochelatin (PC) synthase (Gadd 2010) particularly in the areas of element biotransformations and biogeochemical cycling, metal and mineral transformations, decomposition, bioweathering, and soil and sediment formation. All kinds of microbes, including prokaryotes and eukaryotes and their symbiotic associations with each other and 'higher organisms', can contribute actively to geological phenomena, and central to many such geomicrobial processes are transformations of metals and minerals. Microbes have a variety of properties that can effect changes in metal speciation, toxicity and mobility, as well as mineral formation or mineral dissolution or deterioration. Such mechanisms are important components of natural biogeochemical cycles for metals as well as associated elements in biomass, soil, rocks and minerals, e.g., sulfur and phosphorus, and metalloids, actinides and metal radionuclides. Apart from being important in natural biosphere processes, metal and mineral transformations can have beneficial or detrimental consequences in a human context. Bioremediation is the application of biological systems to the clean-up of organic and inorganic pollution, with bacteria and fungi being the most important organisms for reclamation, immobilization or detoxification of metallic and radionuclide pollutants. Some biominerals or metallic elements deposited by microbes have catalytic and other properties in nanoparticle, crystalline or colloidal forms, and these are relevant to the development of novel biomaterials for technological and antimicrobial purposes. On the negative side, metal and mineral transformations by microbes may result in spoilage and destruction of natural and synthetic materials, rock and mineral-based building materials (e.g., concrete). For instance, the phytochelatin synthase gene (HaPCS) from *H. annuus* overexpressed in *E. coli* and showed the enhanced accumulation of As^{3+} (10.67 μmol/g DCW), As^{5+} (7.58 μmol/g DCW), and Cd^{2+} (25.97 μmol/g DCW) (Singh et al. 2021b).

15.9.2.3.1 Overproducing phytochelatin synthase in transgenic plants

- The gene produced from *Triticum aestivum* is known as Phytochelatin synthase gene TaPCS1. *Nicotiana glauca* uses gene origin. On medium with 800 mM Pb^{2+} or 50 mM Cd^{2+}, genetically modified plant species has 1.6 times longer roots. From polluted soil, the shoots of the altered line NgTP1 collected increased Cd, Cu, Ni, Pb, and Zn (Wang et al. 2012).
- *Ceratophyllum demersum* possesses the Phytochelatin synthase gene CdPCS1. *Arabidopsis* uses the gene origin. Compared to WT, there was a considerable increase in accumulation of Cd, As(III), and As (V) in aerial parts of the plant without any considerable variation in growth parameters (Shukla et al. 2013).
- The phytochelatin synthase gene of *A. thaliana* is AtPCS1. In *N. tabacum* and *B. juncea*, using the gene origin, 1.9 and 1.4 times longer roots and shoots were observed with a concentration of 100 and 500 mM Cd^{2+}, respectively (Zhang et al. 2019).

15.10 Challenges and Future Prospects

The fact that not all substances are biodegradable is a major constraint in bioremediation, which limits the applications of this process. If the material is biodegradable at all, its downstream processing and decomposition may produce harmful compounds as well. There are numerous advantages associated with bioremediation like eco-friendliness, specificity, self-reproducibility, recycling, adaptability, etc. But, it is a very long process, and it takes many years to remediate heavy metals or other pollutants. There are so many other limitations. Furthermore, because all bacteria cannot be cultivated in the laboratory, the microbial population is frequently considered a "black box" when assessing its contribution to bioremediation and impact on the environment. There is more than one metal present in the polluted environment. For remediation, it is crucial to develop multi-metal resistant bacterial strain; again, it is a very challenging task. It is a huge challenge to bring native microbial bioremediation techniques to the commercial level by making them quicker and reusable.

Phytoremediation is also an attractive option for heavy metal remediation, but its practical application is very challenging due to many reasons:

- It takes a long period for remediation.
- The efficacy of phytoextraction is further hampered due to the hyperaccumulator's poor growth rate and low biomass.
- Plants have a limited depth to which they can reach.
- Under increasing metal concentration, the decline in phytoextraction efficiency has also been reported.
- Pest and disease attacks may also affect the metal accumulation capacity in plants.
- Lack of know-how on agronomic practices and control is likewise a vital factor.
- Restricted geographical distribution of metallophytes.
- The climatic conditions and their impact on the plant.
- Risk to indigenous species in case of invasion of hyperaccumulators.
- Metal ions' solubility and bioavailability in soil.
- Currently, most part of the research is still limited to only the laboratory scale.

In the future, maybe it is possible to implicate these researches and techniques in the actual field. Moreover, few methods like metabolomic evaluation can assist in the identification of the metabolites related to heavy metallic and metalloid stresses that could similarly map its metabolic pathways to identify the associated candidate genes (Kumar et al. 2014). Laser ablation inductively coupled plasma mass spectrometry (LA-ICP-MS), Matrix-Assisted Laser Desorption and Ionization (MALDI), and Fourier Transform Ion Cyclotron Resonance Mass Spectrometry (FT-ICR-MS) have all been used to achieve this goal (Jones et al. 2015).

Future efforts should focus on developing techniques to improve metal tolerance, its uptake by the plant, its hyperaccumulation inside the cell, and its translocation to different tissues by using genomic and metabolic engineering techniques. CRISPR/Cas9-mediated targeted mutagenesis of poplar has recently been proven effective and successful (Kosicki et al. 2018). Genomics, transcriptomics, proteomics, and metabolomics should help identify participating genes for plant repair, using unique strategies to combine trans-genes, cis-gene, gene-stacking, genome editing, and metabolic engineering.

15.11 Conclusion

Bioremediation is one of the safest and most eco-friendly methods to remove contaminants from the environment or convert them into less toxic forms. It involves the use of various microbes and enzymes to remediate and restore polluted environments. However, due to the formation of hazardous metabolites by microorganisms and the non-biodegradability of heavy metals, bioremediation of heavy metals has limitations.

Plants might be used to clean up heavy metal pollution, which would be an effective pollution management method without producing secondary contaminants. But, it also has certain limitations. Although it is a green and ecofriendly approach, its implication in the actual field is not easy. In phytoremediation, mainly phytoextraction is becoming a more helpful method for heavy metal(loid) s remediation. In the coming days, microbe-assisted phytoremediation can be a rising technology to resolve heavy metal(loid)s problems without much adverse impact. The various bacteria enhance the solubility and bioavailability of heavy metals, thus, promoting phytoremediation through a variety of methods, including biomethylation, microbial biosurfactants, redox processes, siderophores, and organic acids. PGPB play an important role in phytoremediation by generating growth elevating substance. Transgenic plants play a significant role in making the plant more tolerant by increasing accumulation capacity and, in another way, by overexpressing modified genes. Now it is more important to shift from know how to do how because currently, maximum research is only limited to the laboratory.

References

Akbari, M., A. Hallajisani, A. R. Keshtkar, H. Shahbeig and S. Ali Ghorbanian. 2015. Equilibrium and kinetic study and modeling of Cu(II) and Co(II) synergistic biosorption from Cu(II)-Co(II) single and binary mixtures on brown algae *C. indica*. J. Environ. Chem. Eng. 3: 140–149.

Ali, H., E. Khan and M. A. Sajad. 2013. Phytoremediation of heavy metals—concepts and applications. Chemosphere 91(7): 869–881.

Ali, H. and E. Khan. 2018. What are heavy metals? Long-standing controversy over the scientific use of the term 'heavy metals'—proposal of a comprehensive definition. Toxicol. Environ. Chem. 100(1): 6–19.

Anju, M. and D. K. Banerjee. 2011. Associations of cadmium, zinc, and lead in soils from a lead and zinc mining area as studied by single and sequential extractions. Environ. Monit. Assess. 176(1): 67-85.

Anju, M. 2017. Biotechnological strategies for remediation of toxic Metal(loid)s from environment. pp. 315–360. *In*: Gahlawat, S. K. et al. (eds.). Plant Biotechnology: Recent Advancements and Developments, Springer, ISBN: 978-981-10-4731-2.

Awa, S. H. and T. Hadibarata. 2020. Removal of heavy metals in contaminated soil by phytoremediation mechanism: a review. Water Air Soil Pollut. 231: 1–15.

Awofolu, O. R. 2005. A survey of trace metals in vegetation, soil and lower animal along some selected major roads in metropolitan city of Lagos. Environ. Monit. Assess. 105: 431–447. https://doi.org/10.1007/s10661-005-4440-0.

Bano, A., J. Hussain, A. Akbar, K. Mehmood, M. Anwar, M. S. Hasni, S. Ullah, S. Sajid and I. Ali. 2018. Biosorption of heavy metals by obligate halophilic fungi. Chemosphere 199: 218–222.

Basha, S. A. and K. Rajaganesh. 2014. Original research article microbial bioremediation of heavy metals from textile industry dye effluents using isolated bacterial strains. Int. J. Curr. Microbiol. Appl. Sci. 3: 785–794.

Bienert, G. P., M. Thorsen, M. D. Schüssler, H. R. Nilsson, A. Wagner, M. J. Tamás and T. P. Jahn. 2008. A subgroup of plant aquaporins facilitate the bi-directional diffusion of $As(OH)_3$ and $Sb(OH)_3$ across membranes. BMC biol. 6(1): 1–15.

Brunetti, P., L. Zanella, A. D Paolis, D. Di Litta, V. Cecchetti, G. Falasca and M. Cardarelli. 2015. Cadmium-inducible expression of the ABC-type transporter AtABCC3 increases phytochelatin-mediated cadmium tolerance in *Arabidopsis*. J. Exp. Bot. 66(13): 3815–3829.

Chaffai, R. and H. Koyama. 2011. Heavy metal tolerance in *Arabidopsis thaliana*. Adv. Bot. Res. 60: 1–49.

Chandran, H., M. Meena and K. Sharma. 2020. Microbial biodiversity and bioremediation assessment through omics approaches. Front. Environ. Chem.1: 9.

Cheng, J., W. Yin, Z. Chang, N. Lundholm and Z. Jiang. 2017. Biosorption capacity and kinetics of cadmium(II) on live and dead Chlorella vulgaris. J. Appl. Phycol. 29: 211–221.

Chojnacka, K. 2010. Biosorption and bioaccumulation–the prospects for practical applications. Environ. Int. 36(3): 299–307.

Costa, F. and T. Tavares. 2016. Biosorption of nickel and cadmium in the presence of diethylketone by a *Streptococcus equisimilis* biofilm supported on vermiculite. Int. Biodeterior. Biodegrad. 115: 119–132.

Cristaldi, A., G. O. Conti, E. H. Jho, P. Zuccarello, A. Grasso, C. Copat and M. Ferrante. 2017. Phytoremediation of contaminated soils by heavy metals and PAHs. A brief review. Environ. Technol. Innov. 8: 309–326.

Cui, Z., X. Zhang, H. Yang and L. Sun. 2017. Bioremediation of heavy metal pollution utilizing composite microbial agent of *Mucor circinelloides, Actinomucor* sp. and *Mortierella* sp. J. Environ. Chem. Eng. 5: 3616–3621.

Curie, C., J. M. Alonso, M. L. Jean, J. R. Ecker and J. F. Briat. 2000. Involvement of NRAMP1 from *Arabidopsis thaliana* in iron transport. Biochem. J. 347(3): 749–755.

Czarnecki, S. and R. A. Düring. 2015. Influence of long-term mineral fertilization on metal contents and properties of soil samples taken from different locations in Hesse, Germany. Soil. 1(1): 23–33.

Daghan, H., M. Arslan, V. Uygur and N. Koleli. 2013. Transformation of tobacco with ScMTII gene-enhanced Cadmium and Zinc accumulation. CLEAN–Soil, Air, Water. 41(5): 503–509.

Daghan, H. 2019. Transgenic tobacco for phytoremediation of metals and metalloids. pp. 279–297. *In*: Prasad, M. N. V. (ed.). Transgenic Plant Technology for Remediation of Toxic Metals and Metalloids. Academic Press. ISBN: 978-0-12-814389-6

Deniz, F. and A. Karabulut. 2017. Biosorption of heavy metal ions by chemically modified biomass of coastal seaweed community: Studies on phycoremediation system modeling and design. Ecol. Eng. 106: 101–108.

Dixit, A. R. and O. P. Dhankher. 2011. A novel stress-associated protein 'AtSAP10'from *Arabidopsis thaliana* confers tolerance to nickel, manganese, zinc, and high temperature stress. PLoS one 6(6): e20921.

Gadd, G. M. 2010. Metals, minerals and microbes: geomicrobiology and bioremediation. Microbiol. Soc. 156: 609–643.

Gola, D., P. Dey, A. Bhattacharya, A. Mishra, A. Malik, M. Namburath and S. Z. Ahammad. 2016. Multiple heavy metal removal using an entomopathogenic fungi *Beauveria bassiana*. Bioresour. Technol. 218: 388–396.

Guarino, F., A. Miranda, S. Castiglione and A. Cicatelli. 2020. Arsenic phytovolatilization and epigenetic modifications in *Arundo donax* L. assisted by a PGPR consortium. Chemosphere 251: 126310.

Gunarathne, V., S. Mayakaduwa, A. Ashiq, S. R. Weerakoon, J. K. Biswas and M. Vithanage. 2019. Transgenic plants: benefits, applications, and potential risks in phytoremediation. pp. 89–102. *In*: Prasad, M. N. V. (ed.). Transgenic Plant Technology for Remediation of Toxic Metals and Metalloids. Academic Press. ISBN: 978-0-12-814389-6.

Hassan, A., A. Periathamby, A. Ahmed, O. Innocent and F. S. Hamid. 2020. Effective bioremediation of heavy metal–contaminated landfill soil through bioaugmentation using consortia of fungi. J. Soils Sediments 20: 66–80.

He, Z., J. Shentu, X. Yang, V. C. Baligar, T. Zhang and P. J. Stoffella. 2015. Heavy metal contamination of soils: sources, indicators and assessment. J. Environ. Indic, 9: 17–18.

Hirschi, K., V. D. Korenkov, N. L. Wilganowski and G. J. Wagner. 2000. Expression of Arabidopsis CAX2 in tobacco. Altered metal accumulation and increased manganese tolerance. Plant Physiol. 124(1): 125–134.

Huang, J., Y. Zhang, J. S. Peng, C. Zhong, H. Y. Yi, D. W. Ow and J. M. Gong. 2012. Fission yeast HMT1 lowers seed cadmium through phytochelatin-dependent vacuolar sequestration in Arabidopsis. Plant Physiol. 158(4): 1779–1788.

Imran, M., A. Rehim, N. Sarwar and S. Hussain. 2016. Zinc bioavailability in maize grains in response of phosphorous–zinc interaction. J. Plant Nutr. Soil Sci. 179: 60–66.

Jaiswal, A., A. Verma and P. Jaiswal. 2018. Detrimental effects of heavy metals in soil, plants, and aquatic ecosystems and in humans. J. Environ. Pathol. Toxicol. Oncol. 37: 183–197.

Jiang, L., W. Zhou, D. Liu, T. Liu and Z. Wang. 2017. Biosorption isotherm study of Cd2+, Pb2+ and Zn2+ biosorption onto marine bacterium *Pseudoalteromonas* sp. SCSE709-6 in multiple systems. J. Mol. Liq. 247: 230–237.

Jin, Y., Y. Luan, Y. Ning and L. Wang. 2018. Effects and mechanisms of microbial remediation of heavy metals in soil: a critical review. Appl. Sci. 8: 1336.

Jones, O. A. H., D. A. Dias, D. L. Callahan, K. A. Kouremenos, D. J. Beale and U. Roessner. 2015. The use of metabolomics in the study of metals in biological systems. Metallomics 7: 29–38.

Joutey, N., H. Sayel, W. Bahafid and N. El Ghachtouli. 2015. Mechanisms of hexavalent chromium resistance and removal by microorganisms. Rev. Environ. Contam. Toxicol. 233: 45–69.

Kaewdoung, B., T. Sutjaritvorakul, G. M. Gadd, A. J. S. Whalley and P. Sihanonth. 2016. Heavy metal tolerance and biotransformation of toxic metal compounds by new isolates of wood-rotting fungi from Thailand. Geomicrobiol. J. 33: 283–288.

Kang, C. H., S. J. Oh, Y. J. Shin, S. H. Han, I. H. Nam and J. S. So. 2015. Bioremediation of lead by ureolytic bacteria isolated from soil at abandoned metal mines in South Korea. Ecol. Eng. 74: 402–407.

Kawalec, M., A. P. Dove, L. Mespouille and P. Dubois. 2013. Morpholine-functionalized polycarbonate hydrogels for heavy metal ion sequestration. Polym. Chem. 4: 1260–1270.

Khalid, S., M. Shahid, N. K. Niazi, B. Murtaza, I. Bibi and C. Dumat. 2017. A comparison of technologies for remediation of heavy metal contaminated soils. J. Geochem. Explor. 182: 247–268.

Kim, Y. O., H. Kang and S. J. Ahn. 2019. Overexpression of phytochelatin synthase AtPCS2 enhances salt tolerance in *Arabidopsis thaliana*. J. Plant Physiol. 240: 153011.

Kosicki, M., K. Tomberg and A. Bradley. 2018. Repair of double-strand breaks induced by CRISPR–Cas9 leads to large deletions and complex rearrangements. Nat. Biotechnol. 36: 765–771.

Kotrba, P. 2013. Transgenic approaches to enhance phytoremediation of heavy metal-polluted soils. In Plant Based Remediation Processes (239–271). Springer, Berlin, Hiedelberg.

Kumar, A., U. Kage, K. Mosa and D. Dhokane. 2014. Metabolomics: a novel tool to bridge phenome to genome under changing climate to ensure food security. Med. Aromat. Plants 3 (e154). https://doi.org/http://dx.doi.org/10.4172/2167-0412.1000e154.

LeBlanc, M., A. Lima, P. Montello, T. Kim, R. B. Meagher and S. Merkle. 2011. Enhanced arsenic tolerance of transgenic eastern cottonwood plants expressing gamma-glutamylcysteine synthetase. Int. J. Phytoremediat. 13: 657–673.

Li, C., K. Zhou, W. Qin, C. Tian, M. Qi, X. Yan and W. Han. 2019. A review on heavy metals contamination in soil: effects, sources, and remediation techniques. Soil Sediment Contam.: An Int. J. 28(4): 380–394.

Li, M., X. Cheng and H. Guo. 2013. Heavy metal removal by biomineralization of urease producing bacteria isolated from soil. Int. Biodeterior. Biodegrad. 76: 81–85.

Mani, D. and C. Kumar. 2014. Biotechnological advances in bioremediation of heavy metals contaminated ecosystems: an overview with special reference to phytoremediation. Int. J. Environ. Sci. Technol. 11: 843–872.

Merlot, S., L. Hannibal, S. Martins, L. Martinelli, H. Amir, M. Lebrun and S. Thomine. 2014. The metal transporter PgIREG1 from the hyperaccumulator *Psychotria gabriellae* is a candidate gene for nickel tolerance and accumulation. J. Exp. Bot. 65: 1551–1564.

Mohamed, M. S. M., N. I. El-Arabi, A. El-Hussein, S. A. El-Maaty and A. A. Abdelhadi. 2020. Reduction of chromium-VI by chromium-resistant *Escherichia coli* FACU: a prospective bacterium for bioremediation. Folia Microbiol. 65: 687–696.

Mohanty, M. 2021. Proteomics and bioinformatics as novel tools in phytoremediation technology-an overview. J. Bot. Res. 3(3): 49–54. DOI: https://doi.org/10.30564/jbr.v3i3.3380.

Mosa, K. A., K. Kumar, S. Chhikara, J. Mcdermott, Z. Liu, C. Musante, J. C. White and O. P. Dhankher. 2012. Members of rice plasma membrane intrinsic proteins subfamily are involved in arsenite permeability and tolerance in plants. Transgenic Res. 21: 1265–1277.

Mosa, K. A., I. Saadoun, K. Kumar, M. Helmy and O. P. Dhankher. 2016. Potential biotechnological strategies for the cleanup of heavy metals and metalloids. Front. Plant Sci. 7: 303.

Muthusaravanan, S., N. Sivarajasekar, J. S. Vivek, T. Paramasivan, M. Naushad, J. Prakashmaran et al. 2018. Phytoremediation of heavy metals: mechanisms, methods and enhancements. Environ. Chem. Lett. 16: 1339–1359.

Oladipo, O. G., O. O. Awotoye, A. Olayinka, C. C. Bezuidenhout and M. S. Maboeta. 2018. Heavy metal tolerance traits of filamentous fungi isolated from gold and gemstone mining sites. Brazilian J. Microbiol. 49: 29–37.

Ozyigit, I. I., H. Can and I. Dogan. 2020. Phytoremediation using genetically engineered plants to remove metals: a review. Environ. Chem. Lett. 19: 669–698.

Pan, A., M. Yang, F. Tie, L. Li, Z. Chen and B. Ru. 1994. Expression of mouse metallothionein-I gene confers cadmium resistance in transgenic tobacco plants. Plant Mol. Biol. 24(2): 341–51. doi: 10.1007/BF00020172.

Pandey, A. K., V. K. Gaur, A. Udayan, S. Varjani, S.-H. Kim and J. W. C. Wong. 2021. Biocatalytic remediation of industrial pollutants for environmental sustainability: Research needs and opportunities. Chemosphere 272: 129936.

Podder, M. S. and C. B. Majumder. 2016. Fixed-bed column study for As(III) and As(V) removal and recovery by bacterial cells immobilized on Sawdust/$MnFe_2O_4$ composite. Biochem. Eng. J. 105: 114–135.

Popova, L., L. Maslenkova, R. Y. Yordanova, A. P. Ivanova, A. P. Krantev, G. Szalai et al. 2009. Exogenous treatment with salicylic acid attenuates cadmium toxicity in pea seedlings. Plant Physiol. Biochem. 47: 224–231.

Pratush, A., A. Kumar and Z. Hu. 2018. Adverse effect of heavy metals (As, Pb, Hg, and Cr) on health and their bioremediation strategies: a review. Int. Microbiol. 21(3): 97–106. doi: 10.1007/s10123-018-0012-3.

Raffa, C. M., F. Chiampo and S. Shanthakumar. 2021. Remediation of metal/metalloid-polluted soils: a short review. Appl. Sci. 11: 4134.

Raj, A., A. Yadav, A. P. Rawat, A. K. Singh, S. Kumar, A. K. Pandey et al. 2021. Kinetic and thermodynamic investigations of sewage sludge biochar in removal of Remazol Brilliant Blue R dye from aqueous solution and evaluation of residual dyes cytotoxicity. Environ. Technol. Innov. 23: 101556.

Rangabhashiyam, S., E. Suganya, A. V. Lity and N. Selvaraju. 2015. Equilibrium and kinetics studies of hexavalent chromium biosorption on a novel green macroalgae *Enteromorpha* sp. Res. Chem. Intermed. 42: 1275–1294.

Roane, T. M., C. Rensing, I. L. Pepper and R. M. Maier. 2015. Microorganisms and metal pollutants. pp. 415–439. *In*: Maier, R. M., I. L. Pepper and C. P. Gerba (Eds.). Environmental Microbiology. Academic Press, ISBN 978-0-12-370519-8.

Rodríguez, A., M. L. Castrejón-Godínez, E. Salazar-Bustamante, Y. Gama-Martínez, E. Sánchez-Salinas, P. Mussali-Galante et al. 2020. Omics Approaches to pesticide biodegradation. Curr. Microbiol. 77(4): 545–563.

Roy, M. and L. M. Mcdonald. 2015. Metal uptake in plants and health risk assessments in metal-contaminated smelter soils. L. Degrad. Dev. 26: 785–792.

Rudakiya, D. M., A. Tripathi, S. Gupte and A. Gupte. 2019. Fungal bioremediation: a step towards cleaner environment. Adv. Front. Mycol. Mycotechnology Basic Appl. Asp. Fungi 229–249.

Rudakiya, D. M. and Y. Patel. 2021. Bioremediation of metals, metalloids, and nonmetals. Microb. Rejuvenation Polluted Environ. 26: 33–49.

Salem, I. A. and M. S. El-Maazawi. 2000. Kinetics and mechanism of color removal of methylene blue with hydrogen peroxide catalyzed by some supported alumina surfaces. Chemosphere 41: 1173–1180.

Saran, A., L. Fernandez, F. Cora, M. Savio, S. Thijs, J. Vangronsveld and L. J. Merini. 2020. Phytostabilization of Pb and Cd polluted soils using *Helianthus petiolaris* as pioneer aromatic plant species. Taylor Fr. 22: 459–467.

Sargin, I., G. Arslan and M. Kaya. 2016. Microfungal spores (Ustilago maydis and U. digitariae) immobilised chitosan microcapsules for heavy metal removal. Carbohydr. Polym. 138: 201–209.

Sarwar, N., M. Imran, M. R. Shaheen, W. Ishaque, M. A. Kamran, A. Matloob et al. 2017. Phytoremediation strategies for soils contaminated with heavy metals: Modifications and future perspectives. Chemosphere 171: 710–721.

Schmid, M. H., C. Meuli-Simmen and J. Hafner. 2002. Repair of cutaneous defects after skin cancer surgery. Cancers Ski. 225–233.

Sharma, P., A. K. Pandey, S.-H. Kim, S. P. Singh, P. Chaturvedi and S. Varjani. 2021a. Critical review on microbial community during in-situ bioremediation of heavy metals from industrial wastewater. Environ. Technol. Innov. 101826.

Sharma, P., A. K. Pandey, A. Udayan and S. Kumar. 2021b. Role of microbial community and metal-binding proteins in phytoremediation of heavy metals from industrial wastewater. Bioresour. Technol. 326: 124750.

Shazia, I., G. Rukh Sadia and A. Talat. 2013. Bioremediation of heavy metals using isolates of filamentous fungus *Aspergillus fumigatus* Collected from Polluted Soil of Kasur, Pakistan. Int. Res. J. Biol. Sci. 2: 1–6.

Sheng, J., K. Liu, B. Fan, Y. Yuan, L. Shen and B. Ru. 2007. Improving zinc content and antioxidant activity in transgenic tomato plants with expression of mouse Metallothionein-I by mt-I Gene. J. Agric. Food Chem. 55: 9846–9849.

Shukla, A., S. Srivastava and S. F. D'Souza. 2018. An integrative approach toward biosensing and bioremediation of metals and metalloids. Int. J. Environ. Sci. Technol. 15: 2701–2712.

Shukla, D., R. Kesari, M. Tiwari, S. Dwivedi, R. D. Tripathi, P. Nath and P. K. Trivedi. 2013. Expression of Ceratophyllum demersum phytochelatin synthase, CdPCS1, in *Escherichia coli* and *Arabidopsis* enhances heavy metal(loid)s accumulation. Protoplasm 250: 1263–1272.

Singh, D., S. K. Singh, V. K. Singh, H. Verma, M. Mishra, K. Rashmi and A. Kumar. 2021a. Plant growth-promoting bacteria: application in bioremediation of salinity and heavy metal–contaminated soils. pp. 73–78. *In*: Kumar, A., V. K. Singh, P. Singh and V. K. Mishra (Eds.). Microbe Mediated Remediation of Environmental Contaminants. Woodhead Publishing, ISBN: 978-0-12-821199-1.

Singh, S., V. Kumar, S. Datta, D. S. Dhanjal, S. Singh, S. Kumar et al. 2021b. Physiological responses, tolerance, and remediation strategies in plants exposed to metalloids. Environ. Sci. Pollut. Res. 28: 40233–40248.

Song, W. Y., E. J. Sohn, E. Martinoia, Y. J. Lee, Y. Y. Yang, M. Jasinski and Y. Lee. 2003. Engineering tolerance and accumulation of lead and cadmium in transgenic plants. Nat. Biotechnol. 21: 914–919.

Sriprang, R., M. Hayashi, M. Yamashita, H. Ono, K. Saeki and Y. Murooka. 2002. A novel bioremediation system for heavy metals using the symbiosis between leguminous plant and genetically engineered Rhizobia. J. Biotechnol. 99: 279–293. https://doi.org/10.1016/S0168-1656(02)00219-5.

Suman, J., O. Uhlik, J. Viktorova and T. Macek. 2018. Phytoextraction of heavy metals: A promising tool for clean-up of polluted environment? Front. in Plant Sci. 871.

Sumiahadi, A. and R. Acar. 2018. A review of phytoremediation technology: heavy metals uptake by plants. IOP Conf. Ser. Earth Environ. Sci. 142: 012023.

Taleei, M. M., N. K. Ghomi and S. A. Jozi. 2019. Arsenic removal of contaminated soils by phytoremediation of vetiver grass, chara algae and *Water Hyacinth*. Bull. Environ. Contam. Toxicol. 102: 134–139.

Thomas, J. C., E. C. Davies, F. K. Malick, C. Endreszl, C. R. Williams, M. Abbas et al. 2003. Yeast metallothionein in transgenic tobacco promotes copper uptake from contaminated soils. Biotechnol. Prog. 19: 273–280.

Tiwari, S. and C. Lata. 2018. Heavy metal stress, signaling, and tolerance due to plant-associated microbes: an overview. Front. Plant Sci. 9: 452.

Tomar, P. R., A. R. Dixit, P. K. Jaiwal and O. P. Dhankher. 2016. Engineered plants for heavy metals and metalloids tolerance. Genet. Manip. Plants Mitig. Clim. Chang. 143–168.

Tripathi, R. D., S. Srivastava, S. Mishra, N. Singh, R. Tuli, D. K. Gupta and F. J. M. Maathuis. 2007. Arsenic hazards: strategies for tolerance and remediation by plants. Trends Biotechnol. 25: 158–165.

Tripathi, S. and K. M. Poluri. 2021. Heavy metal detoxification mechanisms by microalgae: Insights from transcriptomics analysis. Environ. Pollut. 285: 117443.

Umar, A. 2017. Phytohyperaccumulator-AMF (arbuscular mycorrhizal fungi) interaction in heavy metals detoxification of soil. Acta Biológica Paranaense 46.

van der Zaal, B. J., L. W. Neuteboom, J. E. Pinas, A. N. Chardonnens, H. Schat, J. A. C. Verkleij et al. 1999. Overexpression of a novel Arabidopsis gene related to putative zinc-transporter genes from animals can lead to enhanced zinc resistance and accumulation. Plant Physiol. 119: 1047–1056.

Vazquez-Nunez, E., J. M. Pena-Castro, F. Fernandez-Luqueno, E. Cejudo, M. G. De La Rosa-Alvarez and M. C. García-Castañeda. 2018. A review on genetically modified plants designed to phytoremediate polluted soils: biochemical responses and international regulation. Pedosphere 28: 697–712.

Velkova, Z., G. Kirova, M. Stoytcheva, S. Kostadinova, K. Todorova and V. Gochev. 2018. Immobilized microbial biosorbents for heavy metals removal. Eng. Life Sci. 18: 871–881.

Verma, A., S. Kumar and S. Kumar. 2017. Statistical modeling, equilibrium and kinetic studies of cadmium ions biosorption from aqueous solution using S. filipendula. J. Environ. Chem. Eng. 5: 2290–2304.

Wang, J., X. Feng, C. Anderson, Y. Xing and L. Shang. 2012. Remediation of mercury contaminated sites–a review. J. Hazard. Mater. 221–222: 1–18.

Wilkin, R. T., T. R. Lee, D. G. Beak, R. Anderson and B. Burns. 2018. Groundwater co-contaminant behavior of arsenic and selenium at a lead and zinc smelting facility. Appl. Geochem. 89: 255–264.

Xu, J., Y. S. Tian, R. H. Peng, A. S. Xiong, B. Xiong Zhu, X. L. Hou and Q. H. Yao. 2010. Cyanobacteria MT gene SmtA enhance zinc tolerance in Arabidopsis. Mol. Biol. Rep. 37: 1105–1110.

Yan, C., F. Wang, H. Geng, H. Liu, S. Pu, Z. Tian, H. Chen, B. Zhou, R. Yuan and J. Yao. 2020. Integrating high-throughput sequencing and metagenome analysis to reveal the characteristic and resistance mechanism of microbial community in metal contaminated sediments. Sci. Total Environ. 707: 136116.

Yan, K., P. Yuan and Q. Wei. 2019. Effects of different inducer on the accumulation of essential oil from endophytic fungi of *Cinnamomum longepaniculatum*. OALib 06: 1–12.

Yang, Z., W. Shi, W. Yang, L. Liang, W. Yao, L. Chai et al. 2018. Combination of bioleaching by gross bacterial biosurfactants and flocculation: A potential remediation for the heavy metal contaminated soils. Chemosphere. 206: 83–91.

Zhang, D., T. Yamamoto, D. Tang, Y. Kato, S. Horiuchi, S. Ogawa et al. 2019. Enhanced biosynthesis of CdS nanoparticles through *Arabidopsis thaliana* phytochelatin synthase-modified *Escherichia coli* with fluorescence effect in detection of pyrogallol and gallic acid. Talanta. 195: 447–455.

Zhao, F., J. Ma, A. Meharg and S. P. McGrath. 2009. Arsenic uptake and metabolism in plants. New Phytol. 181: 777–794. doi: 10.1111/j.1469-8137.2008.02716.x.

Chapter 16

Synthetic Biology Approaches for Bioremediation of Metals

Rohit Ruhal and *Rashmi Kataria**

16.1 Introduction

Bioremediation can use biological ways to completely eliminate environmental pollutants (Azubuike et al. 2016). Based on this, it can be defined as *in situ* bioremediation if the bioremediation process occurs on the same spot, while it may be defined as *ex situ* if contaminated materials are transferred to a different place (Azubuike et al. 2016). Nature has processes of recycling pollutants with the help of microbes (bacteria, fungi, algae), plants with specific metabolic activity involved in breaking xenobiotic chemicals, and hazardous metals (Verma and Kuila 2019). Microbes are now widely used. In one example, heavy metals (cadmium, copper and zinc) from wastewater were removed using rotating biological contactor and multiple sorption and desorption options (Omokhagbor et al. 2020). Plants may also help in bioremediation, and the process is known as phytoremediation. But, plants do not have particular enzymes for degradation, so researchers engineered genes in plants for naphthalene degradation. This also made the plants pollutant stress resistant (Rayu et al. 2012). In the cases of extensive land contamination, it is difficult and very costly to bioremediate back to non-pollutants. In order to restore polluted places, bioremediation based on biological methods like plants and bacteria may be helpful. The main inorganic pollutants, the heavy metals viz. Pb, Cd, Cu, Hg, Sn, and essential micronutrients can be toxic as they bioaccumulate in tissues and living organisms (Kapahi and Sachdeva 2019). Occasionally, the source of metals are petrochemicals, agrochemicals, coal, and the mining industry (Zhang et al. 2020). The metal pollutants in fertilizers and sewage may be washed off to lakes and rivers leading low quality of water for plants and animals (Dixit et al. 2015).

In order to develop bioremediation strategies based on the application of microorganisms, there is a need for deep knowledge of the biology of microorganisms' ability to degrade contaminants (Rylott and Bruce 2020). For this, a number of points must be kept in mind, for example, the metabolic capabilities of microorganisms in contaminated environments (Dvořák et al. 2017). So, the physical and chemical processes of these micro-organisms for bioremediation need to be understood (Diep et al. 2018). This may lead to the development of models to predict the best metabolic pathways helping bioremediation. The implementation of such strategies is only possible by using synthetic biology, which includes genome sequencing to metabolic engineering of relevant

Department of Biotechnology, Delhi Technological University, Delhi, India
* Corresponding authors: rohitmetabolomics@gmail.com, rashmikataria@dtu.ac.in

identified microorganisms (probably by metagenomics) (Rylott and Bruce 2020). The beginning is from the isolation of the pure culture organism and characterization for molecular analysis of ecology.

In the current scenario, sequencing of genomes involved in bioremediation has been done. Analysis of the genome sequences of microbes has displayed a number of genes with a possible role in biodegradation and bioremediation. For instance, genome sequencing of *Pseudomonas* spp. represented genes with properties like oxidoreductase, dehydrogenase, oxygenase, cytochromes, efflux pumps, membrane transporters, and metal specific transcriptional regulators (Miyazaki et al. 2015, García-Hidalgo et al. 2020). Further, the sequencing has represented how heavy metals and other pollutants may be bioremediated. Additionally, increased use of genetically modified microorganisms to clean hazardous waste, including heavy metals were reported. There are genes in microbes that help in metal resistance, and they are designated as microbial metal resistance genes (Shi et al. 2019, Xavier et al. 2019). These are used for genetic engineering with the help of synthetic biology. These genes are studied in detail with the help of omics studies, including metagenomics, transcriptomics, and proteomics. These can be further tailored in genetically modified microorganisms, which leads to novel metabolic pathways. The genetically modified organism must be studied for metabolic burden in order to get most efficient bioremediation. In addition, there is a need for a detailed investigation of transcriptional regulation for engineering accordingly. For instance, regulated expression of metallothionein and kinase led to efficient mercury bioremediation. Different research groups have used numerous bacteria like *E. coli* JM109 to remove mercury from polluted water (Zhao et al. 2005), and *Ralstonia metallidurans*, *Caulobacter* spp. for Cd contaminated wastewater. The generation of mutants by shuffling nucleotides in genes of DNA can help to create enzymatic and degradative pathways as well as generate efficient enzymes.

A detailed overview of synthetic biology and its applications in bioremediation, especially metals, are discussed in the book chapter. The first section describes the basics of synthetic biology, followed by sections describing the role of synthetic biology for metal bioremediation.

16.2 Synthetic Biology Techniques

The applications of synthetic biology in bioremediation are presented in Fig. 16.1. Synthetic biology is a multidisciplinary approach that includes biotechnology, molecular biology, genetic engineering, omics, biophysics, chemical engineering, and computational approaches (Hartwell et al. 1999, Karoui et al. 2019). The main purpose of synthetic biology is to create new biological systems based on information collected from above all fields. In the current scenario, synthetic biology tools are used to construct and control microbes by manipulating communication networks, regulating gene expression, and engineering interaction pathways (Sharma and Shukla 2020a, Bhattacharjee et al. 2020). During constructing new organisms, a need for a low metabolic burden is significant. One of the important tools of synthetic biology is omics approaches (Rodríguez et al. 2020, Ufarté et al. 2015). The genomic approach includes genome sequencing and refers to sequencing the entire genome of an organism. Many high throughput sequencing technologies and their data handling tools and technologies are available. In the beginning, Sanger sequencing was used, which was later made robust and automated leading to the sequencing of *Haemophilus influenzae*. This was known as shotgun sequencing, in which longer sequences are subdivided into smaller fragments and later on assembled together (Oyewusi et al. 2021, Ekblom and Wolf 2014). The advantage of sequencing is to identify potential causative variants to follow up studies of gene expression. At present, next generation sequencing is most frequently used as it is much efficient and error free. It is much faster than classic and rapidly sequence whole genomes and deeply sequence targets. Illumina also does RNA sequencing for transcriptomics and identifies novel genes (Sato et al. 2019, Cacho et al. 2016, Stoler and Nekrutenko 2021). Similarly, proteomics has developed enough to study large-scale study proteomes. Further, structural and functional characterization of proteins may lead to details

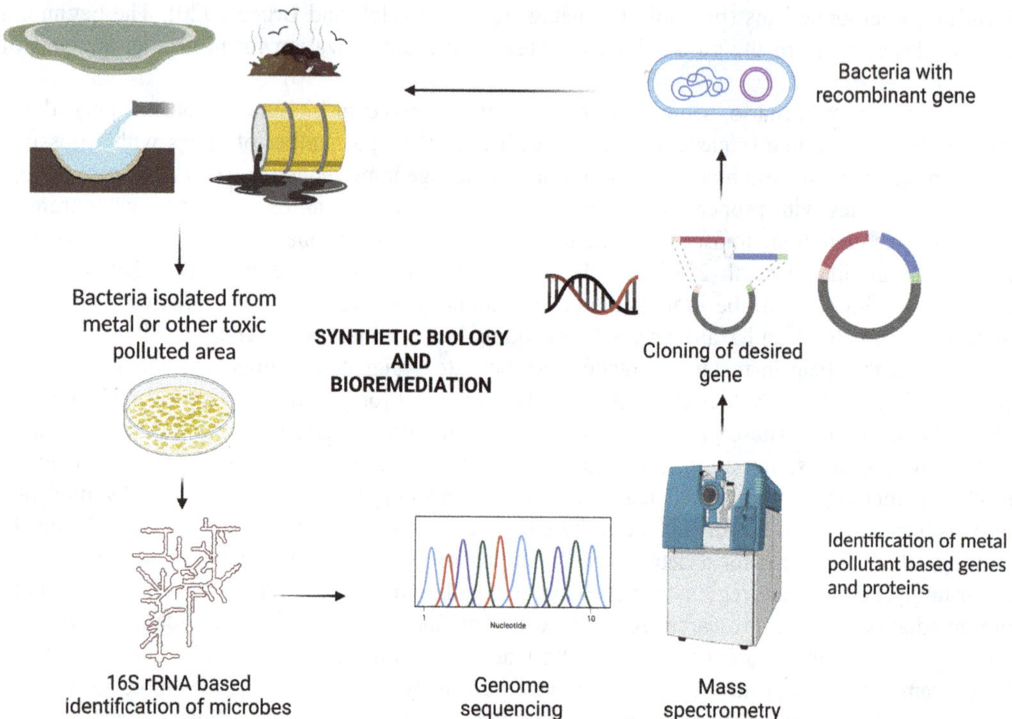

Fig. 16.1. Schematic representation of metal bioremediation with the knowledge of synthetic biology. The microbial DNA is isolated and identified by 16S rRNA sequencing. The genome sequencing help in identification of relevant desired genes involved in bioremediation. The desired genes and protein expressed are identified by MS. Genetic engineered strains with these genes may be used for enhanced metal bioremediation. Figure was created on biorender.com.

required for applications. In order to characterize, we need to purify the protein and proceed with biochemical characterization. For in detail study of protein active site, crystallization of proteins and analysis of the active site by x-ray diffraction is needed (Chen et al. 2020).

For engineering microbes, several molecular biology technologies are used. The host cell used for genetic modification is known as chassis and is referred to as a model organism for which abundant characterized genetic parts are available. Molecular cloning is the process of transforming chassis with recombinant DNA molecules (Clasen et al. 2018, Lebuhn et al. 2015, Xu et al. 2020). Molecular cloning has two major components, the desired DNA fragment (gene) and the vector (plasmid). The vector contains all components for replication within the host. The gene of interest or regulatory elements or operons are generally prepared by cloning and excised out with restriction enzymes. It is further copied by polymerase chain reaction (PCR). The plasmid vector and gene of interest are cut with similar restriction enzymes (restriction enzymes are DNA specific enzymes). Plasmids are circular DNA replicated in the host separately from the host chromosome. The gene of interest and plasmid are allowed to ligate using DNA ligase enzyme and transformed into host cells. A more advanced technique known as Gibson assembly is more efficient. In this technique, the reaction is carried out in isothermal conditions with three enzymes—exonuclease, polymerase, and DNA ligase. The process is fast, efficient, and low error prone.

Another revolutionary technique used is CRISPR (Clustered Regularly Interspaced Short Palindromic Repeats) and Cas9 protein (Bodapati et al. 2020). These are essential adaptive immunity selected in bacteria and archaea to eliminate invading DNA, especially of bacteriophages. How these technologies were used for metal bioremediation are described in the following sections.

16.3 Application of Microbial Genome Sequencing for Bioremediation of Metals

Complete genome sequencing helps in the identification of novel genes related to bioremediation (Das et al. 2015, Hong et al. 2017). Some of the important genes studied related to metal bioremediation are shown in Table 16.1. Especially, in cases based on physiology data, a bacteria is considered to play a role in particular degradation, but genome sequencing may indicate the presence of other relevant genes, and this gives chances to explore it further (Muccee and Ejaz 2020, Das et al. 2015, Wang et al. 2020). Many research articles have shown how a bacterial strain isolated from metal pollutant area was studied in detail. Analyzing genome sequencing led to the discovery of several metal bioremediation pathways. In one of the studies, a strain isolated from a mining area was identified based on 16S rRNA sequencing belonging to *Bacillus* spp. (Ayangbenro and Babalola 2020). It was observed that bacteriocin production was the reason for metal resistance. On analyzing genome sequencing, different types of bacteriocins genes were found in this strain. Thus, 16S rRNA sequencing and genome sequencing made the project more helpful in understanding the biology of metal bioremediation. Similarly, it was observed that biosurfactant was capable of removing Pb, Cd, and Cr (69%, 54%, and 43%) (Ayangbenro and Babalola 2020).

In another study, strains were isolated from manganese mines and studied further. One of the strains identified *Pseudaminobacter manganicus* (Xia et al. 2018). This strain has multiple heavy metal resistance and effectively removed Mn^{2+} and Cd^{2+}. Considering its capability and potential for metal bioremediation, it was further studied for genome sequencing. Several putative genes related to heavy metal resistance and exopolysaccharide synthesis are found in the genome. With the world of the genomic era, it is now possible to sequence the genes available in the contaminated zone (Cabello et al. 2018). There are possibilities of identifying key genes responsible for degradation,

Table 16.1. Representative genes and transporters used to engineer bacteria for enhanced bioremediation.

Gene	Function/Application	References
nahA	Naphthalene degrading gene	Fleming et al. 1993
merA/merR	Mercury transcriptional regulator/Application in biosensor	Cai et al. 2018
ArsR	Arsenic transcription factors/Biosensor applications	Tchounwou et al. 1999
gfp or lux or luc	Reporter genes used for developing fluorescent or luminescence based biosensors	Inda and Lu 2020
ZntR	Molecular reporters used in whole cell based assay	Fernandez-López et al. 2015
ArsR	Molecular reporters used in whole cell based assay	Fernandez-López et al. 2015
copA	Deleting *copA* led to increased copper sensitivity. The *copA* is exporter of copper	Kang et al. 2018
arsM	Gene from *Rhodopseudomonas palitris* over expressed in *Bacillus* assisted in removing arsenic	Liu et al. 2011
SpPCS	Phytochelatin synthase from Schizosaccharomyces pombe overexpressed in *E. coli* for cadmium bioremediation	Seung et al. 2007
CrR	Gene in *Methylococcus capsulatus* removes Cr^{6+}	Hasin et al. 2010
ChrR/nfsA/YieF	Chromate reductase removes chromium	Ackerley et al. 2004
NiCoT permease	Transporter used for bioaccumulation of nickel	Hebbeln and Eitinger 2004
GlpF	Glycerol facilitator in *E. coli* used channels to improve antimonite accumulation	Sanders et al. 1997
Fps1	Major facilitator protein superfamily from *Saccharomyces* used for metal uptake	Wysocki et al. 2001

and their relative abundance may be correlated to the type of metalloids or contaminants present (Heidelberg et al. 2002).

In case high concentration of mRNA of gene involved in degradation is observed than that means the area is associated with high degradation activity of contamination. One of the best example of such case is presence of mRNA of *nahA* (naphthalene degrading gene) in correlation of naphthalene degradation (Fleming et al. 1993). Similarly, the reduction of soluble ionic mercury to volatile mercury from water correlated to the concentration of mRNA *merA* in mercury contaminated waters (Naguib et al. 2018). Although, these observations were not consistent in other studies, which indicate that degradation is a complicated process and involve many regulatory factors (Das et al. 2016). For instance, *Geobacter* was considered to be non-motile and is regarded as an important organism for metal bioremediation (Hu et al. 2013, Lu et al. 2016). After genome sequencing, it was determined that these bacteria possess flagellar genes. A related bacteria, *Geobacter metallireducens*, produced flagella when the organism was growing on insoluble Fe (III) or Mn (IV) oxides. Numerous bacterial genome sequencing is available, which play a role in metal bioremediation. For instance, *Dephalococcoids ethanogens* is known for dichlorination of chlorinated solvents to ethylene, and many of its genes indicate its capability to degrade chlorinated solvents on the subsurface. Similarly, *Geobacter sulfurreducens* and *Geobacter metallireducens* are known for the anaerobic oxidation of aromatic hydrocarbons, and reductive precipitation of uranium and hence may be explored for further investigation. Thus, genome sequencing is not only helpful to identify potential metal resistant bacteria but also the mechanism of bioremediation as well.

16.4 Developing Metal Pollutant Biosensors using Synthetic Biology

Another application of synthetic biology in bioremediation is the development of biosensors for particular contaminant metals. For example, MerR transcriptional regulator was developed for mercury biosensors (Liu et al. 2019). When Hg^{2+} ions bind to MerR, it gets derepressed, and *mer* operon gene expression starts, so the *mer* operon was replaced by the GFP reporter gene (or luciferase) (Cai et al. 2018, Ndu et al. 2015). Thus, this system became mercury induced biosensors.

Biosensors denote a kind of device which is self-contained and is capable of providing specific quantitative analytical data by using a receptor, which has a biological origin and may contain an electrochemical transduction element that helps to represent observation (Fig. 16.2) (Mehrotra 2016). Heavy metals include all metals except Al, Na, K, Mg and Ca, etc.; in other words, metals with a density higher than 5 g/cm^3 are considered heavy metals. These can include both toxic (As, Pb, Hg, Cd) and non-toxic metals (Au, Ag) (Jia et al. 2020, Yang et al. 2018). The heavy metal can be a threat to human health as well as the environment at very low concentrations (Tchounwou et al. 2012). This is due to their inability to get degrade, which causes serious pollution problems. The toxicity of metals or metalloids may differ and can be divided into groups based on level of toxicity, e.g., some metals are toxic at very low concentrations like lead, mercury, or cadmium, while arsenic, indium thallium have low toxicity and need certain concentration to be toxic (Bruins et al. 2000, Bhat et al. 2019). Similarly, copper, zinc, cobalt and selenium, and iron have certain values above which they could be toxic. All these informations are required for developing biosensors. In general, bioremediation can be monitored by measuring oxidation and reduction potential and other indirect parameters as pH, temperature, and oxygen content (Naresh and Lee 2021). One of the methods which can be applied using recombinant biotechnology includes reporter gene based system (Jones et al. 2019). In order to utilize and create such reporter system, we need to understand the detailed biology of metal resistance genes known in the microbial world (Kim et al. 2018). Plasmids carrying these resistance genes specifically expressed or induced as they bind with these metals are significant for such systems (Fig. 16.2). All these are tightly regulated and are specific to a particular metal. The regulatory proteins and promoters from these operons are used in such biosensors, and by using metal specific biosensors, one can regulate metal specific contamination

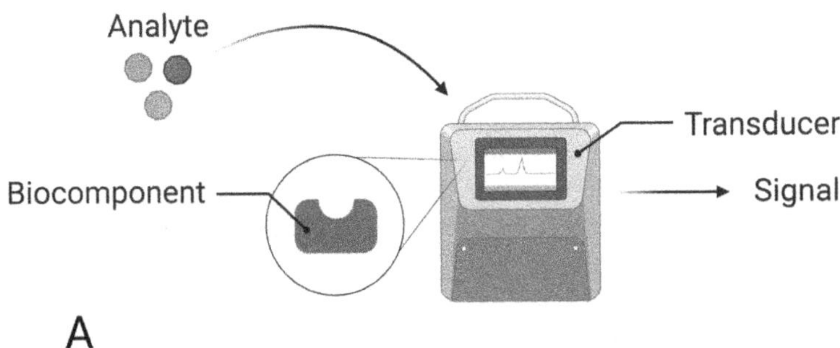

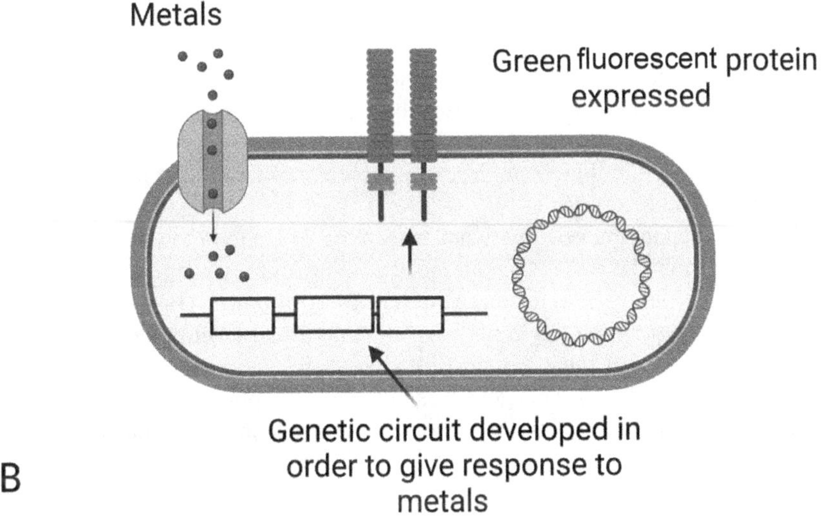

Fig. 16.2. Representative figure of biosensor and its applications. (A) The basic principle of biosensor is presented where biocomponent and analyte are put together in machine which can detect signal. (B) Similar concept is applied in whole cell microbe biosensor, synthetic biology may help in development of genetic circuit which gives response to metal. The response is in the form of luminescence or fluorescent protein (as shown in B part). Figure was created on biorender.com.

(Jones et al. 2019). Further, there are chances to distinguish the concentration of metals in a given environment. In addition, transcriptional regulator-based biosensors are also developed. As described in the above section, the MerR includes a family of transcriptional regulator proteins which get activated in response to mercury metals. Looking in detail at MerR structure, it is characterized by three domains-N terminal DNA binding domain (DBD), linker and effector binding domain (EBD) at C-terminal (Zeng et al. 1998, Guo et al. 2010). In DBD, first 44 amino acids represent the helix turn helix motif, while the EBD region has metal binding domains consisting of three cysteines (Utschig et al. 1995). In MerR, the cys center can bind to Hg^{2+}, but other families of proteins may bind to other metals as well. The ArsR transcription factors (TF) family is known for metal sensing. The biosensors developed based on ArsR for sensing arsenic contamination in water have been successful, especially in low-income countries (Tchounwou et al. 1999). The mechanism of these TF includes controlling operons to protect bacteria against toxic metals. These factors occur in Gram-positive and Gram-negative bacteria. Thus, based on all studies, such kinds of genes are noted and are used for developing biosensors. The reporter genes are selected which have the property of measurement, and generally, luminescent or fluorescent genes were selected, e.g., luciferase, gfp, or lux (Fig. 16.2). There are examples using microbial sensors using reporter genes. For instance, metal antimony was detected using reporter gene *luc* in *Staphylococcus aureus*, *Bacillus subtilis*, and *E. coli* (Bilal and Iqbal 2019, Inda and Lu 2020). Similarly, arsenic used reporter *luc* in *Pseudomonas flourscens* and *lux* in *E. coli* (Xiaoqiang Jia et al. 2019). Other metals like cadmium, cobalt, copper, lead, mercury, and nickel were detected using *Lux*, *luc*, and *gfp* genes.

Some authors use the term whole cell biosensor while using any prokaryotic or eukaryotic cell (Wu et al. 2021). These are like single reporter, which incorporates both bioreceptor and transducer in the same cells (Fig. 16.2). Thus, these organisms need to be experimentally modified and increase their sensitivity (Berepiki et al. 2020). These whole cell biosensors can be used as turn off and turn on. For turn off, it is more like microbial toxicological bioassay, and toxicity can be measured by the degree of inhibition of growth or motility or any physiological activity of bacteria. In these bioassays, quantitative data is measured by any cellular function inhibition and correlated to the degree of toxicity of metal. The quantifiable molecular reporter is fused to a specific gene promoter activated by the metal pollutant. In general, 85% are of bacterial origin genetically modified, and 15% are eukaryotic, including bacteria and fungus. A number of genetic components of microbes recognizing external stress are exploited for this purpose (Fernandez-López et al. 2015). The popular ones are metal responsive operon systems and metalloproteins, e.g., ZntR, ArsR. These proteins play transcriptional factors regulating the expression of genes in the presence of metalloids. Therefore, during the designing of bacteria, part of these operons is taken to increase the selectivity and sensitivity of metalloids. For example, promoter regions of znt-operon and gfp may be combined to detect cadmium and mercury.

In order to modulate the selectivity and sensitivity of bacteria towards metalloids, sometimes one can modulate amino acid sequence for metal binding loops. Changing the amino acid sequence (gene sequence accordingly) may increase the selectivity of metals (Plyuta et al. 2020). There may be cases where overexpression might not have an impact, and deleting genes results in the desired effect. It was observed that the *copA* gene deleted in *E. coli* resulted in increased sensitivity in copper ions (Kang et al. 2018). This gene was involved in exporting copper ions outside cells, and deletion led to the accumulation of copper ions in cells, which led to enhanced sensitivity. In this manner, it is significant to understand the biology of microbes in response to metals and modulate accordingly. The methodology may be applied not only to metallic pollutants but also to many other organic pollutants. The advantage of using microbes is reflected as an economic tool for bioremediation as they are easy to manipulate. But limitations may not be denied; for example, it is challenging to measure bioavailability. Moreover, there is a possibility that a portion of metal ions accumulated in the bacterial cells could only be measurable within bacteria. There is a possibility of detection limitation, especially due to the transport of certain metals inside the cells (Jarque et al. 2016, Carpenter et al. 2018). A detailed study of the biology of the metabolic pathway of metal detection and transporters is needed. Consequently, a lot of biological data is required to achieve any conclusion.

16.5 Engineering Metabolic Pathways for Metal Bioremediation

Genetic engineering may be advantageous for several bacterial strains isolated from metal polluted environments (Rosenberg 2013). Genetic modification may lead to highly efficient multiple degradative pathways. The degradation could be partial or highly efficient compared to wild strain (Kaur et al. 2020). In most cases, the genetic targets may include transcriptional regulation or the genes encoding enzymes for pollutant degrading pathways. Using molecular biology tools, it is possible to modify coding regions or gene sequence transcribing into mRNA (Malik et al. 2021). These modifications could be at the level where enzyme substrate interaction is increased. Further, enzyme active sites may be modified (Sharma et al. 2018). This can lead to the development of robust microbes targeting specific pollutants. There is a possibility of operon including genes which together may play a role in the degradation of pollutants (Dangi et al. 2019). The operons may be controlled by positively acting regulatory proteins and are activated by substrate or metabolites (Sharma and Shukla 2020b). Especially, if these operons operate in the presence of a specific substrate, there is a possibility of no wastage of carbon and energy. In bacteria *Rhodopseudomonas palutris*, a gene *arsM* encodes the homolog of mammalian Cyt19 As(III)-S-adenosylmethionine methyltransferase regulated by arsenicals. An expression of *arsM* introduced into the strains of methylation of arsenic, especially in *Bacillus idriensis* and in soil system about 2–5% of arsenic was removed by bio volatilization in 30 days (Liu et al. 2011). Thus, expressing arsM in bacteria helped in the removal of arsenic by volatilization from contaminated soil. This makes the possibility of the development of inexpensive, efficient arsenic removing system. Similarly, plasmid pTP6 that provides novel genes *merB*, *merG*, and other *mer* genes involved in mercury resistance and volatilization were studied and transformed to *Cupriavidus metallidurans*, which was genetically and biochemically characterized for mercury resistance (Rojas et al. 2011). The strain was able to maintain plasmid over 70 generations when no selective conditions were given. It was observed that organomercurial lyase protein (merB) and mercuric reductase (MerA) were expressed in the presence of Hg^{2+}. Bioremediation was possible when used in mercury contaminated aqueous systems, and polluted waters showed a decrease in mercury. Phytochelatin synthase (SpPCS) from *Schizosaccharomyces pombe* overexpressed in *E. coli* for cadmium bioremediation. The accumulation achieved was 7.5 times higher (Seung et al. 2007). *Methylococcus capsulatus* (Bath) have shown the capability to bioremediate chromium(VI) pollution in the range of concentrations (1.4–1000 mg L-1 of Cr6+) (Hasin et al. 2010). Some of the genes were observed in this bacteria, like CrR, which, if engineered in bacteria, may help in bioremediation.

Another study displayed that ChrR protects *Pseudomonas putida* against chromate toxicity, and it was shown that NfsA or YieF overproduction led to tolerance in *E. coli* to this compound (Ackerley et al. 2004). NfsA is the major oxygen-insensitive nitroreductase, and YieF is a divalent chromate reducer of *Escherichia coli.*

Another important aspect of using the genetically modified system is that the stability of mRNA of desired genes may influence the expression of the desired trait. There is a need to look at other organisms, e.g., mRNA of gene 32 of bacteriophage T4 is considered stable due to the formation of particular structural determinants at the 5' region (Qiu et al. 1996, Morgan et al. 1983). This is advantageous as this can be fused to the native promoter of gene 32 to several heterologous genes, and it was observed that stable mRNA was formed. For instance, fusion to the xylE gene responsible for catechol 2,3 dioxygenase was reported to result in the expression of high levels of mRNA. Further, efforts were made to introduce DNA cassettes in 5 prime untranslated regions, which stabilized mRNA. It was due to the formation of a hairpin structure, which is known to increase the half-life of mRNA. The increased stability of mRNA led to more protein production and hence enhanced bioremediation. Thus, methods applied to increase transcription and translation result in increased production of the enzymes and proteins. The poor stability of protein, and less specificity of proteins may restrict the efficiency of bioremediation. The stability and activity of proteins may be enhanced by altering the sequence of their amino acids. For instance, changing even a few amino acids in the protein sequence can make it more tolerant to pollutants with stable activity. Thus, efforts are made to develop a hybrid gene cluster, which might have the advantage of enhanced activity and substrate specificity. These hybrids may be developed by gene clusters producing different enzyme clusters to make a single enzyme with super transforming activity. This can be applied for metal bioremediation as shown in the degradation of organic pollutants; for example, *E. coli* was engineered to express a hybrid gene cluster with toluene metabolic tod operon and biphenyl metabolic *bph* operon. It was able to degrade trichloroethylene much faster (Furukawa et al. 1993).

Overall, detailed biology of metabolic pathway responses in bacteria against metals is required. This can be determined by different omics studies, proteomics and transcriptomics, in response to the metal pollutant (Rawat and Rangarajan 2019, Chandran et al. 2020, Giovanella et al. 2020).

Different studies have suggested strategies for microbial intracellular accumulation of metals. There are various import-storage systems amongst microbial world known for bioaccumulation. Some of them are commonly used, like NiCoT permease for Ni bioaccumulation (Hebbeln and Eitinger 2004). Channels of Mer operon are popularly utilized for mercury bioaccumulation. In order to discover more of such a system, lab studies need to be carried out by exposing bacteria to different metals and determining changes in their physiology. We can do transcriptomic or proteomics. The LC/MS was used to determine proteomes of *Acidithiobacillus ferrooxidans* exposed to copper and discovered that enzymes with histidine and cysteine biosynthesis pathways were upregulated (Vargas-Straube et al. 2020, Almárcegui et al. 2014). Altogether, there is a need to explore more microbial strains in the metal contaminated zones and study their physiology.

16.6 Membrane Proteins Targets for Metal Bioaccumulation using Bioengineering

One of the strategies to develop microbes for bioremediation is by enhancing their increased ability of bioaccumulation (Sharma and Fulekar 2009). Bioaccumulation is a metabolically active process in bacteria and other eukaryotic microbes involving heavy metals uptake through intracellular space (Massoud et al. 2019). The importer includes different membrane transport complex and translocation may involve outer membrane, periplasmic and inner membrane (Fig. 16.3). As soon as these metals reach intracellular space, these can be further processed and sequestered by proteins and peptide ligands (Ruiz et al. 2011). Consequently, the capacity of bioaccumulation will be calculated with respect to biomass. The limitation of bioaccumulation concerns the inner architecture of the cell; there is a possibility of gene and protein expression related to stress due to toxic effluents. Synthetic biology may help to increase the bioaccumulation capacity as it can optimize the trade-off between genetically engineered protein machinery and cellular growth (Njoku et al. 2020).

Biosorption is the process of particles to a biological matrix, and the interaction may be based on electrostatic, chemical, or chelation (Beni and Esmaeili 2020, Redha 2020). Several researchers engineered recombinant metal binding proteins and peptides on the outer surface of bacteria. These can act as microbial bio sorbents, although, several challenges arise in adsorption methods which may include the slight variation in pH and ionic strength. Bio-adsorbents might have lifespans and, with time, the possibility of degradation (Torres 2020). Bioaccumulation is a metabolically active process in which microorganisms can uptake metal complex and accumulate intracellularly (Gobas et al. 2016, Franke et al. 1994). The bioaccumulative capacity of a biomass for a target may differ and be measured as µmol/dry biomass. After studying the biology of gene expression levels and stress response to toxic composition in response to metal pollutant, proteins required may be identified. This information may be utilized for synthetic biology to optimize trade-offs between genetic engineering protein machinery dedicated to the import of metals and cellular growth. The genetically engineered strains are more robust as their regulatory elements are modulated and may be better controlled by external stimuli. Especially, superior control over the import and export system of metals makes the strains ideal for efficient bioremediation.

In order to improve the metal uptake system of bacteria, transporters must be modulated. In order to understand the strategy of overexpressing transporters, differential expression of membrane proteins is important to study in response to metals. Most frequently, the focus is on Gram-negative bacteria using inner membrane importers like major transporters, channels, secondary carriers, and active transporters (Tanaka et al. 2018, Long et al. 2012, Delmar et al. 2015). Channels are single components and are generally helical proteins that facilitate the passive diffusion of metals based on the concentration gradient across the membrane. Normally, ATP is not required, although, they also do not require proton motive force. A number of examples are available exploiting channels for different metals accumulation within bacteria. For instance, researchers used channels to improve As^{3+}, the homo-tetramer GlpF glycerol facilitators in *E. coli* (Sanders et al. 1997). This was used in

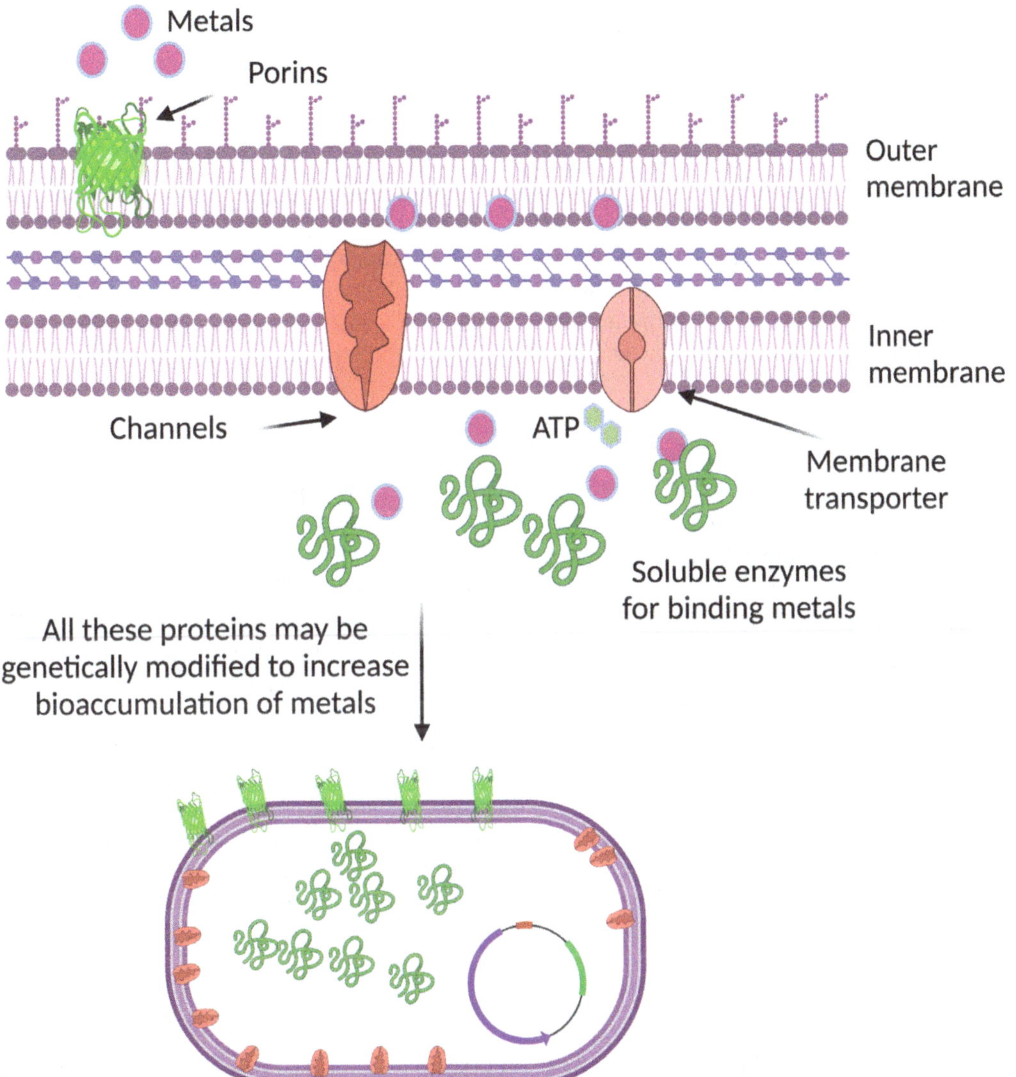

Fig. 16.3. The figures represent how membrane protein including porins, channels act as transporter of metals inside bacterial cells leading to bioaccumulation. Many soluble enzymes in cytoplasm may help in sequestering metals. These can be engineered for metal bioremediation by bioaccumulation. Figure was created on biorender.com.

other bacteria like *Corynebacterium diptheriae*. A homologous protein Fps1 from Saccharomyces has also been known for uptake of metals (Wysocki et al. 2001). These proteins come in the category of the major facilitator protein superfamily; some other examples are MerT/P transporter in *Serratia* spp. and *Pseudomonas* spp. The Mer transporters are known for their mercury transporter and known as Mer transporter. Since these channels have a low energy burden, they are considered the best choice for bioaccumulation. The passive uptake has its own limitation, and even if the bacteria are genetically engineered, it could limit bioaccumulation. In such cases, storage capacity and mechanism of accumulation must be determined. Another class of proteins involved in transport is known as porins (especially in Gram negative bacteria). The FrpB4 channel from *Helicobacter pylori* is involved in Ni uptake. Overexpressing these porins has led to an enhanced uptake rate, and porins also may be manipulated according to our needs. Single component proteins are termed uniporters, symporters, and many of them were reported for bioaccumulation. There are possibilities of the different topology of these transporters, e.g., symporters like NiCoR family have transmembrane topology. Similarly, many sugar porter families and phosphate family are known for metal accumulation. On the other hand, the active transporter consists of a multicomponent protein complex with transmembrane and a cytoplasmic energy coupling ATPase. Thus, these transporters can actually transport against the concentration gradient by hydrolyzing ATP, GTP, and other nucleotide sugars. Thus, huge reserves of ATP may help these proteins to store and accumulate metals.

In order to develop recombinant strains with a high capability of metal binding, there is a need to engineer membrane transporters with enhanced capability of binding metals. In this respect, functional characterization of proteins and their amino acids sequence need to be done. This strategy has led to the development of strains helping in the bioaccumulation of Zn, Cd, and copper.

16.7 Conclusion

Synthetic biology collects in detail information about microbial physiology concerning metal bioremediation, which may be utilized for bioremediation strategies. Biodegradation of organic pollutants is less complicated but challenging with inorganic pollutants like metals. The reason for the complication is due to the requirement of the system for trapping metals and biological chelators (metallothionein). All these complications generate a requirement for engineered microbial strains with enhanced ability of bioaccumulation. This led to the need for the detailed physiological study of strains capable of surviving in metal polluted areas. In order to increase bioaccumulation, engineering of outer membrane pores, ABC transporters, and histidine based storage proteins is required. Since many extremophiles have to face similar stressful conditions in their natural environment, they may be studied in detail for harsh wastewater treatment with metal contamination. Which kind of adaptation mechanism is applied, especially transporters, membrane channels, and metal chelating enzymes, must also be studied in detail. Based on these, genetically modified organisms are made, which can be tested in the lab and with careful procedures on sites. The use of synthetic biology for bioremediation is still in infancy but full of exciting possibilities for using engineered organisms, providing a cleaner and safer environment.

Acknowledgement

The authors would like to thank CSIR-Govt of India for providing salary under the CSIR Scientist Pool Scheme to Rohit Ruhal and DBT-Govt of India to provide Ramalinga Swamy fellowship Rashmi Kataria. We acknowledge Biorender.com for all figures.

References

Ackerley, D. F., C. F. Gonzalez, M. Keyhan, R. Blake and A. Matin. 2004. Mechanism of chromate reduction by the *Escherichia coli* protein, NFSa, and the role of different chromate reductases in minimizing oxidative stress during chromate reduction. Environ. Microbiol. 6: 851–860.

Ali Redha, Ali. 2020. Removal of heavy metals from aqueous media by biosorption. Arab J. Basic Appl. Sci. 27: 181–193.

Almárcegui, R. J., Claudio A. Navarro, A. Paradela, J. P. Albar, D. von Bernath and C. A. Jerez. 2014. Response to copper of *Acidithiobacillus ferrooxidans* ATCC 23270 grown in elemental sulfur. Res. Microbiol. 165(9): 761–772.

Ayangbenro, A. S. and O. O. Babalola. 2020. Genomic analysis of *Bacillus cereus* NWUAB01 and its heavy metal removal from polluted soil. Sci. Rep. 10(1): 1–12.

Azubuike, C. C., C. B. Chikere and G. C. Okpokwasili. 2016. Bioremediation techniques–classification based on site of application: principles, advantages, limitations and prospects. W. J. Microbiol. Biotechnol. 32: 1–18.

Beni, A. A. and A. Esmaeili. 2020. Biosorption, an efficient method for removing heavy metals from industrial effluents: a review. Environ. Technol. Innov.17: 100503.

Berepiki, A., R. Kent, L. F. Machado and N. Dixon. 2020. Development of high-performance whole cell biosensors aided by statistical modelling. ACS Synth. Biol. 9(3): 576–589.

Bhat, S. A., T. Hassan and S. Majid. 2019. Heavy metal toxicity and their harmful effects on living organisms. Int. J. Med. Sci. Diagn. Res. 3(1).

Bhattacharjee, G., N. Gohil and V. Singh. 2020. Synthetic biology approaches for bioremediation. pp. 303–312. *In*: Pandey, V. C. and V. Singh. (eds.). Bioremediation of Pollutants. Elsevier.

Bilal, M. and H. M. N. Iqbal. 2019. Microbial-derived biosensors for monitoring environmental contaminants: recent advances and future outlook. Process Safety and Environmental Protection. 124: 8–17.

Bodapati, S., T. P. Daley, X. Lin, J. Zou and L. S. Qi. 2020. A benchmark of algorithms for the analysis of pooled crispr screens. Gen. Biol. 21: 1–3.

Bruins, M. R., S. Kapil and F. W. Oehme. 2000. Microbial resistance to metals in the environment. Ecotoxic. Environ. Safety. 45: 198–207.

Cabello, P., V. M. Luque-Almagro, A. Olaya-Abril, L. P. Sáez, C. Moreno-Vivián and M. Dolores Roldán. 2018. Assimilation of cyanide and cyano-derivatives by *Pseudomonas pseudoalcaligenes* CECT5344: from omic approaches to biotechnological applications. FEMS Microbiol. Lett. 365: p.fny032.

Cacho, A., E. Smirnova, S. Huzurbazar and X. Cui. 2016. A comparison of base-calling algorithms for illumina sequencing technology. Briefings in bioinformatics 17(5): 786–95.

Cai, S., Y. Shen, Y. Zou, P. Sun, W. Wei, J. Zhao and C. Zhang. 2018. Engineering highly sensitive whole-cell mercury biosensors based on positive feedback loops from quorum-sensing systems. Analyst 143(3): 630–634.

Carpenter, A. C., I. T. Paulsen and T. C. Williams. 2018. Blueprints for biosensors: design, limitations, and applications. Genes. 9: 375.

Chandran, H., M. Meena and K. Sharma. 2020. Microbial biodiversity and bioremediation assessment through omics approaches. Front. Environ. Chem. 1: 9.

Chen, C., J. Hou, J. J. Tanner and J. Cheng. 2020. Bioinformatics methods for mass spectrometry-based proteomics data analysis. Int. J. Mol. Sci. 2873.

Clasen, B. E., A. De O. Silveira, D. B. Baldoni, D. F. Montagner, R. J. S. Jacques and Z. I. Antoniolli. 2018. Characterization of ectomycorrhizal species through molecular biology tools and morphotyping. Scient. Agric. 75(3): 246–54.

Dangi, A. K., B. Sharma, R. T. Hill and P. Shukla. 2019. Bioremediation through microbes: systems biology and metabolic engineering approach. Crit. Rev. Biotechnol. 39: 79–98.

Das, S., B. M. F. Pettersson, P. R. K. Behra, M. Ramesh, S. Dasgupta, A. Bhattacharya and L. A. Kirsebom. 2015. Characterization of three *Mycobacterium* spp. with potential use in bioremediation by genome sequencing and comparative genomics. Gen. Biol. Evol. 7(7).

Das, S., H. R. Dash and J. Chakraborty. 2016. Genetic basis and importance of metal resistant genes in bacteria for bioremediation of contaminated environments with toxic metal pollutants. Appl. Microbiol. Biotechnol. 100: 2967–2984.

Delmar, J. A., C. C. Su and E. W. Yu. 2015. Heavy metal transport by the C us CFBA efflux system. Prot. Sci. 24: 1720–1736.

Diep, P., R. Mahadevan and A. F. Yakunin. 2018. Heavy metal removal by bioaccumulation using genetically engineered microorganisms. Front. Bioengg. and Biotechnol. 6: 157.

Dixit, R., D. Malaviya, K. Pandiyan, U. B. Singh, A. Sahu, R. Shukla et al. 2015. Bioremediation of heavy metals from soil and aquatic environment: an overview of principles and criteria of fundamental processes. Sustainability 7: 2189–2212.

Dvořák, P., P. I. Nikel, J. Damborský and V. de Lorenzo. 2017. Bioremediation 3.0: engineering pollutant-removing bacteria in the times of systemic biology. Biotechnol. Adv. 35: 845–866.

Ekblom, R. and J. B. W. Wolf. 2014. A field guide to whole-genome sequencing, assembly and annotation. Evol. Appl. 7: 1026-1042.

Fernandez-López, R., R. Ruiz, F. de la Cruz and G. Moncalián. 2015. Transcription factor-based biosensors enlightened by the analyte. Front. Microbiol. 6: 648.

Fleming, J. T., J. Sanseverino and G. S. Sayler. 1993. Quantitative relationship between naphthalene catabolic gene frequency and expression in predicting pah degradation in soils at town gas manufacturing sites. Environ. Sci. Technol. 27(6): 1068–1074.

Franke, C., G. Studinger, G. Berger, S. Böhling, U. Bruckmann, D. Cohors-Fresenborg and U. Jöhncke. 1994. The assessment of bioaccumulation. Chemosphere 29(7): 1501–1514.

Furukawa, K., J. Hirose, A. Suyama, T. Zainki and S. Hayashida. 1993. Gene components responsible for discrete substrate specificity in the metabolism of biphenyl (bph operon) and toluene (tod operon). J. Bacteriol. 175(16): 5224–5232.

García-Hidalgo, J., D. P. Brink, K. Ravi, C. J. Paul, G. Lidénb and M. F. Gorwa-Grauslund. 2020. Vanillin production in pseudomonas: whole-genome sequencing of *Pseudomonas* sp. strain 9.1 and reannotation of *Pseudomonas putida* cala as a vanillin reductase. Appl. Environ. Microbiol. 86(6):02442–19.

Giovanella, P., G. A. L. Vieira, Igor v. R. Otero, E. P. Pellizzer, B. de J. Fontes and L. D. Sette. 2020. Metal and organic pollutants bioremediation by extremophile microorganisms. J. Hazard. Mat.: 382: 121–24.

Gobas, F. A. P. C., L. P. Burkhard, W. J. Doucette, K. G. Sappington, E. M. J. Verbruggen, B. K. Hope, M. A. Bonnell, J. A. Arnot and J. v. Tarazona. 2016. Review of existing terrestrial bioaccumulation models and terrestrial bioaccumulation modelling needs for organic chemicals. Integr. Environ. Assess. Manag.. 12: 123–134.

Guo, H. B., A. Johs, J. M. Parks, L. Olliff, S. M. Miller, A. O. Summers et al. 2010. Structure and conformational dynamics of the metalloregulator MERr upon binding of Hg(ii). J. Mol. Biol. 398(4).

Hartwell, L. H., J. J. Hopfield, S. Leibler and A. W. Murray. 1999. From molecular to modular cell biology. Nat. 402: C47–52.

Hasin, A. A. L., S. J. Gurman, L. M. Murphy, A. Perry, T. J. Smith and P. H. E. Gardiner. 2010. Remediation of chromium(vi) by a methane-oxidizing bacterium. Environ. Sci. Technol. 44(1): 400–405.

Hebbeln, P. and T. Eitinger. 2004. Heterologous production and characterization of bacterial nickel/cobalt permeases. FEMS Microbiol. Lett. 230(1): 129–35.

Heidelberg, J. F., I. T. Paulsen, K. E. Nelson, E. J. Gaidos, W. C. Nelson, T. D. Read, J. A. Eisen et al. 2002. Genome sequence of the dissimilatory metal ion-reducing bacterium *Shewanella oneidensis*. Nat. Biotechnol. 20(11): 1118–1123.

Hong, Y. H., C. C. Ye, Q. Z. Zhou, X. Y. Wu, J. P. Yuan, J. Peng, H. Deng and J. H. Wang. 2017. Genome sequencing reveals the potential of *Achromobacter* sp. hz01 for bioremediation. Front. Microbiol. 8: 1507.

Hu, H., H. Lin, W. Zheng, S. J. Tomanicek, A. Johs, X. Feng, D. A. Elias, L. Liang and B. Gu. 2013. Oxidation and methylation of dissolved elemental mercury by anaerobic bacteria. Nat. Geosci. 6(9): 751–754.

Inda, M. E. and T. K. Lu. 2020. Microbes as biosensors. Ann. Rev. Microbiol. 74: 337–359.

Jarque, S., B. Michal, B. Ludek and H. Klara. 2016. Yeast biosensors for detection of environmental pollutants: current state and limitations. Trends Biotechnol. 34: 408–419.

Jia, X., T. Fu, B. Hu, Z. Shi, L. Zhou and Y. Zhu. 2020. Identification of the potential risk areas for soil heavy metal pollution based on the source-sink theory. J. Hazard. Mat. 393. 122424.

Jia, X., R. Bu, T. Zhao and K. Wu. 2019. Sensitive and specific whole-cell biosensor for arsenic detection. Appl. Environ. Microbiol. 85(11). e00694-19.

Jones, K. A., J. Zinkus-Boltz and B. C. Dickinson. 2019. Recent advances in developing and applying biosensors for synthetic biology. Nano Fut. 3(4): 042002.

Kang, Y., W. Lee, S. Kim, G. Jang, B. G. Kim and Y. Yoon. 2018. Enhancing the copper-sensing capability of *Escherichia coli*-based whole-cell bioreporters by genetic engineering. Appl. Microbiol. Biotechnol. 102(3): 1513–1521.

Kapahi, M. and S. Sachdeva. 2019. Bioremediation options for heavy metal pollution. J. Health Pollut. 9: 24.

Karoui, E., M., M. Hoyos-Flight and L. Fletcher. 2019. Future trends in synthetic biology—a report. Front. Bioeng. Biotechnol. 7: 175.

Kaur, D., A. Singh, A. Kumar and S. Gupta. 2020. Genetic engineering approaches and applicability for the bioremediation of metalloids. *In*: Plant Life Under Changing Environment, 207–235.

Kim, H. J. and S. J. Lee. 2018. Synthetic biology for microbial heavy metal biosensors. Anal. Bioanal. Chem. 410: 1191–1203.

Lebuhn, M., S. Weiß, B. Munk and G. M. Guebitz. 2015. Microbiology and molecular biology tools for biogas process analysis, diagnosis and control. Biogas Sci. Technol. 1–40.

Liu, S., F. Zhang, J. Chen and G. Sun. 2011. Arsenic removal from contaminated soil via biovolatilization by genetically engineered bacteria under laboratory conditions. J Environ. Sci. 23(9): 1544–1550.

Liu, T., Z. Chu and W. Jin. 2019. Electrochemical mercury biosensors based on advanced nanomaterials. J. Mat. Chem B. 7: 3620–3632.

Long, F., C. C. Su, H. T. Lei, J. R. Bolla, S. v. Do and E. W. Yu. 2012. Structure and mechanism of the tripartite CusCBA heavy-metal efflux complex. Philosophical Transactions of the Royal Society B: Biological Sciences. 1592: 1047–1058.

Lu, X., Y. Liu, A. Johs, L. Zhao, T. Wang, Z. Yang, H. Lin et al. 2016. Anaerobic mercury methylation and demethylation by *Geobacter bemidjiensis* bem. Environ. Sci. Technol. 50(8): 4366–4373.

Malik, G., R. Arora, R. Chaturvedi and M. S. Paul. 2021. Implementation of genetic engineering and novel omics approaches to enhance bioremediation: A Focused Review. Bulletin of Environ. Contam. Toxicol. 1–8.

Massoud, R., M. R. Hadiani, P. Hamzehlou and K. Khosravi-Darani. 2019. Bioremediation of heavy metals in food industry: application of saccharomyces cerevisiae. E. J. Biotechnol. 37: 56–70.

Mehrotra, P. 2016. Biosensors and their applications—a review. J. Oral Biol. Craniofacial Res. 6: 153–159.

Miyazaki, R., C. Bertelli, P. Benaglio, J. Canton, N. de Coi, W. H. Gharib, B. Gjoksi et al. 2015. Comparative genome analysis of *Pseudomonas knackmussii* b13, the first bacterium known to degrade chloroaromatic compounds. Environ. Microbiol. 17(1): 91–104.

Morgan, W. D., D. G. Bear and P. H. von Hippel. 1983. Rho-dependent termination of transcription. i. identification and characterization of termination sites for transcription from the bacteriophage lambda pr promoter. J. Biol. Chem. 258(15): 9553–64.

Muccee, F. and S. Ejaz. 2020. Whole genome shotgun sequencing of pops degrading bacterial community dwelling tannery effluents and petrol contaminated soil. Microbiol. Res. 238.126504.

Naguib, M. M., A. O. El-Gendy and A. S. Khairalla. 2018. Microbial diversity of mer operon genes and their potential rules in mercury bioremediation and resistance. The Open Biotechnol. J. 12(1).

Naresh, V. and N. Lee. 2021. A review on biosensors and recent development of nanostructured materials-enabled biosensors. Sensors. 21,1109.

Ndu, U., T. Barkay, R. P. Mason, A. T. Schartup, R. Al-Farawati, J. Liu and J. R. Reinfelder. 2015. The use of a mercury biosensor to evaluate the bioavailability of mercury-thiol complexes and mechanisms of mercury uptake in bacteria. PLoS ONE 10(9): e0138333.

Njoku, K. L., O. R. Akinyede and O. F. Obidi. 2020. Microbial remediation of heavy metals contaminated media by *Bacillus megaterium* and *Rhizopus stolonifer*. Sci. Afr. 10.

Omokhagbor, A., Godleads, P. T. Fufeyin, S. E. Okoro and I. Ehinomen. 2020. Bioremediation, Biostimulation and Bioaugmention: A Review. Int. J. Environ. Biorem. Biodegrad. 3(1): 28–39.

Oyewusi, H. A., R. A. Wahab and F. Huyop. 2021. Whole genome strategies and bioremediation insight into dehalogenase-producing bacteria. Mol. Biol. Rep. 1–15.

Plyuta, V. A., D. E. Sidorova, G. B. Zavigelsky, V. Yu Kotova and I. A. Khmel. 2020. Effects of volatile organic compounds synthesized by bacteria on the expression from promoters of the znta, copa, and arsr genes induced in response to copper, zinc, and arsenic. Mol. Genet. Microbiol. Virol. 35(3): 152–8.

Qiu, H., K. Kaluarachchi, Z. Du, D. W. Hoffman and D. P. Giedroc. 1996. Thermodynamics of folding of the rna pseudoknot of the t4 gene 32 autoregulatory messenger rna. Biochem. 35(13): 4176–4186.

Rawat, M. and S. Rangarajan. 2019. Omics approaches for elucidating molecular mechanisms of microbial bioremediation. In: Smart bioremediation technologies: 191–203. Academc Press.

Rayu, S., D. G. Karpouzas and B. K. Singh. 2012. Emerging technologies in bioremediation: constraints and opportunities. Biodegradation 23(6): 917–26.

Rodríguez, A., M. L. Castrejón-Godínez, E. Salazar-Bustamante, Y. Gama-Martínez, E. Sánchez-Salinas, P. Mussali-Galante, E. Tovar-Sánchez and M. L. Ortiz-Hernández. 2020. Omics approaches to pesticide biodegradation. Curr. Microbiol. 77: 545–63.

Rojas, L. A., C. Yáñez, M. González, S. Lobos, K. Smalla and M. Seeger. 2011. Characterization of the metabolically modified heavy metal-resistant cupriavidus metallidurans strain msr33 generated for mercury bioremediation. PLoS ONE 6(3): e17555.

Rosenberg, E. 2013. Biotechnology and applied microbiology. pp. 315–328. *In*: Rosenberg, E., E. F. DeLong, S. Lory, E. Stackebrandt and F. Thompson (eds.). The Prokaryotes: Prokaryotic Biology and Symbiotic Associations.

Ruiz, O. N., D. Alvarez, G. Gonzalez-Ruiz and C. Torres. 2011. Characterization of mercury bioremediation by transgenic bacteria expressing metallothionein and polyphosphate kinase. BMC Biotechnol. 11: 82.

Rylott, E. L. and N. C. Bruce. 2020. How synthetic biology can help bioremediation. Curr. Opin. Chem. Biol. 58: 86–95.

Sanders, O. I., C. Rensing, M. Kuroda, B. Mitra and B. P. Rosen. 1997. Antimonite is accumulated by the glycerol facilitator glpf in *Escherichia coli*. J. Bacteriol. 179(10): 3365–3367.

Sato, M. P., Y. Ogura, K. Nakamura, R. Nishida, Y. Gotoh, M. Hayashi, J. Hisatsune et al. 2019. Comparison of the sequencing bias of currently available library preparation kits for illumina sequencing of bacterial genomes and metagenomes. DNA Res. 26(5): 017.

Seung, H. K., S. Singh, J. Y. Kim, W. Lee, A. Mulchandani and W. Chen. 2007. Bacteria metabolically engineered for enhanced phytochelatin production and cadmium accumulation. Appl. Environ. Microbiol. 73(19): 07.

Sharma, B., A. K. Dangi and P. Shukla. 2018. Contemporary enzyme based technologies for bioremediation: a review. J. Environ. Manag. 210: 10–22.

Sharma, B. and P. Shukla. 2020a. Designing synthetic microbial communities for effectual bioremediation: a review. Biocatal. Biotransfor. 38: 405–414.

Sharma, B. and P. Shukla. 2020b. Futuristic avenues of metabolic engineering techniques in bioremediation. Biotechnol. Appl. Biochem. 51–60.

Sharma, J. and M. H. Fulekar. 2009. Potential of *Citrobacter freundii* for bioaccumulation of heavy metal - copper. Biol. Med 1(3): 7–14.

Shi, B., J. Chen, Y. Liu, H. Zhang and J. Li. 2019. Spatial distribution of heavy metal and the microbial metal-resistance genes in copper tailing. Acta Sci Circumstant. 39(8).

Stoler, N. and A. Nekrutenko. 2021. Sequencing error profiles of illumina sequencing instruments. NAR Genom. and Bioinform. 3(1). Iqab019.

Tanaka, K. J., S. Song, K. Mason and H. W. Pinkett. 2018. Selective substrate uptake: the role of atp-binding cassette (abc) importers in pathogenesis. Biochim. Biophysic. Acta Biomemb. 1860: 868–877.

Tchounwou, P. B., B. Wilson and A. Ishaque. 1999. Important considerations in the development of public health advisories for arsenic and arsenic-containing compounds in drinking Water. Rev. Environ. Health. 14: 211–29.

Tchounwou, P. B., C. G. Yedjou, A. K. Patlolla and D. J. Sutton. 2012. Heavy metal toxicity and the environment. Mol. Clin. environ. Toxic. 133–164.

Torres, E. 2020. Biosorption: A review of the latest advances. Processes. 8: 1584.

Ufarté, L., É. Laville, S. Duquesne and G. Potocki-Veronese. 2015. Metagenomics for the discovery of pollutant degrading enzymes. Biotechnol. Adv. 33: 1845–54.

Utschig, L. M., J. W. Bryson and T. v. O'Halloran. 1995. Mercury-199 NMR of the metal receptor site in merr and its protein-dna complex. Sci. 268(5209): 380–385.

Vargas-Straube, M. J., S. Beard, R. Norambuena, A. Paradela, M. Vera and C. A. Jerez. 2020. High copper concentration reduces biofilm formation in *Acidithiobacillus ferrooxidans* by decreasing production of extracellular polymeric substances and its adherence to elemental sulfur. J. Proteomics 225: 103874.

Verma, S. and A. Kuila. 2019. Bioremediation of heavy metals by microbial process. Environ. Technol. Innov. 14: 100369.

Wang, B., D. Zhang, S. Chu, Y. Zhi, X. Liu, P. Zhou and R. Kant. 2020. Genomic analysis of *Bacillus megaterium* nct-2 reveals its genetic basis for the bioremediation of secondary salinization soil. Int. J. Genome.

Wu, Y., C. W. Wang, D. Wang and N. Wei. 2021. A Whole-cell biosensor for point-of-care detection of waterborne bacterial pathogens. ACS Synth. Biol. 10(2): 333–44.

Wysocki, R., C. C. Chéry, D. Wawrzycka, M. van Hulle, R. Cornelis, J. M. Thevelein and M. J. Tamás. 2001. The glycerol channel fps1p mediates the uptake of arsenite and antimonite in saccharomyces cerevisiae. Mol. Microbiol. 40(6): 1391–1401.

Xavier, J. C., P. E. S. Costa, D. C. Hissa, V. M. M. Melo, R. M. Falcão, V. Q. Balbino, L. A. R. Mendonça, M. G.S. Lima, H. D. M. Coutinho and L. C. L. Verde. 2019. Evaluation of the microbial diversity and heavy metal resistance genes of a microbial community on contaminated environment. Appl. Geochem. 105: 1–6.

Xia, X., Jiahong Li, Zijie Zhou, Dan Wang, Jing Huang and Gejiao Wang. 2018. High-quality-draft genome sequence of the multiple heavy metal resistant bacterium *Pseudaminobacter manganicus* jh-7.stand. Genomic Sci. 13(1): 1–8.

Xu, Wei, Evaldas Klumbys, Ee Lui Ang and Huimin Zhao. 2020. Emerging molecular biology tools and strategies for engineering natural product biosynthesis. Metabol. Eng. Commun.10: e00108.

Yang, Q., Z. Li, X. Lu, Q. Duan, L. Huang and J. Bi. 2018. A review of soil heavy metal pollution from industrial and agricultural regions in china: pollution and risk assessment. Sci Total Environ. 15: 690–700.

Zeng, Q., C. Stålhandske, M. C. Anderson, R. A. Scott and A. O. Summers. 1998. The core metal-recognition domain of merR. Biochem. 37(45): 15885–95.

Zhang, H., X. Yuan, T. Xiong, H. Wang and L. Jiang. 2020. Bioremediation of co-contaminated soil with heavy metals and pesticides: influence factors, mechanisms and evaluation methods. Chem. Eng. J. 398: 125657.

Zhao, X. W., M. H. Zhou, Q. B. Li, Y. H. Lu, N. He, D. H. Sun and X. Deng. 2005. Simultaneous mercury bioaccumulation and cell propagation by genetically engineered *Escherichia coli*. Process Biochem. 40(5): 06–14.

Section IV
Nanotechnology and Metal(loid)s Remediation

CHAPTER 17

Bioremediation of Heavy Metals from Ecosystem
Nanotechnological Perspectives

Ruma Ganguly,[1,*] *Anshu Mathur*[2] and *R.P. Singh*[2]

17.1 Introduction

The industrial revolution has changed the face of mankind. Once considered a boon, it is now turning into a bane owing to the pollution it is causing. This pollution has severely affected all forms of life in the ecosystem. While other forms of pollution are easier to control and can be subjected to bioremediation to some extent, this does not hold true in the case of metalloids. The metals, on the other hand, enter the food chain and lead to their biomagnification. This leads to deleterious effects on biological systems at cellular levels, resulting in metal-induced carcinogenesis, neuromuscular defects, mental retardation, renal malfunction, and other disastrous health issues (Wang and Shi 2001).

To tackle this problem and mitigate its side effects, scientists all over the world have tried their hands on physical, chemical, and biological remediation processes. The commonly used physicochemical processes are thermal, adsorption, chlorination, chemical extraction, ion-exchange, membrane separation, electrokinetics, bioleaching, and many more. Though these techniques can be efficiently used to remove heavy metals from the ecosystem, they are reported to cause many aggravating new problems like sludge disposal and generation of secondary pollutants (Selvi et al. 2019, Thomas et al. 2020).

The biological remediation mechanisms include processes like biosorption, metal-microbe interactions, bioaccumulation, biostimulation, biotransformation, and bioleaching. Though bioremediation does not pose problems of secondary pollutants and is eco-friendly in nature, it has its shortcomings of having very slow kinetics, thus rendering the process unfeasible (Akhtar et al. 2020).

The use of materials with sizes ranging from 1 to 100 nm, i.e., nanomaterials, which exhibit an exceptional capacity to remediate metal(loids), are being extensively attempted for environmental decontamination processes. Broadly, nanomaterials can be divided into two types: (i) carbon-

[1] Department of Biotechnology, Mata Gujri College of Professional Studies, Indore-452012, India.
[2] Department of Biosciences and Bioengineering, Indian Institute of Technology Roorkee, Roorkee-247667, India.
* Corresponding author: ruma_biot@yahoo.co.in

based nanomaterials (such as carbon nanotubes and graphenes) and (ii) inorganic nanomaterials (also known as metallic nanoparticles) (Ju-Nam and Lead 2008, Tang et al. 2015). Among them, the most commonly used for nanoremediation process are nanoscale zero-valent iron (nZVI), stabilizer-modified nZVI, nano apatite-based materials including nano-hydroxyapatite particles (nHAp) and stabilized nanochlorapatite (nCLAP), carbon nanotubes (CNTs), and titanium dioxide nanoparticles (TiO2 NPs) (Cai et al. 2019).

Nano bioremediation is a term used by Cecchin et al. (2017), where nanomaterials integrated with biological processes are further used to accelerate the remediation process (Kumari and Singh 2016). Thereafter, nano bioprocesses are categorized according to the type of organism used, such as phyto-nanoremediation for plant-assisted nanoremediation, microbial nanoremediation for microorganism assisted nanoremediation, and zoo-nanoremediation for animal-assisted nanoremediation (El-Ramady et al. 2017, Vázquez-Núñez et al. 2020). Nanoparticles owe a special place in the bioremediation process due to advantages such as an increased surface area for reactivity, the relatively lower activation energy required for chemical reactions, i.e., quantum dot effect, surface plasmon effect for detection of toxic materials, and ease of penetration in the contaminated zone (Rizwan et al. 2014).

17.2 Nanomaterials for Remediation of Metal(Loid)S

17.2.1 Carbon Nanotubes

A term coined by Lijma in 1919, it is basically a graphite sheet rolled up into tubes by varying the rolls; the carbon nanotubes can be bifurcated into two types, i.e., Single-wall carbon nanotubes (SWCNTs) and Multi-walled carbon nanotubes (MWCNTs) having the strength of 100 times higher than steel (Baby et al. 2019, Yu et al. 2021a).

The three methods of preparation of CNTs (Carbon Nano Tubes), i.e., chemical vapor deposition, laser ablation, and arc discharge methods, administer to a large extent to their adsorption capacity. Apart from adsorption, there are also other pathways for heavy metal removal by carbon nanotubes, such as electrostatic interaction, reduction, and ion-exchange methods (Yu et al. 2021b).

(a) SWCNTs—They have a diameter ranging from 0.3 nm to 3 nm (Soni et al. 2020). The surface allows for extensive adsorption capacity, intensive porosity, and user-friendly operations, thus rendering them highly pertinent to the remediation process (Baby et al. 2019). The carbon nanotubes have four adsorption sites that facilitate the adsorption of heavy metals-1. Internal Site 2. Interstitial Site 3. External Groove Site 4. External Surface (Gadhave et al. 2014).

(b) MWCNTs—The nanotubes with multiple rolls are called multi-walled carbon nanotubes having a diameter up to 100 nm (Iijima and Ichihashi 1993). The multi-walled concentric structure of nanotubes can be described by two models. (1) Russian doll model having sheets of graphite arranged in concentric rings. (2) Parchment model having one sheet of graphite rolled among itself like that of parchment of a newspaper (Alsharefa et al. 2017).

17.2.1.1 Functionalization of carbon nanotubes

The π–π electron interactions in the molecular structure of CNTs leads to intratubular aggregation of the CNTs. This leads to their poor solubility in solvents and decreased applicability. Two different methods to overcome this drawback are the mechanical and chemical methods. Due to abrasive actions (mixing, grinding, rubbing, ultrasonication), the mechanical methods can lead to the breakage of the nanotube. Also, these processes have been found to be ineffective in some instances. The chemical methods, which increase the adsorption capacity of CNTS, have led to increased applicability (Mallakpour and Soltanian 2016). The functional groups like carboxyl, hydroxyl and amide, and other functional groups are added to the surface by various methods to increase their adsorptive capacity (Anitha et al. 2015). Functionalization is important for changing the hydrophobic nature of CNTs to hydrophilic, thus increasing their adsorptivity. The functionality of CNTs can be reversed by heating them at 2200°C (Gadhave et al. 2014). The functionalization of CNTs can be achieved either by exohedral or endohedral mechanisms. Exohedral functionalization involves the grafting of molecules on the surface of CNTs by either covalent or non-covalent bonds, whereas the endohedral functionalization involves impregnating molecules in the cavities of CNTs (Tîlmaciu and Morris 2015).

Covalent functionalization: Covalent bond formation for functionalization of carbon CNTs can be performed either at the end caps of the carbon nanotubes or at the sidewalls (having defects). The direct functionalization of side walls leads to a change of hybridization from sp^2 to sp^3, resulting in the loss of the p-conjugation system on the graphene layer. The other method is defect functionalization which involves grafting functional groups on the already existing defects in CNTs (Jeon et al. 2011).

Non-covalent functionalization: The non-covalent functionalization does not change or disturb the electronic arrangement of CNTS. This type of functionalization is facilitated by forces such as Van-der Walls, hydrogen bonds, electrostatic interactions, and π-stacking interactions (Mallakpour and Soltanian 2016).

17.2.1.2 Nanoremediation for heavy metal and metal(loids) removal by functionalized CNTs

The functionalized MWCNTs exhibited increased adsorption of heavy metals and metalloids due to the formation of a number of multiple adsorption sites on the tube surface. The bioremediation of heavy metals from contaminated sites by functionalized CNTs is due to interactions between heavy metal and the functional group on the carbon nanotubes. These interactions lead to adsorption and thus bioremediation of heavy metals from contaminated sites. In a further study of the relation between adsorption of metals and the functional group present, Ntim and Mitra (2011) observed higher arsenic adsorption by zirconia functionalized nanocomposites in comparison to CNTs functionalized with iron oxides. Therefore, further studies on functionalizing of nanotubes by various groups are required for increasing the potential of remediation of heavy metal pollutants from water effectively (Aslam et al. 2021).

17.2.2 Graphene and Graphene Oxide Nanomaterials

The graphene oxide, owing to its high hydrophilic nature, large surface area, and presence of functional group (inbuilt), makes its applicability for use in the water detoxification process a success. The bioremediation of heavy metal technically is based on the adsorption of heavy metal through the reaction between the oxygen functional group on graphene and the metal contaminants present (Tahoon et al. 2020).

In comparison to CNT_S, graphenes, particularly graphene-based nanosheets, are preferred for heavy metal removal owing to the presence of innate oxygen functional group present (Park et al. 2005, Baniamerian et al. 2009, Azzaza et al. 2017).

In a study on multilayered graphene oxide nanosheets fabricated for removing cadmium and chromium from aqueous solution by adsorption mechanism (Hummers et al. 1958), the remediation efficiency was found to be largely dependent on pH and functional groups present, etc. (Yang et al. 2019a,c).

17.2.3 Metal Based Nanomaterials

Metallic nanomaterials (oxides of manganese, copper, magnesium, titanium, iron, silver) are commonly used to remove metal pollutants from water. Among the various metallic nanomaterials, it was observed that the Fe-based nanomaterials (iron sulfide nanoparticles, bimetallic Fe nanoparticles, and nanosized FeO), particularly the nano zero-valent iron, is most commonly used for heavy metal removal (Kumar and Gopinath 2016, Yang et al. 2019a).

Nanoscale zero-valent iron (nZVI) is a nanocomposite of Fe(0) with ferric oxide coating, which had been used for adsorption of various heavy metals like Hg (II), Cr (VI), Cu (II), Ni (II), Cd from contaminated areas. Owing to high reducing capacity and large surface area, nanoscale zero-valent iron (nZVI) is the first choice for the removal of metal contaminants (Yang et al. 2019a). The adsorptivity of the nZVI could be exploited further by metal doping (d, Cu, Ni, Pt) or by surface modifications (Fu et al. 2014). The study was further supported by an experiment conducted by Huang et al. (2015), where 98.919 percent removal of chromium could be achieved when nZVI conjugated with SDS (sodium dodecyl sulfate) was used. Further materials like bentonite had also been used to increase the adsorption capacity of nanoscale zero-valent iron (nZVI). They were used to remove lead, copper, cadmium, cobalt, zinc, and nickel from water (Zarime et al. 2018).

17.2.3.1 Magnetic nanoparticles for metal remediation

Magnetic nanoparticles are basically composed of two parts, one part is made up of a magnetic material (iron, nickel, and cobalt), and the second part is formed by a functionally active component. They can be functionally maneuvered using a magnetic field (Bossmann and Wang 2017).

The magnetic nanomaterials have been efficiently used to remove heavy metals from an aqueous medium. This is because of a few factors like a large surface-to-volume ratio, increased adsorption sites, and easy manipulation using a magnetic field. Mostly the remediation mechanisms involve the use of chelators (EDTA) attached to carbon nanomagnets which result in the formation of magnetic reagents. These magnetic reagents aid in the removal of metallic contaminants to concentrations as low as a few micrograms from aqueous solution (Koehler et al. 2000, Azzaza et al. 2017). Some of the applications of magnetic nanoparticles for metal and metalloids removal are outlined henceforth.

Magnetic nanoparticles modified with NH_2/PEI-EDTA had been successfully used by Zhang et al. (2011) to achieve 98.8% removal of lead from an aqueous solution up to three cycles of remediation. Surface engineered (with carboxyl, amine, and thiol group) magnetic nanoparticles were used by Singh et al. (2011) to remove chromium, cobalt, nickel, cadmium, lead, and arsenic, as well as bacterial pathogens from water samples (Carlos et al. 2013). Almomani et al. (2020) fabricated iron oxide magnetic nanoparticles (MNPs) grafted on hyperbranched polyglycerol (HPG) for removal of heavy metals like nickel, copper, and aluminum from industrial wastewater and observed a highly efficient process independent of organic matter and phosphorous content with operational efficiency for many recurrent cycles.

Iron-containing sludge (iMNP) and chemical agent (cMNP) were used to synthesize magnetic nanoparticles for the removal of arsenic from water. Magnetic nanoparticles prepared using two chemical agents could remove about 90 percent of arsenic within 60 min of initiation of the process. Maximum adsorption capacity and subsequent arsenic removal were observed when using magnetic nanoparticles synthesized by iMNP (12.74 mg/g) in comparison to magnetic nanoparticles of cMNP (11.76 mg g^{-1}). Moreover, iMNP fabricated magnetic nanoparticles proved to be more environment friendly in contrast to chemically synthesized ones (Zeng et al. 2020).

The magnetic nanoparticles prepared for the removal of heavy metals have a large surface area, resulting in nanomaterials with considerable magnetic power. This feature (a decrease in surface to volume ratio) leads to a non-superparamagnetic nature, thus leading to their agglomeration and failed adsorptivity in future cycles of operation. To overcome this problem, Pardo et al. (2021) prepared iron oxide magnetic nanoparticles (8.50 nm) with a high specific area, facilitating efficient adsorption of metals. A proper controlled synthetic procedure was adopted to retain the super paramagnetic feature with high magnetization and no aggregation. An adsorption value of *ca.* 400 mg of arsenic oxide and 500 mg of mercury oxide per gram of MNPs were obtained after

24 hr of incubation of metal solutions. The pH for maximal adsorption for arsenic was 10, while mercury favored a lower pH of 5.5. The kinetics of the process pointed to a chemisorption mechanism of remediation (Pardo et al. 2021).

In another study, calcium alginate beads' adsorbents were prepared by entrapping them on magnetic nanoparticles. For increasing their arsenic remediation efficiency, the beads were functionalized with methionine (MFMNABs). Among the various variables studied for maximal adsorption, a pH of 7.0–7.5 (at 110 min) was found to be the most optimum and led to 99.56 percentage removal of contaminants. The process was found to be economical and eco-friendly (Lilhare et al. 2021). Thao et al. (2021) synthesized magnetic magnetite (Fe_3O_4) nanoparticles of varying sizes (5.11, 10.53, and 14.76 nm). The different sized nanoparticles resulted in variation of arsenic remediation, thus pointing to the fact that adsorption is dependent on the size of the nanomaterial. An adsorption max of 99.02 percentage was achieved in 30 min with an adsorption capacity of 14.46 mg/g, with nanoparticles having a size of about 5nm. Thus, it can be concluded that smaller nanoparticles lead to a high remediation efficiency.

17.2.4 Photocatalytic Nanomaterials

The removal of heavy metals by photocatalytic process involves reduction of high valence toxic heavy metals to low or zero valence, nontoxic metals. Among the various semiconductor metals for photocatalytic removal of heavy metals, titanium dioxide (TiO_2) and zinc oxide (ZnO) are the most commonly used (Ren et al. 2021). Chromium Cr(VI) is toxic to humans while Cr (II) is not; this conversion of chromium can be achieved by photocatalytic process rendering the contaminants harmless (Subramaniam et al. 2019).

TiO_2-based nanomaterials have the dual functionality of being photocatalytic and high adsorption efficiency, thus favoring the easy removal of heavy metals. The activity of TiO_2, i.e., the adsorption efficiency, depends to a large extent on particle size, structure, crystal morphology, and surface area. The crystalline TiO_2 was reported to exhibit low adsorption compared to the amorphous TiO_2 for metals like manganese, iron, copper, and lead (Azzaza et al. 2017). In recent times, photocatalytic conversion of toxic Cr(VI) to nontoxic Cr(II) was achieved by using oxide (II) bismuth carbonate hybrid photocatalyst (Kar and Equeenuddin 2019). A hybrid WO3/reduced graphene oxide (rGO) was fabricated by Kumar and Gopinath (2016) that could photocatalytically reduce Cr(VI) efficiently.

To increase the photocatalytic efficiency and increase the visible light sensitivity, the photocatalyst was doped with metallic and nonmetallic materials, resulting in a nanocomposite photocatalyst with high metal removing efficiency (Subramanyam et al. 2019). The efficiency of metal remediation by the photocatalytic process was found to be influenced by the presence of other contaminants. In this context, Liu et al. (2020) had reported that photoreduction of Cr(VI) in the presence of Rhodamine B (RhB) was better in comparison to the solution containing Cr(VI) alone as a pollutant. Similar results were observed when the photocatalytic conversion of Cr(VI) in the presence of a few organic pollutants occurred (Ruan et al. 2019). The development of binary and tertiary photocatalysts has gained momentum recently, stressing the need to develop photocatalysts having high surface reactive sites and increased sensitivity to visible lights (Subramanyam et al. 2019).

17.2.5 Nanocomposites

Nanomaterials having polymeric substances integrated with them are termed nanocomposites. They can be classified based on the presence or absence of polymeric material present in them. They are basically composites (materials formed by blending two or more materials having specific properties) in which one part has nanoscale structural morphology, i.e., like nanotubes or nanoparticles, etc. The nanocomposites can be categorized into two types depending on the presence or absence of polymers, known as polymer-based nanocomposite and non-polymer-based nanocomposites, respectively. The nanopolymer based nanocomposite can be further categorized into (a) metal nanocomposite, (b) ceramic nanocomposite, (c) ceramic-ceramic nanocomposite, whereas the polymer-based are subcategorized into (a) polymer/ceramic-based, (b) inorganic/organic polymer-based, (c) inorganic/organic hybrid (d) polymer/layered silicate-based nanocomposites (Sen 2020).

17.2.5.1 Nanocomposites in remediation of heavy metal contaminants

Nanocomposites assisted remediation process for heavy metal removal can be categorized based on the constituents of nanocomposites. The different processes are briefly discussed below.

17.2.5.1.1 Remediation by inorganic based nanocomposite

Inorganic supports of nanocomposites which are used for heavy metals removal are mainly made up of activated carbon (AC), CNTs, natural materials like bentonite, zeolites, etc. (Tounsadi et al. 2016). Glass column packed with CNTs/chitosan nanocomposite exhibited substantial remediation efficiency of Cu(II), Zn(II), Cd(II), and Ni(II) from aqueous solution (Salam et al. 2011). Bioremediation of Cr(VI), Pb(II), Cd(II) from the water had been efficiently done by AC-supported nanocomposites (Kang et al. 2015, Parlayici et al. 2015, Fernando et al. 2018, Jayaweera et al. 2018, Yang et al. 2019a). A hyper metal accumulator bentonite, when combined with NZVI, magnetite, hexadecyltrimethylammonium bromide (CTMAB), ethylene diamine tetraacetic acid (EDTA), 2-mercaptobenzothiazole (MBT), cellulose, resulted in increased metal remediation potential (Yang et al. 2019a). Zeolite, another nanoparticle stabilizer, could remediate Pb(II) and Cd(II) by adsorption in the range of Pb (II) 55.55 mg g^{-1} and Cd (II) 40.16 mg g^{-1}, whereas when another form of zeolite, i.e., Hydroxyapatite/zeolite nanocomposite (HAp/NaP) was used, an increase in bioremediation efficiency (93% for lead and 89 percent for arsenic) was observed (Zendehdel et al. 2016, Wangi et al. 2020).

17.2.5.1.2 Remediation by organic polymer-based nanocomposite

Organic polymers like Polystyrene (PS), polyaniline, etc., based nanocomposites have been used efficiently for the heavy metal contaminant remediation process (Yang et al. 2019a). Polypyrrolepolyaniline/Fe_3O_4 magnetic nanocomposite was observed to have 100 percentage removal of lead at pH 8–10, with the initial concentration of lead being 20 mgL^{-1} (Afshar et al. 2015). A novel composite synthesized by treating FE-BTC with dopamine which polymerized to form polydopamine (PDA) and further formed the nanocomposite Fe-BTC/PDA was found to exhibit appreciable remediation of Pb and Hg^{2+} (99.8%) and yielded drinkable water in seconds. The nanocomposite was found to be resistant to fouling and was fully functional for many cycles (Sun et al. 2018). Biopolymers (cellulose, chitin, alginate, etc.) were also used for the fabrication of nanocomposites (Yang et al. 2019b). Cellulose, a plant-based biopolymer, has ample coordination sites for heavy metals due to its metal binding affinity of the hydroxyl group of the glucose ring (Cai et al. 2019, Yu. et al. 2021). In an experimental setup developed by using nanocellulose (NC)-Ag nanoparticles (AgNPs) embedded pebbles-based composite material, a high efficiency (99.48 percent) of Pb(II) and (98.30 percent) of Cr(III)) remediation was observed (Yang et al. 2019b).

A layered cellulose nanocomposite membrane having two layers, one support (cellulose microfiber sludge as a support) layer and the other functional layer (cellulose nanocrystals of $CNSL_{sl}$, CNC_{BE}, or PCN_{SL} in a gelatin matrix), was fabricated for bioremediation of mirror industry waste having metals like Ag^+ and $Cu^{2+}/Fe^{3+}/Fe^{2+}$ as the contaminant. The nanocomposite membrane showed remarkable metal bioremediation efficiency; the highest remediation efficiency of up to 100 percent was achieved by PCN_{SL} followed by CNC_{BE} and CNC_{SL}, respectively. The high remediation efficiency for the nano biopolymers was due to interactions between negatively charged nanocomposite (nanocellulose) and the positively charged contaminants (metal ions) (Karim et al 2016).

Two superior biopolymers used for the bioremediation process due to their larger bio adsorption efficiency are chitosan and alginate. Chitosan is a deacetylation product of chitin (found mostly in crustaceans) and alginate is derived from sea algae (Gokila et al. 2017, Spoiala et al. 2021). The adsorption potentials of chitosan and alginate nanocomposites were studied by Gokila et al. (2017) under batch conditions by varying physicochemical parameters. They reported pH 5.0 to be the most suitable for maximal adsorption. The adsorption capacity was further studied using Langmuir and Freundlich adsorption isotherms, and the performed experiments were found to be well fitted in the Freundlich isotherm specifications. Moreover, the adsorption capacity of both the nanocomposites was found to be quite high, and the adsorption mechanism of multilayered adsorption was found to be more effective for remediation. Thus, the nanocomposites proved to be super absorbers for the removal of chromium from wastewater.

A Thiol-lignocellulose sodium bentonite (TLSB) nanocomposite, fabricated by Zhang et al. (2020), was found to exhibit appreciable adsorption capacities of metal (357.29 mg g^{-1} for Zn(II), 458.32 mg g^{-1} for Cd(II), and 208.12 mg g^{-1} for Hg(II)) from aqueous solutions. Further, the metals ions were recovered back efficiently from TLSB using acids (HCl was used for Zn(II) and Cd(II), HNO_3 for Hg(II), making the nanocomposite pollutant-free and ready for use for the further cycle of remediation

17.2.5.1.3 Remediation by magnetic nanocomposite

Magnetic nanocomposites are materials having one part of the inorganic magnetic component and the other being organic polymer compounds with an efficient metal bioremediation efficiency (Zhu et al. 2013). Hierarchically structured manganese oxide-coated magnetic nanocomposites developed by Kim et al. (2015) showed high metal removal efficiency of Cd(II), Cu(II), Pb(II), and Zn(II) compared to plain Fe_3O_4 nanoparticles. Geetha and Belagali (2021) had fabricated magnetic nanocomposites for the removal of metal ions and dye from water. The nanocomposite exhibited excellent bioremediation efficiency for both metal and dye, with adsorption for Pb(II) being 1476.4 mg g^{-1} and 6947.9 mg g^{-1} for methyl orange (MO), respectively, from aqueous solution.

Magnetic guanidinylated chitosan (MGC) nanobiocomposite was synthesized by Eivazzadeh-Keihan (2020) for evaluation of its metal adsorption and cancer treatment functions. The maximal adsorption of copper, lead, and chromium was observed to be, respectively, 100, 98.64, and 33.76 percent. Metal and dye removal efficiency was detected by *in situ* synthesized $MnFe_2O_4$/GO nanocomposite and showed maximal adsorption of pollutants (pH 6.0) at an increased concentration of pollutants (625 mg g^{-1} for Pb_2^+ ions and 90 mg g^{-1} for NR dye). Also, the $MnFe_2O_4$/GO nanocomposite was reported to show competent bioremediation potential up to 5 cycles of operation (Katubi et al. 2021). Though nanocomposites exhibited high potential for remediation of metal pollutants, their applicability can further be exploited by exploring their biocompatibility studies with the environment (Yang et al. 2019a).

17.2.6 Other Nanomaterials in Remediation of Metals

Among other nanomaterials, zeolite nanomaterials showed great potential in the removal of heavy metals owing to their high porosity (Tahoon et al. 2020). Similarly, a comparative study was made between zeolite immobilized and alumina nanoparticles for the removal of cobalt and chromium contaminants. It was observed that the zeolite nanoparticles exhibited higher remediation potential (i.e., 17% increased efficiency of cobalt removal and 31.77% removal of chromium) as compared to alumina nanoparticles (Deravanesiyan et al. 2015).

Also, in another study, it was reported that clinoptilolite zeolite nanoparticles modified with pentetic acid exhibited a favorable adsorption capacity of cadmium (138.9 mg/g of cadmium) from water (Shafiof and Nezamzadeh-Ejhieh 2020).

17.3 Nanomaterials Assisted Bioremediation

Nanobioremediation or nanomaterial-assisted bioremediation is a technology wherein nanoparticles or nanomaterials are used for increasing the efficiency of the biological (plant, bacteria, algae, and fungi) remediation process (Yadav et al. 2017) (Fig. 17.1). Biogenic nanoremediation is mainly

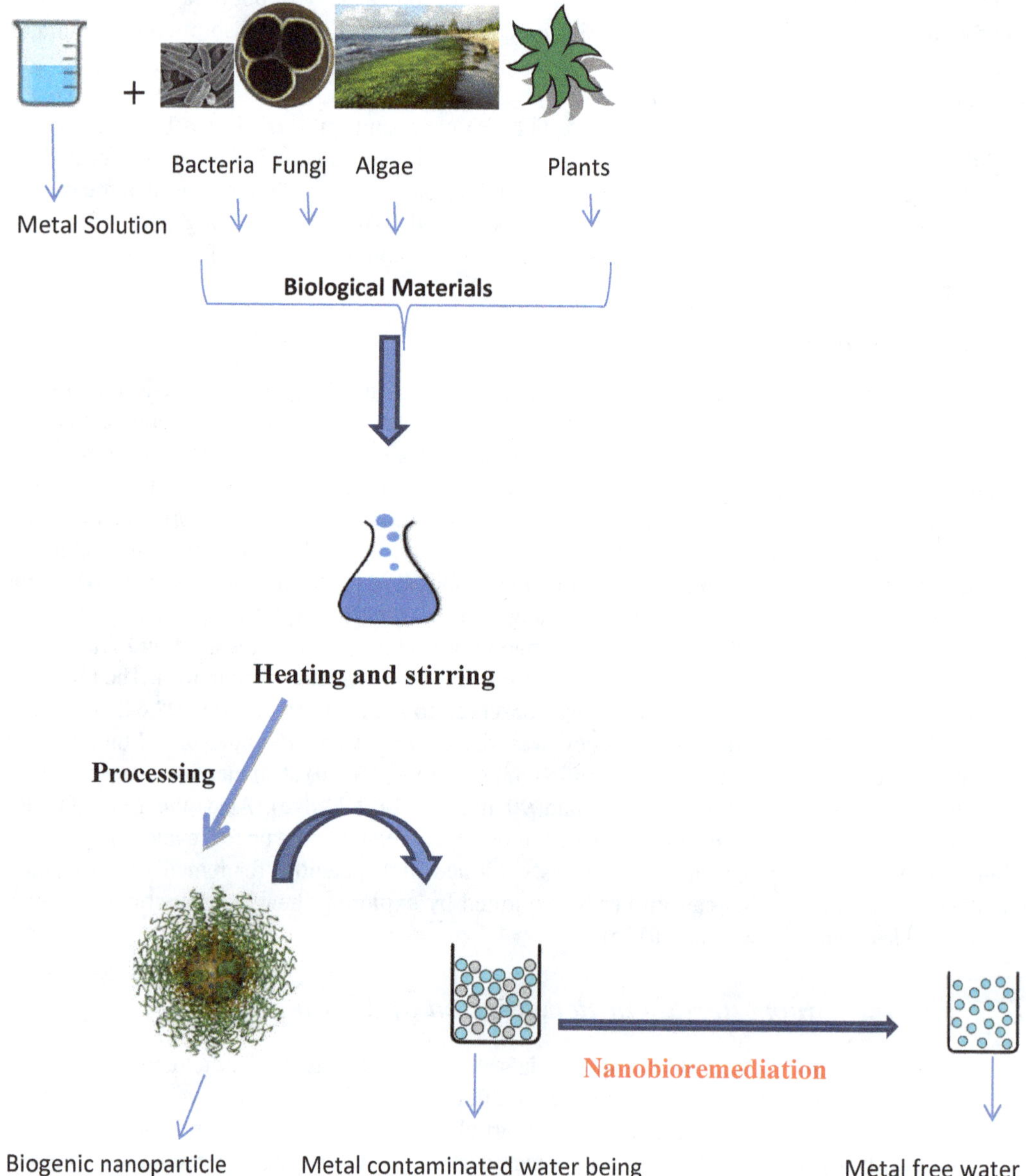

Fig. 17.1. Synthesis and application of biogenic nanoparticles in heavy metal removal.

Table 17.1. Nanobiocomposite in metal(loid)s removal.

Sr. No.	Nanoparticle	Plant used/ Microrganism used	Metal contaminant removed	References
1.	Plant biomass nanoparticle	*Noaea mucronata*	Cu, Zn, Pb and Ni	Mohsenzadeh et al. 2012
2.	nZVI (nanobiocomposite)	Oak and mulberry leaf extract	Nickel and Copper	Poguberović et al. 2016
3.	Nanobiocomposite of iron oxide	Orange peel pith	Cr(IV)	López-Téllez et al. 2011
4	nZVI nanoparticle (nanobiocomposite)	Orange peel	Cadmium	Gupta and Nayak 2012
5.	Iron oxide nanoparticles (nanobiocomposite)	*Bacillus subtilis*	Arsenic	Khan et al. 2021
6.	Biogenic nanoparticle	*Aspergillus tubingensis* (STSP 25)	Pb (II), Ni (II), Cu (II), and Zn (II)	Mahanty et al. 2020
7.	Chitosan nanocomposite	Chitosan of *Cunninghamella elegans*	Cu2+, Zn2+ and Pb2+)	Tayel et al. 2016
8.	Biogenic magnetite nanoparticles (BMNs)	Metal-reducing bacteria (Geocha-1, a mixed culture)	Chromium	Kim and Roh 2019
9.	nZVI particles	Leaf extracts of *Eucalyptus globules*	Cr(VI)	Madhavi et al. 2013
10.	Iron-based nanoparticles	*Syzygiumjambos* (L.) *Alston* leaf extract	Cr(VI)	Xiao et al. 2016
11.	Polymer-nano zero valent iron (nZVI)	Sulfate reducing bacteria	Cr(VI)	Ravikumar et al. 2017

concerned with utilizing biologically synthesized nanomaterials, whereas the process involving nanomaterial of physical or chemical origin in conjunction with a biological system (for remediation) is known as nanomaterial-assisted biological remediation or nanobioremediation (Karthik 2020) (Table 17.1).

17.3.1 Phytoremediation

Phytoremediation, also known as plant-mediated bioremediation (Chaney et al. 1997), is a cost-effective, efficient, environment-friendly, solar-driven soil and water detoxification strategy (Ali et al. 2013, Shehata et al. 2019, Shah and Davarey 2020, Shikha et al. 2021). Plants have an ability to accumulate all kinds of essential (Ca, Co, Cu, Fe, K, Mg, Mn, Mo, Na, Ni, Se, V, and Zn) as well as non-essential (Al, As, Au, Cd, Cr, Hg, Pb, Pd, Pt, Sb, Te, Tl, and U) metals from the soil for their biological and some unknown functions, and need different concentrations for growth and development (Djingova and Kuleff 2000, Jadia and Fulek 2009).

The plants capable of accumulating heavy metals or metalloids in their body tissues in huge amounts are known as hyperaccumulator (Reeves 2003). Plants of the *Brassica* family, such as *Brassica juncea* L., *Brassica napus* L. and *Brassica rapa* L., were found to show Zn and Cd accumulation ability moderately. Species of *Pistia* (water lettuce) were found to be good absorbers for Cd, while Canola is of Cu, Cd, Pb, Zn, and Sunflower is of lead. Plants having transgenic varieties have shown an increased heavy metal absorption capacity as compared with the normal ones. The transgenic variety of *Arabidopsis thaliana* possessing an over-expressive MT gene has been reported to show an increased Cu accumulating efficiency than wild type. *Nicotiana glauca* (shrub tobacco) is a hyperaccumulator of Pb and Cd, which was found to show heightened efficiency of metal absorption in both shoot (50 percentage) and the root system (85%) (Sarkar et al.

2021*)*. The phytoremediation technology (by hyperaccumulators) involves either of the processes phytoextraction (or phytoaccumulation), phytofiltration, phytostabilization, phytovolatilization, and phytodegradation (Tanghu et al. 2011, Ali et al. 2013), some of which are explained henceforth.

Phytofilteration: The removal of contaminants from water or wastewater by the plant is known as phytofilteration (Mukhopadhyay and Maiti 2010). If the removal of pollutants is through plant root, it is known as rhizofiltration, and if through seedling as blastofiltration and caulofiltration wherein the cut plant shoots is used (Mesjasz-Przybylowicz et al. 2004). In phytofiltration, the contaminants are absorbed or adsorbed, and thus their movement to underground water is minimized. This process minimizes the seepage of pollutants into groundwater (Ali et al. 2013).

Phytostabilization: Also known as phytoimmobilization, it is a property of some plant species to immobilize the contaminants in soil or water primarily by adsorption on roots or precipitation within the root zone, thus preventing their entry into the food chain. The root exudates various enzymes, which helps in the conversion of toxic metals to their less toxic counterparts (Wu et al. 2010).

Phytovolatilization: It involves the volatilization of pollutants from soil to the air via the plants. This process is commonly seen in a few heavy metals and metalloids. This technique can be used for organic pollutants and some heavy metals like mercury and selenium. In a way, this method transports the pollutants from one part of the ecosystem to the other (Jadia et al. 2008, Ali et al. 2013).

Phytodegradation: Owing to their ability to biodegrade organic pollutants, a feature known as phytodegradation, plants are also known as green livers of the biosphere. This mechanism is not applicable for heavy metals as they are non-biodegradable (Yan et al. 2020, Ali et al. 2013).

Phytoextraction: It is also known as phytoaccumulation, phytoabsorption, or phytosequestration. It involves the extraction of metals from the soil by plant roots and further transportation of them in the shoots. Moreover, the plants covering the soil protect it from erosion. The metal pollutants in this way are removed from the soil and transported to shoots (Jadia et al. 2009, Ali et al. 2013).

Out of all of the phytoremediation strategies, phytoextraction is found to be aptly suitable for the extraction of heavy metals and metalloids from contaminated soil and water and can be used for commercial applications as well (Ali et al. 2013). Phytoextraction suitability of the plants is determined by a few characteristics, which are as follows: (a) high efficiency of translocation of metals (metalloids) from root to aerial parts, (b) extensively branched roots, (c) very fast growth rate, (d) heavy metal toxicity tolerance ability, (e) easy cultivability, (f) high adjustability to varied environmental conditions, (g) unpalatability to herbivores and subsequently non entrance in the food chain (Mejáre and Bülow 2001, Tong et al. 2004, Adesodun et al. 2010, Li et al. 2010, Sakakibara et al. 2011, Shabani and Sayadi 2012, Ali et al. 2013, Wang et al. 2019, Arnao and Hernández-Ruiz 2019). In elucidating the mechanism for metal uptake by plants, it was observed that the metal ions were first actively absorbed by root cells through plasmalemma and further were transported on the cell walls by passive diffusion or absorbed and moved by proteins of the membrane transport system (Gomes et al. 2016).

17.3.1.1 Nanotechnology assisted phytoremediation

The bioremediation efficiency of plants for toxic metals had been found to increase when nanoparticles are used in conjunction with the phytoremediation process (Song et al. 2019) (Fig. 17.2). Apart from being synthesized by physical and chemical means, the nanoparticles can also be synthesized with plant parts for the bioremediation process. These are known as biogenic nanoparticles (Yadav et al. 2017). In this context, *Noaea mucronata*, a metal hyperaccumulator plant, was used in

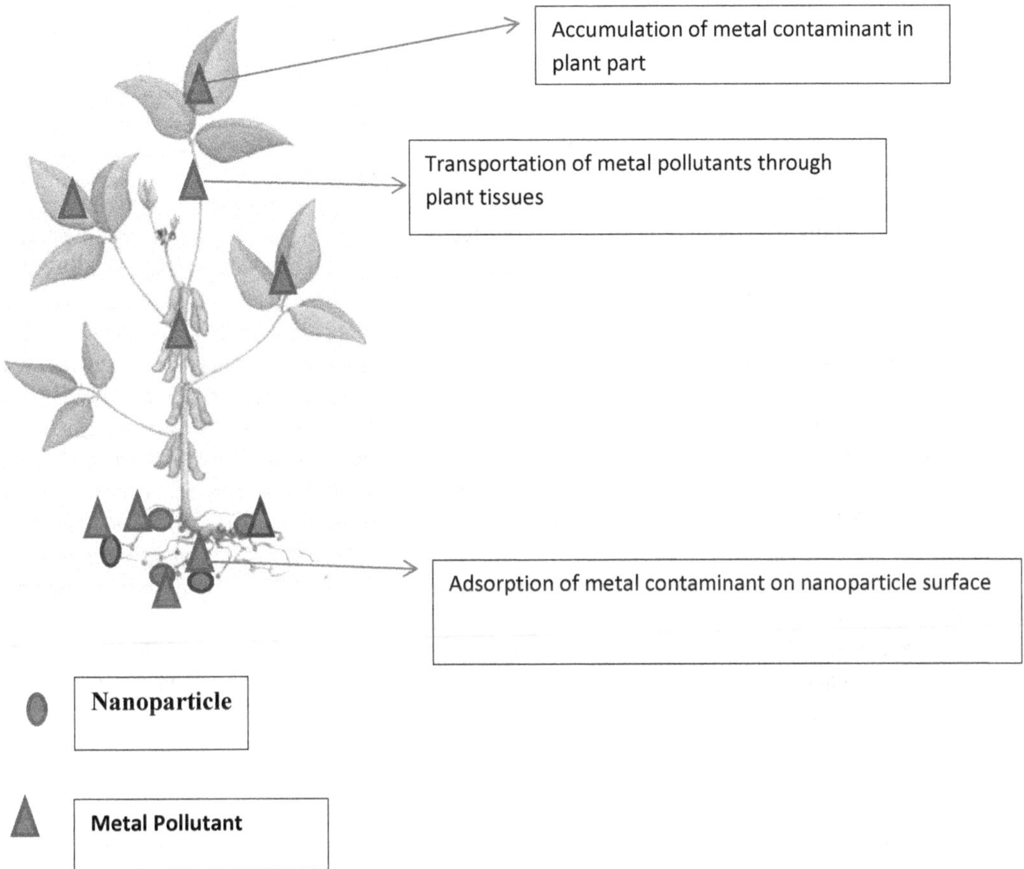

Fig. 17.2. Nanoparticle assisted phytoremediation of metal (loid)s.

field studies by Mohsenzadeh et al. (2012) for the removal of Cu, Pb, Zn, and Ni from the soil. The experiment was conducted using nanoparticles prepared from *Noaea mucronata* biomass. It was observed that bionanoparticles showed an appreciable decrease in metal ion concentrations (Pb-92%, Zn-76.05%, Cu- 74.66%, Cd-69.08%, Ni-31.50%) with Pb being the maximally remediated. The data showed that the concentrations of all the metals subjected to the bioremediation process decreased substantially (Pb-92%, Zn-76.05%, Cu-74.66%, Cd-69.08%, Ni-31.50%), among which Pb showed the highest decrease. The authors had earlier also used *Euphorbia macroclada* biomass for nanoparticle formation and bioremediation efficiency. They observed high bioremediation efficiency, with the removal of lead being 92% and zinc 76.05%.

Poguberović et al. (2016) used biogenic zero-valent nanoparticles (nZVI) synthesized from oak and mulberry leaf extracts to effectively remove Ni(II) and Cu(II) from aqueous solutions. They observed the process to be dependent on pH, with adsorption increasing with an increase of pH from 7.0 to 8.0.

In another study, it was observed that the nano biocomposite of iron oxide embedded in orange peel pith could effectively adsorb and reduce Cr(IV) in solution leading to increased metal bioremediation (71% removal) as compared to plain orange peel (34%) (López-Téllez et al. 2011). The biogenic nanoparticles synthesized using orange peel had also been used to remove cadmium from water (Gupta and Nayak 2012). Similarly biosynthesized nanoparticles such as zerovalence nanoparticle with mango peel and nZVI nanoparticle using medicinal plant leaves of *Rosa damascene*, *Thymus vulgaris* and *Urtica diocica* and Fe biosynthesized nanomaterial using loquat, i.e., *Eriobotrya japonica* were used for removal of metal pollutants like chromium (Yirsaw et al. 2016, Fazlzadeh et al. 2017, Zhu et al. 2018, Onal et al. 2019, Yi et al. 2019) and also Cd and Ni from contaminated biosystem (Mohsenzadeh 2011).

After studying the various nanoparticles used for the bioremediation process, Song et al. (2019) concluded nZVI to be most effective for facilitating the phytoremediation process owing to its low phytotoxicity. Cadmium metal had been found to affect plant growth and development adversely. An application of nano-Tio2 had been found to reduce the toxicity of metal to plant while supporting metal uptake by the plant and increasing uptake from 128.5 to 507.6 μg/plant (Singh and Lee 2016). The phytoextraction efficiency of Pb from soil was studied by Liang et al. (2017), ranging over a 12 months period. They observed that with an addition of only 0.2% (w/w) nano-hydroxyapatite, the rate of accumulation of Pb in the shoot system showed an appreciable increase within 1.5 months. The Pb uptake in the control group fluctuated between 16.74% to 31.76%, while that with nano assistance peaked over 30 percent after one month. The remediation level reached 44.39% in three months and after 12 months reached 46.55%. Bioremediation efficiency of Rye grass (*Lolium perenne*) assisted with nZVI nanoparticles was studied at varying concentrations (0, 100, 200, 500, 1000, and 2000 mg kg^{-1}) by Huang et al. (2018). They observed that a low concentration of nanoparticle (100 mg kg^{-1}) was more favorable for the bioremediation process, as high concentration (1000 and 2000 mg/kg) was toxic to the plant (Song et al. 2019).

Singh and Lee (2016), in their study of nano phytoremediation efficiency of Cd by soybean plants with TiO_2 nanoparticles at varying concentrations (100, 200, and 300 mg kg^{-1}) of the soil, observed that the accumulation of Cadmium metal in shoot increased to 1.9, 2.1, and 2.6 times, while in the underground system, it increased to 2.5, 2.6, and 3.3 times, respectively, thus remediating the soil of heavy metals. Increased phytoextraction of Cd from the soil by *Boehmeria nivea* (L.) Gaudich (ramie) was shown when used in conjunction with nZVI particles, as reported by Gong et al. (2017). They had added starch stabilized nZVI nanoparticles (100, 500, and 1000 mg kg^{-1}) to the Cd polluted soil before sowing the seedlings. An appreciable increase in Cd concentration in the roots, stems, and leaves were observed from 16 to 50%, 29 to 52%, and 31 to 73%, respectively. By incorporating salicylic acid nanoparticles to phytoextraction of arsenic by *Isatis cappadocica*, Souri et al. (2017) found increased tolerance of arsenic by the plant. Prior to the phytoextraction process, the authors treated the seedlings with 250 mM of salicylic acid nanoparticles for a period of 10 days. They observed a positive effect on nanoparticles both in terms of plant growth and metal tolerance with arsenic concentration in root reaching about 1188 mg/kg while that of the shoot to 705 mg kg^{-1} (Song et al. 2019).

Harikumar et al. (2019) had enumerated the bioremediation efficiency of lead by some hyperaccumulator plants in the combination of nanoscale zero-valent iron (nZVI). They observed an increase in metal content in plant biomass in the combined process compared to plain phytoremediation only. In their study of translocation factors (TF) and bioconcentration factor (BCF), which suggested the ability of plants to transport and concentrate the metal, they pointed to the fact that metal accumulation was high in roots in comparison to shoots, thereby suggesting the metal immobilizing potential of plants. Three different aquatic plants' (*Lemna minor* L., *Azolla filiculoides Lam.* and *Pistia stratiotes*) powders after modifications with interfacial layer synthesized silver nanoparticle showed an appreciable increase in heavy metals' absorption compared to control (Awady et al. 2021). In the plant nano combination system, nanomaterials are found to play a role in the phytoremediation system by directly removing pollutants, promoting plant growth, and increasing the phyto-availability of pollutants. Nano-zero-valent iron (NZVI) is currently the most studied nanomaterial to promote phytoremediation (Ding et al. 2021). An investigative study on the phytoremediation efficiency of *S. bicolor* in association with TiO_2 nanoparticles on antimony polluted soil was carried out by Zand and Heir (2020). They studied not only the nano-phytoremediation efficiency but also the plant growth and chlorophyll content. They observed that a high concentration of nanoparticles (1000 mg/kg) had a deleterious effect on seed germination. A study of varied concentration of antimony (43.11, 57.33, 73.88, 44.38 and 29.45 mg kg^{-1}) in phytoremediation efficiency of *S. bicolor* with different concentration of nanoparticles (50, 100, 250, 500, and 1000 mg/kg) compared to control having Antimony in concentration of 31.78 mg kg^{-1} was also done. TiO_2 nanoparticles at a concentration of 250 mg kg^{-1} NPs led to an increased (2.23 times = 319.16 mg kg^{-1}) antimony accumulation by *S. bicolor*. The authors also observed insignificant bioremediation efficiency when TiO_2 NPs concentration dipped down to 50 mg kg^{-1}, whereas a very high concentration of 1000 mg kg^{-1} had a toxic effect on plant growth. So the most favorable Tio2 concentration ranged from 50–1000 mg kg^{-1}, with 250 mg kg^{-1} being most favorable for Sb bioremediation. In their bioremediation study using both plain ZVI (B-nZVI) and starch-stabilized nZVI (S-nZVI) nanoparticles for Cd, Pb, and Zn, using sweet sorghum, Cheng et al. (2021) observed that both the nanoparticles enhanced phytoextraction of heavy metals considerably. Some of the suitable characteristics for the nano phytoremediation process in terms of suitability of plant and nanoparticle are characterized as follows (Srivastav et al. 2018).

Suitable characteristics of plants for bioremediation of heavy metals include an extensive deep and branched rooting system, tremendous toxicity tolerance levels for contaminants, high capacity of accumulation of metal(loid)s on aerial parts of plants, i.e., high translocation efficiency, non-palatability of plant parts by both man and animals and the ease of genetic manipulation. The selected nanoparticles used in the nanobioremediation process are required to possess specific characteristics like the non-toxic nature of nanoparticles, having a positive effect on growth, germination, and weight of the plant, a stimulatory effect on plant enzyme (or hormones) and the most important being the tremendous capability to bind pollutants together and increasing their availability for bioremediating plants (Srivastav et al. 2018). Nano phytoremediation studies have shown a promising future for the bioremediation of heavy metals and metalloids. But most of the work is at a preliminary level, and a further investigative study is the need of the hour before the technology could be made available to the scientifically unskilled farm dwellers.

In order to study the effect of absorbed nanoparticles on the physiology of *Boehmeria nivea*(L), and Gaudich(ramie), Gong et al. (2017) devised an experiment set up using starch stabilized nanoscale zerovalent iron(S-nZVI) to bioremediate cadmium at different concentrations (100, 500 and 1000 mg kg^{-1}) added to the soil. The accumulation of cadmium during nanoremediation was higher inside the cell organelle as compared to the cell wall, suggesting the metal accumulation being favored by nanomaterials' lower concentration of cadmium (100 mg kg^{-1}), which was helpful in easing the oxidative injury to seedling while a higher concentration (500 and 1000 mg kg^{-1}) was detrimental to plant growth and also magnified the oxidative damage to the plant. Thus, a lower concentration of starch stabilized nanoscale zerovalent iron had a positive effect on phytoremediation.

17.3.2 Microbial Bioremediation of Heavy Metals and Metalloids

Bioremediation is a technique of using microorganisms for removing toxic contaminants from the ecosystem either by *ex situ* or *in situ* process. While the *ex situ* process may disturb the ecological balance since it involves carrying polluted materials to bioremediation sites, this is not the case for *in-situ* bioremediation (Kapahi and Sachdeva 2019). The microbial bioremediation process can be brought about by bacteria, fungi, and algae. Among the various pollutants, heavy metals are the most problematic owing to their non-biodegradable nature and easy seepage in the groundwater (Usman et al. 2020), while the organic pollutants are bio converted through the various metabolic processes to nontoxic forms by the microorganism. The microorganisms carry out bioremediation of metals primarily by the one of the processes viz. biosorption, metal-microbe interactions, bioaccumulation, biomineralization, biotransformation, bioleaching, precipitation, intracellular and extracellular metal sequestration, exclusion by permeability barrier, and bioproduction of metal chelators (Shefali and Chetan 2021, Tayang and Songachan 2021).

Strains of *Bacillus* sp., *Pseudomonas* sp., *Rhodobacter* sp., *Sporosarcina pasteurii*, *Enterobacter cloacae* have shown the ability to decontaminate metal(loid)s such as lead, chromium, cadmium, and mercury. Apart from bacterial species, a large number of fungal species, i.e., *Aspergillusniger, Aspergillus fumigatus, Aspergillus flavus, Acremonium, Cellulosimicrobium* sp. *Termitomyces clypeatus, Penicillium chrysogenum, Trichoderma* sp., *Trichoderma viride, S. cerevisiae* had exhibited high metal bioremediation efficiency. Investigations on the bioremediation efficiency of metals are not only limited to bacterial and fungal species but also have been carried for the photosynthetic algal (*Cladophora fascicularis, Anabaena sphaerica, Fucus vesiculosus, Chlamydomonas reinhardtii, Spirogyra* sp., *Sargassum* sp., *Saccharina japonica*) systems (Deng et al. 2007, Christofordis et al. 2015, Poo et al. 2016, Tayang and Songachan 2021, Volaric et al. 2021, Ummalyma and Singh 2021).

17.3.2.1 Bioremediation mechanisms

The microorganisms use the metals for their growth and development and, in return, remediate the soil and water of metal pollutants. The immobilization of metals occurs in the biosphere with different mechanisms like biosorption, bioconversion, and precipitation. The microorganisms also facilitate bioremediation by mobilizing the pollutants into soluble forms, thus increasing their bioavailability for microbes. Also, the polymeric substance exudates by bacterial species have been found to bind metals effectively (Shaifali and Kumar 2021). After entering into the cell, the metals are taken up by a class of protein called metallothionein and sequestered, which are then stored in intracellular granules or transported out of the microbial cell. By performing oxidation and reduction reactions, the toxic form of metal is converted into its less toxic form, Hg(II) to its volatile form Hg(0), Fe(III) to Fe(II), As(V) to As(III), Cr(VI) to Cr (III), thus facilitating bioremediation. Bacterial catecholate siderophores and fungal hydroxamate siderophores are involved in the chelation of metal ions, thus enabling their removal from the soil (Volarić et al. 2021, Ojuederie and Babalola 2017).

17.3.2.2 Nanotechnology assisted microbial remediation of heavy metals and metalloids

The impetus on nanotechnology-assisted microbial remediation (including nano mycoremediation and bacterial assisted nano remediation) is considerable because the heavy metal detoxification processes like biosorption, bioconversion, and biodegradation, clean the environment without generating toxic end products (Geetha and Belagali 2021). Biosynthesis of biogenic nanoparticles using microorganisms makes the process more eco-friendly. Similarly, numerous microorganisms had been used for the biogenic synthesis of nanoparticles, which had been further used for the bioremediation process owing to its increased absorptivity (Goutam and Saxena 2021). *Aspergillus tubingensis* STSP 25 biogenic nanoparticles were used by Mahanty et al. (2020) for their metal bioremediation studies. The authors observed 90 percent and higher bioremediation efficiency for Pb (II), Ni (II), Cu (II), and Zn (II) from wastewater with stability for 5 cycles of application (Mandeep and Shukla 2020). Nano adsorption capacity of *Bacillus subtills* synthesized iron oxide nanoparticles was found to be effective in reducing arsenic content in paddy fields, thus providing favorable conditions for plant growth (Khan et al. 2021). Iwahori et al. (2014) observed increased efficiency of metal detoxification by biogenic magnetite nanoparticles (30- to 40-fold) in comparison with plain magnetite nanoparticles.

In order to remove metal pollutants from water, Mohammed et al. (2021) had developed a novel technique of converting the metal pollutants into nanoparticles using culture filtrate of *Fusarium solani* YMM20. This process served the dual purpose of removing metal pollutants as well as their bioconversion into nanoparticles. Bionano-Met nanoparticles were produced by the precipitation of metals (transition series) like gold, iron, and palladium on the bacterial surface of *Shewanella oneidensis* and resulted in the removal of palladium by the formation of nanoparticles on the bacterial surface (White et al. 20016, Batool et al. 2014). Chitosan of *Cunninghamella elegans* was found to have antimicrobial properties and metal absorbent (Cu^{2+}, Zn^{2+}, and Pb^{2+}) potential (Tayel et al. 2016). Further nanoparticles synthesized from chitosan, i.e., chitosan nanoparticles of *Cunninghamella elegans*, showed effective absorption of heavy metal (Pb and Cu at a range of 100–200 μg) from both liquid and solid matrices. In comparison to pure chitosan, the chitosan nanoparticles were found more effective in the bioremediation of heavy metals, especially Pb and Cu (Alsharari et al. 2018).

Nanoparticles coated with PVP (Polyvinylpyrrolidone) enhanced the heavy metal bioremediation efficiency of *Halomonas* sp., a gram-negative bacteria. The conjugated efforts had a positive effect on bioremediation efficiency both in terms of decrease in metal toxicity and the time taken. It was observed that about 100 percent removal of Pb after 24 hr and of Cd after 48 hr occurred, which was far better either with only nanoparticles or bacteria. The ability to produce exopolysaccharides was found to be important for metal removal efficiency as both the metals were found to be transported intracellularly by exopolysaccharides (Cao et al. 2020).

Kim and Roh (2019) experimentally compared the chromium removal efficiency of biogenic magnetite nanoparticles (BMNs) with chemically synthesized magnetite nanoparticles (CMNs) and found that the Cr(VI) bioremediation efficiency of BMN was far higher (100 percent) as compared to CMNs (82%) following 14 days of incubation. A pH variation study (pH 2–12) denoted that lower pH coupled with increased exposure to contaminant led to increased bioremediation efficiency (94 percent) of Cr(VI) from groundwater. Metal(loid) bioremediation efficiency was explored using plant *Arbuscular Mycorrhiza* (AM) in association with nano zero-valent iron (nZVI) nanoparticles assisted process. The nanoparticles (0.5%) were incorporated in the experimental site (metalloid contaminated Litavka river soil, Czech Republic). Physiochemical and biological studies pointed out that AM had an inhibitory effect on plant metal(loid) uptake. The nZVI had also been found to have an inhibitory effect on AM growth and development (Wu et al. 2018).

Fungi like *Phanerochaete chrysosporium* had been a preferred organism for the bioremediation process owing to their immense absorption and degradation ability. In spite of all the positive attributes, the bioremediation process by *P. chrysosporium* suffers from a few limitations, such as prolonged biodegradation process and susceptibility to pollutant toxicity. To alleviate this problem, *P. chrysosporium* coated with nitrogen-doped TiO_2 nanoparticles was structured and studied for bioremediation efficiency. The bioremoval of metal Cd(II) was in the range of 84.2% while that for 2,4-dichlorophenol (2,4-DCP) was 78.9%, while the pH varied between 4 to 7 following 60 hr of incubation. The initial metal and DCP concentrations had strongly affected the mycogenic nanoremediation efficiency. It was observed that the carboxyl, amino, and hydroxyl groups on the surface of PTNs were mainly involved in the biosorption process, further supporting the fact that PTNs have an efficient metal removal (2.4 DCP also) capability (Chen et al. 2012). *Saccharomyces cerevisiae* immobilized on chitosan coated magnetic nanoparticles (SICCM) was observed to have 96.8 percent Cu(II) removal efficiency from aqueous solution. The optimal conditions for bioremediation were pH 4.5 and initial Cu(II) dose 60 mg L^{-1}. The maximal Cu(II) adsorbed was observed to be 144.9 mg g^{-1}. The competent biosorption capacity of the bio nanomaterial was due to, e.g., the hydroxyl group of glucose units leading to increased hydrophobicity, functional groups having metal adsorbing potential, reactive functional group, and flexible polymeric structure (Peng et al. 2010, Batool et al. 2014). Synergistic effect of AM fungi (*Glomus macrocarpum*) and ZnO nanoparticles on lead toxicity studies conducted on wheat plants showed promising results for the growth of wheat plants on Pb contaminated soils (Raghib et al. 2020). *Nannochloropsis* sp. and *Chlorella vulgaris* were utilized for the synthesis of silver (AgNPs) and gold nanoparticles (AuNPs)

Table 17.2. Nanoparticle assisted bioremediation of metal(loid)s.

Sr. No.	Nanoparticle	Plant/Microrganism used	Metal contaminant	References
1.	Crystalline iron oxide nanoparticles (IO-NPs)	*Aspergillus tubingensis* (STSP 25)	90% of heavy metals [Pb (II), Ni (II), Cu (II), and Zn (II)]	Mahanty et al. 2020
2.	Iron oxide NPs coated with polyvinylpyrrolidone (PVP)	*Halomonas* species	Cd and Pb	Cao et al. 2020
3.	TiO_2 nanoparticles	Soyabean plant	Cadmium	Singh and Lee 2016
4.	Nanoscale zero valent iron (nZVI)	*Bacillus subtills*	Arsenic	Khan et al. 2021
5.	Silicon nanoparticles	Pea (*Pisum sativum* L.)	chromium	Tripathi et al. 2015
6.	ZnO nanoparticles	White popinac [*Leucaena leucocephala* (Lam.) de Wit	Cd and Pb	Venkatachalam et al. 2015

for metal bioremediation of pharmaceutical industry effluents. *Nannochloropsis* sp. and *C. vulgaris* derived nanoparticles exhibited 70.35% and 74.62% bio removal of zinc, respectively, while gold nanoparticles of *Nannochloropsis* sp. and *C. vulgaris* had respectively shown appreciable reduction of zinc from effluents (60.32% and 66.83%, respectively). The study from Ogunsona et al. (2020) helped in the development of an algal-based nanobioremediation technique for heavy metal removal from wastewater (Table 17.2).

17.4 Conclusions and Future Perspectives

The present century has been perceiving the diverse use of nanomaterials in a number of areas. Its use in the remediation of heavy metals has been observed to be progressively gaining momentum. The nanoparticles assisted remediation process had been found showing sturdiness even to the most corrosive contaminants. The technology can be further tailored to the needs of the process and the contaminants. Also, the added advantage of increased surface-to-volume ratio, high adsorptivity, low environmental toxicity, and economic feasibility had further added to its choice in the remediation process. Nanotechnology can be implicated to provide an ecofriendly approach to environmental bioremediation with minimum damage to the ecosystem. Nanoparticles and heavy metal accumulators together have opened a newer dimension of the nanobioremediation process. This innovative approach of bioremediation has proved to be a cost-effective and more environment friendly and effective remediation process. As compared to the physicochemical approaches, the nanobioremediation process is still in its early stages. In addition, more efforts are essential to develop cost-effective routes for nanoparticles synthesis and practice them for remediation of effluents. Furthermore, scaling up of the process is required for the operational field application of the nanomaterials. As biological materials are environmentally compatible, the study of biogenic nanoparticles or nano-assisted biological remediation process needs to be vigorously explored. The application of recombinant DNA technology in developing microorganisms having a wide-ranging capacity for metal tolerance and its use in nanobioremediation needs to be a priority area for future efforts.

Acknowledgments

The authors wish to thank all the researchers who had provided their research papers.

References

Adesodun, J. K., M. O. Atayese, T. Agbaje, B. A. Osadiaye, O. Mafe and A. A. Soretire. 2010. Phytoremediation potentials of sunflowers (*Tithonia diversifolia* and *Helianthus annuus*) for metals in soils contaminated with zinc and lead nitrates. Water Air Soil Pollut. 207: 195–201.

Adenigba, V. O., I. O. Omomowo, J. K. Oloke, B. A. Fatukasi, M. A. Odeniyiand and A. A. Adedayo. 2020. Nanotechnology Applications in Africa: Opportunities and Constraints. IOP Conf. Series: Materials Science and Engineering 805:012030. IOP Publishing.

Akhtar, Nahid and M. Amin-ul Mannan. 2020. Mycoremediation: Expunging environmental pollutants. Biotech Rep. 26: e00452.

Ali, Hazrat. E. K. Khan and M. A. Sajad. 2013. Phytoremediation of heavy metals—Concepts and applications. Chemosphere 91(7): 869–881.

Almomani, F., R. Bhosale, K. Majeda, K. Anand and T. Almomani. 2020. Heavy metal ions removal from industrial wastewater using magnetic nanoparticles (MNP). Appl. Surf. Sci. 506: 144924.

Alsharari, S. F., A. A. Tayel and S. H. Moussa. 2018. Soil emendation with nano-fungal chitosan for heavy metals biosorption. Int. J. Biol. Macromol. 118: 2265–2268.

Alsharefa, J. M. A., Mohd. R. Tahaa and T. A. Khana. 2017. Physical dispersion of nanocarbons in composites—A review. J. Teknologi. 79. 10.11113/jt.v79.7646.

Afshar, A., S. Seyed, M. Afsaneh and E. Mohammad. 2015. Polypyrrole-polyaniline/Fe3O4 magnetic nanocomposite for the removal of Pb(II) from aqueous solution. Korean J. Chem. Eng. 33(2).

Anitha, K., S. Namsani and J. K. Singh. 2015. Removal of heavy metal ions using a functionalized single-walled carbon nanotube. A molecular dynamics study. J. Phys. Chem. 119(30): 8349–8358.

Arnao, M. B. and J. H. Ruiz. 2019. Role of melatonin to enhance phytoremediation capacity. Plants. 8: 295.

Aslam, M. M. A., H. W. Kuo, W. Den, M. Usman, M. Sultan and H. Ashraf. 2021. Functionalized Carbon Nanotubes (CNTs) for water and wastewater treatment: preparation to application. Sustain. 13: 5717.

Awady, F. R. E., M. A. Abbas, A. M. Abdelghany and Y. A. E. Amir. 2021. Silver modified hydrophytes for heavy metal removal from different water resources. Biointerface Res. Appl. Chem. 11(6): 14555–14563.

Azzaza, S., R. T. Kumar, J. J. Vijaya and M. Bououdina. 2016. Nanomaterials for heavy metal removal. *In*: Hussain, C. M. and B. Kharisov (eds.). Advan Environ Ana: Appli of Nano. 1.

Baniamerian, M. J., S. E. Moradi, A. Noori and H. Salahi. 2009. The effect of surface odification on heavy metal ion removal from water by carbon nanoporous adsorbent. Appl. Surf. Sci. 256. 5: 1347–1354.

Baby, R., B. Saifullah and M. Z. Hussein. 2019. Carbon nanomaterials for the treatment of heavy metal-contaminated water and environmental remediation. Nanoscale Res. Lett. 14: 341.

Batool, S., S. Akib, M. Ahmad, K. S. Balkhair and M. A. Ashraf. 2014. Study of modern nano enhanced techniques for removal of dyes and metals. J. Nano. ID 864914.

Bossmann, S. H. and H. Wang. 2017. Magnetic nanomaterials. (eds.). Royal Society of Chemistry, Cambridge.

Cai, C., M. Zhao, Yu. Zhen, H. Rong, C. Zhang. 2019. Utilization of nanomaterials for *in-situ* remediation of heavy metal(loid) contaminated sediments: A review. Sci. Total Environ. 662: 205–217.

Cao, X., A. Alabresm, Y. P. Chen, A. W. Decho and J. Lead. 2020. Improved metal remediation using a combined bacterial and nanoscience approach. Sci. Total Environ. 704: 0048–9697. 135378.

Carlos, L., F. S. G. Einschlag, M. C. González and D. O. Mártire. 2013. Applications of magnetite nanoparticles for heavy metal removal from wastewater. Chapter 3. pp. 63–76. *In*: Fernando S. García Einschlag and Luciano Carlos (eds.). Waste Water - Treatment Technologies and Recent Analytical Developments. Intechopen.

Cecchin, I., K. R. Reddy, A. Thome, E. F. Tessaro and F. Schnaid. 2017. Nanobioremediation: Integration of nanoparticles and bioremediation for sustainable remediation of chlorinated organic contaminants in soils. Int. Biodeterior. Biodegrad. 119: 419–428.

Chaney, R. L., M. Malik, Y. M. Li, S. L. Brown, S. A. J. Brewer and A. J. M. Baker. 1997. Phytoremediation of soil metals. Curr. Opin. Biotechnol. 8(3): 279–284.

Chen, B., M. Yuan and L. Qian. 2012. Enhanced bioremediation of PAH-contaminated soil by immobilized bacteria with plant residue and biochar as carriers. J. Soils Sediments. 12: 1350–1359.

Cheng, P., S. Zhang, Q. Wang, X. Feng, S. Zhang, Y. Sun and F. Wang. 2021. Contribution of nano-zero-valent iron and *Arbuscular mycorrhizal* fungi to phytoremediation of heavy metal-contaminated soil. Nanomaterials. 11: 1264.

Christoforidis, A., S. Orfanidis, S. Papageorgiou, A. Lazaridou and E. Favvas. 2015. Mitropoulos, study of Cu (II) removal by *Cystoseira crinitophylla* biomass in batch and continuous flow biosorption. Chem. Eng. J. 277: 334–340.

Deng, L., Y. Su, H. Su, X. Wang and X. Zhu. 2007. Sorption and desorption of lead (II) from wastewater by green algae *Cladophora fascicularis*. J. Hazard. Mater. 143(1-2): 220–5.

Deravanesiyan, M., M. Beheshti and A. Malekpour. 2015. Alumina nanoparticles immobilization onto the NaX zeolite and the removal of Cr (III) and Co (II) ions from aqueous solutions. J. Ind. Eng. Chem. 21: 580–586.

Ding, Na., H. Mengxuan, H. Ya, W. Xinshuai, P. Yuxuan, L. Hua and Y. Guo. 2021. Advances in application of nanomaterials in remediation of heavy metal contaminated soil. 7th International conference on energy and environmental engineering (ICEMME 2021), Zhangiajie, China. *In*: Mostafa, M. M. H. and S. Manickam (eds.). E3S Web of Conferences, 261. id04027.

Djingova, R. and I. Kuleff. 2000. Instrumental techniques for traceanalysis. pp. 137–185. *In*: Vernet, J. P. (ed.). Trace Elements: Their Distribution and Effects in the Environment. Elsevier, London,UK.

El-Ramady, H., T. Alshaal, M. Abowaly, N. Abdalla, H. S. Taha, A. H. Al-Saeedi, T. Shalaby, M. Amer, M. Fári and É. Domokos-Szabolcsy. 2017. Nanoremediation for sustainable crop production. pp. 335–363. *In*. Ranjan, S., N. Dasgupta and E. Lichtfouse (eds.). Nanoscience in Food and agriculture, 1st edition. Springer. Singapore.

Fazlzadeh, M., K. Rahmani, A. Zarei, H. Abdoallahzadeh, F. Nasiri and R. Khosravi. 2017. A novel green synthesis of zero valent iron nanoparticles (NZVI) using three plant extracts and their efficient application for removal of Cr(VI) from aqueous solutions. Adv. Powder Technol. 28(1): 122–130.

Fernando, E., K. Tajalli and K. Godfrey. 2018. The use of bio-electrochemical systems in environmental remediation of xenobiotics: a review. J. Chem. Technol. Biotechnol. 94(7).

Gadhave, W. A. and J. Waghmare. 2014. Removal of metal ions from waste water by carbon nanotubes (CNTs). Int. J. Chem. Sci. Appl. 5(2): 66–77.

Geetha, K. S. and S. L. Belagali. 2021. Removal of heavy metals and dyes using lowcost adsorbents from aqueous medium. a review. IOSR J. Environ. Sci. Toxicol. Food Technol. 4: 56–68.

Gokila, S., T. Gomathi, P. N. Sudha and S. Anil. 2017. Removal of the heavy metal ion Chromium (VI) using Chitosan and Alginate nanocomposites. Int. J. Biol. Macromol. 104(Pt B): 1459–1468.

Gomes, M. A., R. A. Hauser-Davis, A. N. de-Souza and A. P. Vitória. 2016. General strategies, genetically modified plants and applications in metal nanoparticle contamination. Ecotoxicol. Environ. Saf. 134: 133–147.

Gong, X., D. Huang., Y. Liu, G. Zeng., R. Wang, J. Wan, C. Zhang, M. Cheng, X. Qin and W. Xue. 2017. Stabilized nanoscale zerovalent iron mediated cadmium accumulation and oxidative damage of *Boehmeria nivea* (L.) Gaudich Cultivated in Cadmium Contaminated Sediments. Environ. Sci. Technol. 51: 11308−11316.

Goutam, S. P. and Gaurav Saxena. 2021. Biogenic nanoparticles for removal of heavy metals and organic pollutants from water and wastewater: advances, challenges, and future prospects. pp. 623–636. Chapter 25. *In*: Saxena, G., V. Kumar and M. P. Shah (eds.). Bioremediation for Environmental Sustainability. Elsevier. ISBN. 9780128205242.

Gupta. V. K. and A. Nayak. 2012. Cadmium removal and recovery from aqueous solutions by novel adsorbents prepared from orange peel and Fe_2O_3 nanoparticles. Chem. Eng. J. 180: 81–90.

Harikumar, P. S., T. V. Lamya and T. Shalna. 2019. Enhanced phytoremediation efficiency of lead contaminated soil by zero Valent Nano Iron. Int. J. Innov. Eng. Technol. 12(2): 44–5.

Huang, D., X. Qin, Z. Peng, Y. Liu, X. Gong, G. Zeng, C. Huang, M. Cheng, W. Xue, X. Wang and Z. Hu. 2018. Nanoscale zero-valent iron assisted phytoremediation of Pb in sediment: Impacts on metal accumulation and antioxidative system of *Lolium perenne*. Ecotoxicol. Environ. Saf. 153: 229–237.

Hummers, W. S. and R. E. Offeman. 1958. Preparation of graphitic oxides. J. Am. Chem. Soc. 80(6): 1339.

Iijima, S. and T. Ichihashi. 1993. Single-shell carbon nanotubes of 1-nm diameter. Nature. 363: 603–605.

Iwahori, K., J. I. Watanabe, Y. Tani, H. Seyama and N. Miyata. 2014. Removal of heavy metal cations by biogenic magnetite nanoparticles produced in Fe(III)-reducing microbial enrichment cultures. J. Biosci. Bioeng. 117(3): 333–335.

Jadia, C. D. and M. H. Fulekar. 2008. Phytoremediation: The application of vermicompost to remove zinc, cadmium, copper, nickel and lead by sunflower plant. Environ. Eng. Manag. J. 7: 547–558.

Jadia, C. D. and M. H. Fulekar. 2009. Phytoremediation of heavy metals: Recent techniques. Afr. J. Biotechnol. 8(6): 921–928.

Jayaweera, H. D. A. C., S. K. M. N. de-Siriwardane, K. M. N. de-Silva and R. M. de-Silva. 2018. Synthesis of multifunctional activated carbon nanocomposite comprising biocompatible flake nano hydroxyapatite and natural turmeric extract for the removal of bacteria and lead ions from aqueous solution. Chem. Cent. J. 12(1): 18.

Jeon, In-Yup., D. W. Chang, A. K. Nanjundan and J. B. Baek. 2011. Functionalization of carbon nanotubes. pp. 91–110. *In:* Siva Yellampalli (ed.). Carbon nanotubes - polymernanocomposites. IntechOpen, Rijeka, Croatia.

Ju-Nam, Y. and J. R. Lead. 2008. Manufactured nanoparticles: An overview of their chemistry, interactions and potential environmental implications. Sci. Total Environ. 400(1-3): 396–414.

Kang, C. H., S. J. Oh, Y. J. Shin, S. H. Han, I. H. Nam and J. S. So. 2015. Bioremediation of lead by ureolytic bacteria isolated from soil at abandoned metal mines in South Korea. Ecol. Eng. 74: 402–407.

Kapahi, M. and S. Sachdeva. 2019. Bioremediation options for heavy metal pollution. J. Health Pollut. 9(24): 191203.

Karthik, C., N. Swathi, P. S. Pandi and D. G. Caroline. 2020. Green synthesized rGO-AgNP hybrid nanocomposite—An effective antibacterial adsorbent for photocatalytic removal of DB-14 dye from aqueous solution. J. Environ. Chem. Eng. 8: 103577.

Kar, S. and S. K. Equeenuddin. 2019. Adsorption of hexavalent chromium using natural goethite: isotherm, thermodynamic and kinetic study. J. Geol. Soc. India. 93(3): 285–292.

Karim, Z., A. P. Mathew, V. Kokol, J. Weid and M. Grahne. 2016. High-flux affinity membranes based on cellulose nanocomposites for removal of heavy metal ions from industrial effluents. RSC Adv. 6: 20644.

Katubi, K. M. M., N. S. Alsaiari, F. M. Alzahrani, M. S. Siddeeg and A. M. Tahoon. 2021. Synthesis of manganese ferrite/graphene oxide magnetic nanocomposite for pollutants removal from water. Process. 9: 589.

Khan, S., N. Akhtar., S. U. Rehman, S. Shujah, E. S. Rha and M. Jamil. 2021. Biosynthesized iron oxide nanoparticles (Fe3O4 NPs) Mitigate arsenic toxicity in rice seedlings. Toxics. 9(1): 2.

Kim, Y. and Y. Roh. 2019. Environmental application of biogenic magnetite nanoparticles to remediate Chromium(III/VI)-Contaminated Water. Miner. 9: 260.

Kohler, R., S. R. Raghavan and E. W. Kaler. 2000. Microstructure and dynamics of wormlike micellar solutions formed by mixing cationic and anionic surfactants. J. Phys. Chem. B.104(47): 11035–11044.

Kumar, S. and P. Gopinath. 2016. Nano-bioremediation applications of nanotechnology for bioremediation. Chapter 2. pp. 27–48. *In*: Remediation of Heavy Metals in the Environment.

Kumari, B. and D. P. Singh. 2016. A review on multifaceted application of nanoparticles in the field of bioremediation of petroleum hydrocarbons. Ecol. Eng. 97: 98–105.

Li, J. T., B. Liang Liao, C. Y. Lan, Z. H. Ye, A. J. M. Baker and W. S. Shu. 2010. Cadmium tolerance and accumulation in cultivars of a high-biomass tropical tree (*Averrhoa carambola*) and its potential for phytoextraction. J. Environ. Qual. 39: 1262–1268.

Liang, S. X., Y. Jin, W. Liu, X. Li, S. G. Shen and L. Ding. 2017. Feasibility of Pb phytoextraction using nano-materials assisted ryegrass: Results of a one-year field-scale experiment. J. Environ. Manag. 190: 170–175.

Lilhare, S., S. B. Mathew, A. K. Singh and S. A. C. Carabineiro. 2021. Calcium alginate beads with entrapped iron oxide magnetic nanoparticles functionalized with methionine-a versatile adsorbent for arsenic removal. Nanomate. 11: 1345.

Liu, Enzhou., Y. Du, X. Bai, J. Fan and X. Hu. 2020. Synergistic improvement of Cr(VI) reduction and RhB degradation using RP/g-C3N4 photocatalyst under visible light irradiation. Arab. J. Chem. 13: 3836–3848.

López-Téllez, G., C. E. Barrera-Díaz, P. Balderas-Hernández, G. Roa-Morales and B. Bilyeu. 2011. Removal of hexavalent chromium in aquatic solutions by iron nanoparticles embedded in orange peel pith. Chem. Eng. J. 173: 480–485.

Madhavi, V., T. N. V. K. V. Prasad, A. V. B. Reddy, B. Ravindra Reddy and G. Madhavi. 2013. Application of phytogenic zerovalent iron nanoparticles in the adsorption of hexavalent chromium. Spectrochim. Acta - Part A Mol. Biomol. Spectrosc. 116: 17–25.

Mahanty, S., S. Chatterjee, S. Ghosh, P. Tudu, T. Gaine, M. Bakshi, S. Das, P. Das, S. Bhattacharyya, S. Bandyopadhyay and P. Chaudhuri. 2020. Synergistic approach towards the sustainable management of heavy metals in wastewater using mycosynthesized iron oxide nanoparticles: Biofabrication, adsorptive dynamics and chemometric modeling study. J. Water Process Eng. 37: 101426.

Mallakpour, S. and S. Soltanian. 2016. Surface functionalization of carbon nanotubes: fabrication and applications. RSC Adv. 6: 109916–109935.

Mandeep and P. Shukla. 2020. Microbial nanotechnology for bioremediation of industrial wastewater. Front Microbiol. 11: 590631. DOI: 10.3389/fmicb.2020.590631.

Mejáre, M. and L. Bülow. 2001. Metal-binding proteins and peptides in bioremediation and phytoremediation of heavy metals. Trends Biotechnol. 19: 67–73.

Mesjasz-Przybylowicz, J., M. Nakonieczny, P. Migula, M. Augustyniak, M. Tarnawska, W. U. Reimold, C. Koeberl, W. Przybylowicz and E. Glowacka. 2004. Uptake of cadmium, lead, nickel and zinc from soil and water solutions by the nickel hyperaccumulator *Berkheya coddii*. Acta Biol. Cracov. Bot. 46: 75–85.

Mohammed, Y. M. M. and Y. I. Khedr. 2021. Applications of *Fusarium solani* YMM20 in bioremediation of heavy metals via enhancing extracellular green synthesis of nanoparticles. Water Environ. Res. 93(9): 1600–1607.

Mohsenzadeh, F. and A. C. Rad. 2012. Bioremediation of heavy metal pollution by nano-particles of *Noaea Mucronata*. Int. J. Biosci. Biochem. Bioinf. 2(2): 85–89.

Mukhopadhyay, S. and S. K. Maiti. 2010. Phytoremediation of metal enriched mine waste: a review. Global J. Environ. Res. 4: 135–150.

Ntim, A. and S. Mitra. 2011. Removal of trace Arsenic to meet drinking water standards using tron oxide coated multiwall carbon nanotubes. J. Chem. Eng. Data. 56: 2077–2083.

Ojuederie, O. B. and O. O. Babalola. 2017. Microbial and plant-assisted bioremediation of heavy metal polluted environments: a review. Int. J. Environ. Res. Public Health. 14(12): 1504.

Onal, E. S., Y. Tolga, T. Aslanov, M. Ergut and A. Ozer. 2019. Research Article Biosynthesis and characterization of iron nanoparticles for effective adsorption of Cr (VI). Int. J. Chem. Eng. Article ID 2716423, 13 pages.

Pardo, A., B. Pelaz, P. D. Pino, A. Al-Modlej, A. Cambón, B. Velasco, R. Domínguez-González, A. M. Piñeiro, P. B. Barrera, S. Barbosa and P. Taboada. 2021. Monodisperse superparamagnetic nanoparticles separation adsorbents for high-yield removal of arsenic and/or mercury metals in aqueous media. Nanomater. 11: 1345.

Park, S. J. and Y. M. Kim. 2005. Adsorption behaviours of heavy metal ions onto electrochemically oxidized activated carbon fibers. Mater. Sci. Eng. Process. A. 391: 121–123.

Parlayici, S, V. Eskizeybek, A. Ahmet and E. Pehlivan. 2015. Removal of Chromium (VI) using activated carbon-supported functionalized carbon nanotubes. J. Nanostruct. Chem. 5: 255–263.

Peng, Q., L. Yunguo, G. Zeng, W. Xu, C. Yang and J. Zhang. 2010. Biosorption of Copper (II) by immobilizing *Saccharomyces cerevisiae* on the surface of chitosan-coated magnetic nanoparticles from aqueous solution. J. Hazard. Mater. 177(1-3): 676–82.

Poguberović, S. S., D. M. Krčmar, S. P. Maletić, Z. Kónya, D. D. T. Pilipović, D. V. Kerkez and S. D. Rončević. 2016. Removal of As (III) and Cr(VI) from aqueous solutions using "green" zero-valent iron nanoparticles produced by oak, mulberry and cherry leaf extracts. Ecol. Eng. 90: 42–49.

Poo, K. M., E. B. Son, J. S. Chang, X. Rez, Y. J. Choi and K. J. Chae. 2018. Biochars derived from wasted marine macro-algae (*Saccharina japonica* and *Sargassum fusiforme*) and their potential for heavy metal removal in aqueous solution. J. Environ. Manag. 206: 364–372.

Raghib, F., M. I. Naikoo, F. A. Khan, M. N. Alyemeni and P. Ahmad. 2020. Interaction of ZnO nanoparticle and AM fungi mitigates Pb toxicity in wheat by upregulating antioxidants and restricted uptake of Pb. J. Biotechnol. 323: 254–263.

Ravikumar, K. V. G., S. Argulwar, S. Sudakaran, M. Pulimi, N. Chandrasekaran and A. Mukherjee. 2017. Nano-Bio sequential removal of Cr(VI) using polymer-nano zero valent iron composite film and sulfate reducing bacteria under anaerobic condition. Environ. Technol. Innov. 9(24).

Reeves, R. D. 2003. Tropical hyperaccumulators of metals and their potential for phytoextraction. Plant Soil. 249: 57–65.

Ren, G., H. Han, Y. Wang, S. Liu, J. Zhao, X. Meng and Z. Li. 2021. Recent advances of photocatalytic application in water treatment: a review. Nanomater. 11: 1804.

Rizwan, Md., M. Singh, C. K. Mitra and R. K. Morve. 2014. Review article-ecofriendly application of nanomaterials. nanobioremediation. J. Nanopart. Article ID 431787, 7 pages.

Ruan. X., Y. Sun., W. Du, Y. Tang, Q. Liu, Z. Zhang, W. Doherty, R. L. Frost, G. Qiana and D. C. W. Tsang. 2019. Formation, characteristics, and applications of environmentally persistent free radicals in biochars: A review. Bioresour. Technol. 281: 457–468.

Salam, M. S., M. S. I Makki and M. Abdelaal. 2011. Preparation and characterization of multi-walled carbon nanotubes/ chitosan nanocomposite and its application for the removal of heavy metals from aqueous solution. J. Alloys Compd. 509: 2582–2587.

Sakakibara, M., Y. Ohmori, N. T. H. Ha, S. Sano and K. Sera. 2011. Phytoremediation of heavy metal contaminated water and sediment by *Eleocharis acicularis*. Clean: Soil Air Water. 39: 735–741.

Sarakar, S., M. K. Enamala, M. Chavali, G. V. S. S. Sarma, K. M. Mannam, A. Kadier, A. Veeramuthu, K. Chandrasekhar, K. Ponvel and R. K. Kandikonda. 2020. Nanophytoremediation: An Overview of Novel and Sustainable Biological Advancement. DOI: 10.5772/Intechopen.93300.

Selvi, A., A. Rajasekar, J. Theerthagiri, A. Ananthaselvam, K. Sathishkumar, J Madhavan and P. K. S. M. Rahman. 2019. Integrated remediation processes toward heavy metal removal/recovery from various environments-A Review. Front. Environ. Sci. 7: 66.

Sen, M. (ed.). 2020. Nanotechnology and the Environment. IntechOpen, London. DOI: 10.5772/intechopen.87903.

Shabani, N. and M. H. Sayadi. 2012. Evaluation of heavy metals accumulation by two emergent macrophytes from the polluted soil: an experimental study. Environmentalist. 32: 91–98.

Shah, V. and A. Daverey. 2020. Phytoremediation: A multidisciplinary approach to clean up heavy metal contaminated soil. Environ. Technol. Innov. 18.

Shafiof, M. al. S. and A. Nezamzadeh-Ejhieh. 2020. A comprehensive study on the removal of Cd(II) from aqueous solution on a novel pentetic acid-clinoptilolite nanoparticles adsorbent: Experimental design, kinetic and thermodynamic aspects. Solid State Sci. 99: 106071.

Shaifali and G. Chethan Kumar. 2021. Microbial remediation of heavy metals. Int. J. Chem. Studies. 9(2): 92–107. DOI: 10.22271/chemi.2021.v9.i2b.11706.

Shehata, S. M., R. K. Badawy and Y. I. E. Aboulsoud. 2019. Phytoremediation of some heavy metals in contaminated soil. Bull. Nat. Res. Cent. 43: 189.

Shikha, D. and P. K. Singh. 2021. *In situ* phytoremediation of heavy metal–contaminated soil and groundwater: a green inventive approach. Environ. Sci. Polllut. Res. 28: 4104–4124.

Singh, J. and B.-K. Lee. 2016. Influence of nano-TiO2 particles on the bioaccumulation of Cd in soybean plants (*Glycinemax*): A possible mechanism for the removal of Cd from the contaminated soil. J. Environ. Manag. 170: 88–96.

Singh, S., K. C. Barick and D. Bahadur. 2011. Surface engineered magnetic nanoparticles for removal of toxic metal ions and bacterial pathogens. J. Hazard. Mater. 192: 1539–1547.

Song, B., P. Xu, M. Chen, W. Tang, G. Zeng, J. Gong, P. Zhang and S. Ye. 2019. Using nanomaterials to facilitate the phytoremediation of contaminated soil. Crit. Rev. Environ. Sci. Technol. 49(9): 791–824.

Soni, R., A. K. Pal, P. Tripathi, J. A. Lal, K. Kesari and V. Tripathi. 2020. Review Article An overview of nanoscale materials on the removal of wastewater contaminants. Appl. Water Sci. 10: 189.

Souri, Z., N. Karimi, M. Sarmadi and E. Rostami. 2017. Salicylic acid nanoparticles (SANPs) improve growth and phytoremediation efficiency of *Isatis cappadocica Desv.*, under Arsenic stress. IET Nanobiotechnol. 11(6): 650–655.

Spoiala, A. Ilie, C.-I. Illie, D. Ficai, A. Ficai and E. Andronescu. 2021. Chitosan-based nanocomposite polymeric membranes for water purification—a review. Mater. 14: 2091.

Srivastav, A., K. K. Yadav, S. Yadav, N. Gupta, J. K. Singh, R. Katiyar and V. Kumar. 2018. Nano-phytoremediation of pollutants from contaminated soil environment: current scenario and future prospects: management of environmental contaminants. 6. Chapter 16: 383–401. In book: Phytoremediation. Publisher: Springer Nature.

Subramaniam, M. N., P. S. Goh, W. J. Lau and A. F. Ismail. 2019. The roles of nanomaterials in conventional and emerging technologies for heavy metal removal: a state-of-the-art review. Nanomater. 9: 625.

Sun, D. T., L. Peng, W. S. Reeder and S. M. Moosavi. 2018. Rapid, selective metal removal from water by a metal-organic framework/polydopamine composite. ACS Cent. Sci. 4(3).

Tahoon, M. A., S. M. Siddeeg, N. S. Alsaiari., W. Mnif, F. Ben and Rebah. 2020. Effective heavy metals removal from water using nanomaterials: A review. Process. 8(6): 645.

Tang, L., Y. Wang and J. Li. 2015. The graphene/nucleic acid nanobiointerface. Chem. Soc. Rev. 44 6954–6980.

Tangahu, B. V., S. R. S. Abdullah, H. Basri, M. Idris, N. Anuar and Mohd. Mukhlisin. 2011. A Review on Heavy Metals (As, Pb, and Hg) Uptake by Plants through Phytoremediation. Int. J. Chem. Eng. Article ID 939161. 31 pages.

Tayang, A. and L. S. Songachan. 2021. Microbial bioremediation of heavy metals. Curr. Sci. 120(6): 1013–1025.

Tayel, A. A., M. M. Gharieb, H. R. Zaki and N. M. Elguindy. 2016. Bio-clarification of water from heavy metals and microbial effluence using fungal chitosan. Int. J. Biol. Macromol. 83: 277–81.

Thomas, P., P. R. Nelson, P. J. George, Lai, Chin. Wei, P. Tyagi, Mohd. R. B. Johan and M. P. Saravanakumar. 2020. Remediation of heavy metal ions using nanomaterials sourced from wastewaters. pp. 255–296. Chapter 12. *In*: Nanotechnology for Food, agriculture and Environment. © Springer Nature Switzerland AG 2020.

Tîlmaciu, C. M. and M. C. Morris. 2015. Carbon nanotube biosensors. Front. Chem. 3: 59.

Tong, Y. P., R. Kneer and Y. G. Zhu. 2004. Vacuolar compartmentalization: a second-generation approach to engineering plants for phytoremediation. Trends Plant Sci. 9: 7–9.

Tounsadi, H., A. Khalidi, M. Abdennouri and N. Barka. 2016. Activated carbon from *Diplotaxis harra* biomass: optimization of preparation conditions and heavy metal removal. J. Taiwan Inst. Chem. Eng. 59: 348–358.

Tripathi, D. K., V. P. Singh, S. M. Prasad, D. K. Chauhan and N. K. Dubey. 2015. Silicon nanoparticles (SiNp) alleviate chromium (VI) phytotoxicity in *Pisum sativum* (L.) seedlings. Plant Physiol. Biochem. 96: 189–98.

Ummalyma, S. B. and A. Singh. 2021. Importance of algae and bacteria in the bioremediation of heavy metals from wastewater treatment plants. pp. 343–357 Chapter-12. *In:* Maulin, P. Shah, Susana Rodriguez Couto, Vineet Kumar. (eds.). New Trends in Removal of Heavy Metals from Industrial Wastewater. Copyright © 2021 Elsevier Inc. All rights reserved.

Usman, K., H. A. Jabri, M. H. Abu-Dieyeh and M. H. S. A. Alsaf. 2020. Comparative assessment of toxic metals bioaccumulation and the mechanisms of Chromium (Cr) tolerance and uptake in *Calotropis procera*. Front. Plant Sci. 11: 883.

Vázquez-Núñez, E., C. E. Molina-Guerrero, J. M. Peña-Castro, F. Fernández-Luqueño and Ma. G. de la Rosa-Álvarez. 2020. Use of nanotechnology for the bioremediation of contaminants: A Review. Process. 8: 826 17 pages.

Venkatachalam, P., M. Jayaraj, R. Manikandan, N. Geetha, E. R. Rene, N. C. Sharma and S. V. Sahi. 2015. Zinc oxide nanoparticles (ZnONPs) alleviate heavy metal-induced toxicity in *Leucaena leucocephala* seedlings: A physiochemical analysis. Plant Physiol. Biochem. 110: 59–69.

Volarić, A., Z. Svirčev, D. I. Tamindžija and D. Radnovic. 2021. Microbial bioremediation of heavy metals. Hemijska Ind. 75: 103–115.

Wang, B., H. L. Xie, H. Y. Ren, X. Li, L. Chen and B. C. Wu. 2019. Application of AHP, TOPSIS, and TFNs to plant selection for phytoremediation of petroleum-contaminated soils in shale gas and oil fields. J. Clean. Prod. 233: 13–22.

Wang, S. and X. Shi. 2001. Molecular mechanisms of metal toxicity and carcinogenesis. Mol. Cell Biochem. 222: 3–9.

Wangi, G. M., P. W. Olupot, J. K. Byaruhanga, R. N. Kulabako, E. d. martínez, a. prado and m. gonzalez. 2020. A review for potential applications of Zeolite-based nanocomposites in removal of heavy metals and *Escherichia coli* from drinking water. Nanotechnol. Russia. 15: 686–700.

White, G. F., M. J. Edwards, L. Gomez-Perez, D. J. Richardson, J. N. Butt and T. A. Clarke. 2016. Mechanisms of bacterial extracellular electron exchange. pp. 87–138. *In:* Robert K. Poole (ed.). Advances in Microbial Physiology. Academic Press. Volume 68.

Wu, G., H. Kang, X. Zhang, H. Shao, L. Chu and C. Ruan. 2010. A critical review on the bio-removal of hazardous heavy metals from contaminated soils: issues, progress, eco-environmental concerns and opportunities. J. Hazard. Mater. 174: 1–8.

Wu, S., M. Vosátka, K. Vogel-Mikus, A. Kavčič, M. Kelemen, L. Šepec, P. Pelicon, R. Skála, A. R. V. Powter, M. Teodoro, Z. Michálková and M. Komárek. 2018. Nano Zero-Valent Iron Mediated Metal(loid) Uptake and Translocation by *Arbuscular Mycorrhizal* Symbioses. Environ. Sci. Technol. 52(14): 7640–7651.

Xiao, Z., H. Zhang, M. Yuan, X. Jing, J. Huang, Q. Li and D. Sun. 2016. Ultra-efficient removal of chromium from aqueous medium by biogenic iron based nanoparticles. Sep. Purif. Technol. 174: 466–473.

Yadav, K. K., J. K. Singh, N. Gupta and V. Kumar. 2017. A review of nanobioremediation technologies for environmental cleanup: a novel biological approach. J. Mater. Environ. Sci. 8(2): 740–757.

Yan, A., Y. Wang, S. N. Tan, M. L. Mohd. Yusof, S. Ghosh and Z. Chen. 2020. Phytoremediation: a promising approach for revegetation of heavy metal-polluted land. Front. Plant Sci. 11: 359.

Yang, J., B. Hou, J. Wang, B. Tian, J. Bi, N. Wang, X. Li. and X. Huang. 2019a. Review nanomaterials for the removal of heavy metals from wastewater. Nanomater. 9: 424.

Yang, J., C. Dahlström, H. Edlund, B. Lindman and M. Norgren. 2019b. pH-responsive cellulose–chitosan nanocomposite films with slow release of chitosan. Cellulose. 26: 3763–3776.

Yang, W., X. Deng, W. Huang, X. Qing and Z. Shao. 2019c. The physicochemical properties of graphene nanocomposites influence the anticancer effect. J. Oncol. ArticleID 7254534.

Yi, Yunqiang, G. Tu, P. E. Tsang, S. Xiao and Z. Fang. 2019. Green synthesis of iron-based nanoparticles from extracts of *Nephrolepis auriculata* and applications for Cr (VI) removal. Mater. Lett. 234: 388–391.

Yirsaw, B. D., S. Mayilswami and M. Megharaj. 2016. Effect of zero valent iron nanoparticles to *Eisenia fetida* in three soil types. Environ. Sci. Pollut. Res. 23: 9822–9831.

Yu, G., X. Wang, J. Liu, P. Jiang, S. You, N. Ding, Q. Guo and F. Lin. 2021a. Applications of nanomaterials for heavy metal removal from water and soil: a review. Sustain. 13.713.

Yu, Z., L. Gao, K. Chen, W. Zhang, Q. Zhang, Q. Li and K. Hu. 2021b. Nanoparticles: a new approach to upgrade cancer diagnosis and treatment. Nanoscale Res. Lett. 16: 88.

Zand, A. and A. Heir. 2020. Phytoremediation: data on effects of titanium dioxide nanoparticles on phytoremediation of antimony polluted soil. Data Brief. 31: 105959.

Zarime, A., Y. W. Zuhairi and J. Habibah. 2018. Removal of heavy metals using bentonite supported nano-zero valent iron particles. AIP Conference Proceedings. 1940. 020029. 10.1063/1.5027944.

Zendehdel, M., B. Y., Shoshtari and G. Cruciani. 2016. Removal of heavy metals and bacteria from aqueous solution by novel hydroxyapatite/zeolite nanocomposite, preparation, and characterization. J. Iran Chem. Soc. 13: 1915–1930.

Zeng, H., L. Zhai, T. Qiao, Y. Yu, J. Zhang and D. Li. 2020. Efficient removal of As(V) from aqueous media by magnetic nanoparticles prepared with Iron-containing water treatment residuals. Sci. Rep. 10(1): 9335.

Zhang, F., Z. Zhu, Z. Dong, Z. Cui, H. Wang, W. Hu, P. Zhao, P. Wang, S. Wei, R. Li and J. Ma. 2011. Magnetically recoverable facile nanomaterials: synthesis, characterization and application in remediation of heavy metals. Microchem. J. 98: 328–333.

Zhang, W., Y. An and S. Li. 2020. Enhanced heavy metal removal from an aqueous environment using an eco-friendly and sustainable adsorbent. Sci. Rep. 10: 16453.

Zhu, J., C. Minjiao, H. Qingliang, L. Shao, S. Wei and Z. Guo. 2013. An overview of the engineered graphene nanostructures and nanocomposites. RSC Adv. 3: 22790–22824.

Zhu, Nan., H. Ji, P. Yu, J. Niu, M. U. Farooq, M. W. Akram, I. O. Udego, H. Li and X. Niu. 2018. Review surface modification of magnetic iron oxide nanoparticles. Nanomater. 8: 810.

Chapter 18

Mitigation of Arsenic Pollution by using Iron-based Nano-adsorbents

R. Suresh, Saravanan Rajendran* and *Lorena Cornejo Ponce*

18.1 Introduction

Elemental arsenic (As) is a gray colored metalloid that acquired the twentieth rank among the most abundant elements in the Earth's crust (Siddiqui and Chaudhry 2017). It commonly occurs in organic (e.g., monomethylarsonic acid, dimethylarsinous acid, etc.) and inorganic (e.g., H_3AsO_3, As_2S_3, etc.) forms in the soil, air, and aquatic ecosystems (Siddiqui and Chaudhry 2017, Vishwakarma et al. 2021). There are nearly two hundred kinds of arsenic minerals occurring in the biosphere (Siddiqui and Chaudhry 2017). In those compounds, arsenic acquired variable oxidation states (–3, 0, +3, and +5); among these, As(III) and As(V) as oxyanions are naturally found in water bodies (Asere et al. 2019). Through the various natural processes, arsenic minerals release arsenic oxyanions in aquatic ecosystems (Cornejo et al. 2019, Bowell et al. 2014). Nonetheless, the occurrence of a higher concentration of arsenic species in water ecosystems, including groundwater, is governed by anthropogenic sources (Asere et al. 2019). Anthropogenic sources of arsenic species are the usage of feed additives in poultries, pesticides in agricultural land, mining activities, and discharge of untreated arsenic containing effluents from industries like electronics, copper, and hydrometallurgical industry (Bundschuh et al. 2021, Rathi and Kumar 2021). Arsenic polluted water causes several acute health effects, including diabetes, heart, skin, neurological diseases, and cancers in humans (Rahaman et al. 2021). Due to the strong affinity with membranes of muscles, lungs, and kidneys, As(III) species are accumulated in these body parts (Abdul et al. 2015).

As per the guidelines of WHO, the permissible limit of arsenic in water is 0.01 mg L^{-1} (Asere et al. 2019). But, water in many regions of the world contains arsenic exceeding this limit (Asere et al. 2019, Cornejo et al. 2020, Chen et al. 2017). In the case of arsenic poisoning in humans, the chelating agent is administered for their treatment (Kim et al. 2019). However, this treatment cannot be afforded for a huge population. Hence, it is better to provide arsenic-free pure water for human consumption to avoid arsenic related health problems. Besides, due to the physicochemical behaviors (structure, dissociation capability, affinity, and toxicity) of arsenic ions, appropriate treatment methods are required for arsenic contaminated water. Table 18.1 summarizes available techniques for the treatment of arsenic contaminated waters (Alka et al. 2021).

Laboratorio de Investigaciones Medioambientales de Zonas Áridas, Departamento de Ingeniería Mecánica, Facultad de Ingeniería, Universidad de Tarapacá, Avda. General Velásquez 1775, Arica, Chile.

* Corresponding author: sureshinorg@gmail.com

Table 18.1. Purification methods available for the treatment of arsenic polluted water.

Treatment methods	Principle	Merits	Demerits
Precipitation	Precipitating reagent reacts with arsenic species to form insoluble solid that can be separated from solution via filtration	Simple and effective	High costs
Membrane technology	Membrane selectively separates arsenic species from water	• High efficiency • Low energy consumption	• High costs • High rejection of water
Phytobial remediation	Interactions between plants and microbes and/or nanoparticles facilitate arsenic absorption behavior	• Inexpensive • Eco-friendly • Control phytopathogens	• Secondary pollution may be created • Plant's growth will be affected
Phyto-remediation	Plants are used to absorb arsenic species from water and soil medium	• Inexpensive • Eco-friendly	• Very slow process • Depends on climate • Secondary pollution may be created • Plant's growth will be affected
Ion-exchange process	Ions from solid exchanged with equal arsenic anions from polluted water	• Highly efficient • Less toxic sludge production	• Regular maintenance is essential • Selectivity of ion-exchanger is less
Electro-coagulation	Precipitation of arsenic ions by metal ions that are produced directly by electricity	• Efficient • Relatively cheaper	• Not suitable for removing trivalent arsenic ions • A huge amount of sludge is produced • Electrical supply is required
Adsorption	Solid particles (adsorbent) adsorb arsenic species from aqueous solutions via physical or chemical interactions.	• Highly efficient • Easy operation • Low cost • No sludge generation	• Adsorbent will be exhausted • Not automatic • It is suitable for wastewater containing traces of arsenic
Electrodialysis	Under electrical field, dialysis bag selectively allows arsenic ions to pass faster	Simple	• High cost • Less effective
Reverse osmosis	Under external pressure, arsenic ions move from less concentrated solution to high concentration solution	• Chemicals are not required • No toxic wastes • Easy to monitor	• High cost • Require pre-oxidation of As(III) ions

Figure 18.1 reveals that the number of publications related to arsenic adsorption increased from 2000 onwards (Scopus database). It means that attention to the adsorption method for arsenic remediation by researchers has progressively increased. This is due to the following important reasons. (a) The adsorption method does not require any costlier instruments; (b) it does not involve critical experimental procedures; (c) it is effective for contaminated water containing even trace amount of arsenic species (Awual et al. 2011), (d) it does not produce toxic sludge, (e) it is possible to prepare efficient and recoverable adsorbents with selectivity towards targeted pollutant (Pincus et al. 2021, Li et al. 2021), and (f) exhausted adsorbents can also be regenerated by various desorption processes.

The basic principle of the adsorption method for wastewater treatment is a physical or chemical attraction between adsorbent material (solid phase) and adsorbate (dissolved substance) in an

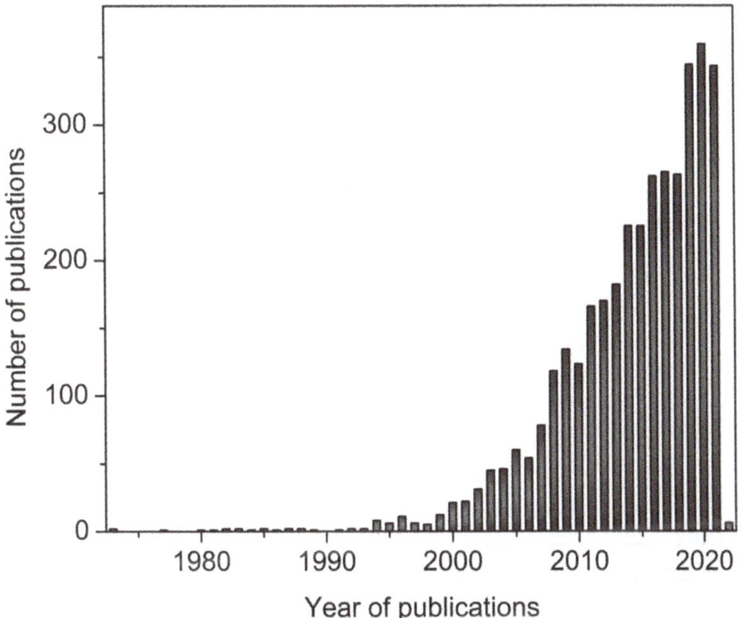

Fig. 18.1. Plot of years versus the number of publications (peer-reviewed journals) related to adsorptive removal of arsenic species from water (Source: Scopus database; keyword: Adsorption + Arsenic removal, searched on 30 August 2021).

aqueous solution. The solid phase material on which adsorption occurs is termed an adsorbent. The substance which accumulates on adsorbent is known as adsorbate (like arsenic species) (Fig. 18.2). Adsorptive removal of arsenic occurs by the following steps: (a) migration of arsenic species toward the adsorbent's surface, (b) physical and/or chemical adsorption of arsenic species takes place, and (c) desorption of arsenic species from the adsorbent's surface by suitable methods (Lata and Samadder 2016).

The nature of adsorbents and their properties strongly influence the decontamination of arsenic from water. Generally, adsorbent particles should possess a smaller size, unique morphology, greater specific surface area, and amorphous character (Seynnaeve et al. 2021). Also, adsorbents should be

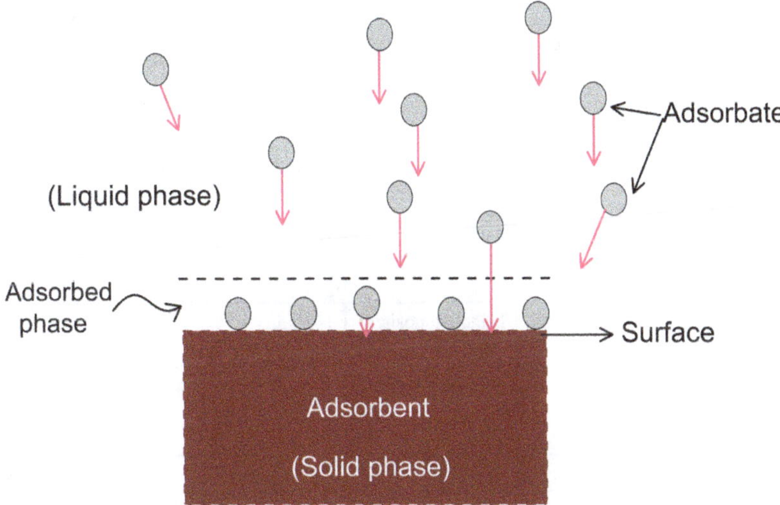

Fig. 18.2. Diagrammatic illustration of the adsorption process. The arrow (→) mark indicates the migration of pollutants towards the adsorbent's surface.

biocompatible, cheap, and easy to synthesize on a large scale. Table 18.2 displays different types of adsorbents and their arsenic adsorption capacity.

From Table 18.2, it can be inferred that the application of nano-sized adsorbent materials is promising in arsenic adsorption capacity than other adsorbents such as polymer and minerals, etc.

Table 18.2. Different types of adsorbent materials used for capturing arsenic species (As(III) and (As(V)) from aqueous solution.

Type of adsorbents	Example	Morphology	Adsorption capacity ($mg\ g^{-1}$)		References
			As(III)	As(V)	
Metal oxide	TiO_2 (Transition metal oxide)	Nanoparticles	39.86 μmol g^{-1}	40.85 μmol g^{-1}	Gupta et al. 2013
	MgO (Alkaline earth metal oxide)	Hollow spheres	892	-	Purwajanti et al. 2016
	CeO_2 (Inner transition metal oxide)	Nanoparticles	80%[a]	-	Pang et al. 2019
Doped metal oxide	Fe-modified ZrO_2	Powder	80	45	Sun et al. 2009
Spinel ferrite	$MgFe_2O_4$	Nano-platelets	-	13.33	Wu et al. 2018a
Metal	Fe^0	Nanoparticles	97%[b]	-	Adio et al. 2017
Binary metal oxide	Aluminum magnesium oxide	Composite particles	813	912	Li et al. 2016
Metal oxyhydroxide	MnOOH	Nanorods	432.1	-	Guo et al. 2015a
Mineral	Pyrite (FeS_2)	Microparticles	4.9	42.6	Fu et al. 2021
Metal organic frameworks	EB-COF:Br[c]	Spheres	-	53.1	Yang et al. 2020
Waste material	Alum sludge[d]	-	-	22.2	Jo et al. 2021
Mesoporous material	Natural zeolite	Microparticles		1.495	Zamudio et al. 2013
Fibers	Ligand exchange fibers	-	-	0.057 mmol g^{-1}	Awual et al. 2012
Polymer	Chitosan	-	1.83	1.94	Chen and Chung 2006
Layered double hydroxides	Mg/Al–NO_3 layered double hydroxides	-	-	1.56 mmol g^{-1}	Wang et al. 2009
Carbon material	Activated carbon	-	-	0.083 mmol g^{-1}	Amano et al. 2014
Biomass	*Melia azedarach* biomass	Aggregated particles	> 90%[e]	> 90%[f]	Sarwar et al. 2015
Metal compound/ Metal compound composite	Fe_3O_4/TiO_2	Core-shell microspheres	18.22	7.74	Deng et al. 2021
Metal compound/ Carbon material	FeO_x/graphene oxide	Nanocomposites	147	113	Su et al. 2017
Metal compound/ Biopolymer composite	Cellulose/Zn/Al layered double hydroxide	Nanocomposites	30.15	42.00	Bessaies et al. 2020
Ternary composite	$Au/ZnO/ZrO_2$	Nanocomposites	-	33.23	Hua 2021

[a,b,e,f]Removal efficiency; [c]EB-COF: Br – ethidium bromide-covalent organic frameworks: Br; [d]Alum sludge-mixture of amorphous $Al(OH)_3$ and organic matters

Especially, iron-based adsorbents are most attractive for arsenic adsorption from water due to their favorable properties, which are described later in this chapter. Importantly, the prior oxidation step is a must for adsorbing As(III) species. Chlorine and permanganate are generally used as oxidizing agents for As(III) → As(V) conversion (Wang et al. 2021). But, this oxidation step makes the process cost-ineffective and leads to the generation of secondary pollutants. For instance, chlorine may react with natural organic matter present in arsenic contaminated water to give toxic chlorinated organic compounds (Song et al. 2021). Notably, iron-based adsorbents avoid additional oxidants because iron compounds act as catalysts for As(III) oxidation (Singh et al. 2021). With this regard, this book chapter provides the application of nanostructured iron-based adsorbent materials for decontamination of arsenic species (As(III) and (As(V) ions) from the aqueous phase.

18.2 Iron-based Materials

Iron (Fe) is one of the most abundant elements with variable oxidation states (Fe(III) and Fe(II)). It combines with oxygen (O) to form iron oxides. Depending on the content of Fe ions, iron oxides are broadly classified as iron(III) oxides and iron(III, II) oxide. The most stable α-Fe_2O_3, β-Fe_2O_3, and γ-Fe_2O_3 are typical examples of iron(III) oxides, while Fe_3O_4 is an example of iron(III, II) oxide. Iron (oxy)hydroxides like α-FeOOH, β-FeOOH, γ-FeOOH, δ-FeOOH, ferrihydrite and $Fe(OH)_3$ have also been reported (Cornell and Schwertmann 2003). Siderite ($FeCO_3$), a carbonate form of iron ore, has also been studied in detail (Guo et al. 2007). Nanostructures of these iron-based adsorbents can be prepared by various synthesis methods. They have the following favorable edge over other adsorbent nanomaterials, making them superior in the adsorption of arsenic from water.

(i) Abundance in Earth's crust

Many iron compounds, including Fe_2O_3, Fe_3O_4, FeOOH, $FeCO_3$, and $Fe(OH)_3$, naturally occur in the earth's crust. Since the availability of these compounds is high, their production cost will be less.

(ii) Magnetic nature

Numerous iron compounds have magnetic properties. For example, superparamagnetic γ-Fe_2O_3 with saturation magnetization of 54.197 emu g^{-1} was reported by Lin et al. (2012). Similarly, superparamagnetic Fe_3O_4 particles (saturation magnetization = 52 emu g^{-1}) were also synthesized by Kumar et al. (2016). In another study, superparamagnetic $CuFe_2O_4$ with high saturation magnetization (62.52 emu g^{-1}) was also achieved (Tu et al. 2012). In general, superparamagnetic nanoparticles will be highly dispersed in water; while applying an external magnetic field, they accumulate rapidly near the magnet. Hence, these materials could be easier to separate from water.

(iii) High surface area

Iron-based adsorbents with a great active surface area can be achieved by chemical methods. For instance, γ-Fe_2O_3 nanoparticles with 168.73 $m^2\ g^{-1}$ of specific surface area were reported by Lin et al. (2012). As mentioned earlier, a promising adsorbent must possess a high surface area.

(iv) Surface functional groups

Surface functional groups play a vital role in the adsorption process. Different surface groups functionalized on iron-based adsorbents can be easily prepared. For example, diethylamine functionalized Fe_3O_4 particles were synthesized by the precipitation method (Kumar et al. 2016).

(v) Inherent affinity with arsenic

The functional groups (e.g., –OH in FeOOH or $Fe(OH)_3$) of iron-based adsorbents can readily react with arsenic oxyanions to form mono or bi-dentate complexes (Pham et al. 2020).

(vi) Catalytic activity

Generally, Fe(III)/Fe(II) redox couple can effectively mediate the catalytic oxidation process. In the arsenic removal process, prior oxidation of As(III) ions is also required for effective decontamination. Iron redox couple could be useful for As(III) oxidation (Singh et al. 2021); in this way, iron-based adsorbents have the potential for arsenic remediation.

(vii) Easy synthesis

Iron-based adsorbents with different shapes like nanosheets, nanoflowers, nanotubes, rods, etc., can be easily synthesized by facile and cheaper chemical methods.

18.3 Iron Oxide Adsorbents for Arsenic Removal

Iron oxides, hydroxides, oxyhydroxides, carbonates, and zero-valent iron and their composites with polymer, carbon materials have been fabricated for adsorption of arsenic anions from water. The details are elaborately discussed in this section.

18.3.1 Fe_2O_3 Based Adsorbents

18.3.1.1 Pure Fe_2O_3

Generally, hexagonal structured α-Fe_2O_3 and cubic structured γ-Fe_2O_3 nanoparticles are employed as adsorbents in wastewater treatment. For example, Lin et al. (2012) have prepared γ-Fe_2O_3 nanoparticles with high surface area (168.73 $m^2\ g^{-1}$) for adsorption of As(III) and As(V) ions from water. Under the optimized experimental conditions (temperature 50°C; contact time 30 min), adsorption capacities for As(III) and As(V) were reached as 74.83 and 105.25 mg g^{-1}, respectively. Chemical bonding with the surface hydroxyl group of γ-Fe_2O_3 is accountable for the adsorption of As(III) and As(V) ions. The arsenic adsorption capacities were not affected by pH in the range 3 to 11 and chloride, nitrate, and sulfate ions. But, phosphate ions severely affected the arsenic adsorption capacity of γ-Fe_2O_3 nanoparticles. In another study, α-Fe_2O_3 adsorbent was prepared using $FeCl_3$ and Aloe vera leaves viscous gel as the iron source and precipitating agent, respectively (Mukherjee et al. 2016). This α-Fe_2O_3 adsorbent was exclusively used to adsorb As(V) specie from synthetic arsenic contaminated water, and the observed adsorption capacity was 38.48 mg g^{-1}. Adsorption experiments were carried out at 20°C ([As(V)] concentration 2–30 mg L^{-1}), and the adsorbent was nearly saturated within 45 minutes. This α-Fe_2O_3 powder (1 g L^{-1}) has also been tested for As(III) adsorption (2–30 mg L^{-1}) at a pH of 6.5. The initial concentration of As and adsorbent dosage has the foremost influence on the adsorption capacity, while pH did not have any significant effect within the experimental range (pH 6 to 8). About 41.98 mg of As(III) ions were adsorbed on one gram of adsorbent. Further, As(V) adsorbed α-Fe_2O_3 adsorbent was added as an ingredient in the preparation of silicate glass, which can be used to make bottles and window glasses. Compared to commercial Fe_2O_3 based glasses, glass prepared using As(V) adsorbed α-Fe_2O_3 showed high stability.

Mesostructured α-Fe_2O_3 adsorbent, synthesized from MIL-100(Fe) (MIL: Materials of Institute Lavoisier), has been reported for abatement of arsenic species (Liu et al. 2018). α-Fe_2O_3 showed adsorption capacities of 109.89 and 181.82 mg g^{-1} for As(III) and As(V), respectively (contact time = 30 minutes). This study revealed that As(V) adsorption efficiency is high at very low pH, while solution pH didn't affect As(III) adsorption capacity. This is due to the nonionic form of As(III) (H_3AsO_3) up to pH 9.1. The ionic interaction has a key role in the As(V) adsorption process, while in the case of As(III), ligand exchange plays a major role (Banerjee et al. 2008). The orderly mesoporous structure of α-Fe_2O_3 powder is also one of the major reasons for the greater adsorption capacity of As(III) and As(V) species.

18.3.1.2 Doped Fe_2O_3

Cation/anion doped Fe_2O_3 nanostructures could also show improved arsenic adsorption behavior. For example, at pH 5.0 (temperature = of 20°C), Ce(IV) doped Fe_2O_3 with greater As(V) ions adsorption capacity (70.4 mg g^{-1}) was reported (Zhang et al. 2003). It was also determined that interfering ions like chloride, nitrate, sulfate, and fluoride did not affect As(V) adsorption by Ce(IV) doped Fe_2O_3 adsorbent. Also, surface hydroxyl groups are found to be important for As(V) adsorption process, as they form bidentate surface complex with As(V) ions. Further, aluminum (Al)-substituted ferrihydrite ($Fe_2O_3.0.5H_2O$) adsorbents, considered for the elimination of arsenic species from water, have also been examined. For example, Adra et al. (2016) have studied the influence of Al on As(III)/As(V) removal ability (pH 6.5) of ferrihydrite. When Al/Fe molar ratio in ferrihydrite increased, sorption density for As(III) decreased due to the poor binding ability of $As(OH)_3$ with Al incorporated ferrite. On the other hand, sorption density increased for As(V) when Al/Fe molar ratio increased. This is due to the formation of the surface complex between As(V) with Al-ferrihydrate. Later, Souza et al. (2021) have found that Al substituted ferrihydrite showed superior As(V) adsorption behavior due to the surface hydroxyl group. Importantly, 15 mol% Al substituted sample displayed higher adsorption capacity than other samples, indicating that optimum Al substitution level is necessary.

18.3.1.3 Surface modified Fe_2O_3

Surface groups play a key role in the capturing of arsenic ions from water. For example, stearic acid capped Fe_2O_3 nanoparticles exhibited nearly complete adsorption of As(III) ions (Goswami et al. 2011). It is because functionalized Fe_2O_3 nanoparticles have $-COO^-$ groups, which effectively interact with H_3AsO_3 species (pH = 6.5). In another study, starch functionalized γ-Fe_2O_3 adsorbent was reported for adsorption of As(III) ions (Siddiqui et al. 2020). Compared to pure γ-Fe_2O_3 (7.46 mg g^{-1}), starch functionalized γ-Fe_2O_3 showed a relatively higher adsorption capacity (8.88 mg g^{-1}) at 35°C.

18.3.1.4 Fe_2O_3 based composite

Different support materials are composited with Fe_2O_3 nanoparticles to enhance arsenic removal through the adsorption process. Sahu et al. (2017) have prepared Fe_2O_3/jute fiber (C: O: Fe is 41.65: 40.82: 17.53 wt.%) composites for As(V) removal from water through the batch adsorption process. At pH 3, Fe_2O_3/jute fiber composite exhibited 48.06 mg g^{-1} of As(V) adsorption capacity. Based on the Fourier transform infrared spectroscopic and zeta potential measurements, it was observed that As(V) adsorption occurs on Fe_2O_3/jute fiber through ionic interaction and surface complex formation (mono and bidentate). Moreover, Fe_2O_3/jute fiber has superparamagnetic nature, which makes it easy to recover through the external magnet. Alternatively, γ-Fe_2O_3/covalent triazine framework was prepared and tested for adsorptive removal of As(III) and As(V) from synthetic and real waters (Leus et al. 2018). The adsorption capacity of γ-Fe_2O_3/covalent triazine framework towards As(III) and As(V) were 198.0 and 102.3 mg g^{-1}, respectively.

18.3.2 Fe_3O_4 Based Adsorbents

Fe_3O_4 is a black colored iron oxide with favorable magnetic properties. Pure and modified Fe_3O_4 materials have been applied extensively as adsorbents in arsenic remediation (Hao et al. 2018).

18.3.2.1 Pure Fe_3O_4

Cuboid/pyramid shaped Fe_3O_4 particles with saturation magnetization of 6.9 emu g^{-1} were prepared using tea waste and used for adsorption of As(III) and As(V) ions from aqueous solution (Lunge et al. 2014). Fe_3O_4 nanoparticles showed 188.69 and 153.8 mg g^{-1} for As(III) and As(V), respectively. The recyclability of this adsorbent was also found to be good. On the other hand, Fe_3O_4 nanoparticles (size = 10 nm) with surface area (100 m^2 g^{-1}) were utilized for As(V) removal (Iconaru et al. 2016). The observed adsorption capacity of Fe_3O_4 nanoparticles was 66.53 mg g^{-1}, which is greater than that for commercial Fe_3O_4 (39.24 mg g^{-1}). These investigations indicate that the morphology and particle size of adsorbents influence their adsorption capacity.

18.3.2.2 Surface functionalized Fe_3O_4

Monodispersed diethylamine functionalized Fe_3O_4 and yeast cross-linked Fe_3O_4 powder was prepared by a self-assembly process. The prepared powder was tested as an adsorbent for As(V) from water (Kumar et al. 2016). It was found that yeast cross-linked Fe_3O_4 adsorbent (99%) has complete As(V) adsorption efficiency than diethylamine functionalized Fe_3O_4 (83.33%).

18.3.2.3 Fe_3O_4 based composites

Numerous Fe_3O_4 based composites synthesized for abatement of inorganic arsenic contaminated water are tabulated in Table 18.3.

From Table 18.3, the following conclusions can be extracted: (a) surface area is not important, while active surface sites are necessary for arsenic ions adsorption. (b) Nature and amount of counterparts (e.g. metal oxide, polymer, carbon material, layer double hydroxides, etc.) remarkably influence iron-based composites' arsenic adsorption capacity/efficiency. (c) Improved adsorption performance of iron-based composites is due to enhanced catalytic activity, binding capacity, and pH sensitive surface charges.

18.3.3 MFe_2O_3 Based Adsorbents

Nanostructured spinel ferrites (MFe_2O_4, where M represents divalent transition or alkaline earth metal ions) are emerging as potential adsorbents for arsenic decontamination from waste water. This is due to their unique structure, surface properties, magnetic and redox properties (Wu et al. 2018b). In this section, the arsenic removal ability of pure ferrites and ferrite based composite adsorbents is described.

18.3.3.1 Pure spinel ferrites

Parsons et al. (2009) have prepared pure Fe_3O_4, Mn_3O_4, and $MnFe_2O_4$ nanoparticles for As(III) and As(V) adsorption from water. These samples showed pH independent adsorption capacity in pH range 2 to 6, when the initial concentration of As(III)/As(V) was 100 ppb. Compared to Fe_3O_4, and Mn_3O_4, $MnFe_2O_4$ showed enhanced arsenic adsorption capacities (µg g^{-1}): $MnFe_2O_4$ (718) > Fe_3O_4 (32.2) > Mn_3O_4 (8.9) for As(III) and $MnFe_2O_4$ (2125) > Fe_3O_4 (1575) > Mn_3O_4 (212) for As(V). $CuFe_2O_4$ nanoparticles synthesized from printed circuit board sludge were used for the removal of As(V) from water (Tu et al. 2012). The As(V) adsorption capacity of recycled $CuFe_2O_4$ was 45.66 mg g^{-1} at pH 3.7. Further, As(V) did not reduce as highly toxic As(III) at the surface of $CuFe_2O_4$. Moreover, this adsorbent can be recovered within 20 s by using an external magnet. Later, Wu et al. (2018b) have deposited $CuFe_2O_4$ on Fe-Ni foam for decontamination of water containing As(III) and As(V). $CuFe_2O_4$ foam showed adsorption capacity of 44.0 and 85.4 mg g^{-1} for As(III) and As(V), respectively, with outstanding recyclability.

Sulfite activated $MnFe_2O_4$ for arsenic oxidation (As(III) to As(V)) followed by adsorption was also demonstrated by Ding et al. (2021). As(III) adsorption capacity of $MnFe_2O_4$/sulfite was 26.257 mg g^{-1}, which was greater than $MnFe_2O_4$ alone (9.491 mg g^{-1}). Also, As(III) adsorption is more effective than a pre-oxidation of As(III) ions followed by adsorption. Mn(III) was the

Table 18.3. Fe_3O_4 based composite adsorbents for adsorption of arsenic ions from the liquid phase.

Fe_3O_4 based adsorbents	Morphology	Surface area ($m^2 g^{-1}$)	Species	Contact time	pH	Adsorption capacity ($mg g^{-1}$)	Significance	References
Fe_3O_4/TiO_2	Sheets	~ 89.4	As(V) As(III)	45 min	3–9	36.36 30.96	• Photocatalysis of arsenite followed by adsorption • UV illumination increased adsorption capacity	Deng et al. 2019
Fe_3O_4/CeO_2	Nanoparticles	174.69	As(V) (from p-arsanilic acid)	40 min	4–8	122.19	Photocatalytic degradation of p-arsanilic acid (28 min) followed by adsorption of As	Sun et al. 2019
Fe_3O_4/CaO_2	Nanoparticles	-	As(III)	24 h	6	56.35	• Oxidation of As(III) (15 min) followed by adsorption • Phosphate interference was less	Song et al. 2019
Fe_3O_4/biochar Fe_3O_4/activated biochar	Microporous	28.9 482.4	As(V)	1 h	7	457 868	Activated biochar provides more surface area for adsorption of As(V) species	Alchouron et al. 2020
Fe_3O_4/zeolitic imidazolate framework-8	Nanoparticles	245.5	As (III)	180 min	6	21.12	Adsorption was due to hydrogen bonding and electrostatic interactions	Wu et al. 2021
Fe_3O_4/porous carbon	Yolk-shell	251.36	As (V)	120 min	7	100% (in water) and 98.3% (in mud)	Role of porous carbon on As(V) adsorption	Li et al. 2019
Fe_3O_4/polydopamine/ reduced graphene oxide	Microstructures	13.22$m^3 g^{-1}$	As(III) As(V)	24 h	7 4	250 $\mu g g^{-1}$ 240 $\mu g g^{-1}$	Polydopamine strengthens graphene-and enhanced binding ability of Fe_3O_4	Guo et al. 2015b
Fe_3O_4/Al_2O_3/Zn-Fe layer double hydroxides	Nanoparticles	156.67	As specie	12.8 min	4	67.57	Adsorbent based on ultrasound-assisted dispersive magnetic solid phase extraction	Adlnasab et al. 2019
Fe_3O_4/Polyaniline/silver diethyldithiocarbamate	Nanostructures	-	As specie	45 min	6	175.4	A ternary composite used to remove As and bacteria from water	Hashemi et al. 2019

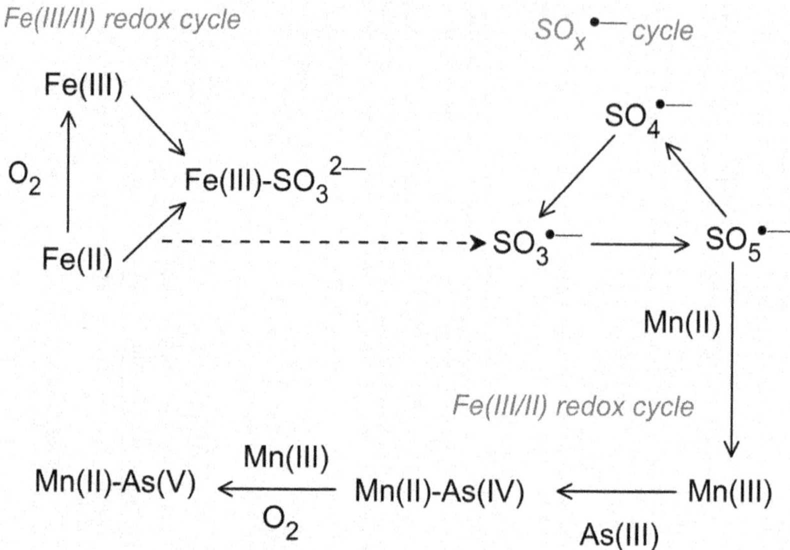

Fig. 18.3. Pathway of As(III) oxidation followed by adsorption on $MnFe_2O_4$/sulfite system (Redrawn with modifications from Ding et al. 2021).

predominant oxidant for As(III) oxidation, while sulfite acted as a precursor of oxysulfur radicals. Pathway of As(III) oxidation at $MnFe_2O_4$/sulfite system is displayed as Fig. 18.3.

Defects rich $MgFe_2O_4$ nano-platelets were prepared and found as potential adsorbents for adsorption of As(V) ions even from very low concentrated solutions (Wu et al. 2018a). $MgFe_2O_4$ nano-platelets contain an abundance of surface –OH groups which could form tridentate hexanuclear and monodentate mononuclear complexes with As(V) ions easily. Due to the above-said reasons, As(V) concentration decreased to 4.9 μg L^{-1} from 1 mg L^{-1} in polluted water after adsorption.

18.3.3.2 Ferrite based composites

Besides the application of pure spinel ferrites nanoparticles as an adsorbent for arsenic species, they have disadvantages like aggregation and dissolution in acidic conditions. Therefore, ferrite should be coupled with other materials such as carbon materials, natural minerals, and biomass. A few examples of spinel ferrite based composite adsorbents and their arsenic removal behavior have been listed in Table 18.4.

Table 18.4. Spinel ferrites based adsorbents for arsenic removal from water.

Ferrite based composites	**Species**	**Contact time (min)**	**pH**	**Adsorption capacity (mg g^{-1})**	**References**
$Ni_{0.5}Zn_{0.5}Fe_2O_4$/carbon nanotubes	As(V)	30	2	66.0	Ahangari et al. 2019
$MnFe_2O_4$/granular activated carbon/*Bacillus arsenicus*	As(III) As(V)	180	7	2584.668 2651.675	Podder and Majumder 2015
$CoFe_2O_4$ aggregated schwertmannite	As(III)	240	5.3	1011 μg g^{-1}	Dey et al. 2014
$MnFe_2O_4$/Neem leaves/*Bacillus arsenicus*	As(III) As(V)	240	7	53.175 56.239	Podder and Majumder 2016
$CuFe_2O_4$/graphene oxide	As(III) As(V)	180	7	51.64 124.69	Wu et al. 2018c
$ZnFe_2O_4/Zn_5(OH)_6(CO_3)_2$	As(V)	24 h	7	18.4 μg mg^{-1}	Tresintsi et al. 2018

The following important inferences can be derived from the above mentioned studies in Table 18.4. (a) Composite adsorbents have improved As(III)/As(V) ion adsorption efficiency than their pure counterparts due to strong affinity with arsenic ions. (b) Magnetic property could also be improved by composite formation. (c) Composite adsorbents show pH sensitive arsenic adsorption behaviors.

18.3.4 FeOOH Based Adsorbents

Iron oxyhydroxide (FeOOH) nanostructures were applied as adsorbents for arsenic removal from water. For example, iron oxyhydroxide nanoparticles were synthesized in the presence of sulfate, nitrate, and chloride ions at various pH (Zhang et al. 2021). The effect of the drying process and media on As(V) ions adsorption behavior of these samples were tested. Adsorption capacity decreased in the case of freeze-dried samples, while suspension showed high adsorption capacity. The low adsorption capacity of the dried sample is due to aggregation and high crystallinity. At pH 6, reaction media didn't have any influence, while at pH 4, As(V) adsorption was more for iron oxyhydroxide prepared in the presence of sulfate and chloride than nitrate. It can be attributed to changes in the microstructure of the sample in the presence of sulfate and chloride ions.

In another study, sulfide (S^{2-}) modified α-FeOOH was effectively used for removing As(V) (384.6 mg g^{-1}) ions from water (Li et al. 2020). Importantly, As(V) adsorption was not affected by interfering species like chloride, sulfate, nitrate, silicate, and phosphate ions. Based on the X-ray photoelectron spectroscopic results, it was identified that a reduction of As(V) to As(III) occurred via hydroxyl groups of S^{2-} modified α-FeOOH during the adsorption process.

Like iron oxides and ferrites, FeOOH nanoparticles were also composited with other adsorbent materials. For example, FeOOH/root powder of *Eichhornia crassipes* composite was reported for adsorptive removal of arsenic ions from an aqueous solution (Lin et al. 2018). FeOOH modified root powder showed adsorption capacities of 9.43 and 5.78 mg g^{-1} for As(III) and As(V), respectively, at pH 9. The complexation of arsenic with the hydroxyl group of FeOOH was the major reason for observed adsorption performance. On the other hand, Safi et al. (2019) have developed (53.7%) FeOOH/cationic polymer (monomers: N, N-dimethylamino propyl acrylamide and methyl chloride quaternary; crosslinker: N, N′-Methylene bisacrylamide) composite for adsorption of arsenic species. This composite has a good adsorption capacity (123.4 mg g^{-1}) and selectivity towards As(V) in the presence of SO_4^{2-} ions at pH 7. As(V) species adsorption follows chemisorption via ion exchange process (replacement of chloride ions by arsenate ions) with amino group cationic polymer adsorption with FeOOH particles (Fig. 18.4).

FeOOH based ternary composite adsorbents were also prepared. For instance, FeOOH/CuO/water bamboo cellulose was applied for As(III) adsorption from water (Liu et al. 2020). At pH 3.5 and As(III) concentration 150 mg L^{-1}, the adsorption capacity was determined as 76.1 mg g^{-1}. The mechanism of arsenic adsorption on FeOOH/CuO/water bamboo cellulose followed both physical and chemical reactions.

18.3.5 $Fe(OH)_3$ Based Adsorbents

Attention has been paid to ferric hydroxide ($Fe(OH)_3$) based adsorbent on purifying water containing arsenic ions. For example, Hlavay and Polyák (2005) synthesized $Fe(OH)_3/Al_2O_3$ adsorbent for selective abatement of As(III) and As(V) ions from water. Arsenic adsorption process is mainly dependent on solution pH. For this adsorbent, As(V) adsorption was more than As(III) ions. Later, Pham et al. (2020) have prepared $Fe(OH)_3$ based adsorbent (surface area = 22.69 m^2 g^{-1}) from iron ore. The synthesized granular $Fe(OH)_3$ adsorbent also contained silica, hematite, and maghemite phases. It was used to remove As(V) ions from an aqueous solution through batch experiments at pH 3. The maximum adsorption capacity for powdered $Fe(OH)_3$ and granular $Fe(OH)_3$ was

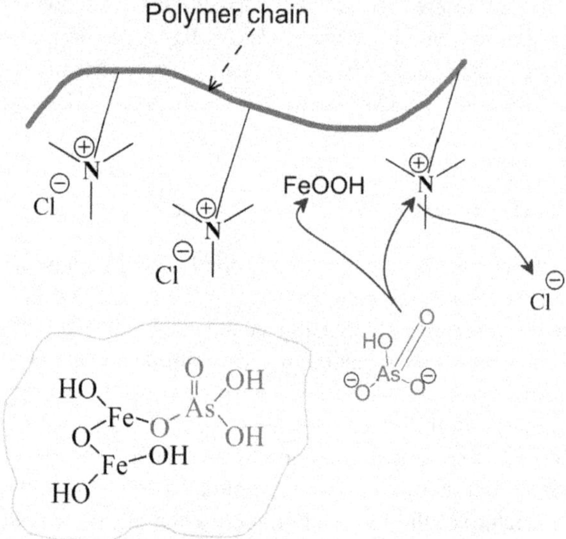

Fig. 18.4. Adsorption mode of As(V) species ($HAsO_4^{2-}$) on FeOOH/polymer composite (Redrawn with modifications from Safi et al. 2019).

2896.9 and 250 μg g^{-1}, respectively. As(V) adsorption on $Fe(OH)_3$ based adsorbent is attributed to the formation of monodentate complexation (Eq. 18.1 – 18.3) (Pham et al. 2020).

$$\equiv FeOH + H_3AsO_4 \rightarrow \equiv FeH_2AsO_4 + H_2O \quad \text{(Eq. 18.1)}$$

$$\equiv FeOH + H_2AsO_4^- \rightarrow \equiv FeHAsO_4^- + H_2O \quad \text{(Eq. 18.2)}$$

$$\equiv FeOH + HAsO_4^{2-} \rightarrow \equiv FeAsO_4^{2-} + H_2O \quad \text{(Eq. 18.3)}$$

18.3.6 Siderite ($FeCO_3$)

Siderite ($FeCO_3$) is one of the naturally occurring iron ores commonly found in the Earth's crust. Due to its abundance on Earth and cheapness, it is considered an ideal adsorbent for arsenic removal. For instance, Wang et al. (2018) have used a mixture of $FeCO_3$ (particle size 1–2 mm) and river sands as adsorbents for the removal of arsenic from groundwater (247.57 mg L^{-1} arsenic concentration). The adsorption capacity was 0.5233 mg g^{-1}. Arsenic forms complex with surface –OH of $FeCO_3$, which leads to the decrease in the concentration of arsenic in water (Shi et al. 2007). At the same time, under the experimental conditions, $FeCO_3$ undergoes hydrolysis to form iron oxide hydroxide and HCO_3^- ions (Eq. 18.4) (Wang et al. 2018). Hence, arsenic removal has also likely occurred through adsorption on iron oxide hydroxide particles. It was also observed that the co-existence of sulfide ions favored the adsorption of arsenic from water.

$$FeCO_3 + H_2O \leftrightarrow \equiv Fe\text{–}OH + HCO_3^- \quad \text{(Eq. 18.4)}$$

Su et al. (2021) have studied As(III) species removal mechanism in waste sulfuric acid by $FeCO_3$ adsorbent. In this process, arsenic is precipitated as scorodite ($FeAsO_4{\cdot}2H_2O$) by the following three steps (Su et al. 2021): (a) neutralization of waste sulfuric acid with $FeCO_3$ and production of Fe(II) ions (Eq. 18.5) which are subsequently oxidized by molecular oxygen to Fe(III) ions (Eq. 18.6),

(b) oxidation of As(III) species in waste sulfuric acid by 28% H_2O_2 and (c) precipitation of As(V) species with Fe(III) ions as $FeAsO_4{\cdot}2H_2O$ (Eq. 18.7).

$$FeCO_3 + H_2SO_4 \rightarrow FeSO_4 + H_2O + CO_2 \qquad \text{(Eq. 18.5)}$$

$$4Fe^{2+} + O_2 + 4H^+ \rightarrow 4Fe^{3+} + 2H_2O \qquad \text{(Eq. 18.6)}$$

$$H_3AsO_4 + Fe^{3+} + H_2O \rightarrow FeAsO_4{\bullet}2H_2O + 3H^+ \qquad \text{(Eq. 18.7)}$$

Nearly 99.99% of arsenic removal efficiency was obtained at 10 hours of contact time (Fe/As = 2, 95°C).

18.3.7 $Fe^{(0)}$ Based Adsorbents

Metallic iron (Fe^0) nanoparticles can be applied effectively in arsenic removal due to their huge surface area, high dispersibility in water, electropositive character, low cost, and high removal efficiency. For example, Morgada et al. (2009) have used commercial Fe^0 iron nanoparticles to remove As(V) species in the presence of oxygen. After 150 minutes of contact time, 90% of As(V) was removed from water by Fe nanoparticles (0.05–0.1 g L^{-1}). The high arsenic removal efficiency was attributed to high surface area and intrinsic activity. However, in the dark, the addition of humic acids decreased (50% of removal efficiency), while ultraviolet light doubled arsenic removal efficiency even in the presence of humic acid. Green synthesized Fe^0 nanoparticles (using Aloe vera plants extract) were also synthesized and used for removal of As(III) species from water (Adio et al. 2017). Around 95% removal efficiency was obtained with 100 mg green synthesized Fe^0 nanoparticles (pH 3). The optimum contact time, shaker speed, and initial concentration were fixed to be 120 minutes, 120 rpm, and 2 mg L^{-1}, respectively.

Kang et al. (2018) have investigated As(III) removal performance of Fe^0/oxidizing agent system. H_2O_2, persulfate, and peroxymonosulfate were used as oxidizing agents. As(III) removal was dependent on oxidant dosage and adsorbent dosage. In presence of Fe^0/persulfate (pH 3), maximum As(III) removal (k_1 = 0.0189 min^{-1}; q_e = 115.27 mg g^{-1}) was achieved. The order of As(III) removal efficiency for various oxidants was persulfate > H_2O_2 > peroxymonosulfate. As(III) removal happened via oxidation followed by adsorption and co-precipitation processes. Oxidation of As(III) occurs by *in situ* generated hydroxyl ($^{\bullet}OH$) and sulfate ($SO_4^{\bullet-}$) radicals produced from the Fe^0/persulfate (Eq. 18.8 and 18.9) (Kang et al. 2018).

$$SO_4^{\bullet-} + H_2O \rightarrow SO_4^{2-} + {}^{\bullet}OH + H^+ \qquad E_0 = 1.8\text{–}2.7\ V \qquad \text{(Eq. 18.8)}$$

$$Fe^0 + S_2O_8^{2-} + 2H_2O \rightarrow Fe^{2+} + 2{}^{\bullet}OH + 2SO_4^{2-} + 2H^+ \qquad E_0 = 2.5\text{–}3.1\ V \qquad \text{(Eq. 18.9)}$$

Recently, Singh et al. (2021) have synthesized sulfur modified Fe^0 nanoparticles, which showed enhanced arsenic (As(III) and As(V) ions) adsorption efficiency under optimized conditions (S/Fe ratio = 0.1; adsorbent dose = 0.5 g L^{-1}). The obtained adsorption capacity for As(III) and As(V) was 89.29 and 79.37 mg g^{-1}, respectively. After two weeks of aging, the activity of pure Fe^0 decreased, while sulfur-modified Fe^0 nanoparticles remained unchanged. It was observed that synthesized pure Fe^0 nanoparticles have a smooth outer shell of iron oxide while sulfur-modified Fe^0 nanoparticles contained distorted flaky shells of FeS and iron oxide. Based on this result, it can be concluded that the arsenic removal mechanism of sulfur modified Fe^0 is entirely different from pure Fe^0 nanoparticles. In the presence of pure Fe^0, arsenic undergoes adsorption and reduction. In the case of sulfur-modified Fe^0, arsenic is removed via the adsorption and precipitation process. The detailed mechanism was explained by Singh et al. (2021), shown in Fig. 18.5.

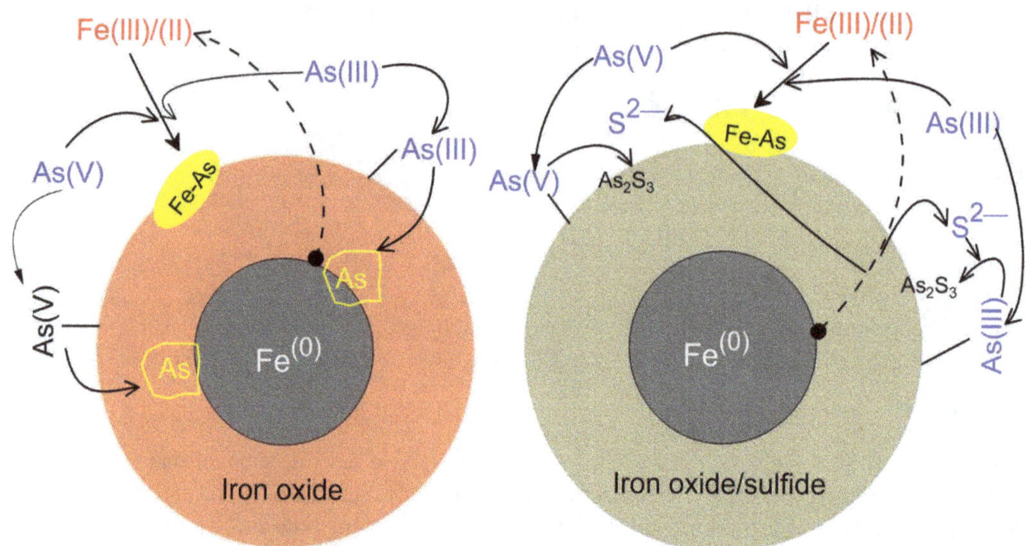

Fig. 18.5. Schematic illustration of adsorptive removal of As(III) and As(IV) by pure zero-valent iron and sulfur modified iron nano-adsorbents (Redrawn with modifications from Singh et al. 2021).

18.4 Separation and Regeneration of Iron-based Adsorbents

Depending on the nature of adsorbents, separation procedures like sedimentation, filtration, centrifugation, and external magnetic field are applied. Magnetic iron-based adsorbents such as γ-Fe_2O_3, Fe_3O_4, $CoFe_2O_4$, etc., can be simply recovered by a magnet from aqueous solutions. The magnetic separation method consists of a magnetic field stainless steel column separator. Non-magnetic iron-based adsorbents such as α-Fe_2O_3, FeOOH, etc., can be separated by centrifugation (20,000 to 50,000 rpm) or filtration using membrane or Whatman filter paper with great separation efficiency and free from an aggregation of nanoparticles. In order to reduce the additional cost of the separation process, iron-based adsorbents containing polymer gels (e.g., FeOOH/cationic polymer) were also fabricated (Safi et al. 2019). Such polymer gel can be readily separated after the adsorption process.

To reduce cost and increase reusability, spent adsorbents should be regenerated. Thus, they can be further used for subsequent adsorptive removal processes for the purification of arsenic contaminated water. Besides, arsenic should also be desorbed from adsorbent for safe discard. Examples of regeneration techniques are pressure swing, thermal and electrochemical methods, and pH change (Jain et al. 2021). Especially, pH change is widely used for the regeneration of iron-based adsorbents. After the regeneration process, iron-based adsorbents could be utilized again to remove arsenic species from water, as regenerated adsorbents retain their adsorption capacity. Generally, NaOH has been widely used to elute adsorbed arsenic species from iron-based adsorbents. For example, Leus et al. (2018) have used 0.1 M NaOH solution as regenerating agent followed by hydrogen peroxide to desorb arsenic ions from the adsorbents. The temperature of the regeneration process was 70°C. Alternatively, the regeneration process was also performed with an aqueous solution of NaCl. Adsorbent like FeOOH/cationic polymer gel was regenerated using 0.5 M NaCl aqueous solution, and the obtained regeneration efficiency was 87.6% (Safi et al. 2019).

18.5 Conclusions

In this book chapter, adsorptive removal of arsenic species (As(III) and As(V)) from polluted water using iron-based nanoparticles such as Fe_2O_3, Fe_3O_4, ferrite, FeOOH, $Fe(OH)_3$, Fe^0, ferrihydrates, and $FeCO_3$ were explained. Finally, separation and regeneration of spent iron-based adsorbents after arsenic adsorption experiments were also highlighted. Overall, the following conclusions have arrived from the literature survey:

(a) Iron-based adsorbents have gained much attention in the field of arsenic remediation owing to their abundance, redox couple, and inherent affinity with arsenic ions.

(b) Arsenic adsorption capacity of pristine iron-based adsorbents depends on the active surface area, morphology, surface functional groups, and drying process of adsorbents.

(c) Addition of an external oxidant can also activate the adsorption of arsenic ions on iron-based adsorbents. Nature and dosage of external oxidants influence adsorption features of iron-based adsorbents.

(d) Arsenic removal efficiency can be further enhanced by doping with a suitable amount of metal ions, surface modification, and composite formation.

(e) Solution pH, temperature, nature and amount of adsorbents, co-existing ions and their nature, light and initial concentration of arsenic ions also govern iron-based adsorbents' removal efficiency.

(f) Adsorption of arsenic ions on iron-based adsorbents occurs through physical adsorption, mono and bidentate surface complexes, ion-exchange, and precipitation.

(g) Catalytic oxidation/reduction of arsenic ions happens at the iron-based adsorbents' surface, which further enhanced adsorption removal efficiency.

Acknowledgments

The authors wish to thank Solar Energy Research Center, SERC-Chile (FONDAP/ANID/15110019).

References

Abdul, K. S., S. S. Jayasinghe, E. P. Chandana, C. Jayasumana and P. M. De-Silva. 2015. Arsenic and human health effects: A review. Environ. Toxicol. Pharmacol. 40: 828–846.

Adio, S. O., M. H. Omar, M. Asif and T. A. Saleh. 2017. Arsenic and selenium removal from water using biosynthesized nanoscale zero-valent iron: A factorial design analysis. Process Saf. Environ. Prot. 107: 518–527.

Adlnasab, L., N. Shekari and A. Maghsodi. 2019. Optimization of arsenic removal with Fe_3O_4@Al_2O_3@Zn-Fe LDH as a new magnetic nano adsorbent using Box-Behnken design. J. Environ. Chem. Eng. 7: 102974.

Adra, A., G. Morin, G. O. Nguema and J. Brest. 2016. Arsenate and arsenite adsorption onto Al-containing ferrihydrites. Implications for arsenic immobilization after neutralization of acid mine drainage. Appl. Geochemistry 64: 2–9.

Ahangari, A., S. Raygan and A. Ataie. 2019. Capabilities of nickel zinc ferrite and its nanocomposite with CNT for adsorption of arsenic (V) ions from wastewater. J. Environ. Chem. Eng. 7: 103493.

Alka, S., S. Shahir, N. Ibrahim, M. J. Ndejiko, D. V. N. Vo and F. A. Manan. 2021. Arsenic removal technologies and future trends: A mini review. J. Clean. Prod. 278: 123805.

Alchouron, J., C. Navarathna, H. D. Chludil, N. B. Dewage, F. Perez, E. B. Hassan et al. 2020. Assessing South American *Guadua chacoensis* bamboo biochar and Fe_3O_4 nanoparticle dispersed analogues for aqueous arsenic(V) remediation. Sci. Total Environ. 706: 135943.

Amano, Y., Y. Matsushita and M. Machida. 2014. Arsenic adsorption by activated carbon with different amounts of basic sites under different solution pH and coexistent ions. Sep. Sci. Technol. 49: 345–353.

Asere, T. G., C. V. Stevens and G. D. Laing. 2019. Use of (modified) natural adsorbents for arsenic remediation: A review. Sci. Total Environ. 676: 706–720.

Awual, M. R., S. A. El-Safty and A. Jyo. 2011. Removal of trace arsenic(V) and phosphate from water by a highly selective ligand exchange adsorbent. J. Environ. Sci. 23: 1947–1954.

Awual, M. R., M. A. Shenashen, T. Yaita, H. Shiwaku and A. Jyo. 2012. Efficient arsenic(V) removal from water by ligand exchange fibrous adsorbent. Water Res. 46: 5541–5550.

Banerjee, K., G. L. Amy, M. Prevost, S. Nour, M. Jekel, P. M. Gallagher et al. 2008. Kinetic and thermodynamic aspects of adsorption of arsenic onto granular ferric hydroxide (GFH). Water Res. 42: 3371–3378.

Bessaies, H., S. Iftekhar, B. Doshi, J. Kheriji, M. C. Ncibi, V. Srivastava et al. 2020. Synthesis of novel adsorbent by intercalation of biopolymer in LDH for the removal of arsenic from synthetic and natural water. J. Environ. Sci. 91: 246–261.

Bowell, R. J., C. N. Alpers, H. E. Jamieson, D. K. Nordstrom and J. Majzlan. 2014. The environmental geochemistry of arsenic—An overview. Rev. Mineral. Geochem. 79: 1–16.

Bundschuh, J., J. Schneider, M. A. Alam, N. K. Niazi, I. Herath, F. Parvez et al. 2021. Seven potential sources of arsenic pollution in Latin America and their environmental and health impacts. Sci. Total Environ. 780: 146274.

Chen, C. C. and Y. C. Chung. 2006. Arsenic removal using a biopolymer chitosan sorbent. J. Environ. Sci. Health A. 41: 645–658.

Chen, X. M., X. C. Zeng, J. N. Wang, Y. M. Deng, T. Ma, E. Guoji et al. 2017. Microbial communities involved in arsenic mobilization and release from the deep sediments into groundwater in Jianghan plain. Central China. Sci. Total Environ. 579: 989–999.

Cornejo-Ponce, L., P. Vilca-Salinas, H. Lienqueo-Aburto, M. J. Arenas, r. Pepe-Victoriano, E. Carpio and J. Rodríguez. 2020. Integrated Aquaculture Recirculation System (IARS) supported by solar energy as a circular economy alternative for resilient communities in arid/semi-arid zones in Southern South America: A Case Study in the Camarones Town. Water. 12: 3469.

Cornejo, L., H. Lienqueo and P. Vilca. 2019. Hydro-chemical characteristics, water quality assessment and water relationship (HCA) of the Amuyo Lagoons, Andean Altiplano, Chile. Desalination Water Treat. 153: 36–45.

Cornell, R. M. and U. Schwertmann. 2003. The iron oxides: structure, properties, reactions, occurrences and uses. Wiley V.C.H, ISBN 3-527-30274-3.

Deng, M., X. Wu, A. Zhu, Q. Zhang and Q. Liu. 2019. Well-dispersed TiO_2 nanoparticles anchored on Fe_3O_4 magnetic nanosheets for efficient arsenic removal. J. Environ. Manag. 237: 63–74.

Deng, M., M. Chi, M. Wei, A. Zhu, L. Zhong, Q. Zhang et al. 2021. A facile route of mesoporous TiO_2 shell for enhanced arsenic removal, Colloids Surf. A: Physicochem. Eng. Asp. 627: 127138.

Dey, A., R. Singh and M. K. Purkait. 2014. Cobalt ferrite nanoparticles aggregated schwertmannite: A novel adsorbent for the efficient removal of arsenic. J. Water Process. Eng. 3: 1–9.

Ding, W., H. Zheng, Y. Sun, Z. Zhao, X. Zheng, Y. Wu et al. 2021. Activation of $MnFe_2O_4$ by sulfite for fast and efficient removal of arsenic (III) at circumneutral pH: Involvement of Mn(III). J. Hazard. Mater. 403: 123623.

Fu, D., T. A. Kurniawan, L. Lin, Y. Li, R. Avtar, M. H. D. Othman et al. Arsenic removal in aqueous solutions using FeS_2. J. Environ. Manag. 286: 112246.

Goswami, R., P. Deb, R. Thakur, K. P. Sarma and A. Basumallick. 2011. Removal of As(III) from aqueous solution using functionalized ultrafine iron oxide nanoparticles. Sep. Sci. Technol. 46: 1017–1022.

Guo, H., D. Stüben and Z. Berner. 2007. Adsorption of arsenic(III) and arsenic(V) from groundwater using natural siderite as the adsorbent, J. Colloid Interface Sci. 315: 47–53.

Guo, L., P. Ye, J. Wang, F. Fu and Z. Wu. 2015b. Three-dimensional Fe_3O_4-graphene macroscopic composites for arsenic and arsenate removal, J. Hazard. Mater. 298: 28–35.

Guo, S., W. Sun, W. Yang, Q. Li and J. K. Shang. 2015a. Superior As(III) removal performance of hydrous MnOOH nanorods from water. RSC Adv. 5: 53280–53288.

Gupta, K. K., N. L. Singh, A. Pandey, S. K. Shukla, S. N. Upadayay, V. Mishra et al. 2013. Effect of anatase/rutile TiO_2 phase composition on arsenic adsorption. J. Dispers. Sci. Technol. 34: 1043–1052.

Hao, L., M. Liu, N. Wang and G. Li. 2018. A critical review on arsenic removal from water using iron-based adsorbents, RSC Adv. 8: 39545–39560.

Hashemi, S. A., S. M. Mousavi and S. Ramakrishna. 2019. Effective removal of mercury, arsenic and lead from aqueous media using polyaniline-Fe_3O_4-silver diethyl dithiocarbamate nanostructures. J. Clean. Prod. 239: 118023.

Hlavay, J. and K. Polyák. 2005. Determination of surface properties of iron hydroxide-coated alumina adsorbent prepared for removal of arsenic from drinking water. J. Colloid Interface Sci. 284: 71–77.

Hua, J. 2021. Synthesis and characterization of gold nanoparticles (AuNPs) and ZnO decorated zirconia as a potential adsorbent for enhanced arsenic removal from aqueous solution. J. Mol. Struct. 1228: 129482.

Iconaru, S. L., R. Guégan, C. L. Popa, M. M. Heino, C. S. Ciobanu and D. Predoi. 2016. Magnetite (Fe_3O_4) nanoparticles as adsorbents for As and Cu removal. Appl. Clay Sci. 134: 128–135.

Jain, A., S. Kumari, S. Agarwal and S. Khan. 2021. Water purification via novel nano-adsorbents and their regeneration strategies. Process Saf. Environ. Prot. 152: 441–454.

Jo, J. Y., J. H. Choi, Y. F. Tsang and K. Baek. 2021. Pelletized adsorbent of alum sludge and bentonite for removal of arsenic. Environ. Pollut. 277: 116747.

Kang, Y. G., H. Yoon, W. Lee, E. Kim and Y. S. Chang. 2018. Comparative study of peroxide oxidants activated by nZVI: Removal of 1,4-Dioxane and arsenic(III) in contaminated waters, Chem. Eng. J. 334: 2511–2519.

Kim, J. J., Y. S. Kim and V. Kumar. 2019. Heavy metal toxicity: An update of chelating therapeutic strategies. J. Trace Elem. Med. Biol. 54: 226–231.

Kumar, S. R., V. Jayavignesh, R. Selvakumar, K. Swaminathan and N. Ponpandian. 2016. Facile synthesis of yeast cross-linked Fe_3O_4 nanoadsorbents for efficient removal of aquatic environment contaminated with As(V). J. Colloid Interface Sci. 484: 183–195.

Lata, S. and S. R. Samadder. 2016. Removal of arsenic from water using nano adsorbents and challenges: A review, J. Environ. Manag. 166: 387–406.

Leus, K., K. Folens, N. R. Nicomel, J. P. H. Perez, M. Filippousi, M. Meledina et al. 2018. Removal of arsenic and mercury species from water by covalent triazine framework encapsulated γ-Fe_2O_3 nanoparticles. J. Hazard. Mater. 353: 312–319.

Li, Q., R. Li, X. Ma, B. Sarkar, X. Sun and N. Bolan. 2020. Comparative removal of As(V) and Sb(V) from aqueous solution by sulfide-modified α-FeOOH, Environ. Pollut. 267: 115658.

Li, S., H. Xu, L. Wang, L. Ji, X. Li, Z. Qu et al. 2021. Dual-functional sites for selective adsorption of mercury and arsenic ions in $[SnS_4]^{4-}$/MgFe-LDH from Wastewater. J. Hazard. Mater. 403: 123940.

Li, W., D. Chen, F. Xia, J. Z. Y. Tan, P. P. Huang, W. G. Song et al. 2016. Extremely high arsenic removal capacity for mesoporous aluminium magnesium oxide composites. Environ. Sci. Nano. 3: 94–106.

Li, Y., M. Xu, H. Yin, W. Shi, G. I. N. Waterhouse, H. Li et al. 2019. Yolk-shell Fe_3O_4 nanoparticles loaded on persimmon-derived porous carbon for supercapacitor assembly and As (V) removal. J. Alloys Compd. 810: 151887.

Lin, S., D. Lu and Z. Liu. 2012. Removal of arsenic contaminants with magnetic γ-Fe_2O_3 nanoparticles. Chem. Eng. J. 211-212: 46–52.

Lin, S., H. Yang, Z. Na and K. Lin. 2018. A novel biodegradable arsenic adsorbent by immobilization of iron oxyhydroxide (FeOOH) on the root powder of long-root *Eichhornia crassipes*, Chemosphere 192: 258–266.

Liu, H., P. Li, F. Qiu, T. Zhang and J. Xu. 2020. Controllable preparation of FeOOH/CuO@WBC composite based on water bamboo cellulose applied for enhanced arsenic removal. Food Bioprod. Process. 123: 177–187.

Liu, Z., J. Chen, Y. Wu, Y. Li, J. Zhao and P. Na. 2018. Synthesis of magnetic orderly mesoporous α-Fe_2O_3 nanocluster derived from MIL-100(Fe) for rapid and efficient arsenic(III,V) removal, J. Hazard. Mater. 343: 304–314.

Lunge, S., S. Singh and A. Sinha. 2014. Magnetic iron oxide (Fe_3O_4) nanoparticles from tea waste for arsenic removal. J. Magn. Magn. Mater. 356: 21–31.

Morgada, M. E., I. K. Levy, V. Salomone, S. S. Farıas, G. Lopez and M. I. Litter. 2009. Arsenic (V) removal with nanoparticulate zerovalent iron: Effect of UV light and humic acids. Catal. Today 143: 261–268.

Mukherjee, D., S. Ghosh, S. Majumdar and K. Annapurna. 2016. Green synthesis of α-Fe_2O_3 nanoparticles for arsenic(V) remediation with a novel aspect for sludge management. J. Environ. Chem. Eng. 4: 639–650.

Pang, J. H., Y. Liu, J. Li and X. J. Yang. 2019. Solvothermal synthesis of nano-CeO_2 aggregates and its application as a high-efficient arsenic adsorbent. Rare Met. 38: 73–80.

Parsons, J. G., M. L. Lopez, J. R. Peralta-Videa and J. L. Gardea-Torresdey. 2009. Determination of arsenic(III) and arsenic(V) binding to microwave assisted hydrothermal synthetically prepared Fe_3O_4, Mn_3O_4, and $MnFe_2O_4$ nanoadsorbents. Microchem. J. 91: 100–106.

Pham, T. T., H. H. Ngo, V. S. Tran and M. K. Nguyen. 2020. Removal of As(V) from the aqueous solution by a modified granular ferric hydroxide adsorbent. Sci. Total Environ. 706: 135947.

Pincus, L. N., P. V. Petrovic, I. S. Gonzalez, E. Stavitski, Z. S. Fishman, H. E. Rudel et al. 2021. Selective adsorption of arsenic over phosphate by transition metal cross-linked chitosan. Chem. Eng. J. 412: 128582.

Podder, M. S. and C. B. Majumder. 2015. Removal of arsenic by a *Bacillus arsenicus* biofilm supported on GAC/$MnFe_2O_4$ composite. Groundw. Sustain. Dev. 1: 105–128.

Podder, M. S. and C. B. Majumder. 2016. Study of the kinetics of arsenic removal from wastewater using *Bacillus arsenicus* biofilms supported on a neem leaves/$MnFe_2O_4$ composite. Ecol. Eng. 88: 195–216.

Purwajanti, S., H. Zhang, X. Huang, H. Song, Y. Yang, J. Zhang et al. 2016. Mesoporous magnesium oxide hollow spheres as superior arsenite adsorbent: synthesis and adsorption behavior. ACS Appl. Mater. Interfaces. 38: 25306–25312.

Rahaman, M. S., M. M. Rahman, N. Mise, M. T. Sikder, G. Ichihara, M. K. Uddin et al. 2021. Environmental arsenic exposure and its contribution to human diseases, toxicity mechanism and management. Environ. Pollut. 289: 117940.

Rathi, B. S. and P. S. Kumar. 2021. A review on sources, identification and treatment strategies for the removal of toxic Arsenic from water system. J. Hazard. Mater. 418: 126299.

Safi, S. R., T. Gotoh, T. Iizawa and S. Nakai. 2019. Development and regeneration of composite of cationic gel and iron hydroxide for adsorbing arsenic from ground water, Chemosphere 217: 808–815.

Sahu, U. K., S. S. Mahapatra and R. K. Patel. 2017. Synthesis and characterization of an eco-friendly composite of jute fiber and Fe_2O_3 nanoparticles and its application as an adsorbent for removal of As(V) from water. J. Mol. Liq. 237: 313–321.

Sarwar, A., Q. Mahmood, M. Bilal, Z. A. Bhatti, A. Pervez, A. N. S. Saqib et al. 2015. Desalination Water Treat. 53: 1632–1640.

Seynnaeve, B., K. Folens, C. Krishnaraj, I. K. Ilic, C. Liedel, J. Schmidt et al. 2021. Oxygen-rich poly-bisvanillonitrile embedded amorphous zirconium oxide nanoparticles as reusable and porous adsorbent for removal of arsenic species from water. J. Hazard. Mater. 413: 125356.

Shi, R., Y. Jia and C. Wang. 2007. A review of arsenic adsorption onto mineral constitutions in the soil. Chin. J. Soil Sci. 38: 584–589 (in Chinese).

Siddiqui, S. I. and S. A. Chaudhry. 2017. Iron oxide and its modified forms as an adsorbent for arsenic removal: A comprehensive recent advancement, Process Saf. Environ. Prot. 111: 592–626.

Siddiqui, S. I., P. N. Singh, N. Tara, S. Pal, S. A. Chaudhry and I. Sinha. 2020. Arsenic removal from water by starch functionalized maghemite nanoadsorbents: Thermodynamics and kinetics investigations. Colloids Interface Sci. Commun. 36: 100263.

Singh, P., P. Pal, P. Mondal, G. Saravanan, P. Nagababu, S. Majumdar et al. 2021. Kinetics and mechanism of arsenic removal using sulfide-modified nanoscale zerovalent iron. Chem. Eng. J. 412: 128667.

Song, X., P. Chen, X. Luo, Y. Zhang and J. Liu. 2019. A novel laminated Fe_3O_4/CaO_2 composite for ultratrace arsenite oxidation and adsorption in aqueous solutions. J. Environ. Chem. Eng. 7: 103427.

Song, Z. M., L. L. Yang, Y. Lu, C. Wang, J. K. Liang, Y. Du et al. 2021. Characterization of the transformation of natural organic matter and disinfection byproducts after chlorination, ultraviolet irradiation and ultraviolet irradiation/chlorination treatment. Chem. Eng. J. 426: 131916.

Souza, T. G. F., E. T. F. Freitas, N. D. S. Mohallem and V. S. T. Ciminelli. 2021. Defects induced by Al substitution enhance As(V) adsorption on ferrihydrites. J. Hazard. Mater. 420: 126544.

Su, H., Z. Ye and N. Hmidi. 2017. High-performance iron oxide–graphene oxide nanocomposite adsorbents for arsenic removal, Colloids and Surfaces A: Physicochem. Eng. Aspects 522: 161–172.

Su, R., X. Ma, X. Yin, X. Zhao, Z. Yan, J. Lin et al. 2021. Arsenic removal from hydrometallurgical waste sulfuric acid via scorodite formation using siderite ($FeCO_3$). Chem. Eng. J. 424: 130552.

Sun, T., Z. Shi, X. Zhang, X. Wang, L. Zhu and Q. Lin. 2019. Efficient degradation of p-arsanilic acid with released arsenic removal by magnetic CeO_2-Fe_3O_4 nanoparticles through photo-oxidation and adsorption. J. Alloys Compd. 808: 151689.

Sun, X., C. Hu and J. Qu. 2009. Adsorption and removal of arsenite on ordered mesoporous Fe-modified ZrO_2. Desalination Water Treat. 8: 139–145.

Tresintsi, S., E. Kokkinos, A. Kamou, K. Simeonidis, G. Kyriakou, A. Zouboulis et al. 2018. One step preparation of $ZnFe_2O_4/Zn_5(OH)_6(CO_3)_2$ nanocomposite with improved As(V) removal capacity. Sep. Sci. Technol. 53: 1457–1464.

Tu, Y. J., C. F. You, C. K. Chang, S. L. Wang and T. S. Chan. 2012. Arsenate adsorption from water using a novel fabricated copper ferrite. Chem. Eng. J. 198-199: 440–448.

Vishwakarma, Y. K., S. Tiwari, D. Mohan and R. S. Singh. 2021. A review on health impacts, monitoring and mitigation strategies of arsenic compounds present in air. Clean. Eng. Technol. 3: 100115.

Wang, S. L., C. H. Liu, M. K. Wang, Y. H. Chuang and P. N. Chiang. 2009. Arsenate adsorption by Mg/Al–NO_3 layered double hydroxides with varying the Mg/Al ratio, Appl. Clay Sci. 43: 79–85.

Wang, Z., T. Ma, Y. Zhu, O. K. Abass, L. Liu, C. Su et al. 2018. Application of siderite tailings in water-supply well for As removal: Experiments and field tests. Int. Biodeterior. Biodegrad 128: 85–93.

Wang, Z., Y. Fu and L. Wang. 2021. Abiotic oxidation of arsenite in natural and engineered systems: Mechanisms and related controversies over the last two decades (1999–2020). J. Hazard. Mater. 414: 125488.

Wu, C., J. Tu, C. Tian, J. Geng, Z. Lin and Z. Dang. 2018a. Defective magnesium ferrite nano-platelets for the adsorption of As(V): The role of surface hydroxyl groups. Environ. Pollut. 235: 11–19.

Wu, L. K., H. Wu, Z. Z. Liu, H. Z. Cao, G. Y. Hou, Y. P. Tang et al. 2018b. Highly porous copper ferrite foam: A promising adsorbent for efficient removal of As(III) and As(V) from water. J. Hazard. Mater. 347: 15–24.

Wu, L. K., H. Wu, H. B. Zhang, H. Z. Cao, G. Y. Hou, Y. P. Tang et al. 2018c. Graphene oxide/$CuFe_2O_4$ foam as an efficient absorbent for arsenic removal from water. Chem. Eng. J. 334: 1808–1819.

Wu, Q., D. Wang, C. Chen, C. Peng, D. Cai and Z. Wu. 2021. Fabrication of Fe_3O_4/ZIF-8 nanocomposite for simultaneous removal of copper and arsenic from water/soil/swine urine. J. Environ. Manag. 290: 112626.

Yang, C. H., J. S. Chang and D. J. Lee. 2020. Covalent organic framework EB-COF:Br as adsorbent for phosphorus (V) or arsenic (V) removal from nearly neutral waters. Chemosphere 253: 126736.

Zamudio, F. M., J. V. Garcia, A. G. Alvarez, D. M. Figueroa and W. P. Ela. 2013. Adsorption of arsenic on pre-treated zeolite at different pH levels. Chem. Speciat. Bioavailab. 25: 280–284.

Zhang, D., R. Cao, Y. Wang, S. Wang and Y. Jia. 2021. The adsorption of As(V) on poorly crystalline Fe oxyhydroxides, revisited: Effect of the reaction media and the drying treatment. J. Hazard. Mater. 416: 125863.

Zhang, Y., M. Yang and X. Huang. 2003. Arsenic(V) removal with a Ce(IV)-doped iron oxide adsorbent. Chemosphere 51: 945–952.

Section V

General Aspects/Case Studies on Bioremediation of Metal(loid)s

CHAPTER 19

Restoration of Old Mining Sites Polluted by Metal(loid)s by using Various Amendments

Manhattan Lebrun,[1,2,3,*] *Sylvain Bourgerie*[1] and *Domenico Morabito*[1]

19.1 Introduction

Mining activities are key economic activities (Karaca et al. 2018) and participate in the development of many countries. Mining is defined as the extraction of geological materials and valuable materials from Earth's soil (Worlanyo and Jiangfeng 2021); these materials can then be used in various industrial processes. However, due to the lack of legislation, especially in the past and now in developing countries, intensive mining activities have led to the degradation of the environment. Indeed, after the extraction activity, a huge amount of tailing wastes are generated, which still contain elevated metal(loid) concentrations. But, it turns out that metal(loid)s are an important environmental and health issue due to their non-degradability and thus their accumulation in the soil. They can enter the food chain, causing deleterious effects on human health, as most of the metal(loid)s are carcinogenic (Ashraf et al. 2019). In addition to metal(loid) pollution, mine tailings are also characterized by an extreme pH (usually highly acidic), an absence of soil structure, and reduced fertility, as can be deduced from Table 19.1, which gives some characteristics viz. pH, organic matter content, cation exchange capacity, C, H and N content and metal(loid) concentrations, of soils sampled from mine sites or sites in the vicinity and impacted by the adjacent mining activities. Due to these extreme conditions, after extraction has ended, most mine sites lack vegetation. These bare soils are thus subjected to wind erosion and water leaching. Especially, acid mine drainage (AMD) is an important issue on former mine sites. Although, AMD is generated by the oxidation of minerals, and occurs naturally, the mining activities accelerate its generation. AMD is more important after the mine exploitation than during (Simate 2014, Skousen 2019). AMD is characterized by low pH and high dissolved metal(loid) concentrations (Rodríguez-Galán 2019). The transportation of metal(loid) pollution, through wind erosion, water leaching, and AMD, endangers the surrounding areas. Indeed, it has been shown that areas around the mine sites also present high metal(loid) concentrations and low fertility (Table 19.1).

[1] INRA USC1328, LBLGC EA1207, University of Orléans, Orléans, France.
[2] Université Paris-Saclay, INRAE, AgroParisTech, UMR EcoSys, 78850 Thiverval-Grignon.
[3] AGHYLE (SFR Condorcet FR CNRS 3417), UniLaSalle, 19 Rue Pierre Waguet, 60026 Beauvais, France.
* Corresponding author: manhattan.lebrun@univ-orleans.fr

Table 19.1. Physico-chemical properties and metal(loid) contents of former mining soil or soils affected by adjacent mine activities. WHC = water holding capacity, CEC = cation exchange capacity.

Soil	pH	WHC (%)	Organic matter (%)	CEC (cmol. kg^{-1})	C (%)	H (%)	N (%)	Element concentrations ($mg.kg^{-1}$)		References
	4.82							As	1068	Lebrun et al. 2021e
								Fe	6325	
								Pb	23387	
	7.05		2.6	14.9			1.77	As	95.6	Li et al. 2018b
	6.7			7.6			0.04	As	3.8	Derakhshan Nejad et al. 2021
								Cd	1	
								Cu	54.9	
								Pb	74.4	
								Zn	291	
Cu mine	2.62							Cu	636	Forján et al. 2016
								Pb	8.21	
								Zn	63.7	
					29	3.6	1.2	Cd	6.9	Jain et al. 2014
								Co	7.3	
								Cr	11.7	
								Cu	3.7	
								Fe	130	
								Ni	7.5	
								Pb	8.2	
								Zn	10.1	
	5.33		0.76	7.03				Cd	7.21	Meng et al. 2018
								Cu	219	
								Pb	32673	
								Zn	792	

	6.5				0.3	0.3	0.01	Pb	3642.7	Huang et al. 2018
								Zn	981	
								Cu	70.5	
								Cd	31.8	
								As	1587.1	
	3.9		0.49	10.2			0.05	As	203	Alvarenga et al. 2014
								Cd	2.6	
								Cr	22	
								Cu	301	
								Ni	15	
								Pb	243	
								Zn	836	
	3.86							As	211	Madejón et al. 2006
								Cd	4.44	
								Cu	119	
								Pb	471	
								Zn	381	
	3.06				0.6		0.1	Al	42500	Touceda-González et al. 2017
								Fe	105000	
								Mn	823	
								Cu	596	
								Zn	123	
								Cr	96	
								Ni	43	
								Cd	1.6	
								Co	12.2	

Table 19.1 contd. ...

...Table 19.1 contd.

Soil	pH	WHC (%)	Organic matter (%)	CEC (cmol. kg^{-1})	C (%)	H (%)	N (%)	Element concentrations ($mg.kg^{-1}$)		References
Gold mine	8.38							Cd	16.8	Al-Wabel et al. 2019
								Cu	1251	
								Fe	18920	
								Mn	775.9	
								Pb	465.4	
								Zn	4276	
Farmland close to coal mine								Cu	61.89	Dai et al. 2018
								As	23.91	
								Zn	129.36	
								Cd	0.58	
Soil adjacent to a mine	7.05	39.76	1.1					As	1458.97	El Naggar et al. 2018, 2020
								Cd	16.03	
								Cr	40.17	
								Cu	55.77	
								Ni	22.27	
								Pb	2224.43	
								Zn	1277.7	
								Mn	1600.6	
								Co	11.83	
								Sr	11.1	
Pb/Zn mine	8.13							Al	230	Fellet et al. 2011
								Cd	25.7	
								Cr	3.44	
								Cu	21.6	
								Ni	7.56	
								Pb	2873	
								Tl	130	
								Zn	11519	

Soil contaminated by wastewater of mine								Pb	2050	He et al. 2019
								Zn	495	
								Cu	654	
								Cd	0.705	
								As	1177	
Mine	5.44							Cd	32.9	Ippolito et al. 2017
								Pb	102	
								Cu	4370	
								Zn	5080	
Mine	5.33							Cd	103	
								Pb	1230	
								Cu	4840	
								Zn	18700	
Mine	5.12							Cd	40.8	
								Pb	1430	
								Cu	4490	
								Zn	7900	
Mine	3.97							Cd	12.8	
								Pb	96.2	
								Cu	5530	
								Zn	1120	
Gold mine	8.06		2.84	19.5				Zn	230.4	Ali et al. 2019
								Pb	393.2	
								Cd	1.58	
								Cu	141.36	
								Al	40,529	
								Fe	19,215	
								Mg	8592	
								Na	10,574	

Table 19.1 contd. ...

...Table 19.1 contd.

Soil	pH	WHC (%)	Organic matter (%)	CEC (cmol. kg^{-1})	C (%)	H (%)	N (%)	Element concentrations (mg.kg^{-1})		References
Mine	7.76			1.52				Al	76.1	Fellet et al. 2014
								Cd	74.4	
								Cu	2.24	
								Cr	55.9	
								Fe	22600	
								Mn	90.2	
								Ni	6.8	
								Pb	9235	
								Zn	20976	
Gold mine	7.15							As	1028	Lomaglio et al. 2017
								Pb	321	
								Sb	52	
Zn mine	6.1			9.01				Cd	0.9	Puga et al. 2015
								Zn	60	
								Pb	141.1	
Soil affected by mine sludge spill	3.5							As	250	Ciadamidaro et al. 2017
								Cd	1.3	
								Cu	180	
								Mn	520	
								Pb	600	
								Zn	370	
Soil affected by mine sludge spill	6.5							As	112	Ciadamidaro et al. 2017
								Cd	3.82	
								Cu	166	
								Mn	135	
								Pb	236	
								Zn	507	

In this context, due to their physico-chemical properties as well as the threats they pose to the surrounding environment, mine sites are in need of remediation.

One environment-friendly remediation process is phytoremediation, whose success depends on plant development, which can require the application of amendments.

This chapter will focus on the use of amendments in phytoremediation by describing: (i) the definitions of the different phytoremediation techniques and some examples will be given, (ii) the effects of amendment application will be detailed, with a focus on carbon-based amendments, compost and redmud, and (iii) the mechanisms by which amendments can improve contaminated mine soil will be explained. Finally, the process of assisted phytoremediation and its advantages will be concluded.

19.2 Phytoremediation—A Promising Remediation Technique

Following the fact that former mine sites are highly polluted and pose a threat to the environment and human health, their remediation has become a priority. Therefore, many remediation processes have been developed. Mainly physical and chemical processes have been used for decades, such as soil washing, vitrification, electrokinetic, etc. (Liu et al. 2018). However, these techniques are expensive and difficult to implement on large scales. They are disruptive to the soil, leaving it improper for vegetation, and can even lead to secondary pollution, notably by chemicals. On the contrary, phytoremediation is a more environment-friendly and sustainable technique. The term phytoremediation was first introduced in 1983 and has been defined as the use of plants, and their associated microorganisms, to remove or render less harmful metal(loid)s (Ashraf et al. 2019). The principle of phytoremediation is to establish a plant cover that will stabilize the soil and possibly accumulate metal(loid)s. Since it utilizes the plant as a solar-driven pump, it is cost-effective; it is also aesthetically pleasing and finally publically accepted (Awa and Hadibarata 2020). Based on the localization of the metal(loid)s inside the plants, phytoremediation is mainly divided into two processes: phytoextraction and phytostabilization.

In the phytoextraction process, plants are grown with the purpose of extracting the metal(loid)s completely from the soil. Plants mainly take up the metal(loid)s through their roots and translocate them to their aerial biomass. In general, metal(loid) concentrations in the aerial tissues of the plants are much higher than in the roots. Such plants are called (hyper)accumulators (Ali et al. 2013). Following the metal(loid)s accumulation, the aerial biomass has to be harvested and properly disposed of. Phytoextraction is especially used when elements that are economically interesting, such as Ni, have to be extracted. This is called phytomining (Ashraf et al. 2019). However, most (hyper)accumulator plants are herbaceous species and thus produce low biomass, reducing the rate of metal(loid) extraction. Moreover, when soils have very high contamination levels, the extraction of the whole contamination will take centuries. That is why the phytostabilization process can be a better option.

Phytostabilization does not correspond to the removal of the contamination from the soil but rather corresponds to stabilization of it. In this process, the plants either do not translocate metal(loid) s to their upper parts or translocate very poorly (Ashraf et al. 2019). In this case, plants are qualified as metal(loid) excluders (Ali et al. 2013), given that metal(loid)s are mainly immobilized at the root zone, including the interior compartment of the roots, the root surface, and the soil in the vicinity. Indeed, compounds exuded by the roots can complex or precipitate metal(loid)s in the rhizosphere zone (Liu et al. 2018). Due to the low metal(loid) concentrations in the aboveground plant parts, phytostabilization does not require the harvesting and disposal of the aerial biomass, reducing the process cost. Moreover, if metal(loid) concentrations are below the limits for industrial use, this biomass can be recovered for energy production, adding an economic benefit to the process (Tack and Meers 2010).

Several studies evaluated the potential of diverse plant species for the phytoremediation of former mine sites. For instance, Banerjee et al. (2019) evaluated the potential of vetiver (*Chrysopogo zizaniodes* L.) on a Fe mine. They showed that vetiver could grow on this mine and thus could help remediate the soil. In another study, Patra et al. (2020) tested two wild species, *Sesbania sesban*, and *Brachiara mutica*, for the remediation of a chromite site. The authors demonstrated that *S. sesban* could be suitable for phytostabilization. Lebrun et al. (2021b,c) demonstrated that alder and birch, sampled in the surrounding of an abandoned silver-lead extraction mine site, as well as the endemic *Agrostis capillaris*, were tolerant to the contaminants encountered on the site and thus could grow on the site and allow its revegetation.

However, due to the improper growing conditions of the mine sites, as shown in the introduction and Table 19.1, plant development is often reduced. For instance, in the study of Lebrun et al. (2021b) evaluating the potential of five varieties of flax (*Linum usitatissimum* L.) for the remediation of a former mine site highly polluted by As and Pb, the plants showed reduced growth on the contaminated soil, insufficient for further analysis. Similarly, on the same abandoned mine soil, germination of *Populus nigra* L. was impossible (Nandillon et al. 2019c). To overcome that, the soil conditions have to be improved, i.e., the soil acidity and metal(loid) toxicity need to be reduced, while nutrient and organic matter contents need to be increased. Such ameliorations can be done through the application of amendments.

19.3 The Potential of Amendments to Improve Phytoremediation Success

Although several studies demonstrated the potential of various plants to remediate abandoned mine technosols (soils resulting from mining extraction activities), these sites have extreme conditions (Table 19.1), which reduces plant growth and thus phytoremediation success. Therefore, to improve plant growth and development, it can be necessary to apply amendments to the soil (Fig. 19.1). These amendments will serve three purposes: (i) supply the nutrients necessary for plants, (ii) improve soil physico-chemical properties and (iii) immobilize metal(loid)s.

Many different amendments can be applied to the soil, either organic or inorganic (Rizwan et al. 2017). However, this chapter will focus on three types of amendments, viz. carbon-based, compost, and redmud, due to their increasing interest and use in research studies.

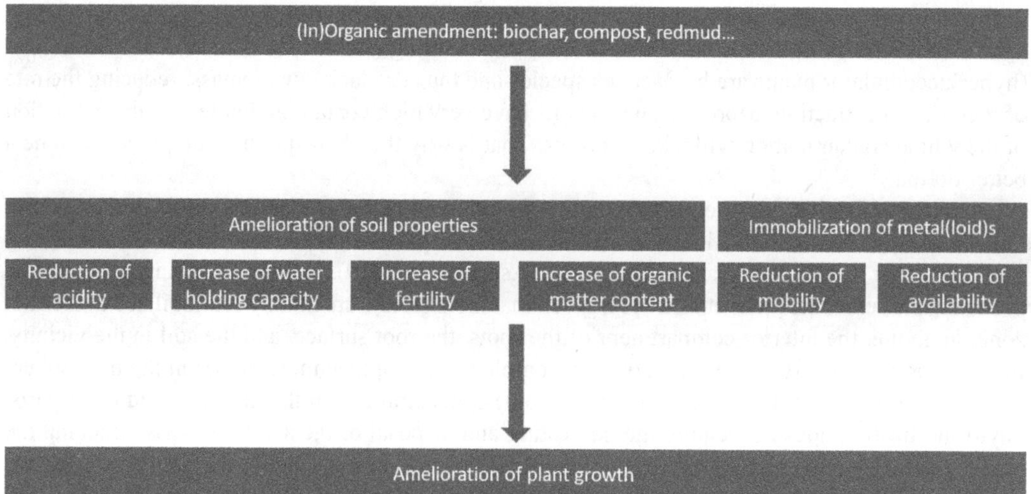

Fig. 19.1. Scheme of the effects of (in)organic amendment application on soil properties, metal(loid) behavior, and plant development.

19.3.1 Carbon-based Amendments

The term carbon-based amendments mainly refer to biochar, which has gathered much research attention over the last decades. It is obtained by the pyrolysis of biomass. Pyrolysis is a thermochemical transformation of biomass in the absence of oxygen (Tack and Egene 2019). It can be modulated mainly according to three parameters: (i) temperature (usually in the range of 200°C to 1000°C) (Yu et al. 2019), (ii) the residence time, and (iii) the heating rate, which allows to obtain biochars with different characteristics. Moreover, biochar can be produced from many diverse biomasses, i.e., vegetable waste, manures, sludges, etc., which also influences biochar properties. Thus, biochar characteristics are diverse, depending on pyrolysis and feedstock. Moreover, biochar is characterized by a porous structure, and it is also defined by its pH, cation exchange capacity, surface area, and carbon, hydrogen, and nitrogen contents. In general, biochar pH is alkaline to neutral. This is due to the decomposition of organic substances during pyrolysis (Huang et al. 2017). Its pH has been shown to be affected by feedstock and pyrolysis temperature. For instance, apple tree branches were pyrolyzed at five temperatures (400°C, 500°C, 600°C, 700°C, and 800°C), and the pH of the resulting biochars increased from 7 to 10 with increasing pyrolysis temperature, until 600°C (Jindo et al. 2014). Similarly, the pH of canola straw biochars was higher for the biochar produced at 700°C (pH 11) than the ones produced at 500°C (pH 9) or 300°C (pH 7) (Kwak et al. 2019, Yuan et al. 2011). In the study of Singh et al. (2010), biochar pH ranged from 6.9 to 10.3 depending on the feedstock (eucalyptus wood, eucalyptus leaves, paper sludge, poultry litter, and cow manure). However, few studies also showed that, in some cases, biochar could have an acidic pH. For instance, when bamboo biomass was pyrolyzed at 450°C, biochar pH was 5.2 (Yao et al. 2012), while wildfire biochar produced at 350–500°C had a pH of 3.1 (Zhang et al. 2016). Another important point about biochar is its cation exchange capacity (CEC). It corresponds to the capacity of a material to hold cations (Dai et al. 2017), which is an important parameter to assess the immobilization of metal(loid)s. Generally, biochar has a high CEC, from a few cmol to several hundred cmol per kilogram (Higashikawa et al. 2016, Kelly et al. 2014, Lu et al. 2017). Biochar is also known for its elevated surface area, going from a few meters per gram (Jindo et al. 2014, Sun et al. 2014) to several hundred meters per gram (Lu et al. 2017, Yao et al. 2012). It mainly contains carbon, between 40 % and 80 % (Yuan et al. 2019), with low levels of hydrogen and nitrogen (Tan et al. 2017a). Finally, biochar is characterized by the presence of specific functional groups on its surface. Infra-red analyses demonstrated that biochar surface was covered by OH, CH, C≡N, C≡C, C=C=C, N=C=O groups (Angın 2013, Lebrun et al. 2018c, Samsuri et al. 2013, Stella Mary et al. 2016).

All of these characteristics will affect how biochar influences soil properties and plant growth.

The main soil properties affected by biochars are pH, organic matter (OM) content, water holding capacity (WHC), and nutrient contents. For instance, the application of a rice straw biochar @ 10 % to 50 % to farmland close to a mine increased soil pH from 0.1 to 0.3 units (Dai et al. 2018). A Pb/Zn mine was amended with 1% to 10% prune residue biochar, which induced an increase in WHC (from 1.3 to 3.9-fold), pH (from 0.6 to 0.9 units), and nutrient contents (phosphorus from 2.8 to 5.4-fold and potassium from 4.2 to 28.8-fold) (Fellet et al. 2011). The amendment of an abandoned silver-lead extraction mine technosol by pinewood biochar also led to an increase in soil pH by 2.2-units when added at 2 % and 2.9-units when applied at 5% (Lebrun et al. 2017), while the application of 5 % hardwood biochar to the same increased pH by 3.7-units (Lebrun et al. 2019).

Moreover, as stated above, the effects of biochar will depend on several parameters: feedstock, pyrolysis temperature, particle size, and application rate are the main ones. For instance, the study of El-Naggar et al. (2018) evaluated the effects of three biochars, one made from *Amur silvergrass* residues, one made from *Oryza sativa*, and one made from *Maesopis eminii*, applied to a soil sampled from two sites adjacent to a mine. It showed that two biochars increased pH, while the third one decreased it, and the reverse effect was observed for electrical conductivity. All three biochars increased OM content, between 32% to 45%, and exchangeable K concentration, from 1.6 to 4.8 times. In another study, the amendment of a gold mine area by date palm biochar produced at different temperatures (300, 500, and 700°C) and added at several rates (5, 15 and 30 g kg^{-1}) led to various modifications in soil parameters (Al-Wabel et al. 2019). In more detail, no variation in soil pH was measured except for the decreases of 0.2 and 0.3 units with the addition of 15 and 30 g.kg^{-1} of the 300°C date palm biochar, while electrical conductivity was increased in all cases. The highest rises were observed with the biochars produced at 700°C and with the higher application rate. Soil organic carbon (SOC) content was also increased by all the biochar treatments. The rise was related to the application rate and among the three temperatures, the biochar produced at 300°C led to higher SOC content increase than the other two biochar temperatures (Al-Wabel et al. 2019). Finally, Lebrun et al. (2018a) showed that biochars with finer particle sizes were able to increase more importantly pH and electrical conductivity than the coarser particle size biochar.

Additionally, biochar amendment has been shown to immobilize metal(loid)s by reducing their mobility and availability. For instance, Forján et al. (2016) showed that biochar was able to reduce Cu availability when applied at 20%, 40%, and 60% to a former mine. Similarly, He et al. (2019) studied topsoil affected by acidic mine wastewaters, to which biochars made from kenaf core or sewage sludge, both pyrolyzed at 350°C and 550°C, were applied at 4%. Both biochar feedstocks reduced Pb and Cu labile fractions, while only the kenaf core biochars immobilized Zn, Cd, and As. Four soils collected from abandoned mine sites were amended with 5%, 10%, and 15% of two biochars made of beetle-killed lodgepole and tamarisk. Following this amendment, concentrations in $CaCl_2$-extractable Pb, Cu, Cd, and Zn were reduced (Ippolito et al. 2017). The extent to which metal(loid) availability was reduced depended on the biochar feedstock, temperature, and application rate (Ippolito et al. 2017). Moreover, in previous studies (Lebrun et al. 2018a,b, 2017, Nandillon et al. 2019b,c), diverse biochars were applied to a former silver-lead extraction mine, which induced a reduction in soil pore water Pb and extractable ($CaCl_2$ And NH_4NO_3) Pb concentrations. In congruence to the study of Ippolito et al. (2017), these studies showed that the efficiency of the biochars to reduce Pb mobility and availability depended on the feedstock, particle size, and application rate. However, in general, they were inefficient for As or had negative effects. Similarly, Beesley et al. (2013) observed an increase in As mobility following the addition of 30% orchard prune residue biochar to a land impacted by mining activities.

Finally, plants were shown to be affected by biochar amendment in terms of growth and metal(loid) accumulation. In Alhar et al.'s (2021) study, mine exposed materials were sampled from five disused mines and amended with rice husk and wheat straw biochars. Following the application of these biochars at 5% and 10%, ryegrass yield was increased, and the accumulation of Cd, As, and Sb in plants decreased. Similarly, ramie plants were grown on a Cu mine site amended with 2.5% to 10% rice straw biochar. Compared to the control, plant biomass was higher in the amended conditions, while Cu concentrations were lower (Rehman et al. 2019). Finally, studies showed that *Salicaceae* plants had a poor growth on non-amended mine technosol and the biochar addition increased it (Lebrun et al. 2018a,b, 2017), while metal(loid) accumulation was differently affected, either decreasing or increasing depending on the biochar feedstock, particle size, and application rate.

More examples of the effects of biochar on soil, metal(loid)s, and plants are given in Table 19.2. This table shows that in general, following biochar amendment, soil physico-chemical properties are improved, metal(loid)s are immobilized, and plant growth is enhanced. Moreover, from Table 19.2, it can be seen that biochar effects are dependent on feedstock, pyrolysis temperature, particle size, and application rate. Finally, the studies summarized in Table 19.2 showed that biochars are very efficient for metal cations immobilization but much less efficient for anions, especially As.

One way to improve biochar's beneficial effects on metal(loid) immobilization is to modify its surface. The process is called functionalization or activation, and the resulting product is called functionalized biochar or activated carbon/biochar. There are two main types of modifications:

Table 19.2. Effect of biochar on soil, metal(loid)s and plants. EC = electrical conductivity, OM = organic matter content, WHC = water holding capacity, NS = non significant effect.

Soil	Feedstock	Pyrolysis temperature (°C)	Particle size	Time	Application rate	Plant species	pH	EC	OM	WHC	[Cd]	[Zn]	[Cu]	[Pb]	[As]	[Al]	[Cr]	[Ni]	[Sb]	Biomass	References
Gold mine	Date palm	300	1 mm	30 days	5 g/kg		NS	+9%													Al-Wabel et al. 2019
					15 g/kg		–0.20	+15%			–18.1%	–23%									
					30 g/kg		–0.34	+31%			–32.2%	–35.2%	–66.6%	–66%							
		500	1 mm		5 g/kg		NS	+8%													
					15 g/kg		NS	+13%													
					30 g/kg		NS	+31%													
		700	1 mm		5 g/kg		NS	+10%													
					15 g/kg		NS	+19%													
					30 g/kg		NS	+34%													
Farmland close to coal mine	Rice straw	550	2 mm	30 days	10%		+0.1				–42.04%	–51.37%	–57.26%		–6.94%						Dai et al. 2018
					20%		+0.2														
					30%		+0.2														
					40%		+0.3														
					50%		+0.3														
Soil adjacent to a mine	Amur silvergrass residues	500		90 days	30 t/ha		+0.07	–16%	+32%												El Naggar et al. 2018
	Paddy straw						–0.07	+34%	+45%												
	Umbrella tree residues						+0.04	–7%	+48%												
Soil adjacent to a mine	Amur silvergrass residues	500		473 days	30 t/ha									–58.3%	NS						El Naggar et al. 2020
	Paddy straw							0.3-fold						–36.8%	2-fold						
	Umbrella tree residues													–82.6%	NS						

Table 19.2 contd. ...

...Table 19.2 contd.

Soil	Feedstock	Pyrolysis temperature (°C)	Particle size	Time	Application rate	Plant species	pH	EC	OM	WHC	[Cd]	[Zn]	[Cu]	[Pb]	[As]	[Al]	[Cr]	[Ni]	[Sb]	Biomass	References
Pb/Zn mine	Prune residues from orchard	500	2 mm	15 days	1%		+0.6	1.6-fold		1.3-fold	+8%	NS	+17%	NS		NS	NS	NS			Fellet et al. 2011
					5%		+1.6	3.8-fold		2.5-fold	NS	NS	+75%	NS		NS	NS	NS			
					10%		+1.9	7.2-fold		3.9-fold	NS	NS	2.7-fold	NS		3.7-fold	+76%	NS			
Mine	Lodgepole pine	300-550			5%		+1.65				50%	85%	NS	73%							Ippolito et al. 2017
					10%		+0.92				50%	85%	NS	90%							
					15%		+0.95				50%	85%	NS	90%							
	Tamarisk	300-550			5%		+0.83				30%	33%	NS	66%							
					10%		+1.61				57%	84%	NS	83%							
					15%		+1.73				71%	92%	NS	83%							
Mine	Lodgepole pine	300-550			5%		+0.53				80%	70%	94%	NS							
					10%		+0.66				80%	70%	98%	NS							
					15%		+0.89				80%	70%	98%	NS							
	Tamarisk	300-550			5%		+0.68				32%	37%	69%	NS							
					10%		+1.21				32%	37%	97%	NS							
					15%		+1.39				61%	67%	97%	NS							
Mine	Lodgepole pine	300-550			5%		+0.79				60%	58%	93%	NS							
					10%		+1.09				60%	58%	97%	NS							
					15%		+1.14				60%	58%	97%	NS							
	Tamarisk	300-550			5%		+0.91				20%	NS	NS	NS							
					10%		+1.59				50%	58%	93%	NS							
					15%		+1.78				50%	58%	98%	NS							
Mine	Lodgepole pine	300-550			5%		+1.23				57%	50%	94%	85%							
					10%		+2.43				60%	65%	98%	100%							
					15%		+2.36				85%	75%	98%	99%							
	Tamarisk	300-550			5%		+1.85				NS	50%	89%	94%							
					10%		+3.16				NS	75%	94%	100%							
					15%		+3.52				NS	100%	94%	100%							
Gold mine	Wood				0.5%	Wheat	NS													NS	Ali et al. 2019
					1%		NS													NS	
					2%		+0.24													+35.77%	

Mine	Pruning residues		2 mm		1.5%		+1.01	NS			NS	NS	NS	NS		2.5-fold	NS	-41%			Fellet et al. 2014
					2%		+1.53	2.3-fold			NS	NS	NS	NS		3.8-fold	NS	-49%			
	Fir tree pellets				1.5%		+0.23	NS			NS	NS	NS	NS		NS	NS	NS			
					2%		+0.18	NS			NS	NS	NS	NS		NS	NS	NS			
	Manure + fir tree pellets				1.5%		+1.14	2.8-fold			NS	NS	3-fold	NS		NS	NS	NS			
					2%		+1.4	3.4-fold			NS	NS	5.1-fold	NS		2.9-fold	NS	+36%			
Farmland near mine	Rice straw	500			2.5%	Rice	+0.57	2.9-fold	1.7-fold											NS	Li et al. 2018a
					5%		+0.99	3.5-fold	2.3-fold											+39%	
Gold mine	Pinewood	500			2%	Bean	+1.09	1.5-fold						−33%	+34%				+72%	NS	Lomaglio et al. 2017
					5%		+1.48	1.8-fold						−78%	+71%				+77%	NS	
Ag/Pb mine	Hardwood	500			2%	None	+2.2	3.1-fold						>99%	NS						Norini et al. 2019
					5%		+3.1	16.9-fold						>99%	NS						
					2%	Willow	+2.2	NS						>99%	NS					14.4-fold	
					5%		+2.9	11.5-fold						>99%	NS					15.2-fold	
					2%	Ryegrass	+2.3	4.4-fold						NS	NS					2.4-fold	
					5%		+3	14.3-fold						NS	NS					2.2-fold	
Ag, Au, Cu mine	Miscanthus	700			1%	Blue wildrye	NS	NS				-32%	NS							NS	Novak et al. 2018
					2%		NS	NS				-29%	NS							NS	
					5%		+0.29	NS				-40%	NS							NS	

Table 19.2 contd. ...

...Table 19.2 contd.

Soil	Feedstock	Pyrolysis temperature (°C)	Particle size	Time	Application rate	Plant species	pH	EC	OM	WHC	[Cd]	[Zn]	[Cu]	[Pb]	[As]	[Al]	[Cr]	[Ni]	[Sb]	Biomass	References
Zn mine	Sugar cane straw	700			1.5%	Jack Bean	NS		NS											+22%	Puga et al. 2015
					3%		NS		+14%											+27%	
					5%		NS		+29%											+35%	
					1.5%	Mucuna aterrima	-0.2		+14%											+8%	
					3%		-0.3		+29%											+12%	
					5%		-0.1		+43%											+18%	
Cu mine	Rice straw	300		40 days	2.5%	Ramie														1.8-fold	Rehman et al. 2019
					5%															2.4-fold	
					10%															3.2-fold	
Ag/Pb mine	Pinewood	500			2%	None	+2.2	2.1-fold						–69%	NS						Lebrun et al. 2017
					5%	None	+2.9	2.9-fold						–97%	NS						
					2%	Salix viminalis														3-fold	
					5%															3-fold	
					2%	Salix alba														6-fold	
					5%															11-fold	
					2%	Salix purpurea														5-fold	
					5%															3-fold	
Ag/Pb mine	Lightwood	500	< 0.1 mm	45 days	2%	Salix viminalis	+1.8	1.9-fold	2-fold	NS				–97%	-68%					3.7-fold	Lebrun et al. 2018b
						Populus euramericana			1.8-fold											NS	
					5%	Salix viminalis	+2.5	5-fold	3.6-fold	NS				–99%	NS					2.3-fold	
						Populus euramericana			3-fold											5.9-fold	
			0.2–0.4 mm		2%	Salix viminalis	+1.5	NS	2.1-fold	NS				–97%	-78%					2.3-fold	
						Populus euramericana			2-fold											3.1-fold	
					5%	Salix viminalis	+2.3	2.6-fold	3.6-fold	1.2-fold				–99%	NS					3.6-fold	
						Populus euramericana			3.6-fold											6-fold	

	Pinewood		< 0.1 mm		2%	Salix viminalis	+1.1	1.1-fold	2.1-fold	NS				–81%	–78%					2.1-fold	
						Populus euramericana			2.1-fold											2.1-fold	
					5%	Salix viminalis	+1.4	1.5-fold	3.7-fold	-15%				–89%	NS					3.2-fold	
						Populus euramericana			3.7-fold											2.1-fold	
			0.2-0.4 mm		2%	Salix viminalis	+0.3	NS	2.9-fold	NS				–60%	NS					3.1-fold	
						Populus euramericana			1.7-fold											2-fold	
					5%	Salix viminalis	+0.8	1.3-fold	4-fold	1.4-fold				–77%	NS					3.3-fold	
						Populus euramericana			3.3-fold											4-fold	
Ag/Pb mine	Hardwood	500	< 0.1 mm	45 days	2%	Salix viminalis	+2.4	3.3-fold		NS				–98%	NS					2.7-fold	Lebrun et al. 2018a
					5%		+2.6	4.9-fold		NS				–98%	NS					2.3-fold	
			0.2–0.4 mm		2%		+1.3	1.5-fold		NS				–88%	-67%					3.9-fold	
					5%		+2.4	2.7-fold		1.2-fold				–97%	NS					3.2-fold	
			0.5–1 mm		2%		+0.4	NS		NS				–29%	NS					1.6-fold	
					5%		+0.9	NS		1.1-fold				–65%	NS					3.3-fold	
			1–2.5 mm		2%		+0.4	NS		NS				–35%	NS					1.6-fold	
					5%		+0.5	NS		1.2-fold				–32%	NS					2.8-fold	
Ag/Pb mine	Hardwood	500	0.2–0.4 mm		5%	Salix viminalis	+3.7	4.3-fold	2.7-fold	1.3-fold				–91%	> –99%					3-fold	Lebrun et al. 2019
Ag/Pb mine	Hardwood	500	0.5–1 mm		1%	Agrostis	+0.9	NS						–39%	NS					NS	Lebrun et al. 2021d
Ag/Pb mine	Hardwood	500	0.5–1 mm		2%	Alder	+1.7	3.1-fold						–93%	NS					1.6-fold	Lebrun et al. 2021e
						Birch	+3.2	1.7-fold						–94%	NS					NS	

Table 19.2 contd. ...

...Table 19.2 contd.

Soil	Feedstock	Pyrolysis temperature (°C)	Particle size	Time	Application rate	Plant species	pH	EC	OM	WHC	[Cd]	[Zn]	[Cu]	[Pb]	[As]	[Al]	[Cr]	[Ni]	[Sb]	Biomass	References
Ag/Pb mine	Hardwood	500	0.2–0.4 mm		2%	Salix dasyclados	+2.1							–66%	NS					4.3-fold	Lebrun et al. 2021f
Ag/Pb mine	Bark oak	500	0.2–0.4 mm		2%	Phaseolus vulgaris	+2.8	13-fold						–89%	NS					4.1-fold	Lebrun et al. 2020
			0.5–1 mm				+2.9	8.6-fold						–90%	–50%					3.8-fold	
			1–2.5 mm				+2	3.9-fold						–68%	–50%					2.9-fold	
	Sapwood oak		0.2–0.4 mm				+0.9	NS						–51%	–83%					NS	
			0.5–1 mm				+0.9	NS						–44%	–83%					NS	
			1–2.5 mm				+0.4	NS						–41%	–67%					NS	
	Heart wood oak		0.2–0.4 mm				+0.4	NS						–38%	–67%					NS	
			0.5–1 mm				+0.3	NS						–30%	–50%					NS	
			1–2.5 mm				+0.2	NS						–28%	NS					NS	
Ag/Pb mine	Hardwood	500	0.2–0.4 mm		2%	Trifolium repens	+1.3	2.2-fold						–62%	NS					1.9-fold	Lebrun et al. 2021a
			0.5–1 mm				+1.7	2.2-fold						–67%	NS					2.5-fold	
			1–2.5 mm				+1.2	1.7-fold						–50%	NS					NS	
	Oak bark and sapwood		0.2–0.4 mm				+1.3	2.7-fold						–48%	NS					1.6-fold	
			0.5–1 mm				+1.3	2.7-fold						–47%	NS					NS	
	Pine bark		0.2–0.4 mm				NS	NS						–29%	–97%					NS	
			0.5–1 mm				NS	NS						–37%	–85%					NS	
			1–2.5 mm				NS	NS						NS	> –99%					NS	

	Coconut		< 2 mm				NS	2.2-fold						–31%	NS					NS	
	Bamboo						+1.4	3.2-fold						–69%	–99%					NS	
	Pseudotsuga		2–3 mm				NS	NS						–49%	–96%					NS	
Tin mine	Hardwood	500	0.5–1 mm		2%	Phaseolus vulgaris	NS	NS						NS	NS					NS	Lebrun et al. 2021c
Ag/Pb mine	Hardwood	500	< 0.1 mm	45 days	2%	Populus euramericana	+0.9	1.2-fold	2.2-fold	NS				–98%	NS					4-fold	Lebrun et al. 2021g
					5%		+1.2	1.4-fold	3.4-fold	NS				–99%	2.5-fold					6.4-fold	
			0.2–0.4 mm		2%		+1.1	NS	1.6-fold	NS				–99%	NS					2-fold	
					5%		+1.1	NS	3.3-fold	NS				–99%	2-fold					3.6-fold	
			0.5–1 mm		2%		+1	NS	1.8-fold	NS				–97%	2-fold					2.8-fold	
					5%		+1	1.1-fold	3.2-fold	NS				–98%	2.5-fold					4.8-fold	
			1–2.5 mm		2%		+1.2	1.1-fold	1.7-fold	NS				–97%	NS					2.8-fold	
					5%		+1.2	1.4-fold	3.4-fold	NS				–99%	NS					4.8-fold	
Ag/Pb mine	Hardwood	500	0.2–0.4 mm		5%	Phaseolus vulgaris	2.7	1.9-fold						–98%	–87%					1.8-fold	Nandillon et al. 2019a
Ag/Pb mine	Hardwood	500	0.2–0.4 mm		5%	Trifolium repens	+1.3	3.2-fold		1.3-fold				–89%	NS						Nandillon et al. 2019b

(i) physical, such as magnetization, steam activation, and gas activation, and (ii) chemical, such as amino modification, methanol modification, acid and base treatments (Rajapaksha et al. 2016, Tan et al. 2017b). Generally, steam activation and gas activation are processed during pyrolysis. However, magnetization and chemical modifications are carried out before or after the pyrolysis, as shown in Fig. 19.2. When it is done before pyrolysis, the feedstock is put in contact with the activation element and then pyrolyzed, while when it is done post-pyrolysis, the feedstock is first pyrolyzed, and then the resulting biochar is treated with the activating material. The purpose of biochar activation/functionalization is to improve biochar properties such as porosity, surface area or to modify its surface structure, for instance, to fix charged ions (Tan et al. 2017b). Such modifications are intended to ameliorate biochar effects. For instance, bamboo biochar was modified by chitosan application after the pyrolysis step, which has allowed to increase Cd, Cu, and Pb removal from a liquid solution (Zhou et al. 2013). Similarly, biogas residue was mixed with $ZnCl_2$ before the pyrolysis; the positive effect has been to increase the resulting biochar adsorption capacity towards As (Xia et al. 2016). On the other hand, steam activation of canola straw biochar increased its Pb adsorption capacity (Kwak et al. 2019). Finally, the study of Wu et al. (2016) evaluated the effects of three modes of biochar activation by addition of H_2O_2, NH_3, or HNO_3 on coconut biochars and found that depending on the modification type and pyrolysis temperature, activated biochar had either a higher or a lower sorption capacity than pristine biochar.

However, although these modifications are much studied in sorption tests to evaluate their adsorption capacity, the application of such modified biochars to mine sites is still scarce. Only a few studies evaluated the effect of activated/functionalized biochars on mine soil properties and plant growth. For instance, Lebrun et al. (2018c) showed that Fe-functionalized biochar was able to sorb As from a liquid solution, contrary to the pristine hardwood biochar. However, the positive results were lost once it was applied to the mine technosol. Furthermore, Lebrun et al. (2021c) found that activated carbon (made from coconut husk) had a higher Pb sorption capacity than the two non-activated biochars (made from hardwood and coconut husk), applied at the same rate (2%). However, when this activated carbon was applied to the mine technosol, it induced a lower pH increase and had a similar Pb immobilization than the non-activated biochars and did not improve *Salix dasyclados* growth. Further, several activated carbons (from vegetal and mineral biomass, chemically or physically activated) were applied to a former technosol, highly contaminated with

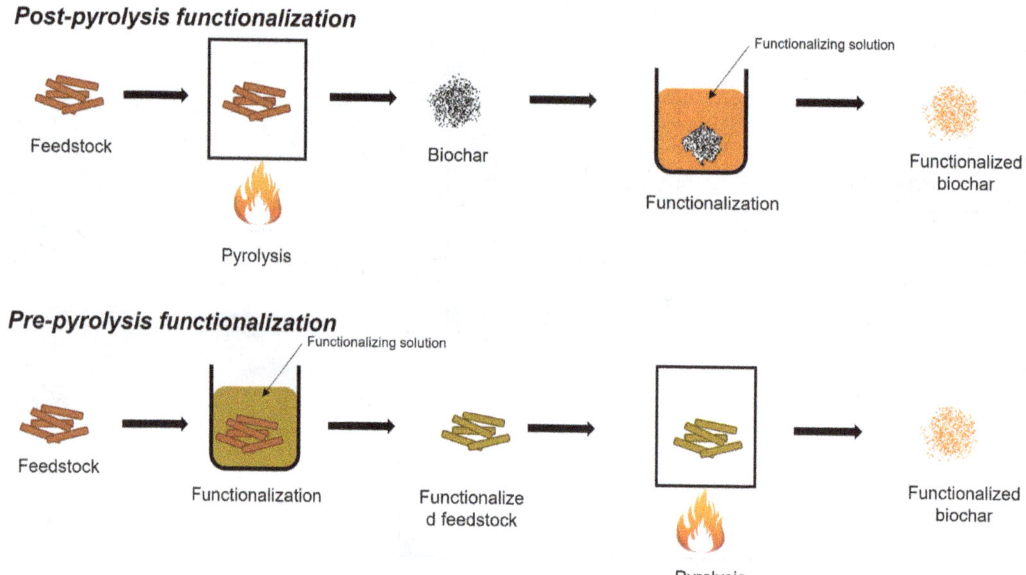

Fig. 19.2. Biochar functionalization process: pre-pyrolysis and post-pyrolysis functionalization.

As and Pb (Lebrun et al. 2021a). Results showed that they had no effect on soil pH and metal(loid) mobility, while *Trifolium repens* growth was improved in a few cases.

Based on these studies, it can be concluded that the modifications of biochar properties following functionalization/activation could improve its metal(loid) sorption capacity, while results in field conditions in mine soils showed contradictory effects, with either positive, neutral, or negative outcomes. Studies on biochar activation/functionalization in soil need to be made on a more regular basis to confirm or invalidate the results of sorption experiments.

To conclude, biochars, and their activated materials, showed beneficial effects towards soil properties and plant growth, which could have a positive effect on acid mine drainage through reduction of metal(loid) mobility and leaching, thanks to the immobilization capacity of carbon-based amendments and stabilization of the soil by plant establishment.

19.3.2 Compost

Compost is the product of the microbial degradation of organic materials (Diacono and Montemurro 2010). It is widely used in agriculture due to its high nutrient content. Moreover, compost is characterized by its richness in humic substances, such as high molecular weight substances and microorganisms (Fischer and Glaser 2012, Huang et al. 2016). In general, composts have alkaline pH, elevated EC, and high OM content. Due to these properties, and because former mine technosols have poor fertility, compost has also been studied for its potential beneficial effect when used in phytoremediation.

For instance, Alvarenga et al. (2014) applied two different composts—a mixed municipal solid waste compost and a green-waste derived compost—to a mine soil, at a rate of 50 t ha^{-1}. Following the addition of these composts, soil pH increased by 2 and 3 units, while OM content rose by 4- and 3.3-fold. The contents of N, P, and K also increased between 2.5- and 15-times. Finally, *Agrostis tenuis* was able to grow on the amended soil compared to the non-existent growth on the contaminated soil (Alvarenga et al. 2014). Similarly, Ciadamidaro et al. (2017) studied two mines, to which they applied 50 t ha^{-1} of a biosolid compost or an alperujo compost and grew *Medicago polymorpha* or *Poa annua*. Results showed that, on both soils, metal availability decreased following compost addition. However, soil pH and organic carbon content only increased in the acid mine soil by 3 units and 3-fold on average, respectively. Finally, *M. polymorpha* was not affected in the neutral soil, while its growth was improved on the acidic soil, and *P. annua* growth increased in all cases. In their study published in 2016, Gil-Loaiza et al. observed an increase in the fertility of a mine tailing pile following the application of 15% to 20% of compost, determined by an increase in pH (3.3 to 5 units), organic carbon content, and nitrogen content (Gil-Loaiza et al. 2016). In a study, Madejón et al. (2006) observed that adding 30 t ha^{-1} of a biosolid compost to a mine led to an increase in soil pH (+1.5 units), organic C content (1.7-fold), and a rise in vegetation cover (Madejón et al. 2006). In the subsequent study, the mine was amended with 30 t ha^{-1} of either a biosolid compost or an alperujo compost, and *Paulownia fortune* plants were grown. Following these amendment applications, soil pH, EC, and organic C content increased (Madejón et al. 2014). These two compost types were applied to another abandoned mine, and they induced a rise in soil pH and *Poa annua* and *Medicago polymorpha* growth. However, no effect on organic C content was observed, and even As and Cu availability was increased (Montiel-Rozas et al. 2015). Alam et al. (2020) applied three different composts to an agricultural field affected by mine activities. Composts were vermicompost, spent mushroom compost, and leaf compost. Compost application increased soil pH; it also increased the survival of radish plants, as well as their height, leaf number, root length and weight, while they decreased metal(loid) concentrations in radish. Finally, the amendment of a former silver lead technosol by different composts showed different results. In more details, the application of 5% compost made of animal manure and vegetal materials increased soil pH (+3 units), soil EC (1.7 to 4.3-fold), and As mobility (17 to 18-fold), while Pb mobility decreased by 90 to 97%. On these amended soils, the growth of *Salix viminalis* and *Trifolium repens* increased by 4.1-fold and 6.5-fold, respectively (Lebrun et al. 2019, Nandillon et al. 2019b). Also, a compost made of peat moss, softwood bark, green compost, and seaweed was applied @ 2% and 5% and induced a rise in soil pH (2.2-units on average), and in As mobility (2-fold), and reduction in Pb mobility by 92% to 97%, an improvement in alder (2.1-fold) and bean (2-fold) growth, and no effect on birch (Lebrun et al. 2021d, Nandillon et al. 2019a).

A detailed account of the effects of compost amendments on soil properties, metal(loid)s, and plants is given in Table 19.3.

Based on these studies, it can be concluded that, in general, compost application improved soil fertility and plant growth. However, it had contrasting effects on metal(loid)s, immobilizing most metal(loid)s, especially cations, but mobilizing As and Cu.

19.3.3 Redmud

Another amendment that has gathered research attention over the last years is redmud. It is a by-product of the Bayer process, which produces alumina from Bauxite (Bhatnagar et al. 2011, Hua et al. 2017). Due to its elevated metal(loid) content, redmud is considered a waste (Hua et al. 2017) and poses a problem for its disposal due to the large amount of redmud produced each year, 90 to 120 million tons (Hua et al. 2017, Liu et al. 2011). In addition to its high metal(loid) content, redmud is characterized by a very fine texture, a high surface area (10–30 m^2 g^{-1}), and an alkaline pH, generally around 10 to 13 (Bhatnagar et al. 2011, Hua et al. 2017, Khairul et al. 2019). Moreover, redmud has a sorption affinity towards metal(loid)s, making it a potential amendment for metal(loid) remediation. Such application on the mine sites will have the additional advantages of reducing the need for redmud disposal.

The use of redmud for metal(loid) phytoremediation of mine site started to have a research interest in the last couple of decades. For instance, redmud was applied @ 1%, 2%, 4%, 5%, and 7% to a mine soil and led to an increase in soil pH (between 0.2 and 2 units) and a decrease in soil Pb and Zn concentration by 95% and 94% at the highest application rate (Argyraki et al. 2017). Similarly, Clemente et al. (2019) added 2 g kg^{-1} redmud to soil affected by former mining activities. They observed an increase in soil pH (1.1 unit in the presence of *Silybum manrianum* and 1.4 units in the presence of *Piptatherum miliaceum*), organic C content (3.4- and 4.9-fold, respectively), N and P contents, and As availability (between 1.2 and 1.8-fold), as well as an improvement in *Silybum masianum* and *Piptatherum miliaceum* growth. Feigl et al. (2012) also applied redmud to a former Pb/Zn mine and found that such amendment increased soil pH (0.2 to 0.3 units) and decreased metal availability; Cd availability decreased by 19% and 41%, Zn availability by 37% and 62%, Pb availability was not affected and As mobility decreased by 47 and 72%, following the addition of 2% or 5% redmud, respectively. Garau et al. published three papers in which they applied redmud to a disused mine. In these studies, they observed that redmud increased soil pH (1.9 to 2.9 units) and EC (1.3 to 3-fold) and improved *P. vulgaris* and *T. vulgare* growth, while the available concentrations of Cd, Zn, Pb, and As showed contrasting responses to redmud amendment, either increasing, decreasing and being not affected (Garau et al. 2014, 2011, 2007). Similarly, the studies of Lee et al. published in 2011 and 2014 evaluated the effects of the redmud amendment on a gold mine soil and a Pb/Zn mine. Results given in these studies showed that redmud increased pH (0.5 to 4.1 units) and EC (1.6 to 4.2-fold) and decreased metal(loid)s availability (Cd 77 to 98%, Zn 20 to 99%, Cu 56 to 58%, Pb 70 to 98%, and As 12 to 87%) (Lee et al. 2014, 2011).

More detailed effects of redmud amendment on soil, metal(loid)s, and plant are given in Table 19.4, revealing that redmud improved soil conditions, immobilized both metals and metalloids, and improved plant growth.

Table 19.3. Effect of compost on soil, metal(loid)s and plants. EC =electrical conductivity, OM = organic matter content, OC = organic carbon content, NS = non-significant effect.

Soil	Compost type	Application rate	Plant species	pH	EC	OM	OC	[C]	[N]	[P]	[K]	[Cd]	[Zn]	[Cu]	[Pb]	[As]	[Al]	[Cr]	[Ni]	[Sb]	Height	Biomass	Reference
Pyrite mine	Mixed municipal solid waste	50 t.ha-1	*Agrostis tenuis*	+2	NS	4-fold			5-fold	5-fold	15-fold		NS	–96%	NS	NS						Allow growth	Alvarenga et al. 2014
	Green waste	50 t.ha-1		+3	NS	3.3-fold			2.5-fold	3.8-fold	12-fold		–99%	> –99%	NS	6.4-fold						Allow growth	
Soil affected by mine sludge spill	Alperujo	50 t.ha-1	*M. polymorpha*	NS			NS															NS	Ciadamidaro et al. 2017
			P. annua	NS			NS															1.5-fold	
	Biosolid	50 t.ha-1	*M. polymorpha*	–0.3			NS															NS	
			P. annua	NS			NS															2.2-fold	
Soil affected by mine sludge spill	Alperujo	50 t.ha-1	*M. polymorpha*	+3.2			NS															Allow growth	
			P. annua	+3.2			3.9-fold															Allow growth	
	Biosolid	50 t.ha-1	*M. polymorpha*	+2.9			2.3-fold															Allow growth	
			P. annua	+3.4			2.3-fold															Allow growth	
							3.2-fold																
Soil affected by a mine	Biosolid	30 t.ha-1		+1.5			1.7-fold					NS	NS	NS	NS	NS							Madejón et al. 2006
Ag/Pb mine	Animal manure + vegetable material	5%	*Salix viminalis*	+3.3	4.3-fold	1.5-fold									–90%	18-fold						4.1-fold	Lebrun et al. 2019

Table 19.3 contd. ...

...Table 19.3 contd.

Soil	Compost type	Application rate	Plant species	pH	EC	OM	OC	[C]	[N]	[P]	[K]	[Cd]	[Zn]	[Cu]	[Pb]	[As]	[Al]	[Cr]	[Ni]	[Sb]	Height	Biomass	Reference
Ag/Pb mine	Peat moss, softwood bark and green compost and seaweed	2%	Alder	NS	NS										–92%	2-fold						2.1-fold	Lebrun et al. 2021e
			Birch	+1.8	NS										–97%	2-fold						NS	
Ag/Pb mine	Peat moss, softwood bark and green compost and seaweed	5%	*Phaseolus vulgaris*	+2.6	1.8-fold										–97%		1.6-fold					2-fold	Nandillon et al. 2019a
Ag/Pb mine	Animal manure + vegetable material	5%	*Trifolium repens*	+2.7	1.7-fold										–97%	17-fold						6.5-fold	Nandillon et al. 2019b

Table 19.4. Effect of redmud on soil, metal(loid)s availability and plants. EC = electrical conductivity, OM = organic matter content, OC = organic carbon content, NS = non-significant.

Soil	Time	Application rate	Plant species	pH	EC	OM	OC	N	P	K	Cd	Zn	Cu	Pb	As	Biomass	References
Soil affected by mine activities		2.2 g.kg-1	None	NS			NS					−29%	NS	NS	1.8-fold		Clemente et al. 2019
			S. manrianum	+1.1			3.4-fold					−58%	−82%	−75%	1.6-fold		
			P. miliaceum	+1.4			4.9-fold					−68%	−91%	−90%	1.2-fold		
Mine		2%		+0.2							−19%	−37%		NS	−47%		Feigl et al. 2012
		5%		+0.3							−41%	−62%		NS	−72%		
Mine		4%		+2.9	3-fold		NS	NS			2-fold	2.1-fold		−15%			Garau et al. 2007
Mine		4%		+1.9	1.3-fold		−37%	−27%	−24%		−43%	NS		NS			Garau et al. 2011
Mine		3%		+2.0	1.6-fold		NS	NS			NS	NS	NS	NS	NS		Garau et al. 2014
Agricultural soil next to a mine	40 days	2%		+2.2	2.7-fold						> −99%	−37%		4-fold			Lee et al. 2009
		5%		+3.2	4.2-fold						−54%	−27%		11.8-fold			
Gold mine		2%	Lettuce	+2.9	3-fold						−88%	−98%		−96%	−68%	5.1-fold	Lee et al. 2011
		5%		+4.1	4.2-fold						−98%	−99%		−98%	−29%	2.3-fold	
Pb/Zn mine		2%	*M. sinensis*	+0.5	1.7-fold						−77%		−58%	−70%	−86%	NS	Lee et al. 2014
			P. aquilinum	+0.5	1.6-fold						−79%		−56%	−71%	−87%	NS	
Gold mine	30 days	0.05%		+0.6		NS			1.2-fold	+4%		−20%	1.2-fold		−12%		
	60 days	0.05%	Brachiara													NS	Lopes et al. 2016

Table 19.4 contd. ...

...Table 19.4 contd.

Soil	Time	Application rate	Plant species	pH	EC	OM	OC	N	P	K	Cd	Zn	Cu	Pb	As	Biomass	References
			Crotalaria													NS	
			Stylosanthes													NS	
Ag/Pb mine		2%	*Salix dasyclados*	+2.2										–85%	ns	5.1-fold	Lebrun et al. 2021f
Ag/Pb mine		2%	*Trifolium repens*	+1.6	2.6-fold									–64%	–85%	NS	Lebrun et al. 2021a
		2%		+1.4	2.8-fold									–57%	–87%	1.6-fold	

19.3.4 Amendment Association

As shown in the previous sections, many studies evaluated the effects of biochar, compost, and redmud amendments on the soil characteristics, the metal(loid) mobility and availability, plant growth, and metal(loid)s accumulation. These studies showed that these amendments were effective in ameliorating soil properties and fertility, immobilizing metal(loid)s, and improving plant growth when applied to abandoned mine sites or on soils affected by mining activities. However, in some cases, these amendments were not capable of such positive effects. For instance, biochar is very effective in immobilizing cation metals but not anions like As. It can even induce its mobilization, as shown in the previous section and Table 19.2. Biochar is not a fertilizer *per se*, while compost is very rich in nutrients and capable of immobilizing cations but was shown to mobilize arsenic. Finally, redmud effectively immobilizes both the cations and anions but does not contain nutrients and has a high salt content, which could induce salinity issues. Moreover, the examples cited in the previous sections used the amendments singularly. But, due to the different properties of these amendments and their differing effects on cations and anions as well as soil nutrients, they could be used in combination for better efficiency through additive or synergetic effects (Fischer and Glaser 2012). Such amendment combination in the context of mining soils has been studied in a few studies.

For instance, Beesley et al. (2014) tested a compost and a biochar mixture to a soil collected from a mining area. These amendments applied alone or combined increased As mobility while decreased mobility of Cd, Cu, and Pb. They also increased pH. Based on a measure-monitor-model, the authors concluded that the combination of biochar and compost was more effective than their single-use. Several studies evaluated different combinations of biochar, compost, and redmud for the stabilization of As and Pb on a former silver-lead extraction mine technosol. For instance, in the study of Nandillon et al. (2019a), a phytotoxicity test was performed to assess the effect of biochar, combined with organic amendments. Among the combinations tested, biochar and compost association was tested. These amendments increased soil pH, EC, and dissolved organic carbon content, decreased As and Pb mobility, and improved *Phaseolus vulgaris* growth, demonstrating that they reduced the phytotoxicity of the soil. Moreover, compared to the single amendments, the combination of biochar and compost induced a higher increase in soil pH and decreased Pb mobility, but plant growth was not different between the single and the combined amendments, while the original soil was totally unsuitable for plant growth. Similarly, in another study (Lebrun et al. 2019), the combination of biochar and compost was tested, and the results showed that this combination led to better improvements than the single amendment application on several parameters such as WHC, OM content, soil pH, and Pb immobilization. Following this, the combination of redmud with biochar was also tested. This study showed that soil pH increase was higher in the combined treatment than the single treatment, while Pb immobilization was the same between the single biochar and redmud amendment than their combined application. Similarly, plant dry weight and metal(loid) accumulation were not different in the single amendment than the combined treatment (Lebrun et al. 2021e).

19.4 The Mechanisms Involved in Soil Properties Improvement and Metal(Loid) Immobilization

The previous section showed that biochar, compost, and redmud amendments improved soil properties, and in particular, they reduced soil acidity, immobilized metal(loid)s, and improved nutrient content. All these soil ameliorations had a positive effect on plant growth.

Generally, nutrient and organic matter contents are increased through a direct addition by the amendment, especially compost for nutrients and OM and biochar for OM, although nutrient availability is also governed by soil pH. However, the mechanisms by which the amendments affect soil pH and metal(loid) behavior are more diverse and can be direct or indirect. These mechanisms will be summarized in this section.

Soil pH is one of the most common parameters affected by amendment application, which induces a rise in soil pH in most cases. The first explanation for such a pH increase is the alkalinity of the amendments. Indeed, these three amendments are usually characterized by their alkaline pH (Ippolito et al. 2017, Lopes et al. 2016, Rossini-Oliva et al. 2017). In addition, these amendments contain functional groups (COO- and O-for instance) on their surface, especially biochar, which can consume protons and thus pH increases (Ciadamidaro et al. 2017, Meng et al. 2018). Moreover, biochar and compost contain carbonates, which can dissolve into the soil solution, raising its pH (Alam et al. 2020, Meng et al. 2018). Finally, redmud contains high levels of sodium, which induces alkalinization of the soil (Lopes et al. 2016).

Finally, these three amendments were shown efficient to immobilize metal(loid)s. This immobilization has been attributed to the direct sorption of the metal(loid)s on the amendments and indirect effects through the modification of soil properties.

The sorption capacity of the amendments has been particularly demonstrated in the case of biochar, which can sorb metal(loid)s through five main mechanisms, illustrated in Fig. 19.3 (Ding et al. 2014, 2017, 2016, Li et al. 2017):

(i) Complexation: Biochar has a lot of functional groups on its surface, especially oxygen-containing ones. Metal(loid)s can complex with these functional groups, which induces the liberation of H^+ ions.

(ii) Cation exchange: Cations, such as Ca^{2+}, Na^{+} and K^+, present on the biochar can be released and replaced by metal(loid) ions.

(iii) Precipitation: Instead of replacing the cations on the surface of the biochar, metal(loid)s can also precipitate with them.

(iv) Intra-particle diffusion: Biochar surface is highly porous, and thus metal(loid)s can diffuse inside the pores.

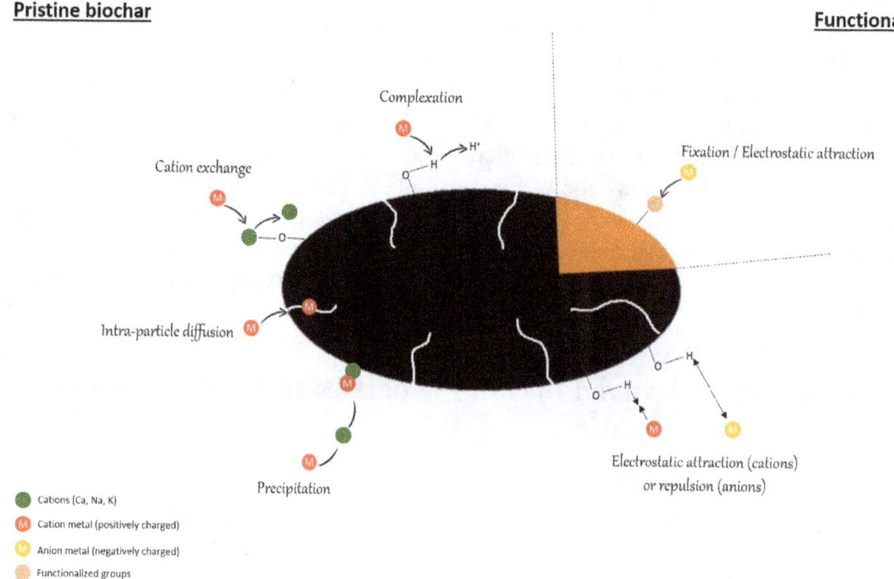

Fig. 19.3. Mechanisms of metal(loid) sorption on biochar surface.

(v) Electrostatic attraction: This attraction occurs between negatively charged functional groups and positively charged metal(loid)s.

These mechanisms have been mostly demonstrated for cation contaminants, especially electrostatic attraction. But, when the pollutant is negatively charged, such as arsenic, repulsion occurs with the biochar functional groups. To overcome this, as discussed previously, the biochar can be functionalized, which modifies its surface and can make it partially positively charged. Consequently, negatively charged metal ions can be fixed on these new surface groups, or they can be electrostatically attracted, increasing the sorption capacity of the biochar towards As.

The sorption capacity of compost and redmud has also been demonstrated (Lebrun et al. 2021e, Soner Altundoğan et al. 2000), although, the precise mechanisms involved need to be further studied.

In addition to direct sorption of metal(loid)s on the amendment surface, metal(loid)s can be immobilized indirectly through the effect of amendments on the soil properties, such as pH and OM content. Soil pH is an important parameter influencing metal(loid) mobility and availability (Forján et al. 2016), and it usually increases following amendment application. In general, cation mobility decreases with pH increase, while it is the reverse for anions. This can also explain why biochar and compost tend to induce mobilization of metal(loid)s like arsenic, while they are efficient for cations (Al-Wabel et al. 2015, Ciadamidaro et al. 2017, Madejón et al. 2006, Meng et al. 2018). The mobilization of As was not observed following the redmud amendment, even though it increased soil pH. Contrary to biochar, redmud has the capacity to sorb arsenic on its surface due to its Fe and Al oxide contents (Castaldi et al. 2010, Derakhshan Nejad et al. 2017). The content of organic matter in soil also has an influence on metal(loid) mobility since it can form complexes with them, leading to either their solubilization or immobilization depending on the state of the organic matter (Clemente et al. 2019, Touceda-González et al. 2017).

19.5 Conclusion

This chapter demonstrated the environmental issues posed by mining extraction activities, which led to colossal metal(loid) pollution levels. Such pollution prevents plant cover establishment. Thus, the derelict mine extraction sites are subjected to wind erosion, water leaching, especially acid mine drainage that solubilizes metal(loid)s and transports them to the surrounding environment (Fig. 19.4). One solution to reduce the negative environmental impact of mining activities is phytoremediation, defined as the use of plants to reduce the toxicity of metal(loid)s. However, due to the extreme soil conditions, i.e. acidic pH, low organic matter and nutrient contents, high metal(loid) concentrations, plant germination, growth, survival, and reproduction are reduced, affecting long term phytoremediation success. To overcome this, amendments can be applied. Biochar, compost, and redmud are amendments that showed positive effects on soil properties, metal(loid)s remediation, and plant growth and development. Therefore, following amendment application, soil properties are improved, and metal(loid)s get immobilized, and thus plants can be established on the site. Such vegetation cover, together with amendments, reduces contamination spreading, protecting the surrounding area (Fig. 19.4). In addition, the development of a plant cover restores biodiversity and soil functionality.

The effects of these three amendments, alone or in combination with others, have been much studied in laboratory and greenhouse conditions. However, field studies need to be implemented on a more regular basis to confirm laboratory results. Moreover, the implementation of field studies will allow the assessment of the long-term effects of amendments on the soil and metal(loid) behavior, which is deeply needed.

Moreover, the impact of amendments on the microorganisms, in terms of activity and diversity, has not been detailed in this chapter, but it is a parameter that deserves further investigation. In addition to plant biomass and metal(loid) accumulation, amendments can also affect plant

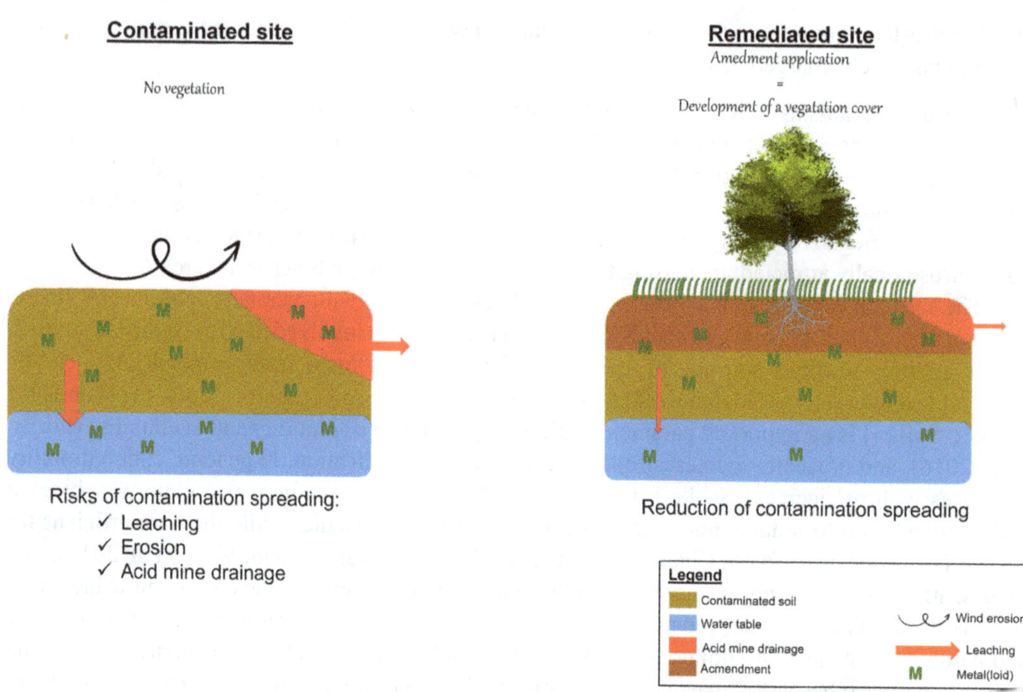

Fig. 19.4. Overview of the phytoremediation process and its benefits.

physiology, such as root exudation, proteomic profiles, and oxidative stress responses, which still lack evaluation.

Finally, the mechanisms involved in biochar metal(loid) immobilization and amendment soil alkalinization have been deeply analyzed. Still, the mechanisms through which amendments affect the other soil properties need to be investigated. Especially, the relation between the amendment and soil properties and their effects on soil and plant have to be assessed, which may help build models for easier selection of amendments depending on soil and contamination types.

References

Alam, M., Z. Hussain, A. Khan, M. A. Khan, A. Rab, M. Asif, M. A. Shah and A. Muhammad. 2020. The effects of organic amendments on heavy metals bioavailability in mine impacted soil and associated human health risk. Sci. Hortic. 262: 109067. https://doi.org/10.1016/j.scienta.2019.109067.

Alhar, M. A. M., D. F. Thompson and I. W. Oliver. 2021. Mine spoil remediation via biochar addition to immobilise potentially toxic elements and promote plant growth for phytostabilisation. J. Environ. Manage. 277: 111500. https://doi.org/10.1016/j.jenvman.2020.111500.

Ali, A., D. Guo, P. G. S. Arockiam Jeyasundar, Y. Li, R. Xiao, J. Du, R. Li and Z. Zhang. 2019. Application of wood biochar in polluted soils stabilized the toxic metals and enhanced wheat (*Triticum aestivum*) growth and soil enzymatic activity. Ecotoxicol. Environ. Saf. 184: 109635. https://doi.org/10.1016/j.ecoenv.2019.109635.

Ali, H., E. Khan and M. A. Sajad. 2013. Phytoremediation of heavy metals—Concepts and applications. Chemosphere 91: 869–881. https://doi.org/10.1016/j.chemosphere.2013.01.075.

Alvarenga, P., A. de Varennes and A. C. Cunha-Queda. 2014. The effect of compost treatments and a plant cover with *Agrostis tenuis* on the immobilization/mobilization of trace elements in a mine-contaminated Soil. Int. J. Phytoremediat. 16: 138–154. https://doi.org/10.1080/15226514.2012.759533.

Al-Wabel, M. I., A. R. A. Usman, A. H. El-Naggar, A. A. Aly, H. M. Ibrahim, S. Elmaghraby and A. Al-Omran. 2015. Conocarpus biochar as a soil amendment for reducing heavy metal availability and uptake by maize plants. Saudi J. Biol. Sci. 22: 503–511. https://doi.org/10.1016/j.sjbs.2014.12.003.

Al-Wabel, M. I., A. R. A. Usman, A. S. Al-Farraj, Y. S. Ok, A. Abduljabbar, A. I. Al-Faraj and A. S. Sallam. 2019. Date palm waste biochars alter a soil respiration, microbial biomass carbon, and heavy metal mobility in contaminated mined soil. Environ. Geochem. Health 41: 1705–1722. https://doi.org/10.1007/s10653-017-9955-0.

Angın, D. 2013. Effect of pyrolysis temperature and heating rate on biochar obtained from pyrolysis of safflower seed press cake. Bioresour. Technol. 128: 593–597. https://doi.org/10.1016/j.biortech.2012.10.150.

Argyraki, A., Z. Boutsi and V. Zotiadis. 2017. Towards sustainable remediation of contaminated soil by using diasporic bauxite: Laboratory experiments on soil from the sulfide mining village of Stratoni, Greece. J. Geochem. Explor. 183: 214–222. https://doi.org/10.1016/j.gexplo.2017.03.007.

Ashraf, Sana, Ali, Q., Z. A. Zahir, S. Ashraf and H. N. Asghar. 2019. Phytoremediation: Environmentally sustainable way for reclamation of heavy metal polluted soils. Ecotoxicol. Environ. Saf. 174: 714–727. https://doi.org/10.1016/j.ecoenv.2019.02.068.

Awa, S. H. and T. Hadibarata. 2020. Removal of heavy metals in contaminated soil by phytoremediation mechanism: a review. Water. Air. Soil Pollut. 231: 47. https://doi.org/10.1007/s11270-020-4426-0.

Banerjee, R., P. Goswami, S. Lavania, A. Mukherjee and U. C. Lavania. 2019. Vetiver grass is a potential candidate for phytoremediation of iron ore mine spoil dumps. Ecol. Eng. 132: 120–136. https://doi.org/10.1016/j.ecoleng.2018.10.012.

Beesley, L., M. Marmiroli, L. Pagano, V. Pigoni, G. Fellet, T. Fresno, T. Vamerali, M. Bandiera and N. Marmiroli. 2013. Biochar addition to an arsenic contaminated soil increases arsenic concentrations in the pore water but reduces uptake to tomato plants (*Solanum lycopersicum* L.). Sci. Total Environ. 454–455: 598–603. https://doi.org/10.1016/j.scitotenv.2013.02.047.

Beesley, L., O. S. Inneh, G. J. Norton, E. Moreno-Jimenez, T. Pardo, R. Clemente J. J. C. Dawson. 2014. Assessing the influence of compost and biochar amendments on the mobility and toxicity of metals and arsenic in a naturally contaminated mine soil. Environ. Pollut. 186: 195–202. https://doi.org/10.1016/j.envpol.2013.11.026.

Bhatnagar, A., V. J. P. Vilar, C. M. S. Botelho and R. A. R. Boaventura. 2011. A review of the use of red mud as adsorbent for the removal of toxic pollutants from water and wastewater. Environ. Technol. 32: 231–249. https://doi.org/10.1080/09593330.2011.560615.

Castaldi, P., M. Silvetti, S. Enzo and P. Melis. 2010. Study of sorption processes and FT-IR analysis of arsenate sorbed onto red muds (a bauxite ore processing waste). J. Hazard. Mater. 175: 172–178. https://doi.org/10.1016/j.jhazmat.2009.09.145.

Ciadamidaro, L., M. Puschenreiter, J. Santner, W. W. Wenzel, P. Madejón and E. Madejón. 2017. Assessment of trace element phytoavailability in compost amended soils using different methodologies. J. Soils Sediments 17: 1251–1261. https://doi.org/10.1007/s11368-015-1283-3.

Clemente, R., E. Arco-Lázaro, T. Pardo, I. Martín, A. Sánchez-Guerrero, F. Sevilla and M. P. Bernal. 2019. Combination of soil organic and inorganic amendments helps plants overcome trace element induced oxidative stress and allows phytostabilisation. Chemosphere 223: 223–231. https://doi.org/10.1016/j.chemosphere.2019.02.056.

Dai, S., H. Li, Z. Yang, M. Dai, X. Dong, X. Ge, M. Sun and L. Shi. 2018. Effects of biochar amendments on speciation and bioavailability of heavy metals in coal-mine-contaminated soil. Hum. Ecol. Risk Assess. Int. J. 24: 1887–1900. https://doi.org/10.1080/10807039.2018.1429250.

Dai, Z., X. Zhang, C. Tang, N. Muhammad, J. Wu, P. C. Brookes and J. Xu. 2017. Potential role of biochars in decreasing soil acidification - A critical review. Sci. Total Environ. 581-582: 601–611. https://doi.org/10.1016/j.scitotenv.2016.12.169.

Derakhshan Nejad, Z., J. W. Kim and M. C. Jung. 2017. Reclamation of arsenic contaminated soils around mining site using solidification/stabilization combined with revegetation. Geosci. J. 21: 385–396. https://doi.org/10.1007/s12303-016-0059-0.

Derakhshan Nejad, Z., S. Rezania, M. C. Jung, A. A. Al-Ghamdi, A. E.-Z. M. A. Mustafa and M. S. Elshikh. 2021. Effects of fine fractions of soil organic, semi-organic, and inorganic amendments on the mitigation of heavy metal(loid)s leaching and bioavailability in a post-mining area. Chemosphere 271: 129538. https://doi.org/10.1016/j.chemosphere.2021.129538.

Diacono, M. and F. Montemurro. 2010. Long-term effects of organic amendments on soil fertility. A review. Agron. Sustain. Dev. 30: 401–422. https://doi.org/10.1051/agro/2009040.

Ding, W., X. Dong, I. M. Ime, B. Gao and L. Q. Ma. 2014. Pyrolytic temperatures impact lead sorption mechanisms by bagasse biochars. Chemosphere 105: 68–74. https://doi.org/10.1016/j.chemosphere.2013.12.042.

Ding, Y., Y. Liu, S. Liu, Z. Li, X. Tan, X. Huang, G. Zeng, Y. Zhou, B. Zheng and X. Cai. 2016. Competitive removal of Cd(II) and Pb(II) by biochars produced from water hyacinths: performance and mechanism. RSC Adv. 6: 5223–5232. https://doi.org/10.1039/C5RA26248H.

Ding, Y., Y. Liu, S. Liu, X. Huang, Z. Li, X. Tan, G. Zeng and L. Zhou. 2017. Potential benefits of biochar in agricultural soils: a review. Pedosphere 27: 645–661. https://doi.org/10.1016/S1002-0160(17)60375-8.

El-Naggar, A., S. S. Lee, Y. M. Awad, X. Yang, C. Ryu, M. Rizwan, J. Rinklebe, D. C. W. Tsang and Y. S. Ok. 2018. Influence of soil properties and feedstocks on biochar potential for carbon mineralization and improvement of infertile soils. Geoderma 332: 100–108. https://doi.org/10.1016/j.geoderma.2018.06.017.

El-Naggar, A., M. H. Lee, J. Hur, Y. H. Lee, A. D. Igalavithana, S. M. Shaheen, C. Ryu, J. Rinklebe, D. C. W. Tsang and Y. S. Ok. 2020. Biochar-induced metal immobilization and soil biogeochemical process: An integrated mechanistic approach. Sci. Total Environ. 698: 134112. https://doi.org/10.1016/j.scitotenv.2019.134112.

Feigl, V., A. Anton, N. Uzigner and K. Gruiz. 2012. Red mud as a chemical stabilizer for soil contaminated with toxic metals. Water. Air. Soil Pollut. 223: 1237–1247. https://doi.org/10.1007/s11270-011-0940-4.

Fellet, G., L. Marchiol, G. Delle Vedove and A. Peressotti. 2011. Application of biochar on mine tailings: Effects and perspectives for land reclamation. Chemosphere 83: 1262–1267. https://doi.org/10.1016/j.chemosphere.2011.03.053.

Fellet, G., M. Marmiroli and L. Marchiol. 2014. Elements uptake by metal accumulator species grown on mine tailings amended with three types of biochar. Sci. Total Environ. 468–469: 598–608. https://doi.org/10.1016/j.scitotenv.2013.08.072.

Fischer, D. and B. Glaser. 2012. Synergisms between compost and biochar for sustainable soil amelioration. *In*: Kumar, S. (ed.). Management of Organic Waste. InTech. https://doi.org/10.5772/31200.

Forján, R., V. Asensio, A. Rodríguez-Vila and E. F. Covelo. 2016. Contribution of waste and biochar amendment to the sorption of metals in a copper mine tailing. CATENA 137: 120–125. https://doi.org/10.1016/j.catena.2015.09.010.

Garau, G., P. Castaldi, L. Santona, P. Deiana and P. Melis. 2007. Influence of red mud, zeolite and lime on heavy metal immobilization, culturable heterotrophic microbial populations and enzyme activities in a contaminated soil. Geoderma 142: 47–57. https://doi.org/10.1016/j.geoderma.2007.07.011.

Garau, G., M. Silvetti, S. Deiana, P. Deiana and P. Castaldi. 2011. Long-term influence of red mud on As mobility and soil physico-chemical and microbial parameters in a polluted sub-acidic soil. J. Hazard. Mater. 185: 1241–1248. https://doi.org/10.1016/j.jhazmat.2010.10.037.

Garau, G., M. Silvetti, P. Castaldi, E. Mele, P. Deiana and S. Deiana. 2014. Stabilising metal(loid)s in soil with iron and aluminium-based products: Microbial, biochemical and plant growth impact. J. Environ. Manage. 139: 146–153. https://doi.org/10.1016/j.jenvman.2014.02.024.

Gil-Loaiza, J., S. A. White, R. A. Root, F. A. Solís-Dominguez, C. M. Hammond, J. Chorover and R. M. Maier. 2016. Phytostabilization of mine tailings using compost-assisted direct planting: Translating greenhouse results to the field. Sci. Total Environ. 565: 451–461. https://doi.org/10.1016/j.scitotenv.2016.04.168.

He, E., Y. Yang, Z. Xu, H. Qiu, F. Yang, W. J. G. M. Peijnenburg, W. Zhang, R. Qiu and S. Wang. 2019. Two years of aging influences the distribution and lability of metal(loid)s in a contaminated soil amended with different biochars. Sci. Total Environ. 673: 245–253. https://doi.org/10.1016/j.scitotenv.2019.04.037.

Higashikawa, F. S., R. F. Conz, M. Colzato, C. E. P. Cerri and L. R. F. Alleoni. 2016. Effects of feedstock type and slow pyrolysis temperature in the production of biochars on the removal of cadmium and nickel from water. J. Clean. Prod. 137: 965–972. https://doi.org/10.1016/j.jclepro.2016.07.205.

Hua, Y., K. V. Heal and W. Friesl-Hanl. 2017. The use of red mud as an immobiliser for metal/metalloid-contaminated soil: A review. J. Hazard. Mater. 325: 17–30. https://doi.org/10.1016/j.jhazmat.2016.11.073.

Huang, D., L. Liu, G. Zeng, P. Xu, C. Huang, L. Deng, R. Wang and J. Wan. 2017. The effects of rice straw biochar on indigenous microbial community and enzymes activity in heavy metal-contaminated sediment. Chemosphere 174: 545–553. https://doi.org/10.1016/j.chemosphere.2017.01.130.

Huang, L., Y. Li, M. Zhao, Y. Chao, R. Qiu, Y. Yang and S. Wang. 2018. Potential of *Cassia alata* L. coupled with biochar for heavy metal stabilization in multi-metal mine tailings. Int. J. Environ. Res. Public. Health 15: 494. https://doi.org/10.3390/ijerph15030494.

Huang, M., Y. Zhu, Z. Li, B. Huang, N. Luo, C. Liu and G. Zeng. 2016. Compost as a soil amendment to remediate heavy metal-contaminated agricultural soil: mechanisms, efficacy, problems, and strategies. Water. Air. Soil Pollut. 227: 359. https://doi.org/10.1007/s11270-016-3068-8.

Ippolito, J. A., C. M. Berry, D. G. Strawn, J. M. Novak, J. Levine and A. Harley. 2017. Biochars reduce mine land soil bioavailable metals. J. Environ. Qual. 46: 411–419. https://doi.org/10.2134/jeq2016.10.0388.

Jain, S., B. P. Baruah and P. Khare. 2014. Kinetic leaching of high sulphur mine rejects amended with biochar: Buffering implication. Ecol. Eng. 71: 703–709. https://doi.org/10.1016/j.ecoleng.2014.08.003.

Jindo, K., H. Mizumoto, Y. Sawada, M. A. Sanchez-Monedero and T. Sonoki. 2014. Physical and chemical characterization of biochars derived from different agricultural residues. Biogeosciences 11: 6613–6621. https://doi.org/10.5194/bg-11-6613-2014.

Karaca, O., C. Cameselle and K. R. Reddy. 2018. Mine tailing disposal sites: contamination problems, remedial options and phytocaps for sustainable remediation. Rev. Environ. Sci. Biotechnol. 17: 205–228. https://doi.org/10.1007/s11157-017-9453-y.

Kelly, C. N., C. D. Peltz, M. Stanton, D. W. Rutherford and C. E. Rostad. 2014. Biochar application to hardrock mine tailings: Soil quality, microbial activity, and toxic element sorption. Appl. Geochem. 43: 35–48. https://doi.org/10.1016/j.apgeochem.2014.02.003.

Khairul, M. A., J. Zanganeh and B. Moghtaderi. 2019. The composition, recycling and utilisation of Bayer red mud. Resour. Conserv. Recycl. 141: 483–498. https://doi.org/10.1016/j.resconrec.2018.11.006.

Kwak, J.-H., M. S. Islam, S. Wang, S. A. Messele, M. A. Naeth, M. G. El-Din and S. X. Chang. 2019. Biochar properties and lead(II) adsorption capacity depend on feedstock type, pyrolysis temperature, and steam activation. Chemosphere 231: 393–404. https://doi.org/10.1016/j.chemosphere.2019.05.128.

Lebrun, M., C. Macri, F. Miard, N. Hattab-Hambli, M. Motelica-Heino, D. Morabito and S. Bourgerie. 2017. Effect of biochar amendments on As and Pb mobility and phytoavailability in contaminated mine technosols phytoremediated by Salix. J. Geochem. Explor. 182: 149–156. https://doi.org/10.1016/j.gexplo.2016.11.016.

Lebrun, M., F. Miard, R. Nandillon, N. Hattab-Hambli, G. S. Scippa, S. Bourgerie and D. Morabito. 2018a. Eco-restoration of a mine technosol according to biochar particle size and dose application: study of soil physico-chemical properties and phytostabilization capacities of Salix viminalis. J. Soils Sediments 18: 2188–2202. https://doi.org/10.1007/s11368-017-1763-8.

Lebrun, M., F. Miard, R. Nandillon, J. C. Léger, N. Hattab-Hambli, G. S. Scippa, S. Bourgerie and D. Morabito. 2018b. Assisted phytostabilization of a multicontaminated mine technosol using biochar amendment: Early stage evaluation of biochar feedstock and particle size effects on As and Pb accumulation of two Salicaceae species (Salix viminalis and Populus euramericana). Chemosphere 194: 316–326. https://doi.org/10.1016/j.chemosphere.2017.11.113.

Lebrun, M., F. Miard, S. Renouard, R. Nandillon, G. S. Scippa, D. Morabito and S. Bourgerie. 2018c. Effect of Fe-functionalized biochar on toxicity of a technosol contaminated by Pb and As: sorption and phytotoxicity tests. Environ. Sci. Pollut. Res. 25: 33678–33690. https://doi.org/10.1007/s11356-018-3247-9.

Lebrun, M., F. Miard, R. Nandillon, G. S. Scippa, S. Bourgerie and D. Morabito. 2019. Biochar effect associated with compost and iron to promote Pb and As soil stabilization and *Salix viminalis* L. growth. Chemosphere 222: 810–822. https://doi.org/10.1016/j.chemosphere.2019.01.188.

Lebrun, M., F. Miard, N. Hattab-Hambli, G. S. Scippa, S. Bourgerie and D. Morabito. 2020. Effect of different tissue biochar amendments on As and Pb stabilization and phytoavailability in a contaminated mine technosol. Sci. Total Environ. 707: 135657. https://doi.org/10.1016/j.scitotenv.2019.135657.

Lebrun, M., F. Miard, R. Nandillon, N. Hattab-Hambli, J. C. Léger, G. S. Scippa, D. Morabito and S. Bourgerie. 2021. Influence of biochar particle size and concentration on Pb and As availability in contaminated mining soil and phytoremediation potential of poplar assessed in a mesocosm experiment. Water. Air. Soil Pollut. 232: 3. https://doi.org/10.1007/s11270-020-04942-y.

Lebrun, M., S. Bourgerie and D. Morabito. 2021a. Effects of different biochars, activated carbons and redmuds on the growth of *Trifolium repens* and As and Pb stabilization in a former mine technosol. Bull. Environ. Contam. Toxicol. https://doi.org/10.1007/s00128-021-03271-y.

Lebrun, M., F. Miard, S. Drouet, D. Tungmunnithum, D. Morabito, C. Hano and S. Bourgerie. 2021b. Physiological and molecular responses of flax (*Linum usitatissimum* L.) cultivars under a multicontaminated technosol amended with biochar. Environ. Sci. Pollut. Res. https://doi.org/10.1007/s11356-021-14563-5.

Lebrun, M., F. Miard, R. Nandillon, D. Morabito and S. Bourgerie. 2021c. Effect of biochar, iron sulfate and poultry manure application on the phytotoxicity of a former tin mine. Int. J. Phytoremediation 1–9. https://doi.org/10.1080/15226514.2021.1889964.

Lebrun, M., R. Nandillon, F. Miard, L. Le Forestier, D. Morabito and S. Bourgerie. 2021d. Effects of biochar, ochre and manure amendments associated with a metallicolous ecotype of *Agrostis capillaris* on As and Pb stabilization of a former mine technosol. Environ. Geochem. Health 43: 1491–1505. https://doi.org/10.1007/s10653-020-00592-5.

Lebrun, M., R. Nandillon, F. Miard, G. S. Scippa, S. Bourgerie and D. Morabito. 2021e. Application of amendments for the phytoremediation of a former mine technosol by endemic pioneer species: alder and birch seedlings. Environ. Geochem. Health 43: 77–89. https://doi.org/10.1007/s10653-020-00678-0.

Lebrun, M., R. Van Poucke, F. Miard, G. S. Scippa, S. Bourgerie, D. Morabito and F. M. G. Tack. 2021f. Effects of carbon-based materials and redmuds on metal(loid) immobilization and growth of *Salix dasyclados* Wimm. on a former mine Technosol contaminated by arsenic and lead. Land Degrad. Dev. 32: 467–481. https://doi.org/10.1002/ldr.3726.

Lebrun, M., F. Miard, R. Nandillon, N. Hattab-Hambli, J. C. Léger, G. S. Scippa, D. Morabito and S. Bourgerie. 2021g. Influence of biochar particle size and concentration on Pb and As availability in contaminated mining soil and phytoremediation potential of poplar assessed in a Mesocosm Experiment. Water, Air, & Soil Pollution, 232(1): 1–21. https://doi.org/10.1007/s11270-020-04942-y.

Lee, S.-H., J. S. Lee, Y. Jeong Choi and J. G. Kim. 2009. *In situ* stabilization of cadmium-, lead-, and zinc-contaminated soil using various amendments. Chemosphere 77: 1069–1075. https://doi.org/10.1016/j.chemosphere.2009.08.056.

Lee, S.-H., E. Y. Kim, H. Park, J. Yun and J. G. Kim. 2011. *In situ* stabilization of arsenic and metal-contaminated agricultural soil using industrial by-products. Geoderma 161: 1–7. https://doi.org/10.1016/j.geoderma.2010.11.008.

Lee, S.-H., W. Ji, W. S. Lee, N. Koo, I. H. Koh, M. S. Kim and J. S. Park. 2014. Influence of amendments and aided phytostabilization on metal availability and mobility in Pb/Zn mine tailings. J. Environ. Manage. 139: 15–21. https://doi.org/10.1016/j.jenvman.2014.02.019.

Li, H., X. Dong, E. B. da Silva, L. M. de Oliveira, Y. Chen and L. Q. Ma. 2017. Mechanisms of metal sorption by biochars: Biochar characteristics and modifications. Chemosphere 178: 466–478. https://doi.org/10.1016/j.chemosphere.2017.03.072.

Li, H., H. Xu, S. Zhou, Y. Yu, H. Li, C. Zhou, Y. Chen, Y. Li, M. Wang and G. Wang. 2018a. Distribution and transformation of lead in rice plants grown in contaminated soil amended with biochar and lime. Ecotoxicol. Environ. Saf. 165: 589–596. https://doi.org/10.1016/j.ecoenv.2018.09.039.

Li, L., C. Zhu, X. Liu, F. Li, H. Li and J. Ye. 2018b. Biochar amendment immobilizes arsenic in farmland and reduces its bioavailability. Environ. Sci. Pollut. Res. 25: 34091–34102. https://doi.org/10.1007/s11356-018-3021-z.

Liu, L., W. Li, W. Song and M. Guo. 2018. Remediation techniques for heavy metal-contaminated soils: Principles and applicability. Sci. Total Environ. 633: 206–219. https://doi.org/10.1016/j.scitotenv.2018.03.161.

Liu, Y., R. Naidu and H. Ming. 2011. Red mud as an amendment for pollutants in solid and liquid phases. Geoderma 163: 1–12. https://doi.org/10.1016/j.geoderma.2011.04.002.

Lomaglio, T., N. Hattab-Hambli, A. Bret, F. Miard, D. Trupiano, G. S. Scippa, M. Motelica-Heino, S. Bourgerie and D. Morabito. 2017. Effect of biochar amendments on the mobility and (bio) availability of As, Sb and Pb in a contaminated mine technosol. J. Geochem. Explor. 182: 138–148. https://doi.org/10.1016/j.gexplo.2016.08.007.

Lopes, G., P. A. A. Ferreira, F. G. Pereira, N. Curi, W. M. Rangel and L. R. G. Guilherme. 2016. Beneficial use of industrial by-products for phytoremediation of an arsenic-rich soil from a gold mining area. Int. J. Phytoremediation 18: 777–784. https://doi.org/10.1080/15226514.2015.1131240.

Lu, K., X. Yang, G. Gielen, N. Bolan, Y. S. Ok, N. K. Niazi, S. Xu, G. Yuan, X. Chen, X. Zhang, D. Liu, Z. Song, X. Liu and H. Wang. 2017. Effect of bamboo and rice straw biochars on the mobility and redistribution of heavy metals (Cd, Cu, Pb and Zn) in contaminated soil. J. Environ. Manage. 186: 285–292. https://doi.org/10.1016/j.jenvman.2016.05.068.

Madejón, E., A. P. de Mora, E. Felipe, P. Burgos and F. Cabrera. 2006. Soil amendments reduce trace element solubility in a contaminated soil and allow regrowth of natural vegetation. Environ. Pollut. 139: 40–52. https://doi.org/10.1016/j.envpol.2005.04.034.

Madejón, P., J. Xiong, F. Cabrera and E. Madejón. 2014. Quality of trace element contaminated soils amended with compost under fast growing tree Paulownia fortunei plantation. J. Environ. Manage. 144: 176–185. https://doi.org/10.1016/j.jenvman.2014.05.020.

Meng, J., M. Tao, L. Wang, X. Liu and J. Xu. 2018. Changes in heavy metal bioavailability and speciation from a Pb-Zn mining soil amended with biochars from co-pyrolysis of rice straw and swine manure. Sci. Total Environ. 633: 300–307. https://doi.org/10.1016/j.scitotenv.2018.03.199.

Montiel-Rozas, M. M., E. Madejón and P. Madejón. 2015. Evaluation of phytostabilizer ability of three ruderal plants in mining soils restored by application of organic amendments. Ecol. Eng. 83: 431–436. https://doi.org/10.1016/j.ecoleng.2015.04.096.

Nandillon, R., F. Miard, M. Lebrun, M. Gaillard, S. Sabatier, S. Bourgerie, F. Battaglia-Brunet and D. Morabito. 2019a. Effect of biochar and amendments on Pb and As phytotoxicity and phytoavailability in a technosol. CLEAN - Soil Air Water 47: 1800220. https://doi.org/10.1002/clen.201800220.

Nandillon, R., O. Lahwegue, F. Miard, M. Lebrun, M. Gaillard, S. Sabatier, F. Battaglia-Brunet, D. Morabito and S. Bourgerie. 2019b. Potential use of biochar, compost and iron grit associated with Trifolium repens to stabilize Pb and As on a multi-contaminated technosol. Ecotoxicol. Environ. Saf. 182: 109432. https://doi.org/10.1016/j.ecoenv.2019.109432.

Nandillon, R., M. Lebrun, F. Miard, M. Gaillard, S. Sabatier, M. Villar, S. Bourgerie and D. Morabito. 2019c. Capability of amendments (biochar, compost and garden soil) added to a mining technosol contaminated by Pb and As to allow poplar seed (*Populus nigra* L.) germination. Environ. Monit. Assess. 191: 465. https://doi.org/10.1007/s10661-019-7561-6.

Norini, M.-P., H. Thouin, F. Miard, F. Battaglia-Brunet, P. Gautret, R. Guégan, L. Le Forestier, D. Morabito, S. Bourgerie and M. Motelica-Heino. 2019. Mobility of Pb, Zn, Ba, As and Cd toward soil pore water and plants (willow and ryegrass) from a mine soil amended with biochar. J. Environ. Manage. 232: 117–130. https://doi.org/10.1016/j.jenvman.2018.11.021.

Novak, J. M., J. A. Ippolito, T. F. Ducey, D. W. Watts, K. A. Spokas, K. M. Trippe, G. C. Sigua and M. G. Johnson. 2018. Remediation of an acidic mine spoil: Miscanthus biochar and lime amendment affects metal availability, plant growth, and soil enzyme activity. Chemosphere 205: 709–718. https://doi.org/10.1016/j.chemosphere.2018.04.107.

Patra, D. K., C. Pradhan, J. Kumar and H. K. Patra. 2020. Assessment of chromium phytotoxicity, phytoremediation and tolerance potential of Sesbania sesban and Brachiaria mutica grown on chromite mine overburden dumps and garden soil. Chemosphere 252: 126553. https://doi.org/10.1016/j.chemosphere.2020.126553.

Puga, A. P., C. A. Abreu, L. C. A. Melo and L. Beesley. 2015. Biochar application to a contaminated soil reduces the availability and plant uptake of zinc, lead and cadmium. J. Environ. Manage. 159: 86–93. https://doi.org/10.1016/j.jenvman.2015.05.036.

Rajapaksha, A. U., S. S. Chen, D. C. W. Tsang, M. Zhang, M. Vithanage, S. Mandal, B. Gao, N. S. Bolan and Y. S. Ok. 2016. Engineered/designer biochar for contaminant removal/immobilization from soil and water: Potential and implication of biochar modification. Chemosphere 148: 276–291. https://doi.org/10.1016/j.chemosphere.2016.01.043.

Rehman, M., L. Liu, S. Bashir, M. H. Saleem, C. Chen, D. Peng and K. H. M. Siddique. 2019. Influence of rice straw biochar on growth, antioxidant capacity and copper uptake in ramie (*Boehmeria nivea* L.) grown as forage in aged copper-contaminated soil. Plant Physiol. Biochem. 138: 121–129. https://doi.org/10.1016/j.plaphy.2019.02.021.

Rizwan, M., S. Ali, F. Abbas, M. Adrees, M. Zia-ur-Rehman, M. Farid, R. A. Gill and B. Ali. 2017. Role of organic and inorganic amendments in alleviating heavy metal stress in oilseed crops. *In*: Ahmad, P. (ed.). Oilseed Crops. John Wiley & Sons, Ltd, Chichester, UK, pp. 224–235. https://doi.org/10.1002/9781119048800.ch12.

Rodríguez-Galán, M. 2019. Remediation of acid mine drainage. Environ. Chem. Lett. 10.

Rossini-Oliva, S., M. D. Mingorance and A. Peña. 2017. Effect of two different composts on soil quality and on the growth of various plant species in a polymetallic acidic mine soil. Chemosphere 168: 183–190. https://doi.org/10.1016/j.chemosphere.2016.10.040.

Samsuri, A. W., F. Sadegh-Zadeh and B. J. Seh-Bardan. 2013. Adsorption of As(III) and As(V) by Fe coated biochars and biochars produced from empty fruit bunch and rice husk. J. Environ. Chem. Eng. 1: 981–988. https://doi.org/10.1016/j.jece.2013.08.009.

Simate, G. S. 2014. Acid mine drainage: Challenges and opportunities. J. Environ. Chem. Eng. 19.

Singh, B., B. P. Singh and A. L. Cowie. 2010. Characterisation and evaluation of biochars for their application as a soil amendment. Soil Res. 48: 516. https://doi.org/10.1071/SR10058.

Skousen, J. G. 2019. Acid mine drainage formation, control and treatment_ Approaches and strategies. Extr. Ind. Soc. 9.

Soner Altundoğan, H., S. Altundoğan, F. Tümen and M. Bildik. 2000. Arsenic removal from aqueous solutions by adsorption on red mud. Waste Manag. 20: 761–767. https://doi.org/10.1016/S0956-053X(00)00031-3.

Stella Mary, G., P. Sugumaran, S. Niveditha, B. Ramalakshmi, P. Ravichandran and S. Seshadri. 2016. Production, characterization and evaluation of biochar from pod (Pisum sativum), leaf (Brassica oleracea) and peel (Citrus sinensis) wastes. Int. J. Recycl. Org. Waste Agric. 5: 43–53. https://doi.org/10.1007/s40093-016-0116-8.

Sun, Y., B. Gao, Y. Yao, J. Fang, M. Zhang, Y. Zhou, H. Chen and L. Yang. 2014. Effects of feedstock type, production method, and pyrolysis temperature on biochar and hydrochar properties. Chem. Eng. J. 240: 574–578. https://doi.org/10.1016/j.cej.2013.10.081.

Tack, F. M. G. and E. Meers. 2010. Assisted Phytoextraction: Helping Plants to Help Us. Elements 6: 383–388. https://doi.org/10.2113/gselements.6.6.383.

Tack, F. M. G. and C. E. Egene. 2019. Potential of biochar for managing metal contaminated areas, in synergy with phytomanagement or other management options. pp. 91–111. *In*: Yong, O. k., Daniel Tsang, Nanthi Bolan and Jeffrey Novak (eds.). Biochar from Biomass and Waste. Elsevier. https://doi.org/10.1016/B978-0-12-811729-3.00006-6.

Tan, X., S. B. Liu, Y. Liu, Y. Gu, G. Zeng, X. Hu, X. Wang, S. H. Liu and L. Jiang. 2017b. Biochar as potential sustainable precursors for activated carbon production: Multiple applications in environmental protection and energy storage. Bioresour. Technol. 227: 359–372. https://doi.org/10.1016/j.biortech.2016.12.083.

Tan, Z., C. S. K. Lin, X. Ji and T. J. Rainey. 2017a. Returning biochar to fields: A review. Appl. Soil Ecol. 116: 1–11. https://doi.org/10.1016/j.apsoil.2017.03.017.

Touceda-González, M., V. Álvarez-López, Á. Prieto-Fernández, B. Rodríguez-Garrido, C. Trasar-Cepeda, M. Mench, M. Puschenreiter, C. Quintela-Sabarís, F. Macías-García and P. S. Kidd. 2017. Aided phytostabilisation reduces metal toxicity, improves soil fertility and enhances microbial activity in Cu-rich mine tailings. J. Environ. Manage. 186: 301–313. https://doi.org/10.1016/j.jenvman.2016.09.019.

Worlanyo, A. S. and L. Jiangfeng. 2021. Evaluating the environmental and economic impact of mining for post-mined land restoration and land-use: A review. J. Environ. Manage. 279: 111623. https://doi.org/10.1016/j.jenvman.2020.111623.

Wu, W., J. Li, N. K. Niazi, K. Müller, Y. Chu, L. Zhang, G. Yuan, K. Lu, Z. Song and H. Wang. 2016. Influence of pyrolysis temperature on lead immobilization by chemically modified coconut fiber-derived biochars in aqueous environments. Environ. Sci. Pollut. Res. 23: 22890–22896. https://doi.org/10.1007/s11356-016-7428-0.

Xia, D., F. Tan, C. Zhang, X. Jiang, Z. Chen, H. Li, Y. Zheng, Q. Li and Y. Wang. 2016. ZnCl 2 -activated biochar from biogas residue facilitates aqueous As(III) removal. Appl. Surf. Sci. 377: 361–369. https://doi.org/10.1016/j.apsusc.2016.03.109.

Yao, Y., B. Gao, M. Zhang, M. Inyang and A. R. Zimmerman. 2012. Effect of biochar amendment on sorption and leaching of nitrate, ammonium, and phosphate in a sandy soil. Chemosphere 89: 1467–1471. https://doi.org/10.1016/j.chemosphere.2012.06.002.

Yu, H., W. Zou, J. Chen, H. Chen, Z. Yu, J. Huang, H. Tang, X. Wei and B. Gao. 2019. Biochar amendment improves crop production in problem soils: A review. J. Environ. Manage. 232: 8–21. https://doi.org/10.1016/j.jenvman.2018.10.117.

Yuan, J.-H., R. K. Xu and H. Zhang. 2011. The forms of alkalis in the biochar produced from crop residues at different temperatures. Bioresour. Technol. 102: 3488–3497. https://doi.org/10.1016/j.biortech.2010.11.018.

Yuan, P., J. Wang, Y. Pan, B. Shen and C. Wu. 2019. Review of biochar for the management of contaminated soil: Preparation, application and prospect. Sci. Total Environ. 659: 473–490. https://doi.org/10.1016/j.scitotenv.2018.12.400.

Zhang, J., Q. Chen and C. You. 2016. Biochar effect on water evaporation and hydraulic conductivity in sandy soil. Pedosphere 26: 265–272. https://doi.org/10.1016/S1002-0160(15)60041-8.

Zhou, Y., B. Gao, A. R. Zimmerman, J. Fang, Y. Sun and X. Cao. 2013. Sorption of heavy metals on chitosan-modified biochars and its biological effects. Chem. Eng. J. 231: 512–518. https://doi.org/10.1016/j.cej.2013.07.036.

CHAPTER 20

Bioremediation of Mining Waste and Other Copper-containing Effluents by Biosorption

Javier I. Ordóñez,[1,*] *Ana Mercado*,[2] *Liey-si Wong-Pinto*[1] and *Sonia I. Cortés*[1]

20.1 Introduction

In the first part, this chapter presents a brief revision about the processes to obtain primary metal resources, focusing on the massive wastes generated and the environmental concern related to acid mine drainage (AMD) and heavy metal spread. In that context, the main methods to mitigate and treat AMD are commented, and subsequently, the bioremediation as an alternative to passively treat effluents loaded with heavy metals is discussed. In terms of the mechanisms involved in the bioremediation techniques, biosorption is highlighted. The last part of the chapter highlights a review of cases where bioremediation has been validated to treat actual (and complex) solutions for copper removal, unlike most of the available investigations, which demonstrate the capability to recover/remove metals with biomasses from synthetic or pure metal solutions.

20.2 Mineral Processing and Mining Waste

20.2.1 Copper and Mineral Processes

Worldwide, copper occurs in many mineralogical forms and may be classified as sulfide, oxidized, and native. Approximately 80% of the copper produced from minerals is obtained from sulfides, with chalcopyrite ($CuFeS_2$) being the one containing about 50% of all copper deposits. Other secondary sulfide copper minerals include bornite (Cu_5FeS_4), enargite (Cu_3AsS_4), covellite (CuS), and chalcocite (Cu_2S). On the other hand, the oxide ores are as cuprite (Cu_2O), brochantite $Cu_4SO_4(OH)_6$, atacamite ($Cu_2Cl(OH)_3$), malachite ($Cu_2CO_3(OH)_2$), and azurite ($Cu_3(CO_3)_2(OH)$). Copper from sulfide deposits is concentrated using flotation, whilst native and oxide ores are treated by hydrometallurgical processes. The remaining copper supply comes from recycling scrap (around 10%) (Gentina and Acevedo 2016).

[1] Department of Chemical Engineering and Mineral Processes, Universidad de Antofagasta, Antofagasta, Chile.
[2] Department of Biotechnology, Universidad de Antofagasta, Antofagasta, Chile.
[*] Corresponding author: javier.ordonez@uantof.cl

Copper has an indispensable role in current technologies, being important in the generation, transmission, and electricity distribution, which are increasingly necessary for developing an economy based on a low-carbon energy system. It is estimated that the energy transition, from fossil to renewable sources, will increase the copper demand about three times in the period 2000–2050 for implementing new electricity technologies in fields of generation and transportation (de Koning et al. 2018, Henckens and Worrell 2020).

There are currently around 250 copper mines in operation in nearly 40 countries, with a global production estimated at 20 million tons in 2020, which is 30% higher than a decade ago (US Geological Survey (USGS) 2021). The forecast for 2050 is that the global market will require an additional amount of copper equivalent to between two and ten years of the current global copper production (Hertwich et al. 2015, Vidal et al. 2018).

The main copper-producing countries in 2017 were Chile, Peru, and China, and the worldwide reserves stand at around 870 million tons, being a quarter located in Chile. Mining operations in Peru produced 2.2 million tons of copper in 2020, which is expected to increase to 3.1 million tons in 2024 after the production decrease associated with COVID-19 pandemics. As for China, the Asian country produced 1.7 million tons of copper (primary production) in 2020, which complements the production of refined copper (secondary) and the importation of large volumes of copper to supply its demand, being the largest global metal consumer (US Geological Survey (USGS) 2021).

Mining in Chile is currently one of the main economic activities, contributing about 9.8% of GDP (COCHILCO 2019), more than 50% of exports, and the leading recipient of foreign direct investment: accounting for one in every three dollars entering the country. The mining sector has steadily grown in the last 60 years: it tripled between 1960 and 1990; and trebled again during 1990–2016, reaching 5.5 million tons in 2016, causing Chile to be the world's leading producer, accounting for 30% of total production (CNP 2017). However, the relevance that the mining industry exhibits entails significant challenges from both a social and environmental point of view since it is an activity that relies on the intensive use of natural resources and the generation of large amounts of potentially harmful materials for the health of people and the environment (Ordóñez et al. 2021).

As mentioned before, Chile is the leading copper mine producer in the world. In the past, oxide ores with copper content over 5% were abundant and easy to exploit. However, the mineralogical composition of ores has changed over the years. Currently, copper oxides have become scarce, and sulfide minerals have become the main source, with contents of 1% Cu approximately, and complex to extract. In accordance with this, the copper production in Chile is based mainly on the exploitation of copper-sulfide ores. The reserves of copper in Chile consist mainly of chalcopyrite, a refractory sulfide ore, which in 2012 was estimated to be 190 million tons of copper, representing between 20–30% worldwide reserves (Gentina and Acevedo 2016, CNP 2017, Lutter and Giljum 2019).

Mineral processes are related to extracting, concentrating, and recovering mineral species, metallic, non-metallic, and fuels. Copper beneficiation operations include comminution by crushing and grinding and separation, such as flotation. On the other hand, chemical treatment operations are related to the dissolution of the species from the mineral to an intermediate solution through leaching, concentration using solvent extraction, and recovery of solid materials through electrowinning and smelting. In all these processes, along with obtaining the enriched metal streams, waste streams are produced with no commercial value because they are depleted of the target metal or because their chemical characteristics make them difficult or expensive to treat (Ordóñez et al. 2021) (Fig. 20.1).

Of the Chilean copper production, approximately 80% is obtained from sulfide minerals, which are treated firstly by crushing and grinding. Then, the fraction of mineral particles rich in copper is concentrated by froth flotation, reaching 20–30% copper from sulfide ores (0.5–2.0% Cu) (Schlesinger et al. 2011). Subsequently, the concentrate is processed by smelting and refining (pyrometallurgical steps).

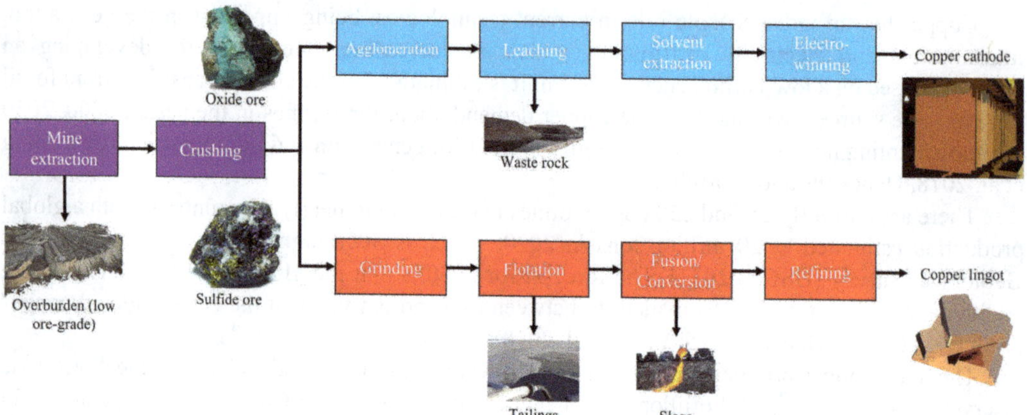

Fig. 20.1. Mineral processing and main products and waste streams.

Froth flotation entails conditioning a water-ore mixture (pulp) to make its Cu sulfide mineral particles repellent to water while leaving its non-Cu minerals (gangue) 'wetted'. The Cu minerals attach to rising air bubbles, 'floating' them out of a water-ore mixture. The flotation is made selective using reagents that wish to make the Cu sulfide minerals water repellent while leaving the other minerals' 'wetted' (Parekh and Miller 1999). The Cu ore concentrating process requires a large amount of water, with a unit consumption of 0.5 m^3 t^{-1} ore, about seven times greater than water consumption of hydrometallurgical processing (0.08 m^3 t^{-1}). The tailings generated by this process are in the form of liquid slime. These tailings can be transported to the tailings pond or dam, solids can be sedimented, and water recovered and recycled back to the ore enrichment process. When recirculation is not possible, part of the water is sent for industrial use, and the rest is returned to the environment under controlled conditions (Schlesinger et al. 2011).

Copper concentrate by flotation is smelted to produce blister copper. An interesting aside is that for every ton of copper produced, 1.5 t of slag and 2 t of sulfur dioxide are also produced as waste products. Sulfur dioxide can be captured and converted to sulfuric acid. Nevertheless, there are other by-products from copper concentrate by flotation. Arsenic occurs at varying levels in some copper ore bodies and is a significant environmental hazard in the copper smelting process when emissions are released into the atmosphere. Mining companies delivering copper concentrates containing high levels of arsenic to smelters are subject to penalties, making some copper ore deposits economically unviable. Also, the dumping of copper slag may cause economic, environmental, and space problems.

On the other hand, heap leaching is responsible for approximately 21% of copper production worldwide (3.9 million tons) (Marsden and Botz 2017). When the ore is exploited, vast quantities of overburden are mined, which is separated and either hauled to the waste site or stockpiled for future site reclamation. The overburden can produce acid mine drainage (AMD). Subsequently, the extracted ore is crushed and stacked into pads, which are lined with synthetic or natural materials such as high-density polyethylene (HDPE) or compacted clay. Sulfuric acid is fed onto copper-bearing ore, seeping through, and dissolving copper and other metal species. The liquid that passes through the bed, known as pregnant leach solution (PLS), is collected and sent to a solvent extraction plant to increase the copper content and remove impurities. The PLS is mixed with an organic solvent that binds to copper and chemically separates the copper from the rest of the species. In the last stage, electrowinning, an electrical field deposits the copper from PLS onto a metal sheet, producing a cathode. The remaining liquid is known as raffinate, a waste product that is returned to the leach pile for reuse, completing the solution stream cycle (USEPA 2021).

20.2.2 Mineral Residues and Effluents

The generation of waste is inherent in the production of minerals and metal raw materials (Diaby et al. 2007). The amount of waste generated by mining makes this industry one of the most intensive in the consumption of raw materials and the use of land to dispose of its waste (Ghorbani and Kuan 2017). Among the main massive mining wastes are overburden, low-grade ores, tailings, and slags, which, as mentioned above, are produced in the different stages of mining and mineral processing (Krishna et al. 2020).

In terms of tons of mineral waste, the coal mining industry is by far the most generating, with values that for 2002 were estimated at almost 22 Gt per year, since operations are intensive in the extraction of overburden (Rankin 2011). However, when analyzed concerning the specific generation of residues, i.e., the mass of residues per mass of product, mining activities of less abundant species have the highest indicators. Thus, for example, the specific value handled in gold mining is about 260,000 kg kg^{-1} (Chen et al. 2018). On the other hand, from the perspective of the mine type, open-pits remove about ten times more mineral than underground operations, with stripping rates between 1 and 4 t residue per t ore extracted, compared to 0.05 0.3 t t^{-1} for underground operations (Rankin 2011).

Tailings disposal without a management plan or insufficient physicochemical stability control can severely impact the environment. In the world, various events of failures in tailings facilities have been reported over time, affecting nearby territories and hydrographic basins (Owen et al. 2020). The critical incidents in tailings dams have raised as the reservoirs' number and size increase. In 2008, 32 events were registered, which grew to 43 in 2017 (Bowker and Chambers 2017). However, the potential impact of tailings deposits on human health and agricultural productivity depends not only on the catastrophes associated with the failure of dams but also on the simple drainage to surface water or groundwater, wind dispersal, absorption by vegetation, and bioaccumulation in food chains. This effect is enhanced by the granulometric characteristics of tailings, where the material has particle sizes smaller than 1 mm (Tutu et al. 2008, Kuter 2013).

It is important to mention that, although heavy metals are present in the environment naturally, anthropogenic activities, such as mining, tannery, and electroplating industries, may significantly increase their presence to toxic levels. Thus, in the case of mining, the extraction and processing of base metal sulfide minerals (such as copper) have been an important source of environmental pollution of rivers and soils (Zhuang et al. 2009, Punia 2021).

Environmental oxidation of sulfur minerals results in the generation of sulfuric acid, a phenomenon known as acid mine drainage (AMD) (Anju and Banerjee 2010). The AMD has three critical phases: in the first stage, the minerals are oxidized by the action of water and oxygen, producing the solubilization of the metallic species and the acidification of the medium. In the second stage, the products generated, particularly the Fe(II) and S(0) species, are transformed to Fe(III) and S(VI) by the action of microorganisms, such as *Acidithiobacillus ferrooxidans*, *A. thiooxidans*, and *Leptospirillum ferrooxidans* (Gan et al. 2018). Finally, the microbial action supposes a significant enhancement of the generation of AMD since the resulting species, particularly Fe(III), is a powerful oxidant that attacks sulfur minerals to follow the chain of associated reactions. Once the reaction has started, it is not easy to reverse, as the ferric ion acts as a powerful oxidizing agent. For example, in the general reactions for the oxidation of pyrite (Eq. 20.1 and 20.2), a sulfide ore commonly found in mining operations, and for chalcopyrite (Eq. 20.3 and 20.4), the main copper sulfide mineral, show that dissolution of these species produces acid in a high proportion, which results not only in a progressive decrease in pH but also in the release of other metallic species present in the mineral matrix, which are dispersed when reaching nearby watercourses (Katiyar and Randhawa 2020). Eq. 20.1–20.4 represent the initial and advanced oxidation reactions for pyrite and chalcopyrite. In contrast, Eq. 20.5 and 20.6 correspond to the bio-oxidation reactions mediated by microorganisms that provide more acid and ferric to continue sulfide dissolution reactions.

$$FeS_2\,(s) + 3.5O_2 + H_2O \rightarrow Fe^{2+} + 2SO_4^{2-} + 2H^+ \qquad \text{(Eq. 20.1)}$$

$$FeS_2\,(s) + 14Fe^{3+} + 8H_2O \rightarrow 15Fe^{2+} + 2SO_4^{2-} + 16H^+ \qquad \text{(Eq. 20.2)}$$

$$CuFeS_2\,(s) + O_2 + 4H^+ \rightarrow Cu^{2+} + Fe^{2+} + 2S^0 + 2H_2O \qquad \text{(Eq. 20.3)}$$

$$CuFeS_2\,(s) + 16Fe^{3+} + 8H_2O \rightarrow Cu^{2+} + 17Fe^{2+} + 2SO_4^{2-} + 2H^+ \qquad \text{(Eq. 20.4)}$$

$$4Fe^{2+} + 4H^+ + O_2 \rightarrow 4Fe^{3+} + 2H_2O\ (A.\ ferrooxidans) \qquad \text{(Eq. 20.5)}$$

$$2S^0 + 3O_2 + 2H_2O \rightarrow 2SO_4^{2-} + 4H^+\ (Acidithiobacillus) \qquad \text{(Eq. 20.6)}$$

The contribution of metallic pollutants through the generation of AMD depends on the composition of the minerals and environmental conditions in which oxidative weathering of sulfides occurs. Naidu et al. (2019) recently carried out a geochemical cadaster of 45 mining drains from mining operations worldwide. It is possible to observe a significant variability of pH and concentration of metallic species (Fig. 20.2). For example, typical pH values for AMD are between 2 and 5, depending on the stage of drainage formation. Mature drains have a more neutral pH, with a high suspended solids content, while young drains have very low pH, high levels of oxygen and dissolved metals (Skousen et al. 2017).

Also, within the so-called heavy metals, the concentrations are high compared to the guideline suggested by the World Health Organization, where the recommended limits for metals in drinking water are vastly exceeded in drains (WHO 2017). Although many drains are not discharged into direct sources of human supply, some are in danger of affecting channels used for human, livestock, and agricultural consumption. On the other hand, considering the metal toxicity (Zwolak et al. 2019), it is important to note that mining operations frequently produce effluents with the main toxic metals (Table 20.1).

Efficient technologies that minimize the negative impacts of AMD are required, which guarantee the environmental sustainability of mining operations in the long term (Zwolak et al. 2019). Metals are persistent and easily assimilated in food chains until they accumulate to toxic levels in plants, animals, and humans; for this reason, the control of AMD becomes a necessary downstream process.

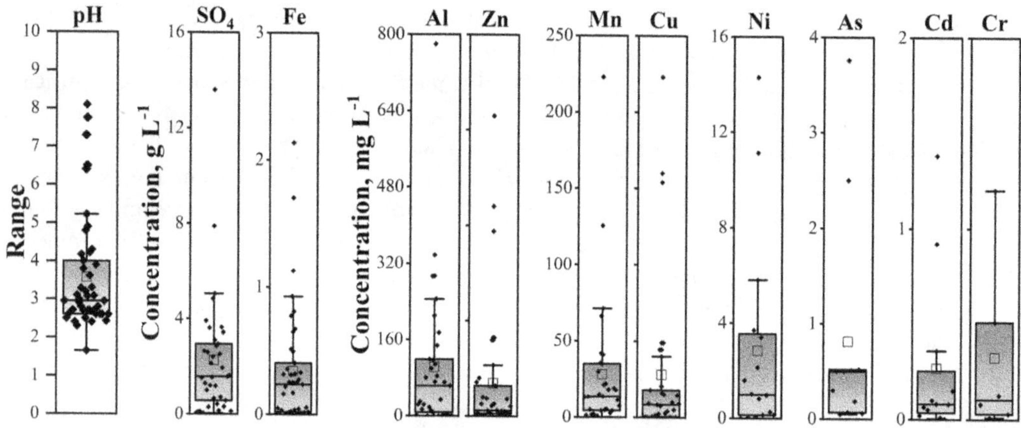

Fig. 20.2. Statistical analysis of AMD reported by Naidu et al. (2019).

Table 20.1. Guidelines for drinking-water quality (WHO 2017).

Species	Al	Mn	SO_4	Cl	Cd	Cr	As	Zn	Cu	Ni
WHO Guideline concentration, mg L^{-1}	0.9*	0.4*	500**	5	0.003	0.05	0.01	3**	2	0.07

* Health-based value
** Advised

Traditionally, industrial effluents loaded with heavy metals, such as AMD, have been treated by physicochemical abatement methods, classified according to the requirements for adding chemical reagents during the purification process. On the one hand, active methods require a constant supply of chemical inputs that neutralize the pH of the AMD. In contrast, passive methods employ strategies that reduce the frequency or substitute chemicals for treatment, using biomass or other biological materials as addenda.

Active treatment methods include the application of alkaline chemicals to precipitate metals and other techniques such as adsorption, ion exchange, and membrane technologies. The treatments based on membranes comprise microfiltration, nanofiltration, and reverse osmosis (Gaikwad et al. 2010, Fu and Wang 2011, Motsi et al. 2011, Ricci et al. 2015, Pino et al. 2018). Among the conventional active chemical options for the AMD treatment, the neutralization by using alkaline chemicals such as calcium hydroxide or limestone is a common method widely applied for metal removal as hydroxides and sulfates precipitate (Tolonen et al. 2014). One of the challenges that alkaline neutralization has is the generation of sludge that can be a secondary waste from effluent remediation. Some methods have achieved a low level of sludge generation, such as partial desalination, but its energy requirement is high, becoming an expensive technique.

Passive methods generally consider treatment under a biological strategy; wetlands, bioreactors with sulfate-reducing bacteria (SRB), and permeable reactive barriers have been the most used for the AMD treatment (Tolonen et al. 2014, Skousen et al. 2017). The passive technologies are more appropriate for treating abandoned mines than for continuous active AMD generation systems since they benefit from low operating costs and maintenance (Lukacs and Ortolano 2015). Passive treatments are less effective than active methods and require a longer time to remediate effluents and AMD, making them less used by mining operations (Iakovleva et al. 2015). The choice of treatment options for AMD depends on several factors such as environmental, chemical, and economical.

Among the impacts that mining produces, beyond the eventual generation of acid drainage, is the dispersion of heavy metals to the soils that surround the mineral waste deposits, being a critical aspect in those cases where tailings and waste-rock facilities are located in the vicinity of populated areas. Thus, throughout the world, the contaminating effect of soils by potentially toxic elements has been reported in mining regions, such as India, Morocco, Iran, China, Australia, and Chile (Anju and Bancrjee 2011, Khalil et al. 2013, Liu et al. 2013, Solgi and Parmah 2015, Reyes et al. 2021). A case of study is the city of Taltal, Chile, whose mining history is long. For many years, gold-copper tailings were collected without environmental protection on the beach and inserted in the urban radius. It has recently been shown that these abandoned tailings have substantially raised the geochemical levels of heavy metals in the surrounding soils of the city. Concentrations of Cd, Cu, Zn, Cr, and Pb have exceeded the recommended levels for evaluating ecological risk and/or human health (Reyes et al. 2021).

20.3 Biosorption of Metals in the Bioremediation

As mentioned before, mining activities may severely impact the environment by releasing heavy metals and other species to soils and watercourses that destroy vegetation, crops, and ecological equilibrium. In addition, the polluting effect and accumulation of metals in the human body have been deeply reported, including anemia, respiratory, gastrointestinal, nervous diseases, and cancer

(Dodson et al. 2012). The mining industry, as other activities, has widely adopted conventional active chemical treatments in order to mitigate the contamination events, mainly by neutralization. This technique involves adding alkaline reagents such as limestone, calcium oxide, caustic soda, and sodium carbonate, among others, to precipitate dissolved heavy metals by increasing the pH (Balladares et al. 2018, Carneiro Brandão Pereira et al. 2020). Other methods, such as ion-exchange and activated carbon adsorption, solvent extraction, and chemical precipitation, have been used by similar industries. However, the high amount of energy and chemical agents required, and the generation of large quantities of secondary wastes, are drawbacks that limit their massive application. In addition, the efficiency for very dilute solutions is generally low.

Other techniques based on a biological approach have been gradually applied to treat mine tailings and other industrial wastewater due to their environmental friendliness and cost-efficiency at low heavy metals concentrations (Leong and Chang 2020). Some examples of these bioremediation techniques correspond to phytostabilization, where plants stabilize metallic contaminants in mine effluents, and bioimmobilization, where microorganisms promote the biogenesis of insoluble species like jarosite (Ciarkowska et al. 2017, Piervandi et al. 2020). In a novel approach, the biorecovery of metals from industrial wastewaters, i.e., obtaining critical metals or metal nanoparticles through bacteria and other biomass, has been emergently studied (Yin et al. 2019, Wong-Pinto et al. 2021).

The mechanisms governing the different bioremediation procedures include bio-weathering, microbial reduction, biomineralization, bioaccumulation, and biosorption (Pollmann et al. 2018). Biosorption can be classified using different criteria, such as the metabolism dependence of metal-biosorbent interaction and the location. The nature of interaction determines if the biosorption is active or passive, i.e., if the metal uptake is performed by living or dead biomass, which is related to metabolites or structural biomolecules (Alwaleed et al. 2021). On the other hand, the place where biosorption occurs indicates if the process is intra or extracellular. It has been observed that intracellular biosorption requires longer contact times since metals must be entered into the cell through specialized pumps. The biosorption carried out extracellularly, in the media or on the cell walls, can bioadsorb metals faster because it is independent of metabolism, implying that enzymatic reactions do not govern the interactions. Electrostatic and chemical interaction mechanisms play an important role in independent-metabolism biosorption, including van der Waals forces, displacement of metal cations by ion exchange, complexation, diffusion, surface adsorption, and precipitation (Kotrba et al. 2011).

Biosorption is a clean and straightforward alternative for recovering metals from dilute solutions that require low capital investment costs. It employs sorbents that are renewable, inexpensive, and that can be obtained from cultures and secondary sources such as bacteria, algae, fungi, agro-industrial, and aquaculture wastes (Davis et al. 2003, Beni and Esmaeili 2020). A special advantage of using biomasses for biosorption is related to the afterward processing of residual sorbent. Once the biosorption has been carried out, the spent biosorbent is biodegradable, limiting the pressure on the environment due to the minimization of the use of chemical reagents. The sludge and effluents obtained are potentially integrable to industrial processes.

The fast and reversible ion binding process could be metabolism-independent; therefore, dead biomass can be used (Vijayaraghavan and Yun 2008). Other advantages of biosorption are the possibility to store the biosorbents for prolonged times, no nutrient limitations, ease in the metal recovery, and flexibility to be used in a wide range of pH (3–9) and temperature (4–90°C) (Cortés et al. 2020). The accumulated experience in different perspectives of the biosorption process such as mechanisms, chemistry, physics, and engineering gives essential information on the natural interaction between biomass and metals. It provides the opportunity to develop ecological and low-cost techniques to treat industrial effluents (Gadd 2009).

The main biosorbents employed in the biosorption investigations are microorganisms (bacteria and cyanobacteria), fungi, and macroalgae (and their extracts) (Fig. 20.3a). There are other biomasses

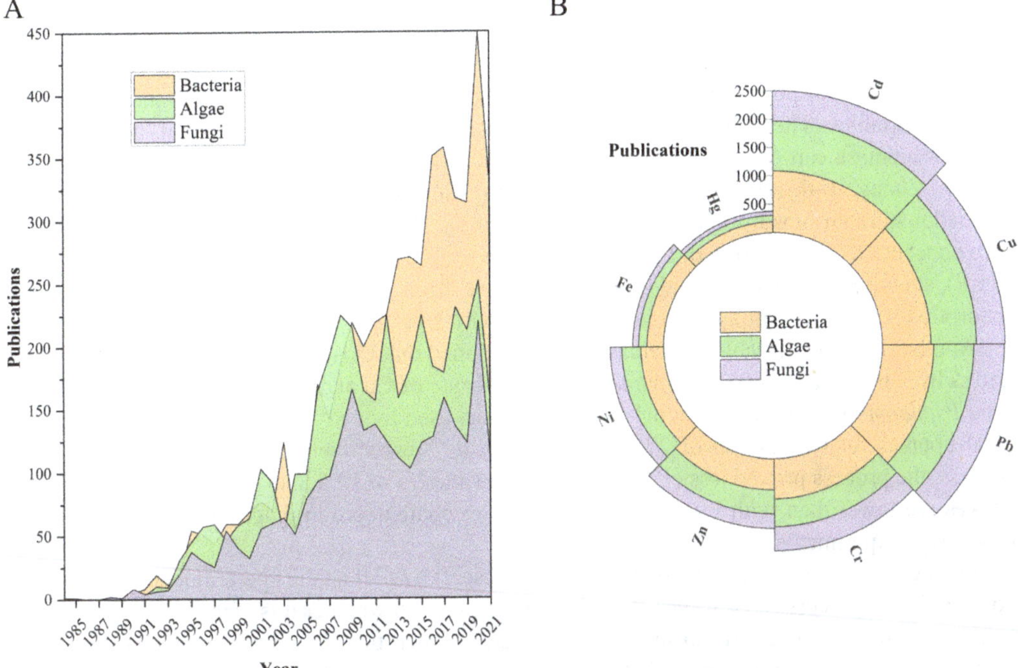

Fig. 20.3. Bibliometric analysis associated with the publications reported in the WOS database on biosorption of heavy metals with the most used biomasses. (a) Evolution of the investigations on biosorption of heavy metals by type of biomass and (b) Profile of total biosorption investigations by metal and type of biomass. A publication can address more than one metal and/or biomass, so the number of articles should be taken as a reference and not as exact for each category.

as plants that also contribute to the knowledge about biosorption, but still to a lesser extent. Metal biosorption has been extensively approached from a mechanistic perspective and on the basis of demonstrating the capacity of the biomass to be able to carry out the removal/recovery of various metals. For this purpose, the literature strongly shows the use of monometallic solutions of defined concentration or in mixtures of metals with established proportions. These experiments have made it possible to elucidate many of the phenomena that occur during biosorption and those operating conditions that are best for biosorption, and those that are specific to metalbiosorbent interaction.

Biosorption is a subject of increasing research. Since the mid-80s, with the first reports on biosorption in bacteria and fungi, and since the 90s, the incursion of seaweeds as biosorbents appeared and consolidated in a sustained manner. By 2020, historical maximums were observed in terms of publications on biosorption. It is expected that the cumulative contribution of biosorption may significantly impact bioremediation activities, especially for those metals whose treatment is difficult or expensive.

It is observable that the biomasses most evaluated for metal biosorption correspond to bacteria and algae, including biopolymers and metabolites derived from these biomasses. When evaluating the metals most studied for biosorption, cadmium has the largest number of publications with about 21% of reports, which is based on how toxic it is in the environment and how critical its removal is (Fig. 20.3b). Likewise, it is an indicator that biotechnological methods represent a competitive alternative to traditional processes. Subsequently, copper and lead are found (19% each), which are heavy metals highly spread by mining and industrial operations. The biosorption of more abundant metals, such as iron (5%) and aluminum, has been less studied because the precipitation methods by neutralizing agents are quite effective and inexpensive, which does not support a sustained motivation for their research.

20.3.1 Bacteria

Different authors have studied the biosorption of metals and the effect that operating parameters have on the removal performance. Thus, the pH and metal concentration are the most frequently analyzed variables. The interactions in biosorption are specific to biomass-metal, and therefore, the same biomass can exhibit different affinity levels and efficiencies with different metals. For instance, Zhang et al. (2017) used *Acinetobacter* sp. to remove Ni and Cu from a solution and obtained maximum biosorption capacities of 56.7 mg g^{-1} for Ni and 157.2 mg g^{-1} for Cu. Using a more concentrated Ni solution did not reflect a high metal uptake (75 mg L^{-1} for Ni and 50 mg L^{-1} for Cu), showing that this bacterial biomass has more affinity for Cu than Ni, which indicates that biosorption is a process implying interesting selectivity levels.

Another relevant variable is biomass; it is mentioned that biosorption can be done by living or dead cells. Chen et al. (2005) compared the removal rate of Cu and Zn from aqueous solutions by using *Pseudomonas putida CZ1* at pH 5. When authors used living bacterium culture, they obtained metal uptakes for Cu and Zn of 27.6 and 27.4 mg g^{-1}, respectively. However, using non-living biomass, the process proved less efficient with removal rates of 15.8 and 17.7 mg g^{-1}. Although the uptakes are lower than with live cells, the values are comparable in magnitude to other reported biosorption experiments.

On the other hand, the use of dead biomass represents a relief on the operational requirements in bioremediation processes. However, there is no consensus on whether it is better to use live or dead biomass for biosorption. Several investigations have shown that dead biomasses result in higher metal biosorption performances, as is the case of the recent work of Hu et al. (2020). They studied the Pb biosorption using *Rhodococcus* sp. *HX-2* with live and dead cells. The maximum uptakes were 88.7 and 125.5 mg g^{-1} for live and dead biomass, respectively. The results reaffirm the idea that biosorption is a process that is complex to generalize because the nature of molecular interaction among biomolecules and cations is specific.

The study of chemical groups involved in metal biosorption has been addressed in many articles. The results obtained from FTIR (Fourier transformed Infrared Spectroscopy) and XPS (X-ray photoelectron spectroscopy) are the techniques used to investigate the presence of certain groups and biomolecules. Amines, phosphates, and carboxyl groups are recurrently reported as responsible biochemicals in biosorption for all types of biomasses. These groups are present in proteins, carbohydrates, nucleic acids, and more complex substances.

20.3.2 Fungi

In the case of fungi, both yeasts and filamentous fungi have demonstrated a metal biosorption activity. The bread yeast *Saccharomyces cerevisiae* was used to treat four different solutions containing Cu, Fe, Ni, and Zn. The best biosorption condition was obtained at an initial copper concentration of 4.6 mg g^{-1}, pH 3-4. It was determined that carboxyl and amino groups play a fundamental role and that the process was governed by ion exchange and surface complexation (Zinicovscaia et al. 2020). Before, Veit et al. (2005) worked with dead biomass of two fungal species, *Pleurotus pulmonius CCB019* and *Schizophyllum commune*, demonstrating that both organisms were able to remove copper from metal solutions with uptakes of 6.20 and 1.52 mg g^{-1}, respectively, which were obtained at pH 4, 3 g L^{-1} biomass concentration, and 100 mg L^{-1} initial copper concentration. Other studies have shown the removal of copper and other heavy metals by other fungal biomasses, such as *Aspergillus niger* (2.66 mg g^{-1}) and basidiomycetes (2.61–4.77 mg g^{-1}).

20.3.3 Macroalgae

On the other hand, macroalgae started to acquire attention as metal biosorbents in the 90s since their high binding affinity, an abundance of binding sites, and large surface area (Vijayaraghavan et al. 2005, Romera et al. 2007, Leong and Chang 2020). Seaweeds do not form a homogeneous group within macroalgae; they are divided into three classes according to physical and physiological differences: red (*Rhodophyta*), brown (*Chromophyta*), and green (*Chlorophyta*) algae. These differences are presented in the cell wall structure, pigments, and phycocolloids that they produce. Brown alga walls generally contain cellulose, which is the structural support; alginic acid, a polymer of mannuronic and glucuronic acids; and sulfated polysaccharides. Red algae also contain cellulose and a matrix of sulfated polysaccharides based on galactans (agar and carrageenans). Green algae cell walls are mainly composed of cellulose and a high percentage of proteins bonded to polysaccharides as glycoproteins (Volesky 2001). It should be noted that brown and red seaweeds are the only source for agar, alginate, and carrageenan, natural polymers highly demanded by the cosmetic and food industry.

The biomolecules related to phycocolloids contain several functional groups, such as amino, carboxyl, sulfate, and hydroxyl, which play an important role in the biosorption of metals (Romera et al. 2007). Within marine algae, brown algae have the highest sorption uptakes due to the presence of alginates in the cell walls, followed by red algae, especially those with high content of carrageenan in their cell walls. Green algae exhibit the lowest adsorption uptakes because they are poor in fucoidal substances.

The use of seaweeds for biosorption has been validated for various metal solutions. Plaza Cazón et al. (2012a) used *Macrocystis pyrifera* to remove zinc and cadmium from mono and bimetallic solutions, demonstrating in both cases goods uptake capacities, which were similar to other studies performed with *Sargassum filipendula*, *Gymnogongrus torulosus*, and *Fucus vesiculosus*, other species reported in previous works (Mata et al. 2008, Luna et al. 2010). In addition, the use of *M. pyrifera* and *Undaria pinnatifida* for chromium and mercury biosorption from aqueous solutions revealed that the carboxylic and amino groups are strongly involved in chromium binding, while amino and sulfhydryl for mercury uptake, establishing that the interaction would be specific between metal and functional group (Plaza Cazón et al. 2011, 2012b).

Brown alga *Sargassum* sp. was used to remove Pb and Cu from stormwater, resulting in biosorption capacities of 196.1 mg g^{-1} and 84.0 mg g^{-1} for Pb and Cu, respectively. The analysis of the functional groups of the algae using FTIR showed that the carboxyl was the main group responsible for the biosorption (Perumal et al. 2007). Fawzy (2020) used *Codium vermilara* to remove copper, with an efficiency of about 85% under an alga dosage of 0.75 g L^{-1}, pH 5.3, contact time 70.5 min, and copper concentration of 48.8 mg L^{-1}. The green microalga *Chlorella vulgaris* biosorbed copper with a recovery of 90.3% under pH 7, 105 min of contact time, and 20 mg L^{-1} of initial copper concentration. It was determined that the chemical groups involved were amines and carboxyl. The presence of copper was extracellular on the surface of the cell wall (Indhumathi et al. 2018).

20.4 Biosorption of Copper From Mining and Industrial Wastewater

In the previous section, studies conducted to elucidate the mechanisms and types of biomasses that have demonstrated the ability to bioadsorb heavy metals were reviewed. Although these investigations are the fruit of rigorous experimental works, they have evaluated the removal of

Table 20.2. Copper biosorption investigations developed in real mining and industrial effluents.

Wastewater type	Biosorbent	Operating conditions	Copper uptake, mg g^{-1}	Removal efficiency, %	References
Leached metallurgical sludge	Treated sugarcane bagasse	21 mg L^{-1} Cu, pH 5, 3 h	16.6	99.5	Xie et al. 2018
Acid-leached tailings	*Gracilaria chilensis*	200 mg L^{-1} Cu, 1 h, pH 1.5, 0.5 g L^{-1} biomass	0.31 mmol g^{-1}	-	Cortés et al. 2020
Industrial effluent	*Kappaphycus alvarezii*	16 mg L^{-1} Cu, pH 7, 13 h	-	81.5	Pandya et al. 2017
Treated AMD	Ca alginate beads	3 mg L^{-1} Cu, pH 5.8	6.7	86	Park and Lee 2017
Mining effluent	*Escherichia coli*-zeolite	53 mg L^{-1} Cu, pH 3.3	-	51	Khosravi et al. 2020
Neutralized mining effluent	*Escherichia coli*-zeolite	3.1 mg L^{-1} Cu, pH 6.2	-	95	Khosravi et al. 2020
Acid-leached slags	Hydrolyzed olive cake	5.23 mg L^{1} Cu, pH 6	3.8	97	Fernández-González et al. 2018
Industrial effluent	*Euglena gracilis*	25–100 µmol L^{-1} Cu, 21 d, pH 7	-	80	Jasso-Chávez et al. 2021
Electroplating effluent	Acid-pretreated *Micrococcus* sp.	50 mg L^{-1} Cu, pH 6	30	27	Lo et al. 2003
Tannery effluent	*Bacillus* sp. *FM1*	3 mg L^{-1} Cu, pH 7.9	-	78	Masood and Malik 2011
Paint industry effluent	Kola nut pod powder	35.7 mg L^{-1} Cu	-	93.9	Yahya et al. 2020
Industrial effluent	*Azadirachta indica* (neem leaf) powder	pH 7, 60 min 1 g biomass, 110 mg L^{-1} Cu	-	73	Al Moharbi et al. 2020
Industrial effluent	SRB, *Bacillus cereus*, and *Camellia oleifera* cake (COC)	100 mg L^{-1} Cu, pH 6–9, 60 h	-	60	Wu et al. 2017
Treatment plant discharge	*Chlorella vulgaris*	56.6 mg L^{-1} Cu, pH 7.96, 2 d	-	81.9	Chan et al. 2014
Treatment plant discharge	*Spirulina maxima*	56.6 mg L^{-1} Cu, pH 7.96, 2 d	-	81.7	Chan et al. 2014
Industrial effluent	*Phormidium* sp.	266 mg L^{-1} Cu, 18 d	-	62	Kottangodan et al. 2019
Electroplating effluent	Inactivated *Saccharomyces cerevisiae NCYC 1364*	2.57 mg L^{-1} Cu, pH 6, 18 g L^{-1} biomass, 1 h	-	78	Machado et al. 2010
Electroplating effluent	Sugar beet pulp	15.9 mg L^{-1} Cu, pH 5, 1 g L^{-1} biomass, 2 h	-	50	Castro et al. 2017

Table 20.2 contd. ...

...Table 20.2 contd.

Wastewater type	Biosorbent	Operating conditions	Copper uptake, mg g^{-1}	Removal efficiency, %	References
Electroplating effluent	*Fucus vesiculosus*	15.9 mg L^{-1} Cu, pH 5, 1 g L^{-1} biomass, 2 h	-	69	Castro et al. 2017
Treated electroplating effluent	*Sargassum* sp.	20.9 mg L^{-1} Cu, pH 6.5	55.7	75	Barquilha et al. 2019b
Treated electroplating effluent	Ca alginate beads	23.2 mg L^{-1} Cu, pH 4.5, 13.3 g L^{-1} biomass	1.597 mmol g^{-1}	77	Barquilha et al. 2019a
Electroplating effluent	Native *Spirogyra* sp.	0.6 g L^{-1} biomass, 0.5 h, pH 6	-	82.8	Ilyas et al. 2018
Electroplating effluent	Modified *Spirogyra* sp.	0.6 g L^{-1} biomass, 0.5 h, pH 6	-	96.4	Ilyas et al. 2018
Industrial effluent	Dried cabbage leaves	50 mg L^{-1} and pH 6, 67 min	-	-	Kamar et al. 2016
Electroplating effluent	Peanut shells	20 mg L^{-1} Cu, pH 4–6, 0.5 h	0.028 mmol g^{-1}	-	Oliveira et al. 2010

heavy metals from pure and/or synthetic solutions that simplistically resemble real effluents. It is known that many residual liquid streams have a complex chemical composition, where many metals are mobilized and available to interact with adsorbent matrices, high acidity, and ionic strength. This means an additional step for the validation of biosorbents as a solution to effluent treatment. This section will describe the works that have addressed copper biosorption from real solutions, such as mining, electroplating, industrial, and dying wastewaters. In addition, different biomasses that have been tested, from bacteria and fungi to alga, plant, and agro-industrial waste, will be included (Table 20.2).

20.4.1 Metallurgical Effluents and AMD

A residue that usually results from mineral processes corresponds to the metallurgical sludge, which was also studied for biosorption (Xie et al. 2018). A biosorbent based on sugarcane bagasse recovered copper and other eight metals from a nitric acid-leached sludge. Leaching yielded a copper concentration of 620 mg L^{-1}, which, after a step of iron removal by precipitation, was diluted to 21 mg L^{-1}. Subsequently, the bioadsorption was carried out in a fixed-bed column loaded with the modified biosorbent, achieving a copper uptake at the equilibrium of 16.6 mg g^{-1} and removal efficiency of 99.5%, higher than that of the rest of the metals evaluated and demonstrating that this biosorbent could be used for the selective removal of this metal. After bioadsorption, the column was desorbed with 5 mM EDTA-2Na obtaining a solution with a purity of 95%.

An advantage of copper biosorption with algal biomasses is the possibility to use dead biomass for handling mining effluents because these solutions have unfavorable chemical characteristics for cell growth (high acidity, ionic strength, and heavy metal content). Also, algae can be cultivated and therefore considered a reliable source of supply. In addition, the biomass acts as an ion exchanger and, therefore, the process is carried out quickly, and desorption of the metal is relatively easy. The use of marine algae, predominantly brown algae, to remove heavy metals from solutions through the mechanisms of bioadsorption and bioaccumulation was reviewed by Yadav et al. (2019). Despite the extensive review, the cases are mostly applied in laboratory situations with synthetic solutions. The efficiencies for copper with different seaweed range between 7.2 and 92%.

Recently, the feasibility to bioadsorb copper from tailing derived solutions has been verified using the red algae *Gracilaria chilensis* (Cortés et al. 2020). Tailings, obtained from an abandoned deposit in Chile, were first treated with sulfuric acid to dissolve the metal species. Then, the biosorption experiments were conducted in stirred flasks and columns. It was determined that the copper biosorption is a fast process that occurred in 60 min, and the maximum adsorption (0.31 mmol g^{-1}) was achieved with pH 1.5, 0.5 g L^{-1} of *G. chilensis* dosage, and 200 mg L^{-1} of initial copper concentration, which is in the range of concentrations usually found for this metal in mining effluents. Observations of *G. chilensis* by FTIR demonstrated that the functional chemical groups involved in the biomass-metal interaction were carboxyl, hydroxyl, and amines, which are molecules widely distributed in the basic biomolecules such as proteins, carbohydrates, and their derivative compounds, as glycoproteins. Another red algae evaluated in the biosorption of copper from industrial effluent containing 16 mg L^{-1} Cu has been *Kappaphycus alvarezii*, which achieved a maximum removal rate of 81.5% under pH 7, and after 80 min of contact time in batch conditions (Pandya et al. 2017).

The biosorption using biopolymers derived from seaweed was analyzed by Park and Lee (2017), where spheres of calcium alginate were used to adsorb copper and cadmium from AMD. An advantage of using biomaterials is the ability to separate growth and metabolite production processes from that of biosorption, which provides operational benefits. AMD from the Ilgwang mine, Korea, has copper concentrations of 21.5 mg L^{-1}, which is seven times greater than the Korean limit for groundwater quality in mining areas. The AMD was diluted to 3 mg L^{-1} Cu and its pH fixed to 5.8. Biosorption was conducted in a column packed with alginate spheres. The removal efficiency of Cu achieved was 86%, and a pH neutralization effect was also observed, which allows its use for pH conditions even under 3. The heavy metal removal mechanisms were attributed not only to biosorption, but also to complexation and precipitation (Park and Lee 2017).

The treatment of mine wastewaters has also included the use of bacterial biomasses. Copper and zinc from the Bahonar mining complex effluents, Iran, were removed using *E. coli* immobilized in the zeolite. The hybrid bacterium-zeolite bioadsorption system obtained a removal efficiency from effluents without neutralization pretreatment (copper load of 53 mg L^{-1} and pH 3.3) of 51%, while, with pretreatment, the efficiency increased to 95% (with a Cu content of 3.1 mg L^{-1} and pH 6.2), suggesting the use of this method coupled to a neutralization treatment unit to remove interfering metal species in the bioadsorption (Khosravi et al. 2020).

As part of the pyrometallurgical treatment of ores, slags are generated as a mineral residue after smelting. The abatement of heavy metals from an acid-leached slag solution was performed using hydrolyzed olive cake obtained from oil production. Biosorption from a solution loaded with 5.23 mg L^{-1} Cu was conducted in batch experiments at pH 6, resulting in a copper uptake of 3.8 mg g^{-1} and a removal efficiency of 97% (Fernández-González et al. 2018).

20.4.2 Industrial Wastewater

The recent use of the protist *Euglena gracilis* to treat actual industrial wastewater, and mine tailings was performed to remove copper and other heavy metals such as Cd, Cr, and Fe (Jasso-Chávez et al. 2021). It was demonstrated that under anaerobic and, preferably, microaerophilic conditions, the cells of *E. gracilis* have the maximum metal biosorption. The biofilm formation was critical for metal resistance and for providing the hypoxia conditions. The removal of metals from solutions with copper concentrations between 25–100 μM was achieved in several cases up to 80% after 21 days of culture.

Other evidence related to the application of bacteria on the removal and recovery of copper from wastewater was performed by Lo et al. (2003), who studied *Micrococcus* sp. as a biosorbent. The maximum metal biosorption capacity from a real solution with 50 mg L^{-1} Cu reached 30 mg g^{-1} at pH 6, which markedly decreased when working with more acidic solutions, suggesting that cations and protons compete for the same binding sites in the cell wall. In a similar approach, the Cr and Cu binding capacity of *Bacillus* sp. FM1 was evaluated from a tannery effluent with a Cu load of 3 mg L^{-1}. Although lab experiments determined that the optimum pH for copper biosorption was 5, the employed pH was 7.9 due to operational conditions. It was noticed that *Bacillus* sp. biomass satisfactorily removed metals from actual tannery effluent, with an efficiency of 78%; however, the performance was slightly lower than from synthetic metal solutions (Masood and Malik 2011).

In a more recent approach, the sorption of Ni and Cu from industrial paint effluents was investigated. For this, the kola nut pod was loaded in columns. The operating conditions were optimized and corresponded to a bed height of 10 cm, feed flow rate of 5 mL min^{-1}, and Cu initial concentration of 35.7 mg L^{-1}. The metal uptake reached 93.9%, and it was demonstrated that carboxylic, amide, and hydroxyl groups play a key role in ion capture (Yahya et al. 2020). Additionally, in other research, the use of *Azadirachta indica* (neem leaf) powder was investigated for capturing copper from industrial wastewater. In this case, the chemical groups involved in the biosorption of Cu corresponded to amino and hydroxyl, which led to a maximum copper removal of 73%. The best operational conditions were pH 7, 60 min of stirring, 1 g of *A. indica* biomass, 110 mg L^{-1} Cu concentration, and 125 rpm stirring (Al Moharbi et al. 2020).

The removal of multiple metallic species has also been addressed under the biotechnological approach for decontaminating effluents. The authors used a combination of sulfate-reducing bacteria (SRB), *Bacillus cereus*, and filter cake from the *Camellia oleifera* plant (COC). In this sense, the biodegrading action of *B. cereus* on COC generates suitable anoxic and reducing conditions for SRB, which together with *B. cereus* have an important role in the biosorption and bioaccumulation of metals. The optimal pH was between 6–9, and the initial copper concentration was 100 mg L^{-1}. The removal rate was almost 60%, achieved after 60 h of incubation at 37°C (Wu et al. 2017).

Microalgae such as *Chlorella vulgaris*, *Spirulina maxima*, and other native algae have been used in mixtures and separately to remove heavy metals in secondary effluents of wastewater treatment plants. In terms of copper removal, the highest efficiencies resulted with autoclaved *C. vulgaris* (81.9%) and untreated *S. maxima* (81.7%), confirming that the involved mechanism is metabolism independent (Chan et al. 2014). In a similar approach, the use of marine cyanobacteria has also been evaluated to treat effluents loaded with heavy metals. Copper and other heavy metals were biosorbed using *Phormidium* sp. from a combined industrial effluent from the textile, pharmaceutical, medical, and engineering industries in India. After 18 d of microalgae culture in the effluent, the maximum reduction in the Cu level was obtained, achieving 62%, and going from 0.27 to 0.1 mg L^{-1} (Kottangodan et al. 2019).

20.4.3 Electroplating Effluents

The electroplating industry is also an important source of heavy metals in wastewater and is one of the most investigated effluents in biosorption. Machado et al. (2010) used 18 g L^{-1} of heat-inactivated cells of *S. cerevisiae NCYC 1364* for copper, chromium, and nickel removal from a solution containing 2.57 mg L^{-1} Cu. In the case of copper, the efficiency of biosorption was 78% at pH 6 in 2 batch cycles of 30 min each.

In a scaled study, sugar beet pulp and the brown algae *F. vesiculosus* were used to remove zinc and copper from electroplating wastewater in Valladolid, Spain, with a load of 15.9 mg L^{-1} Cu. Both biomasses were tested individually in fixed beds in a dosage of 1 g L^{-1}. At pH 5-6 and after 2 h, the metal uptake was observed to be maximum, achieving a Cu removal efficiency of up to 50% for sugar beet and 69% for the seaweed. The highlighted performance of algae was confirmed in this study, determining that the main functional group involved in copper biosorption was carboxylate (as alginate) in algae and pectin in the sugar beet (Castro et al. 2017).

Sargassum sp. has been other brown algae evaluated for treating electroplating effluent containing 20.9 mg L^{-1} copper and pH 6.5. The solution was fed into a continuous fixed-bed column, resulting in removal efficiency of 75% and copper uptake of 55.7 mg g^{-1}. In previous observations, the use of the same biomass for synthetic solutions reached 75%, reflecting that ions from real effluents become highly competitive and increase the transfer zone of mass (Barquilha et al. 2019b). Another study by the same authors used an alginate-based biosorbent produced from *Sargassum* sp. to recover Ni and Cu from a real electroplating effluent with 23.2 mg L^{-1} Cu at pH 4.5. It was demonstrated that the biosorption capacity of alginate was 77% for copper using 13.3 g L^{-1} of biosorbent, and the uptake capacity was 1.60 mmol g^{-1} (Barquilha et al. 2019a).

Ilyas et al. (2018) investigated the biosorption behavior of the green alga, *Spirogyra* sp., in its native and modified states, for the removal of copper from an industrial electroplating effluent. In the optimized condition with a sorbent dose of 6 g with 100 ml of effluent for 30 min at pH 6, 82.8% and 96.4% of maximum copper could be adsorbed by the native and modified *Spirogyra* sp., respectively.

The search for cheap biosorbents has led to the exploration of the use of agro-industrial waste. Thus, Kamar et al. (2016) conducted the continuous biosorption of copper, lead, and cadmium from industrial wastewater using a fluidized bed of dried cabbage leaves. The initial concentration of metals was 50 mg L^{-1} and pH 6. The equilibrium time was 67 min, demonstrating a fast biosorption process. In other work, Oliveira et al. (2010) proved peanut shells as biosorbents to recover copper from an actual electroplating effluent with 20 mg L^{-1} Cu. A sorption capacity of 0.028 mmol g^{-1} was obtained at pH 4-6 and after 30 min.

20.5 Conclusions

In this chapter, a comprehensive review of the biosorption approach applied in the bioremediation of heavy metal effluents was performed. Copper mining and other industrial activities as tannery and galvanic are the main anthropogenic sources of copper in wastewaters that may affect the environment and population health.

Treatment of polluted waters requires the understanding of different approaches, among which is bioremediation. The biosorption, i.e., the superficial uptake of metals using reactive biomasses, is an emerging area that explores a broad spectrum of species and metals. This phenomenon has been applied for many years in the research to understand the involved processes; however, in the last decade, the use and application of knowledge on the treatment of actual solutions and wastewater are becoming more intense. It is expected that the number of scientific and technological contributions in this field will be notably increased in the following years.

References

Al Moharbi, S. S., M. G. Devi, B. M. Sangeetha and S. Jahan. 2020. Studies on the removal of copper ions from industrial effluent by *Azadirachta indica* powder. Appl. Water Sci. 10: 1–10.

Alwaleed, E. A., A. A. Abdel Latef and M. El-Sheekh. 2021. Biosorption efficacy of living and non-living algal cells of *Microcystis aeruginosa* to toxic metals. Not. Bot. Horti Agrobot. Cluj-Napoca 49: 12149.

Anju, M. and D. K. Banerjee. 2010. Comparison of two sequential extraction procedures for heavy metal partitioning in mine tailings. Chemosphere 78: 1393–1402.

Anju, M. and D. K. Banerjee. 2011. Associations of cadmium, zinc, and lead in soils from a lead and zinc mining area as studied by single and sequential extractions. Environ. Monit. Assess. 176: 67–85.

Balladares, E., O. Jerez, F. Parada, L. Baltierra, C. Hernández, E. Araneda and V. Parra. 2018. Neutralization and co-precipitation of heavy metals by lime addition to effluent from acid plant in a copper smelter. Miner. Eng. 122: 122–129.

Barquilha, C. E. R., E. S. Cossich, C. R. G. Tavares and E. A. da Silva. 2019a. Biosorption of nickel(II) and copper(II) ions from synthetic and real effluents by alginate-based biosorbent produced from seaweed *Sargassum* sp. Environ. Sci. Pollut. Res. 26: 11100–11112.

Barquilha, C. E. R., E. S. Cossich, C. R. G. Tavares and E. A. da Silva. 2019b. Biosorption of nickel and copper ions from synthetic solution and electroplating effluent using fixed bed column of immobilized brown algae. J. Water Process Eng. 32: 100904.

Beni, A. A. and A. Esmaeili. 2020. Biosorption, an efficient method for removing heavy metals from industrial effluents: A Review. Environ. Technol. Innov. 17: 100503.

Bowker, L. and D. Chambers. 2017. In the dark shadow of the supercycle tailings failure risk & public liability reach all time highs. Environments 4: 75–95.

Carneiro Brandão Pereira, T., K. Batista dos Santos, W. Lautert-Dutra, L. de Souza Teodoro, V. O. de Almeida, J. Weiler, I. A. Homrich Schneider and M. Reis Bogo. 2020. Acid mine drainage (AMD) treatment by neutralization: Evaluation of physical-chemical performance and ecotoxicological effects on zebrafish (*Danio rerio*) development. Chemosphere 253: 126665.

Castro, L., L. A. Bonilla, F. González, A. Ballester, M. L. Blázquez and J. A. Muñoz. 2017. Continuous metal biosorption applied to industrial effluents: A comparative study using an agricultural by-product and a marine alga. Environ. Earth Sci. 76: 491.

Chan, A., H. Salsali and E. McBean. 2014. Heavy metal removal (copper and zinc) in secondary effluent from wastewater treatment plants by microalgae. ACS Sustain. Chem. Eng. 2: 130–137.

Chen, W., Y. Geng, J. Hong, H. Dong, X. Cui, M. Sun and Q. Zhang. 2018. Life cycle assessment of gold production in China. J. Clean. Prod. 179: 143–150.

Chen, X. C., Y. P. Wang, Q. Lin, J. Y. Shi, W. X. Wu and Y. X. Chen. 2005. Biosorption of copper(II) and zinc(II) from aqueous solution by *Pseudomonas putida* CZ1. Colloids Surfaces B Biointerfaces 46: 101–107.

Ciarkowska, K., E. Hanus-Fajerska, F. Gambuś, E. Muszyńska and T. Czech. 2017. Phytostabilization of Zn-Pb ore flotation tailings with *Dianthus carthusianorum* and *Biscutella laevigata* after amending with mineral fertilizers or sewage sludge. J. Environ. Manage. 189: 75–83.

CNP. 2017. Productivity in the Chilean copper mining industry. Maval Spa, Santiago, Chile.

COCHILCO. 2019. Anuario de Estadisticas del Cobre y Otros Minerales 1999–2018. Santiago, Chile.

Cortés, S., E. E. Soto and J. I. Ordóñez. 2020. Recovery of copper from leached tailing solutions by biosorption. Minerals 10: 158.

Davis, T. A., B. Volesky and A. Mucci. 2003. A review of the biochemistry of heavy metal biosorption by brown algae. Water Res. 37: 4311–4330.

de Koning, A., R. Kleijn, G. Huppes, B. Sprecher, G. van Engelen and A. Tukker. 2018. Metal supply constraints for a low-carbon economy? Resour. Conserv. Recycl. 129: 202–208.

Diaby, N., B. Dold, H.-R. Pfeifer, C. Holliger, D. B. Johnson and K. B. Hallberg. 2007. Microbial communities in a porphyry copper tailings impoundment and their impact on the geochemical dynamics of the mine waste. Environ. Microbiol. 9: 298–307.

Dodson, J. R., A. J. Hunt, H. L. Parker, Y. Yang and J. H. Clark. 2012. Elemental sustainability: Towards the total recovery of scarce metals. Chem. Eng. Process. Process Intensif. 51: 69–78.

Fawzy, M. A. 2020. Biosorption of copper ions from aqueous solution by *Codium vermilara*: Optimization, kinetic, isotherm and thermodynamic studies. Adv. Powder Technol. 31: 3724–3735.

Fernández-González, R., M. A. Martín-Lara, I. Iáñez-Rodríguez and M. Calero. 2018. Removal of heavy metals from acid mining effluents by hydrolyzed olive cake. Bioresour. Technol. 268: 169–175.

Fu, F. and Q. Wang. 2011. Removal of heavy metal ions from wastewaters: A review. J. Environ. Manage. 92: 407–418.

Gadd, G. M. 2009. Biosorption: critical review of scientific rationale, environmental importance and significance for pollution treatment. J. Chem. Technol. Biotechnol. 84: 13–28.

Gaikwad, R. W., R. S. Sapkal and V. S. Sapkal. 2010. Removal of copper ions from acid mine drainage wastewater using ion exchange technique: factorial design analysis. J. Water Resour. Prot. 2: 984–989.

Gan, M., J. Li, S. Sun, J. Ding, J. Zhu, X. Liu and G. Qiu. 2018. Synergistic effect between sulfide mineral and acidophilic bacteria significantly promoted Cr(VI) reduction. J. Environ. Manage. 219: 84–94.

Gentina, J. and F. Acevedo. 2016. Copper bioleaching in Chile. Minerals 6: 23.

Ghorbani, Y. and S. H. Kuan. 2017. A review of sustainable development in the Chilean mining sector: past, present and future. Int. J. Mining, Reclam. Environ. 31: 137–165.

Henckens, M. L. C. M. and E. Worrell. 2020. Reviewing the availability of copper and nickel for future generations The balance between production growth, sustainability and recycling rates. J. Clean. Prod. 264: 121460.

Hertwich, E. G., T. Gibon, E. A. Bouman, A. Arvesen, S. Suh, G. A. Heath, J. D. Bergesen, A. Ramirez, M. I. Vega and L. Shi. 2015. Integrated life-cycle assessment of electricity-supply scenarios confirms global environmental benefit of low-carbon technologies. Proc. Natl. Acad. Sci. 112: 6277–6282.

Hu, X., J. Cao, H. Yang, D. Li, Y. Qiao, J. Zhao, Z. Zhang and L. Huang. 2020. Pb2+ biosorption from aqueous solutions by live and dead biosorbents of the hydrocarbon-degrading strain *Rhodococcus* sp. *HX-2*. PLoS One 15: e0226557.

Iakovleva, E., E. Mäkilä, J. Salonen, M. Sitarz, S. Wang and M. Sillanpää. 2015. Acid mine drainage (AMD) treatment: Neutralization and toxic elements removal with unmodified and modified limestone. Ecol. Eng. 81: 30–40.

Ilyas, N., S. Ilyas, Sajjad-ur-Rahman, S. Yousaf, A. Zia and S. Sattar. 2018. Removal of copper from an electroplating industrial effluent using the native and modified *Spirogyra*. Water Sci. Technol. 78: 147–155.

Indhumathi, P., S. Sathiyaraj, J. P. Koelmel, S. U. Shoba, C. Jayabalakrishnan and M. Saravanabhavan. 2018. The efficient removal of heavy metal ions from industry effluents using waste biomass as low-cost adsorbent: thermodynamic and kinetic models. Zeitschrift für Phys. Chemie 232: 527–543.

Jasso-Chávez, R., M. L. Campos-García, A. Vega-Segura, G. Pichardo-Ramos, M. Silva-Flores, M. G. Santiago-Martínez, R. D. Feregrino-Mondragón, R. Sánchez-Thomas, R. García-Contreras, M. E. Torres-Márquez and R. Moreno-Sánchez. 2021. Microaerophilia enhances heavy metal biosorption and internal binding by polyphosphates in photosynthetic *Euglena gracilis*. Algal Res. 58: 102384.

Kamar, F. H., A. A. Mohammed, A. A. H. Faisal, A. C. Nechifor and G. Nechifor. 2016. Biosorption of lead, copper and cadmium ions from industrial wastewater using fluidized bed of dry cabbage leaves. Rev. Chim. 67: 1039–1046.

Katiyar, P. K. and N. S. Randhawa. 2020. A comprehensive review on recycling methods for cemented tungsten carbide scraps highlighting the electrochemical techniques. Int. J. Refract. Met. Hard Mater. 90: 105251.

Khalil, A., L. Hanich, A. Bannari, L. Zouhri, O. Pourret and R. Hakkou. 2013. Assessment of soil contamination around an abandoned mine in a semi-arid environment using geochemistry and geostatistics: Pre-work of geochemical process modeling with numerical models. J. Geochemical Explor. 125: 117–129.

Khosravi, A., M. Javdan, G. Yazdanpanah and M. Malakootian. 2020. Removal of heavy metals by *Escherichia coli* (*E. coli*) biofilm placed on zeolite from aqueous solutions (case study: the wastewater of Kerman Bahonar Copper Complex). Appl. Water Sci. 10: 167.

Kotrba, P., M. Mackova and T. Macek. 2011. Microbial Biosorption of Metals. Springer Netherlands, Dordrecht, Netherlands.

Kottangodan, N., C. Das, A. Ram, R. M. Meena and N. Ramaiah. 2019. Phycoremediation of hazardous mixed industrial effluent by a marine strain of *Phormidium* sp. CLEAN – Soil, Air, Water 47: 1800264.

Krishna, R. S., J. Mishra, S. Meher, S. K. Das, S. M. Mustakim and S. K. Singh. 2020. Industrial solid waste management through sustainable green technology: Case study insights from steel and mining industry in Keonjhar, India. Mater. Today Proc. 33: 5243–5249.

Kuter, N. 2013. Reclamation of degraded landscapes due to opencast mining. *In*: Özyavuz, M. (ed). Advances in Landscape Architecture. IntechOpen.

Leong, Y. K. and J.-S. Chang. 2020. Bioremediation of heavy metals using microalgae: Recent advances and mechanisms. Bioresour. Technol. 303: 122886.

Liu, G., L. Tao, X. Liu, J. Hou, A. Wang and R. Li. 2013. Heavy metal speciation and pollution of agricultural soils along Jishui River in non-ferrous metal mine area in Jiangxi Province, China. J. Geochemical Explor. 132: 156–163.

Lo, W., H. Chua, M. Wong and P. Yu. 2003. Bacterial biosorbent for removing and recovering copper from electroplating effluents. Water Sci. Technol. 47: 251–256.

Lukacs, H. and L. Ortolano. 2015. West Virginia has not directed sufficient resources to treat acid mine drainage effectively. Extr. Ind. Soc. 2: 194–197.

Luna, A. S., A. L. H. Costa, A. C. A. da Costa and C. A. Henriques. 2010. Competitive biosorption of cadmium(II) and zinc(II) ions from binary systems by *Sargassum filipendula*. Bioresour. Technol. 101: 5104–5111.

Lutter, S. and S. Giljum. 2019. Copper production in Chile requires 500 million cubic metres of water. An assessment of the water use by Chile's copper mining industry. FINEPRINT Brief No. 9. Vienna, Austria.

Machado, M. D., H. M. V. M. Soares and E. V. Soares. 2010. Removal of chromium, copper, and nickel from an electroplating effluent using a flocculent Brewer's yeast strain of *Saccharomyces cerevisiae*. Water. Air. Soil Pollut. 212: 199–204.

Marsden, J. O. and M. M. Botz. 2017. Heap leach modeling—A review of approaches to metal production forecasting. Miner. Metall. Process. 34: 53–64.

Masood, F. and A. Malik. 2011. Biosorption of metal ions from aqueous solution and tannery effluent by *Bacillus* sp. *FM1*. J. Environ. Sci. Heal. Part A 46: 1667–1674.

Mata, Y. N., M. L. Blázquez, A. Ballester, F. González and J. A. Muñoz. 2008. Characterization of the biosorption of cadmium, lead and copper with the brown alga *Fucus vesiculosus*. J. Hazard. Mater. 158: 316–323.

Motsi, T., N. A. Rowson and M. J. H. Simmons. 2011. Kinetic studies of the removal of heavy metals from acid mine drainage by natural zeolite. Int. J. Miner. Process. 101: 42–49.

Naidu, G., S. Ryu, R. Thiruvenkatachari, Y. Choi, S. Jeong and S. Vigneswaran. 2019. A critical review on remediation, reuse, and resource recovery from acid mine drainage. Environ. Pollut. 247: 1110–1124.

Oliveira, F., A. Soares, O. Freitas and S. Figueiredo. 2010. Copper, nickel and zinc removal by Peanut hulls: Batch and column studies in mono, tri-component systems and with real effluent. Glob. NEST J. 12: 206–214.

Ordóñez, J. I., L. Wong-Pinto and S. Cortés. 2021. Biotecnología aplicada a la valorización de relaves mineros. pp. 63–91. *In*: Cisternas, L., E. Gálvez, M. Rivas and J. Valderrama (eds.). Economía Circular en Procesos Mineros. RIL Editores, Santiago, Chile.

Owen, J. R., D. Kemp, É. Lèbre, K. Svobodova and G. Pérez Murillo. 2020. Catastrophic tailings dam failures and disaster risk disclosure. Int. J. Disaster Risk Reduct. 42: 101361.

Pandya, K. Y., R. V. Patel, R. T. Jasrai and N. Brahmbhatt. 2017. Biosorption of Cr , Ni & Cu from industrial dye effluents onto *Kappaphycus alvarezii*: Assessment of sorption isotherms and kinetics. Int. J. Eng. Res. Gen. Sci. 5: 137–148.

Parekh, B. K. and J. D. Miller. 1999. Advances in flotation technology. Society for Mining, Metallurgy and Exploration. Inc (SME), Littleton.

Park, S. and M. Lee. 2017. Removal of copper and cadmium in acid mine drainage using Ca-alginate beads as biosorbent. Geosci. J. 21: 373–383.

Perumal, S. V., U. M. Joshi, S. Karthikeyan and R. Balasubramanian. 2007. Biosorption of lead(II) and copper(II) from stormwater by brown seaweed *Sargassum* sp.: Batch and column studies. Water Sci. Technol. 56: 277–285.

Piervandi, Z., A. Khodadadi Darban, S. M. Mousavi, M. Abdollahy, G. Asadollahfardi, V. Funari, E. Dinelli, R. D. Webster and M. Sillanpää. 2020. Effect of biogenic jarosite on the bio-immobilization of toxic elements from sulfide tailings. Chemosphere 258: 127288.

Pino, L., C. Vargas, A. Schwarz and R. Borquez. 2018. Influence of operating conditions on the removal of metals and sulfate from copper acid mine drainage by nanofiltration. Chem. Eng. J. 345: 114–125.

Plaza Cazón, J., M. Viera, E. Donati and E. Guibal. 2011. Biosorption of mercury by *Macrocystis pyrifera* and *Undaria pinnatifida*: Influence of zinc, cadmium and nickel. J. Environ. Sci. 23: 1778–1786.

Plaza Cazón, J., L. Benítez, E. Donati, M. Viera, J. P. H. Cazón, L. Benítez, E. Donati and M. Viera. 2012a. Biosorption of chromium(III) by two brown algae *Macrocystis pyrifera* and *Undaria pinnatifida*: Equilibrium and kinetic study. Eng. Life Sci. 12: 95–103.

Plaza Cazón, J., C. Bernardelli, M. Viera, E. Donati and E. Guibal. 2012b. Zinc and cadmium biosorption by untreated and calcium-treated *Macrocystis pyrifera* in a batch system. Bioresour. Technol. 116: 195–203.

Pollmann, K., S. Kutschke, S. Matys, J. Raff, G. Hlawacek and F. L. Lederer. 2018. Bio-recycling of metals: Recycling of technical products using biological applications. Biotechnol. Adv. 36: 1048–1062.

Punia, A. 2021. Role of temperature, wind, and precipitation in heavy metal contamination at copper mines: A review. Environ. Sci. Pollut. Res. 28: 4056–4072.

Rankin, W. J. 2011. Minerals, Metals and Sustainability. CSIRO Publishing, Collingwood, Victoria.

Reyes, A., J. Cuevas, B. Fuentes, E. Fernández, W. Arce, M. Guerrero and M. V. Letelier. 2021. Distribution of potentially toxic elements in soils surrounding abandoned mining waste located in Taltal, Northern Chile. J. Geochemical Explor. 220: 106653.

Ricci, B. C., C. D. Ferreira, A. O. Aguiar and M. C. S. Amaral. 2015. Integration of nanofiltration and reverse osmosis for metal separation and sulfuric acid recovery from gold mining effluent. Sep. Purif. Technol. 154: 11–21.

Romera, E., F. González, A. Ballester, M. L. Blázquez and J. A. Muñoz. 2007. Comparative study of biosorption of heavy metals using different types of algae. Bioresour. Technol. 98: 3344–3353.

Schlesinger, M., M. King, K. Sole and W. Davenport. 2011. Extractive Metallurgy of Copper. Fifth. Elsevier Ltd, Oxford, UK.

Skousen, J., C. E. Zipper, A. Rose, P. F. Ziemkiewicz, R. Nairn, L. M. McDonald and R. L. Kleinmann. 2017. Review of passive systems for acid mine drainage treatment. Mine Water Environ. 36: 133–153.

Solgi, E. and J. Parmah. 2015. Analysis and assessment of nickel and chromium pollution in soils around Baghejar Chromite Mine of Sabzevar Ophiolite Belt, Northeastern Iran. Trans. Nonferrous Met. Soc. China 25: 2380–2387.

Tolonen, E.-T., A. Sarpola, T. Hu, J. Rämö and U. Lassi. 2014. Acid mine drainage treatment using by-products from quicklime manufacturing as neutralization chemicals. Chemosphere 117: 419–424.

Tutu, H., T. S. McCarthy and E. Cukrowska. 2008. The chemical characteristics of acid mine drainage with particular reference to sources, distribution and remediation: The Witwatersrand Basin, South Africa as a case study. Appl. Geochemistry 23: 3666–3684.

US Geological Survey (USGS). 2021. Mineral Commodities Summaries 2021. 204.

USEPA. 2021. TENORM: Copper Mining and Production Wastes.

Veit, M. T., C. R. G. Tavares, S. M. Gomes-da-Costa and T. A. Guedes. 2005. Adsorption isotherms of copper(II) for two species of dead fungi biomasses. Process Biochem. 40: 3303–3308.

Vidal, O., H. Le Boulzec and C. François. 2018. Modelling the material and energy costs of the transition to low-carbon energy. pp. 1–14. *In*: EPJ Web of Conferences. EDP Sciences, Varenna, Italy.

Vijayaraghavan, K., J. Jegan, K. Palanivelu and M. Velan. 2005. Batch and column removal of copper from aqueous solution using a brown marine alga *Turbinaria ornata*. Chem. Eng. J. 106: 177–184

Vijayaraghavan, K. and Y.-S. Yun. 2008. Bacterial biosorbents and biosorption. Biotechnol. Adv. 26: 266–291.

Volesky, B. 2001. Detoxification of metal-bearing effluents: biosorption for the next century. Hydrometallurgy 59: 203–216.

WHO. 2017. Guidelines for drinking-water quality, 4th edition, incorporating the 1st addendum. Geneva.

Wong-Pinto, L.-S., A. Mercado, G. Chong, P. Salazar and J. I. Ordóñez. 2021. Biosynthesis of copper nanoparticles from copper tailings ore – An approach to the 'Bionanomining'. J. Clean. Prod. 315: 128107.

Wu, M., J. Liang, J. Tang, G. Li, S. Shan, Z. Guo and L. Deng. 2017. Decontamination of multiple heavy metals-containing effluents through microbial biotechnology. J. Hazard. Mater. 337: 189–197.

Xie, Y., W. Xiong, J. Yu, J.-Q. Tang and R. Chi. 2018. Recovery of copper from metallurgical sludge by combined method of acid leaching and biosorption. Process Saf. Environ. Prot. 116: 340–346.

Yadav, P., J. Singh and V. Mishra. 2019. Biosorption-cum-bioaccumulation of heavy metals from industrial effluent by brown algae: deep insight. pp. 249–270. *In*: Tripathi, V., P. Kumar, P. Tripathi and A. Kishore (eds.). Microbial Genomics in Sustainable Agroecosystems. Springer Singapore, Singapore.

Yahya, M. D., K. S. Obayomi, B. A. Orekoya, A. G. Olugbenga and B. Akoh. 2020. Process evaluation study on the removal of Ni(II) and Cu(II) ions from an industrial paint effluent using kola nut pod as an adsorbent. J. Dispers. Sci. Technol. 1–9.

Yin, W., X. Bai, X. Zhang, J. Zhang, X. Gao and W. W. Yu. 2019. Multicolor Light-Emitting Diodes with MoS 2 Quantum Dots. Part. Part. Syst. Charact. 36: 2–6.

Zhang, H., X. Hu and H. Lu. 2017. Ni(II) and Cu(II) removal from aqueous solution by a heavy metal-resistance bacterium: Kinetic, isotherm and mechanism studies. Water Sci. Technol. 76: 859–868.

Zhuang, P., M. B. McBride, H. Xia, N. Li and Z. Li. 2009. Health risk from heavy metals via consumption of food crops in the vicinity of Dabaoshan mine, South China. Sci. Total Environ. 407: 1551–1561.

Zinicovscaia, I., N. Yushin, D. Grozdov, K. Vergel, T. Ostrovnaya and E. Rodlovskaya. 2020. Metal removal from complex copper containing effluents by waste biomass of *Saccharomyces cerevisiae*. Ecol. Chem. Eng. S 27: 415–435.

Zwolak, A., M. Sarzyńska, E. Szpyrka and K. Stawarczyk. 2019. Sources of soil pollution by heavy metals and their accumulation in vegetables: a review. Water. Air. Soil Pollut. 230: 164.

Chapter 21

Remediation of Contaminated Chromite Mine Spoil by Biochar Application

Dipita Ghosh,[1] *Manish Kumar,*[2] *Nabin Kumar Dhal*[2] and *Subodh Kumar Maiti*[1,*]

21.1 Introduction

Heavy metal mining activities cause the dispersion of toxic metals in the air, soil, and water, which causes hazards to human health and the environment. Chromite mining causes severe environmental pollution and degradation due to hexavalent chromium [Cr (VI)], especially in friable ore due to its carcinogenic nature. The Cr (VI) contamination of water bodies is a major issue that requires attention. 0.14% of India is under mining activities and stands 4th in production of chromite ores across the world (Table 21.1). Some of the common sources of Cr [III] include CrB, $CrCl_3.6H_2O$, and $KCr(SO_4)_2$ $12H_2O$. Cr [VI] is obtained from $(NH_4)_2CrO_4$, K_2CrO_4, $PbCrO_4$, $CrCuO_4$, copper–chromium–arsenic (CCA), composites of copper–chromium–boron (CCB), copper–chromium–fluoride (CCF), and copper–chromium–phosphate (CCP). In 2019–2020, the state of Odisha produced about 100% of the total chromite ores of India (IBM 2021). The Sukinda Valley is the most important chromite mining site of Odisha and holds about 183 MT of chromium deposits and spreads over 200 km^2 of area. Hence, the problems associated with chromite mining are quite pronounced in this region (Fig. 21.1).

Chromium is an industrial metal with wide application in metallurgical, steel, and chemical industries. It is an important alloying metal in ferrous metallurgy, which is used in the manufacture of alloys. Opencast chromite mining contaminates the soil and water in the mining vicinity due to acid drainage, soil erosion, and contaminant leaching (Basu et al. 2015, Mandal et al. 2017). Thus, its large-scale contamination of the environment can affect human health and well-being, and the use of proper remediation measures is essential for environmental protection.

[1] Ecological Restoration Laboratory, Department of Environmental Science & Engineering, Indian Institute of Technology (ISM), Dhanbad 826 004, Jharkhand, India.

[2] CSIR-Institute of Minerals and Materials Technology, Environment Sustainability Department, Bhubaneswar, 751013, Odisha, India.

* Corresponding author: subodh@iitism.ac.in; skmism1960@gmail.com

Table 21.1. Chromium production worldwide in 2020 (Statista 2021).

Rank	Country	Million metric tons
1	South Africa	1600
2	Kazakhstan	6700
3	Turkey	6300
4	India	4000
5	Finland	2400
6	Other countries	4800
	Total	**25800**

Fig. 21.1. (a) Open-cast chromite mining site with mine lake formation in Sukhinda mines, Odisha, India; (b) A waste dump generated by chromite mining; (c) Wastewater accumulation from a chromite mine; (d) Phytoremediation of mine dumps by *Cymbopogon flexuosus* (Nees ex Steud.) W. Watson. (Photos taken in July 2020).

Hexavalent chromium can be reduced to less toxic trivalent chromium using various remediation techniques such as physio-chemical and biological techniques. Biochar is one such biological remediation technique which can remediate Cr toxicity by various mechanisms such as reduction, ion-exchange, precipitation, electrostatic attraction, complexation, and physical adsorption (Mandal et al. 2017, Uchimiya et al. 2011). Biochar is a pyrolysis product of biomass in low or no supply of oxygen at temperatures > 250°C and < 750°C (Ghosh and Maiti 2020, Lehmann 2007). Currently, biochar has attracted the attention of researchers due to its ability to contaminate remediation in an economically sound manner. The multiple domains in which biochar can be applied include: an agent for improving soil physico-chemical and biological properties, nutrient sink, an absorbent for heavy metal remediation, and carbon sequestration (Ghosh and Maiti 2021a, Karim et al. 2019,

Mohapatra et al. 2020). Biochar can support soil microbial growth and promote enzymatic activities such as dehydrogenase, amylase, cellulase, and invertase (Ghosh et al. 2021b).

Many studies have reported that biochar can effectively remediate Cr remediation and immobilize it in the substrate. Choudhary and Paul (2018) reported a removal ability of 21.3 mg/g of Cr (VI) by *Eucalyptus* bark biochar application. According to Shakya et al. (2019), the pH of the soil affects the Cr adsorption. Free radicals on biochar surfaces reduce Cr (VI) to less toxic groups (Zhao et al. 2018b). Modification of biochar improves the adsorption capacity of the biochar to remediate Cr (VI) contamination (Mohapatra et al. 2020, Zhao et al. 2017). Abbas et al. (2019) reported that acidified biochar at @3% decreases the mobility of Cr (III) and Cr (VI) and hence reduces its uptake by plants. Similarly, Su et al. (2021) reported an increase in Cr (VI) adsorption by Fe (0) and Fe (II) modification. In conclusion, the current chapter focuses on the scope of biochar application for chromite mine spoil remediation, biochar production, and application conditions and mechanisms involved.

21.2 Common Cr Remediation Techniques

21.2.1 Physico-chemical techniques

Chemical, membrane, ion exchange, and electrocoagulation are some of the common physiochemical techniques used for Cr remediation (Fig. 21.2). Chemical methods include using chemical agents, which reduce the highly toxic Cr (VI) to Cr (III). The common chemical agents include citric acid, hydrochloric acid, oxalic acid, and sulphuric acid (Sun et al. 2019). The membrane method uses semipermeable membrane for the separation of metals; the membrane between the phases restricts ions' and molecules' movement and remediates the Cr toxicity (Malaviya and Singh 2011). In

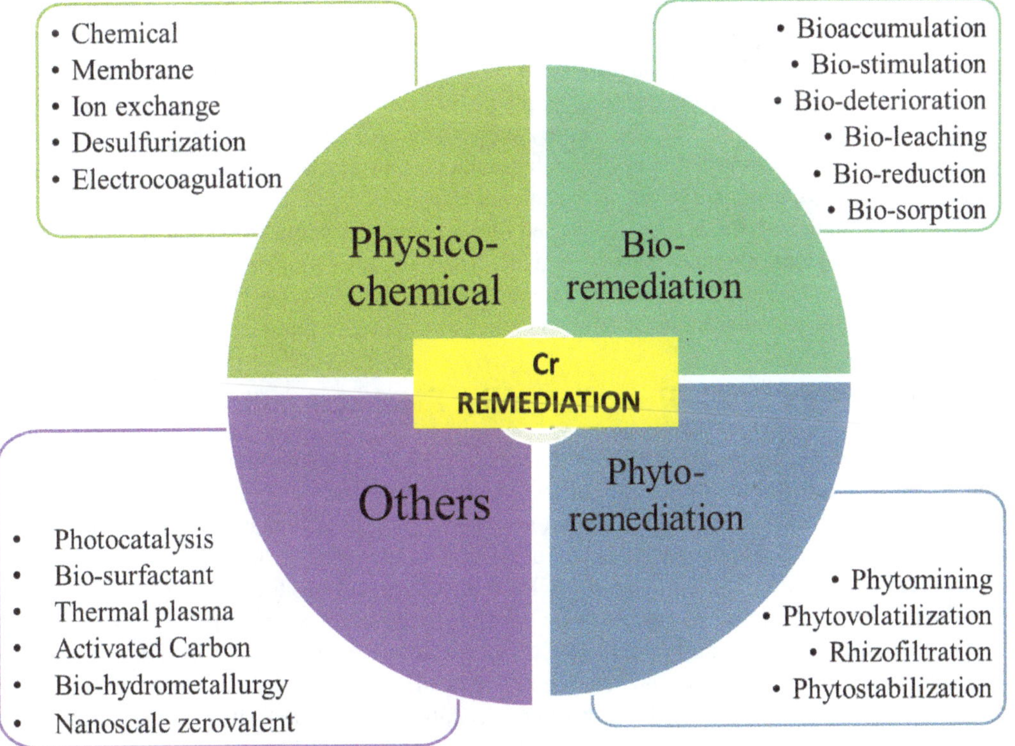

Fig. 21.2. Classification of chromium remediation techniques.

ion-exchange remediation, polymeric resins are usually used for removing Cr (VI) and Cr (III) contamination. This method has fast kinetics, removal efficiency, and high treatment capacity (Nayak et al. 2020). Electrocoagulation of Cr contamination is done by combined effects of chemical precipitation, co-precipitation, coagulation, and adsorption. According to Golder et al. (2007), at higher current density and pH, high removal efficacy of Cr (III) was observed.

21.2.2 Bioremediation

Bioaccumulation, bio-stimulation, bio-deterioration, bio-leaching, bio-reduction, and bio-sorption are some common bioremediation techniques used for Cr contamination (Fig. 21.2). According to Basu et al. (2015), a number of microorganisms catalyze the reduction of Cr (VI) to Cr (V) or Cr (III) in various environmental conditions. Cr (VI) reduction is shown to be metabolic in some species of bacteria but can also be dissimilatory/respiratory when exposed to anaerobic conditions. Although most microbes are sensitive to Cr (VI), some microbes are highly resistant and can tolerate Cr (VI) toxicity in the soil. Metal reductase genes found on plasmids and chromosomes impart the resistance to these microbes for growth in Cr (VI) environment (Patra et al. 2017). Some common microbes that have the potential for Cr remediation include *Acinetobacter*, *Arthrobacter*, *Bacillus* spp., *Cellulomonas* spp., *E. coli*, *Enterobacter cloacae*, *Pseudomonas*, and *Ochrobactrum* (Hossan et al. 2020).

21.2.3 Phytoremediation

Phytoremediation is a bioremediation technique in which hardy and toxicity tolerant vegetation is grown in the contaminated environment to bio-accumulate the toxic metal elements and remediate the soil in which it grows. It is a cost effective and ecologically viable solution for chromite mining remediation. Plant, soil, and microbe interaction forms the basis of phytoremediation and can be broadly divided as phytoextraction (phytoaccumulation), rhizofiltration, phytostabilization, phytovolatilization, and phytodegradation (Fig. 21.2). Rhizoextraction and phytostabilization are some of the most common techniques used by plants for Cr (VI) remediation in contaminated mine spoil. The toxic heavy metals are absorbed by the root system and fixed in the plant body parts. Some common hyperaccumulators used as phytoremediation agents include *Agrostis castellana*, *Brassica napus*, *Melastoma malabathricum*, and *Salix* spp.

Other remediation methods commonly used for Cr toxicity remediation include photocatalysis, biopiles, bio-surfactant, thermal plasma, activated carbon, bio-hydrometallurgy, and nanoscale zerovalent (Fig. 21.2). Biochar is one such remediation method that has immense Cr remediation potential when applied either by itself or in combination with other remediation techniques.

21.3 Mechanism for Cr Remediation by Biochar Application

The most common mechanisms by which biochar can remediate Cr (VI) toxicity include ion-exchange, complexation, electrostatic attraction, oxido-reduction, physical adsorption, and precipitation (Fig. 21.3). In the mechanism involving ion exchange, high cation exchange capacity of biochars causes the release of calcium and magnesium ions, which exchanges with Cr (VI)/Cr (III) on biochar surfaces (Ghosh and Maiti 2021a). The high CEC of biochar increases the adsorption capability for heavy metals. Complexation is a process by which Cr ions form complexes with –COOM and –R–O–M groups. Biochars can immobilize Cr(VI) by surface complexation by certain functional groups on their surface. Biochar contains high content of NH_2- C=O, which causes strong bonds with Cr ions. Minerals present on biochar often precipitate with Cr to form insoluble precipitates (Masto et al. 2013, Uchimiya et al. 2011). According to Sun et al. (2014), biochar has high sorption

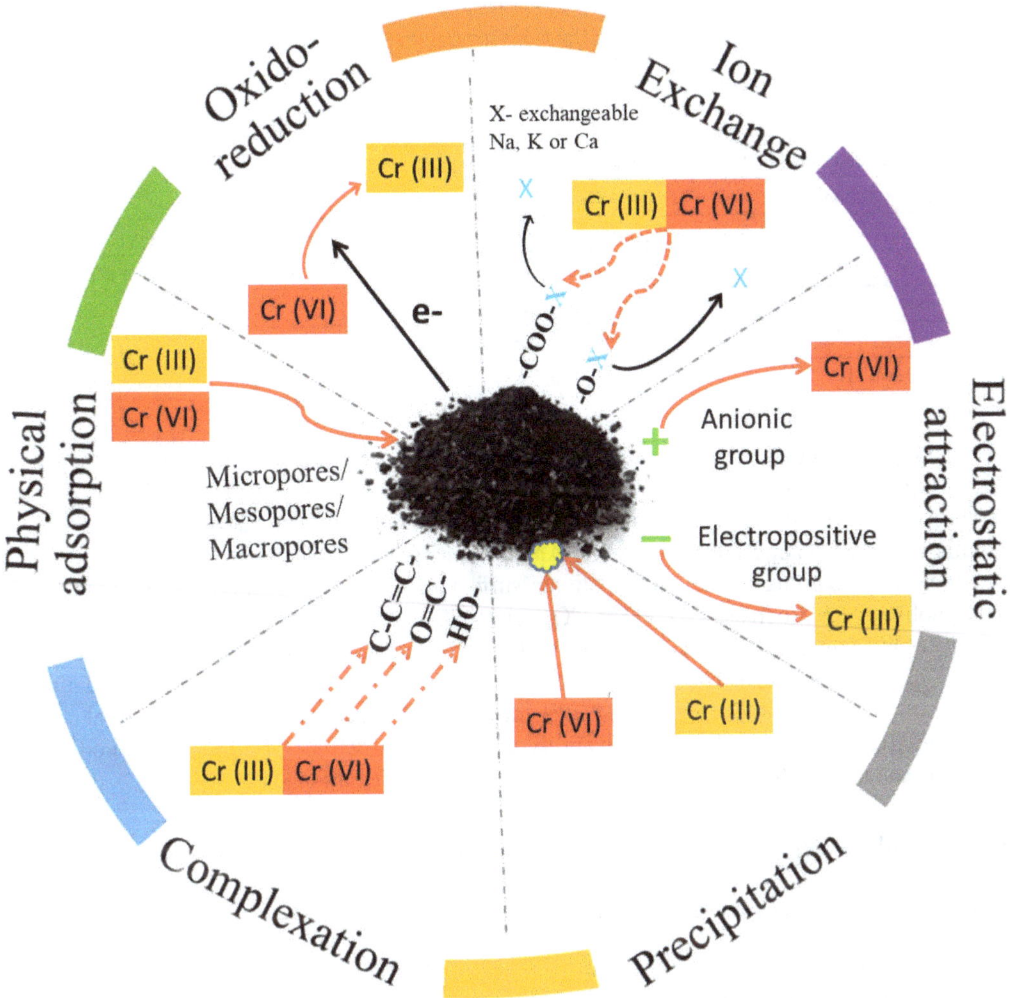

Fig. 21.3. Mechanisms involved in biochar based Cr remediation.

capacity as heavy metals precipitates with minerals such as PO_4^{3-} and CO_3^{2-} present in the biochar surfaces. Similarly, the biochar surface is electronegative in nature and possesses zeta potential, which makes the absorption of positive Cr ions on the biochar surface very easy. The pi-pi e^- donor and acceptor interaction between biochar's pi e^- rich graphene surface and pi e^- deficient positively charged Cr ions immobilize the Cr ion. This method is known as electrostatic attraction. An increase in surface charge is reported in an alkaline medium compared to an acidic medium (Ghosh and Maiti 2020). Thus, an alkaline medium is more suited for Cr(VI) remediation.

21.4 Biochar Modification Techniques

Biochar modification has a better ability to immobilize heavy metals. Modification of biochar improves porosity and changes the surface functional groups. Some methods of biochar modification are loading with minerals, organic functional groups, reductants, and treatment with alkali solution (Table 21.2). While modifying the biochar, certain characteristics such as stability should not be affected as it might alter the basic recalcitrant nature of the biochar. Also, very little research has

Table 21.2. Biochar modification methods for chromite remediation.

Raw Material	**Reagent**	**References**
Eucalyptus	H_3PO_4	Zeng et al. 2021
Sugarcane bagasse	Nickel and nitrogen hybrid carbon nanotubes	Zhu et al. 2021
Auricularia auricular dreg (AAD)	Cetyltrimethyl ammonium bromide (CTAB)	Li et al. 2018
Poultry, cow, sheep manure	Chitosan, ZVI	Mandal et al. 2017
Sugarcane bagasse	Zn nanoparticles	Gan et al. 2015
Bamboo	Zero valent iron	Zhou et al. 2014
Peanut straw	$Na_2SO_3/FeSO_4$	Pan et al. 2014
Rice husk	polyethyleneimine	Ma et al. 2014
Bamboo hardwoods	Sulfur-iron	Wu et al. 2013

been done on producing designer biochar on a large scale for mine soil remediation. Zeng et al. (2021) reported that H_3PO_4 treated Eucalyptus biochar immobilized 99.76% of Cr (VI), which was higher than the pristine biochar removal rate of 25.24%. Li et al. (2018) studied the use of cationic surfactant, namely Cetyltrimethyl Ammonium Bromide, to modify *Auricularia auricula* dreg biochar. According to the study, cationic surfactant modification improved hydrophilic and hydrophobic groups, such as –OH, PO_4^{2-}, CO_3^{2-} and –COO, which helped to improve the adsorption capacity of biochar for Cr(VI). It was observed that surfactant modification increased the adsorption rate and quantity for Cr(VI) by 8.0%. Similarly, Zhu et al. (2021) reported that nickel and nitrogen hybrid carbon nanotube (CNT) modified sugarcane bagasse biochar had a higher removal capacity (824.4 mg/g) for Cr(VI) and a stronger ability to reduce to Cr(III). Thus, a biochar modification technique can be an effective method for Cr contaminant remediation.

21.5 Biochar Application for the Remediation of Cr Contaminated Soil

21.5.1 Cr Immobilization in the Chromite Mine Spoil

Cr mobility and bio-availability are functions of both biochar characteristics and soil in which biochar is applied for amendment (Ghosh and Maiti 2021a). Application of biochar has been reported to show dual interfering impacts on Cr (VI). Biochar improves the humic content in the soil, which in turn reduces the surface area of biochar by clogging the pores present in its surface, which inhibits the adsorption of Cr on the biochar surface. On the contrary, humic acid can catalyze the conversion of Cr (VI) to Cr (III) by transferring electrons from biochar to Cr (VI) ions (Zheng et al. 2021). Biochar has been reported to reduce Cr (VI) more efficiently in acidic soil (Mandal et al. 2017). Biochar applications for remediation of Cr contamination in various contaminated soil are given in Table 21.3. A field-based study by Khan et al. (2020a) reported that hardwood biochar @5% (w/w) decreased the concentrations of Cr by 25.5% in comparison with the control. Another biochar study conducted by Khan et al. (2020b) reported that relative to controls, the poplar wood biochar and sugarcane bagasse biochar reduced Cr uptake in lettuce by 69%, and 73.7%, respectively. A study conducted by Mohapatra et al. (2020) on the application of *Ipomea* and corn cob biochar on chromite contaminated mine spoil reported that modified corn cob biochar adsorbed comparatively higher Cr. It also reduced relatively higher water leachable and enhanced phytoavailable Cr fraction in the overburden. In conclusion, biochar-assisted Cr remediation in the soil is dependent on the types of plants used, microbial growth, soil properties, and biochar applied.

Table 21.3. Application of biochar for remediation of chromite contamination.

Feedstock	Pyrolysis condition	Type of study	Country	Conclusion	References
Sugarcane bagasse	500°C for 2 h	Pot experiment	China	Biochar @ 3% reduced mobility of Cr(III) and Cr(VI) by 44 and 22%, respectively, and decreased its uptake by maize plant.	Abbas et al. 2019
Poplar wood and sugarcane bagasse	550°C for 1 h	Greenhouse experiment	Pakistan	Biochar application reduced Cr uptake in lettuce by 69% to 73.7 %, respectively.	Khan et al. 2020a
Hardwood	500°C for 1 h	Field study	Pakistan	Biochar @5% (w/w) decreased the concentrations of Cr, by 25.5% as compared to control.	Khan et al. 2020b
Corncob and *Ipomoea*	300, 500, and 700°C	Batch absorption experiment	India	Modified corn cob biochar adsorbed Cr (III) (> 90%) and Cr (VI) (> 75%). It reduced water leachable Cr content from 4.88 to 0.87 mg/L and enhanced phytoavailable Cr fraction from 5.73 to 7.41 mg/L.	Mohapatra et al. 2020
Poplar	900°C	Desorption experiments	China	Modified biochar removed aqueous Cr (VI) in a wide pH range. Un-modified biochar application removed 56.8% of Cr (VI) in neutral pH.	Su et al. 2021
Corn straws	300°C for 1 h	Adsorption experiment	China	The adsorption capacity of modified biochar was 476.19 mg g^{-1}. Cr (VI) was remediated by biochar via chemical complexation mechanism.	Zhao et al. 2017
Corn straw	300°C, 500°C, and 700°C for 2 h.	Adsorption experiment	China	Free radicals on biochar surfaces were found to play a role in reducing Cr (VI) to Cr (III).	Zhao et al. 2018a
Sweet lime peel	450°C	Batch adsorption experiment	India	The Cr (VI) adsorption on the biochar surface was favorable at pH of 2; alkaline pH suppressed the adsorption of Cr. The maximum removal efficiency was found to be 95%.	Shakya et al. 2019
Eucalyptus bark	500°C	Adsorption Experiment	India	The maximum removal capacity by biochar application was 21.3 mg/g.	Choudhary and Paul 2018
Barley grass	350°C for 4 h	Column experiment	China	Iron modified biochar amendment removed 71% Cr from contaminated groundwater. The leachability of Cr was reduced by over 81%.	Chen et al. 2021
Bagasse	350°C	Pot experiment	Pakistan	Biochar application reduced Cr (VI) concentration from 18.6 mg/kg to 2 mg/kg and Cr(III) from 64 mg/kg to 54.9 Cr(III) in soil.	Bashir et al. 2020
Walnut	400°C for 2 h	Column Experiment	Iran	Biochar application reduced $CaCl_2$-leachable Cr (VI) by 56.6%	Matin et al. 2020
Sheep manure	450°C for 1 h	Incubation Experiment	South Africa	Biochar application reduced Cr (VI) in soil by 95.28 mg/kg	Mandal et al. 2017

21.5.2 Improvement of Physio – chemical Properties

Application of biochar ameliorates the mine spoil properties by improving the physio-chemical, biological, and nutritive content of the mine spoil (Ghosh et al. 2020, Novak et al. 2019). Application of biochar has been reported to reduce the bulk density (~ 30%), increase porosity (15%–65%), aggregate stability (2%–200%), and decrease particle size density (65%) and tensile strength (42–242%) (Blanco-Canqui 2017). Biochar application can effectively improve soil chemical properties such as pH, cation exchange capacity, and soil organic matter in a mine spoil (Ghosh and Maiti 2020, Liu et al. 2018). Apart from the direct improvement in mine spoil properties, the application of biochar influences microbial activity and plant growth, which in turn influences the immobilization of Cr in soil (Liu et al. 2020a, Mohan et al. 2011). Physicochemical and biological changes by biochar application improve the tolerance of plants to Cr (VI) toxicity. Many studies have reported an increase in the nutritional value of the fruits growing in biochar amended soil (Zheng et al. 2021). The Cr intake and bioaccumulation in plant tissues have also been reported to be reduced by soil ameliorating properties of biochar (Bashir et al. 2020, Khan et al. 2013).

21.5.3 Improvement in Biological Properties

Due to excessive toxicity and degradation of land, a chromite mine spoil is completely derelict and is unable to support microbial activity. This delays the reclamation process in a mine degraded land. The microbial activity ensures the decomposition and physical mixing of organic matter within the soil; the microbial cells act upon the plant litter, liberating plant nutrients and synthesizing soil organic carbons (Maiti 2013). Biochar ameliorates soil acidity, improves soil physio-chemical properties, and provides favorable conditions for microbial growth. The large pore volume and porosity make the microbes more habitable in adverse soil conditions. Biochar is also known to effect intra- and interspecific communication of organisms. N-acyl-homoserine lactone, a soil signaling molecule, is known to be adsorbed by biochar surface and promote its hydrolysis. Biochar application also influences the soil's enzymatic activities (Ghosh and Maiti 2021b). Allosteric regulator/inhibitor and soil physic-chemical properties are affected by biochar application which in turn affects the soil enzymatic and microbial activities.

21.6 Case Studies

21.6.1 Case Study 1: Hardwood Biochar for Remediating Cr Mining Contaminated Agricultural Soil in Pakistan

Khan et al. (2020b) conducted a field based biochar application study to determine the effect of hardwood biochar application on a Cr mine contaminated agricultural soil and analyze its effect on two varieties of *Oryza sativa* growth. Hardwood biochar produced at 500°C was applied at 3% (w/w) mine soil contaminated by (1) chromium; (2) manganese and (3) combination of chromium and manganese to suppress Cr mobility in soils and their bioavailability in two varieties of *O. sativa* plants. The study reported a significant ($P \leq 0.05$) reduction of heavy metal uptake in *O. sativa* plants cultivated in biochar amended soils. Biochar application significantly reduced the estimated daily dose of Cr by 99.1 % for variety 1 grown in the contaminated soil. The highest reduction in Cr accumulation was observed in variety 2 grain (Table 21.4) in chromite mine contaminated soil 14.4%. The highest Cr reduction of 86.2% was observed in manganese mine contaminated soil compared to the control treatments. Thus, the addition of hardwood biochar to mine contaminated soil effectively reduced the bioavailability of Cr and decreased its bioaccumulation in the *O. sativa* plants.

Table 21.4. Chromium concentrations (mg kg^{-1}) in rice grain cultivated in contaminated soil and biochar amended soils (Khan et al. 2020b).

	Chromite mine contaminated soil	**Manganese mine contaminated soil**	**Chromite-manganese mix mine contaminated soil**
Control 1 (variety 1)	76 ± 0.026	81.5 ± 0.023	62.5 ± 0.004
Hardwood biochar (3%)	70.5 ± 0.14	69.7 ± 0.024	55.5 ± 0.005
Control 2 (variety 2)	113 ± 0.01	548 ± 0.015	126 ± 0.023
Hardwood biochar (3%)	57.7 ± 0.024	75.5 ± 0.010	61.7 ± 0.011

21.6.2 Case Study 2: Biochar for Abatement of Cr-polluted Mine Wastewater and Overburden Material in India

Mohapatra et al. (2020) conducted a study on the effect of *Ipomoea* and corncob biochar and modified corncob biochar composites on a chromite mine spoil collected from Sukinda mine spoils, Odisha, India. Biochars were produced at 300, 500, and 700°C, and the modified corncob biochar composites were prepared by 1 M $FeSO_4{\cdot}7H_2O$ at 800°C. A batch adsorption experiment was conducted at different pH (2–10), and the experiment time varied from 5 min to 48 h. Modified corncob biochar composites adsorbed higher concentrations of Cr_{total} (> 90%) and Cr (VI) (> 75%) (Fig. 21.4). Modified corncob biochar composites @5% reduced water leaching to 0.87 mg/L from 4.88 mg/L and enhanced phytoavailable Cr fraction to 7.41 mg/L from 5.73 mg/L. The modified corncob biochar composites had a larger surface area, acidic pH (≤ 3) and resulted in better Cr pollution reduction and remediation potential. The study concluded that modified corncob biochar composites can be used for remediation on highly contaminated chromite mining sites.

21.7 Conclusions

Biochar has the ability to remediate Cr toxicity thereby, remediating a chromite mining degraded soil. In addition, the mobility and toxicity of Cr can also be reduced by its application. The immobilization mechanism implies that Cr (VI) can be reduced to Cr (III). Optimization of biomass feedstock, production conditions, application rate, and background study of soil is critical for effective remediation of mine soil. For successful Cr remediation, the knowledge of the type of contamination and the mechanism involved is vital. In-field biochar production will help to overcome the current cost constraints associated with biochar application. Although biochar is a prospective amendment medium, its field-based application is very limited. Thus, extensive research in this area would popularise its benefits and promote its large-scale applicability for mine restoration.

21.8 Future Recommendations

- Exploring the use of waste biomass and invasive weeds growing in mining sites as feedstock for biochar production and utilizing it as a soil amendment has scope for chromite mine remediation. Also, developing biochar standards for application rate, production conditions, and feedstock needs attention.
- Detailed cost-benefit analysis and life cycle assessment for Cr toxicity remediation, especially for modified biochar application on a large scale, needs to be considered.
- From a mine reclamation point of view, an *in situ*, built-in-place unit would be the most economical as it curtails the transportation and feedstock processing cost. Thus, technological advancement for producing biochar in the chromite mining site needs to be explored.
- Exploring more vegetation as hyperaccumulators for biochar-assisted phytoremediation of chromite mine needs attention.

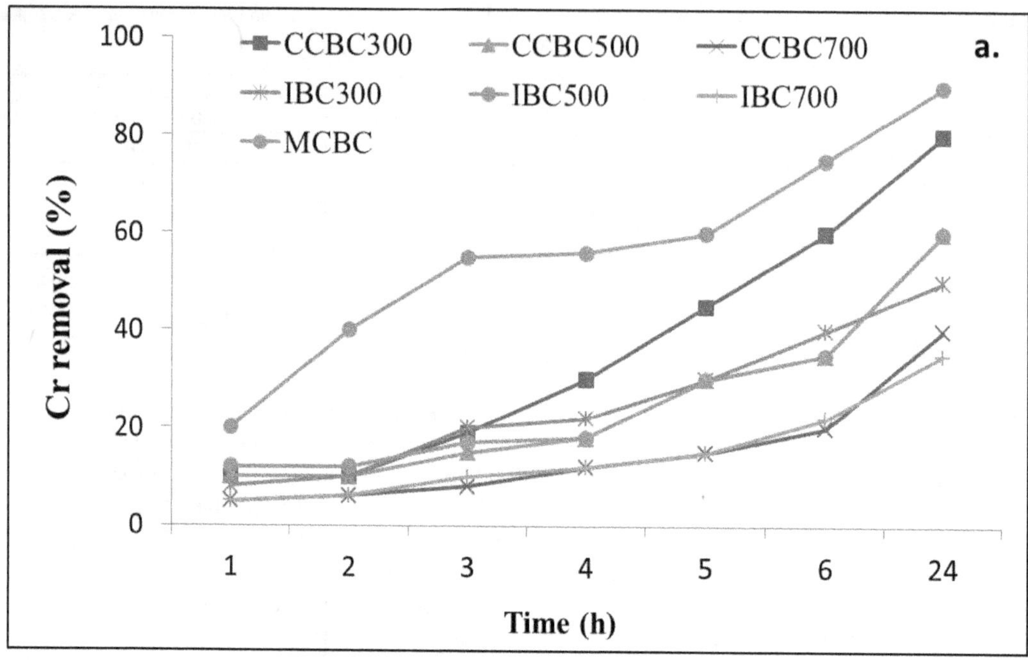

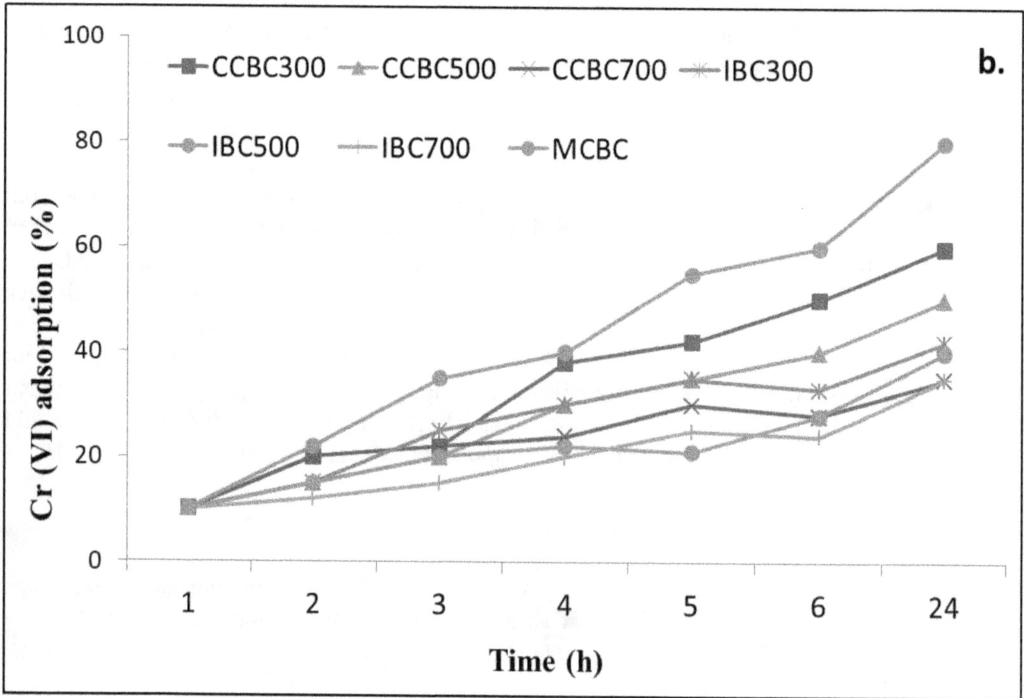

Fig. 21.4. (a) Removal efficacy of Cr (VI) by biochars from mine wastewater samples; (b) Adsorption of Cr (VI) by biochar samples (based on Mohapatra et al. 2020).

References

Abbas, A., M. Azeem, M. Naveed, A. Latif, A. Ali, M. Bilal and L. Ali. 2019. Synergistic use of biochar and acidified manure for improving growth of maize in chromium contaminated soil. Int. J. Phytoremediation 0: 1–10.

Bashir, A. M., M. Naveed, Z. Ahmad and B. Gao. 2020. Combined application of biochar and sulfur regulated growth, physiological, antioxidant responses and Cr removal capacity of maize (*Zea mays* L.) in tannery polluted soils J. Environ. Manage. 259: 110051.

Basu, A., S. Panda and N. K. Dhal. 2015. Potential microbial diversity in chromium mining areas : a review potential microbial diversity in chromium mining areas : a review. Bull. Environ. Pharmacol. Life Sci. 4: 158–169.

Blanco-Canqui, H. 2017. Biochar and soil physical properties. Soil Sci. Soc. Am. J. 81: 687–711.

Chen, X., Y. Dai, J. Fan, X. Xu and X. Cao. 2021. Application of iron-biochar composite in topsoil for simultaneous remediation of chromium-contaminated soil and groundwater : Immobilization mechanism and long-term stability. J. Hazard. Mater. 405: 124226.

Choudhary, B. and D. Paul. 2018. Isotherms, kinetics and thermodynamics of hexavalent chromium removal using biochar. J. Environ. Chem. Eng. 6: 2335–2343.

Gan, C., Y. G. Liu, X. F. Tan, S. F. Wang, G. M. Zeng, B. H. Zheng, T. T. Li, Z. J. Jiang and W. Liu. 2015. Effect of porous zincebiochar nanocomposites on Cr (VI) sorption from aqueous solution. RSC Adv. 5: 35107.

Ghosh, D. and S. K. Maiti. 2020. Can biochar reclaim coal mine spoil? J. Environ. Manage. 272: 111097.

Ghosh, D., R. E. Masto and S. K. Maiti. 2020. Ameliorative effect of *Lantana camara* biochar on coal mine spoil and growth of maize (*Zea mays*). Soil Use Manag. 36(4): 726–739.

Ghosh, D. and S. K. Maiti. 2021a. Biochar assisted phytoremediation and biomass disposal in heavy metal contaminated mine soils : a review. Int. J. Phytoremediation 0: 1–18.

Ghosh, D. and S. K. Maiti. 2021b. Effect of invasive weed biochar amendment on soil enzymatic activity and respiration of coal mine spoil: A laboratory experiment study, Biochar 3(4): 519–533.

Golder, A. K., A. N. Samanta and S. Ray. 2007. Removal of trivalent chromium by electrocoagulation. Sep. Purif. Technol. 53: 33–41.

Hossan, S., S. Hossain and M. R. Islam. 2020. Bioremediation of hexavalent chromium by chromium resistant bacteria reduces phytotoxicity. Int. J. Environ. Res. Public Health 17: 6013.

IBM. 2021. https://mines.gov.in/writereaddata/UploadFile/Mines_AR_2017-18_English_Final%2017052021.pdf.

Karim, A. A., M. Kumar, S. Mohapatra and S. Kumar. 2019. Nutrient rich biomass and effluent sludge wastes co-utilization for production of biochar fertilizer through different thermal treatments. J. Clean. Prod. 228: 570–579.

Khan, A. Z., S. Khan, T. Ayaz, M. L. Brusseau and M. Amjad. 2020a. Popular wood and sugarcane bagasse biochars reduced uptake of chromium and lead by lettuce from mine-contaminated soil. Environ. Pollut. 263: 114446.

Khan, A. Z., S. Khan, M. A. Khan and M. Alam. 2020b. Biochar reduced the uptake of toxic heavy metals and their associated health risk via rice (*Oryza sativa* L.) grown in Cr-Mn mine contaminated soils. Environ. Technol. Innov. 17: 100590.

Khan, S., C. Chao, M. Waqas, M., H. P. H. Arp and Y. G. Zhu. 2013. Sewage sludge biochar influence upon rice (*Oryza sativa* L.) yield, metal bioaccumulation and greenhouse gas emissions from acidic paddy soil. Environ. Sci. Technol. 47: 8624–8632.

Lehmann, J. 2007. A handful of carbon. Nature 447: 10–11.

Li, Y., Y. Wei, S. Huang, X. Liu, Z. Jin, M. Zhang, J. Qu and Y. Jin. 2018. Biosorption of Cr (VI) onto *Auricularia auricula* dreg biochar modified by cationic surfactant : Characteristics and mechanism. J. Mol. Liq. 269: 824–832.

Liu, L., W. Li, W. Song and M. Guo. 2018. Remediation techniques for heavy metal-contaminated soils: Principles and applicability. Sci. Total Environ. 633: 206–219.

Liu, N., Y. Zhang, C. Xu, P. Liu, J. Lv, Y. Y. Liu and Q. Wang. 2020a. Removal mechanisms of aqueous Cr(VI) using apple wood biochar: a spectroscopic study. J. Hazard. Mater. 384: 121371. https://doi.org/10.1016/j.jhazmat.2019.121371.

Liu, S., S. Pu, D. Deng, H. Huang, C. Yan, H. Ma and B. S. Razavi. 2020b. Comparable effects of manure and its biochar on reducing soil Cr bioavailability and narrowing the rhizosphere extent of enzyme activities. Environ. Int. 134: 105277. https://doi.org/10.1016/j.envint.2019.105277.

Ma, Y., W. J. Liu, N. Zhang, Y. S. Li, H. Jiang and G. P. Sheng. 2014. Polyethylenimine modified biochar adsorbent for hexavalent chromium removal from the aqueous solution. Bioresour. Technol. 169: 403–408.

Maiti, S. K. 2013. Ecorestoration of coal mine degraded lands. Springer, New Delhi.

Malaviya, P. and A. Singh. 2011. Physicochemical technologies for remediation of chromium-containing waters and wastewaters crit. Rev. Environ. Sci. Technol. 41: 1111–1172.

Mandal, S., B. Sarkar, N. Bolan, Y. S. Ok and R. Naidu. 2017. Enhancement of chromate reduction in soils by surface modified biochar. J. Environ. Manage. 186: 277–284.

Masto, R. E., M. A. Ansari, J. George, V. A. Selvi and L. C. Ram. 2013. Co-application of biochar and lignite fly ash on soil nutrients and biological parameters at different crop growth stages of *Zea mays*. Ecol. Eng. 58: 314–322.

Matin, H. N., M. Jalali and W. Buss. 2020. Synergistic immobilization of potentially toxic elements (PTEs) by biochar and nanoparticles in alkaline soil. Chemosphere 241: 124932.

Mohan, D., S. Rajput, V. K. Singh, P. H. Steele and C. U. Pittman. 2011. Modeling and evaluation of chromium remediation from water using low cost biochar, a green adsorbent. J. Hazard. Mater. 188: 319–333.

Mohapatra, S., M. Kumar, A. A. Karim and N. K. Dhal. 2020. Biochars evaluation for chromium pollution abatement in chromite mine wastewater and overburden of Sukinda, Odisha, India. Arab. J. Geosci. 13(13): 14.

Nayak, S., S. Rangabhashiyam, P. Balasubramanian and P. Kale. 2020. A review of chromite mining in Sukinda Valley of India : impact and potential remediation measures. Int. J. Phytoremediation 1–15.

Novak, J. M., J. A. Ippolito, D. W. Watts, G. C. Sigua, T. F. Ducey and M. G. Johnson. 2019. Biochar compost blends facilitate switchgrass growth in mine soils by reducing Cd and Zn bioavailability. Biochar 1: 97–114.

Pan, J. J., J. Jiang and R. K. Xu. 2014. Removal of Cr (VI) from aqueous solutions by $Na_2SO_3/FeSO_4$ combined with peanut straw biochar. Chemosphere 101: 71–76.

Patra, J. M., S. S. Panda and N. K. Dhal. 2017. Biochar as a low-cost adsorbent for heavy metal removal : A review. Int. J. Res. Biosci. 6(1): 1–7.

Shakya, A., A. Núñez-delgado and T. Agarwal. 2019. Biochar synthesis from sweet lime peel for hexavalent chromium remediation from aqueous solution. J. Environ. Manage. 251: 109570.

Statista 2021. https://www.statista.com/statistics/1040991/mine-production-of-chromium-worldwide-by-country/.

Su, C., S. Wang, Z. Zhou, H. Wang, X. Xie et al. 2021. Chemical processes of Cr (VI) removal by Fe-modified biochar under aerobic and anaerobic conditions and mechanism characterization under aerobic conditions using synchrotron-related techniques. Sci. Total Environ. 768: 144604.

Sun, Y., B. Gao, Y. Yao, J. Fang, M. Zhang, Y. Zhou et al. 2014. Effects of feedstock type, production method, and pyrolysis temperature on biochar and hydrochar properties. Chem. Eng. J. 240: 574–578.

Sun, Y., F. Guan, W. Yang and F. Wang. 2019. Removal of chromium from a contaminated soil using oxalic acid, citric acid, and hydrochloric acid : dynamics, mechanisms, and concomitant removal of non-targeted metals. Int. J. Environ. Res. Public Health 16: 2771.

Uchimiya, M., K. T. Klasson, L. H. Wartelle and I. M. Lima. 2011. Influence of soil properties on heavy metal sequestration by biochar amendment: 1. Copper sorption isotherms and the release of cations. Chemosphere 82: 1431–1437.

Wu, F., Z. Jia, S. Wang, S. X. Chang and A. Startsev. 2013. Contrasting effects of wheat straw and its biochar on greenhouse gas emissions and enzyme activities in a Chernozemic soil. Biol. Fertil. Soils 49: 555–565.

Zeng, H., H. Zeng, H. Zhang, A. Shahab, K. Zhang, Y. Lu et al. 2021. Efficient adsorption of Cr (VI) from aqueous environments by phosphoric acid activated eucalyptus biochar. J. Clean. Prod. 286: 124964.

Zhao, N., C. Zhao, Y. Lv, W. Zhang, Y. Du, Z. Hao and J. Zhang. 2017. Adsorption and coadsorption mechanisms of Cr (VI) and organic contaminants on H_3PO_4 treated biochar. Chemosphere 186: 422–429.

Zhao, N., Y. Z. Lv, X. X. Yang, F. Huang and J. W. Yang. 2018a. Characterization and 2D structural model of corn straw and poplar leaf biochars. Environ. Sci. Pollut. Res. 25: 25789–25798.

Zhao, N., Z. Yin, F. Liu, M. Zhang, Y. Lv, Z. Hao and G. Pan. 2018b. Environmentally persistent free radicals mediated removal of Cr (VI) from highly saline water by corn straw biochars. Bioresour. Technol. 260: 294–301.

Zheng, C., Z. Yang, M. Si, F. Zhu, W. Yang, F. Zhao and Y. Shi. 2021. Application of biochars in the remediation of chromium contamination : Fabrication, mechanisms, and interfering species. J. Hazard. Mater. 407: 124376.

Zhou, Y. M., B. Gao, A. R. Zimmerman, H. Chen, M. Zhang and X. D. Cao. 2014. Biochar supported zerovalent iron for removal of various contaminants from aqueous solutions. Bioresour. Technol. 152: 538–542.

Zhu, D., J. Shao, Z. Li, H. Yang, S. Zhang and H. Chen. 2021. Nano nickel embedded in N-doped CNTs-supported porous biochar for adsorption-reduction of hexavalent chromium. J. Hazard. Mater. 416: 125693.

Index

About the Editors

Dr. Anju Malik is presently working as Associate Professor and Chairperson at Department of Energy and Environmental Sciences, Chaudhary Devi Lal University, Sirsa, India. She has completed her M.Phil. and Ph.D. from School of Environmental Sciences, Jawaharlal Nehru University, New Delhi. She was awarded with Belgian Government fellowship under bilateral cultural exchange programme at University of Ghent, Belgium. Dr. Anju has also received research scholarships from University Grant Commission (UGC) and Council of Scientific and Industrial Research (CSIR), New Delhi, India. She has also worked in the past as Scientist under Indian Council of Forestry Research and Education, Dehradoon, India. With more than 17 years of research and teaching experience, her research interest is focused on trace elemental contamination, speciation and remediation. She is a life member of The Indian Science Congress Association, Kolkata, India. She has guided several M.Sc., M.Phil. and Ph.D. students. She has published research papers in various peer-reviewed national and international journals of repute and also published book chapters in edited books published by reputed international publishers. She is a reviewer of several international journals.

Dr. Mohd. Kashif Kidwai is currently working as an Associate Professor in the Department of Energy and Environmental Sciences, Chaudhary Devi Lal University, Sirsa, Haryana, India. He has a Ph.D. in Environmental Sciences, University of Lucknow and National Botanical Research Institute (CSIR), Lucknow, India teaching various postgraduate courses and doctoral courses for more than 14 years in the Department of Energy and Environmental Sciences, Chaudhary Devi Lal University, Sirsa, Haryana, India. He has several publications into his credit including research and review articles in reputed national and international journals along with book chapters in edited books of reputed national and international publishers. He has participated and presented in various national and international conferences, etc., and is guiding both post graduate and doctoral students. He has been working on various environmental issues viz. water monitoring studies; agrochemical based phytotoxic studies of agriculturally important plants, development and application of fungal based biopesticides, etc. He is in process of conducting research on diverse environmental aspects.

Prof. (Dr.) Vinod Kumar Garg is presently working as Professor at the Department of Environmental Science and Technology, Central University of Punjab, Punjab, India. He is a well-rounded researcher with more than 30 years of experience in leading, supervising, and undertaking research in the field of water, wastewater management, solid and Hazardous Waste Management. He and his research group are working on Water and Wastewater pollution monitoring and abatement, Solid Waste Management, Pesticide degradation, radioecology and Heavy Metal detoxification. He has published more than 200 research and review articles, 22 proceedings, and 6 editorials in peer-reviewed International and National journals of repute with more than citations 16000 and h-index 67. In addition, he has published 3 books and 15 book chapters and completed 10 sponsored research projects as Principal Investigator funded by various agencies and departments. He was awarded "Thomson Reuters Research Excellence – India Citation Awards 2012". He is an active member of various scientific societies and organizations including, the Biotech Research Society of India, the Indian Nuclear Society, etc.

For Product Safety Concerns and Information please contact our EU representative GPSR@taylorandfrancis.com
Taylor & Francis Verlag GmbH, Kaufingerstraße 24, 80331 München, Germany

www.ingramcontent.com/pod-product-compliance
Lightning Source LLC
Chambersburg PA
CBHW080650220326
41598CB00033B/5154

* 9 7 8 1 0 3 2 1 3 5 7 9 3 *